T0314936

Thermal Systems Design

Thermal Systems Design

Fundamentals and Projects

Second Edition

Richard J. Martin
Martin Thermal Engineering, Inc.
California Polytechnic State University, San Luis Obispo
Santa Clara University, Santa Clara
California, USA

This edition first published 2022

Registered Office
John Wiley & Sons, Inc., 111 River Street, Hoboken, NJ 07030, USA

Editorial Office
111 River Street, Hoboken, NJ 07030, USA

For details of our global editorial offices, customer services, and more information about Wiley products visit us at www.wiley.com.

Wiley also publishes its books in a variety of electronic formats and by print-on-demand. Some content that appears in standard print versions of this book may not be available in other formats.

Library of Congress Cataloging-in-Publication Data applied for:

ISBN: 9781119803478

Cover design by Wiley
Cover images: © AnitaVDB/Getty Images; © Photographee.eu/Shutterstock; © Derek Pinkston/Getty; © leungchopan/Shutterstock; © momente/Shutterstock; © Mint Images/Getty; © Grant Faint/Getty; © Mechanical Engineering/Shutterstock; © Andrii Stepaniuk/Shutterstock; © artpartner-images/Getty; © mikulas1/Getty

Set in 9.5/12.5pt STIXTwoText by Straive, Pondicherry, India

10 9 8 7 6 5 4 3 2 1

Contents

Preface to the First Edition (A Most Practical Guidebook)

The theme and structure of this textbook arose from the author's 16 semesters instructing mechanical and chemical engineering students at the University of Southern California (USC), and much of the specialized content incorporated here arose from investigations the author performed for clients of Martin Thermal Engineering, Inc. The author (in parallel with several colleagues) considered numerous published texts and found that none contained the focus or breadth necessary for a comprehensive class in thermal systems design – hence, the need for this tome was clear.

The intended audience is mechanical or chemical engineering students seeking capstone design guidance for thermal-fluid systems, including heating, drying, boiling, refrigeration, air-conditioning, compression, expansion, combustion, and power generation. Practitioners of thermal engineering design may also find this to be a helpful reference work – one that offers breadth, clarity, and simplicity.

The overarching goal of this textbook is to help students visualize the landscape of a thermal system design project and to equip their intellectual "toolkits" with a wide variety of techniques for applying solid engineering theory toward a useful and successful design, while exposing them to predictable stumbling blocks that will require ingenuity to overcome.

Thermal systems design students should enter this course having successfully completed math and science prerequisites such as advanced calculus, differential equations, chemistry, and physics, as well as intermediate engineering courses in thermodynamics, fluid mechanics, and heat transfer. Prior study of combustion is helpful, but not required. The book adopts use of a technique called "Think Stop," which triggers a pause for reflection whenever the author sees an opportunity for "extracurricular learning" about a subject.

The course can be taught in a 15-week semester with 3-lecture hours per week or in a 10-week quarter with 4-lecture hours per week. Expectations for student projects and homework should be reduced a bit for the 10-week course. Chapters 16–18 may be excluded without a loss of continuity, but students should be encouraged to digest this content in their spare time.

Acknowledgments

Inspiration for much of the material developed here would not be possible without contributions from my colleagues at USC and elsewhere. I am indebted to Larry Redekopp and Geoff Spedding for bringing me into the fold of the Aerospace & Mechanical Engineering Department as a part-time faculty member and for numerous discussions about teaching philosophy and technique. I also thank my colleagues Fokion Egolfopoulos and Paul Ronney for their encouragement and advice about subject matter for which our fondness is kindred. It is with deep gratitude that I acknowledge my classroom colleagues Manny Dekermenjian and Leslie King, without whom the teaching and learning elements that embody the soul of this book would not have been possible. The original illustrations were expertly crafted by Ed Thielen, one of the world's finest creators of clear and compelling courtroom demonstrative exhibits. The editor was Alison Martin, a rising star in the field of composition, rhetoric, and news analysis. The delightful cover art was created by Nik Hallin and Cara Koenig of Motion Squared Design. Brian Martin, John Taber, John McArthur, and Zuhair Ibrahim attended lectures and provided helpful comments, as did numerous students from my Fall 2017 class, who studied from a draft version of this book. Finally, to Dawn Martin, my CFO, my cheerleader, and my more-than-equal partner in all things nonengineering, I give a Tom Hanks-to-Meg Ryan smile of gratitude and love.

Preface to the Second Edition (Fundamentals and Projects)

Where the prior edition claimed to be "most practical," the current edition attempts to earn recognition for having the "coolest" (and the "hottest") projects. The principal changes in this edition relate to the inclusion of many new design projects for the student teams, but it also includes new analytical tools for students to employ as they undertake their design projects. We continue our prior themes of motivating good engineering habits by applying the laws of thermodynamics, fluid mechanics, and heat transfer to create a functional design – while also insisting students demonstrate a higher level of design acumen than required by most other textbooks in the design category. This distinction appears in the projects by the inclusion of phase change (vaporization, condensation, humidity), chemistry (combustion, multiphase thermochemistry), and/or flow in porous media – to go beyondthe more elementary *p*-*T*-*h* analyses found in less-thorough works.

With this edition, the author has adopted an augmented mission: to convince students that they can analyze thermal systems without spending tens of thousands of dollars per year to license a process/flow simulation software package. The outcome we seek is for students to first understand the fundamentals and then approach their design projects with confidence and creativity. If our instruction is successful, students will begin their careers with an unsurpassed breadth of knowledge and a stout collection of engineering design tools in their toolkits.

The most practical way we attempt to further this mission is with a new appendix that contains VBA scripts for customized Excel functions that compute values for thermodynamic properties of fluids and other complex engineering equations. If readers (i.e. students or engineering practitioners) invest a small amount of time copying the scripts and pasting them into VBA modules within Microsoft Excel, their toolkits will be augmented with powerful tools that would cost a small fortune elsewhere. In addition, the property tables from the first edition were relocated to a separate appendix and new property tables for propane, ammonia, and ammonia/water mixtures were added.

Eight projects were presented in the first edition (as end-of-chapter problems in Chapters 5 and 14), and they remain prominent in the second edition, even as new projects are presented via new homework questions. The new projects include: a hot air balloon, an exothermic gas generator, a tenter-frame drying oven, an espresso machine, an ammonia/water/hydrogen absorption refrigeration system with no moving parts, and a thermally assisted air filtration system for destruction of biohazard particles.

The companion website for this book includes a substantial collection of supplementary tools to assist instructors and students. Included in this resource cache are PowerPoint slides with content organized for a 15-week lecture course, "customer specifications" that provide necessary sizing and operating parameters that form the bases for the project designs, and solution keys for most of the homework problems. The solution keys also are repositories for several important derivations and project examples that were too long for the published text.

The new projects are intended to be executed in the same manner as the old projects. First, students create a schematic of the system with major equipment (unit operations) connected by lines (process streams). Then they compute thermodynamic details at each state and prepare a process flow diagram (PFD) withstream table. Next, they determine sizes for pipes/ducts and major equipment using applicable rules for heat transfer, fluid mechanics, phase change, and combustion. And finally, they complete the overall process design by selecting sensors (instruments) and valves (final control elements) and adding feedback control loops and safety interlocks – all of which are communicated via a piping andinstrumentation diagram (P&ID). If the PFD and P&ID drawings are prepared using commercial software, those exercises can provide a "laboratory" experience for design students.

The approach taken here is for students to complete a ***process design*** package, not a ***mechanical design*** package. Consequently, no emphasis is placed on detailed structural or mechanical design of equipment beyond basic sizing (*D*, *L*, # *tubes*) of piping, vessels, and heat exchangers, along with sizing/selection of commodity items such as blowers, pumps, and burners. Economic analyses and cost optimization may be added by instructors, but these topics are excluded from the chapter content here to help ensure the coverage remains manageable for a 45-lecture-hour semester or a 40-lecture-hour quarter.

In addition to the new projects, extensive new content is provided in several chapters: Chapter 4 has a new analytical method for computing equilibrium in fuel-rich combustion and an improved method of estimating destruction efficiency for thermal oxidation based on VOC properties; Chapters 6 and 10 provide a thorough discussion of dew points and bubble points for two-component refrigerant mixtures. Chapter 13 is broadened to include engineering methods for protection of materials against thermal shock and thermal expansion; Chapter 17 contains a more rigorous development of statistical methods for quality and efficiency.

Finally, we wish to call attention to a concern we have encountered regarding the nomenclature of conservation, and why we elected to invent a pedagogically preferred symbology – even though it modestly disrespects scientific orthodoxy. As described in Chapters 1–4, we use a nomenclature shortcut to illuminate the four elements of conservationin a control volume for any arbitrary property $\boldsymbol{B}$: production $\dot{\mathcal{P}}_B$, inflow $\dot{\boldsymbol{B}}_{in}$, outflow $\dot{\boldsymbol{B}}_{out}$, and storage $\dot{\mathcal{S}}_B$.

Strictly speaking, the superdot symbol should apply only to the two flow terms, and the production and storage terms should be expressed as time derivatives (without any superdot). Certainly, this orthodoxy offers the only valid mathematical way to construct a conservation equation and this book carefully embraces these conventions in Chapters 2–4. The orthodox math is valid because (i) it is senseless to think of inflows and outflows as being derivatives of something else (hence the need for a superdot symbol to denote quantity flowing per unit time) and (ii) it is completely sensible to apply differentiation (with respect to time) to extensive fluid properties such as mass, momentum, and energy – because they can and do change with time.

Despite the obvious validity of the long-accepted orthodoxy, we introduce our unorthodox superdot symbology for production and storage in Chapter 1 for its descriptive simplicity, and we believe this approach greatly helps students understand the contrasting concepts of ***production*** and ***storage***. This approach protects the integrity of the physics by employing the mathematical rigor of the Reynolds transport theorem when presenting the detailed conservation equations, while it also highlights the bright line distinguishing production and storage in a simple way.

Acknowledgments

Many thanks are due to Professors Betta Fisher (Cornell University), Mahboobe Mahdavi (Gannon University), and Zuhair Ibrahim (University of Southern California) for identifying minor errors in the first edition and suggesting sections where additional clarity was needed. Special thanks to Brian Kaiser for lengthy discussions about principles of multicomponent, multiphase mixtures, to Jay Hudson for fluid heater sizing and safety know-how, to Craig Schuler for helpful discussions on iteration techniques, and to Gabriel Gundling for a tutorial on ballooning technologies and practices. Special thanks are due again to Ed Thielen for compelling new artwork and to Dawn Martin for side-by-side assistance with table prep and the key word index.

The author is indebted to the Wiley team, especially Lauren Poplawski, Gabby Robles, Jenny Seward, Amudhapriya Sivamurthy, and Becky Cowan, for adopting this second edition textbook as their project and for numerous helpful editorial and pedagogic discussions along the way.

About the Author and the Textbook

Dr. Richard J. Martin played major roles in the design, commissioning, operation, and testing of combustion and heat transfer equipment in the 1980s and 1990s. During this time he became a named inventor on 24 utility patents. In the two decades of the current century, he investigated hundreds of failures (e.g. fires, explosions, thermal equipment failures) – the majority of which originated within thermal systems employed in many fields of commerce. He has been a volunteer leader of technical committees that write safety standards for industrial heating equipment.

In addition to teaching thermal systems design and heat transfer at University of Southern California, he has also taught courses in air pollution, fluid mechanics, heat transfer, programming applications in engineering, and thermal systems design at the California Polytechnic State University (Pomona and San Luis Obispo). Dr. Martin will be starting a new engagement at Santa Clara University shortly after the publication of this textbook, where he will teach Heat Transfer and Thermal Systems Design.

The primary purpose of this textbook is to encourage and exemplify high quality, accurate, well-communicated, engineering design. A design may be amazing, but if it is poorly communicated to those who build and use it, horrible consequences could ensue. Similarly, a design that is communicated with accuracy and clarity may be equally problematic if it was produced using erroneous principles or insufficient forethought.

Another important purpose of this textbook is to motivate engineering students to enhance their: (i) *knowledge* of scientific fundamentals that govern human interactions with the environment; (ii) *habits* of observing, investigating, and analyzing engineering successes and failures; and (iii) *desire* to apply engineering know-how to the service of humankind.

The author's unique background, comprising equal tenures in innovative design and technology failure investigation, gives this textbook a perspective different from most others. Not only are students given valid tools and methods to solve real-world engineering design problems, but they are also cautioned about numerous ways the tools may be misunderstood or accidentally misapplied. By detailing both "right *and* wrong" approaches, students are better equipped to adopt the right, while rejecting the wrong.

The author encourages feedback from readers if any part of this textbook contains inaccurate or confusing information or could benefit from additional technical content that was omitted.

About the Companion Website

This book is accompanied by a companion website:

www.wiley.com\go\Martin\ThermalSystemsDesign2

This website includes:

Instructor site

- Homework Solutions
- All figures from the print book downloadable in color in PowerPoint
- All tables from the print book downloadable in PowerPoint
- Course schedules for instructors
- Project specifications

Student site

- All figures from the print book available in color in PDF

1

Thermodynamics

Thermodynamics is a predictive science that describes interrelations between properties of matter such as temperature, pressure, density, and entropy and how these properties are affected by transfers of mechanical and thermal energy (i.e. work and heat) into and out of the system and by interconversions of energy forms (e.g. kinetic, potential, electrical, chemical, thermal, and mechanical) within a system.

The primary focus of engineering thermodynamics is to characterize the "states" of substances (typically fluids) participating in a thermodynamic cycle and how those states are altered by equipment (e.g. heat exchangers, compressors) incorporated into the cycle. Thermodynamic cycles are often assumed to be steady, in the sense that state variables are in local equilibrium and states are unchanging in time, but they vary spatially. Transient conditions, such as startup and shutdown, are treated separately.

This chapter covers basic thermodynamic topics such as units of measure, thermodynamic laws, control mass control volume, ideal gas law, thermodynamic variables, and availability. Combustion, which is a very important branch of thermochemistry, is covered in Chapter 4. Several advanced thermodynamic topics are covered in Chapter 6.

1.1 Units of Measure

Students must not undervalue the importance of knowing and tracking "units of measure" when performing an engineering calculation. In many cases, erroneous results of analysis or design problem-solving can be traced back to an engineer's failure to ensure that proper units and proper unit conversions are incorporated into a mathematical representation of a physical or chemical process.

Think Stop. Students are encouraged to develop a ruthless habit of labeling all numerical values with the correct units and ensuring that the units on the left side of an equation are identical to the units on the right side. A simple example of this is seen in Equation (1.1), the generalized 1D heat conduction equation:

$$\underbrace{\frac{\partial}{\partial x}\left(k\frac{\partial T}{\partial x}\right)}_{\left(\frac{1}{m}\right)\left[\left(\frac{W}{m\cdot K}\right)\left(\frac{K}{m}\right)\right]} + \underbrace{\dot{e}_{gen}}_{\left(\frac{W}{m^3}\right)} = \underbrace{\rho c\frac{\partial T}{\partial t}}_{\left(\frac{kg}{m^3}\right)\left(\frac{J}{kg\cdot K}\right)\left(\frac{K}{s}\right)} \tag{1.1}$$

This textbook utilizes a unique symbol constituting an "equals" sign superimposed with a "breve" or "crescent" diacritical mark ($\breve{=}$) to mean "has units of." For example, the phrase "heat flux has units of watts per square meter" would be written symbolically as $\dot{q} \breve{=} W/m^2$.

Thermal Systems Design: Fundamentals and Projects, Second Edition. Richard J. Martin.

Companion website: www.wiley.com\go\Martin\ThermalSystemsDesign2

Because thermal processes often involve chemical reactions, students must also have full command of the ability to interrelate molar units and mass units. For example, the molar mass ($\hat{M}$, historically called molecular weight) of carbon dioxide (CO_2), which is comprised of one carbon atom (atomic mass, $A_C = 12.0096\ amu$) and two oxygen atoms (atomic mass, $A_O = 15.999\ 03\ amu$ each), is $\hat{M}_{CO_2} = 44.007\ 66\ (= 1 \times 12.0096 + 2 \times 15.999\ 03)\ \frac{kg}{kmol}$.

Think Stop. Students should individually validate that the following six expressions are equivalent representations of identical quantities of carbon dioxide molecules:

$1.0000\ kmol_{CO_2}$	$1000.0\ \mathrm{mol}_{CO_2}$	$2.204\ 624\ lbmol_{CO_2}$
$44.007\ 66\ \mathrm{kg}_{CO_2}$	$44\ 007.66\ g_{CO_2}$	$97.020\ 36\ lb_{CO_2}$

1.2 Mass/Force Unit Conversion

The use of traditional English units is intentionally kept to a minimum in this textbook because the author prefers to use *SI* units wherever possible, predominantly to avoid errors associated with non-decimal English-to-English unit conversions (e.g. $1.0\ ft = 12.0\ in$, or $1.0\ mi = 5280\ ft$). Despite the *SI* favoritism expressed here, engineering students are nonetheless encouraged to develop skills for working in both sets of units. In keeping with this theme, a minority of the examples and problems given here are expressed in the less-preferred units.

One area where the choice of units can be potentially confounding is the determination of weight (units of force) by multiplying mass times the acceleration of gravity. The author has found it helpful to utilize the concept of a "gravitational conversion constant" (per Newton) as follows: $\boxed{F_w = m(g/g_c)}$, where $F_w(\doteq N)$ is the weight force, $m(\doteq kg)$ is the mass, $g(\doteq m/s^2)$ is the acceleration due to gravity, and $g_c(\doteq [kg \cdot m]/[N \cdot s^2])$ is the gravitational conversion constant. The value of the gravitational conversion constant can be written in many different sets of units, three of which are given in Equations (1.2):

$$g_c = 1.00\ \frac{kg \cdot m}{N \cdot s^2} \tag{1.2a}$$

$$g_c = 1.00\ \frac{slug \cdot ft}{lb_f \cdot s^2} \tag{1.2b}$$

$$g_c = 32.174\ 05\ \frac{lb_m \cdot ft}{lb_f \cdot s^2} \tag{1.2c}$$

The values in Equations (1.2) demonstrate that the standard (earth) acceleration of gravity acting on $1.0\ kg$ mass gives rise to a weight force of $9.806\ 65\ N$, whereas the same gravity acting on $1.0\ lb_m$ gives rise to a weight force of $1.0\ lb_f$. By this logic, it is easy to see that a *slug* of material is equivalent to $32.174\ lb_m$ of the same material.

The gravitational conversion constant becomes vitally important in fluid problems where the density of a fluid (mass per unit volume) affects the computation of pressure (which is force per unit area). In these problems, lb_m must be appropriately converted to lb_f to obtain units of pressure (in *psi* or *psf*). The *SI* calculations are more straightforward in part because mass is expressed in *kg* and force is expressed in *N*, so the ambiguity between lb_m and lb_f is avoided entirely.

Table 1.1 Preferred conversion factors and reference values for fundamental constants.

Conversion factor	Value	Units	Constants	Value	Units	Constants	Value	Units
CF_{mass}*	**0.453 592**	kg/lb_m	$g_{sealevel}$*	**9.806 65**	m/s^2	$\hat{V}_{STP,ideal\ gas}$	24.465 38	$m^3_{STP}/kmol$
CF_{length}*	**3.280 84**	ft/m	$\rho_{Hg\ (0°C)}$	**13 595.08**	kg/m^3	$\hat{M}_{CH_4}$	16.040 96	$kg/kmol$
CF_{press1}*	**760.000**	mm_{Hg}/atm	$\rho_{H_2O,liq\ (0°C)}$	**999.998**	kg/m^3	$\hat{M}_{H_2}$	2.015 68	$kg/kmol$
CF_{press2}	29.921 261	$inch_{Hg}/atm$	$\Gamma_{Hg\ (0°C)}$	13.595 107	—	$\hat{M}_{NH_3}$	17.029 95	$kg/kmol$
CF_{press3}	406.782 75	$inch_{wc}/atm$	A_H*	**1.007 84**	$kg/kmol$	$\hat{M}_{O_2}$	31.998 06	$kg/kmol$
CF_{press4}	101.325 07	kPa/atm	A_C*	**12.0096**	$kg/kmol$	$\hat{M}_{N_2}$	28.012 86	$kg/kmol$
CF_{press5}	14.695 97	psi/atm	A_N*	**14.006 43**	$kg/kmol$	$\hat{M}_{CO_2}$	44.007 66	$kg/kmol$
CF_{energy}*	**1.055 056**	kJ/BTU	A_O*	**15.999 03**	$kg/kmol$	$\hat{M}_{H_2O}$	18.014 71	$kg/kmol$
CF_{force}	4.448 218	N/lb_f	$\hat{R}$*	**8.314 46**	$kJ/(kmol \cdot K)$	$\hat{M}_{CO}$	28.008 63	$kg/kmol$

Source: Martin Thermal Engineering Inc.

A set of preferred conversion factors and fundamental constants are given with moderately high precision in Table 1.1, where reference values (from which all other factors are derived) are indicated with an asterisk. Constants not displayed here (e.g. enthalpy of formation) can be found in the appendices.

1.3 Standard Temperature and Pressure

Thermodynamic systems invariably require a designation of the "standard" state of matter. Chemists typically use the condition $T_{ref,\ chem} = 273.15\ K\ (= 0.0°C)$ and $p_{ref} = 1.00\ atm$. Other practitioners use different reference temperatures (e.g. $T_{ref,\ other1} = 288.15\ K$ or $T_{ref,\ other2} = 293.15\ K$), and recently, a different reference pressure (e.g. $p_{ref,\ other} = 0.100\ MPa$) has been utilized by the National Institute of Standards and Technology (Chase Jr. 1998).

For purposes of this textbook, ***Standard Temperature and Pressure (STP)*** is defined in harmony with the 1971 Edition of the Joint Army Navy Air Force (JANAF) Thermochemical Tables (Stull and Prophet 1971): $\boxed{T_{STP} = 298.15\ K}$ $(= 25.0°C)$ and $\boxed{p_{STP} = 101.325\ kPa}$.

1.4 Control Mass, Control Volume

Thermodynamic laws are applied to ***systems***. Some textbooks exclusively equate the term ***system*** with the unique physical construct known as a control mass. The treatment here affords more flexibility to the term ***system*** such that it can apply to either a control mass or a control volume.

Control Mass. Systems with a known quantity (mass or moles) of material in a known location are usually treated with a control mass approach. In a control mass, the quantity of the material is not explicitly determined by its volume unless the density is also known. Variations in the state of a control mass occur as a function of time and typically involve work or heat crossing the system boundary to alter the system's state. The state of a control mass can also vary when chemical reactions occur inside the system, causing variations in the molecular composition of the mass and interconversions between chemical and thermal energy.

Control Volume. Systems that contain working fluids flowing through pipes and different pieces of equipment are usually treated with a control volume approach. Frequently, the pieces of equipment are connected to each other in a thermodynamic ***cycle***. The term ***cycle*** is used because the working fluid passes through all the equipment in a designated sequence and then returns to its starting point to begin another cycle.

A schematic version of the classic "Rankine cycle" is shown in Figure 1.1.

Analysis of a thermodynamic cycle typically requires an analysis of unique control volumes surrounding one or more pieces of equipment to determine the thermodynamic state of the fluid at the various locations or states (1, 2, 3, 4 as shown). Control volume analysis also can be used to compute the flows of energy and mass across a control volume boundary. For example, if the pump in Figure 1.1 is known to produce a pressure rise of $\Delta p_{pump} = 250\ psi$, then the pressure at State 2 will be 250 *psi* higher than the pressure at State 1. A control volume drawn around the pump would reveal three important boundary crossings – low-pressure liquid water entering at State 1, high-pressure liquid water exiting at State 2, and mechanical power (to pressurize and move the liquid water) entering through the pump shaft.

As will be illustrated several times in later chapters, a cycle analysis can only proceed to completion when enough information is known (or estimated) about the various control volume boundary crossings to determine all the unknown states and flow parameters algebraically. In Figure 1.1, unique control volumes may be established around the boiler, turbine, and condenser to evaluate

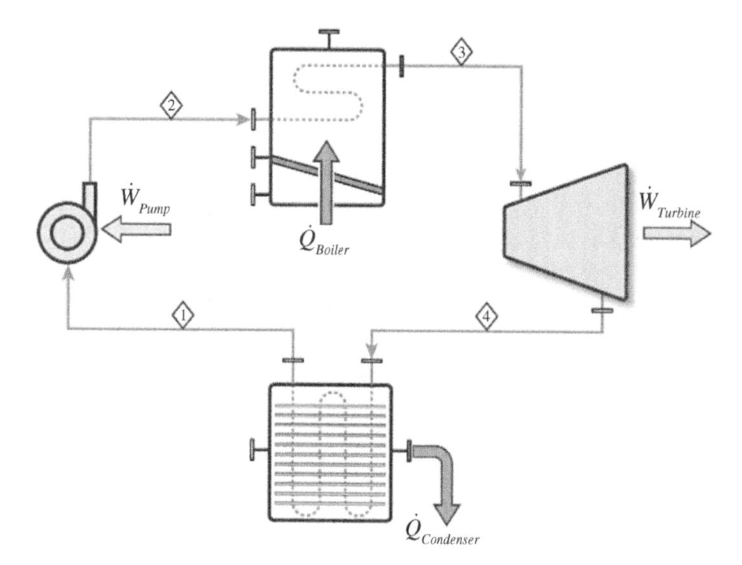

Figure 1.1 Classic Rankine cycle with water as working fluid. *Source:* Martin Thermal Engineering Inc.

state variables upstream or downstream of each of those pieces of equipment. A control volume around the entire system may be useful to determine one of the energy flows if the others are known. Initially, a control volume analysis might involve a trial-and-error methodology to narrow the field of possible equipment choices.

As indicated in the Foreword, this trial-and-error process (placing equipment onto a blank schematic and beginning thermodynamic analyses with little other guidance) is often the most daunting aspect of a thermal system design effort. Over time, students will gain an intuitive feel for where to start an analysis and how to apply control volumes in the most productive manner. When students learn to "embrace the struggle" (especially when helped by judicious questions from the instructor), their tenacity, engineering know-how, and handy bag of tools will propel them forward to many successful outcomes.

1.5 Laws of Thermodynamics

Macroscopic systems (unlike quantum systems, which only obey the same laws as macroscopic systems when they are aggregated together in large ensembles of quantum particles) obey four laws of thermodynamics:

- The ***Zeroth law*** states that when two bodies are simultaneously in thermal equilibrium with a third body, they must be in thermal equilibrium with each other, and all are at the same temperature.
- The ***First law*** states that the total energy of an isolated system does not change, although the energy can be transformed from one type to another (e.g. kinetic energy, potential energy, mechanical energy, thermal energy, and chemical energy).
- The ***Second law*** states that the total entropy of an isolated system must increase or remain constant (it may not decrease).
- The ***Third law*** states that when a pure substance is solidified as a perfect crystal, it has zero entropy at an absolute temperature of zero (0 K or 0 R).

The First and Second laws of thermodynamics are quintessential ***conservation equations*** as covered in the next section. Energy and entropy can flow across the system boundary in one or more of the following forms – shaft work, heat transfer, flows of matter that carry energy or entropy with them.

In the ideal situation, an increment of shaft work δW is ***reversibly*** delivered to or extracted from the fluid in the system, which means that no entropy is ***produced*** by irreversibilities associated with the shaft and impeller motion. Conversely, the ideal situation for heat entering or exiting a system means that the heat flow is delivered reversibly and that zero entropy is ***produced*** by irreversibilities associated with the molecular collisions at the system boundary. Nevertheless, when an increment of heat δQ crosses the system's boundary in (or out), regardless of whether it is delivered ***reversibly***, the working fluid's entropy will increase (or decrease) by an amount dS that is exactly equal to the increment of heat transferred (δQ) divided by the temperature at the boundary (T):

$$dS_{reversible,\ heating/cooling} = \frac{\delta Q_{reversible}}{T_{boundary}}$$

The concept of ***production*** is best illustrated by combining it with the Latin phrase ***ex nihilo***, which means "out of nothing." By using the phrase ***production, ex nihilo***, students can differentiate between ***production*** and ***storage***, which are addressed for several conserved variables in the context of the Reynolds transport theorem in Chapters 2–4.

To clarify here, ***storage*** simply connotes an accumulation of a conserved quantity (e.g. mass or energy) within a control volume, whereas ***production*** connotes something that basically appears out of (or disappears into) thin air. Within this context, the Reynolds transport theorem (see Chapter 2) demonstrates how a change in ***stored*** momentum or entropy can be caused: (i) by a net ***inflow*** or ***outflow*** across the system's boundary or (ii) by ***production***, which is essentially creation or destruction ***ex nihilo***.

The most common ***irreversibilities*** in thermal systems are:

- Friction
- Heat transfer across a finite (i.e. not infinitesimal) temperature difference
- Mass transfer between regions of finite (i.e. not infinitesimal) concentration difference
- Rapid (i.e. not infinitesimally slow) compression/expansion of a fluid to a different pressure
- Rapid (i.e. not infinitesimally slow) accelerations and decelerations of a fluid

Irreversibilities related to shaft work in flowing systems are analytically addressable with a factor called isentropic efficiency (see Chapter 9). The mathematics of this are simple, because the reversible case for shaft work is fully isentropic. On the contrary, irreversibilities caused by other factors (heat, mixing, or friction) involve changes in system entropy even if they are performed reversibly, so the concept of isentropic efficiency is not helpful. The concept of exergy is presented later in this chapter. Its relationship to isentropic efficiency is indirect at best, but it does provide an analytic method for addressing irreversibilities in a system that can transfer heat and work to the environment, which is also called the dead state.

1.6 Conservation Laws

"Conservation laws" are vital to the analysis of thermal systems. The First law is an example of a conservation law wherein energy can be transformed among different types (e.g. thermal, chemical, and mechanical), but the total energy of an isolated system must be conserved.

Conservation laws can be written in the form of Equation (1.3), as follows:

$$\dot{\mathcal{P}}_B + \dot{B}_{in} = \dot{B}_{out} + \dot{\mathcal{S}}_B \tag{1.3}$$

where B is the conserved quantity (e.g. mass, momentum, energy, entropy, and atoms) and:

$\dot{\mathcal{P}}_B$ is the rate of ***production*** of ***B*** (*ex nihilo*) within the system.

$\dot{B}_{in}$ is the rate of ***inflow*** of ***B*** across the system boundary.

$\dot{B}_{out}$ is the rate of ***outflow*** of ***B*** across the system boundary.

$\dot{\mathcal{S}}_B$ is the time rate of change in ***storage*** of ***B*** inside the system.

As will be seen in later chapters, the production rates of mass $\left(\dot{\mathcal{P}}_m\right)$, energy $\left(\dot{\mathcal{P}}_e\right)$, and atoms $\left(\sum \dot{\mathcal{P}}_{i\,[e.g. C,H,O,N]}\right)$ are always zero for conventional thermal systems (i.e. when nuclear reactions are disallowed), but the production rates of momentum $\left(\dot{\mathcal{P}}_{\vec{p}}\right)$, entropy $\left(\dot{\mathcal{P}}_s\right)$, and species

($\sum \dot{\mathcal{P}}_{k\ [e.g. H_2O, CO_2, O_2]}$) are routinely not zero. It is also important to remember that mass, energy, entropy, species, and atoms are nonnegative scalar quantities, but momentum is a vector quantity and can therefore take on negative or positive values. Angular momentum is also a vector quantity that obeys its own conservation equation, however, this book excludes detailed coverage of this phenomenon.

The author notes that use of a ***super-dot*** symbol above each variable in Equation (1.3) is a simplistic way of representing that the units of all four terms include reciprocal time (e.g. sec^{-1}). This simplification is introduced here as a teaching rubric because even though the physics of conservation is well established, the mathematical symbology is not. In Chapters 2–4, we summarize several expressions of the Reynolds transport theorem (Reynolds 1903) for different conserved variables. This classic treatment of the differential mathematics of conservation is simultaneously vital and dense, and in our opinion, it fails to persuasively reinforce physical intuition about the concept of ***production, ex nihilo***, hence the need for Equation (1.3) and its unorthodox symbology.

For a more comprehensive view of the physics of conservation in a control volume, we recommend that students review and compare the treatments by White (2016) and Moran et al. (2018). We feel strongly that Moran's choice to limit usage of the term "conservation" to energy and mass accountings only (and thereby to disallow the Second law and the Momentum equation from being classified as "conservation equations") shortchanges students from a significant learning framework. In contrast, White first introduces the Reynolds transport Theorem as being fully applicable to any property (***B***) of a working fluid and then proceeds to develop the accounting equation for the simplest case – conservation of mass.

Where Moran's treatment of mass and energy as "conserved" quantities is satisfactory as far as it goes, White's treatment is superior because it permits momentum and entropy also to be viewed as "conserved" quantities that obey the Reynolds transport theorem (which they most certainly do). In White's approach, there is one major distinction between mass and momentum – the "production" term is identically zero for the former, and only occasionally nonzero for the latter. Nonetheless, White uses a longer version of Equation (1.3) for both mass and momentum, whereas Moran excludes momentum (and entropy, species, etc.) from the realm of property conservation entirely, because he fails to address the concept of production.

1.7 Thermodynamic Variable Categories

Students are sometimes confused about the nature and use of different thermodynamic variables. The following categories and distinctions are offered to help characterize these variables for gases and some liquids.

For control volume systems (e.g. the Rankine cycle of Figure 1.1), the working fluid spatially experiences different thermodynamic states as the fluid passes through different processing equipment (e.g. pumps, boilers, turbines, etc.).

For control mass systems (e.g. gasoline vapor, air, and combustion products in the cylinder of an automobile engine), the fluid state is often considered to be uniform in space (i.e. throughout the control mass) and varying in time.

Control volume systems, where the fluid flow rate is unchanging in time, are called ***steady-flow*** systems. If the individual fluid states at each point in space do not vary with time, the system is called ***steady state***.

In this textbook, the term "state variable" is applied to intensive properties of substances (e.g. temperature and pressure) that do *not* depend on the quantity of fluid present or rate of fluid flowing through the system. Similarly, the terms "quantity variable" (e.g. mass) and "flow variable" (e.g. enthalpy flow) are applied to extensive properties that express some measure of the size of the system. For example, a dirigible contains about 10 000 000× as much helium as a party balloon, and an ocean cruise liner diesel engine produces about 25 000× as much shaft horsepower as a chainsaw engine.

When a variable can be expressed in both the extensive and intensive forms (e.g. enthalpy H and specific enthalpy h) the lowercase variable represents the intensive property and the uppercase variable represents the extensive property. It should be noted that certain intensive state variables (e.g. temperature, T) historically have used the uppercase form and certain quantity variables (e.g. mass, m) historically have used the lowercase form. This book makes no attempt to overcome these inconsistencies where historical inertia is great.

State Variables (Control Mass). For a control mass, the intensive properties of the system that define its thermodynamic state are temperature and pressure, with volume having relevance as a state variable if the fluid is in a saturated (two-phase) condition. At least one quantity variable (see later) is also required to define the state of a control mass.

$T = Temperature\ (K)$ $\quad$ $p = Pressure\ (Pa)$

$\mathcal{V} = Volume\ (m^3)$ $\quad$ $\hat{M} = Molar\ mass\ (kg/kmol)$

State Variables (Control Volume). For a control volume, there are numerous intensive properties that can be utilized interchangeably to fix or lock the state of the fluid. Intensive thermodynamic state variables such as h, s, and v are tabulated for pure substances (and some mixtures such as air), so that if you know any two state variables, you can find the others.

$T = Temperature\ (K)$ $\quad$ $h = Specific\ enthalpy\ (kJ/kg)$

$p = Pressure\ (Pa)$ $\quad$ $\hat{s} = Molar\ entropy\ (kJ/(kmol \cdot K))$

$v = Specific\ volume\ (m^3/kg)$ $\quad$ $\hat{h} = Molar\ enthalpy\ (kJ/kmol)$

$s = Specific\ entropy\ (kJ/(kg \cdot K))$ $\quad$ $\hat{M} = Molar\ mass\ (kg/kmol)$

Quantity Variables. For a control mass, the quantity variables are necessary to define what constitutes the system's entirety or extent. Control volumes frequently don't need their quantity variables analyzed, but a significant counter example of this is a flowing system whose transient behavior is just as important as its steady-state behavior. The time to reach steady state will depend on the total mass of the working fluid in all parts of the system. Heat (Q) and work (W) are quantity variables for thermal and mechanical energy that cross the control mass boundary from the beginning to end of a transient process. Mole fractions and mass fractions are intensive properties, but they can be used to compute proportions of different constituents in the control mass.

$m = Mass\ (kg)$ $\quad$ $W = Work\ transferred\ (kJ)$

$n = Moles\ (kmol)$ $\quad$ $Q = Heat\ transferred\ (kJ)$

$H = Enthalpy\ (kJ)$ $\quad$ $\chi_i = Mole\ fraction\ i\ (kmol_i/kmol_{tot})$

$S = Entropy\ (kJ/K)$ $\quad$ $Y_i = Mass\ fraction\ i\ (kg_i/kg_{tot})$

Flow Variables. Flow variables are parallel to (and more important than) quantity variables when a control volume system (as opposed to a control mass system) is being analyzed. Typically, a "super-dot" above the variable name signifies a flowing variable that crosses the system boundary, but we reemphasize that the simplistic nomenclature for the production and storage terms in Equation (1.3) ($\dot{\mathcal{P}}_B, \dot{\mathcal{S}}_B$) are shortcuts, and neither involves the flow of anything across a boundary. In Chapters 2–4, the production and storage terms are expressed as time derivatives, because the Reynolds transport theorem dictates that to be the case. For multicomponent systems, mass and mole fractions provide a means for quantifying proportions of flows for the individual constituents.

$\dot{m} = Mass\ flow\ (kg/s)$ $\qquad$ $\dot{W} = Work\ flow\ (kW)$

$\dot{n} = Mole\ flow\ (kmol/s)$ $\qquad$ $\dot{Q} = Heat\ flow\ (kW)$

$\chi_i = Mole\ fraction\ i\ (kmol_i/kmol_{tot})$ $\qquad$ $\dot{H} = Enthalpy\ flow\ (kW)$

$Y_i = Mass\ fraction\ i\ (kg_i/kg_{tot})$ $\qquad$ $\dot{S} = Entropy\ flow\ (kW/K)$

Fluid Transport Property Variables. Fluid transport property variables are not thermodynamic variables at all, but some are listed here to clarify why they should not be included in the four groups aforementioned. Fluid properties such as specific heat, thermal conductivity, and dynamic viscosity are useful for solving flow resistance and heat convection problems, but they are not sufficient to define the state of a substance. For example, if a student was given the dynamic viscosity and the specific heat of steam, no state equation exists to algebraically determine the specific enthalpy and specific entropy of the fluid from those given values, let alone the temperature and pressure.

$k = Thermal\ conductivity\ (W/(m \cdot K))$ $\qquad$ $\mu = Dynamic\ viscosity\ (kg/(m \cdot s))$

$c_v, c_p = Specific\ heat\ (kJ/(kg \cdot K))$ $\qquad$ $\rho = Density\ (kg/m^3)$

Tables of thermodynamic properties of fluids at different temperatures and pressures are available from many different sources, (e.g. Linstrom and Mallard 2017–2021). One older resource that contains properties and equations of state for several important compounds (some of which are not available in Linstrom and Mallard) is thermodynamic properties in SI (Reynolds 1979). Other sources of thermodynamic properties to consider include the JANAF tables (Chase Jr. 1998 or Stull and Prophet 1971) and Abdulagatov et al. (1999).

The thermodynamic and fluid property tables compiled in Appendix A at the end of this book and some of the Visual Basic for Applications (VBA) customized functions given in Appendix B are sourced from the thermophysical properties of fluid systems section of the *NIST (National Institute of Standards and Technology) Chemistry Webbook* (Linstrom and Mallard 2017–2021). Students are encouraged to explore this data source to discover how they, too, can generate their own property tables for compounds not included in Appendix A. A large proportion of the exercises in this book can be solved with the Linstrom fluid property data.

Example 1.1

Determine (a) work input, (b) heat input, (c) change in stored internal energy, and (d) change in stored entropy for a control mass ($m = 1.0\ kg$) of nitrogen in a piston-cylinder apparatus undergoing a constant volume change from an initial state $T_1 = 300\ K, p_1 = 0.02\ MPa$ to a final state $T_2 = 600\ K, p_2 = 0.04\ MPa$.

Answer. Unfortunately, neither state can be obtained from Appendix A.3 because the pressures are too low. However, by using Linstrom's property generating algorithm, we can obtain the precise results in Table 1.2. Based on the table:

(a) $\boxed{W_{12}} = -m \int_{v_1}^{v_2} p \cdot dv \approx \boxed{0\ kJ}$ (Approximately constant volume; slight real gas effects)

(b) First law (control mass): $W_{in} + Q_{in} = \Delta U_{stored}$; Because $W_{in} = 0$, $\boxed{Q_{in} = \Delta U_{stored}} = m(u_2 - u_1) = 1 \cdot (448.87 - 222.34) = \boxed{226.53\ kJ}$

(c) See (b)

(d) $\boxed{\Delta S_{stored}} = m(s_2 - s_1) = 1 \cdot (7.8460 - 7.3239) = \boxed{0.5221\ kJ/K}$

Think Stop. Why is density considered a transport variable and not a state, quantity, or flow variable in the lists aforementioned?

Answer. Specific volume (v) is a thermodynamic state variable, but density (ρ), which is the reciprocal of v, is a transport variable. Density may seem like an intensive variable because it does not depend on the quantity of the substance, but density is neither intensive nor extensive. Unlike true extensive and intensive properties, it is logically nonsensible to divide or multiply density by mass. As will be seen in Chapter 6, classical thermodynamics identifies four state variables (p, V, T, S) and four thermodynamic potentials (U, H, A, G). For these reasons, specific volume is the correct intensive variable for thermodynamic analyses (because it represents the mass quotient of an extensive property – volume).

Table 1.2 Nitrogen data from *NIST Chemistry Webbook*. Properties at two states, defined by pressure and temperature.

T (K)	p (MPa)	v (m3/kg)	u (kJ/kg)	h (kJ/kg)	s (J/g*K)	c_v (J/g*K)	c_p (J/g*K)	μ (Pa*s)	k (W/m*K)	Phase
300	0.02	4.4519	222.34	311.37	7.3239	0.742 97	1.0400	1.79E-05	0.025 834	Vapor
600	0.04	4.4528	448.87	626.98	7.8460	0.778 03	1.0749	2.96E-05	0.043 907	Vapor

Source: Data from Linstrom and Mallard (2017–2021).

1.8 Ideal Gas Law

The ideal gas law is an "equation of state" for gases. It is only valid for gases at conditions where the density approaches zero. It was derived from experiments conducted in the nineteenth century that revealed direct relationships between the macroscopic volume of a gas and its absolute temperature, pressure, and quantity (moles).

The ideal gas law can be expressed in many forms (see Equations 1.4 and 1.5), and thermal engineers should commit them to memory or be able to derive others quickly from a single memorized version. The first two below are for a control mass at a single state, the next three are primarily for a control volume, and the last one is for comparing before and after states of a gas that has undergone a physical process (e.g. expansion, compression, cooling, heating):

For a control mass:

$$pV = n\hat{R}T \quad \left(\text{where } \hat{R} = \text{universal gas constant} = 8.314\,46\ \frac{kJ}{kmol \cdot K}\right) \tag{1.4a}$$

$$pV = mRT \quad \left(\text{where } R = \text{specific gas constant} = \frac{\hat{R}}{\hat{M}} \doteq \frac{kJ}{kg \cdot K}\right) \tag{1.4b}$$

For a control volume:

$$p = \rho RT \quad \left(\text{where } R = \text{specific gas constant} = \frac{\hat{R}}{\hat{M}} \doteq \frac{kJ}{kg \cdot K}\right) \tag{1.4c}$$

$$pv = RT \quad \left(\text{where } v = \text{specific volume} = \frac{\mathcal{V}}{m} = \frac{\mathcal{V}}{n \cdot \hat{M}} \doteq \frac{m^3}{kg}\right) \tag{1.4d}$$

$$\dot{m} = \frac{P\hat{M}}{\hat{R}T}\dot{\mathcal{V}} \tag{1.4e}$$

For an ideal gas density change, as it flows through a piece of equipment:

$$\rho_2 = \rho_1 \cdot \left(\frac{P_2}{P_1} \cdot \frac{T_1}{T_2}\right) \tag{1.5}$$

The universal gas constant can be presented in a variety of different units; students should be comfortable converting values back and forth from the familiar to the unfamiliar, to ensure their calculations are accurate.

$$\hat{R} = 8.314\,46\ \frac{kJ}{kmol \cdot K} \qquad \hat{R} \approx 8.314\ \frac{J}{mol \cdot K} \qquad \hat{R} \approx 0.082\,05\ \frac{L \cdot atm}{mol \cdot K}$$

$$\hat{R} \approx 1545.2\ \frac{ft \cdot lb_f}{lbmol \cdot R} \qquad \hat{R} \approx 10.731\ \frac{ft^3 \cdot psi}{lbmol \cdot R} \qquad \hat{R} \approx 1.986\ \frac{BTU}{lbmol \cdot R}$$

1.9 History of Temperature

Think Stop. Beginning in high school chemistry, students are cautioned repeatedly to remember (i) when absolute temperature units (i.e. Kelvins or Rankines) are mandatory and (ii) when relative temperature units (i.e. degrees Celsius or degrees Fahrenheit) are permissible. This distinction should be indelibly imprinted in the memory banks of third- and fourth-year engineering students, because misuse of relative temperature units (e.g. in the Stefan–Boltzmann equation, where temperature is raised to the fourth power) gives horribly wrong answers.

Table 1.3 provides brief bits of background about how the conceptualization of temperature has varied throughout the history of science. From the time of Aristotle to the time of Charles, temperature was a widely misunderstood thermodynamic variable.

After the invention of the mercury thermometer by Fahrenheit, scientists found that the choice of liquid (e.g. alcohol vs. water or mercury) gave vastly different scale results for the same temperature in the same capillary tube. Seeking a more universal relationship, Charles invented the gas thermometer (essentially a glass tube with a movable plug or piston) and found the gas volume varied linearly with what he called the "adjusted" temperature $T_{adj} = T(°C) + 273.15$.

Table 1.3 A timeline of temperature.

Year	Scientist	Result	Year	Scientist	Result
335 BCE	Aristotle	"Heat" and "cold" are active powers	1821	Seebeck	Bi-Sb thermocouple
1710	Fahrenheit	Mercury thermometer	1824	Carnot	Two-temperature engine
1741	Celsius	Inverted scale 100 to 0	1848	Lord Kelvin	Absolute temperature
1782	Wedgewood	Pyrometric stones	1853	Rankine	Zeroth law of thermodynamics
1800	Herschel	Infrared thermal radiation	1884	Stefan, Boltzmann	Radiation 4th power temperature
1802	Charles	Gas thermometer V, T proportional	1901	Planck	Wavelength, flux, power relationship

Source: Martin Thermal Engineering Inc.

The ideal gas law was derived shortly thereafter by combining Charles's law, Boyle's law, and Avogadro's law and defining the universal gas constant. Referring to Charles's law, it should be obvious that using degrees Celsius instead of Kelvins will give a horribly wrong answer for a 1.0 *mol* of an ideal gas at 1.0 *atm*, as shown in the given example.

$$\textit{Charles's Law}: \frac{V_2}{V_1} = \frac{T_2(K)}{T_1(K)} \quad or \quad \frac{44.8\,L}{22.4\,L} = \frac{546.30\,K}{273.15\,K}\left[=\frac{2}{1}, good\right]$$

$$\textit{Incorrect}: \frac{V_2}{V_1} \neq \frac{T_2(°C)}{T_1(°C)} \quad or \quad \frac{44.8\,L}{22.4\,L} \neq \frac{273.15°C}{0.0°C}\ [\rightarrow \infty, bad]$$

1.10 Thermodynamic States

The concept of a thermodynamic ***state*** is related to the concept of reproducibility. For example, a photocopy (also called a graphic reproduction) may be "functionally identical" to the original document, but it is not "materially identical." Functional identicality relates to "copied" specimens delivering identical effects in some global or macroscopic sense. Material identicality is impossible to achieve if individual atoms and molecules can be identified. From a thermodynamics perspective, a kilomole of gaseous helium atoms in container A is functionally identical to a kilomole of helium atoms in container B if the two gas samples have the same pressure, temperature, and specific volume. For a given collection of helium atoms, the ***state*** of the helium can be changed by changing its thermodynamic variables. Conversely, the ***state*** of the helium can be reproduced by restoring its thermodynamic parameters to their original values.

Josiah Gibbs postulated (Gibbs 1878) that the ***state*** of any collection of molecules in mutual equilibrium could be reproduced by reproducing a fixed number of thermodynamic variables. Because chemical equilibrium was postulated along with mechanical and thermal equilibria, the thermodynamic variables used to track the conditions that would establish the equilibrium were pressure p (mechanical equilibrium), temperature T (thermal equilibrium) and chemical potential for each species μ_i (chemical equilibrium).

A corollary of Gibbs rule is that the entire superset of thermodynamic variables includes other properties (e.g. specific volume, entropy, enthalpy, and internal energy) that are individually unique and meaningful in their own context, but they are dependent on the independent variables (i.e. those controlled by the experimenter). We will discuss the Gibbs phase rule further in Chapter 6.

1.11 Internal Energy, Enthalpy, Entropy

Values for the thermodynamic parameters, enthalpy (h) and internal energy (u) are often found in tables or diagrams, as functions of temperature and pressure (T and p) or temperature and specific volume (T and v). However, these parameters can also be determined by equations of state. For example, if a pure substance is present in a single phase (e.g. helium gas), it has two degrees of freedom. Suppose it is desired to compute the internal energy of helium as a function of temperature and specific volume [i.e. $u = f(T, v)$], the following differential equation (Equation 1.6) could be written:

$$du = \left.\frac{\partial u}{\partial T}\right]_{v = const} dT + \left.\frac{\partial u}{\partial v}\right]_{T = const} dv \tag{1.6}$$

The first partial derivative has a familiar name, specific heat at constant volume (see Equation 1.7):

$$c_v \equiv \left.\frac{\partial u}{\partial T}\right]_{v = const} \tag{1.7}$$

For ideal gases, the partial derivative in the second term was shown experimentally by Joule to be zero, so that *ideal gas* internal energy (u) is a function of temperature (T) only, not specific volume (v). We note here that the ideal-gas-expansion effect observed by Joule should not be confused with the Joule–Thomson effect, which only applies to real gases, and where the second term in the differential equation is decidedly not zero.

Similarly, if the helium's enthalpy is desired to be computed as a function of the two degrees of freedom temperature and pressure [i.e. $h = f(T, p)$], the general differential equation would be written as shown in Equation (1.8):

$$dh = \left.\frac{\partial h}{\partial T}\right]_p dT + \left.\frac{\partial h}{\partial p}\right]_T dp \tag{1.8}$$

The standard convention is used, wherein the variable being held constant is called out in a subscript outside a half-bracket adjacent to the partial derivative.

Since ideal gas internal energy depends only on temperature (not specific volume), one may be tempted to conclude that ideal gas enthalpy depends only on temperature (not pressure). This is confirmed by relying on (i) the relationship between enthalpy and internal energy, (ii) the ideal gas law, and (iii) Joule's experiment, as follows:

$$h = u + pv; \qquad dh = du + d(pv); \qquad dh = du + d(RT) \qquad \text{(1.9a-c) \{IG\}}$$

$$pv = RT \qquad \text{(1.9d) \{IG\}}$$

$$c_p dT = c_v dT + R dT; \qquad c_p = c_v + R \tag{1.9e, f) \{IG\}$$

Thus, for an ideal gas, the second term in Equation (1.8) is zero and the first term is the familiar specific heat at constant pressure (see Equation 1.10)

$$c_p \equiv \left.\frac{\partial h}{\partial T}\right]_p \tag{1.10}$$

A useful relationship for entropy (see Equations 1.11, 1.12, and 1.13) can be derived from either of the Gibbs equations and the *ideal gas law* (discussed further in Chapter 6):

$$dh = Tds + vdp; \qquad \frac{dh}{T} = ds + \frac{v}{T}dp \tag{1.11a, b}$$

$$\underbrace{ds}_{\substack{entropy\\ change}} = \underbrace{\frac{dh}{T}}_{\substack{entropy\\ change\ at\\ const\ p}} - \underbrace{\frac{R}{p}dp}_{\substack{entropy\\ change\ at\\ const\ T}} = c_p\frac{dT}{T} - R\frac{dp}{p} \tag{1.12) \{IG\}$$

$$\int_1^2 ds = \int_1^2 c_p\frac{dT}{T} - \int_1^2 R\frac{dp}{p}$$

$$\boxed{s_2 - s_1 = c_p\, ln\frac{T_2}{T_1} - R ln\frac{p_2}{p_1}} \tag{1.13) \{IG\}$$

Equation (1.13) can be used to determine the change in entropy between two states of an *ideal gas* by summing two terms – a term representing the entropy change at *constant pressure* and a term representing the entropy change at *constant temperature.* Since the thermodynamics of states are not path dependent, the order of performing the two processes is not important. Thus, Equation (1.13) will be seen as useful for "availability" problems described in the next section.

Example 1.2

Using Equation (1.13), illustrate how the pair of processes (one isothermal and the other isobaric) can lead to an isentropic outcome by plotting the three curves on a temperature-entropy diagram.

Answer. See Figure 1.2:

Although arriving at a particular ***state*** can be accomplished by many paths, computing the ***amount of work or heat*** crossing the boundary is path (or process) dependent. For computations of the quantity of work or heat crossing a system boundary, the integration must be carried out along the actual path followed by the working fluid. For the example shown in Figure 1.2, process 1 to 2 illustrates heat addition at constant pressure and process 2 to 3 illustrates compression at constant temperature (which necessitates a smaller amount of heat removal). By contrast, process 1 to 3 illustrates isentropic compression, which is an entirely adiabatic process (i.e. zero heat crosses the boundary) if the compression is carried out reversibly. The path through state 2 involves heat flow (first in, then out) across the boundary, whereas the path that avoids state 2 involves no heat transfer whatsoever.

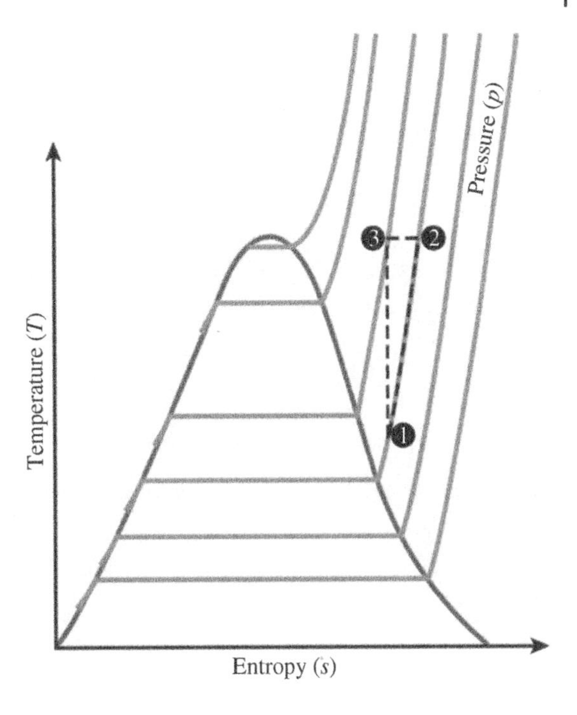

Figure 1.2 Path independence of state variable entropy. Combining an isobaric (heat addition) process 1 to 2 with an isothermal (compression, heat removal) process 2 to 3 leads to same final state value as an isentropic (adiabatic, reversible compression) process 1 to 3. *Source:* Martin Thermal Engineering Inc.

1.12 Availability (Exergy)

Consider a hypothetical control mass in a transient process where heat crosses the boundary at the temperature of the surroundings (T_0) and work leaves the system by a shaft. By definition, no mass is permitted to leave the control mass. It is desired to quantify the available energy in the control volume as it transitions to the "dead state" (i.e. ambient pressure and temperature). In the First law, the production of energy is always zero, but in the Second law, the production of entropy is zero only if all processes are reversible. Hence, the entropy production term is retained in the Second law.

Writing the First law and Second law for a control mass (see Equations 1.14; where subscripts i and f represent the initial and final state of the control mass):

$$\dot{Q}_{in} + [m_i u_i]_{CM} = \dot{W}_{out} + [m_f u_f]_{CM} \tag{1.14a}$$

$$\dot{\mathcal{P}}_s + \frac{\dot{Q}_{in}}{T_0} + [m_i s_i]_{CM} = [m_f s_f]_{CM} \tag{1.14b}$$

Combining Equations (1.14) together by eliminating the heat flow $(\dot{Q})$, and further setting entropy production $(\dot{\mathcal{P}}_s)$ to zero for reversible processes, gives Equation (1.15) for exergy (Ψ) of a control mass.

$$\boxed{\Psi = [m(u_i - T_0 s_i) - m(u_0 - T_0 s_0)]_{CM}} \tag{1.15}$$

The value of available energy or exergy is a function of both the initial state and the dead state.

Example 1.3

Find the availability of 1.0 kg of steam whose initial state is $p_1 = 200\,000\ Pa$, $T_1 = 600.0\ K$, if the dead state is taken to be $p_0 = 101\ 325\ Pa$, $T_0 = 373.13\ K$.

Answer. Find thermodynamic properties of initial state and dead state.

Using the property table in Appendix A.4 for H_2O, the following are obtained for the initial state: $v_1 = 1.379\ m^3/kg$; $h_1 = 3126.6\ kJ/kg$; $s_1 = 7.9871\ kJ/(kg \cdot K)$; $u_1 = 2850.8\ kJ/kg$; (Confirm $u = h - pv = 3126.6 - 1.379 \cdot 200 = 2850.8\ kJ/kg$ ok).

For the dead state: $v_0 = 1.6731\ m^3/kg$; $h_0 = 2675.5\ kJ/kg$; $s_0 = 7.3544\ kJ/(kg \cdot K)$; $u_0 = h_0 - p_0 v_0 = 2675.5 - 1.6731 \cdot 101.325 = 2506.0\ kJ/kg$.

Note, steam is not an ideal gas at either state, which means $pv \neq RT$, so computing $u \approx h - RT$ gives a different value than the exact expression $u = h - pv$. Students should confirm this inequality for themselves as an exercise to refine their intuition about gas conditions that may produce real gas behavior versus ideal gas behavior.

The final answer to the problem is obtained by plugging in these thermodynamic properties into Equation (1.15) to obtain $\Psi = m[(u_i - T_0 s_i) - (u_0 - T_0 s_0)] = 1.0[(2850.8 - 373.13 \cdot 7.9871) - (2506.0 - 373.13 \cdot 7.3544)] = \boxed{108.7\ kJ}$ of exergy for the one kilogram of steam evaluated.

1.13 Homework Problems

1.1 Calculate the change in specific enthalpy of sea level, ambient air heated from $T_i = 300\ K$ to $T_f = 800\ K$.

Answer. 521.3 kJ/kg

1.2 Determine the final temperature of 1.0 kg of liquid water (initially at $p = 101.33\ kPa$ and $T = 298.15\ K$) that undergoes heating at an average rate of $\dot{Q}_{avg} = 17.0\ W$ for a duration $t = 1200\ s$ in a vessel where pressure remains constant.

1.3 For a control mass $m = 1.0\ kg$, in a piston-cylinder apparatus, determine the net input of work, the net input of heat, and the net increase in internal energy from state 1 to state 3. The working fluid is steam, and the conditions are $T_1 = 600\ K$; $T_2 = T_3 = 800\ K$; $p_1 = p_2 = 10.0\ kPa$; $p_3 = 36.5\ kPa$. Assume ideal gas behavior with $\hat{M} = 18.015\ kg/kmol \cdot K$.

Answer. $W_{13,\ net} = +\ 385.8\ kJ$; $Q_{13,\ net} = -\ 61.8\ kJ$; $\Delta U_{13} = +\ 324.0\ kJ$

1.4 A turbine delivers $\dot{W} = 300.0\ kW$ of shaft power to a generator. The inlet stream to the turbine is air at $p_{in} = 1000\ kPa$ and $T_{in} = 600\ K$. The outlet pressure is $p_{out} = 200.0\ kPa$. Determine the mass flow rate $\dot{m} \doteq kg/s$ and the outlet air temperature $T_{out} \doteq K$, if the expansion process is isentropic.

1.5 An engineer claims that she has invented a system that operates according to the ideal Rankine cycle with maximum and minimum temperatures (associated with warm surface water versus colder deep water) equal to $T_h = 26.85°C$ and $T_c = 1.85°C$. The cycle is said to have a net output of shaft power $\dot{W}_{net} = 50\ kW$, and a working fluid flow rate of $\dot{m}_{H_2O} = 1.076\ kg/s$ of (pure, non-ocean) water. Assume State 4 is at saturated vapor conditions and State 1 is at saturated liquid conditions at the cold temperature reservoir. Also assume the two shaft power machines are isentropic (i.e. $s_3 = s_4$ and $s_2 = s_1$).

Answer. No. The Second law is satisfied and the net power out is just right for the mass flow rate.

1.6 A power plant burns natural gas ($\Delta H_{comb} = 49\ 000\ kJ/kg$) at the rate of $\dot{m}_{gas} = 3.5\ kg/s$ and produces electricity at the rate of $\dot{W}_e = 40\ 500\ kW$. Determine the overall cycle efficiency, $\eta_{overall}$.

1.7 Which fluid has greater availability, 1.0 kg of dry air at $T_i = 1000\ K$ and $p_i = 2000\ kPa$ or 1.0 kg of carbon dioxide at $T_i = 1000\ K$ and $p_i = 2000\ kPa$? Assume the dead state is STP.

Answer. Air has greater availability. $\Psi_{air} = 424.5\ kJ/kg$; $\Psi_{CO_2} = 417.0\ kJ/kg$

1.8 3.5 kg of helium is contained in a compressed gas cylinder at $p_i = 1100\ kPa$ and $T_i = 298.15\ K$. Compute the maximum amount of work that helium can do. Assume the dead state is STP.

1.9 A pump collects water ($\rho = 1000\ kg/m^3$) from the top of one reservoir and pumps it uphill to the top of another reservoir with an elevation change of $\Delta z = 800\ m$. The work per unit mass delivered by an electric motor to the shaft of the pump is $W_p = 8200\ kJ/kg$. Determine the percent irreversibility associated with the pump.

Answer. $(1 - \eta_p) = 4.32\%$

Cited References

Abdulagatov, I.M., Rabinovich, V.A., and Dvoryanchikov, V.I.; (1999); *Thermodynamic Properties of Fluids and Fluid Mixtures*; Danbury, CT: Begell House; http://www.begellhouse.com/books/thermodynamic-properties-of-fluids-and-fluid-mixtures.html

Chase, Jr., M.W. (1998). NIST-JANAF Thermochemical Tables, 4e. Monograph No. 9, Vol. I, Introduction: https://janaf.nist.gov/pdf/JANAF-FourthEd-1998-1Vol1-Intro.pdf, Individual compounds by element: https://janaf.nist.gov/janaf4pdf.html.

Gibbs, J.; (1878); "On the equilibrium of heterogeneous substances"; *Am. J. Sci.*; s3–16; **441**.

Linstrom, P.J. and Mallard, W.G., Eds.; (2017-2021); *NIST Chemistry WebBook, SRD 69*; "Thermophysical Properties of Fluid Systems"; National Institute of Standards and Technology, U.S. Dept. of Commerce; http://webbook.nist.gov/chemistry/fluid

Moran, M.J, Shapiro, H.N., Boettner, D.D., and Bailey, M.B.; (2018); *Fundamentals of Engineering Thermodynamics*, 9; Hoboken: Wiley.

Reynolds, Osborne; (1903); *Papers on Mechanical and Physical Subjects. Vol. 3, The Sub-Mechanics of the Universe*. Cambridge: Cambridge University Press.

Reynolds, William C.; (1979); *Thermodynamic Properties in SI*; Stanford University Mechanical Engineering Department

Stull, D.R. and Prophet, H.; (1971); *JANAF Thermochemical Tables*, 2; NBS 37; National Bureau of Standards; https://nvlpubs.nist.gov/nistpubs/Legacy/NSRDS/nbsnsrds37.pdf

White, F.M.; (2016); *Fluid Mechanics* 8th Edition; New York: McGraw-Hill.

2

Fluid Mechanics

A popular synonym for "thermal systems" is "thermal/fluid systems" because fluids are almost always a primary component of thermal systems.

This chapter summarizes many basic concepts from elementary fluid mechanics to provide the reader with the fundamentals required to design fluid conveying systems. Topics covered include viscosity and shear stress; buoyancy; conservation of mass; interconversion between mass, moles, and volume; Reynolds number; Bernoulli's equation; friction factor; Moody charts; and flow in porous media. Chapter 12 provides further details about how to size piping and equipment using these concepts.

2.1 Viscosity, Shear, Velocity

Internal flow of fluids in pipes and ducts involves fluid motion. All internal flows develop profiles where forward velocity varies with radius and the fluid molecules are found to have zero average velocity at the wall (the so-called no-slip condition) and higher average velocities away from the wall, with a maximum at the center of the pipe or duct. Turbulent flows (to be discussed later in this chapter) have velocity profiles that are shaped differently than laminar flows, but both have zero velocity at the wall and maximum velocity at the centerline.

When fluid flows through a pipe, shear forces are exerted on fluid elements because of the no-slip condition (see Figure 2.1). From its own reference frame, the element sees higher-velocity fluid dragging it forward along the element surface closer to the centerline and lower-velocity fluid dragging it backward along the element surface closer to the wall.

Experiments on Newtonian fluids show that the velocity gradient in the y-direction (i.e. starting at the wall and moving radially toward the center) is proportional to the shear stress, and the proportionality constant is the fluid's dynamic viscosity μ, per Equation (2.1). By definition, a Newtonian fluid is one that obeys Equation (2.1). Non-Newtonian fluids behave according to a *nonlinear* relationship between shear force and velocity gradient.

$$\underbrace{\tau}_{\frac{N}{m^2}} = \underbrace{\mu}_{\frac{kg}{m \cdot s}} \underbrace{\frac{\partial u}{\partial y}}_{s^{-1}} \tag{2.1}$$

The no-slip condition is a slightly nonintuitive result of the vast number of fluid molecules that collide with the wall elastically and come into mechanical and thermal equilibrium with it. If one thinks of a billiard ball rebounding elastically off the billiard table's cushion, the ball's velocity

Thermal Systems Design: Fundamentals and Projects, Second Edition. Richard J. Martin.

Companion website: www.wiley.com\go\Martin\ThermalSystemsDesign2

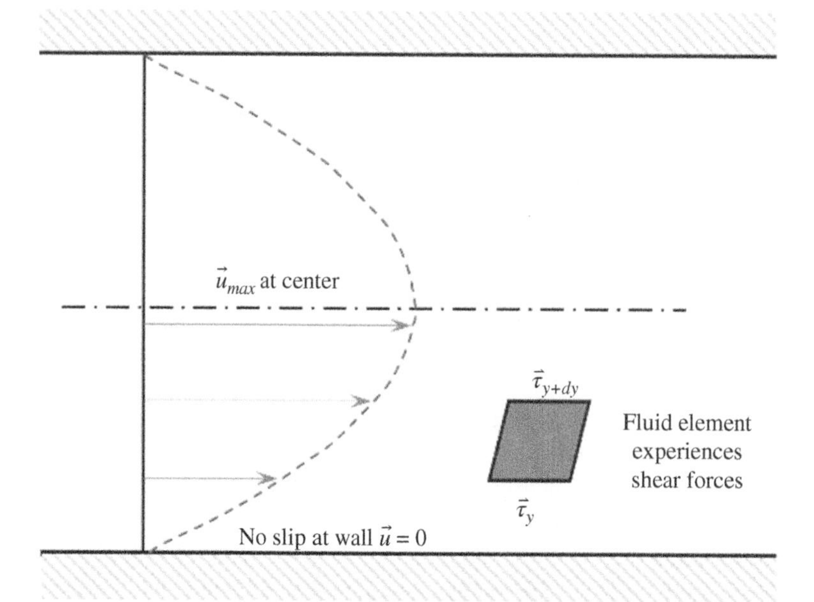

Figure 2.1 Illustration of fluid element under the influence of shear forces arising from a velocity gradient that results from the no-slip condition at a solid wall boundary and a bulk-flow or free-stream velocity imposed at some distance away from the wall. *Source:* Martin Thermal Engineering Inc.

component in the direction parallel to the wall is unchanged after the collision. If one thinks of a stationary fluid contacting a stationary wall, with a large ensemble of fluid molecules colliding and rebounding there with an extraordinary number of times per second, the fluid's average (bulk) velocity is zero, but the average speed of the individual molecules (i.e. the mean value of velocity magnitude, ignoring velocity direction) is very high (hundreds of meters per second at ambient temperatures). This is the fluid's *sound speed.* If one finally supposes that the wall begins moving in a direction parallel to its surface (i.e. exerting a shear force on the adjacent fluid), and one assumes that the wall's speed is slow relative to the sound speed of the molecules, one reaches a slightly nonintuitive conclusion. By vector analysis, one can show that because the collisions with the wall are elastic, the pre- and postcollision velocities still exhibit the same lateral velocity component, and when averaged over the entire ensemble of molecules, the net velocity of the colliding molecules relative to the moving wall is very nearly equal to the wall's velocity.

This *no-slip* analysis is valid whenever the fluid can be considered as a continuum (i.e. a small Knudsen number) and when the wall's speed relative to the fluid is slow relative to the molecular average speed (i.e. a small Mach number). Fluid flows that aren't characterized by these two conditions may exhibit *slip* behavior that should not be ignored. Fortunately, the large majority of important engineering flows can be considered to have *no-slip* behavior at solid wall boundaries.

2.2 Hydrostatics, Buoyancy

Archimedes first observed that objects immersed in water seemed lighter than objects surrounded by ambient air. His experiments led him to conclude that the *buoyancy force* exerted on an immersed object was equal to the weight of the water displaced by the object's immersion.

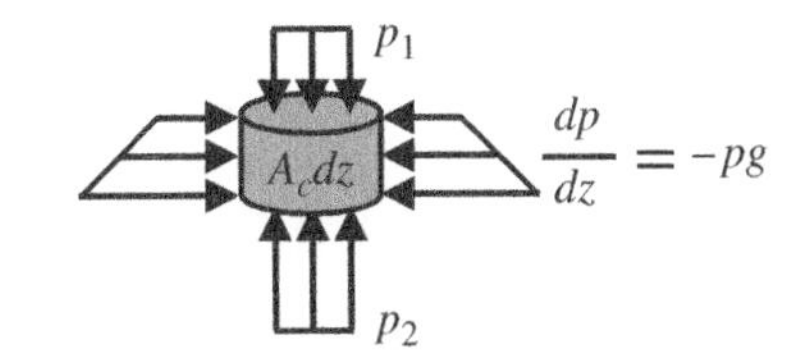

Figure 2.2 Pressure forces exerted on surfaces of a submerged, vertical cylinder. *Source:* Martin Thermal Engineering Inc.

What Archimedes didn't realize was that objects are also *immersed* in air, and the air also exerts a buoyancy force on immersed objects equal to the weight of the air displaced (which is very small compared to the weights associated with displaced liquids such as water or mercury).

Buoyancy forces are important in determining natural convection heat transfer rates and flow rates of heated gases in chimneys. These phenomena will be discussed later in this book after the relationship between the hydrostatic forces on an object's surfaces and the net buoyancy force developed is established.

The hydrostatic pressure of a fluid at a point in space is equal to the weight per unit area of the fluid column above it. In Figure 2.2, which is a free-body diagram of an infinitesimally short (height dz) cylindrical element with fluid pressure acting on top and bottom circular areas, it is seen that $dp/dz = -\rho g$. If constant fluid density is assumed over a column height of practical interest, then $(p_2 - p_1) = -\rho g(z_2 - z_1)$. The buoyancy force discovered by Archimedes is a result of the z-direction pressure gradient.

Consider our *vertical* cylinder immersed in water. The net *horizontal* force due to the water pressure is zero everywhere because equal and opposite fluid pressures exist all around the circumference of the side walls of the cylinder. However, as seen in Figure 2.2, the pressure force p_1 on the upper surface is smaller than the pressure force p_2 on the bottom surface. This is because the bottom surface must also support the weight of the cylinder itself in addition to the force attributable to p_1. This difference in pressure is exactly equal to (the density of the fluid) times (the gravitational acceleration) times (the elevation difference) or $(p_2 - p_1) = \rho g(z_1 - z_2)$, which when multiplied by the upper/lower contact area equates to exactly the weight of fluid displaced, as Archimedes showed.

2.3 The Continuity Equation

Introductory thermodynamics and fluid mechanics textbooks typically provide a sufficient foundation in the fundamentals of mass flow and volume flow through pipes and other types of control volumes. The differential-integral equation that mathematically represents fluid flowing through a control volume is the Reynolds transport theorem (Reynolds 1903). The Continuity equation (Equation 2.2) is the simplest form of the Reynolds transport theorem, where the conserved property is the mass of the flowing fluid. Readers will note that while our treatment of the Reynolds transport theorem uses nomenclature consistent with White (2016), the *captions* below each term follow the simplistic "super-dot" format we established in Chapter 1 for production, inflow, outflow, and storage.

$$\underbrace{0}_{\dot{P}_m} + \underbrace{\int_{C.S.} \rho\left(\vec{V}\cdot\vec{n}\right)dA_{in}}_{\dot{m}_{in}} = \underbrace{\int_{C.S.} \rho\left(\vec{V}\cdot\vec{n}\right)dA_{out}}_{\dot{m}_{out}} + \underbrace{\int_{C.V.} \frac{\partial}{\partial t}(\rho)d\mathcal{V}}_{\dot{S}_m} \tag{2.2}$$

The production term ($\dot{\mathcal{P}}_m$) is identically zero for mass because matter can neither be created nor destroyed (assuming nuclear reactions are excluded). The storage term ($\dot{\mathcal{S}}_m$) may be nonzero (i.e. if the control volume conditions are ***transient***), but it becomes zero when the control volume has reached ***steady state*** (i.e. no changes to velocity or density over time). Under *steady* conditions, the *inflow* of mass must equal the *outflow* of mass, based on Equation (2.2). *Unsteady* flow often implies that fluid (mass) is undergoing positive or negative accumulation (storage) inside the control volume; however, accelerating inflow and outflow is another mode of unsteadiness that does not necessarily result in mass accumulation inside any specific control volume.

When a ***flow*** is considered ***incompressible***, changes in density are assumed negligible, such that the density term can be moved outside the integrals and derivatives. ***Incompressible flow*** does not necessarily mean that the fluid is incompressible (i.e. that the fluid's density is literally constant at all locations and all times). Incompressible flow is a class of flows where the Mach number is small enough (typically $M < 0.3$ is sufficient) so that compressibility effects due to high fluid velocity may be ignored. Natural convection flows are a good example where the flow is driven by small changes in density, yet the Mach number is low enough so that the flow can be considered incompressible. As will be seen in Chapter 3, this counterintuitive finding (i.e. that a buoyancy-driven flow can be called ***incompressible*** even though differences in density comprise the driving force for the fluid motion) is related specifically to the sense that changes in ***velocity*** do not affect the density. When the ***kinetic energy*** (per unit mass) of the flow reaches a magnitude comparable to the ***thermal*** and/or ***mechanical energy*** of the fluid (i.e. when $M > 0.3$ and $M^2 > 0.1$), we call the flow compressible because the velocity has substantially altered the temperature, pressure, and density from what they were when the fluid had little or no kinetic energy.

2.4 Mass, Volume, Mole Flows

Thermal Systems Design students should develop a high level of confidence in their ability to convert flow rate parameters back and forth between units of mass, volume, and moles. The basic interconversion equations are presented here (Equations 2.3), and some practical methodology follows.

$$\hat{M}\left[\stackrel{\vee}{=} \frac{kg}{kmol}\right]$$

$$\rho(T,p)\left[\stackrel{\vee}{=} \frac{kg}{m^3}\right] \qquad \rho_{STP}\left(T_{ref}, p_{ref}\right)\left[\stackrel{\vee}{=} \frac{kg}{m^3_{STP}}\left(=\frac{kg}{Nm^3}\right)\right]$$

$$\dot{m}\left[\stackrel{\vee}{=} \frac{kg}{s}\right]$$

$$\boxed{\dot{n} = \frac{\dot{m}}{\hat{M}}}\left[\stackrel{\vee}{=} \frac{kmol}{s}\right] \tag{2.3a}$$

$$\boxed{\dot{V}\left(=\dot{V}_{act}\right) = \frac{\dot{m}}{\rho}}\left[\stackrel{\vee}{=} \frac{m^3}{s}\right] \tag{2.3b}$$

$$\boxed{\dot{V}_{STP} = \frac{\dot{m}}{\rho_{STP}}}\left[\stackrel{\vee}{=} \frac{m^3_{STP}}{s}\left(=\frac{Nm^3}{s}\right)\right] \tag{2.3c}$$

$$\boxed{\dot{m} = \rho A_c V} \tag{2.3d}$$

where A_c is cross-sectional area of the pipe and $A_c = \pi D^2/4$ for a circular cross section.

Interconversion from mass flow to mole flow is carried out readily by multiplication or division by the molar mass $(\hat{M})$. Interconversion from mass flow to either of the two volume flow variables (***actual*** or ***STP*** volume flow) is also relatively straightforward. The key difference is that the ***actual*** density (ρ) is computed (for ideal gases) using $\rho = (p\hat{M})/(\hat{R}T)$, where p and T are the ***actual*** pressure and temperature of the fluid that is flowing. Conversely, the ***STP*** density is computed using the same formula, but the density is adjusted by applying ***STP*** pressure and temperature ($p_{ref} = 101.325\ kPa$, $T_{ref} = 298.15\ K$) instead of the ***actual*** values.

In a sense, reporting the ***STP*** (also called standard or normal) *volumetric flow rate* of a gas is equivalent to reporting its *mole flow* because Avogadro's law asserts that the ***STP*** volume of an ideal gas is directly and uniquely proportional to the number of moles of the gas.

Here is a practical way of intuitively distinguishing the two formats for volume flow. Suppose the ***actual*** gas density is lighter than its ***standard*** density (e.g. the gas's actual temperature is *higher* than its standard temperature, assuming identical pressures). The consequence of this is that for identical mass flows, the ***actual*** volume flow rate $\dot{V}$ must be *higher* than the ***STP*** volume flow rate.

Example 2.1

Determine the volume flow rate (in both *acfm* and *scfm* units) of warm air $(p = 1.00\ atm, T = 300°F, \hat{M} = 28.85\ lb_m/lbmol)$ if its mass flow rate is $\dot{m} = 1000\ lb_m/h$.

Answer. The actual density of the air is $\rho = (p\hat{M})/(\hat{R}T) = (1.00 \cdot 28.85)/(0.7302 \cdot 759.67) = 0.0520\ lb_m/ft^3$, and the density of the air if it were at STP is $\rho = (1.00 \cdot 28.85)/(0.7302 \cdot 536.67) = 0.073\ 62\ lb_m/ft^3$.

These calculations lead to the two computed volume flow rates as: $\dot{V} = 1000/(0.0520 \cdot 60) = 320.6\ ft^3/min = \boxed{320.6\ acfm}$, and $\dot{V}_{STP} = 1000/(0.073\ 62 \cdot 60) = 226.4\ ft^3_{STP}/min = \boxed{226.4\ scfm}$.

One's intuitive understanding about warm (less dense) air, namely that *acfm* ought to be greater than *scfm*, is thus validated.

2.5 Reynolds Number, Velocity Profiles

For virtually all fluid flow problems, computation of Reynolds number is critical to understand the nature of the flow (i.e. ***laminar*** versus ***turbulent***). Thus, it is not surprising that Reynolds number is one of the first dimensionless parameters introduced to engineering students. Reynolds number is often characterized as the ratio of *momentum flux* (numerator) to *viscous stress* (denominator), as shown in Equation (2.4). Students are cautioned to ensure that the units in the numerator are identical to the units in the denominator, as is required for all dimensionless variables.

$$Re_D = \frac{\overbrace{\rho \cdot D \cdot V}^{Density \cdot Diameter \cdot Velocity}}{\underbrace{\mu}_{\substack{Dynamic \\ Viscosity}}} \sim \frac{\overbrace{\rho V^2}^{\substack{Momentum \\ Flux}}}{\underbrace{\mu \frac{V}{D}}_{\substack{Viscous \\ Stress}}} \tag{2.4}$$

Reynolds discovered the threshold where orderly flows became disorderly and quantified the critical value $Re_{D,\,crit} \approx 2300$ for internal (duct or pipe) flows. For low Reynolds numbers, the flows were called ***laminar*** because they were orderly and layered due to viscous forces overwhelming the momentum flux. For high Reynolds numbers, the flows were called ***turbulent*** because they were disorderly and chaotic due to high momentum overpowering low viscosity. Caution is also urged when the flow geometry does not match the condition under which the dimensionless parameter was studied. Reynolds number relations established for gas flow in a circular pipe will not be useful to predict the transition from laminar to turbulent for gas flow between two computer circuit boards having dozens of geometric irregularities.

Most elementary fluid mechanics textbooks include a derivation from first principles that demonstrates why the ***laminar*** velocity profile as a function of pipe radius $u = f(r)$ in circular pipes (both smooth and rough) is parabolic and the maximum velocity (u_{max}) is exactly twice the average velocity (V), as given in Equation (2.5). The designation {L} after the equation number indicates that the equation should only be used for ***laminar*** flow, not turbulent.

$$u(r) = u_{max}\left(1 - \left[\frac{r}{R}\right]^2\right) \tag{2.5) {L}$$

For steady, incompressible ***laminar*** flows, u is only a function of r and the velocity profile does not vary along the axial direction or with time. The shear stress at the wall can be found by differentiating u with respect to r to obtain Equation (2.6):

$$\tau_{w,lam} = \mu \frac{\partial u}{\partial r}\bigg]_{r=R} = -\frac{2\mu \cdot u_{max}}{R} = \boxed{-\frac{8\mu V}{D}} \tag{2.6) {L}$$

For ***turbulent*** internal flows, some authors postulated a "power-law" relationship for the velocity profile inside the pipe, because the turbulent velocity profile is very steep near the wall, but flatter near the centerline. Unfortunately, these power-law approximations often create more problems than they resolve, and students should avoid the temptation to use them. Demonstrably unrealistic physical consequences introduced by the "power-law" velocity profile for turbulent flow include: (i) velocity gradient independent of wall roughness, (ii) discontinuous velocity gradient at pipe centerline (i.e. velocity profile forms a cusp), (iii) velocity gradient infinite at wall, (i.e. shear stress infinite at wall).

Thanks to Prandtl (1925), Von Karman (1930), and Millikan (1938), a better method for approximating the turbulent velocity profile is available.

The primary elements of the theory are: (i) molecular diffusion that dominates momentum transport near the wall (Prandtl's Law of the Wall), (ii) turbulent eddy motion that dominates molecular momentum transport near the pipe centerline (Von Karman's Velocity Defect Law), (iii) a "logarithmic overlap" formula that accounts for both effects in the intermediate region, where neither dominates (Millikan's logarithmic overlap layer).

The theory is simplified further by the realization that for smooth pipes, the logarithmic law can be extended (with due care) into the two adjoining regions, with nearly insignificant error affecting the parameters of practical interest (average velocity V, maximum velocity u_{max}, and wall shear stress τ_w). Readers who would like a more detailed study of the turbulent velocity profile are referred to sections 6.5 and 6.6 of White (2016).

For the purpose of *plotting* the turbulent velocity profile in a pipe, the more accurate equations for the wall region and the outer region can be ignored, and we adopt the logarithmic overlap velocity profile over the entire radial domain for fully developed turbulent flow in a smooth pipe. Millikan's law is presented here as Equation (2.7), and the designation {T} after the equation number indicates that the equation applies to ***turbulent*** flow only.

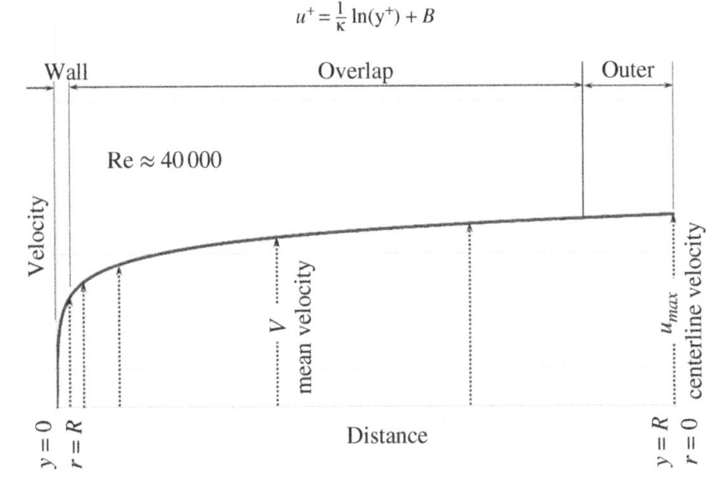

Figure 2.3 General shape of turbulent boundary layer. Wall region is where molecular viscosity dominates, and slope is very steep; overlap region is where molecular viscosity and eddy diffusivity are both important; outer region is where turbulent eddy diffusivity dominates, and slope is very shallow. Relative sizes of regions are not universal. Sizes shown may be relatively larger or smaller for other fluid and flow parameters. *Source:* Martin Thermal Engineering Inc.

$$u^+ = \frac{1}{\kappa} \ln(y^+) + B \tag{2.7) \{T\}$$

Here, the dimensionless parameters are defined as

$$u^+ \equiv u(r)/u^* \qquad y^+ \equiv (R - r)u^*/\nu$$

$$\kappa = 0.41 \qquad B = 5.0$$

and a fictitious velocity parameter (not dimensionless) is defined as

$$u^* \equiv (\tau_w/\rho)^{1/2} \; [\doteq m/s]$$

See Figure 2.3 for a plot of the Logarithmic Overlap Law.

The values for dimensionless parameters κ and B were determined experimentally, and they cover a wide range of turbulent Reynolds numbers. The fictitious velocity u^* has units of velocity, but it is not a physically real velocity in the context of the turbulent boundary layer.

As long as the fluid's viscosity and density, the pipe's diameter, and one of the three "practical parameters" (V, u_{max}, τ_w) are known, the others can be accurately derived from Equation (2.7).

Probably the most important relationship of this analysis is that the wall shear stress is determinable from the other parameters – even though the velocity profile at the wall is not accurately represented by the Logarithmic Overlap Law. If the velocity profile near the wall is known (e.g. via Prandtl's Law of the Wall), the wall shear stress can be determined from Equation (2.1). Unfortunately, attempting to estimate velocity gradient at the wall by using Equation (2.7) is *highly inaccurate*, because for small values of y^+ (i.e. $0 \leq y^+ \leq 5$) the molecular viscosity dominates the momentum transport, and Equation (2.7) badly overpredicts the velocity gradient at the wall. Fortunately, Equation (2.7) can be used to determine the wall shear stress by another method that does not involve computing velocity gradients.

Suppose the centerline velocity $u_{max}(r = 0)$ is known along with the pipe diameter, the fluid density, and the fluid viscosity. If $r = 0$ is entered into the expression for y^+ in Equation (2.7), the fictitious velocity u^* can be determined by guessing an initial value for u^* and iterating on u^* until the left side of Equation (2.7) equals the right side, within an acceptably small error.

Example 2.2

For the following fluid and flow parameters ($D = 0.14\ m$, $\nu = 1.5141 \times 10^{-5}\ m^2/s$, $u_{max} = 5.0\ m/s$), determine the fictitious velocity and the wall shear stress. For additional details, see example 6.5 from White (2016).

Answer. The fictitious velocity is determined by iteration to be $u^* = 0.227\ 563\ m/s$. Once the fictitious velocity is computed, the wall shear stress τ_w is found directly from ρ and u^*.

Similarly, once u^* is known, Equation (2.7) can be integrated analytically to obtain an expression for the average velocity V (see Equation 2.8), where the local velocity $u(r)$ is obtained from Equation (2.7):

$$V = \frac{\dot{\mathcal{V}}}{A_c} = \frac{\int_0^R u(r) 2\pi r \cdot dr}{\pi R^2} = \boxed{\frac{u^*}{2}\left[\frac{2}{\kappa}\ln\left(\frac{Ru^*}{\nu}\right) + 2B - \frac{3}{\kappa}\right]} \qquad (2.8)\ \{T\}$$

For the example parameters given above, average velocity is computed to be $V = 4.167\ m/s$, and the Reynolds number $Re = VD/\nu = 38\ 500$ is most definitely turbulent. The ratio of centerline velocity $u_{cl}(r = 0) = 5.0\ m/s$ to average velocity is approximately 120% at this Reynolds number.

In fact, the ratio of centerline velocity to average velocity is a declining function of increasing Reynolds number, as seen in Figure 2.4.

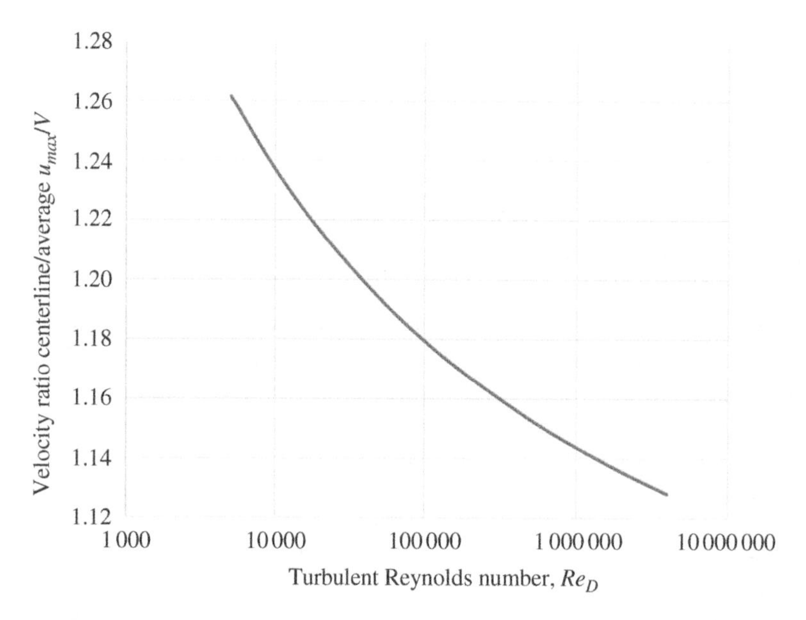

Figure 2.4 Ratio of centerline velocity to average velocity as a function of Reynolds number in turbulent pipe flow. *Source:* Martin Thermal Engineering Inc.

Again, we urge students to memorize the fact that the Logarithmic Overlap law does *not* provide valid results for velocity in the wall region, and thus it would be inappropriate to differentiate Equation (2.7) to find the velocity gradient at the wall. The use of Equation (2.7) across the range $0 < r < R$ is valid only to determine an approximate value for u^*, which indirectly leads to τ_w and $(\partial u/\partial r)_{r=R}$ by using the methods shown here.

2.6 The Momentum Equation

When the Reynolds transport theorem is applied to the momentum of an infinitesimal element of fluid inside the control volume, Equations (2.9a) are obtained. While the form of the momentum equation is similar to Equation (2.2) for mass continuity, the differences are also critically important. We point out that the generic conserved quantity **B** in the transport theorem is now replaced by the fluid momentum $\vec{p}$ (a vector quantity), and we again use the simplistic super-dot variables (i.e. $\dot{\mathcal{P}}_{\vec{p}}$,. $\dot{\vec{p}}_{inflow}$, $\dot{\vec{p}}_{outflow}$, $\dot{\mathcal{S}}_{\vec{p}}$) as parenthetical labels below the terms.

$$\underbrace{\sum \vec{F}}_{\dot{\mathcal{P}}_{\vec{p}}} + \underbrace{\int_{C.S.} \rho\vec{V}\left(\vec{V}\cdot\vec{n}\right)dA_{in}}_{\dot{\vec{p}}_{inflow}} = \underbrace{\int_{C.S.} \rho\vec{V}\left(\vec{V}\cdot\vec{n}\right)dA_{out}}_{\dot{\vec{p}}_{outflow}} + \underbrace{\int_{C.V.} \frac{\partial}{\partial t}\left(\rho\vec{V}\right)d\mathcal{V}}_{\dot{\mathcal{S}}_{\vec{p}}} \tag{2.9a}$$

$$\sum \vec{F} = \sum \vec{F}_{surface} + \sum \vec{F}_{volume} \tag{2.9b}$$

The unique feature of this conservation equation (as compared to the mass equation) is that the ***production*** term, $\dot{\mathcal{P}}_{\vec{p}}$is not necessarily zero. If applied forces exist (acting on infinitesimal areas of the control surface or acting on infinitesimal mass elements inside the control volume), the system (i.e. the fluid inside the control volume) can undergo acceleration. This acceleration is essentially a ***production*** of fluid momentum *ex nihilo*. It can manifest itself in any combination of the six Cartesian directions (up, down, right, left, away, toward), and it adds a new term to the conservation equation so that ***storage*** of momentum $\dot{\mathcal{S}}_{\vec{p}}$ inside the control volume is only equal to ***inflow*** minus ***outflow*** of momentum in the special case of *zero* net applied forces. It is common to replace the generic vector $\vec{p}$ in the convective terms with its respective three-space magnitude values in Cartesian coordinates, along with unit vectors $\vec{i}, \vec{j}, \vec{k}$ in the *x*-, *y*-, and *z*-directions, respectively, but we defer to other authors (e.g. White (2016) or Bergman et al. (2019)) for those details.

2.7 Bernoulli's Equation

Fluids at rest have no kinetic energy – only mechanical energy (in the form of pressure), thermal energy (in the form of temperature), and potential energy (in the form of elevation within a gravitational field). To impart kinetic energy to a stationary fluid (i.e. to cause it to attain a finite bulk velocity in a given direction), the fluid must have work done on it (i.e. from a propeller or impeller attached to a rotating shaft) or it must convert some of its preexisting mechanical, thermal, or potential energy into kinetic energy. The interconversion of energy from one category to another is the essence of Bernoulli's equation.

Bernoulli's equation is typically derived from the momentum conservation equation (with an assumption that effects of friction are small enough to be negligible), but its most intuitive form is as an energy conservation equation. This condition is unrealistic near walls, where friction cannot be ignored, but it is a good assumption along streamlines near the center of ducts and pipes. Bernoulli's equation neither requires the flow to be isothermal nor incompressible, but when applicable, assuming constant fluid density (as we do here) makes the mathematics simpler.

Three formats (***energy***, ***pressure***, ***head***) of the classic Bernoulli's equation are presented next as Equations ((2.10)). The term "classic" is used here as meaning the simplest form of the equation, and one that applies strictly to frictionless flow without shaft work. A "modified" version of the Bernoulli equation (i.e. one that addresses flows undergoing frictional energy losses and shaft-induced energy gains or losses) will be described later in this chapter.

$$\frac{p_1}{\rho} + \frac{V_1^2}{2g_c} + \left[\frac{g}{g_c}\right] z_1 = \frac{p_2}{\rho} + \frac{V_2^2}{2g_c} + \left[\frac{g}{g_c}\right] z_2 \quad \textbf{\textit{Energy}} \left[\doteq \frac{kJ}{kg} \right] \tag{2.10a}$$

$$p_1 + \frac{\rho V_1^2}{2g_c} + \rho \left[\frac{g}{g_c}\right] z_1 = p_2 + \frac{\rho V_2^2}{2g_c} + \rho \left[\frac{g}{g_c}\right] z_2 \quad \textbf{\textit{Pressure}} \left[\doteq kPa \right] \tag{2.10b}$$

$$\frac{p_1}{\rho \left[\frac{g}{g_c}\right]} + \frac{V_1^2}{2g} + z_1 = \frac{p_2}{\rho \left[\frac{g}{g_c}\right]} + \frac{V_2^2}{2g} + z_2 \quad \textbf{\textit{Head}} \left[\doteq m \right] \tag{2.10c}$$

The classic Bernoulli equation applies identically along a streamline. The modified Bernoulli equation applies to pipe flow when velocities are averaged over the cross-sectional area of the pipe.

As written, the Bernoulli equation is most valuable when used to compare two states of the same fluid. The fluid could be stationary (the hydrostatic condition), which drives the kinetic energy term on both sides to zero, and any differences in the elevation terms would be balanced by equivalent differences in the pressure terms. Similarly, if the fluid has bulk motion but no elevation change (e.g. a horizontal, converging, or diverging nozzle), the acceleration of the fluid is "paid for" by a decline in pressure. With the exception of the modified Bernoulli equation, the classic Bernoulli equation is perhaps the most widely useful tool an engineering student can acquire from the study of fluid mechanics.

2.8 Stagnation, Static, Dynamic Pressure

The classic Bernoulli equation can be employed to measure the velocity of a fluid stream by using an instrument called the Pitot-static probe (also called a Prandtl Tube).

The three manometer measurements shown in Figure 2.5 represent the following physical characteristics of the flow:

- Static pressure – the pressure difference (Δp) between the duct pressure (probe transverse to flow direction) and ambient pressure.
- Stagnation pressure – the pressure difference (Δp) between dead-head pressure (probe aligned with flow direction) and ambient pressure.
- Dynamic pressure (also called velocity pressure or $\frac{\rho V^2}{2g_c}$) – the pressure difference (Δp) between the stagnation pressure and the static pressure.

The static pressure is the fluid pressure in the duct where the flow velocity is nonzero (i.e. the pressure remaining after some of the fluid's original mechanical energy was converted to kinetic

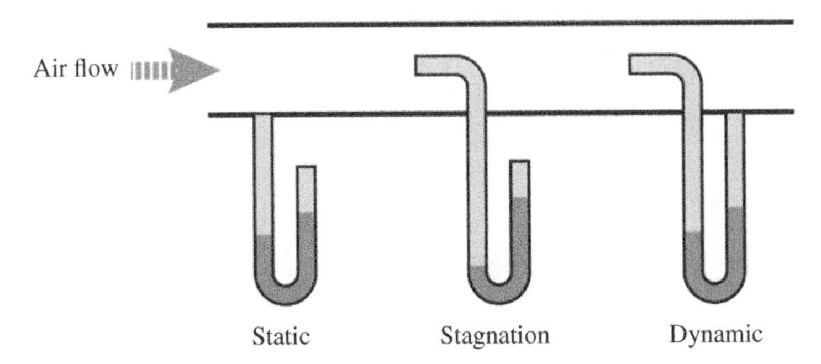

Figure 2.5 Manometers illustrating static, stagnation, and dynamic pressure. *Source:* Martin Thermal Engineering Inc.

energy). The stagnation pressure is the pressure that the flowing fluid would attain if it was decelerated reversibly to zero speed. Alternatively, stagnation pressure can be viewed as the fluid's pressure before it was accelerated (and a portion of its mechanical energy was converted to kinetic energy). Finally, the dynamic pressure is the pressure attained by the kinetic energy alone (without the static pressure component). All three pressures are assumed to be taken at the same elevation, so potential energies are all the same. The flows are assumed to be frictionless, which is a reasonable assumption near the center of the duct.

To determine flow velocity, the pressure format of Bernoulli's Equation (Equation 2.10) is rearranged such that State 1 is where the average fluid velocity is high (static probe) and State 2 is where the average fluid velocity is zero (stagnation probe). The local velocity is thus determined from Equation (2.11):

$$V_1 = \sqrt{\frac{2g_c(p_2 - p_1)}{\rho}} \tag{2.11}$$

If the probe measurement of local velocity (e.g. at the pipe's centerline) can be related to the average velocity in the duct or pipe (i.e. if the laminar or turbulent velocity profile is known), the volume flow rate in the pipe can be determined by multiplying average velocity times cross-sectional area (see Equation (2.12)).

$$\dot{V} = V_{avg} \cdot \pi \frac{D^2}{4} \tag{2.12}$$

For laminar Reynolds numbers, the centerline velocity is exactly twice the average velocity. For low turbulent Reynolds numbers ($Re_D < 5000$), the centerline velocity is approximately 25% higher than the average velocity. For high turbulent Reynolds numbers ($Re_D > 1 \times 10^8$), the centerline velocity is approximately 10% higher than the average velocity. See also Figure 2.4 for velocity comparisons in turbulent flows.

2.9 Friction Factor, Hydraulic Diameter

For flows where friction cannot be ignored (essentially all duct and pipe flows), mechanical energy is dissipated to thermal energy in adherence with the Second law of thermodynamics. The simplest way of understanding this is by drawing a free-body diagram for the fluid inside a horizontal pipe

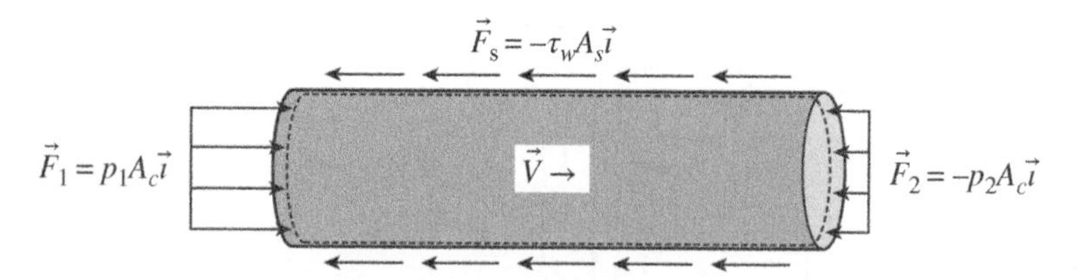

Figure 2.6 Free-body diagram for horizontal flow in a circular pipe. *Source:* Martin Thermal Engineering Inc.

with constant cross section, under steady, incompressible flow conditions (see Figure 2.6). In this example, body forces (e.g. gravity) are zero in the direction of flow (horizontal) so that only surface forces (pressure and shear) are relevant.

For the flow to be steady, there can be no acceleration in time. The incompressible and constant area assumptions imply that there is no acceleration in space either. Hence, the net production term must be zero (i.e. all the forces on the left side of the momentum equation must balance out to a net zero horizontal force). The sole force acting in the forward x–direction is the upstream pressure force arising from the fluid just outside the (imaginary) control volume boundary pressing forward onto the fluid just inside the control volume. This must be balanced by the other two surface forces, namely the downstream pressure force and the shear force along the wall (as shown in Equation 2.13). Because both pressure and shear have units of force per unit area, the force balance must reflect the areas that these forces are acting upon.

$$(p_1 - p_2)\frac{\pi D^2}{4} = \tau_w \cdot \pi DL$$

$$\boxed{(p_1 - p_2) = 4\frac{L}{D}\tau_w} \tag{2.13}$$

The effect of viscosity is nondimensionalized by the Darcy friction factor f, which is defined in Equation (2.14):

$$\boxed{f \equiv \frac{8\tau_w}{\rho V^2}} = 4\left(\frac{\overbrace{\tau_w}^{\text{wall shear stress}}}{\underbrace{\frac{\rho V^2}{2}}_{\text{velocity pressure}}}\right) \tag{2.14}$$

The pressure drop due to friction in a straight pipe run is thus quantified by combining these two expressions (by eliminating τ_w) into what is known as the Darcy–Weisbach Equation (Equation 2.15):

$$(p_1 - p_2)_{friction} = f\frac{L}{D}\frac{\rho V^2}{2g_c} \tag{2.15}$$

For laminar flows in circular cross-section ducts, friction factor is only a function of the Reynolds number per Equation (2.16):

$$\boxed{f_{lam} = \frac{64}{Re_D}} \tag{2.16}$$

For turbulent flow in ducts, the friction factor (f_{turb}) is a function of both the Reynolds number and the relative roughness (ε/D) of the pipe wall. The relationship is frequently plotted on what is called the Moody Chart as discussed in Section 2.9.

For noncircular flow cross sections, the traditional diameter is replaced by the "hydraulic diameter" (Equation 2.17), where $P(\doteq m)$ is the wetted perimeter and $A_c(\doteq m^2)$ is the cross-sectional area of the noncircular tube.

$$\boxed{D_h = \frac{4A_c}{P}} \tag{2.17}$$

When addressing an annulus between concentric circular tubes, the formula above leads to $D_{h,\ annulus} = D_o - D_i$. For a shell-and-tube heat exchanger, the Reynolds number of the flow inside each small tube (called the "tube side") is computed with the inner diameter of those tubes. Conversely, the Reynolds number of the flow on the "shell side" (the annular area surrounding the outside of all the small tubes) is computed starting from the above formula to obtain Equation (2.18). It should be noted that for shell-and-tube heat exchangers with baffles where both flows (shell side and tube side) are not predominantly axial, a very different approach is required compared to the approach above where baffles are assumed absent. Other resources should be consulted for friction losses with baffles and other complex geometries.

$$D_{h,shellside} = \frac{4(A_{i,shell} - n \cdot A_{o,tube})}{(n \cdot P_{o,tube} + P_{i,shell})} \tag{2.18}$$

The relationships between friction factor, wall shear stress, and velocity gradient for both laminar and turbulent flows are summarized in Table 2.1. Some expressions are applicable to both laminar and turbulent flows, whereas others are specific to either the laminar regime or the turbulent regime.

2.10 Moody Chart, Chen Equation

Thermal engineering students have no doubt seen the Moody chart (Moody 1944) on multiple occasions, as it is one of the most widely used charts among engineers of all disciplines. Moody chart became popular instantly with piping engineers and others desiring a visual tool for estimating friction factor. See Figure 2.7 for a representation of a portion of the turbulent regime in the Moody chart.

The accepted formula to compute friction factor used by Moody to produce his diagram was developed by Colebrook (1939). Because the Colebrook formula is implicit (i.e. the desired value f is part of the formula that defines f), and as such it necessitates an iterative solution, it is not presented here.

Instead, an explicit formula that was developed by Chen (1979) is given as Equation (2.19). The Chen Equation correlates very well with the Colebrook results (and Moody #1 in Figure 2.7). Because it is explicit, the Chen equation does not require iteration. Alternative friction factor charts Moody #2 and Moody #3 are discussed in the next section.

Table 2.1 Expressions for friction factors, velocity profiles, and shear stresses for laminar and turbulent flows in circular pipes.

Concept	Applicable to both laminar and turbulent	Applicable to laminar only	Applicable to turbulent only
Friction factor	$f = \dfrac{4\tau_w}{\left(\dfrac{\rho V^2}{2}\right)}$	$f = \dfrac{64}{Re_D}$	Chen Equation (Equation 2.18)
Radial velocity profile	–	$u(r) = u_{max}\left(1 - \left[\dfrac{r}{R}\right]^2\right)$	$u(r) = \dfrac{u^*}{\kappa}\ln(y^+) + B$ $u^* = \left(\dfrac{\tau_w}{\rho}\right)^{1/2}$, $\kappa = 0.41,\ B = 5$
Average velocity	$V(= u_{avg}) = \dfrac{\int_{r=0}^{R} 2\pi r \cdot u(r) \cdot dr}{\int_{r=0}^{R} 2\pi r \cdot dr}$	$V(= u_{avg}) = \dfrac{u_{max}}{2}$	$V(=u_{avg})= \dfrac{u^*}{2}\left[\dfrac{2}{\kappa}\ln\left(\dfrac{Ru^*}{\nu}\right) + 2B - \dfrac{3}{\kappa}\right]$
Wall shear stress	$\tau_w = \mu \left.\dfrac{\partial u}{\partial r}\right]_{r=R}$	$\left.\dfrac{\partial u}{\partial r}\right]_{r=R} = \dfrac{-8V}{D}$	$\left.\dfrac{\partial u}{\partial r}\right]_{r=R} = \dfrac{u^{*2}}{\nu}$

Source: Martin Thermal Engineering Inc.

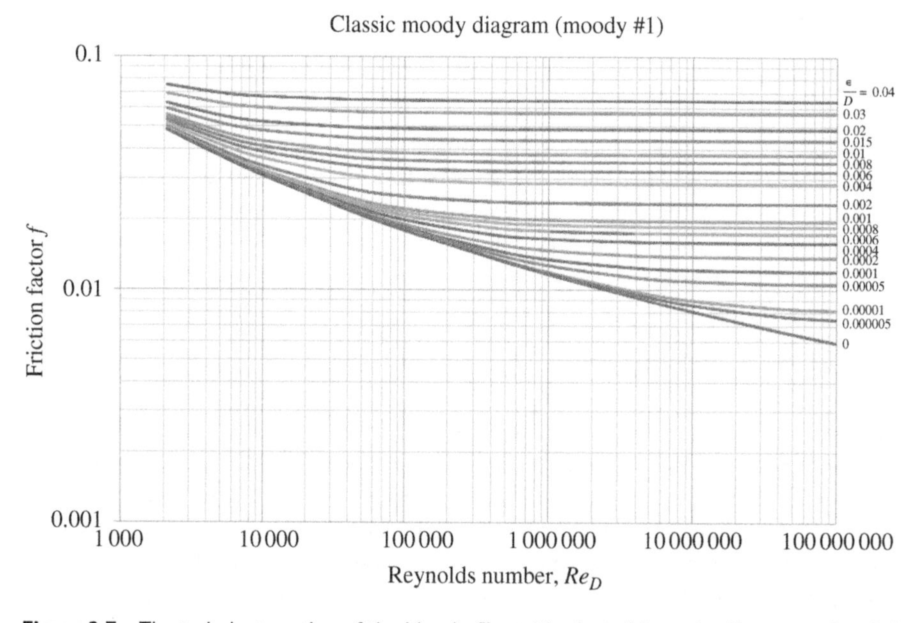

Figure 2.7 The turbulent portion of the Moody Chart #1, plotted from the Chen equation. It is used for determining pressure drop when diameter and volume flow rate are known. *Source:* Adapted from Moody (1944).

$$f = \left[-2\log_{10}\left(\frac{\varepsilon/D}{3.7065} - \frac{5.0452}{Re_D}\cdot \log_{10}\left[\frac{(\varepsilon/D)^{1.1098}}{2.8257} + \frac{5.8506}{Re_D^{0.8981}}\right]\right)\right]^{-2} \tag{2.19}$$

A simpler (and still explicit) expression for turbulent friction factor in smooth pipes (see Equation 2.20) was developed by Petukhov (1970):

$$\boxed{f_{smooth} = (0.79 \ln(Re_D) - 1.64)^{-2}} \tag{2.20}$$

Because Chen equation is numerically too complex to memorize, this textbook provides a customized Excel function to compute friction factor f as function of two arguments (ε/D, Re_D), including a VBA procedure given in Appendix B.1. Instructions for loading VBA scripts into Excel are given in the introduction to Appendix B.

2.11 Modified Bernoulli Equation

The classic Bernoulli equation is not sufficient for thermal systems because most thermal systems involve pipe flows where friction must not be ignored. Therefore, a term for the friction-caused pressure drop is added to make the Bernoulli equation more useful for such problems. The pressure drop term is determined with the friction factor obtained from either the Chen equation or the Moody chart. The modified Bernoulli equation is given as Equation (2.21):

$$\boxed{p_1 + \frac{\rho V_1^2}{2g_c} + \rho\left[\frac{g}{g_c}\right]z_1 = p_2 + \frac{\rho V_2^2}{2g_c} + \rho\left[\frac{g}{g_c}\right]z_2 + f\frac{L}{D}\frac{\rho V_{avg}^2}{2g_c}} \tag{2.21}$$

For steady, incompressible flow in a horizontal pipe with constant cross-sectional area, the pressure drop is equal to the frictional loss term (per the Darcy–Weisbach Equation 2.15):

$$\boxed{\Delta p = (p_1 - p_2) = f\frac{L}{D}\frac{\rho V_{avg}^2}{2g_c}} \tag{2.15}$$

2.12 Alternate Moody Charts

A large majority of all possible piping design problems can be described by one of the following three situations:

1) Determine pressure drop (Δp) when flow rate ($\dot{V}$) and pipe geometry (L, D, ε) are known.
2) Determine volume flow rate when pipe geometry and available pressure drop are known.
3) Determine pipe diameter when pressure drop, flow rate, and pipe length are known.

Use of the traditional Moody chart (or Chen equation) is possible for all three categories, but it requires an iterative solution for categories 2 and 3.

To avoid iteration, two alternate Moody charts are plotted to help evaluate f for the second and third categories. Figure 2.8 is designated Moody chart #2 and Figure 2.9 is designated Moody chart #3. Each of these graphic tools (Janna 2010) offers an exceptionally creative method for avoiding iteration to find the desired unknowns.

The derivation of Moody chart #2 is illustrated in Example 2.3. The derivation for Moody chart #3 is left as an end-of-chapter exercise for the students.

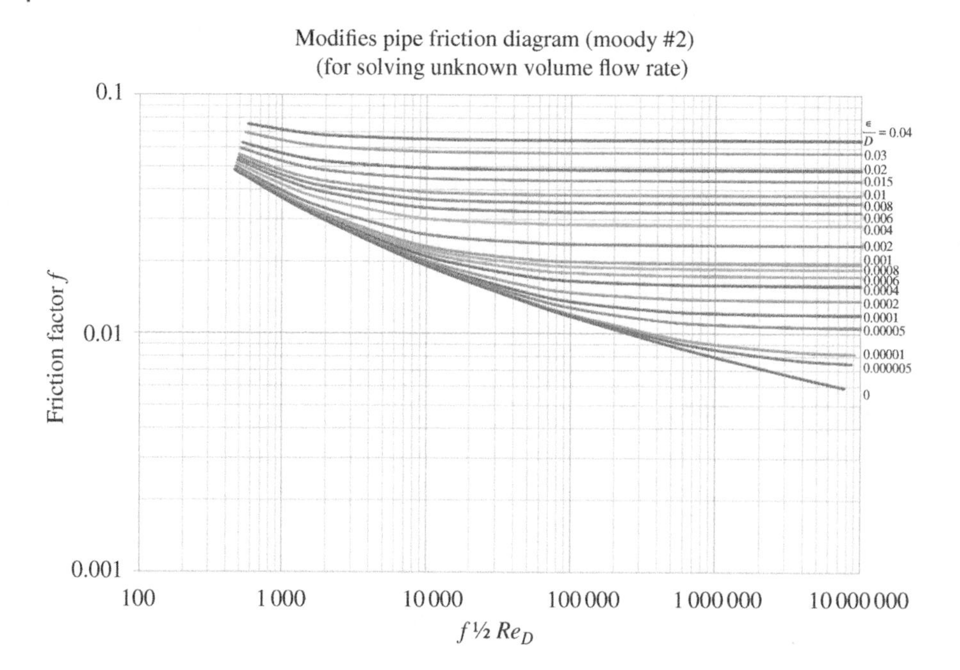

Figure 2.8 Moody chart #2 – for determining volume flow when pipe diameter and pressure drop are known. *Source:* Adapted from Janna (2010)..

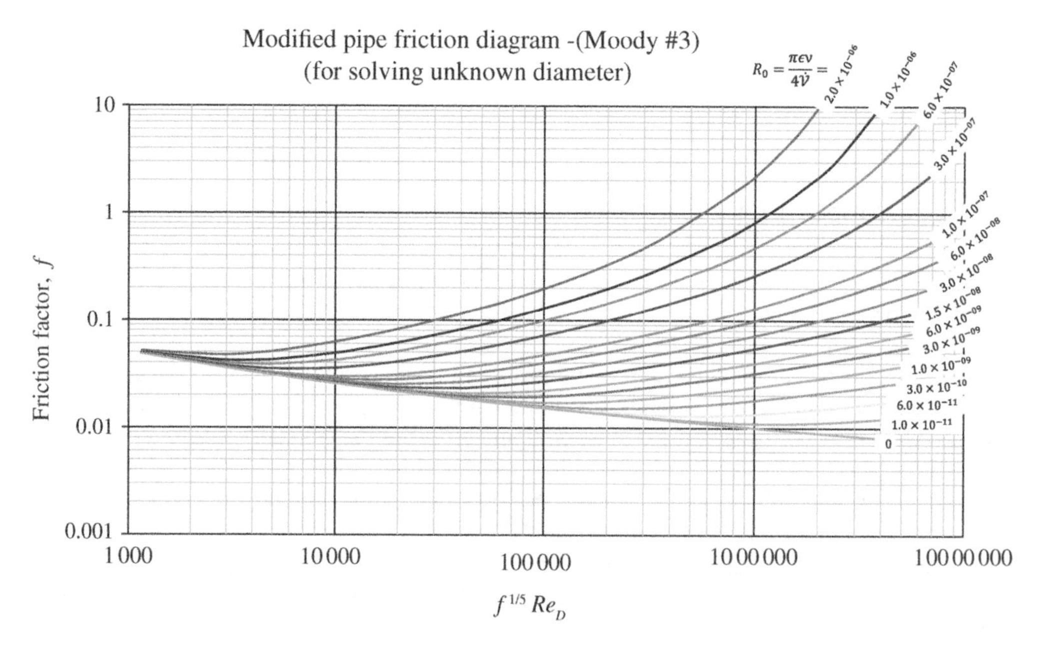

Figure 2.9 Moody chart #3 for determining pipe diameter when pressure drop and volume flow rate are known. The parameter R_0 is not an explicit function of diameter, but it incorporates roughness into a pseudo-reciprocal Reynolds number. *Source:* Adapted from Janna (2010)..

Example 2.3

Consider a horizontal length of pipe ($D = 6.0\ in$, $L = 11\ 270\ ft$, $\varepsilon/D = 0.0017$) with a flow of motor oil ($\nu = 3.49 \times 10^{-2}$). The available pressure at the upstream end of the pipe (i.e. immediately after the pump) is $p_1 = 25.0\ psig$. Derive the parameter plotted on the x-axis of Figure 2.8 and show how it can be computed without knowing volumetric flow rate.

Answer. Solving the Darcy–Weisbach equation for average flow velocity, Equation (2.22) is obtained:

$$V = \sqrt{\frac{2g_c D\Delta p}{fL\rho}} \tag{2.22}$$

Plugging this expression for average velocity into the equation for Reynolds number, Equation (2.23) is derived:

$$Re = \frac{D\sqrt{\frac{2g_c D\Delta p}{fL\rho}}}{\nu} = \sqrt{\frac{2g_c D^3\rho\Delta p}{\mu^2 fL}} \tag{2.23}$$

By isolating the friction factor on the left side along with the Reynolds number, a new parameter ($f^{1/2} \cdot Re$) is derived. Equation (2.24) demonstrates that this new parameter is equal to another parameter comprised of all known quantities.

$$\boxed{f^{1/2} \cdot Re = \sqrt{\frac{2g_c D^3\rho\Delta p}{L\mu^2}}} \tag{2.24}$$

Thus, when the value of the square root formula on the right of Equation (2.24) is used as the x –axis parameter on Moody chart #2, the friction factor can be read directly from the chart.

Think Stop. An equation similar to Equation (2.24) can be derived for Moody chart #3 (see Figure 2.9), which is used for determining pipe diameter when pressure drop and volume flow are known.

The parameter that identifies each distinct plot is a dimensionless number, R_0 (see Equation (2.25)), which is derived from Reynolds number and relative roughness but does not contain diameter (since D is an unknown).

$$R_0 = \frac{\pi\epsilon\nu}{4\dot{V}}\left(= \frac{\epsilon/D}{Re}\right) \tag{2.25}$$

This exercise is given as a Homework Problem at the end of this chapter.

2.13 Entry Effects, Minor Losses

The pressure drop per unit length (dp/dx) in the entry region of any pipe or duct flow is higher than the pressure drop per unit length farther downstream in the pipe (where the velocity profile is fully developed). When the pipe length is long enough compared to its diameter, the entry effects can be ignored because they are small compared to the pressure drop in the fully developed region.

For laminar flow, the approximate ratio of entry length to diameter is linear with the Reynolds number (see Equation 2.26):

$$\frac{L_{entry}}{D} = 0.06 \cdot Re_D \tag{2.26}$$

For low- to mid-range turbulent flow ($5.0 \times 10^3 < Re_D < 1.0 \times 10^7$), the ratio varies as Reynolds number to the ¼ power, as given in Equation (2.27):

$$\frac{L_{entry}}{D} = 1.6 \cdot Re_D^{1/4} \tag{2.27}$$

For low-range turbulent flow exclusively ($5.0 \times 10^3 < Re_D < 5.0 \times 10^4$), it is generally permissible to use a shortcut and approximate the entry length as $L_{entry}/D \approx 10$. This is acceptable due to the weak dependence on Reynolds number in Equation (2.27).

The entry effect is one of several categories of so-called minor losses in pipe flow. Minor losses are attributed to anything that is not a long run of straight pipe. Other examples of minor losses are pipe elbows, tees, contractions, expansions, valves, and flow meters. Often these are truly "minor" in the sense that the sum of all such losses is a relatively small fraction of the total loss due to friction in the straight pipe runs. However, for some of these devices (e.g. orifice plates and partly closed butterfly valves), the pressure loss they induce may be quite major, and the term minor loss is a misnomer.

To quantify the loss of mechanical energy due to minor losses (e.g. valves and fittings), a new term is added to the modified Bernoulli equation, per Equation (2.28):

$$\boxed{p_1 + \frac{\rho V_1^2}{2g_c} + \rho\left[\frac{g}{g_c}\right]z_1 = p_2 + \frac{\rho V_2^2}{2g_c} + \rho\left[\frac{g}{g_c}\right]z_2 + f\frac{L}{D}\frac{\rho V_{avg}^2}{2g_c} + \sum K\frac{\rho V_{avg}^2}{2g_c}} \tag{2.28}$$

where the minor loss coefficient ($K \doteq \langle \text{dimensionless} \rangle$) is defined by the manufacturer of each piping component that induces a minor loss. The value of K is usually a function of its basic geometry (elbow, tee, orifice plate, butterfly valve) and its size. For pipe elbows and wide-open valves, the K values range from about 0.1 to 15.0, with smaller sizes being correlated with larger K values.

Readers are encouraged to obtain K values for devices they employ directly from the technical literature of the device manufacturer. Chapter 6 of White (2016) provides K values for a range of generic valves and fittings, when the manufacturer's information is not available.

2.14 Porous Media Pressure Drop

Thermal system designers are only infrequently tasked with determining pressure drop through a porous medium, but the need does arise in certain types of air pollution control equipment such as packed bed scrubbers and thermal oxidizers. The classic equation by Ergun (1952), given in Equations (2.29) and (2.30), is used frequently by chemical engineers to quantify pressure drop through vessels containing packing materials. See Idelchik (1994) for void fractions and resistances for a wide range of packing materials.

$$\Delta p = \underbrace{f_{pm}}_{\text{porous media}} \cdot \frac{\overbrace{L_v}^{\text{vessel}}}{\underbrace{D_p}_{\text{particle}}} \cdot \frac{\overbrace{\rho}^{\text{fluid density}} \cdot \overbrace{V_{sup}^2}^{\text{superficial velocity squared}}}{2g_c} \cdot \underbrace{\frac{2(1-\sigma)}{\sigma^3}}_{\text{void factor}} \tag{2.29a}$$

$$\boxed{\sigma \equiv \frac{A_{c,void}}{A_{c,vessel}}} = \frac{\mathcal{V}_{void}}{\mathcal{V}_{vessel,tot}} = \left(1 - \frac{\mathcal{V}_{solid\ packing}}{\mathcal{V}_{vessel,tot}}\right) \tag{2.29b}$$

$$\boxed{V_{sup} = \frac{\dot{m}}{\rho \cdot A_{c,vessel}}} \tag{2.29c}$$

$$\boxed{f_{pm} = \frac{150}{Re_{mod,D_p}} + 1.75} \tag{2.30a}$$

$$\boxed{Re_{mod,D_p} = \frac{\rho \cdot V_{sup} \cdot D_p}{\mu(1-\sigma)}} \tag{2.30b}$$

The superficial velocity is what the fluid's average velocity would have been if no packing had been present. Superficial velocity directly contrasted with interstitial velocity, which is the actual velocity of the fluid that passes through the void spaces in the packing. Interstitial velocity is computed from $V_{interstitial} = V_{sup}/\sigma$. The Reynolds number is modified by multiplying the standard Reynolds number by the reciprocal of the packing's solid fraction. The packing's solid fraction is simply one minus the void fraction $(1 - \sigma)$.

Even though random packing materials are sometimes sold by a characteristic dimension (e.g. 1-inch saddles, ¼-inch spheres), their effect on pressure drop is typically characterized by two properties – void fraction $\left(\sigma = V_{void}/V_{packing}[\doteq < dimensionless >]\right)$ and specific surface area $(\alpha = A_{packing}/V_{packing}[\doteq m^{-1}])$. For uniform size spheres, these variables are related to each other through the sphere's radius (see Equation 2.31):

$$\alpha = \frac{3(1-\sigma)}{R_{sphere}} \tag{2.31}$$

It can be shown that the hydraulic diameter of the flow passages in any random packing (e.g. spheres, saddles, other) is given by Equation (2.32):

$$D_{hyd} = \frac{4\sigma}{\alpha} \tag{2.32}$$

Since the Reynolds number in the Ergun equation is calculated with the "particle diameter" instead of the "hydraulic diameter," it is desirable to establish a relationship between the two diameters. The "nominal" diameter for a given packing material (as assigned by the packing manufacturer) should never be used as a direct surrogate for the "particle" diameter unless the packing material comprises uniform spheres. For porcelain and stoneware saddles up to 3 inches nominal, we have found the relationship given in Equation (2.33) to be satisfactory when dimensions are in ***inches***:

$$D_{p,saddles} \approx 0.198 \cdot D_{nom,sadd} + 0.058\ (2.33)\ \{\text{inch}\} \tag{2.33}$$

2.15 Homework Problems

2.1 Determine average and maximum (centerline) velocity for fully developed laminar flow in a smooth pipe ($D = 75\ mm$) with water ($T = 25\ °\ C$) flowing at $\dot{\mathcal{V}} = 0.1\ L/s$.

Answer. $V = 2.26\ cm/s$, $u_{centerline} = 4.53\ cm/s$

2.2 If a liquid enters and exits a pipe/pump combination at $p = 1.0\ atm$ and experiences a total pressure drop greater than $\Delta p = 100\ kPa$ ($\approx 15\ psi$), should the pump be located at (a) the upstream end (prepressurization), (b) the middle (pressure bosting), or (c) the downstream end (repressurization)? Explain.

2.3 Determine average velocity and maximum (centerline) velocity for fully developed turbulent flow ($\dot{V}_{STP} = 955.scfm$) of air ($T = 40\ °\ F$, $p = 14.696\ psia$) in a galvanized steel pipe ($D = 10.0\ inch, \epsilon = 0.006\ inch$).

Answer. $V = 8.28\ m/s$, $u_{centerline} = 9.70\ m/s$

2.4 Determine the total head loss $[\Delta h_{loss} \doteq ft]$ in a horizontal pipe run with $L = 1000\ ft$ of straight, galvanized steel pipe ($D = 6\ inch, sch - 40$) and four 90° long radius elbows. The fluid is water at $T = 85\ °\ F$ and the flow rate is $\dot{V} = 150\ gpm$.

2.5 Liquid water at $T = 20\ °\ C$ flows downward inside an inclined tube of inside diameter $D = 25.0\ mm$. The Reynolds number for the flow is $Re_D = 1800$. Find the angle ($\theta \doteq degrees$) where the gravity force on the fluid is exactly balanced by the viscous shear force all along the inner wall.

Answer. $\theta = 0.021\ 71\ degrees$

2.6 Demonstrate why Moody chart #3 is an appropriate tool for determining pipe diameter, when pressure drop, flow rate, length, and roughness are known.

2.7 Assume the pressure drop in a smooth pipe ($D = 0.025\ m$) is 200 kPa over a length $L = 0.25\ km$. Find the volumetric flow rate if the fluid is (a) liquid water at $T = 25\ °C$ and (b) glycerin at $T = 25\ °C$.

Answer. (a) 658 ml/s, (b) 8.07 ml/s

2.8 Determine the pressure drop of air flowing vertically in a packed bed with the following parameters: $D_{vessel} = 2.0\ m, L_{bed} = 6.0\ m, D_{sphere} = 1.5\ cm, \sigma = 0.55, T = 400\ K, p_{inlet} = 150\ kPa$, $\dot{m}_{air} = 1.1\ kg/s$.

Answer. $\Delta p = 208.7\ Pa \approx 0.84\ inch\ WC$

Cited References

Bergman, T.L., Lavine, A.S.; Incropera, F.P., and DeWitt, D.P.; (2019); *Fundamentals of Heat and Mass Transfer* – 8; Hoboken: Wiley

Chen, N.H.; (1979); "An explicit equation for friction factor in pipe"; *Ind. Eng. Chem. Fundamen.*; **18**; pp.296–297.

Colebrook, C.F.; (1939); "Turbulent flow in pipes, with particular reference to the transition region between the smooth and rough pipe laws"; *J. Inst. Civ. Eng.*; **11**; 133; doi:10.1680/ijoti.1939.13150.

Ergun, S.; (1952); "Fluid flow through packed columns"; *Chem. Eng. Prog.*; **48**; pp. 89–94.

Idelchik, I.E.; (1994); *Handbook of Hydraulic Resistance*, 3; Boca Raton: CRC Press.

Janna, W.S.; (2010); *Design of Fluid Thermal Systems* – 3; Stamford, CT: Cengage Learning.
Millikan, C.M. (1938). A critical discussion of turbulent flows in channels and circular tubes. *Proceedings of the fifth International Congress for Applied Mechanics*. New York: Wiley.
Moody, L.F.; (1944); "Friction factors for pipe flow"; *Trans. ASME*, **66**; 671.
Petukhov, B.S.; (1970); "Heat transfer and friction in turbulent pipe flow with variable physical properties"; In *Advances in Heat Transfer*, Ed. T.F. Irvine & J.P. Hartnett, Vol. **6**; New York: Academic Press.
Prandtl, L.; (1925); *Zeitschr. Angew. Mathematik und Mechanik*; **5**;136; "Bericht uber untersuchungen zur ausgebildeten turbulenz."
Reynolds, O.; (1903); "*Papers on Mechanical and Physical Subjects*"; London: Clay & Sons, Cambridge University Press. https://www.irphe.fr/~clanet/otherpaperfile/articles/Reynolds/N0099463_PDF_1_434.pdf.
Von Karman, T.; (1930), "Mechanische Ähnlichkeit und Turbulenz"; *Mathematik*; **5**: 58 (also as: "Mechanical Similitude and Turbulence", NACA Tech. Mem. 611, https://ntrs.nasa.gov/citations/19930094805).
White, F.M.; (2016); *Fluid Mechanics*, 8, New York: McGraw-Hill.

3

Heat Transfer

For many engineering students, heat transfer seems like the toughest undergraduate subject they have ever undertaken. The theories of conduction and convection are built on partial differential equations, and a vast array of new nomenclature is introduced, including dimensionless constants named after nineteenth- and twentieth-century thermal scientists. Difficulties aside, when students are equipped with a good working knowledge of heat transfer, fluid mechanics, and thermodynamics, many design problems of thermal systems can be solved analytically, and the engineer's ability to predict the behavior of thermal systems can bring satisfying rewards.

The treatment of heat transfer in this chapter tends to be high level – results are often given without mathematical derivation, and assertions of physical intuition call upon students to refine and internalize their basic understanding of the underlying physical phenomena. Readers are encouraged to keep one of the benchmark heat transfer texts by Bergman et al. (2019) or Cengel and Ghajar (2015) on their bookshelves for a more thorough coverage of the subject. The content covered in this chapter includes heat transfer laws by Fourier, Newton, and Stefan and Boltzmann; the electric analogy for heat, film, and bulk temperatures; Nusselt and Prandtl numbers; tube banks; and radiant exchange and heat exchanger calculations.

3.1 Fourier's Law

Jean Baptiste Fourier was a French mathematician and physicist who developed the basic theory for heat conduction in the early part of the nineteenth century (Fourier 1822). The mathematics of heat conduction are similar to those of electricity flow, fluid flow in a pipe, and molecular mass diffusion. In Cartesian coordinates, with the slab wall geometry shown in Figure 3.1, the heat conduction rate $(\dot{Q})$ is proportional to the cross-sectional area (A_c) times the linear temperature slope $\Delta T/\Delta x$ as long as the materials participating in the heat conduction have reached steady state (i.e. no changes in temperature with time anywhere in the domain of interest).

In symbolic terms, Fourier's law is written as shown in Equation (3.1), with the proportionality constant k called the thermal conductivity of the material. The temperature difference ΔT quantifies the ***driving force*** for the transfer of heat, while $\dot{Q}$ is the macroscopic ***heat flow*** through the conducting material. The remaining terms are often lumped together $(kA_c/\Delta x)$ and identified as the reciprocal of heat flow ***resistance***.

Thermal Systems Design: Fundamentals and Projects, Second Edition. Richard J. Martin.

Companion website: www.wiley.com\go\Martin\ThermalSystemsDesign2

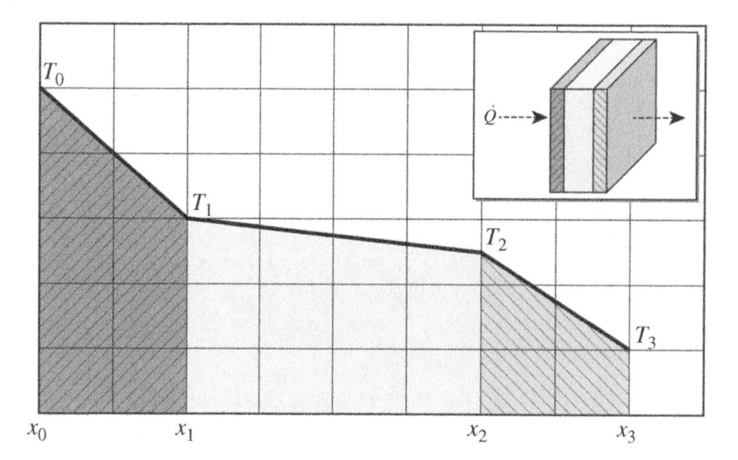

Figure 3.1 Heat conduction through three adjoining slabs with different thermal conductivities. *Source:* Martin Thermal Engineering Inc.

$$\underbrace{\dot{Q}}_{\doteq W} = -\underbrace{k}_{\doteq \frac{W}{m^\circ C}} \underbrace{A_c}_{\doteq m^2} \underbrace{\frac{\Delta T}{\Delta x}}_{\doteq \frac{^\circ C}{m}} \tag{3.1}$$

The negative sign in Fourier's law is a consequence of the Second law of thermodynamics that directs heat to flow from high temperature to low temperature. In Figure 3.1, temperature declines as x-distance increases from left to right. In other words, the slope of temperature in the positive x-direction is negative, but the heat flow is positive from left to right. Hence, the negative sign in Fourier's law is necessary to produce a positive heat flow result whenever the temperature gradient is negative in that same direction.

Also noteworthy is the difference in the steepness of the temperature slope as heat is flowing through the three materials. This is a consequence of variations in the magnitude of thermal conductivity in each material, as elaborated next.

Think Stop. Students should check their intuition about whether a "steeper" slope corresponds to higher or lower thermal conductivity. We have observed about half of new heat transfer students intuitively expect the temperature slope to vary in the direction opposite to what the physics dictates.

For the example considered in Figure 3.1, the steady state assumption implies that heat storage is zero in each slab of material, which also implies (according to the First law) that the inflow of heat into each slab is equal to the outflow of heat from each slab. Since the heat flow is the same through all three slabs, and if the cross-sectional area is assumed to be uniform in the direction of heat flow for all slabs, the product of the thermal conductivity and the temperature gradient must be constant in each slab, per Equation (3.2).

$$\frac{\dot{Q}}{A_c} = -k\frac{\Delta T}{\Delta x} = const \tag{3.2}$$

This implies that greater thermal conductivity leads to a smaller temperature gradient and vice versa for constant heat flux $\dot{q} \equiv \dot{Q}/A_c$. Physically, this makes sense. In order to pass $\dot{q} = 100\ W/m^2$ of heat flux through a slab of copper, you wouldn't need nearly as large a temperature gradient as you would through a slab of ceramic with the same thickness.

3.2 Newton's Law of Cooling

Newton (1701), among his many other contributions to mathematics and physics, first proposed the relationship (Equation 3.3) that the rate of heat loss from a warm object was proportional to the temperature difference between the object and the surroundings:

$$\dot{Q} = hA_s(T_s - T_\infty) \tag{3.3}$$

where the proportionality constant h is called the convective heat transfer coefficient. To be clear, Newton's work preceded both Fourier's and Fahrenheit's, and it only addressed cooling of a warm object in a cooler environment (i.e. both radiative and convective forces were at work), but its utility was obviously not lost on the thermal scientists who succeeded him.

In many cases, convective heat transfer is analyzed by determining the Nusselt number, which is a dimensionless parameter that represents the augmentation of the heat transfer by the fluid's movement over that of conduction in a still fluid.

$$Nu \equiv \frac{\overbrace{h}^{\text{convection coefficient}}}{\underbrace{\left(k/L_c\right)}_{\text{conduction in still fluid}}}$$

The characteristic length (L_c) is related to the fluid's boundary layer thickness that acts like a layer of insulation between the warm object (T_s) and the cool temperature of the surrounding fluid (T_∞). The thermal conductivity (k) is that of the fluid, because it is the boundary layer (not the solid object) that comprises the insulating layer that resists heat flow.

For thermal systems design purposes, a primary goal is to determine the values of h wherever fluid flows along a surface and where heat is being transferred. The most common applications of convection theory are (i) design and analysis of heat exchangers undergoing forced convection across a thin tube from one fluid to another, and (ii) computation of heat loss from a hot surface (furnace, pipe, storage vessel, etc.). A validated series of equations for determining Nusselt number for a given set of parameters will be given later in this chapter.

3.3 The Stefan–Boltzmann Law

The law of thermal radiation heat transfer was developed in the late nineteenth century by Josef Stefan and Ludwig Boltzmann. Stefan first postulated that radiative thermal energy leaving the surface of a hot object was proportional to the fourth power of temperature, and from this relationship, he estimated the surface temperature of the sun to be approximately 5700 K. Boltzmann, who was Stefan's student at University of Vienna, first proved that macroscopic entropy could be derived exactly from a statistical analysis of microscopic entropy. Later they derived the Stefan–Boltzmann proportionality constant (σ) from first principles and estimated its value. The Stefan–Boltzmann law (Equation 3.4) states that the total energy radiated away from a unit area of a black surface is:

$$\frac{\dot{Q}}{A} = \sigma T_s^4 \tag{3.4}$$

On a historical note, Boltzmann's tombstone is inscribed with the equation $S = k \ln \omega$, which was his derivation that the macroscopic entropy S is equal to a fundamental constant k (which is simply the universal gas constant $\hat{R}$ divided by Avogadro's number $\mathcal{N}_{av}$) times the natural logarithm of the number of microstates ω (which is the number of discrete energy states) accessible to all the combinations of molecules in the ensemble. When more microstates become accessible (due to a larger system volume or higher temperature) the macroscopic entropy increases as the logarithm of ω. This is a very exciting outcome, because it bridges the gap between the microscopic behavior of gas molecules (ω) and the macroscopic entropy (S).

Equation (3.4) is typically multiplied on the right side of the equals sign by (i) emissivity ($0 \leq \varepsilon \leq 1$), which diminishes the thermal radiation leaving a nonblack (gray) body, and (ii) view factor ($0 \leq F_{12} \leq 1$) that accounts for radiant exchange between two surfaces at different temperatures. Both factors are dimensionless and represent a reduction from the ideal radiative transfer situation, where surface 1 is a perfect emitter (black) and it sees only surface 2 and nothing else. Several simplified equations for radiant exchange between two bodies are given later in this chapter.

3.4 The Energy Equation

Like the continuity equation was for mass, the energy equation (or First law) is a balance of energy flows across the system boundary and energy storage within it. Like mass, ***production*** of energy is zero because energy can neither be created nor destroyed (absent nuclear reactions).

It is important to track the energy forms (e.g. internal, kinetic, potential, and other) that flow into and out of the control volume along with the fluid that crosses the control surface. Unlike the continuity equation, heat flow ($\dot{Q}$) and shaft work ($\dot{W}_s$) can cross the control surface unaffiliated with matter crossing the boundary, and those energy flows must be accounted for too. The conventional arrangement is to add $\dot{Q}$ on the left side as an inflow of energy and $\dot{W}_s$ on the right as an outflow of energy, even though an individual control volume may have work or heat or both flowing in and/or out (see Equations 3.5). As utilized in the first two chapters, Equation (3.5) is enhanced with parenthetical superdot labels to help students categorize the different terms.

$$\underbrace{0}_{\dot{\mathcal{P}}_E = 0} + \underbrace{\dot{Q}_{in}}_{\substack{Heat\\Inflow\\(net)}} + \underbrace{\int_{C.S.} \rho \mathrm{e} \left(V \cdot \vec{n} \right) dA_{in}}_{\dot{\mathrm{e}}_{in}} = \underbrace{\dot{W}_{s,out}}_{\substack{Shaft\\Power\\Outflow\\Net}} + \underbrace{\int_{C.S.} \rho \mathrm{e} \left(V \cdot \vec{n} \right) dA_{out}}_{\dot{\mathrm{e}}_{out}} + \underbrace{\int_{C.V.} \frac{\partial}{\partial t} (\rho \mathrm{u}) d\mathcal{V}}_{\dot{\mathcal{S}}_E} \tag{3.5}$$

$$\mathrm{e} = \mathrm{e}_{internal} + \mathrm{e}_{kinetic} + \mathrm{e}_{potential} + \mathrm{e}_{other} \tag{3.5a}$$

$$\mathrm{e}_{internal} = u + pv = h \text{ (for the surface integral)} \tag{3.5b}$$

$$\mathrm{e}_{internal} = u \text{ (for the volume integral, no flow work)} \tag{3.5c}$$

$$\mathrm{e}_{kinetic} = \frac{V^2}{2g_c} \tag{3.5d}$$

$$e_{potential} = \left[\frac{g}{g_c}\right] z \tag{3.5e}$$

$$e_{other} = e_{chemical} + e_{electromagnetic} + e_{nuclear} \tag{3.5f}$$

Think Stop. Students should note that the heat flow term $\dot{Q}$ can be further broken down into the three forms of heat flow described in the prior three sections: $\dot{Q} = \dot{Q}_{cond} + \dot{Q}_{conv} + \dot{Q}_{rad}$. The corollary of this subdivision is that none of the previously defined heat transfer laws (Fourier, Newton, and Stefan–Boltzmann) are valid as substitutes for the First law. This is because they only govern specific modes of heat transfer and provide no relevant information about shaft work, inflows/outflows of energy with matter, energy storage, or energy conversion (e.g. electrical or chemical to thermal).

While this distinction may seem obvious when presented in this context, we have observed students forgetting that the First law is an analysis tool that is available to them *in addition to* the laws governing conduction, convection, and radiation. The First law brings another equation to a problem where the number of unknowns exceeds the number of independent relations known.

Students are also advised to refer to Appendix A of this textbook, where "transport properties" (e.g. c_p, k, μ, and ρ) are tabulated for a number of liquids and gases.

3.5 The Entropy Equation

On the first day of each new semester, I often ask new students which thermodynamic law is their favorite, and most students are either baffled by the question or give the answer, "First law." Indeed, energy conservation is a beautiful example of a natural law, and with it, engineers possess a powerful design tool. However, the Second law is unique among all conservation equations. Consider Equation (3.6) and how it contrasts with Equations (2.2) (mass) and (3.5) (energy), but also how it contrasts in a different way to Equation (2.9) (momentum).

$$\underbrace{\frac{dS_{C.M.}}{dt}}_{\dot{P}_S \geq 0} + \underbrace{\int_{C.S.} \left(\dot{Q}_{in}/T_s\right) dA_{\dot{Q}in}}_{\text{Entropy Inflow with Heat}} + \underbrace{\int_{C.S.} \rho s\left(V \cdot \vec{n}\right) dA_{in}}_{\dot{S}_{in}} = \underbrace{\int_{C.S.} \left(\dot{Q}_{out}/T_s\right) dA_{\dot{Q}out}}_{\text{Entropy Outflow with Heat}} + \underbrace{\int_{C.S.} \rho s\left(V \cdot \vec{n}\right) dA_{out}}_{\dot{S}_{out}} + \underbrace{\int_{C.V.} \frac{\partial}{\partial t}(\rho s) dV}_{\dot{S}_S} \tag{3.6}$$

Both contrasting characteristics of the Second law are embodied in the ***production*** term of Equation (3.6). For mass and energy, production is zero; for momentum, production can be positive or negative or zero. However, for entropy, production can only be greater than or equal to zero. The Second law necessitates that entropy can be created but never destroyed. This is the basis for entropy's poetic reputation as the "arrow of time." If entropy in a closed system can only increase or remain constant, then entropy (broadly defined) was smaller in the past and will be larger in the future.

3.6 Electricity Analogy for Heat

Engineering students first see the basic electric circuit (battery + resistor) in a beginning physics course, and they later see the circuit expanded to include capacitors. Both circuits are perfect analogies for heat flow.

Recalling the basic equation for the electric circuit (Equation 3.6):

$$\overbrace{I}^{\substack{Current\\Flow}} = \frac{\overbrace{\Delta V}^{\substack{Voltage\\Drop}}}{\underbrace{R_{elec}}_{\substack{Electrical\\Resistance}}} \tag{3.7}$$

It is notable that analogies for each of these concepts exist in heat transfer, as listed:

- I current flow ~ $\dot{Q}$ heat flow
- ΔV voltage drop ~ ΔT temperature drop
- R_{elec} electrical resistance ~ R_{th} thermal resistance

The analogous relationships for the heat flow circuit are given in Equations (3.7) and (3.8):

$$\overbrace{\dot{Q}}^{\substack{Heat\\Flow}} = \frac{\overbrace{\Delta T}^{\substack{Temperature\\Drop}}}{\underbrace{R_{th}}_{\substack{Heat\ Flow\\Resistance}}} \tag{3.8}$$

$$R_{th} = \frac{\Delta x}{kA_c}\left[\stackrel{\vee}{=} \frac{K}{W}\right] \quad conduction, slab \tag{3.8a}$$

$$R_{th} = \frac{\ln(r_o/r_i)}{2\pi kL}\left[\stackrel{\vee}{=} \frac{K}{W}\right] \quad conduction, hollow\ cylinder \tag{3.8b}$$

$$R_{th} = \frac{1}{hA_s}\left[\stackrel{\vee}{=} \frac{K}{W}\right] \quad convection \tag{3.8c}$$

The rules for summing up heat flows and resistances in series and parallel are precisely identical to the rules for electric resistance circuits:

Specifically, for *series* resistors, 0 —R_1— 1 —R_2— 2 see Equations (3.9):

- Temperature drops add $\Delta T_{01} + \Delta T_{12} = \Delta T_{02}$ (3.9a)
- Resistances add $R_1 + R_2 = R_{tot}$ (3.9b)
- Heat flow is conserved $\dot{Q}_1 = \dot{Q}_2 = \dot{Q}_{tot}$ (3.9c)
- Overall circuit relationship $\dot{Q}_{tot} = \frac{\Delta T_{02}}{R_{tot}}$ (3.9d)

And for *parallel* resistors, 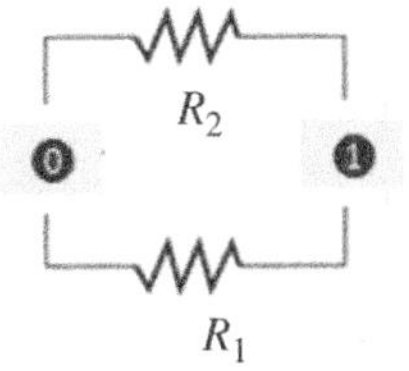

see Equations (3.10):

- Temperature drops are conserved $\Delta T_1 = \Delta T_2 = \Delta T_{01}$ (3.10a)
- Reciprocal resistances add $\frac{1}{R_1} + \frac{1}{R_2} = \frac{1}{R_{tot}}$ (3.10b)
- Heat flows add $\dot{Q}_1 + \dot{Q}_2 = \dot{Q}_{tot}$ (3.10c)
- Overall circuit relationship $\dot{Q}_{tot} = \frac{\Delta T_{01}}{R_{tot}}$ (3.10d)

Thus, the physics of heat flow is fully consistent with (and analogous to) the physics of electricity flow, and the rules for simplifying a resistance circuit are identical in both fields. Although the subject of transient thermal behavior is mostly outside the scope of this textbook, the resistor-capacitor (*RC*) circuit in electricity is also precisely analogous to the equivalent circuit in heat transfer.

3.7 Film, Mean Temperature

For convection problems, many expressions for Nusselt number *Nu* (which relies on *k* for the fluid) require knowledge of Reynolds number, *Re*, and Prandtl number, *Pr* (which both rely, at least in part, on transport properties c_p, μ, and ρ for the fluid). Because these four transport properties are all temperature dependent, a choice must be made as to the temperature used for evaluating fluid properties.

Film Temperature. Fluid properties should be evaluated at ***film temperature*** ($T_f = (T_w + T_\infty)/2$) whenever an ***external convection*** problem is being solved, with one exception – when the formula specifically calls for a property to be evaluated at the wall (usually Pr_w or μ_w).

The following reasons explain why ***film temperature*** should be used to evaluate fluid properties in most external flow cases:

- When the energy equation of the external boundary layer was evaluated, constant properties were assumed. A single temperature must be chosen to evaluate those properties.
- Using the free stream or wall temperature typically overestimates or underestimates the actual heat transfer.
- The film (i.e. boundary layer) is like a layer of insulation that separates the wall from the free stream. Assigning a thermal conductivity to insulation is typically accomplished by selecting the value for *k* that applies at the mean temperature between the hot and cold faces.
- If the convection differential equations are evaluated by integrating the boundary layer velocity and temperature profiles (assuming variable properties), the average value of the fluid properties will be very close to the value of the fluid properties at the average temperature in the boundary layer.

Mean Temperature and Bulk Mean Temperature. For ***internal convection*** problems, fluid temperature typically varies in both the radial and axial directions, and because there is no "free-stream" condition (T_∞, V_∞), film temperature does not exist.

Instead, for internal convection problems, local fluid properties should be evaluated at ***mean temperature*** (see Equation 3.11), which varies along the flow axis:

$$T_m \equiv T(x) = \frac{\int_{r=0}^{R} c_p T(x,r)\rho u(x,r) 2\pi r dr}{\dot{m} c_p} \quad (3.11)$$

Further, for heat exchanger problems, it is usually preferred to compute one ***bulk mean temperature*** for property evaluation of the entire ***hot*** fluid flow path and a different ***bulk mean temperature*** for the entire ***cold*** fluid flow path. In particular, the specific heat of the fluid usually varies from the inlet to the outlet, but the fluid enthalpy carrying capacity ($\dot{m}c_p$) is taken as constant to be able to apply the heat exchanger effectiveness formulas. Furthermore, estimating the bulk mean temperature along a fluid's entire flow path has the advantage of simplifying the determination of the overall heat transfer coefficient (U), as computed from fluid thermal conductivities that vary all along the heat exchanger's length. Bulk mean temperature is defined in Equation (3.12):

$$T_{bm,path1} \equiv \left[\frac{T_{m_{inlet,1}} + T_{m_{outlet,1}}}{2}\right] \quad (3.12)$$

The approximation aforementioned should be used with care. Whenever it is possible to validate the overall results, users of this method should test their result experimentally or numerically (i.e. with a CFD modeling tool).

3.8 Nusselt, Prandtl Numbers

As indicated previously, the Nusselt number is a dimensionless parameter that compares the rate of convection (numerator, h) to the equivalent rate of conduction through a film of stationary fluid of comparable thickness as the specified geometry (denominator, k/δ). For a circular pipe of inside diameter D, Nusselt number is defined in Equation (3.13):

$$Nu_D = \frac{hD}{k} \quad (3.13)$$

For laminar flows, the Nusselt number can be computed from first principles, and the results are somewhat astonishing – the Nusselt number does not depend on diameter, velocity, or even Reynolds number. For ***laminar*** flows, the Nusselt number is simply a constant that depends only on tube geometry. Obviously, the heat transfer coefficient (h) depends on the hydraulic diameter and the fluid's thermal conductivity through the definition of Nusselt number ($Nu = hD/k$).

For ***turbulent*** flows, the Nusselt correlation typically depends on Reynolds number and another dimensionless parameter called the Prandtl number. The Prandtl number (defined in Equation 3.14) is a property of fluids and does not depend on tube geometry or flow velocity.

$$Pr \equiv \frac{\mu}{c_p k} = \frac{\nu}{\alpha} \quad (3.14)$$

Laminar Heat Transfer. Table 3.1 gives exact results for Nusselt number under two different boundary conditions – constant wall temperature, T_s, and constant wall heat flux, $\dot{q}_s$.

Dittus–Boelter (turbulent) Equation. For turbulent internal flow, the Dittus–Boelter equation (Equation 3.15) gives good results for circular tubes. For noncircular tubes, Dittus–Boelter will give reasonable answers by using hydraulic diameter ($D_h = 4A_c/P$), but students may wish to research whether specific correlations are available for other geometries.

Table 3.1 Nusselt number for internal, laminar flow in different tube geometries.

Laminar flow	Circle	Square $b/a = 1$	Rectangle $b/a = 0.5$	Rectangle $b/a = 0.25$	Rectangle $b/a \approx 0$
$Nu_{T\,=\,const}$	3.66	2.98	3.39	4.44	7.54
$Nu_{\dot{q}\,=\,const}$	4.36	3.61	4.12	5.33	8.24

Source: Adapted from Cengel and Ghajar (2015).

$$Nu = 0.023\ Re^{0.8}\ Pr^{n} \qquad \begin{array}{l} n = 0.4\ heating\ of\ fluid \\ n = 0.3\ cooling\ of\ fluid \end{array} \tag{3.15}\{T\}$$

Dittus–Boelter is applicable for flows with $Re > 10\ 000$ and fluids with $0.7 \leq Pr \leq 160$.

Gnielinski Equation. The Gnielinski equation (Equation 3.16) provides better accuracy over a wider range of Reynolds and Prandtl numbers by including the turbulent friction factor (Equation 2.19) as an additional argument in the formula.

$$Nu = \frac{(f/8)(Re - 1000)Pr}{\left[1 + 12.7(f/8)^{0.5}\left(Pr^{2/3} - 1\right)\right]} \tag{3.16}\{T\}$$

The Gnielinski equation is valid for $3 \times 10^{3} < Re < 5 \times 10^{6}$ and $0.5 < Pr < 2000$.

Schleicher and Rouse Equation. For turbulent flow of liquid metals (with very low Prandtl numbers), none of the aforementioned correlations provides satisfactory results. For these conditions, two correlations by Sleicher and Rouse (Equations 3.17) are preferred:

$$Nu_{T\,=\,const} = 4.8 + 0.0156\ Re^{0.85} Pr_{s}^{0.93} \tag{3.17a}\{T, LM\}$$

$$Nu_{\dot{q}\,=\,const} = 6.3 + 0.0167\ Re^{0.85} Pr_{s}^{0.93} \tag{3.17b}\{T, LM\}$$

These correlations provide acceptable results when $10^{4} < Re < 10^{6}$ and $0.004 < Pr < 0.01$.

3.9 Flows Across Tube Banks

Crossflow heat exchangers frequently are equipped with banks of tubes in parallel, carrying a liquid inside and air or a gas mixture flowing across them. The exterior surface of the tubes may be bare or augmented with fins along the external flow direction (i.e. in contact with, but perpendicular to the tubes). The analysis of crossflow with fins depends on too many factors to address in this space, but the case without fins can be addressed briefly. See Figure 3.2 for "in-line" and "staggered" configurations for tube banks.

The most important design factors for tube banks are the spacing between the tubes and the velocity achieved by the air flow as it weaves in and out between them. It is tempting to compute Reynolds number (and by extension, Nusselt number) by using the approach velocity (i.e. upstream of the location where the air stream is accelerated through the smaller gaps between the tubes), but this cannot yield useful results.

Consider two crossflow heat exchangers, both with the same entry cross-sectional area and the same mass flow of air, but one has 100 small tubes (10 rows with 10 tubes per row) with relatively large gaps between adjacent tubes, and the other has 100 larger tubes (also 10 × 10 on the same grid pattern) but with smaller gaps between adjacent tube circumferences. The heat exchanger with the

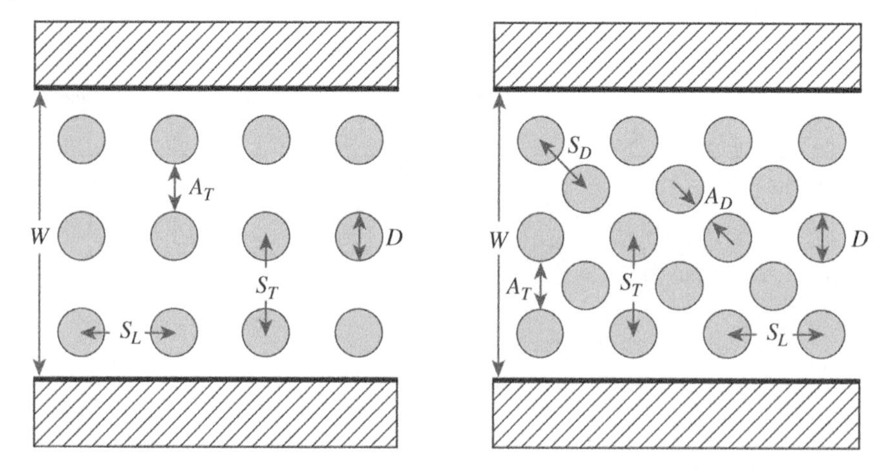

Figure 3.2 In-line versus staggered configurations of tube banks in crossflow heat exchangers. *Source:* Adapted from Zukauskas (1987).

larger tubes will obstruct more of the flow area, and the local velocity will be higher. The increased velocity will enhance the heat transfer and lead to higher convection heat transfer coefficient.

Therefore, the maximum velocity of the air as it weaves through the tube bank must be computed, and V_{max} must be used to compute Reynolds number. The approach by Zukauskas and Ulinskas (1985) and Zukauskas (1987) outlined is preferred. Many of the parameters rely strictly on geometry, and the final Nusselt number correlation has a form similar to Dittus–Boelter. The relevant geometric names (see also Figure 3.2) are:

- D Tube diameter
- H Channel height (overall transverse dimension)
- W Channel width (length of tubes exposed to crossflow)
- L Channel length (number of tube rows times longitudinal spacing)
- A Channel area (flow area width times depth, prior to encountering tubes)
- S_T Transverse spacing of tubes (center to center)
- S_L Longitudinal spacing (center to center)
- S_D Diagonal spacing (center to center), ***staggered only***
- A_T Transverse area (gap between transverse tube walls times depth of channel)
- A_D Diagonal area (gap between diagonal tube walls times depth of channel), ***staggered only***

Equations (3.18) apply to all tube banks, except where designated.

$$V = \frac{\dot{m}}{\rho(H \cdot W)} \quad \textit{upstream velocity} \tag{3.18a}$$

$$V_{max} = \frac{S_T}{(S_T - D)} V \quad \textit{maximum velocity, \textbf{in-line}} \tag{3.18b}$$

$$V_{max} = \frac{S_T}{2(S_D - D)} V \quad \text{max } \textit{velocity, \textbf{staggered}}$$

$$S_D < (S_T + D)/2 \tag{3.18c}$$

Table 3.2 Tube bank relations, where $Nu_D = C\,Re_D^m Pr^n \left(\frac{Pr}{Pr_s}\right)^{1/4}$.

Reynolds range	In-line	Staggered
01 – 00	$C = 0.9$ $m = 0.4$ $n = 0.36$	
0 – 500		$C = 1.04$ $m = 0.4$ $n = 0.36$
100 – 1000	$C = 0.52$ $m = 0.5$ $n = 0.36$	
500 – 1000		$C = 0.71$ $m = 0.5$ $n = 0.36$
1000 – 200 000	$C = 0.27$ $m = 0.63$ $n = 0.36$	$C = 0.35\left(\frac{S_T}{S_L}\right)^{0.2}$ $m = 0.6$ $n = 0.36$
$2 \times 10^5 - 2 \times 10^6$	$C = 0.033$ $m = 0.8$ $n = 0.4$	$C = 0.031\left(\frac{S_T}{S_L}\right)^{0.2}$ $m = 0.8$ $n = 0.36$

Source: Adapted from Zukauskas (1987).

$$Re_D = \frac{V_{max}D}{\nu} \tag{3.18d}$$

$$Nu_D = \frac{hD}{k} = C\,Re_D^m Pr^n \left(\frac{Pr}{Pr_s}\right)^{1/4} \tag{3.18e}$$

Fluid properties are evaluated at the average of the inlet and exit temperatures ($T_{avg} = (T_i + T_e)/2$), with the exception of Pr_s, which is evaluated at the surface temperature of the tubes (T_s). The tube surface temperature will be closer to the temperature of the fluid with the highest value of h (i.e. the lowest resistance). For crossflow heat exchangers with 16 or more tube rows, the correlations in Table 3.2 may be used without correction. If the heat exchanger has fewer than 16 tube rows, the correction factors in Table 3.3 should be applied per Equation (3.19) for cases where $Re_D > 1000$:

$$Nu_{D(N<16)} = F \cdot Nu_D \tag{3.19}$$

Table 3.3 Correction factors for tube banks with fewer than 16 rows of tubes.

N_L	1	2	3	4	5	7	10	13
In-line	0.70	0.80	0.86	0.90	0.93	0.96	0.98	0.99
Staggered	0.64	0.76	0.84	0.89	0.93	0.96	0.98	0.99

Source: Adapted from Zukauskas (1987).

3.10 "Gotcha" Variables

It goes without saying that engineers must take responsibility for determining the context of each variable in every equation they use to solve a problem. Heat transfer equations can be fraught with peril if the engineer fails to double-check the intended use of variables.

We have seen certain "gotcha" variables repeatedly misinterpreted by engineering students, and while no warning can eliminate all such errors, a word to the wise may be sufficient, at least for those who read this list of warnings and internalize them.

- Area. Consider the equation for heat transfer from a hot tube wall to a cooler liquid inside. The First law equates (i) convection into the fluid with (ii) gain in fluid enthalpy:

$$hA\left(T_{wall} - T_{fluid}\right)_{avg} = \underbrace{\rho AV}_{\dot{m}} c_p\left(T_{out} - T_{in}\right)$$

The "gotcha" variables here are the wall surface area $A(=\pi DL)$ and the flow cross-sectional area $A(=\pi D^2/4)$. Using the same A in both places is a very common error. The preferred form is:

$$hA_s\left(T_{wall} - T_{fluid}\right)_{avg} = \underbrace{\rho A_c V}_{\dot{m}} c_p\left(T_{out} - T_{in}\right)$$

- Delta temperature. The symbol ΔT is notorious for being misapplied in heat transfer problems. Using the example aforementioned (where two distinct areas were highlighted in the same equation) it is apparent that two distinct temperature subtraction formulas are present in the same equation. These formulas are clarified as follows: ***temperature change*** $\Delta T_{chg} = (T_{out} - T_{in})$ is the change in temperature of *a single substance* from the inlet to the outlet of a conveying device, whereas, ***temperature difference*** $\Delta T_{diff} = (T_{wall} - T_{fluid})_{avg}$ is the difference in temperature between *two different substances* (e.g. wall and fluid or fluid 1 and fluid 2).
- Degrees of temperature. The importance of distinguishing between absolute temperature (K or R) and nonabsolute temperature (°C or °F) in calculations is well known. One way to cement this distinction is to contemplate the following exchange:

  Q. Why do two of the symbols (K, °C, R, °F) include the degree symbol (°) and the other two exclude it?

  A. The absolute units, R and K, exclude it because temperatures reported in these units are "normal" variables that can be multiplied, divided, and raised to a power. In contrast, temperatures that include the degree symbol, are "not normal" variables, and they must only be used when temperatures are subtracted from one another – hence, the concept of "degrees" of temperature – which portray an image of lines on a thermometer.
- Thermal conductivity. Heat transfer problems often require knowledge of the thermal conductivity of a material to solve the temperature field or determine the total heat loss (or gain). A "gotcha" variable that plagues some engineering students is which thermal conductivity should be used – the solid or the fluid. This mistake often occurs in heat exchanger calculations, where the convection heat transfer coefficient may be derived from one of the Nusselt numbers and that fluid's thermal conductivity (k_{fluid}), but the conduction through a tube wall depends on the solid's thermal conductivity (k_{solid}). Values of k_{fluid} for gases are invariably smaller than values of k_{solid} for steel, copper, or any other conductive pipe material.
- Enthalpy, convection. Enthalpy is often represented by the variable "h" but so is convective heat transfer coefficient. To avoid this conflict, when enthalpy and convection are both needed for a

single analysis, the variable "*i*" is usually substituted for enthalpy. Unlike the first two examples aforementioned, the units are very different for these competing "gotcha" variables, so the tendency to mistake them for one another may be easier to avoid.

- Biot number, Nusselt number. From context, it would seem that these two dimensionless parameters would be nearly impossible to use incorrectly. However, on closer look, they are more similar than they would seem by their common usages: (transient convection to conduction ratio) versus (fluid film convection to conduction ratio). Both parameters comprise a convection parameter, a conduction parameter, and a characteristic length. For the Biot number ($Bi \equiv hL_{char,\ solid}/k_{solid}$), the characteristic length and thermal conductivity are for the solid receiving heat by convection. For the Nusselt number ($Nu \equiv hL_{char,\ film}/k_{fluid}$), the length and thermal conductivity are for the fluid film, which is responsible for the augmentation of conduction by the convective fluid motion.
- Specific heat ratio, thermal conductivity. The variable "*k*" is commonly used for the ratio of specific heats $\left(k = c_p/c_v[\doteq \langle no\ units\rangle]\right)$ and thermal conductivity, $k\ [\doteq W/m°C]$. Again, units and context should be helpful to avoid mixing up these two "gotcha" variables.
- Specific gravity, specific heat ratio. An alternate variable for the ratio of specific heats $\left(\gamma = c_p/c_v \doteq \langle no\ units\rangle\right)$ is sometimes mistaken for the specific gravity of a fluid "*i*" $\left(\gamma_i = \rho_i/\rho_{ref}[\doteq \langle no\ units\rangle]\right)$. These "gotcha" variables are both dimensionless, but the context should be sufficient to avoid confusion and error.

3.11 Radiation and Natural Convection

Many heat transfer textbooks cover the concept of a "radiation" heat transfer coefficient, and thermal engineers may find it useful for certain heat loss computations. The thinking is that for calculating heat losses from a hot vessel to a cold environment, natural convection occurs in parallel to thermal radiation. If the two target temperatures of the cold environment are approximately equal (i.e. $T_\infty \approx T_{surr}$), the total heat loss can be well represented by the sum of (i) the convection coefficient $h_{natconv}$ and (ii) a "convection coefficient" for radiation h_{rad}.

Vertical surfaces. Consider a hot vessel (e.g. furnace) in a large enclosure (e.g. manufacturing building) undergoing heat loss by natural convection (Equations 3.20) and radiation (Equations 3.21), with the combined heat flow given in Equation (3.22). The letter V inside the braces stands for vertical and *Gr* is the Grashof number (defined next).

$$\dot{Q}_{natconv} = \boxed{h_{natconv} A_{vessel}(T_{vessel} - T_\infty)} \qquad (3.20)\{V\}$$

$$\boxed{h_{natconv} = \frac{Nu_{avg} \cdot k_{fluid}}{L_{vess}}} \qquad (3.20a)\{V\}$$

$$Nu_{avg,verticalwall} = \left[0.825 + \left(\frac{0.387 \cdot (Pr \cdot Gr_L)^{1/6}}{\left[1 + \left(\frac{0.492}{Pr}\right)^{9/16}\right]^{8/27}}\right)\right]^2 \qquad (3.20b)\{V\}$$

$$Gr_L = \frac{g\beta(T_{vessel} - T_\infty)L^3}{\nu^2} \qquad (3.20c)\{V\}$$

$$\dot{Q}_{rad} = \sigma\varepsilon_{vess}A_{vess}\left(T_{vess}^4 - T_{surr}^4\right) = \boxed{h_{rad}A_{vess}(T_{vess} - T_{surr})} \quad (3.21a)\{F_{12}=1, F_{22}\approx 1\}$$

$$\boxed{h_{rad} = \sigma\varepsilon_{vess}(T_{vess} + T_{surr})\left[T_{vess}^2 + T_{surr}^2\right]} \quad (3.21b)\{F_{12}=1, F_{22}\approx 1\}$$

$$\dot{Q}_{total} = \dot{Q}_{natconv} + \dot{Q}_{rad} = \boxed{(h_{natconv} + h_{rad})A_{vess}(T_{vess} - T_{\infty})} \quad (3.22)$$

The "radiation heat transfer coefficient" is highly dependent on the two temperatures (vessel, surroundings), so if the vessel exterior temperature is unknown, the solution must be iterative. Nevertheless, the simplicity of the formula is appealing, and a first approximation of the heat loss is easily obtained if the vessel temperature is assumed to be limited by human skin contact safety requirements (see Chapter 13 of this textbook) to $T_{vessel,\ max} \leq 333.15\ K\ (= 60\ °C)$. The design of internal refractory and insulation layers to preclude vessel surface temperatures from exceeding $T_{vessel,\ max}$ may be challenging but possible if material parameters are known.

Where other process considerations preclude the enforcement of a limit on $T_{vessel,\ max}$ ($\leq 60\ °C$), furnace operators should incorporate guards and/or warnings to help prevent employees from inadvertent skin contact burn injuries (see Chapter 13).

Horizontal surfaces. For horizontal surfaces, the natural convection is not as efficient as it is for vertical surfaces, and the convection also depends on which surface (upper or lower) is exposed to the convecting fluid, and whether that fluid is cooler or warmer. Convection is reduced when the fluid tends to stagnate (e.g. when the buoyancy is stabilized by the presence of a horizontal wall). The fluid layers remain stable if (i) a warm plate's lower surface is exposed to a cooling fluid or (ii) a cool plate's upper surface is exposed to a warming fluid. For these less vigorous cases, a different expression (Equation 3.23) is required:

$$Nu_{horiz,stagnant} = 0.27\ (Gr_{L_c} \cdot Pr)^{1/4} \quad (3.23)\{\text{H, stagnant}\}$$

For the more vigorous cases of (i) upper surface cooling of a warm plate or (ii) lower surface heating of a cool plate, Equations 3.20 (for vertical plates) may be used. In both horizontal plate cases (vigorous or stagnant), the characteristic length is not taken to be the plate length along the flow direction (as was applicable to the vertical plate), but rather it should be computed as the ratio of the plate's surface area divided by its perimeter, $L_c = A_s/p$.

3.12 Radiant Exchange

When radiation heat transfer cannot be simplified to the case of a small hot object losing heat to a large cold surrounding, the equations of radiant exchange must be implemented. As will be shown later, the radiant exchange equations will reduce to the "hot object – large room" Equations (3.21), when the view factor of the large room to itself is approximately unity (i.e. the case where $F_{12} = 1.00$; $F_{11} = 0$; $F_{21} \rightarrow 0$; $F_{22} \rightarrow 1.00$).

For a generic two-surface enclosure, the electric circuit analogy can be applied to the radiative heat transfer problem – with resistances for the emissivity of each surface and the view factor between the two surfaces. In the general case (excluding the case where the two surfaces are very large parallel plates), at least one of the two surfaces will have a nonzero "self" view factor. The most general case is where both surfaces have nonzero "self" view factors (i.e. $F_{12} < 1.0$; $F_{21} < 1.0$; $F_{11} > 0.0$, $F_{22} > 0.0$).

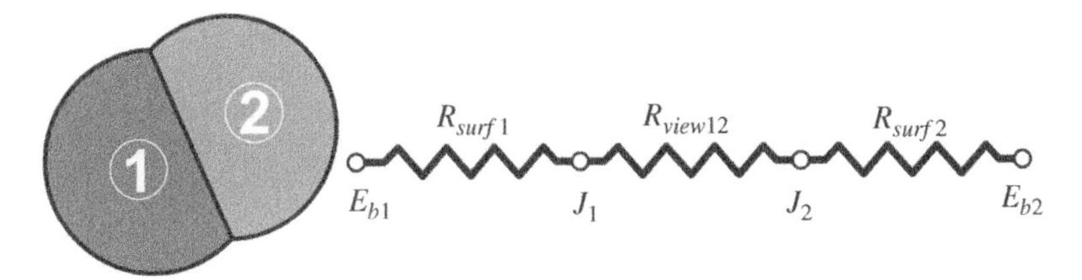

Figure 3.3 Resistance network for radiative exchange between two concave surfaces facing each other and nothing else. *Source:* Martin Thermal Engineering Inc.

Assume that the two surfaces are at different temperatures ($T_1 > T_2$) and have different emissivities, and assume that steady state has been reached. By this definition, the system is not isolated from its surroundings. In order for the two surfaces to have steady and different temperatures while engaged in radiative exchange, each must have an unspecified external heat source or sink available (e.g. via conduction) to maintain its temperature steadily over time.

The resistance network for this problem is illustrated in Figure 3.3:

The "black surface" driving force for radiation is the ***blackbody emissive power*** $\left(E_{b1} = \sigma T_1^4 \doteq W/m^2\right)$. The ***radiosity*** driving force differs from the ***blackbody emissive power*** driving force by the ***surface resistance*** (due to its emissivity being less than 1.0) times the actual heat exchange $\left(\dot{Q}_{12}\right)$ between surface 1 and surface 2, as shown in Equation (3.24):

$$E_{b1} - J_1 = \dot{Q}_{12} \cdot R_{surf1} = \dot{Q}_{12} \cdot ((1-\varepsilon_1)/\varepsilon_1 A_1) \tag{3.24}$$

This equality is given in Equation (3.25) in terms of the heat flow:

$$\dot{Q}_{12} = \frac{\overbrace{E_{b1} - J_1}^{\substack{\textit{Difference}\\ \textit{between}\\ \textit{Black and Gray}\\ \textit{Driving Forces}}}}{\underbrace{R_{surf1}}_{\substack{\textit{Surface}\\ \textit{(Emissivity)}\\ \textit{Resistance}}}} \quad \text{where,} \quad R_{surf1} = \frac{(1-\varepsilon_1)}{\varepsilon_1 A_1} \tag{3.25}$$

It is noteworthy to observe two points: (i) the surface resistance goes up when the area and/or the emissivity go down; (ii) a condition of zero ***surface resistance*** occurs when the emissivity is $\varepsilon = 1$ for a black surface.

Similarly, for surface 2, the same type of relationship exists. However, the heat flow out of surface 2 is negative (i.e. heat inflow occurs), so the ***radiosity*** driving force must be *higher* than the ***blackbody emissive power*** driving force, as seen in Equation (3.25).

$$\dot{Q}_{12} = \frac{J_2 - E_{b2}}{R_{surf2}}; \qquad R_{surf2} = \frac{(1-\varepsilon_2)}{\varepsilon_2 A_2} \tag{3.26a, b}$$

Finally, while addressing the relationship between the two radiosities, their potential values will differ by the view resistance times the heat flow, as seen in Equation (3.26). Here, it is appropriate to use the view factor reciprocity relationship ($A_1F_{12} = A_2F_{21}$).

$$\underbrace{\dot{Q}_{12}}_{\text{Radiative Heat Flow}} = \frac{\overbrace{J_1 - J_2}^{\text{Difference between Radiosities}}}{\underbrace{R_{view12}}_{\text{View Resistance}}}; \qquad R_{view12} = \frac{1}{A_1F_{12}} \tag{3.27a, b}$$

Since all the resistances are in series, the total resistance can be computed as the sum of the three individual resistances. Equation (3.27) is written with the overall driving force in the numerator:

$$\boxed{\dot{Q}_{12} = \frac{E_{b1} - E_{b2}}{R_{tot}}}; \qquad \boxed{R_{tot} = \frac{(1-\varepsilon_1)}{\varepsilon_1 A_1} + \frac{1}{A_1F_{12}} + \frac{(1-\varepsilon_2)}{\varepsilon_2 A_2}} \tag{3.28a, b}$$

One (possibly counterintuitive) result of this relationship is that reducing the emissivity of one of the surfaces will not always cause its ***radiosity*** to decrease. This is due to the fact that the potentials in the circuit must decline monotonically in the direction of heat flow. For the present example, where $T_1 > T_2$, the driving force potential must decline according to the sequence: $\boldsymbol{E_{b1} > J_1 > J_2 > E_{b2}}$.

The first example is relatively intuitive. Suppose the emissivity of surface 1 is decreased (say, for example, from $\varepsilon_1 = 0.9$ to $\varepsilon_1 = 0.1$). This will increase the first resistance (R_{surf1}) and the overall resistance (R_{tot}). Increasing the first resistance causes both radiosities J_1 and J_2 to ***decrease***, while the overall heat transfer rate $\dot{Q}_{12}$ ***decreases***.

The second example may not be intuitive. Suppose the emissivity of surface 2 is decreased (say, from $\varepsilon_2 = 0.9$ to $\varepsilon_2 = 0.1$). This will increase the third resistance (R_{surf2}) and the overall resistance (R_{tot}). Increasing the third resistance causes both radiosities J_1 and J_2 to ***increase***, while the overall heat transfer rate $\dot{Q}_{12}$ ***decreases***.

These results become more intuitive if one thinks of all driving force potentials (E_{b1}, E_{b2}, J_1, J_2) as voltages in a circuit or as temperatures in a conduction path with varying (electrical or thermal) resistors between them.

The urge to think of ***radiosity*** as a measure of *reduced heat flow* when (as compared to ***blackbody emissive power***) is an inappropriate paradigm, even though increased resistance does indeed result in reduced heat flow. When ***radiosity*** and ***blackbody emissive power*** are correctly thought of as *potentials* for heat flow, and when heat flow ($\dot{Q}$) is rightly perceived as depending on both potential difference ($E_{b1} - J_1$) and resistance $[(1 - \varepsilon_1)/\varepsilon_1 A_1]$, a physically correct intuition will be the result.

Small object, large room. Next, this analysis must be circled back to the assertion posited at the beginning of this section that the radiant exchange equation developed for a two-surface enclosure reduces to the ***small hot object – large cold room*** equation whenever $A_2 \gg A_1$ and $F_{12} = 1.0$. The essence of this physical situation is that $A_1 \rightarrow 0$, but the radiant exchange expression must be manipulated (per Equation 3.29) before that limiting condition is implemented.

$$R_{tot} = \frac{(1-\varepsilon_1)}{\varepsilon_1 A_1} + \frac{1}{A_1 F_{12}} + \frac{(1-\varepsilon_2)}{\varepsilon_2 A_2} = \frac{1}{A_1}\left\{\frac{(1-\varepsilon_1)}{\varepsilon_1} + \frac{1}{F_{12}} + \left[\frac{(1-\varepsilon_2)}{\varepsilon_2}\right]\frac{A_1}{A_2}\right\} \tag{3.29}$$

By inspection of the expression aforementioned, it is seen that the middle term $(1/F_{12})$ is unity and the third term $([(1-\varepsilon_2)/\varepsilon_2] \cdot [A_1/A_2])$ is zero, which leads to Equations (3.30) and (3.31), which are equivalent to Equations (3.21) (in which view factors are entirely absent).

$$R_{tot} = \frac{1}{A_1}\left\{\frac{(1-\varepsilon_1)}{\varepsilon_1} + 1 + 0\right\} = \frac{1}{A_1}\left\{\frac{(1-\varepsilon_1)}{\varepsilon_1} + \frac{\varepsilon_1}{\varepsilon_1}\right\} = \frac{1}{A_1}\left\{\frac{(1-\varepsilon_1+\varepsilon_1)}{\varepsilon_1}\right\} \tag{3.30}$$

$$\boxed{R_{tot} = \frac{1}{A_1 \varepsilon_1}} \quad (3.31)\{F_{12}=1, F_{22}\approx 1\}$$

Large parallel plates. Another simplified condition that may be useful for certain heat transfer problems (e.g. adjacent furnaces with hot walls radiating to each other) is the case of ***two very large parallel plates*** that see essentially nothing but each other. The term "essentially nothing" here means that the gap between the two large plates is sufficiently small compared to their areas, that the rate of thermal radiation lost to the surroundings through the gap is negligibly small compared to the rate of thermal radiation transferred from the hot plate to the cold plate. Again, the starting point is the expression for overall resistance in a two-surface enclosure, and the simplifications implemented are: $A_1 = A_2$ and $F_{12} = F_{21} = 1.0$ plus $F_{11} = F_{22} = 0$, with the final result being given as Equation 3.32.

$$R_{tot} = \frac{(1-\varepsilon_1)}{\varepsilon_1 A_1} + \frac{1}{A_1 F_{12}} + \frac{(1-\varepsilon_2)}{\varepsilon_2 A_2} = \frac{1}{A}\left\{\frac{(1-\varepsilon_1)}{\varepsilon_1} + 1 + \frac{(1-\varepsilon_2)}{\varepsilon_2}\right\} = \frac{1}{A}\left\{\left(\frac{1}{\varepsilon_1} - \frac{\varepsilon_1}{\varepsilon_1}\right) + 1 + \left(\frac{1}{\varepsilon_2} - \frac{\varepsilon_2}{\varepsilon_2}\right)\right\}$$

$$\boxed{R_{tot} = \frac{1}{A}\left\{\frac{1}{\varepsilon_1} + \frac{1}{\varepsilon_2} - 1\right\}} \quad (3.32)\{F_{12}\approx 1, F_{21}\approx 1\}$$

Thermal radiation to sky. When computing radiation heat loss from a surface (e.g. a building roof) to the sky, environmental factors play a large role in establishing thermal equilibrium. The most plentiful atmospheric gases N_2, O_2, and Ar are effectively transparent to thermal radiation, but the minor gases H_2O and CO_2 have nonzero absorptivity and provide an effective radiation shield between earth surfaces ($-40\,°C < T_{earth} < 40\,°C$) and deep space ($T_{space} < -270\,°C$).

The concentration of water vapor and the presence of water droplets (clouds) in the atmosphere play the biggest roles in determining the effective sky temperature. For very dry, very cold atmospheric conditions, the night sky temperature depends mostly on the CO_2 concentration in the atmosphere, which is relatively stable compared to the range of possible water vapor concentrations. Under this extreme condition, an assumed night sky temperature of $T_{sky} \approx 230\ K$ is reasonable and represents a conservative design assumption for computing building heat loss in most locations. With this assumption, the sky can be considered to have emissivity $\varepsilon = 1.00$ and the object losing heat to the sky can be analyzed as a small object in a large enclosure (Equations 3.21 and 3.31).

When ambient humidity and cloud cover are significant, the assumed sky temperature should be given a higher value, with the upper limit being the current air temperature at the earth's surface (e.g. $275\ K < T_{sky} < 310\ K$).

3.13 Types of Heat Exchangers

A heat exchanger is a piece of equipment whose purpose is to transfer heat compactly and efficiently from one flowing fluid to another. Some heat exchangers (e.g. boilers, condensers) also effect a phase change because of the heat transfer.

The simplest heat exchanger is a ***parallel-flow, double-pipe*** heat exchanger (Figure 3.4a). The larger pipe forms the outer wall for the isolation and transport of fluid 2 from left to right, and the smaller pipe simultaneously forms the inner wall for the isolation of fluid 2 as well as the outer wall for the isolation and transport of fluid 1. A slightly more complex version of the double-pipe heat exchanger is the ***counterflow*** configuration, where the flow direction is reversed for fluid 2 – entering at the right and exiting at the left.

Slightly more complex is the ***shell-and-tube*** style of heat exchanger (Figure 3.4b). The concept is the same as the double pipe, but one shell contains dozens or hundreds of tubes. This feature permits the heat exchange area to be increased dramatically, which permits greater mass flow rates of both fluids and a much higher overall heat transfer rate.

Another variation is the so-called ***finfan*** (Figure 3.4c, similar to an automobile radiator). In this design, the two fluid streams are perpendicular to each other, and the configuration is "crossflow." Aluminum fins are brazed to the outside of the copper or aluminum hot fluid tubes, and cool air is drawn through the entire arrangement by the fan.

The image in Figure 3.4d is a ***regenerative*** heat exchanger. The unique feature of this design is that the heat transfer surface is a circular disk comprised of a high surface area, porous, thermal storage medium (either metal or ceramic) that rotates between the hot and cold streams. While one (cooler) sector of the disk rotates through the hot stream, it stores thermal energy, and another (hotter) disk sector rotates through the cold steam and rejects thermal energy.

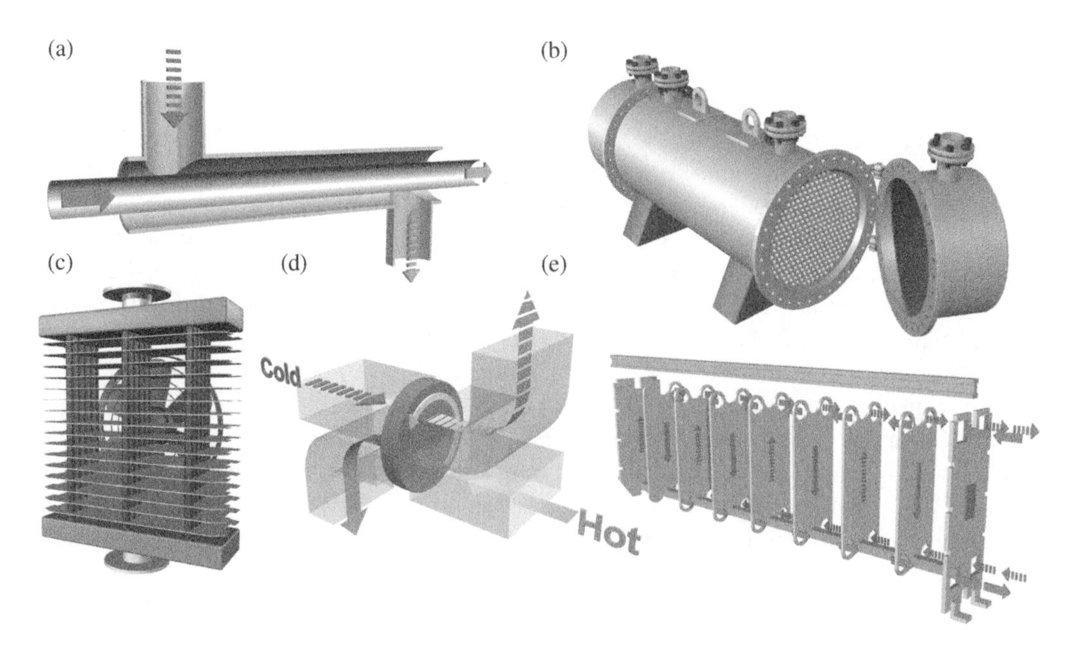

Figure 3.4 Five examples of commercially available heat exchangers: (a) double-pipe, parallel-flow; (b) shell-and-tube; (c) finfan crossflow; (d) regenerative disk; (e) plate-and-frame. *Source:* Martin Thermal Engineering Inc.

Figure 3.4e shows a ***plate and frame*** heat exchanger. The primary selling point of this design is its expandability. The image shows two inlets (hot and cold fluids) and two outlets (warm and cool fluids) at the front end and several parallel flow paths (alternating hot fluid and cold fluid) in counterflow configuration, divided by metallic plates that are clamped together to reduce or eliminate leakage. By adding more plates, both the flow area and exchange area are increased, which increases the heat exchanger's capacity.

3.14 Heat Exchanger Fundamentals

Heat exchanger analysis requires a combination of First law principles and turbulent heat transfer correlations involving flows inside and outside of tubes or across tube banks. (Industrial heat exchangers are rarely designed for laminar flows, but some do exist.) The overall heat transfer is computed as an integration of local heat transfer along the length of the heat exchanger tube(s). Usually, both hot and cold fluid temperatures change along the flow path (x), which causes the driving force for heat transfer $[T_h(x) - T_c(x)]$ to vary logarithmically. The primary exceptions to this logarithmic relationship are boilers and condensers, where one fluid's temperature changes significantly, but the other fluid's temperature remains constant as it absorbs or rejects heat to accomplish the phase change.

The most important relationship in heat exchanger analysis is the First law equality of three phenomena: (i) the loss of heat by the hot fluid, (ii) the gain of heat by the cold fluid, and (iii) the transfer of heat through convection and conduction across the tube wall, as shown in Equations (3.33). This three-way equality is strictly true only when the heat exchanger itself is adiabatic with respect to the surroundings. For shell-and-tube heat exchangers, the likelihood of achieving this condition can be enhanced by ensuring that the fluid whose temperature is closest to the ambient temperature is flowing through the shell-side flow path.

$$\dot{Q}_{h,out} = \dot{Q}_{c,in} = \dot{Q}_{wall(\text{h-to-c})} \tag{3.33a}$$

$$\dot{m}_h c_{p_h}(T_{h_{in}} - T_{h_{out}}) = \dot{m}_c c_{p_c}(T_{c_{out}} - T_{c_{in}}) = \int_{A_w} U[T_h(x) - T_c(x)]dA \tag{3.33b}$$

3.15 Overall Heat Transfer Coefficient

In the expression given in Equation (3.33b), UA is the product of the overall heat transfer coefficient (U) and the total heat transfer surface area (A_s). The UA product is the reciprocal of the overall thermal resistance (R_{tot}) between the hot fluid and the cold fluid, which is comprised of three individual terms (see Equation 3.34):

$$\begin{aligned} R_{tot} &= R_{conv,outside} + R_{cond,wall} + R_{conv,inside} \\ &= \frac{1}{h_o A_o} + \frac{\ln(r_o/r_i)}{2\pi L k} + \frac{1}{h_i A_i} \end{aligned} \tag{3.34}$$

Usually (i.e. when the inside and outside surfaces are clean and free of fouling substances, which can add substantial resistance), the wall resistance can be neglected, which gives Equation (3.35):

$$\boxed{R_{tot} \approx \frac{1}{h_o A_o} + \frac{1}{h_i A_i} = \boxed{\frac{1}{UA}}} \tag{3.35}$$

The inside and outside areas are different $A_i \neq A_o$ (due to nonzero tube wall thickness), but fortunately this minor discrepancy can be remedied fully by always grouping the U and A together and calling the combined term the "UA-product" [$\triangleq W/K$], which is the reciprocal of the overall resistance R_{tot}. When one convection coefficient is much larger than the other (e.g. boiling or condensation), its resistance can be neglected and the UA-product consists of a single term (the less robust convection coefficient times its contact area).

3.16 LMTD Method

As stated previously, the temperature difference between the hot fluid and cold fluid varies logarithmically along the flow path of the two fluids, and the expression for total heat transferred from hot fluid to cold fluid is determined by the integration of temperature difference from one end of the heat exchanger to the other. For an infinitesimal slice through the heat exchanger, an elemental amount of heat transferred from the hot fluid to the cold fluid is illustrated in Figure 3.5. An enlightening treatment of this phenomenon is given in Appendix C of Kays and London (1984) and is discussed in general terms in Section 7.7.

In addition to applying the First law to a control volume that encompasses the entire heat exchanger, the First law can also be applied to the incremental surface area and the incremental amount of heat transferred, per Equation 3.36 and Figure 3.5.

$$d\dot{Q}_{wall} = U[T_h(x) - T_c(x)]dA = -\dot{m}_h c_{p_h} dT_h = \dot{m}_c c_{p_c} dT_c \tag{3.36}$$

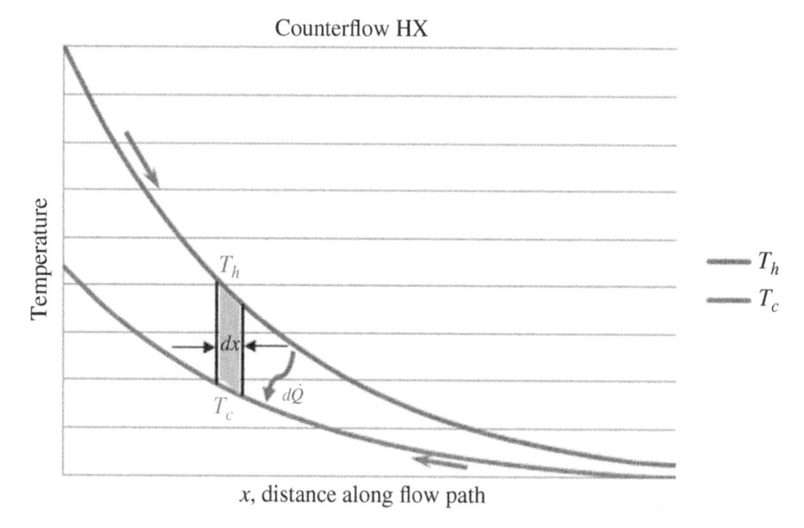

Figure 3.5 Illustration of incremental local heat transfer at an arbitrary location within a counterflow heat exchanger where the hot fluid has the smaller value of $\dot{m}c_p$ product. *Source:* Adapted from Kays and London (1984).

Subtracting the incremental change in the cold fluid temperature from the incremental change in hot fluid temperature, it is seen (Equation 3.37) that this difference is equal to the incremental change in the hot—cold temperature difference in the forward direction along the heat exchanger path.

$$\underbrace{dT_h - dT_c}_{\substack{Difference \\ of\ two\ changes}} = \underbrace{d(T_h - T_c)}_{\substack{Change\ of \\ a\ difference}} \tag{3.37}$$

Then the First law equality between the hot fluid's heat loss and the cold fluid's heat gain is used as a substitute for the difference term on the left (see Equation 3.38):

$$(dT_h - dT_c) = -\frac{d\dot{Q}}{\dot{m}_h c_{p_h}} - \frac{d\dot{Q}}{\dot{m}_c c_{p_c}} = -d\dot{Q}\left(\frac{1}{\dot{m}_h c_{p_h}} + \frac{1}{\dot{m}_c c_{p_c}}\right) = d(T_h - T_c) \tag{3.38}$$

After a moderate amount of nimble algebra, the log mean temperature difference (LMTD) result for a heat exchanger is obtained in Equation (3.39). Students are cautioned that Equation (3.39) is intended solely for counterflow and parallel flow geometries; it does *not* give accurate results for multipass or crossflow heat exchanger geometries (except in the limiting case where $C_{min}/C_{max} = 0$). For crossflow and multipass heat exchanger configurations, an empirical correction factor (usually designated F) must be found from a figure and applied to the counterflow solution. We do not reproduce these figures here because they fail to provide physical intuition, and thus, we encourage students to abandon the LMTD method in favor of the Effectiveness-NTU (number of transfer units) method explained in the next section.

$$\boxed{\dot{Q} = F(UA)_{overall}\Delta T_{lm}} = F(UA)_{overall}\frac{\left[\overbrace{(T_{h2} - T_{c2})}^{\substack{Temp\ Diff \\ at\ Right\ End}} - \overbrace{(T_{h1} - T_{c1})}^{\substack{Temp\ Diff \\ at\ Left\ End}}\right]}{\underbrace{ln\left[\frac{(T_{h2} - T_{c2})}{(T_{h1} - T_{c1})}\right]}_{\substack{Log\ of\ Right\ Temp\ Diff \\ Divided\ by\ Left\ Temp\ Diff}}} \tag{3.39}$$

As stated, $F = 1$ for counterflow and parallel flow heat exchangers, but for crossflow exchangers, F must be determined graphically (e.g. from Cengel and Ghajar 2015 or Bergman et al. 2019). Thankfully, such empiricism is unnecessary with the Effectiveness-NTU method.

3.17 Effectiveness-NTU Method

The Effectiveness-NTU method is derived from First law in the same way as the LMTD method, and both methods produce the same results. Often, the First law needs to be applied in conjunction with one of the Effectiveness-NTU equations, but the combination is very powerful.

Effectiveness is defined in Equation (3.40) as the actual heat transfer in the numerator divided by the maximum possible heat transfer in the denominator.

$$\varepsilon = \frac{\dot{Q}_{act}}{\dot{Q}_{max\,,possible}} \tag{3.40}$$

The actual heat transfer (numerator) can be determined either (i) as the heat flow into the cold fluid or (ii) the heat flow out of the hot fluid, because they are equal if an adiabatic heat exchanger is being analyzed. If the heat exchanger is assumed adiabatic with respect to heat losses to the environment, the equality in Equation (3.41) holds.

$$\dot{Q}_{act} = -\dot{m}_h c_{p_h}(T_{h,out} - T_{h,in}) = \dot{m}_c c_{p_c}(T_{c,out} - T_{c,in}) \tag{3.41}$$

The maximum possible heat transfer (denominator) occurs when the fluid with the minimum value of the mass flow specific heat product $(\dot{m}c_p)_{min}$ undergoes the maximum possible temperature change $(T_{h,\,in} - T_{c,\,in})$. The two mass-flow-specific-heat-product parameters are named C_{min} and C_{max}.

A violation of the First or Second law of thermodynamics would occur if the C_{max} fluid underwent the maximum possible temperature change because the First law would then require the C_{min} fluid to undergo a larger temperature change than the C_{max} fluid, which would necessitate the C_{min} fluid's exit temperature to exceed (negative or positive) the inlet temperature of the C_{max} fluid.

The maximum possible heat transfer $\dot{Q}_{max\,possible}$ is always constrained by the Second law and is defined by the extremes of the hot inlet and cold inlet temperatures as well as the minimum fluid's $\dot{m}c_p$ product. This denominator $(\dot{Q}_{max\,possible})$ remains the same for all geometries – counterflow, parallel flow, cross flow, and for all values of the ratio C_{min}/C_{max}. Rewriting Equation (3.40) in terms of enthalpy-carrying capacity and temperatures, Equation (3.42) is obtained:

$$\varepsilon = \frac{\dot{Q}_{act}}{\dot{Q}_{max\,possible}} = \frac{C_{minfluid}(\Delta T)_{minfluid}}{C_{minfluid}(\Delta T)_{max\,poss}} = \frac{(\Delta T)_{minfluid}}{(\Delta T)_{max\,poss}} \tag{3.42a}$$

$$\varepsilon = \begin{cases} \dfrac{(T_{h,in} - T_{h,out})}{(T_{h,in} - T_{c,in})} \; if\ h = min \\ \dfrac{(T_{c,out} - T_{c,in})}{(T_{h,in} - T_{c,in})} \; if\ c = min \end{cases} \tag{3.42b}$$

Think Stop. Referring back to Equation 3.32 (assuming the integration can be completed), the three terms equated with each other represent a "product" times a "temperature subtraction":

$$\underbrace{\dot{m}_h c_{p_h}}_{mc\ prod}\underbrace{(T_{h_{in}} - T_{h_{out}})}_{\Delta T,\,chg} = \underbrace{\dot{m}_c c_{p_c}}_{mc\ prod}\underbrace{(T_{c_{out}} - T_{c_{in}})}_{\Delta T,\,chg} = \int_{A_w}\underbrace{[T_h(x) - T_c(x)]}_{\Delta T,\,diff}\underbrace{UdA}_{UA\ prod}$$

Like the UA product, the two $\dot{m}c_p$ products have units of $[\doteq W/K]$. As such, the ratio UA/C_{min} is dimensionless, and it is a relative measure of the "size" of the heat exchanger (UA) as compared to the enthalpy-carrying capacity of the minimum fluid (C_{min}). This ratio is called the number of transfer units (NTU).

By integrating $d\dot{Q}$ in various heat exchanger geometries, relationships for heat exchanger effectiveness versus NTU can be derived (see Section 7.7). The expressions in Equations (3.43)–(3.48) constitute the equations deemed most useful for thermal system designers engaged in heat exchanger sizing. For crossflow geometries, the terms ***unmixed*** and ***mixed*** are related to the presence or absence of fins that divide one or both streams into well-defined channels that prevent commingling. Depictions of these geometries can be found in Kays and London (1984), Cengel and Ghajar (2015), or Bergman et al. (2019).

$$\varepsilon_{parallel} = \frac{1 - \exp\left[-NTU \cdot \left(1 + \frac{C_{min}}{C_{max}}\right)\right]}{\left(1 + \frac{C_{min}}{C_{max}}\right)} \tag{3.43}$$

$$\varepsilon_{counter} = \frac{1 - \exp\left[-NTU \cdot \left(1 - \frac{C_{min}}{C_{max}}\right)\right]}{1 - \frac{C_{min}}{C_{max}} \cdot \exp\left[-NTU \cdot \left(1 - \frac{C_{min}}{C_{max}}\right)\right]} \quad \text{for } \frac{C_{min}}{C_{max}} < 1 \tag{3.44a}$$

$$\varepsilon_{counter} = \frac{NTU}{1 + NTU} \quad \text{for } \frac{C_{min}}{C_{max}} = 1 \tag{3.44b}$$

$$\varepsilon_{cross,u-u} = 1 - \exp\left\{\frac{NTU^{0.22}}{\frac{C_{min}}{C_{max}}}\left[\exp\left(-\frac{C_{min}}{C_{max}} \cdot NTU^{0.78}\right) - 1\right]\right\} \tag{3.45}$$

for C_{max} *unmixed* and C_{min} *unmixed*

$$\varepsilon_{cross,\,min\,=\,u} = \left(\frac{C_{min}}{C_{max}}\right)^{-1} \cdot \left(1 - \exp\left\{-\frac{C_{min}}{C_{max}}\left[1 - \exp\left(-NTU\right)\right]\right\}\right) \tag{3.46}$$

for C_{max} ***mixed*** and C_{min} *unmixed*

$$\varepsilon_{cross,\,max\,=\,u} = 1 - \exp\left\{-\left(\frac{C_{min}}{C_{max}}\right)^{-1}\left[1 - \exp\left(-\frac{C_{min}}{C_{max}} \cdot NTU\right)\right]\right\} \tag{3.47}$$

for C_{max} *unmixed* and C_{min} ***mixed***

$$\varepsilon = 1 - \exp\left(-NTU\right) \quad \text{for all HX with } \frac{C_{min}}{C_{max}} = 0 \tag{3.48}$$

Think Stop. Students should ponder the following facts about these NTU relationships:

- For very small heat exchangers ($NTU < 0.25$), all geometries and all values of C_{min}/C_{max} can be approximated by the straight line $\varepsilon \approx NTU$, but the error introduced may be unacceptable for values of $NTU > 0.1$. Students are encouraged to use this approximation only when the magnitude of the error introduced is acceptable.
- For $NTU \to \infty$, counterflow effectiveness approaches 100%, but certain other geometries fall short of 100% (students are tasked with demonstrating this in one of the end-of-chapter homework problems.)

3.18 Porous Media Heat Transfer

Investigative research into porous media heat transfer has been active for many years in the petroleum and chemical industries and has become acutely active in recent years in relation to nanofluids and nanoparticles in many industries. Engineers are continuously seeking new heat transfer solutions that can improve the performance of a heat exchanger through a reduction in the heat flow resistance. One of the primary advantages of porous media is their high value of surface area per unit volume (also called *SSA*, or specific surface area). Because *SSA* is in the denominator of the convection resistance term ($R_{conv} = 1/(hA)$), high *SSA* can lead to low heat transfer resistance (and high heat conductance).

Kays and London (1984) addresses the basic problem of all heat exchangers on its very first page, but this admonishment applies even more acutely to heat exchangers that incorporate porous media:

> The design of a heat exchanger involves a consideration of both the heat transfer rates between the fluids and the mechanical pumping power expended to overcome friction and move the fluids through the exchanger. ... For low density fluids, such as gases, it is very easy to expend as much mechanical energy in overcoming friction power as is transferred in heat.

As the book title suggests, by reducing velocities, and increasing surface areas, a ***compact*** heat transfer solution can be obtained. One of the subjects covered in detail in Kays and London is the pressure drop and heat transfer associated with fluids flowing through a porous medium (see table 3-3, figure 7-10, and table 10-13 of Kays and London 1984). The geometry they explore thoroughly is a bed of randomly packed spheres. (It should be noted that when spheres are closely and regularly packed, like the face-centered-cubic and body-centered-cubic atom geometries from crystallography (see Section 11.5), the void fraction is significantly lower than when they are randomly packed.

As described previously in Section 2.13, the Ergun equation was developed to correlate friction factor to the flow conditions in a bed of randomly packed spheres. The development in Kays and London provides results for both friction and heat as functions of the Reynolds number related to the packing's interstices. (For pressure drop, the Kays and London expression for Reynolds number is different from the Ergun expression for modified Reynolds number, so the two relationships must not be commingled, but the results are nearly identical, and either equation may be used with confidence.)

Think Stop. For porous media heat transfer analyses, the goal is to predict the convection coefficient h as a function of fluid properties, flow properties, and packing properties. Consider the situation where a cylindrical vessel is randomly packed with spheres at some initial temperature T_i. At time t_0^+ a flow of a hot fluid T_{h0} is introduced into the bottom of the vessel and the question to be answered is: what is the instantaneous rate of heat transfer $\dot{Q}$ from the fluid to the solid packing material at that moment?

The heat transfer rate, $\dot{Q}$ will change as a function of time and space, because the temperatures of both the hot fluid and cold bed are changing in both realms. A computation of the convection coefficient h may be valid for longer periods of time and across larger distances, but the initial question of what happens at time t_0^+ is easier to grasp.

The heat transfer rate is a function of the convection coefficient h, the specific surface area SSA of the packing (here relabeled as α), the packed volume of the vessel $\mathcal{V}$, and the fluid to packing temperature difference ΔT (see Equations 3.49).

$$\dot{Q} = hA_s(T_{h0} - T_i) = h \cdot \overbrace{\alpha_{spheres}}^{\left[\doteq \frac{m^2_{spheres}}{m^3_{vessel}}\right]} \cdot \mathcal{V}_{vessel}(T_{h0} - T_i) \tag{3.49a}$$

$$\alpha_{spheres} = \frac{3}{r_{spheres}}\left(1 - \overbrace{\sigma}^{\left[\doteq \frac{m^3_{void}}{m^3_{vessel}}\right]}\right) \tag{3.49b}$$

It is apparent from the second equation that the specific surface area α is a function of only the sphere's radius r and the packing's void fraction σ, where smaller values of both σ and r lead to larger values of α, which enhances the heat transfer rate $\dot{Q}$ and also increases the frictional pressure drop.

The determination of h for a bed of packed spheres is relatively straightforward. Kays and London (1984) provide an equation and a plot of Stanton number as a function Reynolds number and Prandtl number. The following parameters (Equations 3.50 and 3.51) are important to this analysis:

- Hydraulic radius of the interstitial space:

$$r_h = \frac{r_{sphere}}{3} \frac{\sigma}{(1-\sigma)} \tag{3.50a}$$

- Reynolds number:

$$Re = \frac{4\rho V_{interstitial} r_h}{\mu} \tag{3.50b}$$

- Stanton Number:

$$St = \frac{h}{\rho V_{interstitial} c_p} \tag{3.50c}$$

- Interstitial velocity:

$$V_{interstitial} = \frac{\dot{m}}{\rho A_{frontal} \sigma} \left(= \frac{V_{superficial}}{\sigma} \right) \tag{3.50d}$$

The equation for Stanton number is:

$$\boxed{St = 0.23\ Re^{-0.3} Pr^{-2/3}} \tag{3.51}$$

Where gas density changes significantly as its temperature changes from the inlet to the exit of the packed bed, it is usually recommended to take the average density as the reciprocal of the average of the two specific volumes (inlet and exit).

3.19 External Convection to Individual Spheres and Cylinders

While most practical thermal systems can be designed without any assessment of convective flow across the exterior of individual spheres and cylinders, some unique problems may need such analysis. In contrast to packed beds, which must be analyzed by porous media methods given in Sections 2.14 and 3.18, and tube banks, which must be analyzed by the methods given in Section 3.9, the heat and flow analysis of individual cylinders and spheres may be necessary when particulate matter is transported pneumatically or fluidized in a chemical reactor. In such cases, the root problems to be solved are: (i) determining the particle's terminal velocity, where drag force equals gravity force, and (ii) determining the particle's Nusselt number at that velocity.

For individual spheres levitated by an upward flowing fluid, the force balance (where drag is opposed to weight or $\vec{F}_w + \vec{F}_d = 0$) can be rearranged to give the terminal velocity as shown in Equation 3.52:

$$\underbrace{m}_{\substack{\text{particle}\\\text{mass}}} \cdot \underbrace{g}_{\substack{\text{gravity}\\\text{acceleration}}} = \underbrace{C_D}_{\substack{\text{sphere}\\\text{drag}\\\text{coefficient}}} \cdot \frac{1}{2} \cdot \underbrace{\rho_f}_{\substack{\text{fluid (air)}\\\text{density}}} \cdot \underbrace{V_{rel}^2}_{\substack{\text{particle}\\\text{velocity}\\\text{relative}\\\text{to air}}} \cdot \underbrace{A_{fr}}_{\substack{\text{particle}\\\text{frontal area}}}$$

$$V_{rel} = \sqrt{\frac{4\rho_s Dg}{3C_D\rho_f}} \tag{3.52}$$

Unfortunately, the drag coefficient of a sphere varies irregularly with Reynolds number (Schlichting 1960), so the force equation appears to require iteration. However, some researchers have found empirical fit correlations for C_D as a function of Reynolds number, and their results are reasonable. We recommend the piecewise continuous representation given in Tse and Fernandez-Pello (1998) that obeys Stokes' law asymptotically at very small Reynolds numbers, deviates away from the reciprocal Reynolds behavior at moderate Reynolds number, and abruptly imposes a constant value $C_D = 0.4$ for moderate to high Reynolds numbers, as given in Equation 3.53.

$$\boxed{C_{D,sphere} = \begin{cases} \dfrac{24}{Re} \cdot \sqrt{\left(1 + \dfrac{3}{16} Re\right)} & Re < 680 \\ 0.4 & Re \geq 680 \end{cases}} \tag{3.53}$$

This formulation deviates from the empirical profile given in Schlichting (1960) when $Re \geq 2 \times 10^5$, so it should not be used at very high Reynolds numbers unless order-of-magnitude accuracy is acceptable. Designers should also note that the expressions here assume that roughness is zero. The expression for drag coefficient over a sphere is also crudely acceptable for that of a cylinder with aspect ratio $L/D \approx 1$.

For forced convection to a sphere, the Whitaker (1972) relationship is recommended, where fluid properties are taken at either free stream (∞) or surface (s) temperature, as shown in Equation (3.53).

$$Nu_{sph} = \frac{hD}{k_\infty} = 2 + \left[0.4\, Re_\infty^{1/2} + 0.06\, Re_\infty^{2/3}\right] Pr_\infty^{0.4} \left(\frac{\mu_\infty}{\mu_s}\right)^{1/4} \tag{3.54}$$

The drag coefficient for flow across long individual cylinders also follows an irregular profile when a wide enough range of Reynolds numbers is considered (see Schlichting 1960, chapter 1). A piecewise continuous equation can be developed to approximate cylinder drag in the same way we applied the Tse and Fernandez-Pello (1998) relationship for spheres. An approximate relationship that achieves factor-of-two accuracy for long cylinders is given in Equation (3.55). The expression deviates substantially from the empirical profile when $Re \geq 4 \times 10^5$, so it, too, should not be used at very high Reynolds numbers unless order-of-magnitude accuracy is acceptable.

$$C_{D,cylinder} = \begin{cases} 14.5\ Re^{-0.58} & 0.1 \le Re \le 100 \\ 1.0 & Re > 100 \end{cases} \quad (3.55)$$

The recommended expression for Nusselt number addressing heat transfer due to flow over a cylinder is by Churchill and Bernstein (1977) as given in Equation (3.56).

$$Nu_{cyl} = 0.3 + \frac{0.62\,Re^{1/2}Pr^{1/3}}{\left[1 + (0.4/Pr)^{2/3}\right]^{1/4}}\left[1 + \left(\frac{Re}{282\,000}\right)^{5/8}\right]^{4/5} \quad (3.56)$$

3.20 Homework Problems

3.1 Compute the minimum thickness of fiberglass pipe insulation ($k_{insul} = 0.05\ W/m\ °C$) needed on the outside of a steam pipe ($D_o \approx D_i = 15.0\ cm, L = 50\ m$) to ensure that condensation of the steam will not occur inside the pipe, for ambient conditions of STP and steam inlet conditions of $\dot{m} = 0.21\ kg/s, T_{in} = 400\ K, p_{in} = 200\ kPa$. Assume that frictional losses are low enough so that pressure change may be ignored along the pipe length. Assume that both convection resistances are negligible (i) from the ambient air to the outer surface of the pipe insulation and (ii) from the steam to the inner surface of the thin pipe. Estimate the error introduced by the assumption to ignore the convection resistance on the outside of the pipe insulation. Is this a conservative assumption?

Answer. Thickness $t = 5.01\ cm$

3.2 Compute the minimum thickness of polyurethane foam pipe insulation ($k_{insul} = 0.03\ W/m\ °C$) needed on the outside of a copper refrigerant pipe ($D_o = 1.2\ cm$, $D_i = 1.05\ cm$, $L = 40\ m$, $\epsilon = 4.5 \times 10^{-4}\ cm$) to ensure that liquid ammonia inside the pipe will not begin to boil for ammonia inlet conditions of $\dot{m} = 0.024\ kg/s, T_{in} = 250\ K, p_{in} = 200\ kPa$. Compute pressure loss along the pipe length and account for changes in both temperature and pressure to meet the objective.

Answer. Thickness $t = 3.02\ cm$

3.3 Compare all five heat exchanger geometries in Equations (3.43)–(3.46) and determine the geometries that (i) adhere and (ii) do not adhere to the following assertion: For $NTU \to \infty$, effectiveness approaches 100% for $C_{min}/C_{max} = 1.00$. The geometries to compare are counter-flow, parallel-flow, crossflow both unmixed, crossflow min fluid mixed, and crossflow max fluid mixed).

3.4 Demonstrate why, for heat transfer in pipe flow, the mean temperature (T_{avg}) evaluated at a single value of longitudinal location (x) must not be computed as a simple cross-sectional average of local temperature. In other words, show why $T_{avg}(x) \neq (\pi R^2)^{-1}\int_{r=0}^{R} 2\pi r T(r,x)dr$.

3.5 Compute the heat loss $[\doteq kW]$ from a building with a horizontal, flat roof ($L = 30\ m$, $W = 10\ m$) that is losing heat by natural convection ($T_\infty = -20\ °C$) and thermal radiation to the night sky under partly cloudy conditions ($T_{sky} = 253.15\ K$), if the roof insulation comprises a layer ($t = 0.2\ m$) of polyurethane foam ($k_{insul} = 0.03\ W/m\ °C, \varepsilon = 0.97$) and the indoor ceiling temperature is assumed to be $T_s = 22\ °C$.

Answer. $\dot{Q} = 1.837\ kW$

3.6 Determine (i) the UA-product (overall heat transfer coefficient U multiplied by surface area A) and (ii) the rate of heat transfer ($\dot{Q}[\ \hat{=}\ W]$) for a counterflow heat exchanger with the following characteristics: Hot fluid (tube side): mercury (Hg), with $\dot{m}_h = 305\ kg/s, T_{in} = 150°C, \rho =$ $13\ 600\ kg/m^3, \mu = 0.0015\ kg/ms, c_p = 0.14\ kJ/kgK, k = 8.40\ W/m^2K$; cold fluid (shell side): propylene glycol ($C_3H_8O_2$), with $\dot{m}_c = 190\ kg/s, T_{in} = 0°C, \rho = 1036\ kg/m^3, \mu = 0.042\ kg/ms, c_p = 2.727\ kJ/kgK, k = 0.147\ W/m^2K$; shell dimensions of $D_{i,\ shell} = 0.3\ m, L_{shell} = 2.5\ m$; and tube characteristics of $D_{o,\ tube} = 0.020\ m, D_{i,\ tube} = 0.019\ m, L_{tube} = 2.5\ m, N_{tubes} = 109$, $\epsilon = 0.2\ mm$. Assume that conduction resistance through tube wall is negligible.

Answer. $UA = 18\ 320\ W/K$; $\dot{Q} = 2.207\ kW$

3.7 Determine the log mean temperature difference for a parallel-flow, double-pipe heat exchanger with $C_{min}/C_{max} = 0.90$, $NTU = 0.25$, $T_{c,\ in} = 25\ °C$, and $T_{h,\ in} = 1000\ °C$. For this question, it does not matter which fluid is C_{min}.

Answer. $\Delta T_{lm} = 776\ °C$

3.8 Determine (i) the pressure drop, and (ii) the convective heat transfer coefficient from hot air ($T_{in} = 500\ K, p_{in} = 101.3\ kPa, \dot{m} = 3.5\ kg/s$) to a cold-packed bed of spheres ($L_{vessel} = 6.0\ m$, $D_{vessel} = 1.50\ m$, $D_{sphere} = 3.8\ cm$, $\sigma = 0.45$, $T_{init} = 300\ K$, $c_{sphere} = 0.8\ kJ/kgK$, $k_{sphere} = 1.05\ W/mK$).

Answer. (a) $\Delta p = 9.46\ kPa$; (b) $h = 115$

Cited References

Bergman, T.L., Lavine, A.S.; Incropera, F.P., and DeWitt, D.P.; (2019); *Fundamentals of Heat and Mass Transfer*, 8; Hoboken: Wiley

Cengel, Y.A. and Ghajar, A.J.; (2015); *Heat and Mass Transfer: Fundamentals & Applications*, 5; New York: McGraw-Hill.

Churchill, S.W. and Bernstein, M.; (1977); "A correlating equation for forced convection from gases and liquids to circular a cylinder in crossflow"; *J. Heat Trans.*; **99**; 300.

Fourier, J.B.J.; (1822); *Théorie Analytique de la Chaleur*, Cambridge, UK: Cambridge University Press, 2009; https://www.cambridge.org/us/academic

Kays, W.M. and London, A.L.; (1984); *Compact Heat Exchangers*, 3; New York: McGraw-Hill.

Newton, I.; (1701); "Scala graduum caloris, calorum descriptiones & signa" *Philos. T. R. Soc. London*; **22**; 824.

Schlichting, H. (translated by Kestin, J.); (1960); *Boundary Layer Theory*, 4; New York: McGraw-Hill.

Tse, S.D. and Fernandez-Pello, A.C.; (1998); "On the flight paths of metal particles and embers generated by power lines in high winds – a potential source of wildland fires"; *Fire Saf. J.*; **30**; 333

We note that this 1998 article by Tse and Fernandez-Pello is perfectly acceptable for the purpose cited in this chapter (i.e. providing an algebraic formula for drag coefficient over a sphere), but readers who desire to utilize other aspects of Tse and Fernandez-Pello's 1998 article are cautioned to scrutinize a later paper (Anthenien, Tse, and Fernandez-Pello; 2006; *Fire Safety Journal*: 41; 349) that corrects an erroneous value for a combustion rate parameter.

Whitaker, S.; (1972); "Forced convection heat transfer correlations for flow in pipe, past flat plates, single cylinders, single spheres, and for flow in packed beds and tube bundles" *AICHE J.*; **18**; 361–371.

Zukauskas, A.; (1987); "Heat transfer from tubes in cross flow"; In *Handbook of Single-Phase Convective Heat Transfer*; Kakac, S., Shah, R.K. and Aung, W. (Eds.); New York: Wiley.

Zukauskas, A. and Ulinskas, R.; (1985); "Efficiency parameters for heat transfer in tube banks"; *Heat Transf. Eng.*; **6**; 19–25.

4

Introduction to Combustion

Combustion is a chemical reaction where fuel and oxygen combine chemically to form product gases and exothermic energy. This chapter provides fundamental information for engineers to estimate temperatures, flow rates, heat release rates, and product gas composition for burners and other combustion systems. Topics covered in this chapter include balancing chemical equations, equivalence ratio, sensible and chemical enthalpy, property datum states, enthalpy of combustion, adiabatic combustion temperature, fuel-rich combustion, and brief summaries of combustion safety, and pollutant formation and control.

4.1 Fuels for Combustion

Fuels for combustion can be gas, liquid, or solid and may comprise hydrocarbons, carbohydrates, or non-carbonaceous substances. The following list identifies some common combustion fuels.

- Gaseous fuels
  - Natural gas (predominantly methane, CH_4)
  - Liquefied petroleum gas (predominantly propane, C_3H_8)
  - Butane (C_4H_{10})
  - Hydrogen (H_2)
  - Syngas (mixture of carbon monoxide, CO; hydrogen, H_2; plus minor amounts of hydrocarbons, C_xH_y)
- Liquid fuels
  - Gasoline
  - Diesel (also called #2 fuel oil)
  - Alcohol
  - Heavy fuel oils (also called #4, #6 fuel oil)
- Solid fuels
  - Wood (chips, sawdust)
  - Coal (pulverized, lump)
  - Waste (biomass, municipal, and hazardous)

Thermal Systems Design: Fundamentals and Projects, Second Edition. Richard J. Martin.

Companion website: www.wiley.com\go\Martin\ThermalSystemsDesign2

Hydrocarbon and related liquid and gaseous fuels generally fall into one of the following categories, which conveniently, all begin with the letter "A":

- Alkanes – no double bonds (e.g. methane, ethane, propane, butane, octane)
- Alkenes – one double bond (e.g. ethene, propene)
- Aromatics – rings with multiple double bonds (e.g. benzene, toluene)
- Alcohols – hydroxylated hydrocarbons (e.g. methanol, ethanol)
- Aldehydes and ketones – containing a carbonyl group (e.g. formaldehyde, acetaldehyde, acetone, methyl ethyl ketone)

It should be noted that the vast majority of combustion chemistry happens in the gas phase, even if the fuel is liquid or solid at room temperature. For condensed-phase fuels, a portion of the exothermic energy is conducted back into the fuel, causing vaporization of light liquids and pyrolization of heavy liquids and solids. (To pyrolyze a fuel is to thermally decompose it such that large molecules or molecular lattices break apart into fragments that are volatile and are transported away from the fuel's surface into the gas phase combustion zone.) With a few exceptions (e.g. combustible metals that oxidize as solids and generate solid product compounds), the visible flame is where gas phase reactions occur.

4.2 Air for Combustion

In unusual circumstances, combustion can be carried out with oxidizers other than oxygen, but for most combustion applications, the necessary oxygen is provided by the ambient air.

The "standard" composition of dry air can be found in chemistry handbooks or online (e.g. Weast 1974 or PhysLink 2020), and it includes nitrogen, oxygen, argon, carbon dioxide, neon, methane, and helium as the top seven constituents. It is noteworthy that water vapor is the third most prevalent compound in humid air (after nitrogen and oxygen), but its concentration varies so widely by geography, microclimate, and season that the standard composition of air is generally taken to be water-free or "dry."

On relatively dry days ($T_{dew} \leq 7° C$), the atmospheric water vapor concentration falls to the fourth place, behind argon. However, for simplicity this textbook (including all end-of-chapter problems) assumes that ambient, dry air is composed of exactly two components in the proportions shown as follows:

- Oxygen (g): 1 $kmol\ O_2$ in 4.76 $kmol\ air$ (21.01% by volume)
- Nitrogen (g): 3.76 $kmol\ N_2$ in 4.76 $kmol\ air$ (78.99% by volume)

If air is the oxidizing medium for a combustion reaction, the assumption is that for every mole of oxygen gas (O_2) needed to burn the fuel, 3.76 moles of nitrogen gas (N_2) *come along for the ride*. In other words, the nitrogen is essentially an inert diluent gas that absorbs heat from the exothermic reaction, but it does not participate in the chemistry. An important exception to this rule is when *NOx*, a collection of pollutants, forms during combustion, but since these gases typically have product concentrations that are quite low, the decrease in nitrogen (N_2) mole fraction is insignificant and can be ignored.

4.3 Atomic and Molar Mass

Although the atomic mass values of many elements are routinely rounded to 1 or 2 significant digits (e.g. $A_H \approx 1$; $A_{He} \approx 4$; $A_C \approx 12$) in basic chemistry problems, this textbook recommends more precision for atomic mass when performing combustion calculations. The following list (Linstrom and Mallard 2017–2021) includes atomic mass units for the elements most involved in combustion reactions. The standard atomic mass of an element is the arithmetic mean of the masses of all of the element's isotopes weighted by each isotope's abundance on Earth.

- $H = 1.007\,94 \frac{kg}{kmol}$
- $He = 4.002\,602 \frac{kg}{kmol}$
- $C = 12.0107 \frac{kg}{kmol}$
- $N = 14.0067 \frac{kg}{kmol}$
- $O = 15.9994 \frac{kg}{kmol}$
- $F = 18.9984 \frac{kg}{kmol}$
- $Si = 28.085\,5 \frac{kg}{kmol}$
- $P = 30.973\,76 \frac{kg}{kmol}$
- $S = 32.065 \frac{kg}{kmol}$
- $Cl = 35.453 \frac{kg}{kmol}$
- $Ar = 39.948 \frac{kg}{kmol}$
- $Br = 79.904 \frac{kg}{kmol}$

4.4 Balancing Chemical Equations

As stated in Section 4.1, fuels can be gases, liquids, or solids, and most commercial fuels are hydrocarbons or carbohydrates. Five simple (i.e. low molar mass) alkanes and alcohols are shown in Figure 4.1:

Balancing a chemical equation for combustion is a relatively straightforward procedure. The reactants (fuel + air) are on the left side of the equals sign, and the combustion products are on the right side. For a balanced reaction, the total number of each type of atom on the left must be equal to the total number of the same type of atom on the right. If sufficient oxygen is present to react with and consume *all* the fuel molecules, the combustion chemistry is said to have *gone to completion*. Unbalanced chemical reactions violate the atom conservation law and will lead to incorrect mass and energy balances. Students are cautioned *never* to cite or use any chemical equation until it is correctly balanced.

To clarify, a chemical reaction can be ***balanced*** and still have an *excess* or *deficiency* of fuel or oxygen. The term ***balanced*** simply means that numbers of all the atoms on the left side are exactly equal to the numbers of the same atoms on the right side.

If ***insufficient*** (i.e. less than stoichiometric) oxygen is present, products of incomplete combustion (typically carbon monoxide [CO] and hydrogen [H_2], with small amounts of unburned fuel and carbonaceous soot) will be present. This is called ***fuel-rich*** combustion. Determining a ***balanced*** equation for fuel-rich combustion is not possible without additional information (i.e. thermochemical equilibrium constants), which will be discussed in Section 4.12.

If ***sufficient*** (i.e. stoichiometric or excess) oxygen is available, the thermochemical equilibrium dictates that: (i) all C-atoms in the fuel will form carbon dioxide (CO_2, g) and (ii) all H-atoms in the fuel will form water vapor (H_2O, g). If excess air is present (also called ***fuel-lean*** combustion),

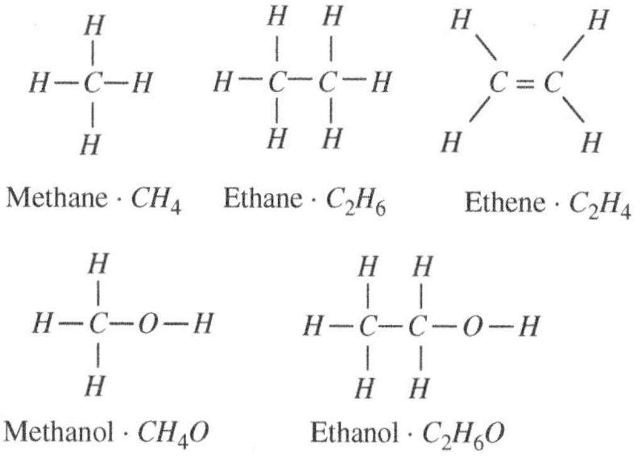

```
    H            H  H           H         H
    |            |  |            \       /
H — C — H    H — C — C — H        C  =  C
    |            |  |            /       \
    H            H  H           H         H
```

Methane · CH_4 Ethane · C_2H_6 Ethene · C_2H_4

```
    H                    H  H
    |                    |  |
H — C — O — H        H — C — C — O — H
    |                    |  |
    H                    H  H
```

Methanol · CH_4O Ethanol · C_2H_6O

Figure 4.1 Five simple fuel molecules, showing structural formulas. *Source:* Martin Thermal Engineering Inc.

oxygen molecules will be left over on the product side of the reaction, and the technique for balancing such reactions is discussed in Section 4.5.

A balanced chemical equation for one mole of a generic hydrocarbon (C_xH_y) reacting with the ***stoichiometric*** amount of oxygen (from air) can be written as follows (Equation 4.1):

$$1C_xH_y + n(O_2 + 3.76N_2) = xCO_2 + \frac{1}{2}yH_2O + (3.76n)N_2 \tag{4.1}$$

For this example, the variable n represents the stoichiometric ratio of oxygen molecules to fuel molecules with zero excess oxygen and zero excess fuel remaining on the right (product) side of the equation. This means that Equation (4.1) is only valid for ***stoichiometric*** combustion. For simplicity, we assume that exactly one fuel molecule (or exactly 1.00 *kmol* of fuel) is present as a reactant.

The process of demonstrating the equation is balanced takes place by counting carbon, hydrogen, oxygen, and nitrogen atoms on the ***left*** side of the equation (i.e. $[1x]$, $[1y]$, $[2n]$, and $[7.52n]$, respectively, in the chemical equation given earlier) and comparing those quantities to the counts of carbon, hydrogen, oxygen, and nitrogen atoms on the ***right*** side of the equation (i.e. $[1x]$, $\left[\frac{2}{2}y\right]$, $\left[2x + \frac{1}{2}y\right]$, and $[7.52n]$, respectively, in the same equation).

To balance the oxygen atoms, the variable $2n$ on the left must be equal to the sum of the variables $2x + \frac{1}{2}y$ on the right. Thus, by choosing n accordingly (i.e. $n = x + \frac{y}{4}$), the equation is ***balanced***. This method for balancing combustion reactions is valid for all values of x and y and can be extended (with modification) to oxygenated fuels (e.g. $C_xH_yO_z$).

4.5 Stoichiometry and Equivalence Ratio

Equation (4.1) gives the general format for a stoichiometric combustion reaction. Under stoichiometric conditions, the fuel and oxygen molecules are exactly in balance, and no excess fuel or oxygen remains in the product gases. An equation for ***fuel-lean*** combustion is given in Equation (4.2), where the value of n is exactly the same as that asserted in the last paragraph of Section 4.4.

$$\overbrace{\phi C_x H_y}^{\phi\, =\, equiv.ratio} \quad + \quad \overbrace{n(O_2 + 3.76N_2)}^{n\, =\, stoichiometric\ air\ for\ x,y}$$
$$= (\phi x)CO_2 + (½\, \phi y)H_2O + ((1 - \phi)n)O_2$$
$$+ (3.76n)N_2 \qquad (4.2)\{\mathrm{FL}\}$$

The variable ϕ is called the "equivalence ratio," and it represents the ratio of actual moles of fuel participating in the combustion reaction being analyzed to the stoichiometric moles of fuel (when no excess fuel or oxygen is present). Note the following definition (Equation 4.3) and terminology:

$$\boxed{\phi \equiv \frac{\left(\dot{n}_f/\dot{n}_{O2}\right)_{actual}}{\left(\dot{n}_f/\dot{n}_{O2}\right)_{stoichiometric}} = \frac{(\phi/n)}{(1/n)}} \qquad (4.3)$$

$\phi < 1$ Fuel-lean combustion
$\phi = 1$ Stoichiometric combustion
$\phi > 1$ Fuel-rich combustion

Although $\phi > 1$ can certainly occur in practical combustion systems, Equation (4.2) (which has the appended parenthetical of {*FL*} for ***fuel-lean***) is not useful for ***fuel-rich*** systems, because it presumes complete combustion, where only excess oxygen and inert nitrogen appear on both sides of the equation. Inspection shows that Equation (4.2) is only useful for ***fuel-lean*** and ***stoichiometric*** combustion because fuel-rich combustion will result in a negative value for oxygen on the right side of the equation: $(1 - \phi) < 0$ when $\phi > 1$.

Equation (4.2) can be balanced by counting carbon, hydrogen, oxygen, and nitrogen atoms on the ***left*** side of the equation (i.e. $[\phi x]$, $[\phi y]$, $[2n]$, and $[7.52n]$, respectively) and comparing those quantities to the counts of carbon, hydrogen, oxygen, and nitrogen atoms on the ***right*** side of the equation (i.e. $[\phi x]$, $[\phi y]$, $\left[2\phi x + \frac{1}{2}\phi y + 2(1-\phi)n\right]$, and $[7.52n]$, respectively). This equation is balanced only if the *O*-atoms are balanced, which is true if $2n = 2\phi x + \frac{1}{2}\phi y + 2(1-\phi)n$. It turns out that the original formula for $n\left(= x + \frac{y}{4}\right)$ still applies here, because the only difference between Equations (4.1) and (4.2) is the inclusion of the equivalence ratio ϕ as a coefficient multiplied by 1 *kmol* of fuel. Verification is achieved by substitution for n:

$$2\left(x + \frac{y}{4}\right) = 2\phi x + \frac{1}{2}\phi y + 2\left(x + \frac{y}{4}\right) - 2\phi\left(x + \frac{y}{4}\right)$$

Algebraic validation of this equality is left to the readers.

Some textbooks and websites prefer using "air-to-fuel ratio" (*A*/*F* ratio), "percent theoretical air" (%TA), or "percent excess air" (%EA) as the proportional measure of richness or leanness for a combustion reactant mixture. Many authors completely exclude discussion of the equivalence ratio, but we caution strongly against blind adoption of the *A*/*F* approach over the equivalence ratio approach.

If performed correctly, the *A*/*F* method is a valid tool to analyze combustion processes. The objective of both methods is to help users properly balance atoms in a combustion process, while simplifying the estimation of richness or leanness into a single number. However, the equivalence ratio method is more intuitive and less susceptible to error (in our opinion), so we avoid utilization of *A*/*F* ratio method in any problems or analyses. Most combustion scientists utilize equivalence ratio for their treatises and technical articles, whereas many burner manufacturers use *A*/*F* ratio, and students should be capable of converting from one to the other.

Example 4.1

For propane combustion in air with an equivalence ratio of $\boxed{\phi = 0.8}$, compute values for A/F, $\%TA$, and $\%EA$.

Answer.

$$C_xH_y = C_3H_8 \quad \therefore x = 3, \qquad y = 8, \qquad n = 5\left(= x + \frac{y}{4}\right)$$

$$\phi C_3H_8 + 5(O_2 + 3.76N_2) = 3\phi CO_2 + 4\phi H_2O + 5(1-\phi)O_2 + 5(3.76)N_2$$

$$\frac{A}{F} = \frac{\dot{m}_{air}}{\dot{m}_{fuel}} = \frac{\widehat{M}_a \dot{n}_a}{\widehat{M}_f \dot{n}_f} = \frac{28.85 \cdot (5 \cdot (1 + 3.76))}{44.097 \cdot 0.8} = 19.5$$

$$\%TA = 100\% \cdot \frac{(\dot{n}_a/\dot{n}_f)_{actual}}{(\dot{n}_a/\dot{n}_f)_{stoichiometric}} = \frac{\frac{28.85\cdot(5\cdot(1+3.76))}{44.097\cdot 0.8}}{\frac{28.85\cdot(5\cdot(1+3.76))}{44.097\cdot 1.0}} = \frac{1.0}{0.8}\left(= \phi^{-1}\right) = 125\,\%TA$$

$$\%EA = \%TA - 100\% = 125 - 100\% = 25\,\%EA$$

4.6 The Atom Equations

The prior two sections have focused on ensuring that the chemical equations are balanced. The conservation equation(s) at the root of this effort are conservation of the flow of C, H, O, and N atoms into and out of the control volume, where combustion occurs (see Equations 4.4). As with previous conservation equations in Chapters 2 and 3, the atom equations are posed here in the broad format of the Reynolds transport theorem with ***production*** of atoms disallowed (because nuclear reactions are forbidden). We also use the same parenthetical super-dot labels to categorize the differential-integral terms for clarity.

$$\underbrace{0}_{\dot{P}_{n_C}=0} + \underbrace{\int_{C.S.} \left(\frac{n_C}{\mathcal{V}}\right)\left(\vec{V}\cdot\vec{n}\right)dA_{in}}_{\dot{n}_{C,in}} = \underbrace{\int_{C.S.} \left(\frac{n_C}{\mathcal{V}}\right)\left(\vec{V}\cdot\vec{n}\right)dA_{out}}_{\dot{n}_{C,out}} + \underbrace{\int_{C.V.} \frac{\partial}{\partial t}\left(\frac{n_C}{\mathcal{V}}\right)d\mathcal{V}}_{\dot{S}_{n_C}} \tag{4.4a}$$

$$\underbrace{0}_{\dot{P}_{n_H}=0} + \underbrace{\int_{C.S.} \left(\frac{n_H}{\mathcal{V}}\right)\left(\vec{V}\cdot\vec{n}\right)dA_{in}}_{\dot{n}_{H,in}} = \underbrace{\int_{C.S.} \left(\frac{n_H}{\mathcal{V}}\right)\left(\vec{V}\cdot\vec{n}\right)dA_{out}}_{\dot{n}_{H,out}} + \underbrace{\int_{C.V.} \frac{\partial}{\partial t}\left(\frac{n_H}{\mathcal{V}}\right)d\mathcal{V}}_{\dot{S}_{n_H}} \tag{4.4b}$$

$$\underbrace{0}_{\dot{P}_{n_O}=0} + \underbrace{\int_{C.S.} \left(\frac{n_O}{\mathcal{V}}\right)\left(\vec{V}\cdot\vec{n}\right)dA_{in}}_{\dot{n}_{O,in}} = \underbrace{\int_{C.S.} \left(\frac{n_O}{\mathcal{V}}\right)\left(\vec{V}\cdot\vec{n}\right)dA_{out}}_{\dot{n}_{O,out}} + \underbrace{\int_{C.V.} \frac{\partial}{\partial t}\left(\frac{n_O}{\mathcal{V}}\right)d\mathcal{V}}_{\dot{S}_{n_O}} \tag{4.4c}$$

$$\underbrace{0}_{\dot{P}_{n_N}=0} + \underbrace{\int_{C.S.} \left(\frac{n_N}{\mathcal{V}}\right)\left(\vec{V}\cdot\vec{n}\right)dA_{in}}_{\dot{n}_{N,in}} = \underbrace{\int_{C.S.} \left(\frac{n_N}{\mathcal{V}}\right)\left(\vec{V}\cdot\vec{n}\right)dA_{out}}_{\dot{n}_{N,out}} + \underbrace{\int_{C.V.} \frac{\partial}{\partial t}\left(\frac{n_N}{\mathcal{V}}\right)d\mathcal{V}}_{\dot{S}_{n_N}} \tag{4.4d}$$

The aforementioned atom conservation equations are presented in terms of mole flows of atoms ($\dot{n}_i$) and molar concentrations of atoms ($n_i/\mathcal{V}$) instead of mass flow of the element and mass density of the element because this format is clearer when chemical transformations are occurring. Not presented is a conservation of species (i.e. molecules) equation, because the production term is the essence of chemical transformations for molecules and the Reynolds transport theorem does not illuminate the concept of production as clearly as can be seen in this and the following sections.

However, it is even more practical to track atom flows by computing the mole flows of the reactant and product compounds and to multiply those mole flows by the formula coefficients for each atom in the compound.

Example 4.2

Considering again the fuel-lean ($\phi = 0.8$) propane combustion example (Example 4.1), write the balanced equation in terms of mole flows of the reactant and product compounds (instead of molar coefficients of the balanced chemical equation). The species flows are shown on the left of the vertical ellipsis and the atom flows are shown on the right of the vertical ellipsis.

Answer.

$$\overbrace{\dot{n}_{C_3H_8,r} + \dot{n}_{O_2,r} + \dot{n}_{N_2,r}}^{reactants} \rightarrow \overbrace{\dot{n}_{CO_2,p} + \dot{n}_{H_2O,p} + \dot{n}_{O_2,p} + \dot{n}_{N_2,p}}^{products}$$

where,

$$\dot{n}_{C\boxed{3}H8,r} = 0.8\frac{kmol_{C_3H_8}}{s} \;\vdots\; \dot{n}_{C,r} = \boxed{3}\cdot 0.8 = \boxed{2.4\frac{kmol_{C\text{-}atoms}}{s}}$$

$$\dot{n}_{C3H\boxed{8},r} = 0.8\frac{kmol_{C_3H_8}}{s} \;\vdots\; \dot{n}_{H,r} = \boxed{8}\cdot 0.8 = \boxed{6.4\frac{kmol_{H\text{-}atoms}}{s}}$$

$$\dot{n}_{O\boxed{2},r} = 5.0\frac{kmol_{O_2}}{s} \;\vdots\; \dot{n}_{O,r} = \boxed{2}\cdot 5.0 = \boxed{10.0\frac{kmol_{O\text{-}atoms}}{s}}$$

$$\boxed{Reactants}\uparrow \quad \dot{n}_{N\boxed{2},r} = 18.8\frac{kmol_{N_2}}{s} \;\vdots\; \dot{n}_{N,r} = \boxed{2}\cdot 5\cdot 3.76 = \boxed{37.6\frac{kmol_{N\text{-}atoms}}{s}}$$

$$\boxed{Products}\downarrow \quad \dot{n}_{\boxed{C}O\boxed{2},p} = 2.4\frac{kmol_{CO_2}}{s} \;\vdots\; \dot{n}_{C,p} = \boxed{1}\cdot 2.4 = \boxed{2.4\frac{kmol_{C\text{-}atoms}}{s}}$$

$$\dot{n}_{H\boxed{2}\boxed{O},p} = 3.2\frac{kmol_{H_2O}}{s} \;\vdots\; \dot{n}_{H,p} = \boxed{2}\cdot 3.2 = \boxed{6.4\frac{kmol_{H\text{-}atoms}}{s}}$$

$$\dot{n}_{O\boxed{2},p} = 1.0\frac{kmol_{O_2}}{s} \;\vdots\; \dot{n}_{O,p} = \boxed{2}\cdot 2.4 + \boxed{1}\cdot 3.2 + \boxed{2}\cdot 1.0 = \boxed{10.0\frac{kmol_{O\text{-}atoms}}{s}}$$

$$\dot{n}_{N\boxed{2},p} = 18.8\frac{kmol_{N_2}}{s} \;\vdots\; \dot{n}_{N,p} = \boxed{2}\cdot 18.8 = \boxed{37.6\frac{kmol_{N\text{-}atoms}}{s}}$$

Think Stop. Students should examine the aforementioned analysis and note the following outcomes and style points:

- The balanced chemical equations are written in terms of mole flows of reactants and mole flows of products, and they use an arrow sign (→) rather than an equals sign (=).

- This is because the sum of the mole flows of all the reactants $\left(\sum_i \dot{n}_{i,reac}\right)$ does not necessarily equal the sum of the mole flows of all the products $\left(\sum_j \dot{n}_{j,prod}\right)$
- When the fuel is methane (CH_4), the mole flow summation of the reactant compounds happens to be precisely equal to the mole flow summation of the product compounds. (Validation of this assertion is given as a homework question at the end of this chapter.)
- However, when the fuel is propane (C_3H_8)., the summation of ***reactant*** mole flows $\left(\sum_i \dot{n}_{i,reac} = 0.8 + 5 + 18.8 = \boxed{24.6}\right)$ clearly does not equal the summation of ***product*** mole flows $\left(\sum_j \dot{n}_{j,prod} = 2.4 + 3.2 + 1 + 18.8 = \boxed{25.4}\right)$

- The rule here is that ***atoms*** must be conserved in *every* combustion reaction, but ***compounds*** are routinely *not* conserved in combustion reactions.
- As the reactant molecules rearrange their bonds to become product molecules, the change in the number of moles, if any, is also a source of increased moles in the ideal gas law.
  - This change can be tracked by computing separate values for: (i) molar mass of the reactant mixture and (ii) molar mass of the product mixture.
  - The total mass flow is always conserved, but the mole flow and volume flow are often not conserved – hence, a "conservation of volume" equation is *not* included in a thermal system designer's tool kit and a "conservation of compound moles" equation has a *nonzero* ***production*** term.

In addition to conservation of atoms, thermal system designers must also ensure their process equipment and flow conveying means are sized in accordance with conservation of momentum and conservation of energy.

Analyzing conservation of fluid momentum in a chemically reacting gas stream requires advanced methods that are not addressed in this textbook. Analyzing conservation of energy in a chemically reacting gas stream is the essence of combustion, which is addressed simply but thoroughly in the next section.

4.7 Sensible and Chemical Enthalpies

The analysis of combustion heat release begins with the energy equation (first given in Section 3.4), and we modify the general form by imposing several simplifications in Equation (4.5). As is our custom, we use the parenthetical super-dot simplifications under each conservation term.

$$\underbrace{0}_{\dot{\mathcal{P}}_E=0} + \underbrace{\dot{Q}_{in}}_{\substack{Net\\Heat\\Flow\\In}} + \underbrace{\int_{C.S.} \rho e\left(\vec{V}\cdot\vec{n}\right)dA_{in}}_{\dot{e}_{in}} = \underbrace{\dot{W}_{s,out}}_{\substack{Net\\Shaft\\Power\\Out}} + \underbrace{\int_{C.S.} \rho e\left(\vec{V}\cdot\vec{n}\right)dA_{out}}_{\dot{e}_{out}} + \underbrace{\int_{C.V.} \frac{\partial}{\partial t}(\rho e)d\mathcal{V}}_{\dot{S}_E} \quad (4.5)$$

The first term to be eliminated is the production term $(\dot{\mathcal{P}}_E)$, which is identically zero for systems where nuclear reactions are excluded. Also, excluded (by assuming adiabatic walls that have no penetrations for rotating shafts) are the heat flow $(\dot{Q}_{in})$ and shaft work $(\dot{W}_{s,out})$ terms that might cross the control surface in a more generic model. In principle, the storage term $\left(\int_{C.V.} \frac{\partial}{\partial t}(\rho e)d\mathcal{V}\right)$ can be accommodated, but for simplicity here, the control volume will be considered to have reached ***steady-flow, steady-state*** conditions (i.e. transients are ignored). Also, changes in kinetic and potential energies of the flowing fluid are ignored, along with other energy forms such as electrical and magnetic. Thus, a simplified form of the steady, adiabatic, one-dimensional, energy equation, including combustion is obtained (Equation 4.6):

$$\underbrace{\int_{C.S.} \rho e\left(\vec{V}\cdot\vec{n}\right)dA_{in}}_{\dot{e}_{in} = \dot{m}_{in}e_{in}} = \underbrace{\int_{C.S.} \rho e\left(\vec{V}\cdot\vec{n}\right)dA_{out}}_{\dot{e}_{out} = \dot{m}_{out}e_{out}} \tag{4.6}$$

where,

$$e_{in} = h_{sensible,in} + h_{chemical,in}\left[\doteq \frac{kJ}{kg}\right]$$

$$e_{out} = h_{sensible,out} + h_{chemical,out}$$

Sensible enthalpy is the familiar type of enthalpy that was tracked for non-reacting flows in Chapters 1 and 3 (i.e. $h_{sensible} = u + pv$). It is the enthalpy of a compound that is associated with its temperature as it enters or leaves the control volume. The name "sensible heat" originated from the idea that when temperature change accompanies internal energy change, the heat can be "sensed."

Chemical enthalpy is the enthalpy associated with the chemical species (i.e. chemical bond arrangements of molecules) that flow into or out of a control volume. Like combustion, phase change (i.e. vaporization or condensation) is also considered as chemical enthalpy change, because the molecules undergo changes in *intermolecular* forces, without any breaking of *intramolecular* bonds. The name "latent heat" of vaporization or condensation originated from the observation that this form of heat transfer was "hidden" since the temperature of the fluid undergoing phase change remained constant during these processes.

Because ***reactants*** flow into the control volume and ***products*** flow out of the control volume, the energy equation for a steady, adiabatic combustor (ignoring frictional losses and changes in kinetic and potential energy) can be summarized by Equations (4.7)

$$\underbrace{\sum_{i\,=\,reactants} \dot{m}_i\left(h_{sensible,i} + h_{chemical,i}\right)}_{\substack{Inflow\ of \\ sensible\ +\ chemical\ enthalpy \\ for\ all\ molecular\ species\ i}} = \underbrace{\sum_{j\,=\,products} \dot{m}_j\left(h_{sensible,j} + h_{chemical,j}\right)}_{\substack{Outflow\ of \\ sensible\ +\ chemical\ enthalpy \\ for\ all\ molecular\ species\ j}} \tag{4.7a}$$

or by using molar format (as is common in many published thermodynamic tables):

$$\sum_{i = reactants} \dot{n}_i \left(\hat{h}_{sensible,i} + \hat{h}_{chemical,i}\right) = \sum_{j = products} \dot{n}_j \left(\hat{h}_{sensible,j} + \hat{h}_{chemical,j}\right) \tag{4.7b}$$

We note that Equations (4.7) are simplified versions of the species conservation equations (with positive and negative ***production*** being the essence of the chemical reactions). To further clarify the practical aspects of using these two forms of enthalpy to predict temperatures and combustion calculations, the following two chemistry definitions are introduced, noting that the subscripts i and j have been temporarily deleted.

Sensible Enthalpy. For ***sensible enthalpy***, because h is not perfectly linear with T, the specific heat must be integrated with temperature, keeping c_p inside the integral. For an indefinite integral, this would generate an arbitrary constant. By scientific convention, to obtain a uniform manner of energy accounting across all chemical species, the constant of integration is set to zero at a single designated temperature (e.g. $T_{STP} = 298.15\ K$). With this understanding, chemical thermodynamicists have generated thermodynamic tables of sensible enthalpies (and other parameters) versus temperature, based on the definite integral given in Equation (4.8), which tracks sensible enthalpy as a change in enthalpy between two temperatures.

$$\hat{\boldsymbol{h}}(\boldsymbol{T}) - \hat{\boldsymbol{h}}_{\boldsymbol{298}} \equiv \left[\hat{h}_{sens}(T) - \hat{h}_{sens}(T_{datum})\right] = \int_{298.15}^{T} c_p dT \tag{4.8}$$

Think Stop. Students should note that publishers of different property tables may designate different datum temperatures (e.g. 25° C, 0° C, or 0 K). One set of property tables maintained by the National Institutes of Standards and Technology (NIST) is called the Joint Army Navy Air Force (JANAF) tables, which are discussed in the next section. The datum temperature for sensible enthalpy in the JANAF tables has always been $T_{datum} = 298.15\ K$.

Chemical Enthalpy. For ***chemical enthalpy***, an accounting of the bond energies in each molecule at some reference state (i.e. standard reference point of temperature and pressure [***STP***]) is required. Conceptually, the simplest way to determine these bond energies would be to measure the enthalpy change that occurs when a compound is assembled from individual atoms.

Unfortunately, this simple approach turns out to be quite challenging experimentally, so the problem is approached differently in the JANAF tables. Instead, scientists have measured the exothermic or endothermic heat that is released or absorbed, when the molecule of interest is assembled isothermally from ***reference elements***.

This thought exercise begins with the precise, stoichiometric quantities of reactant elements in their *natural state* at ***STP*** (e.g. $1H_2$, $\frac{1}{2}O_2$) and ends with 1 *kmol* of product molecules (e.g. $1H_2O$) at ***STP***. For illustration, the ***chemical enthalpy*** (also called ***enthalpy of formation***) of methane is found by forming one mole of $CH_4(g)$ at ***STP*** from reference elements $C(s)$ and $H_2(g)$ at ***STP***, with the energy balance given in Equation (4.9).

$$1.0\ C(s) + 2.0\ H_2(g) \rightarrow 1.0\ CH_4(g)$$

$$\Delta\hat{H}_f^0 = \sum_{j = prod, STP} n_j \hat{h}_j - \sum_{i = reac, STP} n_i \hat{h}_i \tag{4.9}$$

The superscript zero implies that the products and reactants both are at STP, and the enthalpy change can be measured by the amount of heat $[Q \stackrel{\smile}{=} kJ]$ that must be added to or subtracted from the one mole of product to maintain its temperature constant (at T_{STP}) after the reaction has gone to completion. Typically, the reaction is carried out at constant pressure (p_{STP}), so any work performed on or by the environment with respect to the control mass is accounted for by the enthalpy values, and the only measurement needed is the heat input or output that returns the mixture to its initial (standard) temperature.

Think Stop. To reemphasize, the balanced chemical reaction by which ***enthalpy of formation*** is determined must obey the following rules:

- Exactly one mole (*mol*, not *kmol*) of the target compound must be formed.
- The products must not contain any additional species – only the target compound, with no excess reactants, and no added diluents.
- The final products exit at STP.
- The initial reactants enter at STP.
- The reactants are comprised of ***reference elements*** (i.e. pure elements existing in the phase and bond structure of their natural state at STP).

The natural state of an element involves both phases (i.e. solid, liquid, or gas) and bond structures (e.g. solid carbon exists naturally as both diamond and graphite). In the case of the element carbon: (i) the bond structure associated with diamond is unstable (i.e. diamond will revert to graphite if given enough time), and (ii) diamond can be formed endothermically (with heat input to the reactants) from graphite. These factors imply that graphite must be the "reference state" of carbon because it is more stable.

A corollary of these rules is that all the ***reference elements*** themselves are arbitrarily designated to have zero ***enthalpy of formation*** at STP. Thus, $C(s)$ and $H_2(g)$ at *STP* have zero ***total enthalpy*** because ***reference elements*** are assigned zero ***chemical enthalpy***, and all substances at STP have zero ***sensible enthalpy***. Methane, on the other hand, has zero sensible enthalpy at *STP*, but non-zero ***chemical enthalpy***. According to the published thermodynamic property tables (Linstrom and Mallard 2017–2021), the enthalpy of formation of methane is (Equation 4.10):

$$\Delta\hat{h}^0_{f,CH_4} = -74.87\frac{kJ}{mol_{CH_4}} = -74.87\frac{MJ}{kmol_{CH_4}} \tag{4.10}$$

Thus, the formation of one mole of methane at STP from reference elements happens to be an exothermic process. Heat is transferred out of the system to maintain the temperature constant as the product is formed from the reactants.

A graphical aid to the understanding of ***total enthalpy*** (as comprised of ***sensible enthalpy*** plus ***chemical enthalpy***) is given in Figure 4.2. The figure shows enthalpies of oxygen molecules (O_2) and oxygen atoms (O) plotted together on the same axes. The reaction that forms oxygen atoms from oxygen molecules is endothermic (heat input is required to break the oxygen molecule's double bond and drive apart the two oxygen atoms from each other), hence the O-atoms have a positive enthalpy of formation $\left(\Delta\hat{h}^0_{f,O\text{-}atom} = +249.18\ kJ/mol\right)$.

From Figure 4.2, it is apparent that both gases O and O_2 have zero sensible enthalpy at $T_{datum} = 298.15\ K$, and both sensible enthalpies increase with increasing temperature, as expected.

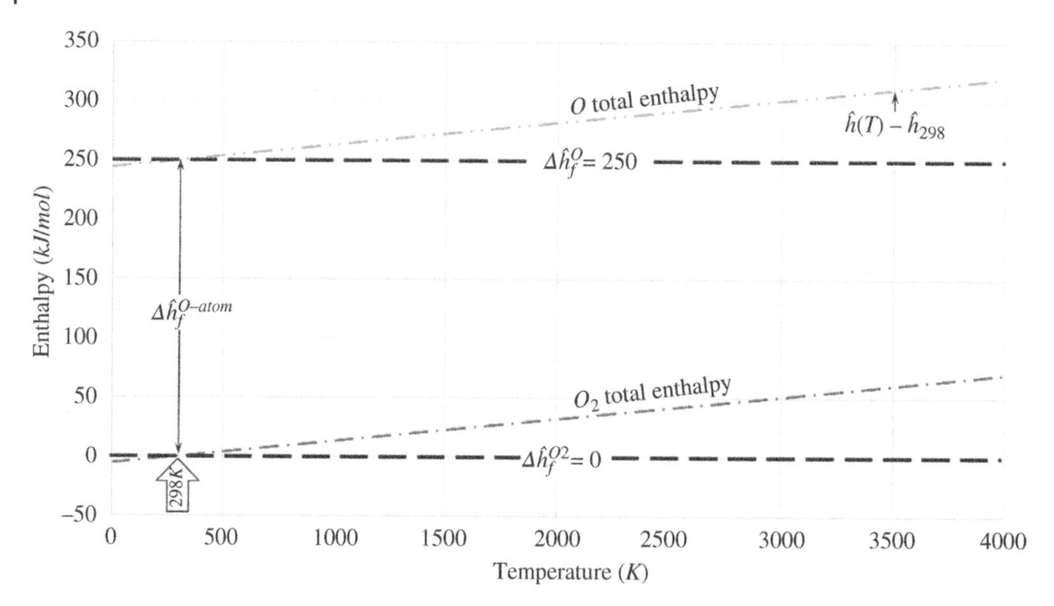

Figure 4.2 Total enthalpy of oxygen atoms and oxygen molecules versus temperature. The total enthalpy of oxygen atoms is found by summing the chemical enthalpy (at 298 *K*) with the sensible enthalpy (from 298 *K* to *T*). *Source:* Martin Thermal Engineering Inc.

However, the total enthalpy of *O* (monatomic) at the datum temperature is approximately +250 *kJ*/*mol*, whereas the total enthalpy of $O_2(g)$ is 0.0 *kJ*/*mol*, because it is a ***reference element***.

4.8 Thermochemical Property Tables

Appendix A of this textbook provides a limited set of thermodynamic data for certain combustion reactants and products, along with data for refrigerants and other fluids of interest. If different compounds are needed, or if an extended range of temperatures is required, the reader must seek data elsewhere. The following resources are recommended:

The most comprehensive source (i.e. most compounds available) are the ***JANAF tables*** (Stull and Prophet 1971 and Chase 1998). Any of the first four editions is acceptable and recommended, although some differences in units and source values exist between them.

The best source for current and easy-to-access thermodynamic data is the ***NIST Chemistry Webbook*** (Linstrom and Mallard 2017–2021). The tables in Appendix A of this textbook were obtained from this information source and students can use this resource to develop their own tables to solve particular problems.

A third source that contains thermodynamic data for some compounds not in the *NIST Chemistry Webbook* is ***Thermodynamic Properties in SI (TPSI)*** (Reynolds 1979). This resource is available from used bookstores and may be available on a "print-on-demand" basis.

Software tools are also available for computing equilibrium conditions under various combustion scenarios. Readers are referred to the Chemical and Biological Engineering Department at Colorado State University (Colorado State University 2020) for an online thermochemical equilibrium calculator, based on STANJAN (Reynolds 1986). Commercially available software (e.g. REI 2020) that computes both equilibrium and kinetics can also be purchased, as can full-service chemical simulation packages (e.g. Chemstations 2021 or ANSYS 2020).

As introduced in Chapter 1 of this textbook, formulas for enthalpy of a limited number of gaseous compounds (sourced from Linstrom and Mallard 2017–2021) have been reduced to Visual Basic for Applications (VBA) scripts in Appendix B. Students are encouraged to employ these customized functions by pasting the scripts into Microsoft Excel's VBA Developer (as shown in the introduction to Appendix A). This technique is developed further in the next several sections of this chapter.

4.9 Enthalpy of Combustion

Enthalpy of combustion ($\Delta h_{comb}\left[\doteq kJ/kg_f\right]$ or $\Delta \hat{h}_{comb}\left[\doteq MJ/kmol_f\right]$) is a property of ***fuels*** that can be computed directly from the enthalpies of formation of reactants and products. It is frequently tabulated for common fuels. Students are cautioned not to equate tabulated values of Δh_{comb} for fuel blends (e.g. natural gas, liquefied petroleum gas) with Δh_{comb} for the dominant constituent (e.g. CH_4, C_3H_8, respectively). The blend's enthalpy of combustion reflects the presence of multiple pure compounds, which invariably causes the blend's Δh_{comb} to be different from that of the dominant constituent. Examples of this are the enthalpies of combustion for natural gas and liquified petroleum gas that have modest differences from those of methane and propane, respectively.

Enthalpy of combustion is a representation of the heat transferred to the environment ($\dot{Q}$) through the walls of a hypothetical chemical reactor, whose inlet and outlet streams are both at STP, but whose reactants (fuel + air) have been completely transformed to products, without any excess of fuel or oxygen remaining from the reactants (see Figure 4.3).

The choice of STP for the states of both reactants and products reduces the problem to one of determining differences in enthalpies of formation for the products and reactants. Sensible enthalpy is by definition zero for both sets of constituents if they are each at STP, and the enthalpies of formation are usually tabulated at STP, which simplifies the exercise further still.

Heating Value. The two common ways to compute enthalpy of combustion are: (i) the lower heating value (LHV) method, which computes exothermic energy of the reaction, assuming water is present in the products as a gas (steam vapor), or (ii) the higher heating value (HHV) method, which computes exothermic energy of the reaction, assuming water is present in the products as a liquid.

Either situation is possible in practice, and the current trend for many combustion devices (particularly home-heating furnaces) is for the combustion-generated water in the products to be (mostly) condensed before the exhaust gas leaves the furnace. If the latent heat of condensation is transferred to the object receiving heat (e.g. the room air being heated for comfort), the waste heat going to the environment is reduced, and the HHV is the legitimate measure of how much heat was delivered by the fuel. If all the water leaves the control volume as a vapor, the energy associated with the latent heat of condensation is lost, and the LHV is the legitimate measure of the amount of heat delivered by the fuel.

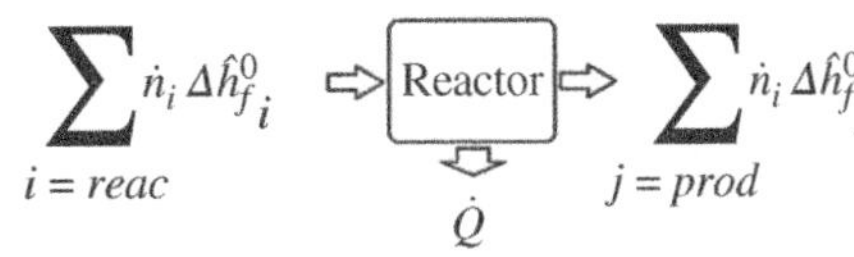

$$\sum_{i=reac} \dot{n}_i \Delta \hat{h}^0_{f_i} \Rightarrow \boxed{\text{Reactor}} \Rightarrow \sum_{j=prod} \dot{n}_i \Delta \hat{h}^0_{f_j} \qquad \dot{Q}$$

Figure 4.3 Schematic of an ideal reactor that converts fuel and oxidizer to combustion products, with all species entering and leaving at STP and the "heat" of combustion exiting the control volume across the walls. *Source:* Martin Thermal Engineering Inc.

Think Stop. If a thermal engineer does not know whether a combustor is designed to condense the combustion generated water vapor to water liquid, it is often safer (more conservative) to assume LHV is the appropriate measure of combustion enthalpy, because that will ensure the burner selected will be large enough to handle that case. Nonetheless, designers should always determine whether other aspects of the design might favor LHV or HHV reporting of combustion enthalpy, every time they perform a new analysis.

Enthalpy of Combustion. An example is provided here for the computation of the enthalpy of combustion for methane. Students should practice this computation with different fuels, such as propane (C_3H_8) and *n*-octane (C_8H_{18}), as given in the homework problems for this chapter.

Example 4.3

Compute the enthalpy of combustion for methane.

Answer. Begin with the balanced, stoichiometric equation for combustion of methane with oxygen. Because the enthalpy change is being normalized per mole of fuel, the procedure works the same whether the oxidizer is pure oxygen or air with 20.9% oxygen. For simplicity, pure oxygen is used in this example.

$$CH_4 + 2O_2 \rightarrow CO_2 + 2H_2O$$

Next, let us assume that products are only carbon dioxide and water vapor. This assumption is not strictly valid for adiabatic methane-oxygen combustion because the products are at such a high temperature that moderate amounts of carbon monoxide (CO) and hydroxyl radical (OH) are formed at equilibrium. However, if the high-temperature combustion products are allowed to cool slowly, the low-temperature equilibrium does favor carbon dioxide and water by a strong margin, and only inconsequential amounts of the chemical fragments CO and OH remain.

Since precisely one mole of fuel is being combusted to form products, the enthalpy of reaction in Equation (4.11) should be computed first, and that result should be divided by the 1 *kmol* of fuel to obtain the enthalpy of combustion. Because the combustion reaction must obey the First law, the total enthalpy of all the reactants should be subtracted from the total enthalpy of all the products and the difference will be the enthalpy of the reaction (which is equal to the enthalpy of combustion in this case of one mole of fuel being burned). Since all reactants and all products are at STP, the total enthalpy values will consist of chemical enthalpy values only. The combustion products will be assumed to be all gases, so the result will be designated as LHV.

$$\Delta\hat{h}_{reaction} = \sum_{j=prod}\left[\dot{n}_j\left(\underbrace{\Delta\hat{h}^0_{f\,j}}_{chemical} + \underbrace{\left[\hat{h}(T) - \hat{h}_{298}\right]_j}_{sensible=0}\right)\right] - \sum_{i=reac}\left[\dot{n}_i\left(\underbrace{\Delta\hat{h}^0_{f\,i}}_{chemical} + \underbrace{\left[\hat{h}(T) - \hat{h}_{298}\right]_i}_{sensible=0}\right)\right] \tag{4.11}$$

$$\Delta\hat{h}^0_{f\,CH_4} = -74.87\frac{MJ}{kmol_{CH_4}} \quad \Delta\hat{h}^0_{f\,O_2} = 0.0\frac{MJ}{kmol_{O_2}}$$

$$\Delta\hat{h}^0_{f\,CO_2} = -393.51\frac{MJ}{kmol_{CO_2}} \quad \Delta\hat{h}^0_{f\,H_2O(g)} = -241.83\frac{MJ}{kmol_{H_2O}}$$

$$\Delta\hat{h}_{comb,CH_4} = \underbrace{[1\cdot(-393.51) + 2\cdot(-241.83)]}_{Total\ enthalpy\ of\ products\ at\ STP} - \underbrace{[1\cdot(-74.87) + 2\cdot(0)]}_{Total\ enthalpy\ of\ reactants\ at\ STP}$$

$$\Delta\hat{h}_{comb,CH_4} = -802.3\frac{MJ}{kmol_{CH_4}}$$

Think Stop. When computing enthalpy of combustion, would not it be better to assume the combustion product H_2O exits from the reactor as a liquid, rather than a gas? Although this may seem logical because we have constrained the product mixture to STP (like the reactant mixture), most academic textbooks assume the exiting mixture to be gaseous, even though the practical outcome is that some (but not all) of the water vapor condenses inside the reactor. On the opposing side, many practical handbooks do not use the term enthalpy of combustion at all – they simply give the results as "heating value" and add a modifier of higher or lower to the phrase, depending on which of the two values is being reported. Engineers are cautioned to be very careful discerning how their source data is defined – especially if a generic term like enthalpy of combustion or heating value is used repeatedly without any LHV or HHV designation.

4.10 Enthalpy Datum States

Think Stop. When solving a thermodynamics problem, students may be tempted to concoct a value for the specific enthalpy of a liquid or gas by multiplying the tabulated value for specific heat at constant pressure ($c_p \doteq kJ/(kg \cdot K)$) by the actual temperature of the substance (i.e. $h \stackrel{?}{=} c_p \cdot T$). This is a terrible practice because it will produce the right answer sometimes and a wrong answer at other times. The preferred practice is to utilize the specific heat as a proportionality that relates differential enthalpy change to differential temperature change (i.e. $dh = c_p dT$).

Undefined Datum. The theoretical reason for this practice being mathematically wrong under most circumstances may be understood by the definition of specific heat (Equation 4.12):

$$c_p \equiv \left.\frac{\partial h}{\partial T}\right]_{p\,=\,const} \tag{4.12}$$

When integrated, the expression becomes Equation (4.13):

$$\int_{h_0}^{h(T)} dh = \int_{T_0}^{T} c_p \cdot dT \; (h - h_0) = c_p(T - T_0) \tag{4.13}$$

where, h_0 is a datum state enthalpy (evaluated at datum temperature T,p T_0). The datum is basically an arbitrary constant of integration. This arbitrariness is well documented in different thermodynamic tables published by different entities, which use different datum states. Based on the aforementioned mathematics, it should be obvious why computing an "absolute" enthalpy based simply on specific heat and temperature (i.e. from "$h \stackrel{?}{=} c_p \cdot T$") is a bad practice. Here are a few examples of how this simple substitution can go horribly wrong.

Errors Due to Temperature Scales. The dangerous equation ($h \stackrel{?}{=} c_p \cdot T$) provides no information about what temperature units belong with the desired values for specific heat and enthalpy. Even if one states (using SI units), the question remains whether T should be entered in degrees Celsius or Kelvin. Clearly, the value of h would go to zero at $T = 0.0°\ C$, but h would be a relatively large number at $T = 273.15\ K$, even though the two numbers represent the same temperature.

Errors Due to Phase Change. If the problem being solved involves a phase change, the dangerous shortcut will not account for the finite difference in enthalpy that occurs during the phase change. For example, the specific heat of a boiling liquid is infinity (∞): finite Δh divided by zero ΔT. Trying to determine the enthalpy change from liquid to vapor requires consultation of a resource that accurately reports discrete values of enthalpy at the beginning and end states (e.g. saturated < or subcooled > liquid versus saturated < or superheated > gas). Using the "$h \stackrel{?}{=} c_p \cdot T$" shortcut is magnificently wrong and would be a huge source of embarrassment for the offending analyst.

Errors Due to Datum State Differences. Potentially, the most crucial reason for avoiding the shortcut equation ($h \stackrel{?}{=} c_p \cdot T$) is that the enthalpy datum state from one table may not be consistent with other tables, even those published by the same source and/or author. Furthermore, as discussed in Chapters 1 and 3, using an improper temperature datum (e.g. $°C$ instead of K) could introduce a significant error into the computed enthalpy, especially if the dependency of c_p on temperature was defined using one temperature scale, but the enthalpy relationship is later used with the other temperature scale.

Presented here (without citation) is a particularly egregious example where one publisher provided two vastly different values for saturated enthalpy of the same refrigerant between their SI table and their English table:

$$h_f(-100°C) = \boxed{+77.3\ kJ/kg}\ h_f(-150°F) = \boxed{-31.2\ BTU/lb_m}$$

We note that the two cited temperatures are slightly different ($-100°\ C = -148°\ F$), but the published enthalpies are miles apart. The reason for this discrepancy is that the publisher uses different values for the reference enthalpy in the different tables: $h_f(-40°C) = +148.4\ kJ/kg$ and $h_f(-40°\ F) = 0.0\ BTU/lb_m$. Using the shortcut expression "$h \stackrel{?}{=} c_p \cdot T$" would not even get consistent signs (+ or −) for enthalpy in the two temperature scales.

4.11 Adiabatic Combustion Temperature

For lean combustion of any fuel in air, a unique adiabatic combustion temperature is associated with every equivalence ratio. Problems of two types are most often faced by thermal engineers – find the adiabatic combustion temperature if the equivalence ratio is known, or the inverse problem – find the equivalence ratio if the adiabatic combustion temperature is known. Both problems usually necessitate iteration to be performed manually using a tabular source of thermodynamic data. As it turns out, the second problem is simpler, because the first problem not only requires iteration, but also interpolation between temperature lines of thermodynamic tables.

For example, suppose the equivalence ratio (ϕ) is to be determined for a lean mixture of methane in air at STP that reaches $T_{ad} = 1200\ K$. For adiabatic combustion, the total enthalpy of the reactants must equal the total enthalpy of the products. Since the primary product gases (CO_2 and H_2O) have such large negative enthalpies of formation, compared to the reactant gases that are largely air (which has zero enthalpy of formation) the sensible enthalpy of the products must be positive and relatively high to offset the large negative chemical enthalpy. The First law (Equation 4.14) is an appropriate tool for this analysis:

The easiest way to complete this exercise is (i) populate all the sensible enthalpies at the specified temperatures (i.e. 298 and 1200 K), (ii) make a first guess of the equivalence ratio, and (iii) compute

Table 4.1 Enthalpy balance worksheet (sensible and chemical enthalpies derived from JANAF tables (1971)). The two values for $\dot{H}$ will be equal when the correct equivalence ratio (ϕ) is found.

	Constituent	$\dot{n}$ (kmol/s)	$\hat{h}_{chem}$ (MJ/kmol)	$\hat{h}_{sens}$ (MJ/kmol)	$\dot{H}$ (MW)	
Reac 298K	CH_4	Guess $\phi = 0.4$	−74.87	0.0		
	O_2	2.0	0.0	0.0		
	N_2	7.52	0.0	0.0		
	TOTAL REAC	$9.52 + \phi$	−3.02	0.0	−29.95	
Prod 1200K	CO_2	ϕ	−393.51	44.484		Error: Product $\dot{H}$ is too negative. Iterate on ϕ
	H_2O	2ϕ	−241.826	34.476		
	O_2	$2(1-\phi)$	0.0	29.765		
	N_2	7.52	0.0	28.108		
	TOTAL PROD	$9.52 + \phi$	−35.369	29.482	−58.40	

Source: Martin Thermal Engineering Inc.

the rates of enthalpy flow from the known mole flow rates and sensible enthalpies, as shown in Table 4.1.

$$\sum_{i=reac}\left[\dot{n}_i\left(\underbrace{\Delta\hat{h}^0_{f_i}}_{chemical}+\underbrace{\left[\hat{h}(T)-\hat{h}_{298}\right]_i}_{sensible=0}\right)\right]=\sum_{j=prod}\left[\dot{n}_j\left(\underbrace{\Delta\hat{h}^0_{f_j}}_{chemical}+\underbrace{\left[\hat{h}(T)-\hat{h}_{298}\right]_j}_{sensible>0}\right)\right] \quad (4.14)$$

Think Stop. The aforementioned problem is given as a homework question at the end of the chapter, as is a related problem where the equivalence ratio is given, and the adiabatic combustion temperature is to be determined. Students are again cautioned to recognize the limitations of the complete combustion method, as compared to the (nearly always valid) thermochemical equilibrium method.

Parenthetically, we point out here that thermochemical equilibrium is reached quickly for virtually all high-temperature gas mixtures, but low-temperature gas mixtures may require analysis with a frozen composition (i.e. one that does not correspond to thermochemical equilibrium). For example, reactant mixtures (e.g. $CH_4 + O_2$) can coexist for exceptionally long periods of time at ambient temperatures without reaching chemical equilibrium. This is because the initiation reactions are too slow to create carrier molecules that participate in chain reactions. Similarly, many frozen compositions can be created artificially by rapidly quenching a high-temperature mixture. For example, a high-temperature combustion product mixture may contain large amount of $CO + OH$, but CO is found to persist after rapid temperature reduction (quenching), because CO is relatively stable at low temperatures and the kinetics to form CO_2 (the favored equilibrium carbon product at lower temperatures) are slow compared with the quenching rate.

A simplified thermochemical equilibrium method for fuel-lean reactant mixtures assumes no products other than the *complete combustion* set CO_2, H_2O, O_2, and N_2. At moderately high combustion temperatures, additional product species are present in moderate concentrations at equilibrium. Even with fuel-lean combustion, sufficiently high temperatures can cause the dissociation of CO_2 to CO and of H_2O to H, O, OH, and H_2.

4.12 Equilibrium and Kinetics

Determination of the composition of a mixture of substances at thermochemical equilibrium is a vitally important subject for combustion scientists and engineers. Similarly, the analysis of combustion kinetics (reaction rates) can be very useful for designers of burners (and other combustion systems) to help minimize pollutant formation (or to maximize destruction of hazardous organic species).

Students who want to specialize in combustion chemistry should familiarize themselves with tools for analyzing reaction rates and equilibrium states of gas mixtures. Several classic textbooks on combustion theory and practice are cited in the references for this chapter, and combustion students should scrutinize such resources and complete advanced coursework in the field.

In this chapter, only one form of equilibrium equations is introduced, and readers are referred to various software resources that compute equilibrium conditions based on thermodynamic property tables. The theory of chemical kinetics is not addressed in this textbook (beyond superficial descriptions of pollutant formation reactions and frozen compositions), and no detailed analytical methods for reaction rates are given. Review of the bibliographic resources is again urged for students who wish to delve further into these subjects.

Equilibrium. In Sections 4.4–4.11, the approach was followed that ***fuel-lean*** and ***stoichiometric*** combustion reactions always go "to completion." While complete combustion is approached very closely for many practical combustion systems, thermochemical equilibrium predicts that tiny amounts of reactants (and intermediates) are present in the final mixture. The concept of "equilibrium" means that forward and reverse reactions are occurring at the same rate, so there is no net change in the composition of the overall mixture. Combustion scientists often utilize the concept of "partial equilibrium," which implies that a subset of the fundamental reactions have reached equilibrium, even though the mixture as a whole is still undergoing a transient combustion process. The partial equilibrium method provides a way to simplify a very complex combustion kinetics problem. Although not discussed further here, one example of the partial equilibrium method is to assume that oxygen molecules (O_2) and oxygen atoms (O) are in partial equilibrium ($O_2 \leftrightarrow 2O$) at the local temperature of the combusting gases in a flame.

Example 4.4

Estimate the equilibrium concentrations of the four compounds in Equation (4.15) for temperatures of 298 and 1000 K, assuming total pressure is 1.0 atm and the $C : H : O$ atom ratio is $1 : 2 : 2$.

$$CO(g) + H_2O(g) \leftrightarrow CO_2(g) + H_2(g) \tag{4.15}$$

Answer. This is the classic *water gas shift* reaction. Ostensibly, the *water gas shift* equilibrium applies to any mixture containing water vapor and carbon monoxide. For simplicity, we assume that molecular oxygen (O_2) and monatomic species (C, H, O) are absent from the reactive mixture. The rightward (or forward) reaction pulls an O-atom away from reactant H_2O (left side) and shifts it to reactant CO (also left side), forming two new compounds on the right (CO_2 and H_2) as the products. The inverse occurs for the leftward (or reverse) reaction. The "shift" associated with the name comes from the tendency of the mixture to shift an oxygen atom back and forth between the carbon and hydrogen species. The leftward shift depletes CO_2 (by transferring O to H_2 and forming H_2O), while the rightward shift depletes H_2O (by transferring O to CO and forming CO_2). The two fuel gases, CO and H_2 are on opposite sides of the equilibrium and the other two participating gases, H_2O and CO_2, are reservoirs for the shifting O-atom.

At ambient temperatures (298 K), the right side of the equilibrium sign (the side with carbon dioxide and hydrogen) is much more thermodynamically stable, so an equilibrium mixture of these four compounds (assuming 1 atm total pressure and a $C:H:O$ atom ratio of 1 : 2 : 2) has relatively small mole fractions of carbon monoxide and water vapor ($\chi_{CO} = \chi_{H_2O} = 0.15\ mol\%$ each). The balance of the product mixture appears as equal mole fractions of carbon dioxide and hydrogen $\chi_{CO_2} = \chi_{H_2} = 49.85\ mol\%$ each.

However, at 1000 K, both sides of the equilibrium have comparable levels of thermodynamic stability, and their concentrations are within 25% of each other. For comparison, at 1000 K, the equilibrium mole fractions are: $\chi_{CO} = \chi_{H_2O} = 22.7\ mol\%$ and $\chi_{CO_2} = \chi_{H_2} = 27.3\ mol\%$. Further temperature increases will push the equilibrium more leftward so that it begins to favor CO and H_2O.

Finding the equilibrium condition is the only viable method to determine the final composition of a ***fuel-rich*** reactant mixture, where the population of oxygen atoms is insufficient to convert all the C-atoms to CO_2 and all the H-atoms to H_2O. The equilibrium rules dictate how much of the oxygen deficiency will appear as carbon monoxide, CO (replacing carbon dioxide, CO_2), and how much will appear as hydrogen gas, H_2 (replacing water, H_2O).

The simplest method for computing the concentration of reacting species at equilibrium uses the ***equilibrium constant***, but for discussion, the starting point for this analysis should be the recognition that at thermochemical equilibrium, the ***Gibbs free energy*** $\hat{g} = \hat{h} - T\hat{s}$ is minimized for the gas mixture in the control volume. Minimization of the ***Gibbs free energy*** is what provides the ***equilibrium constant*** method its fundamental basis for evaluating a mixture of reacting species. Minimization of ***Gibbs free energy*** occurs when the First and Second laws of thermodynamics are adhered to concurrently. Reactions that lead to thermochemical equilibrium must conserve energy ($\hat{h}$) and they must continue in the same direction until they have reached a maximum of entropy ($\hat{s}$) relative to every other mixture of reactant and product gases. Minimizing ***Gibbs free energy*** accomplishes both ends.

The most common format for the ***equilibrium constant*** (for gaseous reactants and products) is K_p, which is published in the JANAF tables (for the formation reaction of each species tabulated there), as $\log_{10}(K_p)$ versus T. With the K_p format, the concentration of each reactant in the balanced equation (i.e. on both sides of the reversible arrow) is given in units of ***absolute*** pressure (kPa, $psia$, or atm). Students are encouraged to experiment with the interconversion of different formats for the equilibrium constant, for example, K_C (where species concentrations have units $C_j \veeeq kmol/m^3$) for concentration. The ideal gas law provides the best tool for performing such interconversions.

For this problem, which uses the water gas shift reaction (Equation 4.15), the JANAF equilibrium constants are $K_p(298\ K) = \boxed{1.042 \times 10^5}$ and $K_p(1000\ K) = \boxed{1.442}$. At equilibrium, the partial pressures of the four constituents must adhere to the equilibrium formula in Equation (4.16), where p_i is the partial pressure of constituent i at equilibrium, and the species on the right side of Equation 4.15 belong in the numerator while the species on the left side belong in the denominator:

$$\frac{\overbrace{p_{CO_2} \cdot p_{H_2}}^{Right\ side}}{\underbrace{p_{CO} \cdot p_{H_2O}}_{Left\ side}} = K_p(T) \tag{4.16}$$

The computation of each of the four partial pressures is tractable when the atom populations are constrained as in our example ($C:H:O$ ratio is 1 : 2 : 2) and the total pressure is also constrained (for example, to comply with $p_{tot} = p_{CO_2} + p_{H_2} + p_{CO} + p_{H_2O} = 1.0\ atm$). As the given values of K_p would indicate, the numerator is very large compared to the denominator for the 298 K example, and the numerator and denominator are nearly equal for the 1000 K case. For memorization purposes, the temperature at which $K_p(T) = 1.00$ for the water gas shift reaction is approximately 1109 K (836° C, 1536° F).

We have developed a reasonably good (±6 % , 200 $K \leq T \leq 2200$ K) analytical expression for the water gas shift equilibrium constant, K_p, based on a curve fit to the JANAF table values. It is given as Equations (4.17) and a VBA function script to compute it automatically from temperature is given in Appendix B.2. Students are warned to use only Kelvin for temperature in Equation (4.17b) and consistent units for partial pressure of each gas on the left-hand side (LHS) term of Equation (4.16).

$$\overbrace{K_p(T)}^{\substack{water\ gas \\ shift\ reaction}} = \exp\left(0.399\,437\,z^0 + 4.113\,18\,z^1 + 0.407\,999\,z^2 - 0.057\,381\,z^3\right) \tag{4.17a}$$

$$z = \left(\frac{1000}{T(K)} - 1\right) \tag{4.17b}$$

Like the Chen equation (Chapter 2), the equation to determine the water gas equilibrium constant numerically is too complex to memorize. To simplify and increase the efficiency of using this formula for ***fuel-rich*** combustion calculations, this textbook provides a VBA script that will create a customized Excel function to compute $K_{p,watgas}$ as a function of one argument T. These VBA instructions are presented in Appendix B.2. Instructions for installing such a VBA script into Excel are given in the Introduction section of Appendix B.

Example 4.5

Compute combustion temperature and estimate product gas composition for fuel-rich combustion of methane and air for the following equivalence ratios: $\phi \in \{1.01, 1.10, 1.5, 2.0\}$. Assume that the reactants are at STP and the combustion is adiabatic. The chemical equation to be balanced is:

$$\phi CH_4 + 2(O_2 + 3.76N_2) \rightarrow aCO_2 + bCO + cH_2O + dH_2 + 7.52N_2$$

Answer. See Figure 4.4 for the computed values of temperature. This problem necessitates an algebraic system of five equations and five unknowns. The unknowns are the molar coefficients of the water gas shift products ($a, b, c, and\ d$) and the adiabatic combustion temperature (T_{ad}), while Equations (4.18a–d) are the rules to be followed:

$$K_p(T) = \frac{(n_{CO_2} \cdot n_{H_2})}{(n_{CO} \cdot n_{H_2O})} = \frac{a \cdot d}{b \cdot c} \quad \text{Equilibrium} \tag{4.18a}$$

$$a + b = \phi \quad \text{Carbon atom balance} \tag{4.18b}$$

$$c + d = 2\phi \quad \text{Hydrogen atom balance} \tag{4.18c}$$

$$2a + b + c = 4 \quad \text{Oxygen atom balance} \tag{4.18d}$$

plus Equation (4.14), the combustion energy balance equation:

$$\sum_{i=reac}\left[\dot{n}_i\left(\underbrace{\Delta\hat{h}^0_{f_i}}_{chemical} + \underbrace{\left[\hat{h}(T_{STP}) - \hat{h}_{298}\right]_i}_{sensible=0}\right)\right] = \sum_{j=prod}\left[\dot{n}_j\left(\underbrace{\Delta\hat{h}^0_{f_j}}_{chemical} + \underbrace{\left[\hat{h}(T_{ad}) - \hat{h}_{298}\right]_j}_{sensible>0}\right)\right] \tag{4.14}$$

Rearranging the equilibrium equation and the three atom balance equations algebraically, we obtain expressions for b (from Eq. 4.18b), d (from Eq. 4.18a), and c (from Eqs. 4.18c and 4.18a). The fourth equation below is a rearrangement of Eq. 4.18d and the others.

$$b = \phi - a$$

$$d = K_p \frac{b \cdot c}{a}$$

$$c = 2\phi - K_p(b \cdot c/a) = 2\phi - K_p((\phi - a) \cdot c/a) = 2\phi/\left[1 + K_p \cdot ((\phi - a)/a)\right]$$

$$2a + (\phi - a) + 2\phi/\left[1 + K_p \cdot ((\phi - a)/a)\right] = 4$$

The additional algebra to further rearrange the equation immediately above (where a is the only unknown) is messy and is not repeated here. After rearrangement, a quadratic equation for a is obtained:

$$\underbrace{(1 - K_p)a^2}_{a_{Qu}} + \underbrace{\left[K_p\phi + (\phi - 4)(1 - K_p) + 2\phi\right]}_{b_{Qu}} a + \underbrace{K_p\phi(\phi - 4)}_{c_{Qu}} = 0$$

The quadratic formula solution is Equations (4.19):

$$a = \frac{-b_{Qu} \pm \sqrt{b_{Qu}^2 - 4a_{Qu}c_{Qu}}}{2a_{Qu}} \tag{4.19a}$$

The expanded form of the right-hand side of the aforementioned quadratic formula is:

$$\frac{-\left[K_p\phi + (\phi - 4)(1 - K_p) + 2\phi\right] \pm \sqrt{\left[K_p\phi + (\phi - 4)(1 - K_p) + 2\phi\right]^2 - 4(1 - K_p)K_p\phi(\phi - 4)}}{2(1 - K_p)} \tag{4.19b}$$

The LHS is simply the stoichiometric coefficient a which represents the moles of carbon dioxide (product) in the balanced chemical equation for fuel-rich combustion. Since two real roots are possible for the quadratic equation, we must decide which is the correct one to choose.

By inspection, the lesser of the two roots $\left[-b_{Qu} - \sqrt{b_{Qu}^2 - 4a_{Qu}c_{Qu}}\right]/2a_{Qu}$ results in a negative value for a or b, and thus it must be ***rejected*** as nonlogical. Consequently, the greater root $\left[-b_{Qu} + \sqrt{b_{Qu}^2 - 4a_{Qu}c_{Qu}}\right]/2a_{Qu}$ must be ***accepted***.

Students should guess an initial value for the adiabatic combustion temperature to temporarily obtain a value for K_p from Equations (4.17). The values of a, b, c, *and* d can then be determined algebraically from the quadratic formula and the three atom balance equations. Finally, the temperature is determined by iteration until the enthalpy of the products precisely equals the enthalpy of the reactants (Equation 4.14), just like the method shown in Table 4.1, only the variable being changed this time is temperature, not equivalence ratio.

As mentioned earlier, it is advantageous to employ the VBA scripts provided in Appendix B for computing sensible enthalpy of the product gases. One way to accomplish this is to employ a single row of Excel to solve for the five unknowns (a, b, c, d, and T_{ad}) associated with any fuel-rich equivalence ratio.

As a further simplification, the author has also created a companion function to go along with the water gas equilibrium constant VBA function described earlier (see Appendix B.2 for details). The

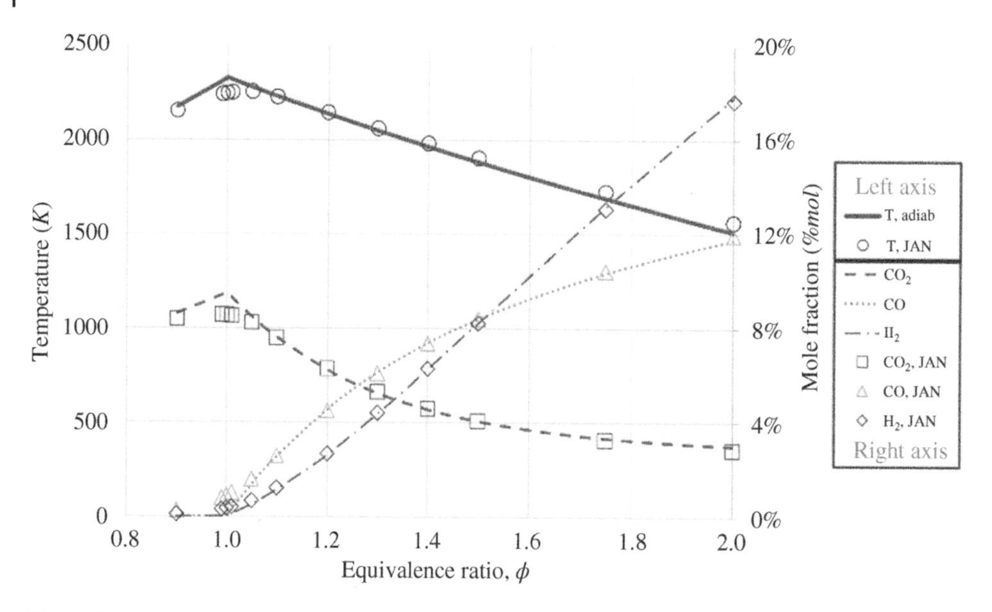

Figure 4.4 Adiabatic combustion temperature and product composition for fuel-rich combustion, based on water gas shift reaction equilibrium. *Source:* Martin Thermal Engineering Inc.

objective is to obtain the critical unknown a (molar coefficient of CO_2) from a pair of inputs ϕ and T_{eq}, by solving the quadratic formula (Equations 4.19) developed earlier. Since the user can input any (reasonable) value for T_{eq}, the process of finding the true value for T_{ad} is not accomplished within the VBA function itself, but the process of obtaining T_{ad} is nonetheless made more efficient by using the additional function in Appendix B.2 to obtain a.

For validation of this algebraic approach to compute the composition of fuel-rich mixtures, Figure 4.4 shows plots of temperature (left axis) and product composition (right axis) as functions of equivalence ratio, ϕ. The lines are predictions using the methodology of this chapter (i.e. Equations 4.14–4.19 and the open symbols are equilibrium computations using STANJAN (Colorado State University 2020) excluding minor constituents such as OH, H, and O. As seen in Figure 4.4, the two sets of values match each other quite well. A slight deviation is evident around $\phi \approx 1$, because the actual transition from fuel-lean equilibrium to fuel-rich equilibrium is smooth, whereas this algebraic method undergoes an abrupt transition (cusp) at $\phi = 1.00$ from "complete combustion" computations on the lean side to "water gas equilibrium" computations on the rich side.

It is important to note that the water gas shift reaction reaches equilibrium very quickly at high temperatures, but the kinetics begin to slow significantly at temperatures below 1200 K. This feature of the chemistry creates a situation where cooling the combustion products from their adiabatic combustion state may cause the composition to become ***frozen*** at values that represent higher equilibrium temperatures than the mixture would experience at room temperature. Readers should be aware of this chemical kinetics artifact and should be careful using this method for mixtures that are grossly depleted of oxygen (i.e. $\phi > 2$).

Based on Figure 4.4, it appears that this equivalence ratio threshold is likely conservative, meaning that $\phi = 2.0$ probably will result in fast enough kinetics that equilibrium would be reached relatively quickly for most applications. However, determining the exact equivalence ratio that fails this kinetics test could be affected by physical parameters such as heat loss due to combustor and burner design, and the extent to which reactants may be preheated before passing through the burner into the furnace enclosure.

4.13 Pollutant Formation and Control

The word "pollutant" is derived from Latin *polluere* (to soil or contaminate) and *lutum* (mud). The sense of pollution as "contamination of the environment" was not common until the late 1940s. The US Congress passed the Federal Water Pollution Control Act (1948) and the Air Pollution Control Act (1955). Subsequent revisions were labeled the Clean Air Act (1970) and the Clean Water Act (1972).

The most basic division among air pollutants are the ***criteria pollutants*** (precisely six categories – particulate matter, ground level ozone, carbon monoxide, sulfur oxides, nitrogen oxides, and lead) versus the *hazardous air pollutants* (a list of 187 specific HAPs). Most of the criteria pollutants are direct or indirect by-products of combustion, and they make their way into the air when combustion exhaust is released to the atmosphere. HAPs are usually not generated by combustion, so their formation will not be a focus of this section.

Pollutants are not harmful to biological life forms when they are present below *de minimis* levels that vary for each compound. In areas where human population is sparse, the local environment is only very rarely burdened with contamination above the *de minimis* levels, hence those areas are deemed to be *unpolluted.* We note that volcanic eruptions and forest fires may create severe "natural pollution" events in remote areas that are otherwise unpolluted. Emissions common to these natural catastrophes are carbon oxides ($CO + CO_2$) acid gases (NO_2, SO_3) and ash (particulate matter).

Particulate Matter. Combustion-generated particulate matter can be either organic (soot) or inorganic (ash). Diesel particulate matter has been found to be a health hazard for people living near locations where heavy trucks are present frequently. Ash from coal combustion is typically captured with a baghouse or electrostatic precipitator, and if pure enough, it may be used as a desirable ingredient in cement.

Ozone. Ozone is too fragile to be directly generated during high-temperature combustion; however, combustion by-products are the primary cause of excessively high ambient ozone in certain urban areas. Automobiles and stationary combustion sources emit nitrogen oxides and volatile organic compounds, which are the primary precursors to tropospheric ozone in the presence of sunlight. This polluted air is called photochemical smog because the ultraviolet photons create excited monatomic oxygen, which then reacts with diatomic oxygen to form ozone. The monatomic oxygen is generated at relatively high levels when nitrogen oxides react with hydrocarbons and oxygen.

Carbon Monoxide. Carbon monoxide is almost exclusively a by-product of combustion processes. CO can be formed from (i) poorly mixed combustion, (ii) very high temperature combustion, and (iii) oxygen-deficient combustion.

Sulfur Oxides. Sulfur is present in coal at sufficiently high concentrations such that emissions of SO_2 and SO_3 can accumulate in the atmosphere, dissolve in clouds, and generate acid rain. Sulfur oxides are typically scrubbed out of the exhausts of industrial combustion processes, and the sulfur is recovered as a sulfate salt.

Nitrogen Oxides. As described earlier, nitrogen oxides ($NO + NO_2$ or NO_x) are formed in combustion and are a key component in the photochemical smog cycle. Three combustion pathways for NO_x are known as: (i) thermal NO_x, where high temperature splitting of N_2 and O_2 molecules to form nitric oxide (NO); (ii) prompt NO_x, where hydrocarbon fragments (typically CH) attack N_2 molecules to initiate the formation of NO; and (iii) fuel NO_x, where fuel-bound N-atoms are released in the flame, and they react with oxygen to form NO. Like ozone, nitrogen dioxide (NO_2) is formed in the atmosphere (and sometimes inside the stack before the gases exit) because the molecule is too fragile to survive at combustion temperatures. In a stack or ambient air, where temperatures are more favorable, NO_2 is formed in three-molecule collisions ($NO + NO + O_2 \rightarrow 2NO_2$), which is a slow, but not inconsequential reaction.

Lead. Up until the 1970s, lead was added to gasoline to prevent premature combustion, or "engine knock." However, the blended organolead compounds would oxidize in the engine to form PbO_2, which is emitted as solid particles. Numerous studies showed that the particles were inhaled by individuals residing near busy streets and highways, causing lead poisoning that would negatively affect the nervous system, brain, and kidneys. Since being regulated by the Clean Air Act (1970), lead particles are no longer a health concern in the urban atmosphere.

Carbon Dioxide. In 2007, the US Supreme Court ruled (SCOTUS 2007) that the Environmental Protection Agency (EPA) was authorized (and in fact required) to declare carbon dioxide a pollutant because it "...cause[s], or contribute[s] to, ***air pollution*** which may reasonably be anticipated to endanger public health or welfare."

We disagree with the designation of the compound carbon dioxide as a ***pollutant***. This disagreement is narrowly focused on the semantics of the word ***pollutant*** because the evidence is clear that excessive emissions of CO_2 by human-originated combustion contributes to global warming, and regulation of those emissions is appropriate. Imposing rules on industrial and automotive sources of greenhouse gas emissions are vital strategies for world nations to employ to reduce the harmful effects of global warming on future generations.

The Supreme Court faced a dilemma because they understood that CO_2 emissions were dangerous to the nation, but they felt powerless to force EPA to regulate it unless the justices declared CO_2 to be a pollutant. With the Clean Act of 1970, Congress had already authorized EPA to regulate pollutants, but they never designated carbon dioxide as either a criteria pollutant or a HAP. Since Congress had not been able to pass a new law that required EPA to regulate CO_2 (either as a pollutant or simply because it can be harmful to humans and the environment), the Court felt it was their duty to give EPA the authorization to regulate it. Their decision to name CO_2 a pollutant is where they got into trouble.

To clarify the conclusion by the Court that CO_2 emissions contributed to ***air pollution*** is illogical on many levels. Consider the following counterarguments: (i) CO_2 is a normal by-product of animal respiration, (ii) the presence of CO_2 in the atmosphere is vital to photosynthesis in plants, (iii) CO_2 is a desirable product of combustion (not a minor by-product), and (iv) by labeling CO_2 gas as a ***pollutant***, regulators theoretically have grounds for restricting the emission of CO_2 from the lungs of animals and people – an undesirable prospect for any sentient organism.

Although the exact range of values is debatable, a desirable concentration of carbon dioxide in the biosphere is critically important to all forms of life. Too little of it and the planet becomes lifeless. Too much of it and the planet is harmed by global warming. Contrary to most other criteria of pollutants, a *de minimis* level cannot be established for carbon dioxide, below which life is ***un***harmed. In fact, the opposite is true – when the concentration of carbon dioxide is too low, life ***is*** harmed.

For a medical analogy, consider the level of potassium ion (K^+) in a human's bloodstream. Humans (and other animals) can perish from either ***too little*** or ***too much*** potassium in their bloodstream. Perhaps readers will concur that it would be foolish to define potassium as a ***pollutant*** or contaminant in blood, and this same logic is appropriately applied to the designation of CO_2 as an atmospheric ***pollutant***.

Although equally controversial from a political standpoint, the Court could have avoided this semantic calamity by issuing a ruling that *forced* Congress to pass a new law regulating CO_2 emissions. Such a new law would certainly have been justified because excessive amounts of greenhouse gases cause global harm and the old law (Clean Air Act of 1970) failed to foresee the possibility that EPA would ever need to regulate gases that were ***not pollutants***. The Supreme Court felt constrained to avoid "legislating from the bench," so they compromised by bending the dictionary definition of ***pollutant*** beyond recognition. They acted as they did because Congress lacked a large enough majority of representatives who wanted to do the right thing – pass a new law regulating greenhouse gases on its own merits.

4.14 Combustion Safety Fundamentals

Thermal systems that utilize fuel combustion to generate heat for a useful purpose must be recognized for the hazardous nature (i.e. the combustibility) of the fuels they use. While there are many aspects to combustion safety (see Sections 7.2, 14.7, and 18.4), the essence of safe combustion is to simultaneously achieve: (i) reactions occurring in the design location and at the design rate and (ii) reactions prevented at other locations and other rates. Clearly, the hazards associated with combustion equipment exceed those associated with pumps, blowers, and heat exchangers.

For clarity, the following four (non-synonymous) terms are defined:

- Deflagration – a subsonic combustion wave propagating through a combustible premixture of fuel and air.
- Detonation – a supersonic combustion wave propagating through a detonable mixture of fuel and oxidizer.
- Explosion – rupture of a vessel or rapid expulsion of burning fuel from a vessel due to pressure buildup associated with rapid combustion. Both deflagrations and detonations can cause explosions.
- Flash fire – a deflagration (indoors or outdoors) that may cause injury or property damage from high temperatures but does not generate enough pressure to cause rupture or structural damage to any containment, including buildings.

Flammability. Fuel gases mixed in air can deflagrate and cause explosions or flash fires. The fuel must be in the *flammable range* (i.e. above its LFL, lower flammability limit, *and* below its UFL, upper flammability limit). The LFL is called the "lean limit" because the concentration of fuel is too low for the combustion chain reactions to deliver enough exothermic energy into the reaction front to preheat the reactant mixture up to its ignition point, which permits the front to propagate. The UFL is called the *rich limit* for the same reason – insufficient exothermicity (due to oxygen deficiency) to permit propagation of a reaction front.

Flammable limits are tabulated in several resources (e.g. Zabetakis 1999; Engineering Toolbox 2003) for different fuels in air at *STP*. For non-*STP* conditions, the limits tend to broaden a bit with increased reactant temperature. For combustion in pure oxygen or oxygen-enriched air blends, the UFL extends to much higher values but the LFL varies only a small amount. For mixtures of two or more fuel gases in air, the mixture LFL and UFL can be evaluated using an approach based on Le Chatelier's principle (NFPA 69 2019).

Flash Point. Liquid fuels do not burn in the liquid phase – they must vaporize in order to deliver fuel to a reaction zone in the air space above the liquid surface. Because liquid volatility (i.e. vapor pressure) is a function of liquid temperature, a temperature threshold (called flash point) exists for each liquid fuel, below which an insufficient amount of vapor is formed to create a flammable mixture of gases.

Fuels such as gasoline ($F.\ P. \approx -45°\ F$) are so volatile that the vapors can be ignited under nearly all practical ambient conditions. Other fuels (e.g. diesel fuel, $F.\ P. \approx +120°\ F$) must be heated above normal ambient temperature to sustain gaseous combustion. Liquid fuels and organics are classified (NFPA 30 2021) according to their degree of flammability as follows:

- Class 1 "Flammable" – Liquids that have a flash point below 100° *F*.
- Class 2 "Combustible" – Liquids that have a flash point at or above 100° *F* and below 140° *F*.
- Class 3A "Combustible" – Liquids that have a flash point at or above 140° *F* and below 200° *F*.
- Class 3B "Combustible" – Liquids that have a flash point at or above 200° *F*.

The foregoing bulleted content was adapted (reproduced) with permission of NFPA from NFPA 30, Flammable and Combustible Liquids Code, 2021 Edition, © 2020, National Fire Protection Association. For a full copy of NFPA 30, please go to www.nfpa.org. Flash points for some common fuels can be found online at Engineering Toolbox (2005) and the Code of Federal Regulations (OSHA 2019).

Transportation Regulations. The US Department of Transportation regulates the transport of dangerous goods, and their classification guidelines are virtually identical to those published by United Nations (2011). Parts of the first five classifications involve combustion hazards and are as follows:

- Class 1 Explosives
- Class 2 Gases
- Class 3 Flammable liquids
- Class 4 Flammable or spontaneously combustible solids
- Class 5 Oxidizers
- Class 6 Poisonous materials
- Class 7 Radioactive materials
- Class 8 Corrosive materials
- Class 9 Environmentally hazardous materials

The technology of combustion safeguards will be discussed in Chapters 7 and 16.

4.15 Other Topics in Combustion

Combustion is such a broad field of study that no single resource can cover the gamut of information available on the subject. We have found the following textbooks to provide helpful information on combustion theory and practice.

For combustion fundamentals that are presented in a manner suitable for an introductory course, see Glassman et al. (2014), Strehlow (1984), or Kuo (2005). For advanced combustion theory, see Williams (1985), and for a high-tech, encyclopedic coverage of the universe of combustion theory and practice, refer to Warnatz et al. (2006). For practical discussions on a wide variety of topics and a storehouse of classic combustion experiments and data, Lewis and Von Elbe (1987) is excellent. For a more recent study of lean combustion technology and its environmental benefits, see Dunn-Rankin (2008). For practical content with an industrial focus, see Fives North American (2010) and IHEA (1994). And finally, for the science of fire ignition, fire spread, and fire protection, see SFPE (2002), Cote (2008), and Babrauskas (2003).

4.16 Homework Problems

4.1 Show that CH_4 is the only normal alkane where the total number of product moles equals the total number of reactant moles for any fuel-lean equivalence ratio. In other words, show that the equality holds for methane, but does not hold for any other n-alkane. Note: n-alkanes are hydrocarbon compounds of the form C_xH_y, where x is a positive integer, and $y = 2(x + 1)$.

4.2 Find or derive an analytical relationship between exhaust oxygen mole fraction ($\%mol$) and equivalence ratio (ϕ) for fuel-lean combustion of CH_4.

4.3 Determine the molar enthalpy of combustion (LHV, $\Delta\hat{h}_{comb}[\doteq kJ/kmol]$) for n-propane C_3H_8. Show work. Why is the value obtained different from the value given by the *NIST Chemistry Webbook*?

Answer. $\Delta\hat{h}_{comb,C_3H_8} = -2\,043\,200\ kJ/kmol_{C_3H_8}$

4.4 Determine the mass enthalpy of combustion (HHV, $\Delta h_{comb}[\doteq kJ/kg]$) for n-octane C_8H_{18}. Show work.

Answer. $-48\,250\ kJ/kg_{C8H18}$

4.5 For fuel-lean methane/air combustion (with all reactants at STP), determine the equivalence ratio that will produce an adiabatic combustion temperature of 1800 K. Assume product gases are comprised of only four compounds: CO_2, H_2O, O_2, and N_2.

Answer. $\phi = 0.673$

4.6 Determine whether a fuel-lean mixture of methane and air ($\phi = 0.44$) at STP is in the flammable range (i.e. above the fuel's *LFL* and below its *UFL*). According to Zabetakis (1999), the LFL for methane is 5$\%vol$ and its UFL is 15$\%vol$. Show work.

4.7 Determine the adiabatic combustion temperature ($T_{ad}[\doteq K]$) of methane/air combustion with equivalence ratio $\phi = 0.6$. Note that multiple iterations are required to obtain this answer. Assume product gases are comprised of only four compounds: CO_2, H_2O, O_2, and N_2.

Answer. 1669 K

4.8 Citing the resource(s) relied upon, find flash points ($°F$ or$°$ C) for the following combustible liquids: ethanol (200 proof, C_2H_5OH), acetone (CH_3CH_3CO), gasoline-a (assume 100% n-pentane, C_5H_{12}), gasoline-b (assume 100% n-octane, C_8H_{18}), benzene (C_6H_6), diesel-c (assume 100% n-dodecane, $C_{12}H_{26}$), linseed oil, and mineral oil.

4.9 For the water gas shift reaction, use Equations (4.14)–(4.19) to validate the numerical answers given in Example 4.4. Begin by using the JANAF tables to determine values for K_p at the two temperatures ($T_1 = 298.15\ K$ and $T_2 = 1000\ K$) and compute the footnote values for the four partial pressures at the two temperatures. Show work.

Answer. For $T_1(=298.15\ K)$, $p_{CO} = p_{H_2O} = 0.4985\ atm$ and $p_{CO_2} = p_{H_2} = 0.0015\ atm$. For $T_2(=1000\ K)$, $p_{CO} = p_{H_2O} = 0.227\ atm$ and $p_{CO_2} = p_{H_2} = 0.273\ atm$.

Cited References

ANSYS Simulation Software (2020). CFD (Computational Fluid Dynamics) code ANSYS-Fluent. http://www.ansys.com/products/fluids (accessed 20 December 2020).

Babrauskas, V.; (2003); *Ignition Handbook*; Issaquah, WA: Fire Science Publishers.

Chase, Jr., M.W.; (1998); *NIST-JANAF Thermochemical Tables, 4*; Monograph No. 9; Volume **I**. Melville, NY: American Institute of Physics. Introduction: https://janaf.nist.gov/pdf/JANAF-FourthEd-1998-1Vol1-Intro.pdf, Individual compounds by element: https://janaf.nist.gov/janaf4pdf.html (accessed 15 August 2021).

Chemstations (2021). CHEMCAD. https://www.chemstations.com/CHEMCAD (accessed 20 February 2021).

Colorado State University (2020). Chemical Equilibrium Calculation (based on STANJAN). *Bioanalytical Microfluidics Program*. http://navier.engr.colostate.edu/code/code-4/index.html (accessed 18 December 2020).

Cote, A.E., ed.; (2008); *Fire Protection Handbook*, 20; Quincy, MA: National Fire Protection Association.

Dunn-Rankin, D., ed.; (2008); *Lean Combustion: Technology and Control*; Burlington, MA: Academic Press.

Engineering Toolbox (2003). Gases – explosion and flammability concentration limits. https://www.engineeringtoolbox.com/explosive-concentration-limits-d_423.html (accessed 1 May 2021).

Engineering Toolbox (2005). Flash points – liquids. https://www.engineeringtoolbox.com/flash-point-fuels-d_937.html.

Fives North American; (2010); *North American Combustion Handbook*, 3; Cleveland: Fives North American

Glassman, I., Yetter R. A., & Glumac, N.G.; (2014), *Combustion*, 5; Cambridge, MA: Academic Press; https://www.elsevier.com/books/combustion/glassman/978-0-12-407913-7.

IHEA; (1994); *Combustion Technology Manual*, 5; Arlington, VA: Industrial Heating Equipment Association.

Kuo, K.K.; (2005); *Principles of Combustion*, 2; Hoboken: Wiley

Lewis B. & Von Elbe, G.; (1987); *Combustion, Flames, and Explosions of Gases, 3*; London: Academic Press. See Appendix C "Limits of Flammability", Table 3 (p. 709) and Figure 353 (p. 712).

Linstrom, P.J. and Mallard, W.G., eds. (2017–2021). "Thermophysical properties of fluid systems"; *NIST Chemistry WebBook, SRD 69*; National Institute of Standards and Technology, U.S. Dept. of Commerce; http://webbook.nist.gov/chemistry/fluid.

NFPA 30; (2021); *Flammable and Combustible Liquids Code*; Classification of liquids; Quincy, MA: National Fire Protection Association.

NFPA 69; (2019); *Standard on Explosion Protection Systems*; Annex B.6 – mixtures of gases; Quincy, MA: National Fire Protection Association.

OSHA (2019). 29 CFR 1910.106. Flammable Liquids. Occupational Safety and Health Administration, U.S. Department of Labor; p. 234 et seq. https://www.govinfo.gov/content/pkg/CFR-2019-title29-vol5/pdf/CFR-2019-title29-vol5.pdf.

PhysLink Physics and Astronomy Online. (2020). Air Composition at Sea Level. https://www.physlink.com/Reference/AirComposition.cfm (accessed 17 December 2020).

REI. (2020). Detailed Chemical Kinetic Modeling. Reaction Engineering International. https://www.reaction-eng.com/detailed-chemical-kinetic-modeling.

Reynolds, W.C.; (1979); *Thermodynamic Properties in SI: Graphs, Tables, and Computational Equations for Forty Substances*; Stanford, CA: Stanford University Mechanical Engineering; https://searchworks.stanford.edu/view/595196

Reynolds, W.C. (1986). The element potential method for chemical equilibrium analysis: implementation in the interactive program STANJAN – Version 3. https://web.stanford.edu/~cantwell/AA283_Course_Material/AA283_Resources/STANJAN_write-up_by_Bill_Reynolds.pdf (accessed 15 August 2021).

SCOTUS. (2007). Massachusetts vs Environmental Protection Agency. U.S. Supreme Court, Docket 05-1120. https://www.oyez.org/cases/2006/05-1120.

SFPE; (2002); *The SFPE Handbook of Fire Protection Engineering*, 3. Society of Fire Protection Engineers; Quincy, MA: National Fire Protection Association.

Strehlow, R.A.; (1984); *Combustion Fundamentals*; New York: McGraw-Hill.

Stull, D.R. and Prophet, H. (1971); *JANAF Thermochemical Tables, 2*; U.S. National Bureau of Standards; NSRDS-NBS 37; U.S. Government Printing Office. For the *4th Edition*, see Chase, M.W. (1998). https://janaf.nist.gov, and see also prior editions published in 1964 (1st) and 1984 (3rd).

UN; (2011); *Recommendations on the Transport of Dangerous Goods – 5th Revised Edition, Amendment 1.* Manual of Tests and Criteria. Report ST/SG/AC.10/11/Rev.5/Amend.1. United Nations. https://www.un-ilibrary.org/content/books/9789210551380.

Warnatz, J., Maas, U., and Dibble, R.W.; (2006); *Combustion: Physical and Chemical Fundamentals, Modeling and Simulation, Experiments, Pollutant Formation, 4*; Berlin: Springer-Verlag

Weast, R., ed. (1974). *CRC Handbook of Chemistry and Physics*, 55; Cleveland: CRC Press.

Williams, F.A.; (1985); *Combustion Theory*, 2; Redwood City, CA: Addison Wesley.

Zabetakis, M.G.; (1999, reaffirmed); "*Flammability Characteristics of Combustible Gases and Vapors*"; U.S. Bureau of Mines, Bulletin 627; ISA–TR12.13.01–1999 (R200X). Research Triangle Park, NC: Instrument Society of Automation. https://studylib.net/doc/18236170/flammability-characteristics-of-combustible-gases-and-vapors and https://www.osti.gov/servlets/purl/7328370,

5

Process Flow Diagrams

The process flow diagram (PFD) is a detailed snapshot of the "circulatory system" of a thermal design. A PFD drawing is typically included as part of the thermal engineer's proposal to a prospective client. With the PFD drawing, an engineer is communicating how the different parts of the thermal system fit together, and how the different pieces of equipment accomplish the overall process goals. In addition to the schematic drawing that lays out the equipment and lines through which the process fluids flow, an important summary of the fluid properties is also incorporated into the PFD drawing – the stream table. Examples of stream tables are given in Section 5.7. For the purpose of introduction, a stream table is a record of the normal operating conditions ($T, p, \dot{m}, \chi_i$, etc.) at all the steady-state process locations (***streams***) in a thermal system.

The stream table imported into the PFD, typically is *not all-inclusive* (i.e. primary thermodynamic properties T, p are always included, but secondary properties v, h, s are usually excluded because space is in short supply, and once the primary state variables and local compositions are defined, all the other properties can be found from tables or state equations). Where additional property values are needed to complete the design process, the engineer will likely create an *all-inclusive* heat and material balance (H&MB) that is kept separate from the stream table embedded into the PFD.

A second highly important drawing will be covered later in this textbook – the piping and instrumentation diagram (P&ID). Both PFD and P&ID are essential parts of a complete design package, but the focus in Chapter 5 is on the PFD, and discussion of the P&ID is deferred until Chapter 14.

5.1 Intelligent CAD

Think Stop. One of the goals of this textbook is to introduce students to tools that will help them to become successful in their engineering design careers. Intelligent CAD (computer-aided design) is one such tool. Instructors and students are advised to investigate and take advantage of the academic licensing options for at least one intelligent CAD program as part of the experience of teaching and learning thermal systems design. While most student outcomes can be achieved with hand drawings, investment of time in getting a CAD package up and running is worth the effort – even more so because the knowledge gained should provide an advantage to the student in their employment search after they complete their degree program.

Thermal Systems Design: Fundamentals and Projects, Second Edition. Richard J. Martin.

Companion website: www.wiley.com\go\Martin\ThermalSystemsDesign2

The many benefits of intelligent CAD have been demonstrated throughout its 40+ year history. A partial list is given here:

- Reduced Errors
  - Correct information embedded into a process drawing can be transferred without transcription errors to structural, mechanical, electrical, and piping drawings, as well as procurement orders.
  - Changes implemented to a process drawing flow automatically to related mechanical and piping drawings.
- Greater Efficiencies
  - Information developed on one project becomes part of the "institutional memory" of the design team – no need to reinvent the wheel.
  - Drawings can be modified and transmitted electronically without physical documents being transported long distances.
- Greater Flexibility
  - Primary work product (i.e. schematics, lists, and specifications) can be sorted, printed, and broadcast in multiple formats.
  - Design parameters in stream tables can be uploaded as setpoints into distributed control system (DCS) software programs for safeguard limits and feedback control processes.
- Enhanced Safety and Quality
  - Incidents of human error in the operation of equipment can be reduced significantly if the design documents are error-free. Operators who are trained with erroneous or incomplete information are more likely to make an operating error, even if the error was corrected during manufacturing or construction.
  - Because human error is at the root of *all* technology failures, safety outcomes depend as much on error-free design and documentation as they do on error-free fabrication, installation, operation, maintenance, and testing.

The citations at the end of this chapter include links to vendors of process schematic software that may be viable for students to use to complete the given exercises. Some of these vendors offer limited duration free software to students or discounted licensing for universities. The cited products are AutoCAD Plant 3D (Autodesk 2021), Visio Plan 2 (Microsoft 2020), SmartDraw (2020), and Lucid (2020).

Citations are also provided to other resources for students who seek additional background content on process flow schematics and stream tables. Information sources for these topics include Thakore and Bhatt (2007), Hipple (2017), and ISA (2009).

5.2 Equipment

When initiating a PFD, the first drawing entities to be laid out on the page are all the major equipment pieces. Equipment (in a thermal system) is defined as anything that changes the thermodynamic state of a process fluid. Here is a partial list of equipment that may be part of a thermal system:

- Pumps increase liquid pressure and motivate liquid flow by performing work on the fluid.
- Blowers and compressors increase gas pressure and motivate gas flow by performing work on the fluid.
- Turbines decrease gas pressure and temperature and extract useful shaft work from the fluid.

- Expanders (also called expansion valves) decrease gas pressure at constant enthalpy but perform no useful shaft work.
- Burners combine fuel and air to produce high-temperature combustion products.
- Heat exchangers change the temperature of one fluid by transferring heat to/from another fluid.
- Boilers and evaporators change the phase of water (boiler) or refrigerant liquid (evaporator) to vapor via a specialized type of heat exchanger.
- Condensers change phase of water or refrigerant vapor to liquid via a specialized type of heat exchanger.
- Tanks provide mass storage of liquids and gases.
- Mixers blend two streams to create a different composition and/or different temperature.
- Separators remove impurities or isolate different constituents or phases to create different compositions.
- Reactors accomplish a composition change via the chemical reaction of reactant species to product species.
- Electrostatic precipitators use electrostatically charged plates to remove particulate matter from an air stream, so it can be disposed or sold.
- Baghouse filters use fabric filter bags to remove particulate matter from an air stream, so it can be disposed or sold.
- Scrubbers remove acid gases (e.g. SO_2) from combustion exhaust by either absorption into liquid water or adsorption into the pores of a reactive particulate solid (e.g. CaO).
- Catalysts reduce pollutant emissions from industrial combustion (and automobile engine combustion) by passing exhaust gas over a noble-metal catalyst in the presence of reducing compounds (for NO_x conversion to N_2) or oxidizing compounds (for CO and HC conversion to CO_2 and H_2O). The selective catalytic reduction (SCR) process utilizes injected ammonia gas (NH_3) in the presence of the catalyst to reduce the NO formed by industrial burners back to N_2.
- Stacks direct an exhaust flow to a safe location, usually a high elevation, above all sources of ignition, and far removed from breathing air intake vents.

Some examples of P&ID symbols used to represent different pieces of equipment are given in Figure 5.1.

Equipment tags are frequently assigned with three parts: **AN-ET-SQ**, which stands for **area number**, **equipment type**, and **sequential number**. A convenient categorization of **area** numbers are as follows:

1) Ambient air
2) Natural gas
3) Exhaust (flue) gas
4) City water
5) Wastewater
6) Brine or glycol
7) Refrigerant, brine
8) Acid/alkaline
9) Other

A useful categorization for ***equipment*** type is:

- B Blower
- C Compressor
- E Exchanger, boiler, and condenser

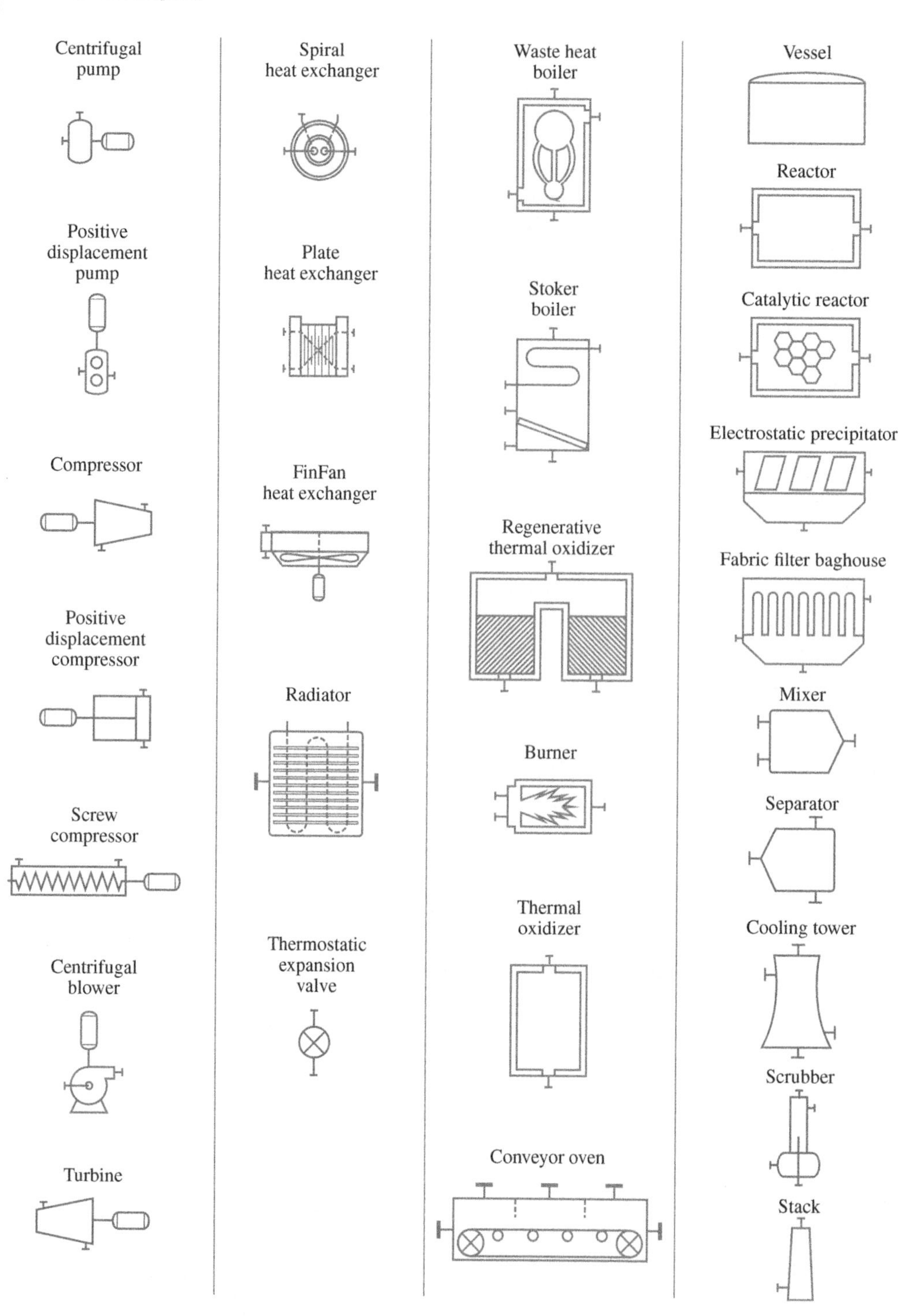

Figure 5.1 Examples of P&ID equipment symbols. *Source:* Selected images/illustrations Images/illustrations from AutoCAD® P&ID 2016 software and used with the permission of Autodesk, Inc. Autodesk and AutoCAD are registered trademarks of Autodesk, Inc., in the United States and other countries.

- F Furnace, burner
- M Mixer
- P Pump
- R Reactor
- S Separator, filter
- V Vessel, tank

  The **sequential** number is self-explanatory. No two pieces of equipment from the same area or type should have the same three-part number. For example, a glycol tank might have the tag 6-V-001 and a brine tag would have a unique sequential number, for example 6-V-002.

Some designers elect to combine the **area** and **sequential** numbers into a single three-digit or four-digit numeral. For example, B-101 might be a blower handling fresh air for a burner, and B-301 might be an induced draft blower handling flue gas.

5.3 Process Lines

The next set of important items to be placed on a PFD are the process fluid lines that interconnect the pieces of equipment. Lines (in a thermal system) are defined as anything that carries a fluid or a control signal. If minor losses of pressure and temperature are ignored, one can assert that the thermodynamic state of a fluid does not change from the inlet to the outlet of the fluid line. Figure 5.2 provides a partial list of lines that may be part of a thermal system:

5.4 Valves and Instruments

After equipment and lines, the next items placed on the P&ID are instruments and valves. However, these items should be excluded from the PFD drawing because their purpose is to maintain control of process parameters, rather than to accomplish chemical or thermodynamic processes. See Chapter 14 for tagging and placing instruments and valves on the P&ID.

5.5 Nonengineering Items

PFD drawings should contain certain nonengineering items that help to communicate clearly the vital process information. A stream tag should be positioned adjacent to each process line, where

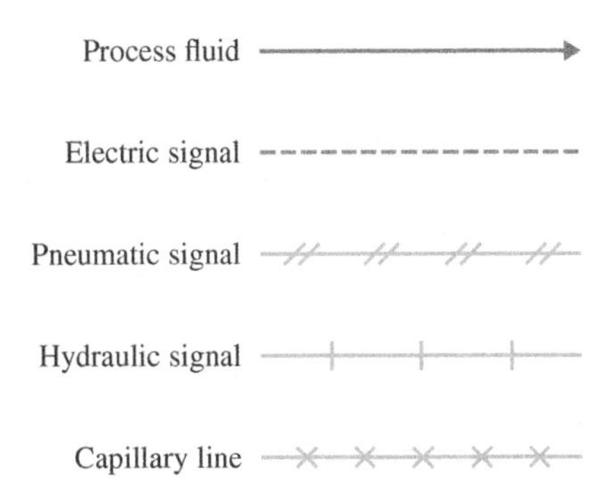

Figure 5.2 Examples of P&ID line symbols. *Source:* Images/illustrations from AutoCAD® P&ID 2016 software and used with the permission of Autodesk, Inc. Autodesk and AutoCAD are registered trademarks of Autodesk, Inc., in the United States and other countries.

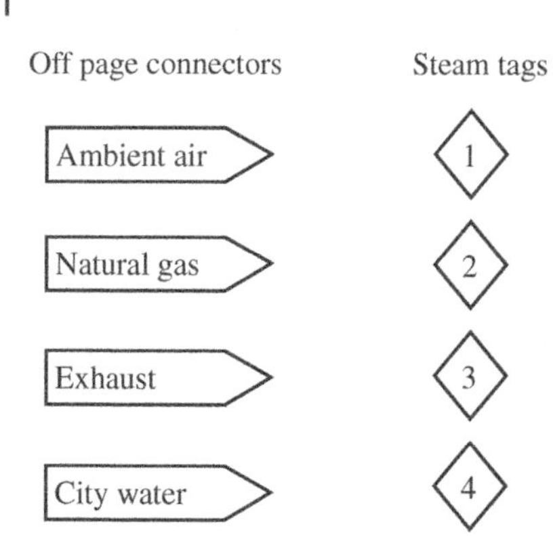

Figure 5.3 Examples of off-page connectors and stream tags. *Source:* Images/illustrations from AutoCAD® P&ID 2016 software and used with the permission of Autodesk, Inc. Autodesk and AutoCAD are registered trademarks of Autodesk, Inc., in the United States and other countries.

each tag has a unique integer value (e.g. 1, 2, 3, ...) that identifies the fluid properties at that location on the PFD. Unlike the PFD, stream tags are purposely excluded from the P&ID drawing, as is the stream table. These customs were developed to reduce clutter; therefore, favored content of the PFD is process data and the favored content of the P&ID is instruments, valves, and control loops.

Another category of nonengineering item that is relevant to the PFD is the off-page connector. These can be box arrows, containing text that describes the fluid entering the process from "off the page" and the direction of flow for that fluid. See Figure 5.3 for examples of off-page connectors and stream tags.

Once the thermal system designer has a vision of the process, including all the equipment and lines, the PFD Schematic should be the first design task. The second design task should be the H&MB, which is the thermodynamic and chemical embodiment of what the designer wishes to accomplish, and it is introduced in the next section. The schematic portion of the PFD can be considered complete when equipment, lines, and tags are all placed, but the PFD is not final until the stream table has been completed.

5.6 Heat and Material Balance

To recapitulate the two primary goals of a thermal systems design – engineering excellence and clear communication, we differentiate the H&MB from the stream table (top-level synopsis). The H&MB is just what its name sounds like – a collection of computations that performs an accounting (like a balance sheet) of all the relevant thermodynamic properties (e.g. enthalpy, mass, and elements) for the given design.

Because processes differ in their working fluids, equipment, and capacity, it is impossible to provide a universal road map for developing the H&MB. A good portion of this book is dedicated to providing examples of how to compute balances of energy flow, mass flow, and atom counts across a spectrum of many different equipment types. Chapter 6 will address preliminary aspects of thermodynamic design, while Chapters 7–14 will provide specific design recommendations for many types of equipment and subsystems. In the remainder of Chapter 5, generic guidance on structuring and populating a stream table is provided, along with tips for arranging all the pieces of the PFD onto the drawing.

Think Stop. Table 5.1 provides an example of what a stream table might look like for a simple process. Students are encouraged to play sleuth and attempt to determine the process being described by the entries in the table given.

Table 5.1 Example stream table for a simple process. PM stands for particulate matter.

Parameter	1	2	3
p (kPa_{abs})	100	94	101
T (K)	298	298	304
$\dot{m}_{air}$ (kg/s)	0.008 23	0.008 23	0.008 23
$\dot{m}_{PM}$ (kg/s)	0.000 01	0.000 00	0.000 00
$\dot{m}_{total}$ (kg/s)	0.008 24	0.008 23	0.008 23

Source: Martin Thermal Engineering Inc.

The structure of a typical stream table is:

- Rows are for process parameters (pressure, temperature, phase, or quality; individual species mass flow rates; and total flow rate).
- Columns are for stream tags (one for each line that interconnects equipment on the PFD drawing).

The number of rows on the stream table is dependent on the nature of the processes captured on the PFD. The variables shown in Table 5.1 (pressure, temperature, individual mass flow rate, and total mass flow rate) are the bare minimum needed to define the process, but sometimes it is also helpful to add additional parameters (e.g. total volume flow rate, mixture molar mass, phase, and saturation quality) and on rare occasions, other properties may be important (e.g. mixture density, enthalpy, and entropy).

The number of columns on the stream table should be equal to the number of process fluid lines (stream tags) on the PFD. For the purpose of the PFD, it is generally assumed that properties do not change substantially along a line, no matter how long or how much pressure drop occurs. If a line is sized with a moderately low velocity, the frictional losses usually can be ignored. See Chapter 12 for examples when pressure drop in a long pipe or duct must not be ignored. Since thermodynamic state is supposed to change across each piece of equipment, it is imperative to have a stream tag associated with every line entering and leaving each piece of equipment. When off-page connectors are placed to identify the flow direction of inlet streams (fuel, fresh air) or outlet streams (exhaust gases), additional stream tags should be included for the streams that tee into or out of the main process stream.

5.7 PFD Techniques

Other than the sequence with which to place PFD equipment along the direction of flow, there are very few prescriptive rules about generating a PFD. However, some practices should be avoided on a PFD, and students learn such lessons best by having their drawings critiqued by an experienced designer.

One convention that we follow regarding stream tag designation is that ***State* 1** is typically assigned to the inlet nozzle of the piece of equipment that increases fluid pressure (i.e. blower, compressor, or pump) and the stream tag numbers increase sequentially in the working fluid's direction of flow. This convention necessitates that the working fluid sometimes moves clockwise and other times counterclockwise around the process cycle.

The benefit of adopting such a convention and applying it consistently within your team gives you the benefit of always knowing which stream number applies to which condition in the cycle – and being able to converse about process conditions using tag numbers – without simultaneously viewing the PFD diagram. For example, using this convention, a simple Rankine cycle would always have the fluid entering the condenser tagged as ***State 4*** and the fluid exiting tagged as ***State 1***, because it is flowing toward the pump. Similarly, for a simple refrigeration cycle, ***State 4*** would always represent the entrance to the evaporator, and ***State 1*** would always be the inlet to the compressor.

Think Stop. Students may refer to the list below for common PFD drawing errors that are *avoidable.*

- Failing to include tag numbers for all relevant streams.
- Entering an insufficient number of rows for the relevant number of process parameters.
- Failing to match text and symbol (i.e. placing text adjacent to a finfan's heat exchanger symbol that describes a shell and tube heat exchanger's characteristics).
- Entering data into the H&MB that does not adhere to conservation laws.
- Drawing lines and arrows that point in the wrong direction relative to the process flow.
- Drawing one or more process lines that enter a vessel with no lines that exit the vessel.
- Using box arrows (off-page connectors) that point opposite to flow arrows connected thereto.
- Utilizing a box arrow without any text (i.e. no off-page information on an off-page connector).
- Resizing equipment icons so they are out of scale (i.e. too small or too large compared to other icons on the drawing).
- Using equipment tags that do not represent the class of equipment shown (e.g. furnace tag number that begins with "P" or a pump tag number that begins with "F"). Stream tag conventions will be addressed in detail in Chapter 14.
- Connecting lines to the wrong piece of equipment (e.g. connecting a water line to a burner).
- Placing equipment that is designed for ambient temperature (e.g. blower) in a line that conveys high-temperature gas (e.g. combustor exhaust).
- Entering text that is too small or illegible.

Example 5.1

Draw a schematic for an ideal Rankine cycle where the fuel is wood waste (i.e. sawdust and wood chips) and the working fluid is water. Also prepare a stream table with the following project "knowns": saturated liquid enters the pump at a temperature of $p_1 = 3.535\,91\ kPa$; water pressure exiting the pump is $p_2 = 5000\ kPa$. maximum steam temperature is $T_3 = 600\ K$; water exiting the condenser is $T_1 = 300\ K$. The stream table should include all of the following rows: p, T, ρ, v, h, s, and X. For simplicity, we have excluded the mass flow of water from the calculations, because that value only affects the size of the system, not the thermodynamic states. Insert values into all cells for which you are sure are correct.

Answer. The schematic should look like Figure 1.1.

A stoker boiler was chosen for Figure 5.4 to illustrate a cycle using a solid fuel such as wood waste, but other boiler symbols would be preferred if the source of fuel was gaseous or liquid. The extra nozzles on the stoker boiler are: (i) bottom = ash out, top = exhaust gas out; (ii) left middle = wood waste in; and (iii) left bottom = fresh air in. For simplicity, we have excluded these non-water flows

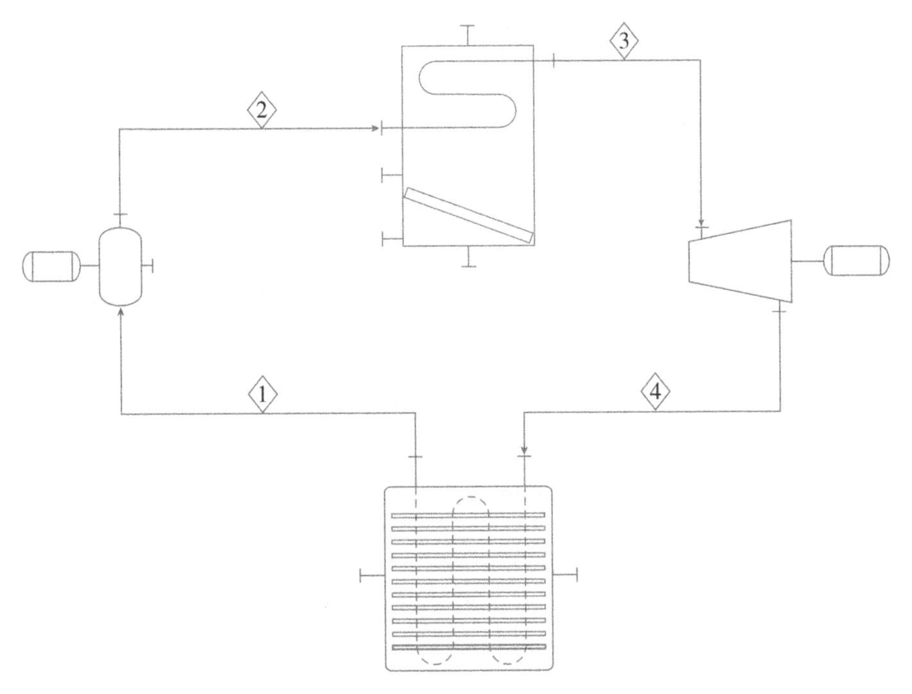

Figure 5.4 Rankine cycle schematic for PFD of Example 5.1. *Source:* Images/illustrations from AutoCAD® P&ID 2016 software and used with the permission of Autodesk, Inc. Autodesk and AutoCAD are registered trademarks of Autodesk, Inc., in the US and other countries.

from this example, but in a complete PFD, all nozzles would have lines (with flow arrows and stream tags) connected to them.

Also, the condensing heat exchanger chosen is a finned air-water device, and it has two nozzles that are shown as being attached to nothing. Those two nozzles would also need lines, connectors, and tags for ambient air inflow and warm air outflow in a completed PFD.

For this example, the stream table is shown in Table 5.2. The known values from the problem statement are given in larger font bold text. Because this was assumed to be an ideal Rankine cycle, we assume no friction (i.e. $p_3 = p_2$ and $p_4 = p_1$) and isentropic shaft work (i.e. $s_2 = s_1$ and $s_4 = s_3$).

Table 5.2 Rankine cycle stream table of Example 5.1. Only the states for working fluid water are shown in this illustration. Large bold values are the known values of problem statements.

Parameter	1	2	3	4
p (kPa_{abs})	**3.535 91**	**5000.**	5000.	3.535 91
T (K)	**300.**	300.0995	**600.**	300.
ρ (kg/m^3)	996.313 59	998.533 05	20.402 405	0.034 913 5
v (m^3/kg)	0.001 003 70	0.001 001 5	0.049 013 8	28.642 20
h (kJ/kg)	111.917 8	116.926 7	3004.934	1898.287
s ($kJ/kg \bullet K$)	0.392 572 6	0.392 572 6	6.344 924	6.344 924
X ($kg_{w,vap}/kg_{w,tot}$)	**0.000 0**	–	–	0.732 537

Source: Martin Thermal Engineering Inc.

To determine the thermodynamic properties of water, we can either use the tables in Appendix A.4 or the customized Excel functions in Appendix B.4. We will choose the VBA functions for this example. The format of the VBA functions is that temperature and density are the two arguments (independent variables) and all the other properties are outputs (dependent variables). For example, we can obtain pressure by employing the VBA function `H2O_p(`T, ρ`)`.

For State 1, since p is already known, we can find ρ by iteration until the VBA function equals the known value $p_1 = 3.535\ 91\ kPa$. Doing so gives a value for $\rho_1 = 996.313\ 59\ kg/m^3$. Next, we will check the quality at this condition with the VBA function `H2O_X(`T, ρ`)` and we can confirm that $X = 0.0000$, which is the saturated liquid. Then, we can compute the remaining properties for State 1 using VBA functions `H2O_h(`T, ρ`)` and `H2O_s(`T, ρ`)`. The computed values are shown in Table 5.2 in normal font.

For State 2, we know pressure (a given value) and entropy (pump assumed isentropic). Since both density and temperature are unknown, we must use the incompressible liquid assumption (Equation 6.2) to estimate Δh across the pump, which will give us an approximate value for $h_2(=h_1 + \Delta h_{pump})$, then temperature T_2 and density ρ_2 can be found by serial iteration.

Ideally, Equation (6.2b) would be integrated ($\Delta h = \int v \cdot dp$), which would address the slight reduction in specific volume during pumping, but it is also acceptable to assume v is constant (taking it outside the integral) and then multiply its average value by Δp. For this problem, we obtain $\Delta h = +\ 5.009\ kJ/kg$, which gives us $h_2 = 116.9267\ kJ/kg$, and after iteration, we obtain $T_2 = 300.0995\ K$ and $\rho_2 = 998.533\ 05$. This is a subcooled liquid, so no value exists for quality. The temperature increases slightly because the pumping was considered to be isentropic, and the liquid was assumed incompressible. Most of the enthalpy change was because of the significantly higher pressure.

For State 3, we know the temperature and pressure, and we know that the water should be a superheated gas after the boiler. We can use the VBA function for pressure to iteratively to find ρ_3, given that $T_3 = 600\ K$. Because the steam is superheated, no quality can be determined. The other properties can then be found from their respective VBA functions and from the known values for T and ρ. Table 5.2 incorporates all these values in small, non-bold font.

For State 4, the exit condition is under the vapor dome, so the quality will be an important property, with $0 < X < 1$. Students should verify that the saturation pressure of water at $T_4 = T_1 = 300\ K$ is $p_4 = p_1 = 3.535\ 91\ kPa$ as given in the problem statement. The density can be found by iteration from known T_4 and $s_4 = s_3$ across the isentropic turbine. The remaining values can be determined from their respective VBA functions.

Think Stop. We suggest to students that the stream table may be thought of as an underconstrained Sudoku puzzle. Consider three types of Sudoku puzzles: (i) an underconstrained puzzle is one where an insufficient number of cells are preassigned, and although many solutions are excluded, more than one valid solution is possible; (ii) a fully constrained puzzle is the type you find in the newspaper, having one and only one solution; and (iii) an overconstrained puzzle is one where the preassigned cells are incompatible with a valid solution.

With the stream table, the puzzle rules are the conservation laws (mass, energy, atoms, entropy, etc.) and other laws (equations of state, Fourier's law, modified Bernoulli's equation, etc.). The rules must be followed as the fluids change properties from one stream identity to the next as they pass through pieces of equipment. The preassigned cells are the known constraints, such as composition, temperature, and pressure of fuel and indoor or outdoor air, customer-defined process values, and final streams that must meet an acceptance criterion. At the end of the game, each individual cell in the stream table must have a value, and all the rules must be followed. However, because the puzzle is underconstrained, more than one solution that meets all the objectives is possible.

The PFD and H&MB are vital elements of a process design package. When facing the blank page of a stream table, student designers are encouraged to "embrace the struggle."

To do so, it requires a sufficient time investment *early* in the design process to differentiate the many solutions that contain weaknesses or errors from the few solutions that meet all the objectives and follow all the rules. Only by "doing the math" these opposite categories of solutions can be differentiated from each other.

5.8 Homework Problems

It is recommended that students utilize a commercially available computer-aided design software package to produce these schematics. Some commercially available software tools are cited in the reference list at the end of this chapter. Instructors may deem manually drawn schematics to be acceptable, but the learning outcomes will be stronger with a CAD software package.

For each project below, the following instructions apply: (i) add fluid movers (pumps, blowers) where appropriate, (ii) label each stream with a stream tag and create a blank stream table with correctly populated row and column headings for this process, and (iii) enter known values for process variables into the appropriate cells (leaving cells blank where values are unknown) and label as DRAFT.

5.1 Draw a schematic that represents a ***lumber kiln*** used to dry batches of green lumber using warm exhaust products from a forced draft, natural gas burner and with the water vapor enriched exhaust gas vented to atmosphere via a stack. Refer to Simpson (1991) for more information.

5.2 Draw a schematic that represents a ***landfill gas thermal oxidizer***, that receives LFG ($50\ \%\ CH_4 + 50\ \%\ CO_2 + ppm - level\ VOC$) and combusts it with air using a burner firing into a horizontal vessel, with exhaust gas vented to atmosphere via a stack. Refer to EPA (2020) for more information.

5.3 Draw a schematic that represents a split system, ***heat pump*** that provides forced air heating in the winter season and forced air cooling in the summer season to a two-story (i.e. two-zone) multistudent ***residence hall***. The split system should have an outdoor compressor unit (including one-way expansion valve, four-way switching valve, and outdoor coil) and indoor forced air unit (FAU) that contains the remaining equipment.

5.4 Draw a schematic that represents a natural gas fired, ***swimming pool water heating and cooling system*** that comprises necessary water piping, a gas burner, a condensing heat exchanger (cross flow) that transfers heat from the hot combustion products to liquid water (without boiling), and an exhaust gas stack. Refer Sections 10.4 and 11.3 for more information and refer Section 11.1 for optional designs.

5.5 Draw a schematic that represents a ***refrigeration system*** capable of freezing and maintaining an appropriate subfreezing temperature of an ***ice rink in an indoor arena***. The components should include all necessary piping (for the refrigerant and the secondary heat transfer fluid which circulates in pipes at the bottom of the ice slab). See ASHRAE (2015) and ASHRAE (2014) in Cited References section for more information.

5.6 Draw a schematic that represents a ***gas-fired, evaporative-distillation-style water desalination process***, including a natural gas, air burner, indirect-heating evaporator, condenser, and storage tank for purified water. Also refer to Sections 11.1 and 11.7 for more information and about optional designs.

5.7 Draw a schematic that represents a ***continuous grain-drying oven***, including a horizontal, air-permeable conveyor belt with two heating zones – each equipped with a forced draft natural gas burner and recirculating air fan, and one cooling zone having a forced draft blower for ambient air. Refer to Section 11.6 for more information.

5.8 Draw a schematic that represents a ***split air conditioning system*** for a 3-zone (i.e. 3-story), 24-inmate ***county jail building*** (a) outdoor finfan condenser coil with compressor and (b) indoor air handler with evaporator and expansion valve. Refer to Section 11.5, ASHRAE (2014), ASHRAE (2015), and ASHRAE (2016) in Cited References section for more information.

5.9 Draw a schematic for a ***hot air balloon*** including pressurized LPG cylinder, burner, balloon envelope, exhaust valve, and passenger basket for a payload of 12 adults with average mass per person $m = 75\ kg$ and a flight duration of $t = 60\ minutes$. The envelope's maximum temperature is $T_{env,\ max} = 120\ °C$. Ensure that the capacity of the fuel storage cylinder is 50% greater than the maximum needed for the mission. Refer to Section 11.10 for more information.

5.10 Draw a schematic for an ***exothermic gas generator for a heat-treating furnace*** to combust methane at fuel rich conditions such that the product gas stream contains approximately $\chi_{H_2} \approx 10\%v$. Assume that the methane supply pressure is higher than the required amount for the desired production rate of special atmosphere gas to the heat-treating furnace, but that air must be pressurized in a blower to combust the fuel in the gas generator chamber. Assume the furnace's demand for special atmosphere gas is $\dot{m}_{generatorgas} = 0.2\ kg/s$. See Section 7.9 for more information.

5.11 Draw a schematic of a ***tenter-frame fabric drying oven*** that removes moisture (H_2O from a dyeing process) from a moving web traveling at $V_{web} = 0.2\ m/s$. The fabric's outlet moisture level should be controlled to $Y_{web,\ out} = 0.05\ kg_w/kg_{tot,\ web}$ for an inlet moisture level of $Y_{web,\ im} = 0.65\ kg_w/kg_{tot,\ web}$. The source of heat are natural gas burners (2×) that produce a dilute stream of hot combustion products at a temperature of $T_{burner} = 140\ °C$. Determine the length of the oven, the number of burners required, and the heat release rating of each burner. Refer to Swanson and Stepner (1976) for more information.

5.12 Draw a schematic of a ***commercial espresso machine*** that produces a batch of hot espresso liquid equivalent to five shots (0.03 $kg/shot$) in the duration of 1 minute. Refer to Section 11.9 for more information.

5.13 Draw a schematic of a ***chemical engineer's refrigerator*** that chills 20 beverage cans ($\mathcal{V} = 500\ mL$, each) from STP to $T_{chill} = 5\ °C$ in $t = 30\ minutes$. Refer to Section 10.10 for more information.

5.14 Draw a schematic of a ***passenger aircraft biohazard destruction system*** that conveys droplets and biohazard particles into a heated metal foam while permitting air to pass through at a rate of $\dot{V}_{air,STP} = 5000\ L/\ min$. The filter temperature should be maintained at $T_{filter} = 200\ °C$ and the recycled exhaust air should be cooled to $T_{returnair} = 20\ °C$ by dilution with cold outside air. Assume a flying altitude of $z_{alt} = 11\ 000\ m$ and determine the volume flow rate of outside air that must be compressed to a density equivalent to $z_{alt,equiv} = 2500\ m$ and blended with the cleaned (warmer) exhaust gas to achieve the desired return air temperature. Refer to Section 7.8 for more information.

Cited References

ASHRAE; (2014); *ASHRAE Handbook: Refrigeration; Chapter 44. Ice rinks*; Atlanta: American Society of Heating, Refrigerating and Air-Conditioning Engineers. (The 2018 Edition should have similar information, but chapters might be numbered differently.)

ASHRAE; (2015); *ASHRAE Handbook: Heating, Ventilating, and Air Conditioning Applications; Chapter 1. Residences; Chapter 5. Places of assembly; Chapter 6. Hotels, motels, dormitories*; Atlanta: American Society of Heating, Refrigerating and Air-Conditioning Engineers.

ASHRAE; (2016); *ASHRAE Handbook: HVAC Systems and Equipment; Chapter 51. Thermal storage*; Atlanta: American Society of Heating, Refrigerating and Air-Conditioning Engineers. (The 2020 Edition should have similar, but updated information.)

Autodesk Inc. (2021). AutoCAD P&ID and Plant 3D Developer Center. https://www.autodesk.com/developer-network/platform-technologies/autocad-p-id-and-plant-3d and "Autodesk Education Community: Design Your Future" https://www.autodesk.com/education/home.

EPA; (2020); *LFG Energy Project Development Handbook.* Landfill Methane Outreach Program. U.S. Environmental Protection Agency; https://www.epa.gov/sites/production/files/2016-11/documents/pdh_full.pdf.

Hipple, J.; (2017); *Chemical Engineering for Non-Chemical Engineers*; Hoboken: Wiley.

ISA 5.1; (2009); "*Instrumentation Symbols and Identification*"; Research Triangle Park: International Society of Automation; https://www.isa.org/products/ansi-isa-5-1-2009-instrumentation-symbols-and-iden

Lucid; (2020); "*P&ID Software*", Lucid Software, Inc.; https://www.lucidchart.com/pages/examples/p-and-id-software

Microsoft; (2020); "*Visio Plan 2: Create Professional Diagrams*"; Microsoft, Inc.; https://www.microsoft.com/en-us/microsoft-365/visio/visio-plan-2

Simpson, W.T., Ed.; (1991); *Dry Kiln Operator's Manual*; USDA Publication AH-188. Madison, WI: U.S. Department of Agriculture, Forest Service, Forest Products Laboratory; https://www.fs.usda.gov/treesearch/pubs/7164.

SmartDraw; (2020); "*Process Flow Diagram Examples*"; SmartDraw, LLC; https://www.smartdraw.com/process-flow-diagram/examples

Swanson, F.K. and Stepner, D.E. (1976). Temperature control system for textile tenter frame apparatus. U.S. Patent No. 3,961,425. Assigned to Measurex Corporation.

Thakore, S.B. and Bhatt, B.I.; (2007); *Introduction to Process Engineering and Design*; New Delhi, Tata McGraw Hill. See Chapter 4. Importance of process diagrams in process engineering.

6

Advanced Thermodynamics

Chapter 1 focused on the fundamentals of thermodynamic system analysis, and Chapter 6 branches into some advanced thermodynamic concepts. The content in this chapter is focused on how the designer can ensure thermodynamic laws are adhered to as fluids enter and leave individual equipment and how to transform a complex problem into a simpler one by modeling.

Advanced macroscopic thermodynamics includes the subjects covered in this section plus chemical thermodynamics, physical gas dynamics, electrodynamics, and more. Advanced microscopic thermodynamics is described by statistical mechanics and to some degree quantum mechanics. Interested readers can refer to Zemansky (1937), Keenan (1941), and Callen (1960) for classical treatises on macroscopic thermodynamics and Kline (1999) for a quick read on engineering implications of entropy. For an extensive coverage of chemical thermodynamics and for state equations of many substances, see Reid et al. (1987), and for properties of water, nothing beats Keenan et al. (1969). For an excellent coverage of physical gas dynamics, see Vincenti and Kruger (1965), and for statistical mechanics, see McQuarrie (1973).

6.1 Equations of State

Beyond the ideal gas law, several other equations of state are often used to predict the new state of a material that has been subjected to a physical or chemical process, or the interrelation between thermodynamic properties of a material that is at a known equilibrium state.

As discussed in Chapter 1, the ideal gas law is an important equation of state for low-density gases; however, it does not provide means for the direct prediction of internal energy or enthalpy of gases and it contains invalid information about gases near the vapor dome and the critical point. Hence, we seek other mathematical forms that can predict the behavior of working fluids at other desired ranges of temperatures and pressures.

Equations (6.1)–(6.5) provide well-known equations of state that may be useful for certain types of analysis problems. None of these are recommended as universally acceptable, but some may prove very helpful for a practicing thermal system engineer and hence their mention here. We recommend the two appendices in this book for their intended scopes, and where more granularity or new materials are needed, the National Institute of Standards and Technologies (NIST) Chemistry Webbook (Linstrom and Mallard 2017–2021) is the best available for the scope it covers. Much of the content in Appendices A and B was derived from Linstrom, and Appendix B.6 is the only section that carries a "not recommended" tag for use in any context other than academic exploration.

Thermal Systems Design: Fundamentals and Projects, Second Edition. Richard J. Martin.

Companion website: www.wiley.com\go\Martin\ThermalSystemsDesign2

- Ideal gas

$$pv = RT \tag{6.1a}$$

$$p\hat{v} = \hat{R}T \tag{6.1b}$$

- Incompressible liquid

$$v = constant \tag{6.2a}$$

with $v_{liq} = f(T, T_{crit}[\text{ and often } v_{crit}])$

$$dh = vdp \tag{6.2b}$$

Useful relations (in English units) for saturated liquid density as a function of temperature for many refrigerants are given in ASHRAE Paper 2313 (Downing 1974) and are reproduced in SI units (Reynolds 1979). The form of these equations (requiring seven empirical constants) is

$$\rho_{satliq} = \sum_{i=1}^{5}\left[D_i\left(1-\frac{T}{T_c}\right)^{(i-1)/3}\right] + D_6\left(1-\frac{T}{T_c}\right)^{½} + D_7\left(1-\frac{T}{T_c}\right)^{2} \tag{6.3}$$

- Clausius–Clapeyron

$$\left.\frac{dP}{dT}\right]_{sat} = \frac{\Delta h_{fg}}{T\Delta v_{fg}} \tag{6.4}$$

where Δh_{fg} is the latent heat of vaporization, Δv_{fg} is the phase transition volume change, and T is temperature. The scope of the Clausius–Clapeyron equation is limited to finding the slope of the saturation curve, but it is derived from elementary thermodynamic principles and applies universally to pure substances in a state of liquid–vapor equilibrium. One of the end-of-chapter problems here asks students to derive it from first principles.

Direct relations for saturation pressure as a function of temperature for many refrigerants are given in ASHRAE Paper 2313 (Downing 1974) and are reproduced in SI units (Reynolds 1979). The form of these equations (requiring six empirical constants) is

$$log_{10}(p_{sat}) = F_1 + \frac{F_2}{T_{sat}} + F_3\, log_{10}T_{sat} + F_4 T_{sat} + F_5\frac{(F_6 - T_{sat})}{T_{sat}}\, log_{10}(F_6 - T_{sat}) \tag{6.5}$$

- Van der Waals

Johannes van der Waals was a self-educated mathematician and physicist who was awarded a PhD at age 36 from the University of Leiden (Netherlands) for the dissertation "On the Continuity of the Gaseous and Liquid States" (Van der Waals 1873).

The equation that bears his name is

$$\left[p + a(\hat{v})^{-2}\right]\cdot(\hat{v} - b) = \hat{R}T \tag{6.6}$$

where a and b are positive constants whose contribution becomes unimportant for high specific volume (ideal gas) conditions. Note that $\hat{v} = V/n = \hat{M}v(\doteq m^3/kmol)$.

The physical significance of the constant b is that the finite volume of the molecules themselves should be subtracted from the volume occupied by the ensemble of gas molecules in space. This fixed volume becomes unimportant compared to $\hat{v}$ when $\hat{v}$ is large.

The physical significance of the constant a is that intermolecular forces become more important when the gas specific volume is small (reciprocal of $\hat{v}$ is large), leading to actual pressures that are smaller than ideal gas pressures for a given $\hat{v}$ and T.

- Real gas
  $pV = Zn\hat{R}T$, where Z is the compressibility factor that is equal to 1.00 for low-density (ideal) gases far removed from the vapor dome but is different from unity at high densities and near the vapor dome. Charts of Z as functions of critical pressure, critical temperature, and critical specific volume can be found online and in some thermodynamics textbooks, under the subject of "Generalized Equation of State."
  Van der Waals theorized that all fluids, when compared at the same reduced temperature ($T_r = T/T_{crit}$) and reduced pressure ($p_r = p/p_{crit}$), have approximately the same compressibility factor (Z). His theory is called the principle of corresponding states and is useful for some temperatures and pressures but overpredicts Z for conditions near the critical point.
- Virial equation for compressibility factor
  A serial expansion for Z has been derived from statistical mechanics and can be used with the specific values of critical pressure and temperature for a given substance to obtain real gas behavior at high pressures or near saturation. The virial expansion provides acceptable results for low-density gases by retaining only the second and third terms, but this gives poorer results near the saturation curve, where more terms are needed.
- Robinson–Peng
  The Robinson–Peng equation relies on the compressibility factor Z and the acentric factor ω to provide good predictions of thermodynamics states for saturated liquids, vapors, and superheated gases. The acentric factor is related to the molecule's deviation from sphericity. The acentric factor is zero for the noble gases, neon, argon, and krypton. This equation of state is beyond the scope of this book but may be of interest to students who require more accuracy than the other equations provide.
- TPSI (Reynolds 1979)

  In addition to detailed p-h and T-s plots and tables for approximately 40 substances, thermodynamic properties in SI (TPSI) also provides a collection of 11 $p(T, v)$ equations, 7 $c_v^0(T)$ equations, 11 $p_{sat}(T)$ equations, and 8 $\rho_f(T)$ equations with unique constants for the same set of 40 different working fluids. Unlike van der Waals, Robinson–Peng, and the other formulations above, the relevance of the Reynolds content is for its engineering practice and not for pioneering theory, although some equations are derived in the Appendix.

6.2 Thermodynamic Property Diagrams

Thermodynamic property diagrams are prepared in several formats, and students are encouraged to become familiar with the general layout of multiple types. They can be a very effective visualization device if one is trying to assess the thermodynamic "landscape" of a cycle or process step.

Mollier Diagram. The p-h diagram is most useful for refrigeration cycle analyses. Isobars (lines of constant pressure) are horizontal. Isenthalps (lines of constant enthalpy) are vertical. Isotherms (lines of constant temperature) coincide with isobars under the vapor dome. In the ideal gas regime, isotherms are nearly coincident with isenthalps, which implies they are nearly vertical. A representative p-h diagram showing isotherms is given in Figure 6.1.

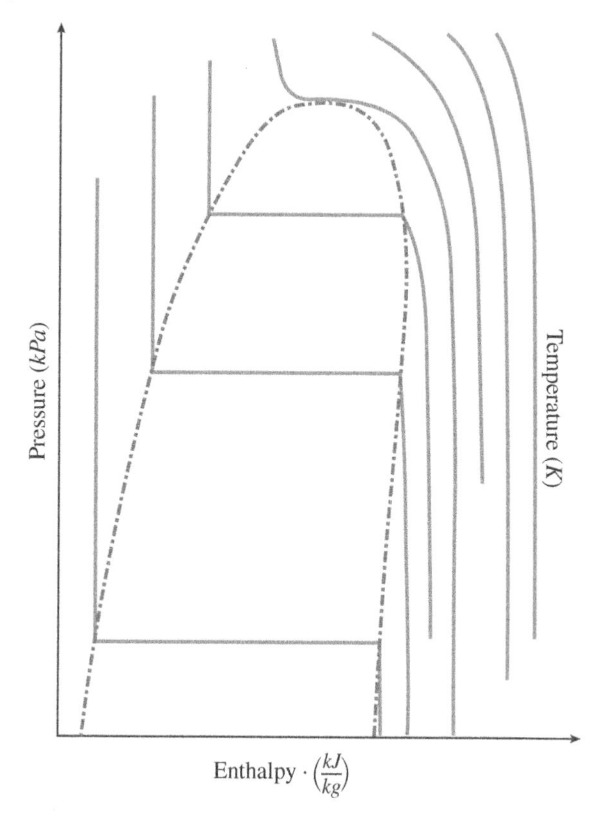

Figure 6.1 Example *p-h* diagram showing a vapor dome (dot-dash line) and isotherms (solid line). *Source:* Martin Thermal Engineering Inc.

Temperature versus Entropy. The *T-s* diagram (see Figure 6.2) is mostly useful for Brayton and Rankine cycle analyses. Isotherms are horizontal. Isentrops (lines of constant entropy) are vertical. Isobars coincide with isotherms under the vapor dome. In the ideal gas regime, isenthalps are nearly coincident with isotherms, which implies they (isenthalps) are nearly horizontal.

Lesser-used diagrams include enthalpy vs entropy and pressure vs specific volume.

6.3 Gibbs, Helmholtz, and Maxwell

Think Stop. The thermodynamics of state can seem a little overwhelming at times, but when students are enthusiastic about the ultimate goal – formulating and utilizing equations that can be used to predict thermodynamic outcomes in practical systems – the extra effort will turn out to be a good investment.

The three scientists whose names adorn this section were all pioneers of the thermodynamics of state. In their day (before Einstein's revolutionary theories of relativity), their findings were thought of as stepping stones on the way to a grand theory of the physics of the universe. And for that time, their contributions were truly remarkable.

In a very simplistic way, the Gibbs and/or Helmholtz equations can be used in conjunction with one of the T, p, and v equations of state to determine entropy s and enthalpy h as functions of

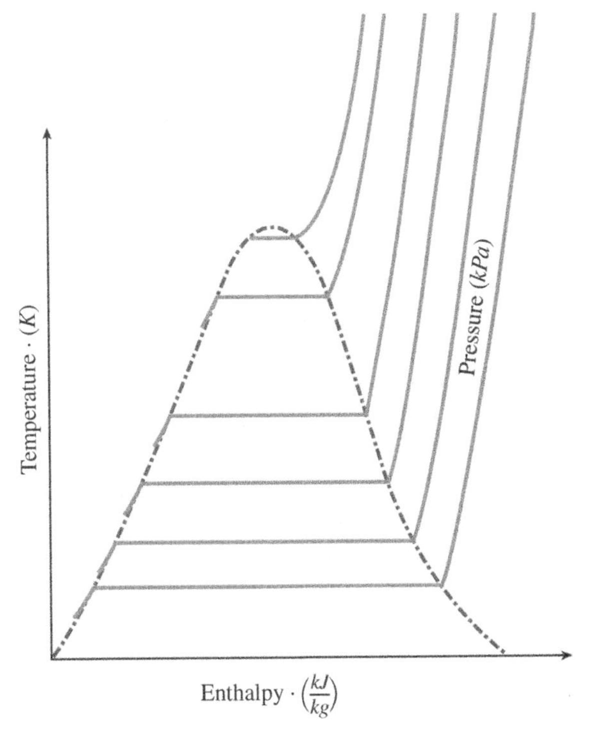

Figure 6.2 Example *T-s* diagram showing a vapor dome (dot-dash line) and isobars (solid line).
Source: Martin Thermal Engineering Inc.

temperature and pressure. To use the Gibbs or Helmholtz equations to track enthalpy and entropy, users must establish an arbitrary datum state (T_{datum}, p_{datum}), where enthalpy and entropy are both set to zero for saturated liquid. For water, this is often taken to be its triple point (0.6117 *kPa*, 273.16 *K*). Values for enthalpy and entropy at different states are determined by clever integration (e.g. first at constant temperature or entropy and then at constant pressure or specific volume).

By comparison, we also want to consider Maxwell's relations. In some elementary thermo classes, Maxwell's relations are introduced as a historical curiosity; a valid way of thinking about thermodynamic variables, but one whose practical value is barely worth the effort to learn. While it is true that Maxwell's relations are simply a set of equalities that flow mathematically from cross-derivatives of thermodynamic potentials with respect to certain thermodynamic state variables, they are exceptional and valuable, nonetheless.

A classic way to approach advanced thermodynamics is to introduce the eight basic thermodynamic parameters as a set of four thermodynamic ***potentials*** (u, h, g, and a) and a set of four ***state variables*** (T, s, p, and v).

The concept of a thermodynamic ***potential*** is analogous to the concepts of electric potential or gravitational potential. A substance's thermodynamic potential is a measure of its ability to perform work (mechanical, electrical, etc.) and/or to transmit heat. These potentials are not directly measurable, but rather are derived from the state variables that are reproducibly quantified by instrumentation.

Engineers are quite familiar with instruments that measure pressure, temperature, mass, and volume. While no engineer has invented an instrument to quantify macroscopic entropy directly and reproducibly, the ability to quantify different microstates in an ensemble of molecules is possible, in principle, although measuring variables such as temperature and specific volume is easier.

The more familiar thermodynamic ***state variables*** constitute intensive properties of fluids that are quantifiable in two very different realms. The macroscopic realm for measuring temperature, pressure, and specific volume is characterized by the comparison of an unknown sample to a standard specimen that is maintained in an unadulterated condition in a controlled environment, within a scientific archive such as NIST.

For example, mass is quantified by comparison of a sample to a bar of platinum on a weight balance subjected to standard earth gravity. Volume is derived from length, which is quantified by counting wavelengths of a standard laser. Temperature is quantified by the voltage generated by a standard thermocouple. Pressure is quantified by a standard manometer using a pure liquid (e.g. mercury) and the previously defined standards for mass and length in the Earth's gravitational field. As stated above, the direct measurement of entropy for a random substance is quantifiable in theory but not in practice.

The basic definitions of the four energy potentials and the identification of each potential's two natural state variables are given in Table 6.1.

Rewriting the four basic definitions from Table 6.1 as differentials, four *universal* (path-independent) equalities are obtained (Equation 6.6).

$$\boxed{du = Tds - pdv} \tag{6.6a}$$

$$du + pdv + vdp = \boxed{dh = Tds + vdp} \tag{6.6b}$$

$$du - Tds - sdT = \boxed{da = -sdT - pdv} \tag{6.6c}$$

$$du + pdv + vdp - Tds - sdT = \boxed{dg = vdp - sdT} \tag{6.6d}$$

These four differential equalities, attributed to Gibbs and Helmholtz, are often quite useful by themselves to correlate changes in the state variables (T, s, p, and v) with changes in the energy potentials (u, h, g, and a), as a fluid's state changes while being processed in a piece of equipment (e.g. heater, compressor). One such relation will be further developed for ideal gases in the next section.

Table 6.1 The four thermodynamic potentials and their natural variables[a].

Thermodynamic potential per unit mass (*J/kg*)	Thermodynamic definition	Natural thermal variable	Natural mechanical variable
Internal energy u	$\underbrace{\int Tds}_{heat\ in} - \underbrace{\int pdv}_{work\ in}$	s	v
Enthalpy h	$\underbrace{u + pv}_{u\ minus\ work}$	s	p
Helmholtz free energy a	$\underbrace{u - Ts}_{u\ minus\ heat}$	T	v
Gibbs free energy g	$\underbrace{u + pv - Ts}_{u\ minus\ work\ minus\ heat}$	T	p

[a] See Callen (1960) and others for additional details.
Source: Martin Thermal Engineering Inc.

Beyond the ***Gibbs*** differential equations (...here, we are using the term Gibbs loosely to mean any of the four versions of Equation 6.6), Maxwell derived four additional equations that are also *universally* true and are potentially just as useful as the Gibbs equations themselves. Maxwell's equations were simply derived from the second derivatives of each thermodynamic potential, taken with respect to both of its natural parameters (see Equation 6.7). It is noteworthy that the natural mechanical and thermal variables identified in Table 6.1 are the same variables expressed as differentials on the right sides of Equations (6.6).

$$\frac{\partial^2 u}{\partial s\,\partial v} = \frac{\partial}{\partial v}\underbrace{\left(\frac{\partial u}{\partial s}\right)_v}_{+T} = \frac{\partial}{\partial s}\underbrace{\left(\frac{\partial u}{\partial v}\right)_s}_{-p} = \boxed{\left.\frac{\partial T}{\partial v}\right]_s = -\left.\frac{\partial p}{\partial s}\right]_v} \tag{6.7a}$$

$$\frac{\partial^2 h}{\partial s\,\partial p} = \frac{\partial}{\partial p}\underbrace{\left(\frac{\partial h}{\partial s}\right)_p}_{+T} = \frac{\partial}{\partial s}\underbrace{\left(\frac{\partial h}{\partial p}\right)_s}_{+v} = \boxed{\left.\frac{\partial T}{\partial p}\right]_s = \left.\frac{\partial v}{\partial s}\right]_p} \tag{6.7b}$$

$$\frac{\partial^2 a}{\partial T\,\partial v} = \frac{\partial}{\partial v}\underbrace{\left(\frac{\partial a}{\partial T}\right)_v}_{-s} = \frac{\partial}{\partial T}\underbrace{\left(\frac{\partial a}{\partial v}\right)_T}_{-p} = \boxed{\left.\frac{\partial s}{\partial v}\right]_T = \left.\frac{\partial p}{\partial T}\right]_v} \tag{6.7c}$$

$$\frac{\partial^2 g}{\partial T\,\partial p} = \frac{\partial}{\partial p}\underbrace{\left(\frac{\partial g}{\partial T}\right)_p}_{-s} = \frac{\partial}{\partial T}\underbrace{\left(\frac{\partial g}{\partial p}\right)_T}_{+v} = \boxed{-\left.\frac{\partial s}{\partial p}\right]_T = \left.\frac{\partial v}{\partial T}\right]_p} \tag{6.7d}$$

The reason an engineer would want to maintain all the Gibbs and Maxwell relations in his or her toolkit is actually a very powerful one – to be able to predict the thermodynamic state of a fluid in order to solve a design problem.

The universal relations given in Equations (6.6) and (6.7) are helpful to solve different types of problems, especially when combined with other equations of state such as the ideal gas law and the incompressible liquid law, as discussed in the next section. Appendix B offers several equations of state converted to Visual Basic for Applications (VBA) instructions that create customized Excel functions. Readers are urged to investigate the potential of these functions to improve both efficiency and accuracy relative to looking up thermodynamic property values in tables or on enthalpy/entropy diagrams.

6.4 Equations of State

As itemized in Section 6.1, several helpful equations of state have been developed for nonideal gases, and saturated liquids and vapors, and these relations should be utilized when the fluids encountered cannot be characterized as "ideal." Nevertheless, many important engineering problems can be tackled successfully by assuming that a gas is ideal or a liquid is incompressible.

It is noteworthy to mention that the term "incompressible liquid" only applies to the effect of pressure on a liquid's density. Liquids are known to expand and contract modestly with changes in temperature, and such variations in liquid density are tabulated with great care in the property tables. From a practical perspective, it is important to apply the "constant density" assumption only to liquids whose pressure may be changing, but whose temperature is being held constant, so that the property of having constant density is a subset of the property of incompressibility.

For example, if an estimate of density is sought for a subcooled liquid, the correct approach is to follow a line of constant temperature toward the saturated liquid line and obtain the saturated density for that temperature. The pressure of the subcooled liquid will be higher than that of the saturated liquid (and consequently, the subcooled density will be marginally higher), but because temperature plays a greater role than pressure, their densities will be approximately equal. This process is akin to drawing a horizontal line leftward from the left side of the dome in a $T-s$ diagram (see Figure 6.2).

Another example would be the ideal gas law, which is considered an equation of state for low-density gases. While it does a great job of predicting unknown ***state variables***, it provides no direct information about the energy ***potentials*** of the substance. The good news is that when combined with the Gibbs equalities and/or Maxwell's relations, the ideal gas law can be used to compute energy potentials (as well as the difficult-to-measure ***state variable*** entropy) mathematically. Three examples of how to use simplified state equations are given here, and another is left to the students to derive as an end-of-chapter problem.

Example 6.1

Find expressions for the enthalpy change and entropy change of an *ideal gas* being heated at constant pressure from an initial temperature T_i to a final temperature T_f.

Answer. Begin with the Gibbs differential equations for enthalpy and entropy of an ideal gas (see Equations 1.11 and 1.13):

$$dh = c_p dT \qquad (1.11)\ \{IG\}$$

$$ds = c_p \frac{dT}{T} - R\frac{dp}{p} \qquad (1.13)\ \{IG\}$$

Since pressure is constant, the second term on the right in Equation (1.13) can be eliminated, and the changes found by direct integration:

$$\int_i^f dh = c_p \int_{T_i}^{T_f} dT \qquad \boxed{\Delta h_{\substack{ideal\\gas}} = c_p\left(T_f - T_i\right)}$$

$$\int_i^f ds = c_p \int_{T_i}^{T_f} \frac{dT}{T} \qquad \boxed{\Delta s_{\substack{ideal\\gas}} = c_p \ln\left(\frac{T_f}{T_i}\right)}$$

Example 6.2

Find an expression for the enthalpy change of an ideal gas being compressed adiabatically and reversibly.

Answer. Processes that are both adiabatic and reversible are also isentropic. Heat addition or removal from a control mass necessitates a change in entropy, regardless of whether it is performed reversibly or not ($dS \geq dQ/T$). Conversely, work input or output does not affect the entropy of a control mass unless it is performed irreversibly.

Begin with Equation (6.6a), which is a statement of the First law, if the heat and work are both transmitted reversibly:

$$\underbrace{du}_{\frac{dU}{m}} = \underbrace{Tds}_{\frac{\delta Q_{in,rev}}{m}} - \underbrace{pdv}_{\frac{\delta W_{in,rev}}{m}} \quad (6.6a)$$

For an adiabatic process, the heat term is zero, and for an ideal gas, the internal energy term can be replaced by a specific heat times temperature differential (Equations 6.8 and 6.9):

$$c_v dT = -pdv \quad (6.8)\ \{IG, Ad\}$$

Next, differentiate the ideal gas law:

$$d(pv) = d(RT) \qquad pdv + vdp = RdT \quad (6.9)\ \{IG\}$$

Isolate dT in both Equations (6.8) and (6.9) and set them equal to each other:

$$\frac{-pdv}{c_v} = \frac{pdv + vdp}{R}$$

Rearrange and simplify by using ($R = c_p - c_v$) and $k = c_p/c_v$:

$$-p(c_p - c_v)dv = pc_v dv + vc_v dp$$

$$-pc_p dv = vc_v dp$$

$$vc_v dp + pc_p dv = 0 \qquad \frac{dp}{p} + \frac{c_p}{c_v}\frac{dv}{v} = 0 \qquad d(\ln p) + kd(\ln v) = 0$$

Integrate to obtain the polytropic equation for a reversible, adiabatic work process (Equation 6.10):

$$\int d(\ln p) + k\int d(\ln v) = \int 0 \qquad e^p + ke^v = const$$

$$\boxed{pv^k = const} \quad (6.10)\ \{IG, Ad, Rev\}$$

Example 6.3

Find relations for the enthalpy and entropy changes of an incompressible liquid being pressurized by an adiabatic, reversible pump.

Answer. For entropy, begin with Equation (6.6a), the first Gibbs equation:

$$du = Tds - pdv \quad (6.6a)$$

By definition, the specific heat at constant volume is only a function of temperature, and for an incompressible liquid, the volume does not change with pressure. These two facts lead to Equation (6.11):

$$c_v dT = Tds \quad (6.11)\ \{IL\}$$

Rearranging and integrating gives Equation (6.12):

$$ds = \frac{c_v dT}{T} \qquad \boxed{s_2 - s_1 = c_v \ln\left(\frac{T_2}{T_1}\right)} \quad (6.12)\ \{IL\}$$

But since the pump is adiabatic and reversible, the entropy change is zero, so the temperature change must also be zero.

For enthalpy, begin with Equation (6.6b), the second Gibbs equation:

$$dh = Tds + vdp \quad (6.6b)$$

If the pumping process is adiabatic and reversible, the entropy does not change. This leads to the expression for enthalpy change while pumping an incompressible liquid (Equation 6.13):

$$\int_1^2 dh = v\int_1^2 dp \qquad \boxed{h_2 - h_1 = v(p_2 - p_1)} \quad (6.13)$$

Think Stop. For a numerical example of this, consider the enthalpy change associated with an adiabatic, reversible pump pressurizing water from saturated liquid at $p_1 = 100\ kPa$, $X = 0.0\%$ to subcooled liquid at $p_2 = 1000\ kPa$. The specific volume of saturated water at p_1 is $v_1(=v_2 = v) = 0.001\ 043\ kg/m^3$.

The isentropic work performed by the pump per unit mass of water is $v(p_2 - p_1) = 0.001\ 043$ $(1000 - 100) = 0.939\ kJ/kg$, which happens to be equivalent to the amount of heat input required to raise the temperature of liquid water by $\Delta T = 0.223\ °C$, or the amount of work input required to isentropically compress a unit mass of steam at $T \approx 374\ K$ from 99.45 to 100.00 kPa. Or put another way, compressing steam by a pressure ratio of 1.0055 requires the same work input as compressing liquid water by a pressure ratio of 10.

6.5 Boiling and Condensation

Phase change processes in practical equipment will be covered in Chapters 8 and 10, but it is useful here to recognize some of the nomenclature and basic thermodynamics of boiling and condensation.

Because the density of the liquid phase is two to three orders of magnitude higher than the density of the gas phase (for most substances when $p \approx 1\ atm$), body forces and elevation changes can drive pressure changes to a greater degree than friction. The separation of vapor from liquid is virtually always a gravity-driven process, in both boilers and condensers.

Quality. For a two-phase mixture of a pure substance, where liquid and vapor coexist under the vapor dome, the measure of the relative progress of the mixture along the isotherm from saturated liquid to saturated vapor is called the ***quality*** (see Equation 6.14):

$$X \veeeq \frac{kg_g}{\left(kg_f + kg_g\right)} \qquad \hat{X} \veeeq \frac{kmol_g}{\left(kmol_f + kmol_g\right)} \quad (6.14)$$

Here, the subscript f indicates saturated liquid and the subscript g indicates saturated vapor, and in this context, the implication is that the two phases are at equilibrium with each other. Our use of f and g instead of *liq* and *vap* is purely due to the inertia of nearly all published resources on the subject and we have decided to surrender without a fight. The use of f (for fluid) is especially weak because both gases and liquids are fluids, and the use of g (for gas) connotes an ideal gas state, which is physically inappropriate for two-phase (liquid + vapor) equilibria.

Example 6.4

Find the quality of an equilibrium mixture of steam and liquid water, where the pressure is $p = 101.325\ kPa$, the temperature is $T = 373.15\ K$, and the specific volume is $v = 1.00\ m^3/kg$.

Answer. A simple way to approach this problem is to examine the saturated steam table in Appendix A.4 to find the specific volume of saturated liquid ($v_f = 0.001\ 043\ m^3/kg$) and saturated vapor ($v_g = 1.673\ m^3/kg$) at the designated pressure. The actual volume of the mixture is the sum of the relative proportions of the vapor and liquid components, as shown in Equation (6.15):

$$\boxed{v_{act} = Xv_g + (1-X)v_f} \tag{6.15}$$

$$\left[\doteq \frac{m^3_{act}}{kg_{act}}\left(=\frac{kg_g}{kg_{act}}\cdot\frac{m_g^3}{kg_g}+\frac{kg_f}{kg_{act}}\cdot\frac{m_f^3}{kg_f}=\frac{m_g^3+m_f^3}{kg_{act}}\right)\right]$$

Expanding and rearranging the variables in the equation above, X is defined in Equations (6.16):

$$v_{act} = Xv_g + (1-X)v_f = v_f + X(v_g - v_f)$$

$$\boxed{X = \frac{(v_{act}-v_f)}{(v_g-v_f)}} = \frac{1.00-0.001\ 043}{1.673-0.001\ 043} = 59.75\% \tag{6.16}$$

Superheating and Subcooling. In heat exchangers where the working fluid undergoes phase change (i.e. boilers and condensers), the initial fluid state is often outside the vapor dome on one side and the final state is outside the vapor on the other side. The generic terms for fluid states that are outside the vapor dome are ***subcooled liquid*** (beyond the dome on the left, where temperatures are low and/or pressures are high) and ***superheated vapor*** (beyond the dome on the right, where temperatures are high and/or pressures are low).

Equations (6.17) provides relationships to quantify how far the fluid is from the saturation condition. These values are usually measured in terms of temperature, assuming a constant pressure heating or cooling process that effects the phase and temperature changes. From Equations (6.17), it may be asserted that dry steam has 50 $°F$ of superheat or liquid water has 10 $°F$ of subcooling:

$$\Delta T_{superheat} = [T - T_{sat}]_{p=const} \tag{6.17a}$$

$$\Delta T_{subcool} = [T_{sat} - T]_{p=const} \tag{6.17b}$$

Example 6.5

Find the degrees of superheat and subcooling for the inlet and exit conditions of a refrigerant condenser operating at constant pressure ($p = 2.0\ MPa$) with the initial state comprising dry $R32$ (CH_2F_2) gas at $T = 320\ K$ and the final state comprising liquid $R32$ at $T = 300\ K$.

Answer. Referring to Appendix A.10, the saturation temperature inside this condenser is found to be $T = 304.58\ K$, which means the entering fluid has $\Delta T_{\text{superheat}} = (320 - 304.58\ K) = 15.42\ °C$ of superheat and the leaving fluid has $\Delta T_{\text{subcool}} = (304.58 - 300\ K) = 4.58\ °C$ of subcooling.

Non-condensible Gases. The presence of non-condensible gases (NCG) in any boiling or condensing flow is a complication at best and a headache at worst. Even if a thermal system is thoroughly leak-tested, gases such as air can enter the piping network through imperfect gaskets or pinholes in tubing and they can interfere with the heat transfer from the tube wall to the two-phase mixture. The engineering analysis of the effect of superfluous gases on the boiling or condensation rate is beyond the scope of this textbook, but readers may wish to independently investigate the effect of NCG on the rate of boiling or condensation if such a level of detail is warranted.

Closed-loop boiler systems often incorporate a deaeration tank between the low-point condensate-return sump and the feedwater supply to the boiler. Condensers sometimes utilize a vacuum pump connected to a port near the top of the vessel to draw out NCG periodically. For facilities that perform steam sterilization of products or tools, a maximum NCG content of 3.5 % *vol* or lower is deemed vital to effective sterilizer performance (BSI 2015). See Chapter 11 for a steam sterilization schematic.

6.6 Psychrometry

Psychrometry is the analysis of the thermodynamic parameters of moist air. Subdivisions of the topic include humidification, dehumidification, and comfort air-conditioning.

Moist air can be modeled as a binary mixture of dry air and water vapor. The ideal gas equation of state is completely valid for the air fraction over a wide range of temperatures, and ideal gas behavior is also acceptable for the water vapor fraction in most air-conditioning applications (ASHRAE 2013). According to this source, ideal gas behavior introduces errors of less than 0.7% for saturated air at standard pressure and temperatures ranging from −60 to 120 °*F*.

Accounting for the variation of ambient pressure with altitude is likely to be important for some design projects at high elevations. The U.S. Standard Atmosphere (NOAA 1976; ASHRAE 2013) as a function of altitude can be approximated by Equations (6.18),

$$p = 101.325(1 - 2.255\,77 \times 10^{-5} \cdot Z)^{5.2559} \tag{6.18a}$$

$$T = 288.15 - 0.006\,500\,07 \cdot Z \tag{6.18b}$$

where $p \doteqdot kPa$, $T \doteqdot K$, and $Z \doteqdot m$ *(above sea level)*. It is obvious that these formulas represent averaged data at a given elevation because both pressure and temperature vary widely with the time of year and geographic location.

The pressure formula (Equation 6.18a) is more useful for psychrometric calculations because the magnitude of the atmosphere's pressure variability with altitude tends to dominate its variability with latitude and season. In other words, if you need to know the ambient pressure at Denver to solve a design problem, it would not be a bad assumption to use Equation (6.18a) with Denver's elevation $Z = 1609\ m$, regardless of whether a low-pressure or a high-pressure front happens to be passing through at any moment.

With temperatures, however, the seasonal and geographic swings are more pronounced than they are with pressures. Imagine using Equation (6.18b) to estimate the temperature difference between Phoenix ($Z = 331\ m$) and Anchorage ($Z = 31\ m$). Your estimates would not only be erroneous in degree, they would also give the false impression that Anchorage's average temperature is hotter than Phoenix's average. For purposes of this textbook, Equation (6.18b) may be most useful for estimating densities at different altitudes, which are needed to compute hot air balloon buoyancy forces (Chapter 11).

Nearly all psychrometric problems can be solved by carefully computing the conservation of mass and energy for each component. Some examples are given after the nomenclature section below.

Wet Bulb and Dry Bulb. A sling psychrometer is a device that contains two mercury-filled glass thermometers side by side. The wet bulb is covered by a small gauze sock saturated with distilled water. The dry bulb has no moisture and no sock – just a bare glass bulb. The device is manually swung around a swivel joint for a sufficient amount of time to allow the water liquid in contact with the wet bulb to come to thermodynamic equilibrium with the humid ambient air, while the dry bulb remains at equilibrium with the actual ambient temperature. In this case, thermodynamic equilibrium is the condition where (i) the rate of vaporization of liquid water away from the sock to the air is exactly equal to (ii) the rate of condensation of water vapor from the air back to the sock.

The time required to reach equilibrium should be checked by successive trials, but 30–60 seconds is often sufficient.

If the ambient air is completely saturated with water vapor (100% relative humidity), the wet sock and humid air are in equilibrium whether the psychrometer is rotating or not, and the wet and dry bulbs will read exactly the same temperature. If the ambient air is unsaturated (less than 100% relative humidity), the temperature of the wet sock declines due to the evaporative cooling effect, until it reaches thermodynamic equilibrium with the unsaturated air. The two temperatures are quickly read, and the values can be checked on a psychrometric chart (discussed below) to determine the relative humidity of the ambient air.

Nomenclature. The following nomenclature is used for psychrometry problems, with the understanding that the underlying conditions are the actual temperature and actual pressure of the humid air:

- Subscripts
  - *da* Subscript for dry air
  - *ha* Subscript for humid air (0 – 100% relative humidity)
  - *w* Subscript for water fraction in humid air
  - *sd* Subscript for difference (Δ) between saturated and dry
  - *f* Subscript for saturated liquid water
  - *g* Subscript for saturated vapor water
- Ratios
  - $W \veeeq \frac{kg_w}{kg_{da}}$ Humidity ratio (mass ratio of water vapor to dry air)
  - $W_s \veeeq \frac{kg_{w,sat}}{kg_{da}}$ Humidity ratio at saturation
  - $\chi_i \veeeq \frac{kmol_i}{kmol_{ha}}$ Mole fraction of i in humid air (i is either w or da)
  - $\gamma \veeeq \frac{kg_w}{kg_{ha}}$ Specific humidity (mass fraction of water in humid air)
- Thermo variables
  - $T \veeeq K$ Absolute temperature in Kelvins (or Rankines)
  - $t \veeeq {}^\circ C$ Gauge temperature in Celsius (or Fahrenheit)
  - $v \veeeq \frac{m^3}{kg}$ Specific volume of actual humid air
  - $v_i \veeeq \frac{m_i^3}{kg_i}$ Specific volume of i in humid air (w or da)
  - $v_s \veeeq \left.\frac{m^3}{kg}\right]_{sat}$ Specific volume of humid air at saturation condition
  - $v_{da} \veeeq \frac{m_{da}^3}{kg_{da}}$ Specific volume of dry air
  - $\rho_w \veeeq \frac{kg_w}{m_{ha}^3}$ Absolute humidity (water density in humid air)
  - $\rho \veeeq \frac{kg_{ha}}{m_{ha}^3}$ Actual density of humid air
  - $\rho_s \veeeq \left.\frac{kg}{m^3}\right]_{sat}$ Actual density of saturated air (reciprocal of v_s)

- $h \triangleq \frac{kJ}{kg}$ Enthalpy of actual humid air
- $h_i \triangleq \frac{kJ_i}{kg_i}$ Enthalpy of constituent i in humid air (w or da)
- $h_s \triangleq \left.\frac{kJ}{kg}\right]_{sat}$ Enthalpy of humid air at saturation
- $s \triangleq \frac{kJ}{kgK}$ Entropy of actual humid air
- $s_i \triangleq \frac{kJ_i}{kg_iK}$ Entropy of constituent i in humid air (w or da)
- $s_s \triangleq \left.\frac{kJ}{kgK}\right]_{sat}$ Entropy of humid air at saturation condition

Table 6.2 Relationships among psychrometric variables.

Formula	Description	Equation number
$W = \frac{\hat{M}_w \chi_w}{\hat{M}_{da} \chi_{da}}$	Conversion from mole fraction ratio to humidity mass ratio	6.19a
$\gamma = \frac{W}{(1+W)}$	Specific humidity, defined above under "Ratios"	6.19b
$\mu = \frac{W}{W_s}$	Saturation degree (actual humidity as a proportion of saturated humidity, by mass)	6.19c
$\phi = \frac{\chi_w}{\chi_{ws}} = \frac{p_w}{p_{ws}}$	Relative humidity (actual humidity as a proportion of saturated humidity, by mole)	6.19d
$\phi_{da} = \mu_{da} = 0.0$	Relative humidity and saturation degree are both zero for dry air	6.19e
$\phi_s = \mu_s = 1.0$	Relative humidity and saturation degree are both unit for saturated air	6.19f
$\Delta v_{sd} = v_s - v_{da}$	Difference between the specific volume of saturated air and that of dry air	6.19g
$\Delta h_{sd} = h_s - h_{da}$	Difference between the enthalpy of saturated vs dry air	6.19h
$\Delta s_{sd} = s_s - s_{da}$	Difference between the entropy of saturated vs dry air	6.19i
$v = v_{da} + \mu \Delta v_{sd}$	Specific volume calculation	6.19j
$h = h_{da} + \mu \Delta h_{sd}$	Enthalpy calculation	6.19k
$s = s_{da} + \mu \Delta s_{sd}$	Entropy calculation	6.19l
$W_s(p, t_{dew}) = W(t, p)$	Solving for t_{dew} gives the dew point of the humid air	6.19m
$p_{da}\hat{v}_{da} = \hat{R}T$	Ideal gas law, where p_{da} is the partial pressure of dry air	6.19n
$p_w\hat{v}_w = \hat{R}T$	Ideal gas law for water vapor, where p_w is the partial pressure of water	6.19o
$p_{da} + p_w = p$	Sum of partial pressures for dry air; water vapor is ambient pressure	6.19p
$\chi_i = \frac{p_i}{p}$	Mole fraction is the ratio of partial pressure to total pressure ($i \in \{w, da\}$)	6.19q
$h/da = h_{da} + Wh_g$	Humid air enthalpy per mass of dry air enthalpy plus the humidity ratio times enthalpy of saturated water vapor	6.19r

Source: Adapted from ASHRAE (2013).

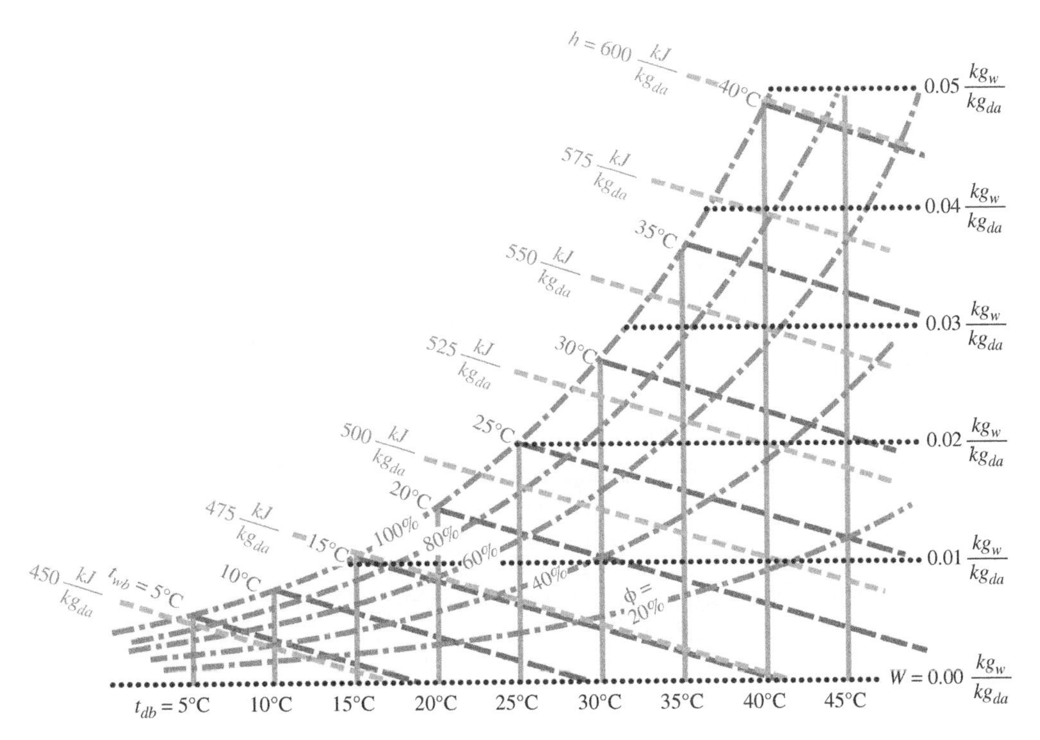

Figure 6.3 A psychrometric chart showing lines of constant dry bulb temperature (**t_{db}**)[vertical and solid], wet bulb temperature (**t_{wb}**)[diagonal and long dashed], humid air enthalpy(**h**) [diagonal and short dashed], humidity ratio (**W**)[horizontal and dotted], and relative humidity (**ϕ**)[curved and dash-dotted]. *Source:* Reynolds (1979). Used with permission by Stanford University Mechanical Engineering Department.

Useful Relations. Table 6.2 contains some useful relations (Equations 6.19) between the variables above and independently for water vapor also.

Figure 6.3 shows a skeleton view of a typical psychrometric chart. The lines shown are for dry bulb, wet bulb, humidity ratio, enthalpy, and relative humidity. Examination of the figure shows that the dry bulb temperature coincides with the wet bulb temperature when the relative humidity is $\phi = 100\%$, as it should.

Think Stop. Readers are challenged to consider what the following four processes would look like on the psychrometric chart: (i) cooling unsaturated, humid air ($\phi < 100\%$) at constant pressure, to reach its dew point; (ii) dehumidifying saturated, humid air ($\phi = 100\%$) at constant temperature, by passing it through a desiccant; (iii) simultaneously humidifying and cooling unsaturated, humid air ($\phi < 100\%$) at constant enthalpy; and (iv) cooling saturated, humid air ($\phi = 100\%$) at constant relative humidity.

Example 6.6

(a) For an air-conditioning system evaporator with the parameters given below, determine the refrigerant mass flow rate required to cool humid air that is initially composed of 0.25 kg/s of dry air plus an unknown mass flow rate of water vapor where the relative humidity is $\phi = 41.9\%$, and where both air and water are initially at $t_{db} = 27\ °C$ and finally at $t_{db} = 13\ °C$, after the cooling. Assume the refrigerant enthalpy rise across the evaporator is $\Delta h_{refr} = +180\ (kJ/kg)_{refr}$. (b) Repeat the calculation to find a new solution for refrigerant mass flow required to cool humid air that is initially composed of 0.25 kg/s of dry air plus an unknown mass flow rate of water vapor where the relative humidity is $\phi = 80.0\%$, and where both air and water are initially at $t_{db} = 27\ °C$ and finally at $t_{db} = 13\ °C$, after the cooling. Continue to use the prior assumption about the refrigerant enthalpy rise. Discuss the results.

Answer. As stated, this problem uses an assumption that the mass flow of dry air remains constant for cases (a) and (b). The problem will first be solved using this assumption and then it will be solved again (see Example 6.5) with a better assumption, and the discrepancy will be estimated.

Students should keep in mind that three different binary comparisons are being made as the problem is addressed from start to finish. The first binary comparison is the ***state*** of the humid air before and after the cooling process. These two states will be labeled *in* and *out*. The second binary comparison is the cooling power required for the two ***cases*** (*a*) and (*b*), which have different values of inlet humidity. The third binary comparison is the ***assumption*** made as to which variable should remain constant when the humidity is changed. The binary assumptions will be labeled *A*1 (dry air mass flow rate remains constant) and *A*2 (volume flow rate of humid air remains constant), respectively.

For the first assumption (*A*1), the problem asks for a comparison of the refrigerant flow needed to cool case (*a*) and case (*b*) when the mass flow of dry air is assumed to be constant for the two cases. The first step is to compute the saturation pressure of water vapor at the initial and final temperatures followed by computing the actual water vapor pressure at the unsaturated initial conditions. For this example, we will use the customized Excel functions for humid air given in Appendix B.5, which have been selectively validated to be accurate to within 1% (for enthalpy) and 0.1% (for saturation pressure) of several of the values given in tables 2 and 3 of chapter 1 of ASHRAE (2013). We note that the ASHRAE tables report enthalpy and entropy values per kg of dry air, whereas the customized Excel functions report enthalpy per kg of humid air.

$$p_{ws}(27\ °C) = 3.5669\ kPa; \qquad p_{ws}(13\ °C) = 1.4976\ kPa$$

$$p_w(27\ °C) = 41.9\% \cdot 3.5669\ kPa = 1.4945\ kPa\ (< 1.4976)$$

As seen above, in case (a), cooling the mixture from the inlet to the outlet (dry bulb) temperature does *not* result in a "supersaturated" outlet condition, and therefore no condensation of liquid water occurs in part (a). The process of cooling a humid air stream from one unsaturated condition to another is reflected by a leftward movement along a constant humidity ratio line ($W =$ const), without reaching the dew point.

From these vapor pressures, the mole and mass fractions of water vapor at the inlet and outlet conditions can be determined.

$$\chi_{w,in} = \frac{p_{w,in}}{p} = \frac{1.4945}{101.325} = 1.475\%$$

$$\chi_{da,in} = \left(1 - \frac{p_{w,in}}{p}\right) = 98.525\%$$

$$W_{in} = W_{out} = \frac{\hat{M}_w \chi_{w,in}}{\hat{M}_{da} \chi_{da,in}} = \frac{18.015 \cdot 1.475\%}{28.85 \cdot 98.525\%} = 0.009\,348\,1 \frac{kg_{w,in}}{kg_{da,in}}$$

$$W_s(27\ °C) = \frac{\hat{M}_w \chi_{w,s}}{\hat{M}_{da} \chi_{da,s}} = 0.022\,783 \frac{kg_{w,s}}{kg_{da}}$$

$$W_s(13\ °C) = \frac{\hat{M}_w \chi_{w,s}}{\hat{M}_{da} \chi_{da,s}} = 0.009\,367\,8 \frac{kg_{w,s}}{kg_{da}}$$

$$\mu_{in} = \frac{W_{in}}{W_s(27\ °C)} = \frac{0.009\,348\,1}{0.022\,783} = 41.03\%$$

$$\mu_{out} = \frac{W_{in}}{W_s(13\ °C)} = \frac{0.009\,348\,1}{0.009\,367\,8} = 99.79\%$$

Again, using the customized Excel functions for humid air given in Appendix B.5, we obtain the values shown below:

$$h_{da,in}(27\,°C) = 27.283\frac{kJ}{kg_{da}};\quad h_{da,out}(13\,°C) = 13.127\frac{kJ}{kg_{da}}$$

$$\Delta h_{sd,in}(27\,°C) = 57.808\frac{kJ}{kg_{da}};\quad \Delta h_{sd,out}(13\,°C) = 23.524\ \frac{kJ}{kg_{da}}$$

$$h_{in/da} = h_{da,in} + \mu_{in}\Delta h_{sd,in} = 27.283 + 41.03\% \cdot 57.808 = 51.001\frac{kJ}{kg_{da}}$$

$$h_{out/da} = h_{da,out} + \mu_{out}\Delta h_{sd,out} = 13.127 + 99.79\% \cdot 23.524 = 36.601\frac{kJ}{kg_{da}}$$

$$\Delta h_{out-in(/da)} = 36.601 - 51.001 = \boxed{-14.40\frac{kJ}{kg_{da}}}$$

Multiplying this humid air enthalpy change (per kg dry air) by the mass flow of dry air gives the total thermal power absorbed in the evaporator for case (a). We note that the cooling rate for case (a) and assumption $A1$ $(\dot{Q}_{aircooling(a,A1)})$ computed below is slightly more than one ton of refrigeration (1.0 TR) as defined in Chapter 10:

$$\dot{Q}_{aircooling(a,A1)} = \dot{m}_{da}\Delta h_{out-in(/da)} = 0.25 \cdot (14.40) = 3.60\ kW_{th}$$

The mass flow rate of refrigerant for case (a) and assumption $A1$ can be obtained by a heat balance, assuming the refrigerant evaporator is adiabatic with regard to any surroundings that are not indoor air (i.e. $\dot{Q}_{refrigant(a,A1)} = \dot{Q}_{aircool(a,A1)}$).

$$\dot{m}_{refr(a,A1)} = \frac{\dot{Q}_{refr(a,A1)}}{\Delta h_{refr}} = \frac{3.6}{180} = \boxed{0.0200\frac{kg_{refr}}{s}}$$

For case (b), two sequential processes occur from inlet to outlet – first is cooling from the initial humid air to saturated air (at constant pressure); second is extracting heat and condensing water vapor out of the saturated air along the saturation curve (at constant pressure). A summary of the key findings for case (b) of this example problem is given in Table 6.3.

The magnitude of the humid air enthalpy change for case (b) $|\Delta h_{out-in(/da)}| = (3.916 + 32.627) = 36.543\ kJ/kg_{da}$ is much greater than the corresponding enthalpy change for case (a) $|\Delta h_{out-in(a)}| = 14.40$ because the initial humidity was much higher in case (b). This initial difference necessitated cooling all the water vapor to the saturation condition and then condensing

Table 6.3 Properties at three states for case (b) of the example psychrometric problem

	State (in)	State (mid)	State (out)
$T(°C)$	27	23.253	13
ϕ	80%	100%	100%
$W(kg_w/kg_{da})$	0.018 095	0.018 095	0.009 368
μ	79.42%	100%	100%
$h_{/da}(kJ/kg_{da})$	73.194	69.278	36.651
$\Delta h_{out-in(/da)}(kJ/kg_{da})$	—	−3.916	−32.627

Source: Martin Thermal Engineering Inc.

a good portion of it to reach the same final dry bulb temperature. Thus, the required refrigerant mass flow rate for case (b) must be higher (by a factor of 2.5×) to accomplish all the required cooling and condensing:

$$\dot{Q}_{evap(b)} = \dot{m}_{da}\left|\Delta h_{out-in(a)}\right| = 0.25 \cdot (3.916 + 32.627) = 9.14kW_{th}$$

$$\dot{m}_{refr(b)} = 0.0508\frac{kg_{refr}}{s}$$

Example 6.7 The original problem (using assumption $A1$) was posed such that the mass flow rate of dry air was assumed the same for both the (a) $\phi = 41.9\%$ and (b) $\phi = 80.0\%$ cases. The problem with the $A1$ assumption is that the ***mass flow*** rate of dry air passing through an air handler does change when its water content is varying – but we surmised that the error was small. We can estimate the level of error by making a better assumption ($A2$) that the ***volume flow*** rate of humid air is the same in both cases. Then, using assumption $A2$, and computing the actual dry air mass flow rates for the two cases (a) and (b), a more correct mass flow rate of dry air can be determined and the final answers for refrigerant mass flow can be adjusted. Because the psychrometry VBA script in Appendix B.5 does not contain customized functions for specific volume, we must utilize the tables from chapter 1 of ASHRAE (2013) or any other trustworthy source:

$$\upsilon_{in} = \upsilon_{da,in} + \mu_{in}\Delta\upsilon_{sd,in}$$

$$\upsilon_{da}(27.0\ °C) = 0.8500\frac{m^3}{kg}; \quad \Delta\upsilon_{sd}(27\ °C) = 0.0311\frac{m^3}{kg}$$

$$\upsilon_{in(a)} = 0.8500 + 41.03\% \cdot 0.0311 = 0.862\,76\frac{m^3}{kg}$$

$$\upsilon_{in(b)} = 0.8500 + 79.42\% \cdot 0.0311 = 0.874\,70\frac{m^3}{kg}$$

$$\underbrace{\dot{V}_{in,(a)A2}}_{\doteq \frac{m^3}{s}} = \dot{V}_{in,(b)A2} = \underbrace{\dot{m}}_{\doteq \frac{kg}{s}} \cdot \underbrace{\upsilon}_{\doteq \frac{m^3}{kg}} = const$$

$$\therefore \frac{\dot{m}_{(b)}}{\dot{m}_{(a)}} = \frac{\upsilon_{(a)}}{\upsilon_{(b)}} = \frac{0.862\,76}{0.874\,70} = 0.9864$$

However, the specific enthalpy values in the ASHRAE tables have units of (kJ/kg_{da}), so the ratio of $\dot{m}_{da,(a)}/\dot{m}_{,da(b)}$ must be determined from the other data available:

$$\dot{m}_{in(a)A2} = \dot{m}_{da,in(a)A2}\left(1 + W_{in(a)A2}\right)$$

$$\dot{m}_{in(b)A2} = \dot{m}_{da,in(b)A2}\left(1 + W_{in(b)A2}\right)$$

$$\frac{\dot{m}_{in,da(b)A2}}{\dot{m}_{in,da(a)A2}} = \frac{\dot{m}_{in(b)A2}}{\dot{m}_{in(a)A2}}\frac{\left(1 + W_{in(b)A2}\right)}{\left(1 + W_{in(a)A2}\right)} = 0.9864\frac{1.018\,095}{1.009\,348} = 0.9779$$

Thus, the correction is made by multiplying $\dot{m}_{da,(b)}$ by 0.9779, which results in $\dot{m}_{refr,(A2)} = 0.0496\ kg_{refr}/s$, which is an error of only 2%. To refine the answer further by accounting for tiny changes in volume flow rate for cases (a) and (b) due to fan curve, effects related to the reduced humid air density would not improve the answer enough to warrant the additional work.

Example 6.8
Using the custom Excel functions created by the VBA script in Appendix B.5, determine the relative humidity (ϕ) associated with a dry bulb temperature of $t_{db} = 24.0\ °C$ and a wet bulb temperature of $t_{wb} = 17.1\ °C$.

Answer. Using Excel's Solver to minimize the error between the computed wet bulb temperature and the desired wet bulb temperature by iteratively changing the relatively humidity, we find $\phi \approx 0.502$.

For validation, it is apparent from examining a psychrometric chart, the intersection of the given wet bulb and dry bulb temperature lines is very close to a relative humidity line of $\phi = 50\%$. For further validation, students may wish to use the ASHRAE tables. Using this method, the following equations must be solved simultaneously:

$$\phi = \frac{\chi_w}{\chi_{ws}}; \qquad \mu = \frac{W}{W_s}$$

$$W_s = \frac{\hat{M}_w \chi_{ws}}{\hat{M}_{da} \chi_{da}} = \frac{\hat{M}_w \chi_{ws}}{\hat{M}_{da}(1-\chi_{ws})}; \quad W = \frac{\hat{M}_w \chi_w}{\hat{M}_{da} \chi_{da}} = \frac{\hat{M}_w \chi_w}{\hat{M}_{da}(1-\chi_w)}; \quad \frac{W}{W_s} = \frac{\hat{M}_w \chi_w \hat{M}_{da}(1-\chi_{ws})}{\hat{M}_{da}(1-\chi_w)\hat{M}_w \chi_{ws}}$$

$$h(t_{db}, \phi) = h(t_{db}) + \mu \cdot \Delta h_{sd}(t_{db})$$

Verification is left for the student.

6.7 Liquid–Vapor Equilibrium for $NH_3 + H_2O$ Mixtures

As discussed in Section 1.10, when considering a system (e.g. refrigerator, heat pump) where thermodynamic properties are desired in the two-phase domain (vapor + liquid), it is appropriate to consider gas and liquid as separate constituents. Often, the ***quality*** $X = \dot{m}_{vap}/(\dot{m}_{vap} + \dot{m}_{liq})$ is the preferred variable to track the relative amounts of vapor and liquid in a one-constituent, two-phase mixture.

When two constituents *and* two phases are present, the Gibbs phase rule calculation results in four degrees of freedom. Clearly, four variables must be set independently to obtain values for mixture internal energy, enthalpy, entropy, and specific volume.

One way to think of the degrees of freedom for a two-constituent mixture is that the state can be fixed by assigning values to the following four independent variables: mixture temperature, mixture total pressure, liquid mole fraction of constituent 1, and vapor mole fraction of constituent 1. The liquid and vapor mole fractions of constituent 2 will be dependent on those of constituent 1 because the combination must add up to 100% in each phase. Alternatively, the same result will be obtained if partial pressure in the gas phase of both constituents 1 and 2 is given in lieu of the mixture total pressure and the vapor mole fraction of constituent 1.

In a flowing (control volume) system, at a given point along the fluid path, the local equilibrium is assumed. ***Mechanical*** equilibrium means all phases and constituents are at the same pressure. Small differences in pressure certainly will occur due to body forces like gravity and surface forces like friction. However, it is frequently OK to assume these variations are small compared to the effects of shaft work on fluid pressure. This may not always be a good assumption where large hydrostatic head or moderately high local Mach numbers are present.

Thermal equilibrium means all phases and constituents are at the same temperature (and thermochemical equilibrium means they are not undergoing chemical reactions). Like mechanical equilibrium, it is frequently acceptable to assume local thermochemical equilibrium in the lengths of pipe that connect the outlets and inlets of equipment such as heat exchangers, reactors, and fluid movers. These connecting pipes and ducts are labeled by stream tags in the process flow diagram (PFD) schematic. Determining the steady-state thermodynamic properties of each stream (i.e. at each tagged state) is the most important task in a thermal system design project.

By convention, when addressing mixtures of two constituents, chemical engineers track the molar proportion of the lighter, more volatile constituent in each phase as their primary intensive variable for phase composition. For example, in a mixture of ammonia and water, the intensive variables for liquid, vapor, and combined composition, respectively, are $\hat{x}$, $\hat{y}$, and $\hat{z}$ for ammonia, and $(1-\hat{x})$, $(1-\hat{y})$, and $(1-\hat{z})$ for water, as defined in Equations (6.20). The subscript ***comb*** in the third equation stands for ***combined***. It will be shown in Section 10.9 why this is designated combined, instead of total:

$$\hat{x} = \frac{\dot{n}_{NH_3,liq}}{(\dot{n}_{NH_3} + \dot{n}_{H_2O})_{liq}} \tag{6.20a}$$

$$\hat{y} = \frac{\dot{n}_{NH_3,vap}}{(\dot{n}_{NH_3} + \dot{n}_{H_2O})_{vap}} \tag{6.20b}$$

$$\hat{z} = \frac{\dot{n}_{NH_3,comb}}{(\dot{n}_{NH_3} + H_2O)_{comb}} \tag{6.20c}$$

Figure 6.4 shows two plots: (a) temperature and (b) pressure as functions of composition for an isobaric system (T vs $\hat{z}$ at $p = 300\ kPa$) and an isothermal system (p vs $\hat{z}$ at $T = 278\ K$), respectively.

The purpose of figures like Figure 6.4 is twofold: (i) with the knowledge of T, p, and $\hat{z}$, determine whether the system is fully ***liquid***, fully ***vapor***, or in the *two-phase* regime; and (ii) if in the ***two-phase*** regime, determine the molar constituent proportions $\hat{x}$ and $\hat{y}$. In each of the figures, the ***solid*** line (comprising the boundary between the fully ***liquid*** and ***two-phase*** regimes) represents a "bubble point" curve, and the ***dashed*** line (comprising the boundary between the fully ***vapor*** and ***two-phase*** regions) represents a "dew point" curve.

Figure 6.4 shows the bubble and dew point curves being approached vertically from below and above, respectively, but these boundaries can also be approached vertically from within the

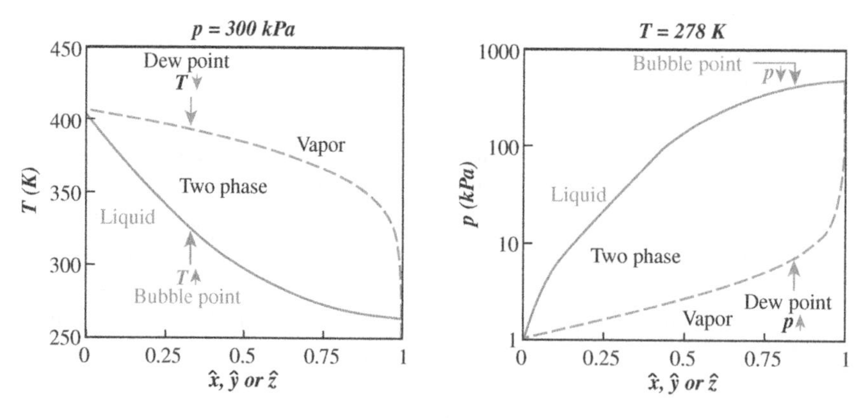

Figure 6.4 Vapor–liquid equilibrium (VLE) diagrams of temperature (left) and pressure (right) versus composition for $NH_3 + H_2O$ mixtures at p = 300 *kPa* and T = 278 *K*, respectively. Bubble point is approached (from the pure liquid regime) along a vertical line (constant $\hat{z}$) by increasing temperature or reducing pressure. Dew point is approached (from the vapor regime) along a vertical line by decreasing temperature or increasing pressure. *Source:* Martin Thermal Engineering Inc.

two-phase regime, where they might be given names such as "fully wet" and "fully dry," respectively. Students should be mindful that each plot of this format represents a single pressure (over a range of temperatures, left image) or a single temperature (over a range of pressures, right image). The ***up*** and ***down*** arrows depict processes that occur at constant $\hat{z}$ for the system, while pressure or temperature is increasing or declining. Students will find it reassuring that a closed system initially in the ***vapor*** region will eventually form dew drops when temperature decreases or pressure increases and that a closed system initially in the ***liquid*** region will eventually form bubbles when the temperature increases or when the pressure decreases.

Determination of the $\hat{x}$ (NH_3 mole fraction in the liquid phase) and $\hat{y}$ (NH_3 mole fraction in the vapor phase) for a specific point within the ***two-phase*** region can be accomplished graphically or analytically. The graphical method for determining these properties is illustrated in Figure 6.5.

The figure (which represents a single pressure of $p = 300\ kPa$) shows a horizontal ***long/short dash*** line that corresponds to a mixture temperature of $T = 390\ K$. For this condition, if the mixture being examined has a combined ammonia mole fraction greater than $\hat{z} > 0.40$, the constituents are entirely vapor. Similarly, if the mixture has $\hat{z} < 0.05$, the constituents are entirely liquid. For the example states shown as **A** and **B**, each with a × on the figure, we see $\hat{z} = 0.30$. For point **A**, the mixture is entirely liquid, and for point **B**, the mixture is in the two-phase region.

A simple calculation for determining the mole fractions of liquid and vapor in the overall mixture is easy to visualize from Figure 6.5. If we imagine an overall mixture (at the given T, p) where

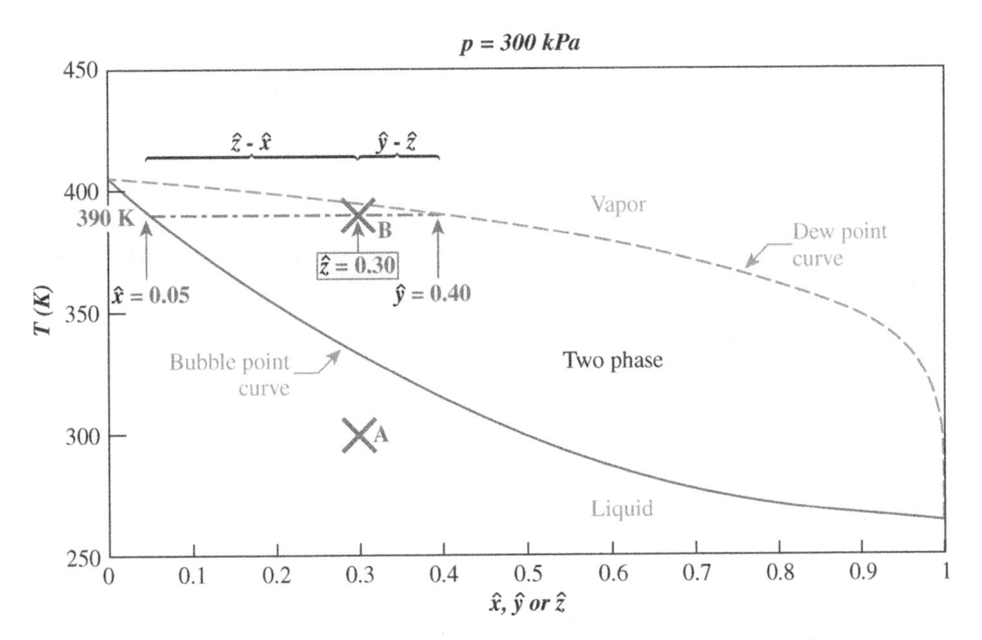

Figure 6.5 Vapor–liquid equilibrium diagram for $NH_3 + H_2O$ mixture at $p = 300\ kPa$. The example shown is for a mixture with $\hat{z} = 0.30$ combined mole fraction of NH_3 and temperature $T = 390\ K$. This diagram shows that for the given values of T and p (horizontal dash-dot line), the mole fraction of NH_3 in the liquid phase is $\hat{x} = 0.05$ and the mole fraction of NH_3 in the vapor phase is $\hat{y} = 0.40$. If a value of liquid or vapor fraction for the combined mixture is desired for any value of $\hat{z}$, it can be determined by the relative lengths of the line segments $[\hat{y} - \hat{z}]$, $[\hat{z} - \hat{x}]$, and $[\hat{y} - \hat{x}]$, as described below. *Source:* Martin Thermal Engineering Inc.

$$\hat{z} = \underbrace{0.05}_{\substack{blue\\solid\\curve}} \; (= \hat{x}\ [\text{at } 390\ K, 300\ kPa])$$

it is plain to see that the entire mixture is at the ***bubble*** point and there is no vapor, so for the combined stream, the vapor mole fraction (i.e. quality) is $\hat{X}_{comb} = 0$ and the combined liquid fraction is $\left(1 - \hat{X}_{comb}\right) = 1$.

Similarly, if we imagine $\hat{z} = 0.4(= \hat{y}\ [\text{at } 390\ K, 300\ kPa])$, then the entire mixture is at the ***dew*** point and there is no liquid, so for the combined stream, the vapor fraction (i.e. quality) is $\hat{X}_{comb} = 1$ and the combined liquid fraction is $\left(1 - \hat{X}_{comb}\right) = 0$.

Finally, if we imagine $\hat{z} = 0.3$, which is the location of the × B, the overall vapor and liquid mole fractions are easily computed from the relative lengths of the two ***dash-dot*** line segments (to the left and right of $\hat{z}$), as per Equations (6.21):

$$\hat{X}_{comb} = \frac{\hat{z} - \hat{x}}{\hat{y} - \hat{x}} \qquad 1 - \hat{X}_{comb} = \frac{\hat{y} - \hat{z}}{\hat{y} - \hat{x}}$$

$$\hat{z} = \hat{x} + (\hat{y} - \hat{x})\hat{X}_{comb} \tag{6.21}$$

These simple expressions for the overall molar vapor quality of the two-constituent, two-phase mixture are algebraic consequences of the mass continuity equation, and they will turn out to be very useful formulas.

Example 6.9

To demonstrate this, consider a vaporizer that receives a mixture of $NH_3 + H_2O$, with ammonia molar flow rate $\dot{n}_{NH3} = 0.3\ kmol/s$ and water molar flow rate $\dot{n}_{H2O} = 0.7\ kmol/s$. Let the inlet conditions of the mixture be $T_{in} = 300\ K$ and $p_{in} = 300\ kPa$, which correspond to the × designated as **A** in Figure 6.4 and is in the fully liquid region. If we assume the vaporizer to be isobaric ($p_{in} = p_{out}$), and that the heat input causes the mixture's temperature to increase such that the evaporator's exit temperature is $T_{out} = 390\ K$, we can compute the exit condition that corresponds to the × designated as **B** in the figure. The vaporization process is modeled as an upward vertical line connecting point **A** to point **B**.

Answer. For simplicity, let us impose the following shorthand to track our six molar flow rates exiting the heater, all of which we assume to have units of *kmol/s*:

$$\hat{a} = \dot{n}_{NH_3,liq} \qquad \hat{b} = \dot{n}_{H_2O,liq} \qquad \hat{c} = \dot{n}_{NH_3,vap}$$

$$\hat{d} = \dot{n}_{H_2O,vap} \qquad \hat{e} = \dot{n}_{NH_3,tot} \qquad \hat{f} = \dot{n}_{H_2O,tot}$$

We can rewrite the proportional variables as functions of these new flow rate variables, shown in Equations (6.22):

$$\hat{x} = \frac{\hat{a}}{\hat{a} + \hat{b}} \quad \hat{y} = \frac{\hat{c}}{\hat{c} + \hat{d}} \quad \hat{z} = \frac{\hat{e}}{\hat{e} + \hat{f}} \quad \hat{X} = \frac{\hat{c} + \hat{d}}{\hat{e} + \hat{f}} \tag{6.22a}$$

We can also rearrange the first two equations for easier substitutions:

$$\hat{b} = \frac{(1 - \hat{x})}{\hat{x}}\hat{a} \qquad \hat{d} = \frac{(1 - \hat{y})}{\hat{y}}\hat{c} \tag{6.22b}$$

And we can also confirm the molar species continuity equations:

$$\hat{a} + \hat{c} = \hat{e} \qquad \hat{b} + \hat{d} = \hat{f} \tag{6.22c}$$

Because of continuity (and the assumption that chemical reactions do not occur between the ammonia and water molecules), the overall molar flow rates of the compounds will not change as the mixture passes through the vaporizer (i.e. $\hat{z}$, $\hat{e}$, and $\hat{f}$ must each stay constant), but the relative amounts of liquid and vapor will vary with mixture temperature (i.e. $\hat{x}, \hat{y}, \hat{a}, \hat{b}, \hat{c}$, and $\hat{d}$ do not remain constant). The inlet stream (point **A**) has only subcooled liquid. When the temperature reaches the bubble point ($T \approx 334\ K$), the following values are obtained: $\hat{X} = 0$, $\hat{x} = \hat{z}$, and $\hat{y} > \hat{z}$.

As the temperature exceeds the bubble point, boiling begins (i.e. $\hat{X}$ increases) and both $\hat{x}$ and $\hat{y}$ decrease (because the slopes of the bubble line and dew line are everywhere negative), while all the equations above (Equations 6.21 and 6.22) continue to be satisfied at every progress point (along the vertical line from **A** to **B**) within the vaporizer.

Now, consider the mixture that exits from the vaporizer at point **B**. Based on the figure, we know the values for $\hat{x}$ and $\hat{y}$, and we can compute the exit molar quality:

$$\hat{X} = \frac{\hat{z} - \hat{x}}{\hat{y} - \hat{x}} = \frac{(0.30 - 0.05)}{(0.40 - 0.05)} = 0.714\,29$$

We will provide here, without derivation, another generally applicable formula and ask the students to derive it in one of the end-of-chapter questions:

$$\boxed{\hat{c} = \frac{\hat{X}\left(\hat{e} + \hat{f}\right)}{1 + \dfrac{(1 - \hat{y})}{\hat{y}}}} \tag{6.23}$$

The utility of Equation (6.23) is easy to see. Once one has determined $\hat{y}$ (possibly from the customized VBA function y_px in B.6), the determination of $\hat{c}$ and $\hat{d}$ is straightforward (when $\hat{X}$ and $\hat{e} + \hat{f}$ are also known) from Equation (6.22b). Ultimately, $\hat{a}$ and $\hat{b}$ are easily determined after $\hat{c}$ and $\hat{d}$ are found from species continuity. Computation of these four molar flow rates is left to the student in the homework problem, as is determination of the relationships between equivalent mass and mole unit formulas.

A table of temperatures, pressures, and enthalpies for bubble and dew points of the two-phase $NH_3 + H_2O$ system is given in Appendix A.12 but in mass units rather than mole units. Specific volumes, enthalpies, and entropies may be computed manually (i.e. by interpolation) from this appendix.

6.8 Efficiency vs Effectiveness

Two measures of how well a system performs are used in thermodynamics and heat transfer – and both begin with the letter "e" – ***effectiveness*** and ***efficiency***. Unfortunately, the two terms are not used consistently across different classes of problems – especially when used for energy conversion systems versus heat exchange devices. Here are several important measures of performance.

Two-temperature Heat Engine. Carnot ***efficiency*** is the rate of useful work output from an ideal, two-temperature, heat engine divided by the rate of heat added from the high-temperature reservoir. The essence of Carnot efficiency is shown in Equation (6.24) and the First law constrains it to remain below 100%:

$$\eta_{carnot} = \frac{\overbrace{\dot{W}_{net,out}}^{\boxed{\textit{shaft power you got}}}}{\underbrace{\dot{Q}_{in}}_{\boxed{\textit{heating you paid for it}}}} < 100\% \tag{6.24}$$

Heat Pump and Refrigerator. As will be shown in Chapter 10, the performance measure for a heat pump, ***HCOP***, has a floor of 100% and no ceiling. The performance measure for a refrigerator, ***COP***, has a floor of 0% and no ceiling. Equations (6.25) illustrates this concept:

$$HCOP_{heat\ pump} = \frac{\overbrace{\dot{Q}_{condenser}}^{\boxed{\textit{heating you got}}}}{\underbrace{\dot{W}_{compressor}}_{\boxed{\textit{electricity you paid for it}}}} \qquad 100\% \leq HCOP < \infty \tag{6.25a}$$

$$COP_{refrigerator} = \frac{\overbrace{\dot{Q}_{evaporator}}^{\boxed{\textit{cooling you got}}}}{\underbrace{\dot{W}_{compressor}}_{\boxed{\textit{electricity you paid for it}}}} \qquad 0\% \leq COP < \infty \tag{6.25b}$$

Heat Exchanger. In contrast to the cycle performance measures above, consider a heat exchanger's measure of performance, its ***effectiveness*** as per Equation (6.26). Here, the Second law constrains the denominator to maximum temperature difference between the hot and cold fluid inlets multiplied by the fluid with the minimum heat capacity product C_{min}:

$$\varepsilon = \frac{\dot{Q}_{act}}{\dot{Q}_{max\,poss}} = \frac{C_{hot}(T_{in} - T_{out})_{hot}}{C_{min}(T_{in,hot} - T_{in,cold})} < 100\% \tag{6.26}$$

Fins. Fin ***efficiency*** and fin ***effectiveness*** (Equations 6.27a and 6.27b) do not fit the pattern of either cycle efficiency or heat exchanger effectiveness:

$$\eta_{fin} = \frac{\dot{Q}_{actual\ fin}}{\dot{Q}_{infinite\ fin}} < 100\% \tag{6.27a}$$

$$\varepsilon_{fin} = \frac{\dot{Q}_{actual\ fin}}{\dot{Q}_{no\ fin}} \qquad 0 \leq \varepsilon_{fin} < \infty \tag{6.27b}$$

In one way, Equation (6.27a) for fin ***efficiency*** appears to be analogous to heat exchanger ***effectiveness*** (actual divided by maximum possible) because it is forbidden from exceeding 100%. On the other hand, Equation (6.27b) for fin ***effectiveness*** is a comparison that resembles ***HCOP*** in that it can theoretically range from zero to infinity, if the denominator is small enough.

In addition to problems addressing psychrometry, equations of state, and thermodynamic property diagrams, the homework questions also ask about thermodynamic performance measures such as efficiency and effectiveness.

6.9 Space vs Time

Perhaps, the most basic question a thermal system designer must answer is ***batch*** or ***continuous***. Is my thermal process better suited to a system where the work piece or working fluid moves physically through different pieces of equipment to achieve the desired result (***continuous***) or is it better suited to stay in one physical location, while its state is altered over time by a managed environment (***batch***)? Sometimes, the best answer to the question is not obvious or intuitive, and either solution may be acceptable.

Think Stop. One way to mentally segregate these categories of space versus time is to consider breakfast rooms at chain hotels and the bread toasting options they provide for their guests. Some offer a pop-up toaster (***batch***) and others offer a moving belt toasting oven (***continuous***). Even though the toasting process requires the slice of bread to remain in each device for approximately the same time duration, one could argue that the continuous oven is more efficient because guests must only wait a few tens of seconds for the conveyor to move far enough that their slices can be added to the lineup at the toaster entry. However, one could also argue that the pop-up toaster is more flexible because each guest is able to customize their degree of toasting by rotating the control knob from lighter to darker, where the conveyor offers no such control. If a thermal design engineer is equipped with tools to solve either problem, their agility and competence will be remarkable.

Batch Processes. By intent, batch processes are transient in time and relatively uniform in space. Batch processes usually rely on mixing blowers (for gases) or agitators (for liquids) to help ensure uniformity is achieved throughout the volume at any given moment, but they are flexible enough to permit a program of different environmental conditions at different times in the batch sequence. A batch process can carry out a physical change (e.g. melting butter, boiling water, mixing yeast and flour, separating milk from cream) or a chemical change (e.g. the Maillard reaction). In cooking, it is the Maillard reaction (Myhrvold and Migoya 2013) that converts the amino acids and sugars in starches and meats into a host of different flavor compounds, depending on the raw materials and the baking environment. Going beyond Maillard is often discouraged because caramelization is the next phase, followed by pyrolysis and charring at higher temperatures.

Continuous Processes. By contrast, continuous processes are nonuniform spatially and steady in time. Continuous processes rely on a conveying means (e.g. moving belts for solids and pipes for fluids) and controls are apportioned spatially with varying setpoints and alarm levels. Like batch processes, continuous processes can carry out physical changes (e.g. condensing steam, evaporating refrigerant, filtering air, blending gasoline, and scrubbing acid gases out of exhaust) or chemical changes (e.g. baking pies, combusting fuel, and synthesizing cement).

Modeling Batch Dehydration. Consider a batch process for drying prunes. Assume the fruit is brought into a drying chamber attached to stems and branches so that each piece of fruit is fully exposed to the drying environment in the chamber. Further, assume the fruit does not fall off during the drying process. The rate of dehydration increases with increasing temperature, but it decreases as the fruit's moisture level declines. Empirical data have demonstrated that only a narrow range of oven temperatures is acceptable because excessively high temperatures can cause explosive boiling at the early drying stages or undesirable chemical changes at the later stages.

Each piece of fruit can be modeled as a sphere, and the batch dehydration chamber can be considered perfectly stirred. A perfectly stirred reactor (PSR) is a theoretical construct that asserts mixing inside the reactor volume is so vigorous that the time required for chemical and physical changes to occur is slow compared to the time necessary for thorough mixing to occur. The PSR model assumes reactant fluids enter the volume through one nozzle and product fluids exit the volume through a different nozzle. The residence time within the volume (see Section 7.3 for more information about residence time) determines the yield of the reaction, with long residence times leading to equilibrium mixtures and short residence times leading to incomplete processes.

While not strictly accurate, the simplest assumption for a dehydration process is to assume a constant rate of moisture loss per unit surface area during the entire process. The actual rate may also be dependent on the temperature or humidity level in the air, but if those parameters are controlled, the rate of drying may be considered constant over a good portion of the overall drying duration. Also, because prunes begin as spheres of a certain size and they ultimately become wrinkled and smaller, their surface area may change over time. The process modeler should consider including a parameter for the changing surface area, but to a first approximation, the surface area can be considered constant throughout the drying process.

If the drying chamber is filled with a certain number of fruit pieces at the beginning of the batch process, the designer must ensure that the inflow of warm, dry air is sufficient to accomplish the steady evaporation of liquid water to water vapor and the removal of the vaporized water out of the dehydration chamber. Furthermore, because all the heat necessary for vaporization to occur must be delivered to each individual fruit sphere, the convective heat transfer coefficient should be high enough to ensure that the necessary heat for vaporization occurs reliably. This is accomplished by ensuring that the mixing velocities are high enough and uniform enough throughout the chamber to attain the required value of h without damaging the fruit by high pressures developed from asymmetric drag forces on the leading and trailing surfaces.

With the assumption of constant moisture loss rate, the overall duration for each batch process can be easily determined from the fruit's initial moisture percent and desired final moisture percent. The use of these simplifications may be adequate to provide approximate values for production rate and energy consumption rate, but designers are urged to rely on test results and model refinements for greater accuracy.

Modeling Continuous Dehydration. For comparison purposes, consider a continuous prune dehydration process. Consider a tunnel dehydrator, with a porous belt holding a single layer of fruit (no stems or branches) spaced far enough apart so that they do not touch. The prunes enter having their moisture content at the freshly picked level, and it is desired for them to exit the dehydrator with a designated dried-fruit moisture percent. Again, it is initially desirable to rely on two simplifying assumptions: (i) the rate of moisture loss per unit surface area is constant throughout the dehydration process and (ii) the surface area for each piece of fruit does not change.

The airflow that passes over the fruit must be able to provide enough convective heat to the surface to facilitate the evaporation of water at the desired rate, and it must also be capable of carrying the water vapor out of the dehydrator tunnel without any moisture condensing in ducts or back onto the fruit itself. Because the fruit is evenly distributed across the belt, the warm airflow past the fruit should be uniform at all locations within the dehydrator. The airflow direction is ideally upward or downward (hence the reason for specifying a porous conveyor belt, which allows vertical air motion over all fruit pieces), rather than horizontal, which could induce undesired variations in air temperature and humidity along the direction of airflow. The belt's ***length L*** is usually its longest dimension, followed by the belt's ***width W*** and followed further by the ***depth D*** of the fruit on the belt.

The tunnel length, width, and speed are determined from the total drying time required and production rate desired. The air velocity and temperature are determined from the limiting parameters that relate to flow uniformity, enthalpy change, evaporation rate, and prevention of fruit damage.

6.10 Homework Problems

6.1 Using a psychrometric table or chart, or the custom Excel functions given in Appendix B.5, estimate the dry bulb and wet bulb temperatures ($t_{db}[\doteq °C], t_{wb}[\doteq °C]$) that correspond to a humidity ratio of $W = 0.0041 kg_w/kg_{da}$ and a relative humidity of $\phi = 0.50$ for humid air pressure of $p = 101.325\ kPa$. Show your work.

Answer. $t_{db} = 11.1\ °C,\ t_{wb} = 6.5\ °C$

6.2 Using a psychrometric table or chart or the custom Excel functions given in Appendix B.5, estimate the change in specific enthalpy associated with a change from state #1 ($t_{db} = t_{wb} = 31\ °C$) to state #2 ($t_{db} = t_{wb} = 30\ °C$) for a humid air pressure of $p = 101.33\ kPa$. Repeat for $\Delta t_{db,\ wb} = 16\ °C \rightarrow 15\ °C$ and compare values. Show your work. Comment on the difference between the enthalpy changes computed for Problem 6.3 and those computed for this problem.

Answer. $\Delta h_{31\,°C\,\rightarrow\,30\,°C} = 5.03\ kJ/kg_{da}$; $\Delta h_{16\,°C\,\rightarrow\,15\,°C} = 2.78\ kJ/kg_{da}$

6.3 Using a psychrometric table or chart or the custom Excel functions given in Appendix B.5, (a) estimate the change in enthalpy associated with a change from state #1 ($t_{db} = 31\ °C$, $\phi = 0.00$) to state #2 ($t_{db} = 30\ °C$, $\phi = 0.00$) for a humid air pressure of $p = 101.33\ kPa$. (b) Repeat for $\Delta t_{db} = 16\ °C \rightarrow 15\ °C$, $\Delta\phi = 0.00 \rightarrow 0.00$ and compare values. Show your work. Comment on the difference between the enthalpy changes computed for Problem 6.2 and those computed for this problem.

Answer. $\Delta h_{31\,°C\,\rightarrow\,30\,°C} = 1.012\ kJ/kg_{da}$; $\Delta h_{16\,°C\,\rightarrow\,15\,°C} = 1.011\ kJ/kg_{da}$

6.4 Using a psychrometric table or chart or the custom Excel functions given in Appendix B.5, estimate the dew point (t_{dew}) for a humid air mixture with $t_{db} = 25\ °C$, $t_{wb} = 22\ °C$. Show your work.

Answer. $t_{dew} \approx 20.75\ °C$

6.5 Starting with the Gibbs equation for internal energy $du = Tds - pdv$ and the definition of Gibbs function $g = h - Ts$, derive the Clapeyron equation. Show your work:

$$\left(\frac{dp}{dT}\right)_{sat} = \frac{\Delta h_{fg}}{T\Delta v_{fg}}$$

Hint. At saturation conditions, where the vapor and liquid are at equilibrium, the Gibbs function for the vapor phase g_g is equal to the Gibbs function for the liquid phase g_f for all values of quality.

6.6 Beginning with Equations (6.21) and (6.22), derive Equation (6.23) and use it to compute individual molar flow rates $\hat{a}, \hat{b}, \hat{c}$, and $\hat{d}$ for Example (6.9) (where $\hat{e} = 0.3$ and $\hat{f} = 0.7$). Verify that all the algebraic relations given in Equations (6.21)–(6.23) are equivalent to the parallel formulas for mass flows and mass proportions.

Answer. $\hat{a} = 0.014\ 286, \hat{b} = 0.271\ 428, \hat{c} = 0.285\ 714$, and $\hat{d} = 0.428\ 572$

6.7 Fill in the blanks (empty boxes) in the heat and material balance table below. The process represented is a heat exchanger performing air dehumidification and cooling. The flow path of the refrigerant ($R32$) is from state 1 to state 2 and the humid airflow path is from state 3 to state 4. Assume the cooler is adiabatic with respect to the environment. For states 3 and 4, the denominators of intensive thermodynamic properties are "per kg dry air" (not per kg total mixture).

Property	State 1	State 2	State 3	State 4
Temperature (°C)	6.62	6.62	28.0	23.0
Pressure (kPa)	1000	1000	101.33	101.33
Density (kg/m^3)	830.97	27.238	1.151 16	1.174 82
Enthalpy (kJ/kg)	272.61		89.761	68.295
Rel humidity	—	—	1.00	
Quality	0.20	1.00	—	—
Dry air (kg/s)			0.15	0.15
Water,liq (kg/s)			0.00	
Water,vap (kg/s)			0.003 631	
R23 (kg/s)			0.00	0.00

Cited References

ASHRAE (2013). *ASHRAE Handbook – Fundamentals*;"Psychrometrics"; p. 1.1; American Society of Heating, Refrigerating, and Air Conditioning Engineering.

BSI. (2015). *BS EN285. Sterilization. Steam Sterilizers. Large Sterilizers*; British Standards Institution.

Callen, H.B.; (1960); *Thermodynamics*; New York: Wiley.

Downing, R.C.; (1974); "Refrigerant equations"; *ASHRAE Transactions;* **81**; p. 158.

Keenan, J.H.; (1941); *Thermodynamics*; New York: Wiley.

Keenan, J.H., Keyes, F.G., Hill, P.C. et al. (1969); *Steam Tables*; New York: Wiley.

Kline, S.J.; (1999); *The Low-Down on Entropy and Interpretive Thermodynamics*; La Canada, CA: DCW Industries.

Linstrom, P.J. and Mallard, W.G., Eds.; (2017–2021); *NIST Chemistry WebBook, SRD 69;* Thermophysical Properties of Fluid Systems. Gaithersburg, MD: National Institute of Standards and Technology, U.S. Dept. of Commerce; http://webbook.nist.gov/chemistry/fluid (accessed 16 August 2021).

McQuarrie, D.A.; (1973); *Statistical Mechanics*; New York: Harper & Row.

Myhrvold, N. and Migoya, F. (2013). Modernist Cuisine. Blog (20 March 2013). https://modernistcuisine.com/mc/the-maillard-reaction (accessed 25 April 2021)

NOAA. (1976). *U.S. Standard Atmosphere, 1976.* National Oceanic and Atmospheric Administration. Report #NOAA-S/T 76-1562. https://apps.dtic.mil/dtic/tr/fulltext/u2/a035728.pdf (accessed 15 January 2021).

Reid, R.C., Prausnitz, J.M. and Poling, B.E.; (1987); *The Properties of Gases and Liquids*, 4; New York: McGraw-Hill.

Reynolds, W. C.; (1979); *Thermodynamic Properties in SI*; Stanford University Mechanical Engineering Department.

Van der Waals, J.D.; (1873); *On the Continuity of the Gaseous and Liquid States*; (2004); Rowlinson, J.S., Ed.; New York: Dover Publications.

Vincenti, W.G. and Kruger, C.H.; (1965); *Physical Gas Dynamics*; New York: Wiley.

Zemansky, M.W.; (1937); *Thermodynamics*; New York: McGraw-Hill.

7

Burners and Heat Recovery

The objective of this chapter is to introduce concepts, terminology, and nuances associated with burners, heat exchangers, and combinations thereof. Also discussed are topics of safeguarding combustion systems and a simple way to discretize a heat exchanger to analyze or model it. Burner and heat exchanger failures are discussed in the concluding section of the chapter.

7.1 Burners

Burners are devices that bring together fuel, oxidizer, and an ignition source in a controlled manner that supports and sustains continuous combustion at a desired rate of heat release. Burners can accommodate gaseous, liquid, or solid fuels. Liquids typically must be atomized by a pressurized nozzle to establish a stable flame within and around the droplet cloud. Solids are frequently pulverized upstream of the burner and pneumatically transported to the flame front by a flow of primary air. Solid particle combustion is typically slower than liquid fuel or fuel gas combustion, so solid-fuel burners have longer flame lengths.

Flame stability can be obtained by (i) the careful commingling of the fuel and air streams at a point where ignition kernels (pockets of high-temperature products and intermediates) are plentiful; (ii) the generation of vortices by small burner obstructions to enhance mixing of cold reactants, hot products, and reactive intermediates; (iii) the introduction of swirl to the combined fuel and airflows to encourage recirculation of products and intermediates into the fuel-burning zone; or (iv) any combination of these techniques.

Flame temperatures are limited on the low side by stoichiometry considerations that affect flame stability. For ***stationary*** or ***propagating*** flames involving *premixtures* of air and fuel, the lower flammability limit (LFL) constrains these mixtures to equivalence ratios of $\phi \approx 0.5$ or greater, because leaner mixtures do not release enough heat to sustain the chain reactions that make flame propagation possible. For flames involving fuels that are *not* premixed with air, flame ***propagation*** is impossible, and ***stationary*** flames burn in space at mixing contours where the equivalence ratio is approximately $\phi \approx 1.0$.

Even premixed burner flames that are controlled to $\phi \approx 1.0$ rarely reach temperatures as high as those calculated thermodynamically ($T \approx 3100\ °F$ $[2000\ K]$). This shortfall is due to convective dilution by excess air and radiation heat loss to the surroundings, and these same phenomena (especially dilution) can cause apparent temperatures of lean flames ($\phi \approx 0.5$) to fall below 1000 $°F$ [$\approx$800 K]. If dilution and heat losses are kept to a minimum, it has been shown that the lowest adiabatic combustion temperature achievable with a burner is that associated with a fuel–air

Thermal Systems Design: Fundamentals and Projects, Second Edition. Richard J. Martin.

Companion website: www.wiley.com\go\Martin\ThermalSystemsDesign2

premixture at its LFL, consistently falls near $T_{LFL} \approx 2400\ °F$ [$\approx$1600 K] for most of the common hydrocarbons (e.g. alkanes, alkenes, and aromatics).

Students should understand and be comfortable using the following terminology related to burners:

- Primary air. The portion of combustion air that is premixed with the fuel (usually applies only to pulverized solid fuels like coal dust and sawdust).
- Secondary air. The portion of combustion air that is provided to the burner in a pipe or duct separate from the fuel. It mixes with the fuel as they enter the flame front together. Secondary air is sometimes called "nozzle mix air."
- Tertiary air. The portion of combustion air that is provided downstream of the burner flame in nozzles that are removed several feet from the burner location. Tertiary air is sometimes referred to as "over fire air" (OFA) because it is introduced over the fire. Tertiary air is typically used to control temperature or fuel richness at the burner zone.
- Theoretical air. The flow rate of air that exactly provides the stoichiometric proportion of oxygen for a given flow rate of fuel. See also discussions in Chapter 4.
- Excess air. The portion of the actual airflow rate that is beyond the required stoichiometric (theoretical) amount for the given flow rate of fuel. See also discussions in Chapter 4.
- On-ratio. A burner firing scenario where the fuel and airflow rates are proportioned exactly at the stoichiometric ratio.
- Turndown. A measure of how much the burner's firing rate $\left(\frac{BTU}{hr}\right)$ can be reduced below its faceplate designated value. Turndown is the quotient of the faceplate firing rate divided by the lowest stable firing rate. Example – if the burner's lowest stable firing rate is 500 000 $\frac{BTU}{hr}$, and its faceplate rating is 1 500 000 $\frac{BTU}{h}$, the burner is sai;d to have a turndown ratio of 3 : 1.
- Flue gas recirculation. Flue gas recirculation (FGR) is a technique for control of NOx emissions from a burner where a moderate amount of exhaust gas (low in O_2 and moderately high in inert gases CO_2 and H_2O) is recirculated from the exhaust duct to the fresh air intake, where it displaces a portion of the fresh air going to the burner. The presence of more inert gas and less oxygen reduces the flame temperature, which reduces thermal-NOx formation. If too much flue gas is recirculated to the burner, the flame can become unstable, which imposes a practical upper limit ($\approx$40% of the combustion airflow) on the amount of FGR addition to maintain flame stability.
- Premix burner. A burner that delivers a premixture of air and gaseous fuel through the interior of the burner and into the flame front. Because the premixture of fuel and air is inherently hazardous (i.e. it can deflagrate), precautions are necessary to ensure that flashback into the premixture conveyance either (i) will not occur or (ii) is not energetic enough to damage the burner or furnace.
- Nozzle mix burner. A burner that delivers air and fuel to the flame front through separate pipes, up until the point where they emerge, mix, react, and release heat. Under normal conditions, nozzle mix burners have essentially no risk of flashback.
- Duct burner. An array of individual burners that are distributed uniformly over the cross section of a large duct. Duct burners are arranged so that the flame envelope of one burner is sufficiently close to all its nearest neighbors that they can all reignite each other rapidly and reliably, whenever an individual burner suffers an unanticipated flame failure. Duct burners are often used to heat large volumes of air (e.g. for multistage gas turbine reheat).
- Afterburner. A synonym for thermal oxidizer, this term describes a burner firing into an open chamber, where hydrocarbons and other combustible gases in a process stream are destroyed so that the stream can be safely directed to the atmosphere.

Procurement of burners requires knowledge of the aforementioned terminology and other nuances associated with burner catalogs and specification sheets. We introduce two new units of measure here: *BTUh* is a single-word acronym (like *kW*) for burner thermal power and *osi* is a measure of pressure like *psi*, but where *o* stands for ounces ($1\ lb_f = 16\ oz_f$).

For one example, a nozzle-mix burner spec sheet presents the following data:

- Air pressure drop across the burner, *osi* (values range from 0.2 to 16.0):
  - The burner is equipped with an internal orifice with pressure taps upstream and downstream, and the pressure drop across the orifice is a function of volume flow rate squared.
  - Pressure drop units are in ounces per square inch (i.e. $16\ osi \approx 28\ inch\ WC$).
  - The given values for air pressure drop are associated with airflow rate values ($\veebar$ *scfh*) for the different burner sizes (each having a different orifice diameter).
- Combustion air capacities, *scfh* (for *BTUh*, multiply by 100):
  - For each burner size represented in the spec sheet, values for combustion airflow rate are given that correspond to each column heading of the air pressure drop.
  - The parenthetical comment about multiplying by 100 to obtain *BTUh* is a simple recognition that for an equivalence ratio of $\phi \approx 1.0$,
    - the volume ratio of fuel to air is approximately $1\ scf_{fuel} : 10\ scf_{air}$;
    - for natural gas, the enthalpy of combustion is approximately $\Delta h_{comb} \approx 1000\ BTU/scf_{fuel}$;
    - when combined, these values lead to an approximate ratio of $100\ BTUh/scfh_{air}$.
- Approximate flame length (self-explanatory)
- Maximum percent excess air at 16 *osi*:
  - The burner is designed to achieve its maximum stable-flame airflow rate at a 16 *osi* pressure drop across the burner orifice.
  - Achieving a 300% excess air under this maximum airflow condition means that the fuel flow rate (and hence, the firing rate) has been ***turned down*** by a factor of four.
  - Recall in Chapter 4 that 300% excess air was shown to be the same as 400% theoretical air, which would be the same as an overall equivalence ratio of $\phi \approx 0.25$.
  - Note, with such a low stoichiometric proportion, this burner would not be stable if it were designed to be operated as a premixed burner. Since it was stated above that the burner style was "nozzle mix," a low overall equivalence ratio can be achieved with tertiary air or other means of downstream dilution.

Selection of a burner can be accomplished after the heat and material balance is completed to the point where the maximum and minimum heat release rates (*BTUh* or kW_{th}) are known for the combustor.

7.2 Combustion Safeguarding

In Section 4.14, we reviewed some fundamentals of combustion safety and here we initiate our review of combustion safeguarding. Further information about combustion safety instruments is given in Section 14.7, and recommended maintenance procedures for combustion safety systems are given in Section 18.4.

A trio of National Fire Protection Association (NFPA) standards that focus on industrial heating equipment provides a wealth of information and best practices for design, construction, installation, operation, maintenance, inspection, and testing of boilers (NFPA 85), ovens (NFPA 86), and fluid heaters (NFPA 87). While the ***burner*** safety requirements are nearly identical across the three standards, other types of safety measures are unique to boilers, furnaces, or heaters exclusively. For example, ovens do not require water level controls, and boilers do not require safety ventilation for evaporated solvents. From this perspective, it is easy to see why differences associated with the ***burner*** system safeguards are minor indeed.

Table 7.1 identifies several key combustion safeguards required for many types of industrial heating systems. In addition to the hardware safeguards shown in Table 7.1, NFPA standards also require users of furnaces, boilers, and fluid heaters to implement procedures for periodic inspection, maintenance, and testing to ensure that safety devices are in place, are functional, and are correctly calibrated.

Beyond the minimum requirements imposed on all systems, NFPA standards also govern the employment of additional safeguards that ensure safety for complex combustion systems that are utilized in ways that would be unsafe for simpler units. One example of this is an industrial oven with multiple burners firing into a single zone. Whenever the zone's safety is proven to be at $T \geq 1400\ ^{\circ}F$ [760 °C], the user can cycle individual burners off and on (to more precisely tailor heat input) without shutting the system down to conduct a preignition purge every time an

Table 7.1 Selected combustion safeguards required for ovens and furnaces by NFPA 86.This table only contains excerpts, and it is not a substitute for a thorough review and application of the standard in its entirety.

Hazard	Causation scenario	Engineering control
Fuel gas accumulation	Residual unburned fuel leftover from prior, unsuccessful ignition trial	Timed preignition purge of at least four furnace volumes of fresh air (**8.5.1**)
<same>	<same>	Proof of minimum purge airflow rate (**8.5.1.2.4**)
<same>	Failure to fully close fuel gas safety shutoff valve (SSOV) during furnace downtime	Redundant SSOVs in fuel line with at least one valve having a "proof-of-closure" interlock (**8.8.2**)
<same>	Failure of SSOV in partially open position due to debris obstruction	Sediment trap and strainer in fuel gas line, upstream of SSOV (**6.2.5**)
<same>	Inability to ignite pilot or main burner	Trial for ignition (TFI) timer (**8.5.2**)
<same>	Flame failure	Flame supervision logic with a sensor having a flame failure response time of 4 seconds or less (**8.10**)
Flame instability and fuel gas accumulation	Combustion air failure	Proof of combustion air blower pressure or flow (**8.7**)
<same>	Ventilation failure	Proof of fresh air or exhaust blower functionality (**8.6**)
<same>	Low fuel gas pressure	Fuel gas low pressure switch (**8.9**)
<same>	High fuel gas pressure	Fuel gas high pressure switch (**8.9**)
Ignition of combustible feed	Excess temperature in the heating zone	Excess temperature interlock (**8.16**)

Source: Reproduced with permission of NFPA from NFPA 86 (2019). © 2018, National Fire Protection Association. For a full copy of NFPA 86, please visit www.nfpa.org.

individual burner is reignited. This exception is allowed because the zone is already hot enough to immediately ignite fuel gas flowing out of the burner, with almost no risk of fuel gas accumulation and subsequent explosion.

7.3 Thermal Oxidizers

Thermal oxidizers (also called fume incinerators and afterburners) are air pollution control devices that are designed to destroy any type of volatile organic compound (*VOC*) that must not be released into the atmosphere because of its toxicity (many such gases are designated as hazardous air pollutants [*HAPs*]), global warming potential (*GWP*), or participation in the photochemical smog cycle (that produces ground-level ozone).

In its simplest form (e.g. Figure 7.1), a thermal oxidizer (*TO*) is an open chamber, where a gaseous stream laden with *VOC* is combined with air, heated (typically by a burner), and given sufficient residence time at a high temperature to carry out the destruction of *VOC*, duly producing CO_2 and H_2O.

The range of *VOC* concentrations that may be continuously admitted to a *TO* traverses the gamut from (i) a very dilute mixture of a single *VOC* in air or in an inert gas (e.g. N_2) to (ii) a mixture of many *VOCs* at very high concentrations and with little or no diluents. If oxygen is lacking in the fume gas, a flow of air must be mixed with the fume to ensure that O_2 molecules are commingled with the VOC molecules. The amount of air required is determined by the stoichiometric ratios and flow rates of each compound present. Typically, any fume actually admitted to a TO chamber should have a minimum of 15% *vol* of O_2 molecules in order to successfully burn the VOC molecules in the residence time available inside the chamber.

Three T's. For a traditional (flamed) thermal oxidizer like the one in Figure 7.1, conventional wisdom instructs a designer to ensure the combustion chamber abides by the rule of three T's – ***temperature, time,* and *turbulence***. The chemical kinetics of VOC destruction are governed at the molecular level by the Arrhenius relationship, which incorporates temperature in two

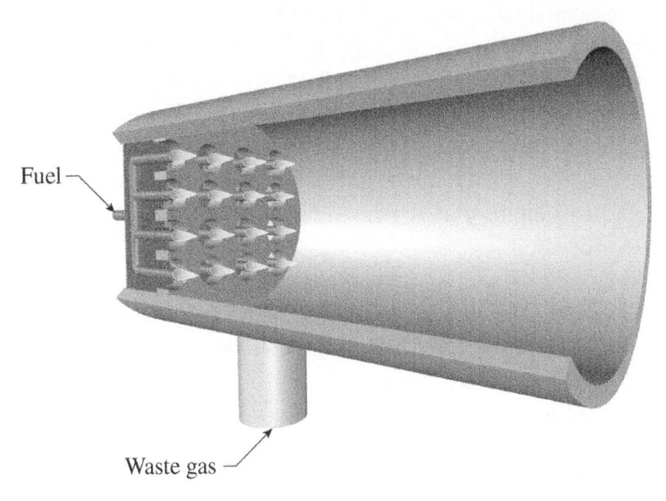

Figure 7.1 Simplified model of a cylindrical, horizontal, straight-through thermal oxidizer, showing a multijet fuel burner on the left, a waste gas entry port at the bottom, and an oxidized gas exit on the right. *Source:* Martin Thermal Engineering Inc.

separate parameters of the same equation: (i) the activation energy term ($exp\left[-E_a/(\hat{R}T)\right]$) and (ii) the power of temperature term (T^b), where E_a and b are the activation energy ($\doteq kJ/kmol$) and temperature exponent (dimensionless), respectively. Detailed discussions of chemical kinetics are beyond the scope of this textbook. Readers are directed to the combustion textbooks cited at the end of Chapter 4 for additional information on that subject.

Residence time at temperature is another vitally important factor, as one would expect. Common residence times in thermal oxidizers are 0.5 and 1.0 s, but other times may be necessary if warranted by a detailed analysis. The computation of residence time (τ) in a vessel is straightforward (see Equation 7.1):

$$\tau = \frac{\mathcal{V}_{vessel}}{\dot{\mathcal{V}}(p, T)} \tag{7.1}$$

where the volume ($\mathcal{V}$) is computed from the inside diameter and inside length of the (refractory-lined) cylindrical vessel, and the volume flow rate ($\dot{\mathcal{V}}$) is computed from the ideal gas law and known total mass flow rate ($\dot{m}$), at the actual temperature and pressure of the reacting gas mixture, which along with the approximately molar mass of the mixture allows the actual gas density (ρ) to be computed. Computing volume flow for standard (STP) conditions will give a very wrong answer, as seen in the example below.

Example 7.1

Compute the maximum flow rate ($scfm$) possible that will achieve $\tau = 1.0\ s$ residence time in a $50\ ft^3$ cylindrical vessel if the fluid's molar mass is $\hat{M} = 30\ kg/kmol$ and the average temperature is $1600\ °F$ (assuming a standard pressure of $1.0\ atm$).

Answer. Determine the maximum volumetric flow rate: $\dot{\mathcal{V}} = 50/1 = 50\ ft^3/s$. Determine the actual mole flow rate equivalent to this volume flow rate, using ideal gas law: $\dot{n} = (p\dot{\mathcal{V}})/(\hat{R}T) = 0.03\ 325\ lbmol/s$. Convert to standard cubic feet by multiplying by standard volume per mole: $\dot{\mathcal{V}}_{stp} = \hat{\mathcal{V}}_{stp}\dot{n} = 391.6\ scf/lbmol \cdot 0.033\ 25\ lbmol/s \cdot 60\ s/min = \boxed{781\ scfm}$. Note the large difference between standard (781 $scfm$) and actual (3000 $acfm$) volumetric flows.

Finally, the emphasis on turbulence is essentially a concern for good mixing. Good mixing can be accomplished readily if the vessel size is large, and the burner velocity is vigorous. The Reynolds number will likely be quite high in a vessel of any practical size, but the Reynolds number alone is not the only factor that determines whether the flow is sufficiently turbulent. For many turbulent duct flows, entry effects may be ignored if the ratio of the vessel's length to its inside diameter is greater than $L/D > 10$. This value designates the point where the turbulent boundary layer has converged at the duct's centerline and all parts of the flow are equally turbulent. It is where the turbulent velocity profile becomes fully developed.

However, even if a thermal oxidation vessel is shorter than the given threshold, a properly designed burner can overcome the wall-related turbulence limitations. When the exit velocity of the burner is high enough, jet mixing occurs, and the turbulent eddies created by the jet usually provide high-intensity mixing from the vessel's inlet to its exit.

To compensate for the possibility of laminar flow near the walls or pockets of stagnant fluid due to obstacles or asymmetries, a high burner velocity is an imperative. While jet mixing design guidelines are beyond the scope of this textbook, interested readers may wish to investigate Schlichting (1960), which covers theories of jet flow and swirling flow, or Uhl and Gray (1966), which provides practical know-how regarding quantifying and attaining good mixing.

Validation of mixing quality can also be performed by experiment (e.g. step-function injection of a tracer gas into the dirty gas stream and measurement of exhaust concentration over time) or by numerical modeling of the subject geometry and flow data (e.g. estimating concentration profiles of a hypothetical tracer gas by using a computational fluid dynamics numerical tool, especially one with large eddy simulation [LES] capabilities.).

7.4 Destruction Efficiency

The recognized performance measure for a thermal oxidizer is its destruction efficiency. Destruction efficiency is defined in Equation (7.2):

$$DE\% \equiv \frac{\sum \dot{m}_{VOC\,in} - \sum \dot{m}_{VOC\,out}}{\sum \dot{m}_{VOC\,in}} \tag{7.2}$$

It is noteworthy that the expression relies on mass flows into and out of the thermal oxidizer, rather than mole or mass fractions. The simple reasoning behind this definition can be ascertained by contemplating the old environmentalist's adage: "The solution to pollution is *not* dilution." One can imagine that by diluting the inflowing stream with enough air (or burner exhaust), the outflowing VOC concentrations could be reduced substantially below the inflowing VOC concentrations, without much destruction occurring at all.

Cooper and Alley (1994) addressed the question of the influence of time, temperature, and chemical identity of the VOC on its level of destruction in a thermal oxidizer. They reported empirically determined equations (Lee et al. 1979, 1982) that incorporate a number of factors associated with the VOC molecule and compute a minimum temperature necessary to achieve a 99.9% destruction efficiency, as follows in Equation (7.3):

$$\begin{aligned} T_{99.9}(\,°F) = {} & 594 - 12.2W_1 + 117.0W_2 + 71.6W_3 + 80.2W_4 + 0.592W_5 - 20.2W_6 - 420.3W_7 \\ & + 87.1W_8 - 66.8W_9 + 62.8W_{10} - 75.3W_{11} \end{aligned} \tag{7.3}$$

W_1 : $\#C$ – atoms	W_2 : *Aromatic* flag
W_3 : $C = C$ *double bond* flag	W_4 : # *N*-atoms
W_5 : *AIT (autoignition temp)*	W_6 : # *O*-atoms
W_7 : $\#S$ – atoms	W_8 : *H/C ratio*
W_9 : *Allyl* flag	W_{10} : $C = C - Cl$ flag
W_{11} : $\ln(\tau)$	

Most of the factors are self-explanatory, with the following clarifications:

W_2: The aromatic flag is given a value of 1.0 when the compound contains one or more benzene rings.

W_3: The $C = C$ double bond flag is given a value of 1.0 when at least one double bond is present in the molecule.

W_9: The allyl flag is given a value of 1.0 if the compound has an allyl group ($-CH{=}CH{-}CH_2$).

W_{11}: Even though the residence time (τ) is not dimensionless, Equation (7.3) was fit to work with the time logarithm having dimensions, so it must be presumed that coefficient W_{11} has units of seconds.

Cooper and Alley (1994) provide autoignition temperatures for more than 30 VOCs that are commonly destroyed in a thermal oxidizer. Some of these values are duplicated here in Table 7.2.

Table 7.2 Autoignition temperatures (°*F*) for selected organic vapors in air.

Substance	AIT (°*F*)	Substance	AIT (°*F*)	Substance	AIT (°*F*)
Acetone (C_3H_6O)	1000	Cyclohexane (C_6H_{12})	514	Isobutane (C_4H_{10})	950
Acrylonitrile (C_3H_3N)	898	1,2-Dichloroethene ($C_2H_2Cl_2$)	775	Methane (CH_4)	999
Ammonia (NH_3)	1200	Ethane (C_2H_6)	986	Methanol (CH_4O)	878
Benzene (C_6H_6)	1075	Ethanol (C_2H_6O)	799	Methyl ethyl ketone (C_4H_8O)	960
Butane (C_4H_{10})	896	Ethene (C_2H_4)	842	Phenol (C_6H_6O)	1319
1-Butanol ($C_4H_{10}O$)	693	Ethyl acetate ($C_4H_8O_2$)	907	Propane (C_3H_8)	871
1-Butene (C_4H_8)	723	Ethylbenzene (C_8H_{10})	870	Propene (C_3H_6)	851
Carbon monoxide (CO)	1205	Hexane (C_6H_{14})	820	Styrene monomer (C_8H_8)	915
Chlorobenzene (C_6H_5Cl)	1245	Hydrogen (H_2)	1076	Toluene (C_7H_8)	1026
Chloroethane (C_2H_5Cl)	965	Hydrogen cyanide (HCN)	1000	Vinyl chloride (C_2H_3Cl)	882
Chloromethane (CH_3Cl)	1170	Hydrogen sulfide (H_2S)	500	Xylene (C_8H_{10})	924

Source: Adapted by permission of Waveland Press from Cooper and Alley (1994). See "Cited References" section at the end of Chapter 7 for full citation.

Previously, Cooper et al. (1982) sought to develop a rigorous (yet simple) analytical method for predicting the destruction efficiency of VOC as a function of the relevant thermal oxidizer parameters (e.g. temperature, residence time, oxygen concentration, and initial VOC concentration). Martin (2019) harmonized their approach to consistent units and extended it with advanced statistical methods for attaining environmental process quality. Details of this new approach are provided in Chapter 17 of this textbook.

7.5 Recuperators and Regenerators

Although it may sound like a violation of the Second law of thermodynamics, recuperators and regenerators are heat exchangers that are designed to transfer heat from a fluid to itself.

More precisely, a recuperator (or a regenerator) is a heat exchanger that admits a hot fluid into its hot side, admits a cold fluid into its cold side, and facilitates the transfer of heat from the hot to the cold fluid – with the unique attribute that the two fluids are indeed the same working fluid, albeit they are flowing through different segments of the overall process path. By this ingenious juxtaposition of a fluid adjacent to itself, recuperators and regenerators are capable of vastly reducing the amount of energy needed to heat a stream from ambient temperature to a desired operating temperature for the equipment between the two heat exchanger paths. Figure 7.2 illustrates this basic concept with a thermal oxidizer placed between the two heat exchanger paths.

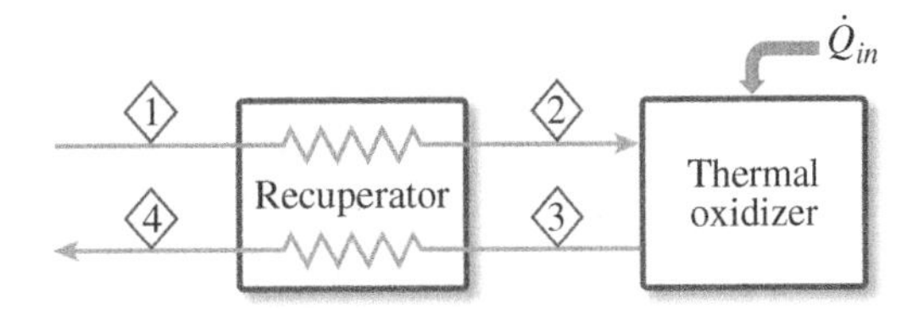

Figure 7.2 Placement of a recuperative heat exchanger between the inlet and outlet paths of a thermal oxidizer to reduce the amount of external energy needed to increase the temperature of the incoming fluid to the operating temperature of the thermal oxidizer. *Source:* Martin Thermal Engineering Inc.

Recuperators. Heat recuperation can be applicable to many types of processes, whenever the cost of adding the heat exchanger is justified by cost savings (e.g. reduced fuel consumption) or some other process benefits (e.g. modification of combustion conditions). The recuperator configuration can be almost anything (e.g. shell-and-tube, pipe-in-pipe, or finned crossflow) with the following flow characteristics posing the most common design constraints: (i) air (or contaminated air) is the most likely working fluid, (ii) high flow rates and low convection coefficients may necessitate large heat-transfer surface areas, and (iii) high temperatures may necessitate refractory metal materials of construction for the heat exchanger. See also Chapter 13 for additional information about refractory materials (that are stable at high temperatures).

The designation $\dot{Q}_{in}$ on the right-hand side of Figure 7.2 is a concession to the Second law, namely, that recuperators are incapable of recovering *all* the heat necessary to accomplish the thermal mission all by themselves (i.e. with no heat entering the control volume across its boundary). The form of the heat input could be a burner firing into a combustion chamber that connects the heat exchanger's cold outlet to its hot inlet, or an electric element adding heat by radiation/convection, or a chemical conversion (and heat release) from the VOC themselves when they reach their ignition temperature.

Recuperator effectiveness defined loosely in Figure 7.2 as:

$$\varepsilon = \frac{\dot{m}(h_2 - h_1)}{\dot{m}(h_3 - h_1)} \approx \frac{(T_2 - T_1)}{(T_3 - T_1)} \approx \frac{(T_3 - T_4)}{(T_3 - T_1)} \tag{7.4}$$

can be as low as 30–40% or as high as 70–80%.

Think Stop. Depending on the specific design, it may or may not be suitable to assume that (i) the mass flow out of the cold side of the recuperator is equal to the mass flow into the hot side of the recuperator (students should ponder the circumstances under which either would be true) or that (ii) the average specific heat on the left side of the equation is equal to the average specific heat on the right side (again, students should ponder why). For some air-to-air applications, these assumptions are excellent, and the temperature-only equations for effectiveness are valid. For others, greater precision may be necessary, and the enthalpy form must be used instead of just temperatures.

Regenerators. In contrast with recuperators, regenerators use a heat-storage medium comprising porous solids to temporarily extract and hold heat from the hot exhaust gas and later transfer it to the cold incoming gas. While recuperators always direct the local fluid flow in a common direction (i.e. from inlet to outlet), regenerators permit the fluid to flow in the reverse direction for approximately half of the operating duration.

Consider a packed-bed style of regenerator, called a regenerative thermal oxidizer (RTO) with two heat-storage beds (see Figure 7.3). In addition to the two packed beds connected by an overhead combustion chamber (with a natural gas burner), the design includes a pair of three-way valves at

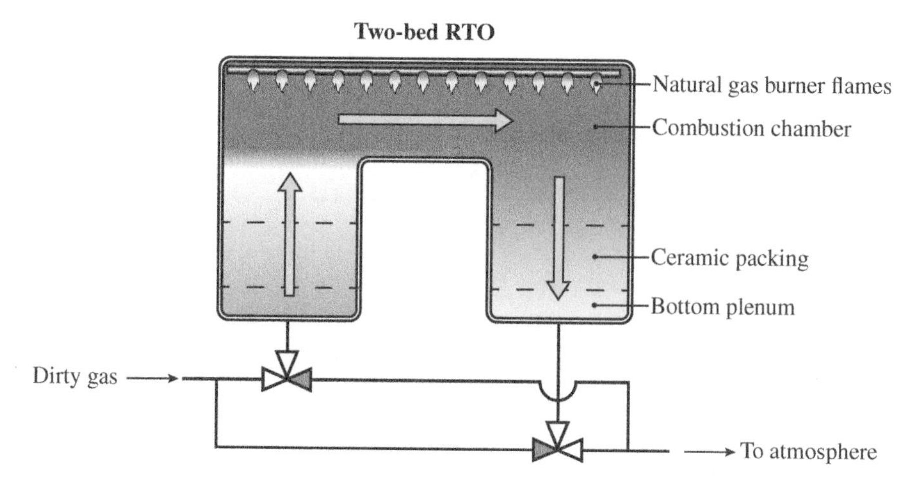

Figure 7.3 Regenerative heat exchanger (specifically designed for VOC destruction). *Source:* Martin Thermal Engineering Inc.

the bottom, which govern which bed (left or right) the dirty inlet gas flows into and which bed the cleaned exhaust gas flows out from.

The image shows the gases flowing clockwise through the system, with the valves arranged to admit cold, dirty (VOC-laden) air into the left bed and to exhaust warm, clean air (i.e. with dilute combustion products and nearly undetectable VOC) out from the right bed. At an appropriate time, the two valves switch positions, and the gas flow moves counterclockwise, with dirty air entering the right bed and clean air exiting the left bed.

The primary value of the two heat regeneration beds is to effectively extract heat from the high-temperature exhaust gases and preheat the ambient temperature incoming gases. If a control volume is drawn around just the two packed beds (effectively, the recuperator in Figure 7.2), the temperature change from the dirty gas inlet and final outlet streams (i.e. $T_4 - T_1$) is very low compared to the difference between the temperature change from the dirty inlet to the combustor exit (i.e. $T_3 - T_1$). In addition to thermal or catalytic oxidation of VOC, the regenerative heat exchange concept is also used in glass-melting furnaces to achieve higher burner exit temperatures than possible with combustion of fuel using room temperature air.

The measured heat recovery effectiveness of many RTOs is greater than 90% (with some systems reaching 95% +). This feature renders RTOs a very good choice for destroying dilute concentrations of VOC in an air stream but also makes them a very poor choice for destroying high concentrations of VOC in an air stream. The combination of the high heat recovery effectiveness, along with a combustion heat release of $\Delta h_{comb} > 100\ BTU/lb_{dirty\ gas}$, runs the risk of melting the ceramic media and compromising the refractory and insulation in an RTO.

Any thermal oxidizer that utilizes ceramic packing for heat transfer and heat storage is very likely to be incompatible with dirty gas streams that either (i) contain solid particulate matter or (ii) form solids as a by-product of combustion. An example of category (i) is a dirty gas stream that has soot or organic particles that may adhere to the cold matrix in the TO's inlet region during the first cycle and then re-vaporize as a new VOC (Martin and Colwell 2003) when the flow direction reverses. An example of category (ii) is a dirty gas stream that contains organometal vapors or metalloid vapors (e.g. tetraethyl lead or silane), which oxidize to form solids (e.g. lead oxide or silicon dioxide). Such oxides have relatively high melting and vaporization points, and they can easily plug the interstices in a ceramic matrix if enough of the material is admitted into the RTO with the dirty gas stream over time.

The first US patent using regenerative technology as a means of destroying air contaminants was issued to Cottrell (1938) and is illustrated in Figure 7.4.

The present author is a named inventor on 24 thermal oxidizer patents that were issued between 1992 and 2016. One such patent (Stilger et al. 1994) shown in Figure 7.5 included the novel concept

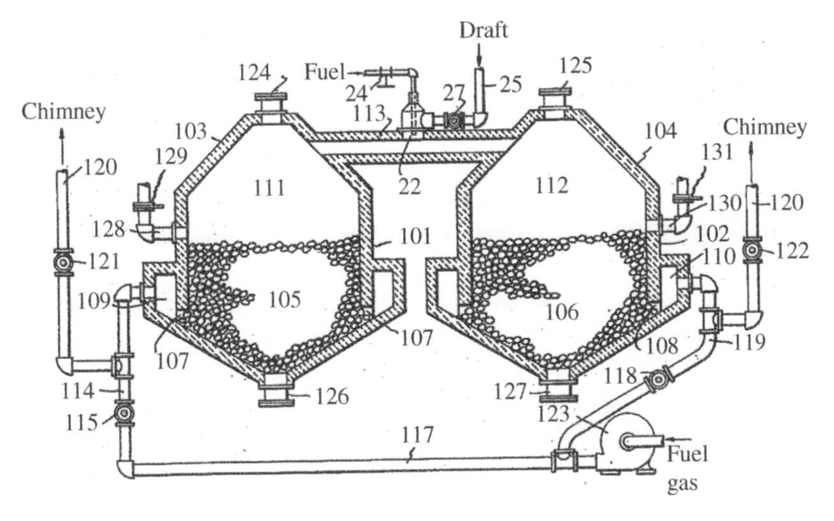

Figure 7.4 Figure from US Patent No. 2,121,733 issued to Cottrell in 1938 – the very first regenerative thermal oxidizer. *Source:* United States Patent and Trademark Office (https://www.uspto.gov) (Public Domain).

Figure 7.5 Figure from US Patent No. 5,320,518 issued to Stilger et al. (1994), the first packed-bed, recuperative, flameless thermal oxidizer. *Source:* United States Patent and Trademark Office (https://www.uspto.gov) (Public Domain).

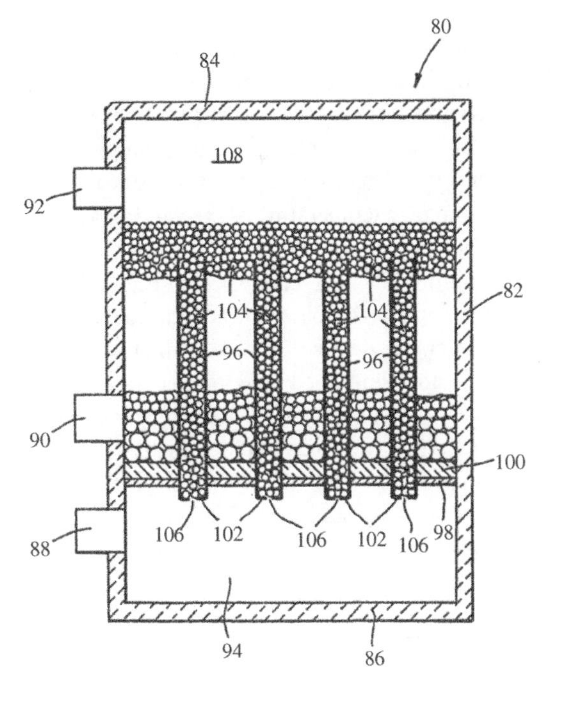

of incorporating large, recuperative heat exchange tubes into the ceramic packing material and employing "radiatively coupled fins" (i.e. the packing) to enhance the heat exchange surface area beyond that of the tubes themselves and thereby to also enhance the overall *UA* product of the heat exchanger.

Think Stop. If a thermal system is being examined for signs of energy inefficiency, what one measurement provides the most immediate information that it might be able to benefit from a recuperator or a regenerator? Temperature is, of course, the only variable to consider, but the bigger question is which temperature sensor (i.e. which location) would provide a telltale sign of inefficiency? The exhaust gas is the hands-down winner because if it is significantly higher than ambient, it represents a source of thermal energy that might be productively utilized to preheat an incoming stream (or alternatively, to run a low-temperature heat engine).

7.6 Packed-bed Heat Storage

Any packed bed used by RTOs (and other chemical reactors) serves as a combined heat-storage and heat-transfer medium. Analysis of the packed-bed heat-storage problem can be likened to an analysis of a cold length of pipe that undergoes gradual heating from a flow of hot gas. The entrance to the pipe begins to heat up first but because the solid pipe is much more massive (per unit volume) than the hot gas, it takes quite some time before the gas leaving the pipe begins to creep above the pipe's initial temperature and even longer for the entire pipe to reach the gas inlet temperature.

For RTOs, the situation is even more pronounced because not only is the hot gas convecting its heat to the pipe wall's mass but it is also losing heat to the solid packing, which can hold 1000× more thermal energy per unit volume (per degree temperature rise) than the exhaust gas alone can hold. Furthermore, because the interstices between packing particles tend to be very small ($D_{hyd} \approx$ ¼ *inch*), the convective heat-transfer area is high, and the convection is very efficient. Thus, depending on geometry and flow rates, it may take minutes or tens of minutes for the gas exiting the packed bed to exceed the initial (cold) temperature of the bed.

Kays and London (1984) solved the packed-bed transient heat-transfer problem numerically. The solution is a table containing a nondimensional temperature field given as a function of the nondimensional distance along the flow path in the bed and the nondimensional time from the initial step change in temperature as the varying parameter. In this geometry, *NTU* is the nondimensional ratio of heat convection to heat storage in the ceramic media (rather than the $\dot{m}c_p$ product for the minimum fluid). In Figure 7.6, the results for a specific case are plotted as time snapshots of temperature profiles in space (the actual bed temperatures versus the actual bed distance), with all unit-bearing values derived from their dimensionless counterparts.

For a more simplified (and less accurate) approach to determining heat transfer to or from a bed of spheres, recall Section 3.17 where an expression for Stanton number was given as a function of the interstitial Reynolds number and the fluid's Prandtl number. If each sphere can be treated as a "one-lump" mass (i.e. where the conduction through the sphere material is large compared to the convection at the surface), the sphere's temperature can be computed from the one-lump transient relationship, where R is sphere radius and rho is solid material density (Equation 7.5).

$$\frac{(T(t)-T_\infty)}{(T_i-T_\infty)} = e^{-bt}, b = \frac{hA_s}{\rho V c_p} = \frac{4\pi R^2 h \cdot 3}{\rho R^3 c_p 4\pi} = \frac{3h}{\rho R c_p} \tag{7.5}$$

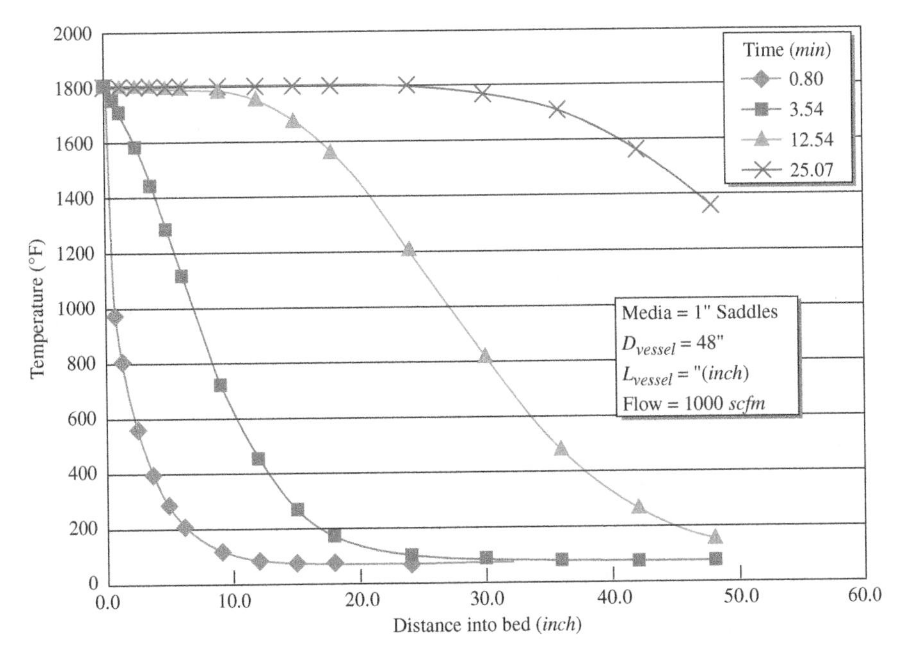

Figure 7.6 Transient packed-bed temperature profiles. The initial bed temperature was 77 °F and the inlet hot gas temperature was 1800 °F. *Source:* Adapted from Kays and London (1984) "Insulated Duct or Porous Matrix Solution" (pp. 90–91).

This condition (where the rate of interior heat conduction is large compared to the surface convection) is known as the small Biot number problem or the lumped capacitance problem, where Biot number is given below and L,c is the sphere's volume to area ratio:

$$Bi = \frac{hL_c}{k} < 0.1$$

7.7 Heat Exchanger Discretization

Kays and London (1984) also provided thorough derivations of several Effectiveness-NTU (number of transfer units) relations in their Appendix C. This appendix developed relations for counterflow, multipass counterflow, and crossflow with one fluid mixed. The details (which were too deep for the main text of Kays and London's book, hence their appearance in the appendix) will not be reproduced here. However, based on the Kays and London's development, six counterflow heat exchanger configurations are possible, and students are encouraged to develop their own discretized models. Briefly, the six configurations (three of which are illustrated in Figure 7.7) are as follows:

1) Hot fluid is the minimum fluid, and it moves forward from left to right. (The extreme subset of this case is where the hot fluid's enthalpy-carrying capacity is negligible compared to the cold fluid's capacity, and the cold fluid's temperature never changes from its inlet value.)

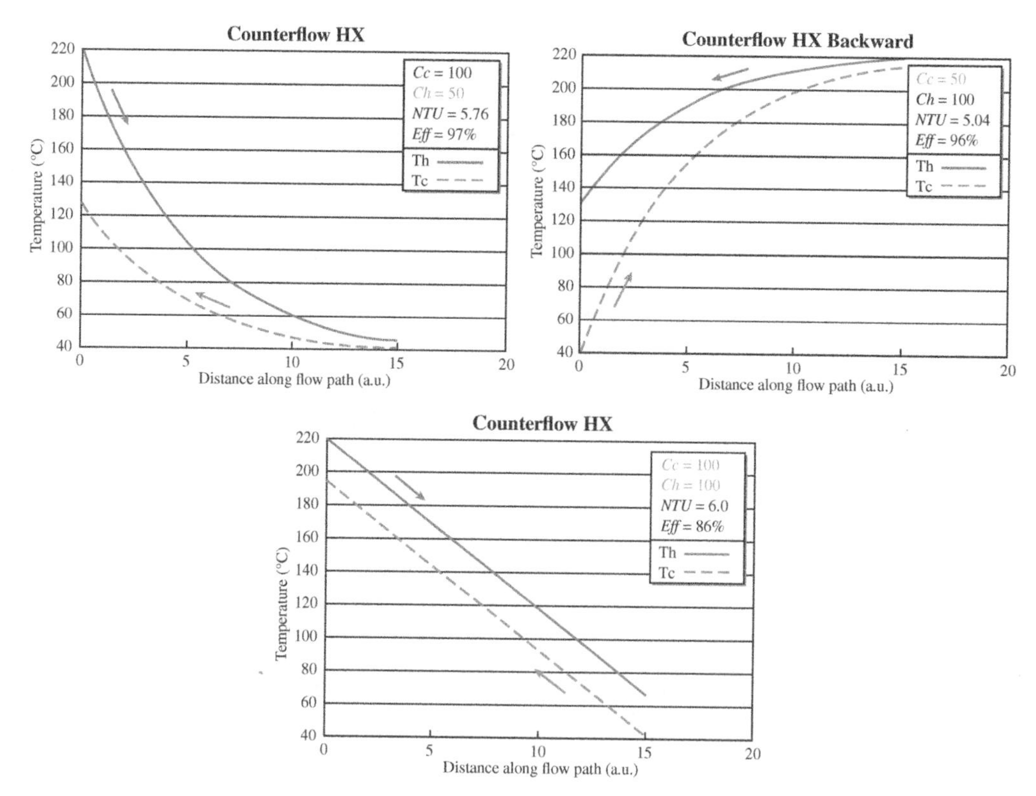

Figure 7.7 Three counterflow heat exchanger configurations. Case 1 has the hot fluid as the minimum (C_h = 50), case 3 has the cold fluid as the minimum (C_c = 50), and case 5 has the two fluids with equal heat-carrying capacities ($C_c = C_h$ = 100). *Source:* Adapted from Kays and London (1984) "Insulated Duct or Porous Matrix Solution" (pp. 90–91).

2) Hot fluid is the minimum fluid, and it moves forward from right to left. No figure is given; it is self-explanatory.
3) Cold fluid is the minimum fluid, and it moves forward from left to right. (The extreme subset of this case is where the cold fluid's enthalpy-carrying capacity is negligible compared to the hot fluid's capacity, and the hot fluid's temperature never changes from its inlet value.)
4) Cold fluid is the minimum fluid, and it moves forward from right to left. No figure is given; it is self-explanatory.
5) Hot and cold fluids have the same enthalpy-carrying capacity, and the hot fluid moves forward from left to right.
6) Hot and cold fluids have the same enthalpy-carrying capacity, and the cold fluid moves forward from left to right. No figure is given; it is self-explanatory.

Think Stop. Students are encouraged to manually sketch profiles 2, 4, and 6 from the configuration list above to develop a feel for the differences among the configurations. Manual or computer sketching of the four configurations where the C_{min}/C_{max} is essentially zero may also be instructive.

Think Stop. For students who have read and understood the Kays and London derivations and want to develop their own discretized, counterflow heat exchanger model, the following pointers may be helpful:

- Set up the model so the computations for both hot fluid and cold fluid march together in a single direction (e.g. forward, left-to-right) even though one of the fluids actually flows in the opposite direction in a counterflow heat exchanger.
- Before setting up the discretization, compute the overall performance of the heat exchanger (UA, C_{min}/C_{max}, NTU, ε) for the given heat exchanger design values.
- Once the effectiveness and the heat-carrying capacities are known, the total heat-transfer rate should be determined.
- Compute the exit temperatures for both streams from the total heat-transfer rate, the heat-carrying capacity values for each fluid, and the inlet temperatures for each fluid.
- Determine the total number of discrete steps desired for the model and create worksheet rows for each location along the path. The author's initial model utilized 16 temperature points for each fluid (including inlet, outlet, and 14 intermediate temperatures), but adaptation for any number of discretized steps is routine once the basic equations are established.
- Use the three-way equality $d\dot{Q} = -C_h dT_h = C_c dT_c$ to compute dT_h and dT_c, and then compute the next temperature in the sequence, regardless of whether the calculations march forward or backward.
- Confirm that $d(T_h - T_c)$ is numerically and logically the same as $dT_h - dT_c$.
- Each of the following columns is likely to be helpful to the overall effort of developing a discretized model:

  - x-position (along the flow axis of the heat exchanger)
  - T_h
  - T_c
  - dT_h (change in hot fluid temperature between adjacent points)
  - dT_c (change in cold fluid temperature between adjacent points)
  - $d\dot{q}$ (increment of heat transferred from hot to cold fluid between adjacent points)

Think Stop. Students should remember that $(T_h - T_c)$ is a temperature ***difference***, whereas dT_h or dT_c represents a temperature ***change***.

7.8 Thermal Destruction of Airborne Pathogens

The 2019–2021 international pandemic and especially the revelation that virus particles can spread and cause infection via the sharing of breathing air motivated many researchers to investigate new ways of removing biohazardous particles from the air. One such team (Yu et al. 2020) tested the components of a hypothetical device that could trap and kill airborne pathogens by heating a porous nickel foam to $T \approx 200\ °C$ and passing the aircraft cabin air through the foam to remove and destroy 99.9% of certain airborne pathogens. We question some of their methods (and the conclusions that rely on them) but are encouraged about the concept. With the exception of the foam temperature, the parameters given in the example derived here are distinctly different from those asserted as useful in Yu's article, and, as a consequence, no logical connection should be drawn between this example and the destruction efficiency or other outcomes cited in Yu's article.

Two of the most important factors for any device that claims to purify the breathing air of a passenger aircraft cabin are safety (ensure that the treatment does not make the breathing air more harmful than the untreated air) and energy efficiency (ensure that the treatment does not consume so much energy that critical power systems for instruments and controls are unreliable). A thermal destruction system for aircraft cabin air runs a high risk of failing both criteria.

If the flow of cabin air must be heated to 200 °C (392 °F) inside the destruction device for it to be effective, then the exhaust must not be released back into the cabin at that temperature because doing so will overheat the cabin atmosphere rapidly and cause great harm to the passengers. One way to avoid this would be to bring in outside air (assume that the exterior air properties are $T \approx -50\ °C$ and $p \approx 0.25\ atm$ for an altitude of $Z = 35\,000\ ft$) to cool the exhaust air coming out of the thermal destruction device or perhaps to replace the dangerously hot exhaust air, which would then be rejected to the exterior. However, this too may be problematic because the exterior pressure is too low for direct breathing and must be pressurized to an equivalent altitude of $Z = 8000\ ft$ ($T \approx 0\ °C$, $p \approx 0.75\ atm$).

Similarly, designers of this process should contemplate the choice of an energy source to maintain the metal destruction filter at its required set point. Electricity is a good option, but if the consumption of electricity is too high, the device would require more generation capacity than is currently available on most aircraft. Fuel combustion most certainly should be ruled out as a ***direct*** source of heat (i.e. combustion products commingled with the breathing air). Unfortunately, using jet fuel combustion as an ***indirect*** heat source (i.e. combustion products separated from breathing air by sealed heat exchanger surfaces) also carries a major drawback (i.e. subtracting a portion of the hot combustion product mass flow from the high-velocity jet exhaust stream reduces the engine's thrust capability).

Readers who recently perused Section 7.5 may be thinking that this is a perfect application for a recuperative heat exchanger, and they are probably correct. But what about the extra weight associated with the recuperator? Aircraft design is a complex enough subject by itself that this book could not pretend to do it justice. However, if we can itemize the relevant constraints and pose the problem in terms that lead to a tractable problem, then the issue becomes one of cost instead of feasibility. Determining the feasibility of this system is posed as an end-of-chapter question, and students who select this system for their design projects should consult with the instructor who will have access to the solution key.

For a recuperative heat exchanger, consider a novel geometry, based on US Patent 6,532,339 (Edgar et al. 2003), where the high-temperature destruction zone is at the center of a spiral heat exchanger (see figure 3 of the Edgar patent). Let the destruction zone be filled with a porous metal foam and assume that a residence time of $\tau = 0.3\ s$ in the core is necessary and sufficient to ensure a thermal destruction of 99.9% of the inflowing pathogens.

7.9 Special Atmosphere Heat Treating

The processing of metals by heat treatment encompasses a wide range of furnace conditions that can be employed to alter the characteristics of the work pieces. Heat treating can soften metals (improves formability), harden them (increases strength), or toughen them (reduces brittleness). It can create a hard or corrosion-resistant skin around a metallic piece whose underlying material has neither of those characteristics. Dossett (2016) and ASM (2015) provide substantial additional details about the field of heat treatment.

The terms *furnace* and *oven* are sometimes used interchangeably, and they are both defined (NFPA 86 2019) as "a heated enclosure used for the processing of materials." Colloquially, the primary difference between a *furnace* and an *oven* is its operating temperature – industrial ovens generally operate at $T < 1000\ °F$ and industrial furnaces generally operate at $T \geq 1000\ °F$.

One of the primary motivators for heat-treating metals in a reducing (i.e. oxygen deficient) atmosphere is to minimize the high-temperature formation of metal oxides on the outermost surfaces of the work piece if oxygen is present. Metal oxides are frequently considered corrosion products because they do not form a stable bond with the substrate metal, and this leads to periodic flaking and substantial loss of material over time. High-temperature oxidation in air can occur at faster rates than water-based corrosion at room temperature, but both processes cause degradation when the oxide flakes away, which exposes a fresh metal substrate to oxygen gas, thereby causing further oxidation.

Heat treating is routinely applied to both ferrous (e.g. steels) and nonferrous (e.g. copper and its alloys) materials. One recipe (Herring 2010) for annealing low-carbon (mild) steel is to combust natural gas in air with a fuel-rich stoichiometry of $\phi \approx 1.46$. Herring states that additional processing (e.g. removal of water by condensation or desiccation) may be necessary for high-carbon steels, but exothermic atmospheric generation using a fuel-rich burner is the simplest recipe for protecting low-carbon steels from surface oxidation during the annealing process (Morris 2013).

One benefit of using a reducing atmosphere instead of an inert gas to protect against surface oxidation is that inert gases are not able to consume stray oxygen molecules by reacting with them. Even if delivered with a sufficiently low (<1 *ppmv*) level of oxygen, a flow of nitrogen, helium, or argon cannot be guaranteed to remain oxygen-free during the entire heat-treating process, which may last several hours. It is always possible for ambient air to leak into the furnace and contaminate the atmosphere, rendering the metal's surface unacceptably oxidized, regardless of how pure the inert gas was to begin with.

See the end-of-chapter questions for a heat treatment furnace design exercise.

7.10 Burner and Heat Exchanger Failures

The following investigation anecdotes provide brief summaries of selected failure investigations that we have performed. The identities of designers, manufacturers, installers, and users have been removed, and the data are generalized so that any similarity between a description below and an actual device system should be purely coincidental. Details of injuries and property damage (which were substantial for some of the cases) are also excluded so that the focus is on design and operational failures, rather than the human aspects of the losses.

Multiburner Heater. A fluid heater equipped with multiple gas-fired burners exploded when a worker failed to shut the hand valve on the gas line to a burner that would not light and then attempted to light a nearby burner while raw gas was still flowing uncombusted out of the adjacent burner. Although the worker inadvertently failed to adhere to his employer's established operating procedures regarding manual valves and ignition sequences, the equipment did not contain common safeguards that likely would have prevented the incident. The subject heater was not covered by any consensus safety standard at the time of the incident, but shortly thereafter, the author became a founding member of the NFPA 87 technical committee and helped to write the first consensus safety document published to cover fluid heaters. The missing safeguards (that NFPA 87 has required since it became a standard) were automatic safety shutoff valves in each of the gas lines leading to the burners. The subject heater had only one safety shutoff valve on the header upstream of the dividing point where the header was split into smaller pipes.

Grain Product Dryer. A belt-conveyor dryer, used for removing moisture from a food product, sustained a major internal fire when substantial quantities of the product accumulated on the floor of the dryer and spontaneously ignited after a 12-hour latency period. Although workers smelled smoke for approximately 15 minutes before seeing visual evidence of fire, they failed to follow established workplace procedures for (i) shutting off burners and the feed conveyor, (ii) opening side doors carefully to find the incipient fire, and (iii) extinguishing it with water spray from a sanitation hose. After the fire, the oven user elected to install inert gas fire suppression systems on their drying ovens because the installation of automatic water sprinklers inside a food-processing oven was considered problematic.

Furnace Exhaust Lines. Two unrelated, poorly maintained, residential heating furnaces began releasing substantial quantities of carbon monoxide into their local environments after accumulation of debris in their exhaust vent lines restricted the exit flow of combustion products from the firebox, thereby limiting the inflow of fresh air to the point where insufficient oxygen was present to burn all the fuel delivered to the burners. For natural draft combustion systems, a functioning chimney is vital to the overall design because it not only permits the combustion product gases to be vented outdoors but also motivates the inflow of fresh air to supply the burners with needed oxygen.

Tunnel Kiln. A multiple-burner tunnel kiln used for firing ceramic work pieces exploded more than a year after workers installed bypass wiring around a flame safeguard that was experiencing what they believed were nuisance shutdowns. The workers took this action despite admonitions in the applicable safety standard (NFPA 86 2019) that safeguards (i) shall not be bypassed and (ii) shall be inspected at least annually to verify that all required safeguards have not been bypassed. Without a functional flame safeguard interlock, one of the burners lost flame, but nothing triggered the safety shutoff valves to close, permitting raw gas to accumulate and migrate to a zone where an operating burner ignited it.

Energy-Efficient Heater. Substantial quantities of carbon monoxide were released when a defectively installed, energy-efficient, water heater exhaust vent line failed. The heater was equipped with advanced heat exchange features that cooled the exhaust gases to below 150 °*F* to recover energy that would otherwise be wasted out the exhaust vent line. The exhaust's low temperature permitted the installer to use a poly vinyl chloride (PVC) exhaust pipe instead of the more common galvanized sheet metal duct. Because the installer failed to cement the plastic pipe joints together, one of the joints inside the heater closet separated, permitting combustion products to fill the space. Eventually, the exhaust gases displaced the air in the room and the burner was starved of oxygen, causing it to generate copious amounts of carbon monoxide, which migrated to nearby spaces designed for human occupancy.

Cement Kiln. An advanced thermal-imaging system, capable of detecting "hot spots" on the exterior steel wall of a 200 *ft*, brick-lined, rotating, cylindrical kiln, failed to alert operators to an internal, high-temperature aberration, which led to rapid failure of the high-density refractory lining and melting of parts of the 4-inch-thick steel shell. The imaging system was equipped with two cameras, one located near the feed end of the kiln (cool zone) and another located near the discharge end (hot zone), along with software that interpreted the imaging system signals to produce a temperature map of the kiln. During a routine calibration of the scanning equipment, a technician accidentally reversed a software toggle for one of the cameras, and the data it was transmitting for the feed end was interpreted to be data for the discharge end. The operators were never aware of any temperature anomalies until the kiln's white-hot product began to dribble out of a breach in the shell just upstream of the discharge point. No safety standards were ignored, and no products were defective, but the technician error led to a very big cleanup and repair exercise.

Batch Drying Oven with Recuperative Thermal Oxidizer. A solvent oven, operated in batch mode, exploded after a burner failure resulted in a safe shutdown, but then was followed by a burner restart without an adequate preignition purge of the combustion chamber. The purge was inadequate for three reasons: (i) for energy- and cost-savings reasons, the system designer failed to adhere to NFPA requirements for safety ventilation and instead equipped the oven with an unsafe hot gas recirculation arrangement that prevented combustion products from being exhausted and fresh air from being admitted at the necessary rates; (ii) the purge timer had been modified by the user to minimize the time delay between each shutdown and start-up; and (iii) exhaust gas from a second batch oven had been recently removed from the thermal oxidizer's infeed, which reduced the flow rate of ventilation air, and thereby necessitated an even longer purge time, which was never undertaken.

7.11 Homework Problems

7.1 Compute the residence time for the gas in a well-insulated thermal oxidizer vessel with inside dimensions $D = 1.4\ m$, $L = 5.5\ m$; a flow rate of $\dot{m} = 1.75\ kg/s$; and a uniform internal temperature of $T = 1050\ K$. Assume the gas has a molar mass $\hat{M} = 29\ kg/kmol$. Show your work.

Answer. $\tau = 1.628\ s$

7.2 Estimate the residence time above a destruction threshold of $T = 1000\ K$ for the gas in a poorly insulated thermal oxidizer vessel with the same dimensions and the same flow rate as in Problem 7.1, but with a temperature profile that declines axially according to $T(x) = (a + bx)$, where $x \doteq m$ and the coefficients are $a = 1050\ K$ and $b = -15\ K/m$. Show your work.

Answer. $\tau = 1.11\ s$

7.3 Using the Cooper and Alley method (Equation 7.3), estimate the minimum temperature required to destroy 99.9% of the inflowing ethylbenzene ($C_6H_5C_2H_5$) in a thermal oxidizer that is designed for a residence time of $\tau = 0.5\ s$. Show your work.

Answer. $T = 1361\ °F$

7.4 Fill in the blanks (empty boxes) in the heat and material balance table given next. The process represented is a counterflow, gas-to-gas recuperator used to recover heat from high-temperature exhaust gas and preheat combustion air going to a burner (see Figure 7.2). The flow path of ambient air being preheated is from state 1 to state 2 and the hot exhaust gas flow path is from state 3 to state 4, with the two flows directed countercurrent to each other through the recuperator. Assume that the recuperator is adiabatic with respect to the environment and pressure drop through both sides of the heat exchanger is insignificant. Also, assume that the properties of the exhaust gas can be represented approximately by the same properties as the inlet air (which is approximated by the composition $\chi_{O_2} = \frac{1}{4.76} \approx 0.210\,08$, $\chi_{N_2} = \frac{3.76}{4.76} \approx 0.789\,92$). Use VBA functions for sensible enthalpy in Appendix B.3 or where necessary, use Appendix A or Linstrom and Mallard (2017–2021) for thermodynamic properties. In addition to completing the blank entries in the table, state which fluid is the minimum fluid and estimate the heat exchanger effectiveness.

Answer. $\varepsilon = 66.15\%$

Property	State 1	State 2	State 3	State 4
Temperature (°C)	25.0	240.0	375.0	
Pressure (kPa)	101.33	101.33	101.33	101.33
Spec volume (m^3/kg)				
Enthalpy (kJ/kg)				
Air (m^3/s)				
Air (kg/s)	1.75		1.75	

7.5 Consider an aircraft that is equipped with one electrically heated, biohazard destruction device for each passenger. Assume that the average human respiration rate is $\dot{V} = 8.0\ L/min$ and the thermal destruction device must process 100% of the air exhaled by the passenger it is assigned to. Also, assume that each passenger occupies a box-shaped space 40 cm × 40 cm × 170 cm and the aircraft's fresh air distribution system provides a one-box volume of fresh air each five minutes. While in flight, assume the inflowing cabin air temperature is $T_{freshair} = -25\ °C$ (after compression to achieve $Z_{equiv} = 8000\ ft$), and the air temperature in the personal box is maintained at $T_{box} = 25\ °C$. Consider using the spiral-shaped recuperative heat exchanger design shown in figure 3 of Edgar et al. (2003) and employ electric resistance heating to maintain the metallic foam at the spiral's center at the desired destruction temperature of $T_{max} = 200\ °C$. Assume the porosity (i.e. void fraction) of the metal foam is $\sigma = 85\%$ and the hydraulic diameter is large enough that the pressure drop across the core is negligible (check the validity of your assumption) but do ensure the core is large enough to obtain a residence time of $\tau = 0.3\ s$ at T_{max}, which will ensure thermal destruction of 99.9% of inflowing pathogens. Let the cold inlet to the spiral heat exchanger be sized to admit ambient cabin air at a velocity of $V = 500\ ft/min$. For the spiral recuperator, initially assume an effectiveness of $\varepsilon \approx 70\%$ is achieved by eight spirally circulating paths in a counterflow arrangement (four alternating paths for inflow and four alternating paths for outflow). Upon computing the actual effectiveness, modify the spiral's configuration if necessary. Assume the spiral device can be made from SS304 stainless steel sheet of thickness 0.9525 mm (i.e. 20-gauge sheet). Also, assume that the two end caps (SS304 sheet) are joined to the opposing edges of the spiral passageways using a proprietary sealing method to prevent cross contamination of dirty into cleaned air. Also, assume an appropriately sized electric blower is positioned at the outlet of the destruction device to draw in the cleaned, cooled air that exits the heated zone of the device.

(a) Determine the actual heat recovery effectiveness of the spiral heat exchanger. (b) Determine if the design is feasible by evaluating whether these objectives are accomplished: (i) total mass (metal foam + spiral walls and end caps) must not exceed $M_{tot} = 20\ kg$ and (ii) electric power consumed must not exceed $\dot{Q}_{elec} \approx 0.10\ kW$.

7.6 Consider a heat-treating furnace for low-carbon (mild) steel parts. An annealing process is proposed for a batch of $n = 10\,000$ hex bolts (1"–8 × 6") with mass per bolt $m_{bolt} = 1.69\ lb_m$ (= 0.7666 kg). Assume the following properties (Cengel and Ghajar 2015) for mild steel:

$$\rho_{bolts} = 7832\ \frac{kg}{m^3}\quad c_{bolts} = 0.434\ \frac{kJ}{kg \cdot K}\quad k_{bolts} = 63.9\ \frac{W}{m \cdot K}$$

The treatment cycle should begin with a rapid temperature ramp upward to reach the austenitizing temperature ($T_{max} = 1150\ K$) and end with a slow downward ramp. For this example, assume (i) all the pieces experience a linear temperature ramp-up over the period $\Delta t_{ramp-up} = 6\ h$ and (ii) the pieces cool through furnace heat loss to the surroundings while the burner is firing at its lowest possible turndown just to ensure that tiny amounts of oxygen that enter the furnace are consumed by the special atmosphere before they can attack and oxidize the bolt surfaces.

Assume the furnace's exterior footprint is $A_{furn} = 2.5\ m \times 2.5\ m$ with its height $H_{furn} = 3.0\ m$. Also, assume the furnace interior is insulated all around with refractory ceramic fiber modules of thickness $t_{RCF,\ modules} = 0.25\ m$ and thermal conductivity (at operating temperature) $k_{RCF} = 0.100\ W/(m \cdot K)$, and a special paint on the furnace walls has a reduced emissivity $\varepsilon = 0.70$. The manufacturing area is indoors with an average air and surrounding temperature of $T_\infty = 300\ K$.

Using the equivalence ratio cited in Herring (2010), determine (i) the adiabatic combustion temperature for the designated equivalence ratio, (ii) the firing rate (kW_{th}) necessary to maintain the designated T_{max} during the ramp-up, and (iii) the maximum chemical enthalpy flow rate of special atmosphere gases ($CO + H_2$) being exhausted from the furnace (that could be a source of exothermicity for some other process).

7.7 Fill in the blanks (empty boxes) in the heat and material balance table given next. The process represented is natural gas burner creating high-temperature combustion products. The air inflow path is state 1; the fuel inflow path is state 2; and the combustion product outflow path is state 3. Assume the burner is firing into a chamber that is adiabatic with respect to the environment and the water of combustion is gaseous (...no liquid water in products). Assume that natural gas is 100 % CH_4 (methane) and air has a composition of $\chi_{O_2} = \frac{1}{4.76} \approx 0.210\,08$, $\chi_{N_2} = \frac{3.76}{4.76} \approx 0.789\,92$). Also, assume complete combustion of the fuel and insignificant pressure drop through the burner (probably not a terrific assumption). Use VBA functions for sensible enthalpy in Appendix B.3 or where necessary, use Appendix A or Linstrom and Mallard (2017–2021) for thermodynamic properties. Note: Where the prepopulated entries in the table given next contain a hyphen (dash), the actual values for chemical and sensible enthalpies should not be assumed zero. The hyphens simply indicate that these values were not provided to the student because the flow rate of the compound is zero at that state, so finding the enthalpies is unnecessary. In addition to completing the blank entries in the table, compute the equivalence ratio of the reactant mixture.

Answer. $\phi = 0.31$

Property	State 1	State 2	State 3
Temperature (K)	300	300	
Pressure (kPa)	101.33	101.33	101.33
Mixture molar mass ($kg/kmol$)	28.8500	16.0410	28.4462
Mixture enthalpy flow (chem + sens) (kW)			
Total mass flow (kg/s)	27.466	0.4973	
Methane CH_4 $\dot{m}$ (kg/s) h_{chem} (kJ/kg) $h - h_{298}$ (kJ/kg)	0.000	0.4973 4667.62 4.19	0.000
Nitrogen N_2 $\dot{m}$ (kg/s) h_{chem} (kJ/kg) $h - h_{298}$ (kJ/kg)	21.066 0.000 1.924	0.000	
Oxygen O_2 $\dot{m}$ (kg/s) h_{chem} (kJ/kg) $h - h_{298}$ (kJ/kg)	6.400 0.000 1.691	0.000	
Water vapor H_2O $\dot{m}$ (kg/s) h_{chem} (kJ/kg) $h - h_{298}$ (kJ/kg)	0.000	0.000	
Carbon dioxide CO_2 $\dot{m}$ (kg/s) h_{chem} (kJ/kg) $h - h_{298}$ (kJ/kg)	0.000	0.000	

Cited References

ASM; (2015); "*Subject Guide: Heat Treating*"; American Society of Materials, International – The Materials Information Society; https://www.asminternational.org/documents/10192/23555666/ASM+Subject+Guide_HeatTreating.pdf.

Cengel, Y.A. and Ghajar, A.J.; (2015); *Heat and Mass Transfer: Fundamentals Applications*, 5; New York: McGraw-Hill.

Cooper, C.D. and Alley, F.C.; (1994); *Air Pollution Control: A Design Approach* 2; Prospect Heights, IL: Waveland Press, Inc. See Chapter 11: VOC incinerators.

Cooper, C.D., Alley, F.C., Overcamp, T.J.; (1982) Hydrocarbon vapor incineration kinetics. *Environ. Prog.*; **1**; 129.

Cottrell, F. (1938). Purifying gases and apparatus therefor. *US Patent* **2**,121,733.

Dossett, J.L.; (2016); *Practical Heat Treating: Basic Principles*; Materials Park, OH: ASM International.

Edgar, B.L., Martin, R.J., and Barkdoll, M.P. (2003). Device for thermally processing a gas stream, and method for same. US Patent 6,532,339, issued 11 March 2003.

Herring, D.H.; (2010); "The annealing process revealed (Part two: Furnace atmosphere considerations)"; *Indus. Htg Mag.*; **78** (18 September 2010).

Kays, W.M. and London, A.L.; (1984); *Compact Heat Exchangers* 3; New York: McGraw-Hill.

Lee, K.C., Hansen, J.L., and Macauley, D.D. (1979). *Predictive Model of the Time Temperature Requirements for Thermal Destruction of Dilute Organic Vapors; 72nd Annual Meeting of Air Pollution Control Association*, Cincinnati, OH (24–29 June 1979).

Lee, K.C., Morgan, N., Hansen, J.L., and Whipple, G.M. (1982). *Revised Model for the Prediction of the Time Temperature Requirements for Thermal Destruction of Dilute Organic Vapors and Its Usage for Predicting Compound Destructability; 75th Annual Meeting of Air Pollution Control Association*, New Orleans, LA (20–25 June 1982).

Linstrom, P.J. and Mallard, W.G., Eds.; (2017–2021); *NIST Chemistry WebBook, SRD 69. Thermophysical Properties of Fluid Systems*. National Institute of Standards and Technology, U.S. Dept. of Commerce.; http://webbook.nist.gov/chemistry/fluid.

Martin, R.J. (2019). Advanced quality methods for thermal oxidizer operation. Paper #71EA-0175; presented at 11th U. S. National Combustion Meeting, Pasadena, CA (24–27 March 2019).

Martin, R.J. and Colwell, J. (2003). De novo VOC from regenerative thermal oxidizers. *Proceedings ASME-HT2003, Summer Heat Transfer Conference*, Las Vegas, NV (22 July 2003).

Morris, A. (2013). Exothermic atmospheres. *Industrial Heating Magazine, BNA Media* (10 June). https://www.industrialheating.com/articles/91142-exothermic-atmospheres.

NFPA 86; (2019); "*Standard for Ovens and Furnaces*"; Quincy, MA: National Fire Protection Association. See chapters 6 "Furnace Heating Systems" and 8 "Safety Equipment and Application.

Schlichting, H.; (1960); *Boundary Layer Theory*, 4 New York: McGraw-Hill Book Company (print by Springer).

Stilger, J., Martin, R., and Holst, M. (1994). Method and apparatus for recuperative heating of reactants in a reaction matrix. *US Patent* **5**,320,518.

Uhl, V.M. and Gray, J.B.; (1966); *Mixing Theory and Practice*, **1**; New York: Academic Press; https://doi.org/10.1002/aic.690130405.

Yu, L., Peel, G.K., Cheema, F.H., Lawrence, W.S., Bukreyeva, N., Jinks, C.W., Peel, J.E., Peterson, J.W., Paessler, S., Hourani, M., and Ren, Z.; (2020); "Catching and killing of airborne SARS-CoV-2 to control spread of COVID-19 by a heated air disinfection system", *Mater. Today Phys.*; **15**; 100249.

8

Boilers and Power Cycles

In earlier days, boilers were the most common centerpiece of home heating systems. Residential boilers would be fired with lump coal (delivered by truck down a chute into the basement) and low-pressure steam would travel through steel pipes to "radiators" that would condense the steam (while heating each room) and circulate the condensate back to the feedwater tank. Alas, the added complexity of heating a working fluid other than air eventually forced boilers into industrial and commercial applications only, and home heating is now predominantly accomplished by gas-fired, forced air units (FAU) and electricity-driven heat pumps. Nevertheless, the industrial market for boilers is still strong, and a large majority of the electricity generated in the United States comes from steam turbines. Table 8.1 gives a breakdown of the electricity-generating capacity among a variety of energy sources in 2016.

The intent of this chapter is to provide insight to students about many of the unique challenges and solutions associated with boiler design and operation, and to solidify their understanding about important components in steam cycles that are associated with power generation and heating.

Among the classic treatises on the topic of boilers are Babcock and Wilcox (2015) and Combustion Engineering (1981). El-Wakil (2002) also provides a good crossover treatment (analytical + practical) of boilers and power generation. For engineering theory of boiling and condensation, we recommend Collier (1981).

8.1 Rankine Cycle

The Rankine cycle is the gold standard for steam power generation. In its simplest form, it consists of four states of water (condensate liquid, high-pressure feedwater, superheated steam, and turbine exhaust steam) and comprises four pieces of equipment (pump, boiler/superheater, turbine, and condenser). A schematic of the basic Rankine cycle is shown in Figure 8.1. The sketch shows the boiler on the top of the layout (because steam rises) and the condenser at the bottom (because condensate falls), with the pump receiving shaft work from a motor (power enters on left) and the turbine delivering shaft work to a generator (power exits on right).

In the ideal form of the basic Rankine cycle, the processes are carried out reversibly, which means the pump and turbine processes are isentropic, and the boiler and condenser processes are isobaric (i.e. frictionless). See Figure 8.2 for a p - H diagram of the ideal Rankine cycle.

Thermal Systems Design: Fundamentals and Projects, Second Edition. Richard J. Martin.

Companion website: www.wiley.com\go\Martin\ThermalSystemsDesign2

Table 8.1 Breakdown of utility-scale and electricity-generating capacity in the United States in 2016.

Energy source	Percent of utility-scale Electricity generation, 2016 (%)	Percent of utility-scale Electricity generation, 2019 (%)
Natural gas	34	38
Coal	30	23
Nuclear	20	20
Renewables, all	15	17
(Nonhydro renewables)	(8.4)	(10.9)
[Hydroelectric]	[6.5]	[6.6]
Petroleum and others	1	*n/a*

Source: Data from EIA (2017–2021).

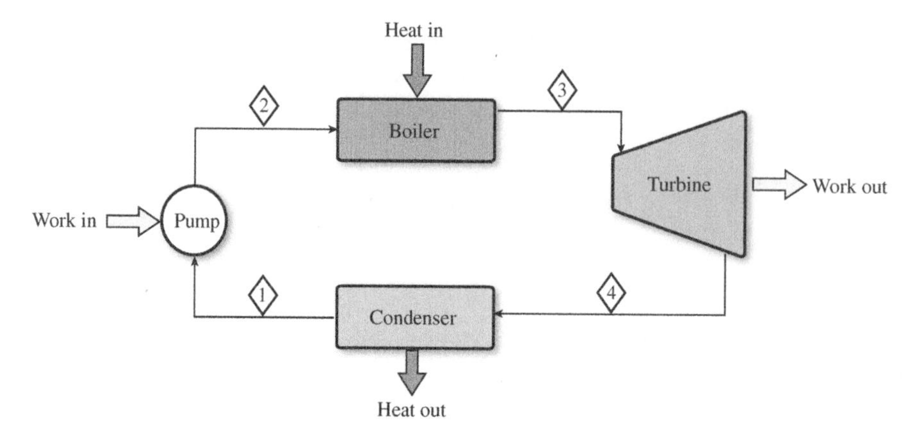

Figure 8.1 The Rankine cycle in its simplest form. *Source:* Martin Thermal Engineering Inc.

In real equipment, none of the shaft work flows are reversible, none of the pipe flows are frictionless, and none of the heating or cooling is performed reversibly. To combat some of these losses, power plant systems attempt to reduce losses by utilizing multiple stages of turbine expansion and recuperation of waste heat into feedwater and combustion air.

For analyzing the thermodynamics of a boiler cycle (with or without heat recovery), readers may utilize the property tables for water in Appendix A.4, which were directly derived from the *NIST Chemistry Webbook* (Linstrom and Mallard 2017–2021) or the Excel-VBA customized functions for water in Appendix B.4, or both by using the tables from Appendix A.4 to QC the results from the VBA functions from Appendix B.4.

Readers are cautioned that the VBA scripts do not compute property values for subcooled liquid water. While the practice of employing saturated liquid properties as an approximation for subcooled liquid properties at the same temperature is reasonable at moderately low pressures, the lack of exact property computations for subcooled liquid water at higher boiler pressures may introduce significant and unacceptable deviations.

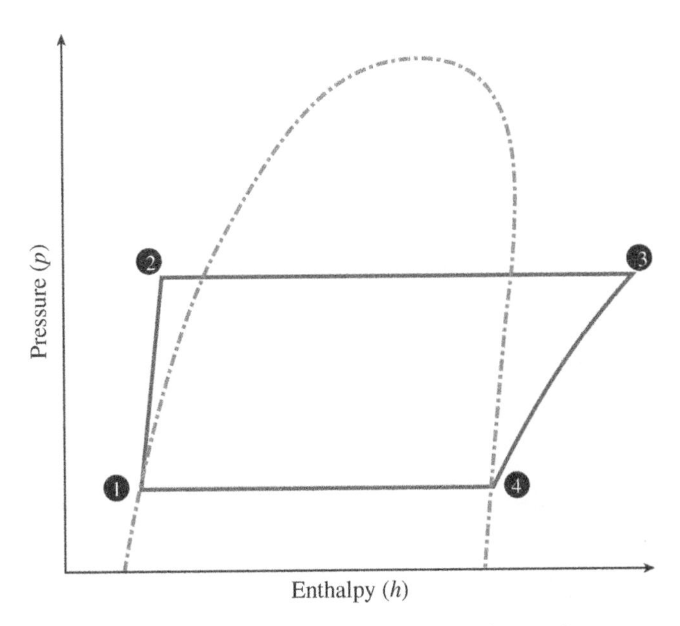

Figure 8.2 Basic Rankine cycle, with reversible $\dot{Q}s$ and $\dot{W}s$. *Source:* Martin Thermal Engineering Inc.

8.2 Boiler Terminology

The following terms of boiler art may be somewhat familiar to readers. The technology has a long and colorful history, which has spawned a number of colloquialisms over more than two centuries since Trevithick's first steam locomotive was built in 1801 (ASME 2021).

Firetube Boiler. The firetube boiler (invented in Great Britain by Richard Trevithick, see Figure 8.3) is typically a large, horizontal, cylindrical drum, with a firebox at one end connected to an exhaust stack at the other end by a large number of small diameter tubes that run the length of the drum. Water is admitted to the large drum, and steam rises to the top of the drum, where it is directed to the point of use. The fuel (originally lump coal or wood logs, but almost exclusively natural gas today) is burned in the firebox with air, and the hot combustion products flow through the small tubes (hence, the name "firetube" because the fire is inside the tubes), where they transfer heat to the water, thereby converting it to steam. The drum is pressurized by a feedwater pump, and the steam is raised at a pressure high enough to do useful work. The earliest firetube boilers (especially locomotives) suffered many failures (typically steam explosions) because the riveted wall joints of the drum would fail catastrophically.

Think Stop. A steam explosion is a form of BLEVE – boiling liquid, expanding vapor explosion. The term BLEVE (pronounced in two syllables, "bleh-vee") is a nonchemical type of explosion and is distinct from chemical explosions (e.g. deflagrations and detonations). A reservoir of water that is pressurized and heated above its ambient boiling point (100 °*C*) contains a lot of stored energy that can perform (destructive) work on the environment if the containment ruptures.

A quick comparison of riveted and welded joints illuminates why BLEVE failures are greatly reduced today: welded joints are typically as strong as the parent material, holes for riveted joints are stress concentration points, rivets do a poor job of sealing curved surfaces, rivets too close to an edge may not have enough material thickness to hold under pressure, etc.

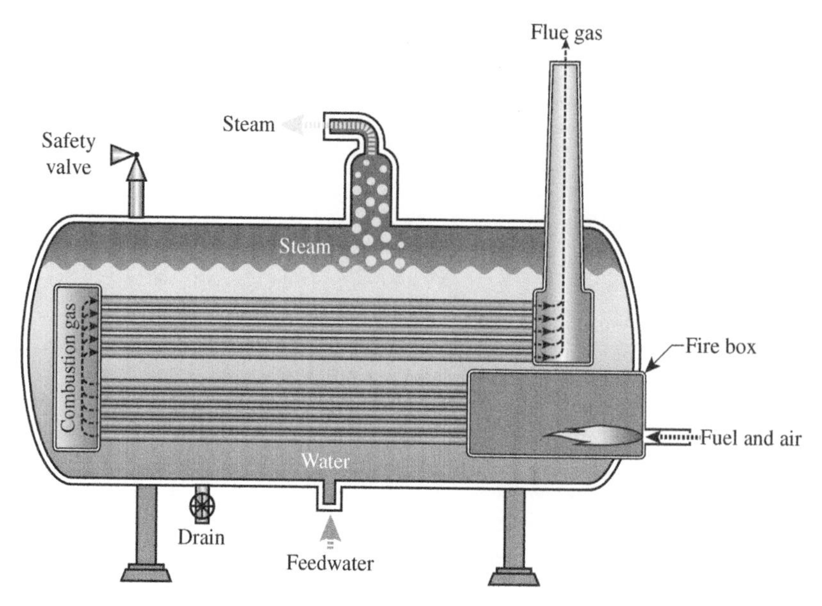

Figure 8.3 Schematic of firetube boiler. *Source:* Martin Thermal Engineering Inc.

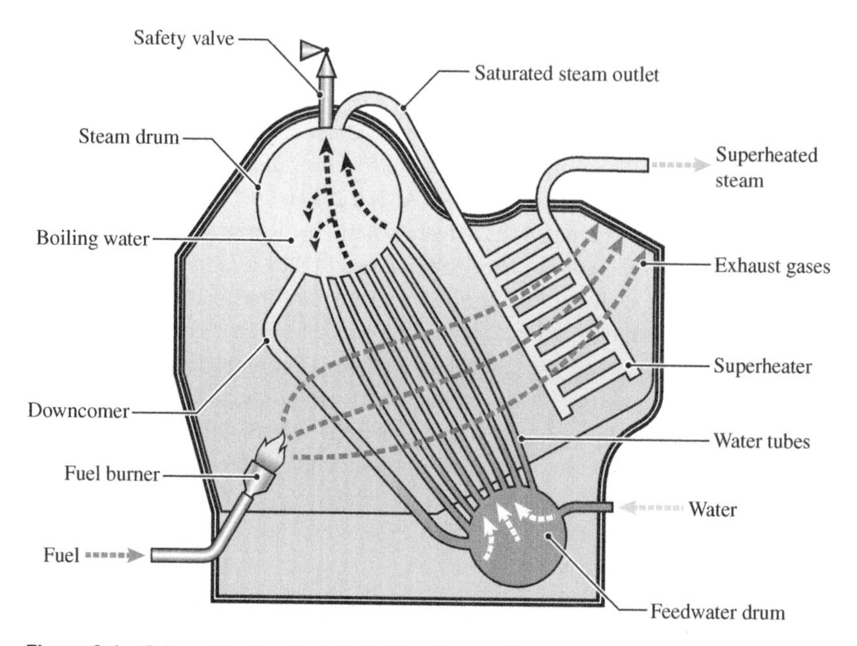

Figure 8.4 Schematic of watertube boiler. *Source:* Martin Thermal Engineering Inc.

Watertube Boiler. A watertube boiler localizes the water (and steam) inside hundreds of steel tubes, while keeping the fire (or combustion products) on the outside (see Figure 8.4). The watertube boiler was invented by Babcock and Wilcox in the late 1800s, after 90 years of market dominance by the firetube style.

By locating the watertubes inside the firebox where it boiled as it was being conveyed upward from the feed [liquid] drum to the steam [vapor] drum, the B&W boiler was inherently safer than the Trevithick design. This was because even if a boiler tube (or three or seven) failed in rapid sequence, the failure(s) would not trigger a massive BLEVE, involving thousands of pounds of rapidly vaporizing water. In fact, Thomas Edison was so impressed with the B&W boiler design; he is credited with the following aphorism (see CA-DIR 2021):

> The Babcock & Wilcox Boiler is "...the best boiler God has permitted man yet to make."
> Thomas Edison (1888)

Stoker Boiler. A stoker is a type of boiler that uses a moving grate to transport solid fuel into the bottom of the boiler while allowing ash to fall through the grates, and into a hopper at the bottom of the boiler. Today, stokers are often used to burn wood waste (e.g. sawdust, scraps).

Horsepower. In the early days of locomotives, the shaft power generated by a steam engine was compared to the power of horses, but the comparison was subject to considerable subjectivity, including how efficiently the steam engine converted mechanical energy (steam pressure) to shaft energy.

Only remotely similar to motors (and horses), the term "boiler horsepower" (or *BHP*) is a measure of the amount of steam that can be generated by a boiler. The exact definition is given in Equation (8.1):

$$\boxed{1.0\ BHP = 33\,475\ \frac{BTU}{h}} \approx \begin{array}{l} Heat\ to\ boil\ 34.5\ \dfrac{lb_{H_2O}}{h}\ from\ sat \\ liq\ at\ 212\,^{\circ}F\ to\ sat\ vap\ at\ 212\,^{\circ}F \end{array} \tag{8.1}$$

The units hp and BTU/h are consistent in the sense that each is a measure of energy per unit time. However, the numerical conversion value is $2544.43\ BTU/h = 1.00\ hp$, which is *not* consistent with the conversion factor from BHP to BTU/h, given above. Furthermore, neither of these units is to be confused with brake-horsepower (bhp), which got its name from the inventor of the Prony brake, a device that was used to measure the power output of an engine (steam, Otto, or Diesel) mounted on a stand (i.e. before experiencing friction losses in the gearbox, differential, and wheels)

These inconsistencies are an unfortunate artifact of the naming history of the term ***horsepower*** because the terms BHP, bhp, and hp are physically unrelated to each other and they were developed and adopted independently.

Steam Demand. Steam demand is the amount (lb/h or kg/s) of steam being drawn out of the boiler (and superheater, if present) by the steam-using devices (e.g. power-generation turbine, sterilization autoclave, or residential heat radiator). When steam demand increases, the boiler must respond by providing more flow (lb/h) of feedwater going into the boiler, and more heat release (BTU/h) from the burner to facilitate the vaporization of the greater inflow of liquid water.

Pinch. Boiler ***pinch*** is the smallest temperature difference between the fireside and the waterside in a counterflow boiler. The pinch is best thought of by considering profiles of gas temperature and water/steam temperature versus distance in a boiler (see Figure 8.5). The pinch may or may not be located at the point shown (i.e. where the water leaves the economizer and enters the boiler) and in fact, pinch could be at the point where the hot gas exits the boiler system entirely. Pinch is a key design consideration for boilers. It is physically impossible for pinch to be zero (because the Second law implies that infinite boiler surface area would be required to do so) but excessively high values of pinch might indicate poor overall energy efficiency for the system.

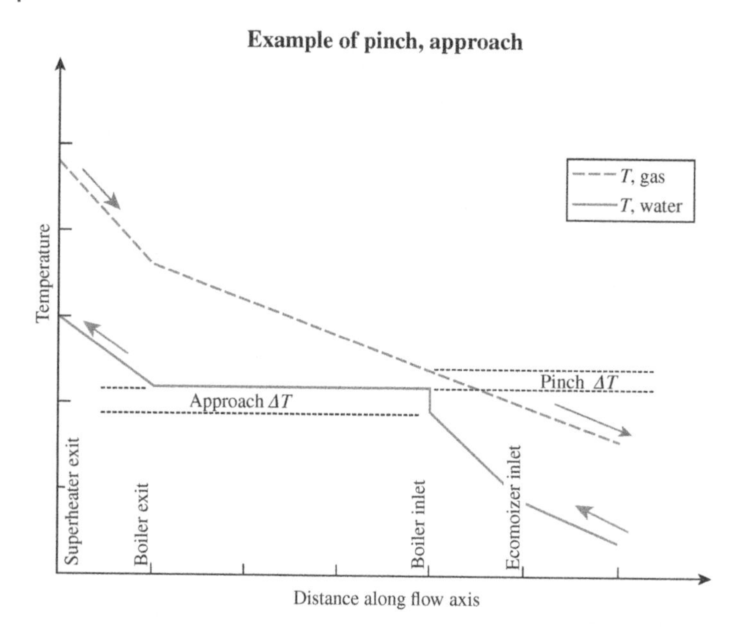

Figure 8.5 Examples of pinch temperature difference and approach temperature change. *Source:* Martin Thermal Engineering Inc.

Think Stop. A typical value for ***pinch*** in large boilers is $\Delta T_{pinch} \approx 20\ °F$, but for student design problems, relaxing this value to a higher number (e.g. 25 °C) is acceptable.

Approach. Boiler ***approach*** is the temperature change (jump) on the waterside between the economizer exit and boiler inlet (see Figure 8.6). ***Approach*** is simply a recognition that economizers are not designed to permit boiling, so the approach provides a safety margin to help prevent it. The purpose of an ***economizer*** is to extract leftover heat from the combustion gases by preheating the boiler's feedwater to a temperature as close as reasonably possible to the water's boiling temperature.

Think Stop. In practice, values for approach temperature are similar to those for pinch; however, for student design problems, we recommend setting the approach to $\Delta T_{approach} \approx 0\ °F$ to simplify the exercise. Unlike pinch, this learning simplification affects heat exchanger surface area in a marginal way only.

8.3 Efficiency Improvement

The ability to capture energy that would otherwise be lost can result in significant efficiency improvements for Rankine cycle systems. Innovative methods to improve energy efficiency are nearly always installed first on large utility boilers because the payback is easier to demonstrate. Figure 8.6 shows an artist's rendering of a full-scale, utility power plant based on the Rankine cycle that contains many of the efficiency improvements described as follows.

The improvements fall into the following basic categories:

- Capture and use waste heat before it leaves the system.
- Reduce mechanical friction by effective lubrication.

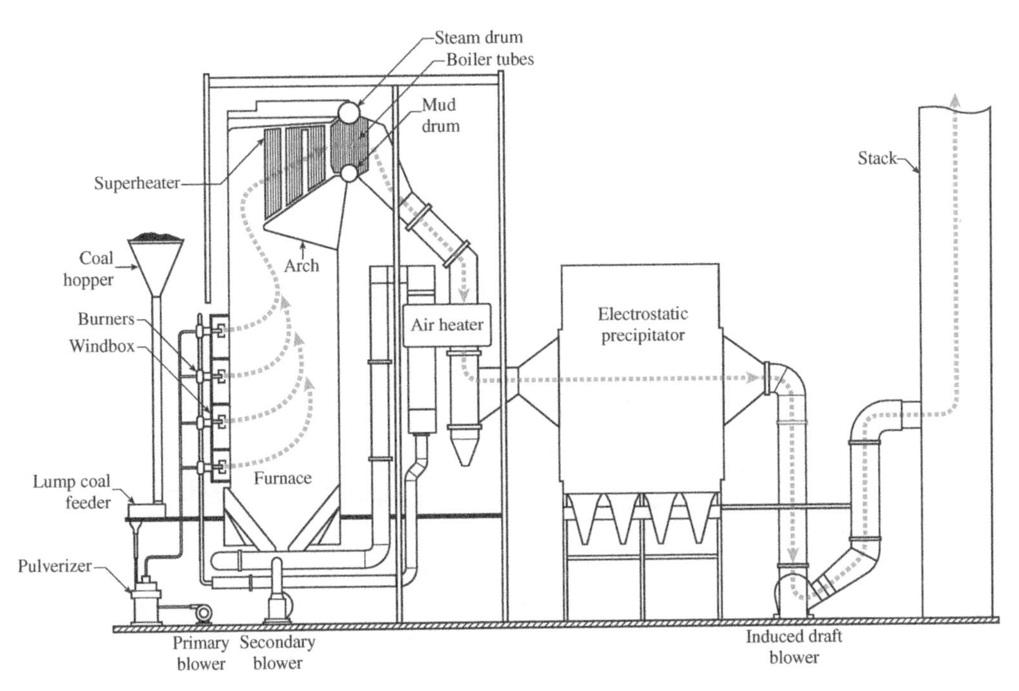

Figure 8.6 Artist's rendering of a coal-fired utility boiler. *Source:* Babcock and Wilcox (1978). Used with permission.

- Reduce fluid friction by sizing and maintaining pipes properly.
- Reduce fluid friction by replacing air dampers with variable speed blower drives.
- Reduce entropy production by making compression and expansion more reversible.
- Improve combustion efficiency to reduce fuel waste.
- Improve combustion processes to reduce operating costs for pollution abatement.
- Improve electrical generating processes to reduce losses.

Steam Superheating. Large boilers are often equipped with a "convective" pass of tubes that superheats the saturated steam after it exits the steam drum. Superheaters in utility boilers sometimes add 300–400 $°F$ of superheat past the water's saturation temperature. Although superheat may be undesirable for some applications, superheating improves cycle efficiency for many others (e.g. Rankine cycle). When shaft work from a turbine is not desired (e.g. when a residential boiler serves radiators and air handlers), superheating reduces installed cost (boiler is smaller) and reduces energy efficiency (fuel is wasted). When the steam is wet, it can transfer heat at the same rate ($\dot{Q} = UA \cdot \Delta T_{lm}$) in a smaller package (smaller A) because condensation imparts a higher U than dry steam convection.

In contrast, steam turbines can more efficiently convert fluid mechanical energy to shaft work when the temperature is higher (more working fluid enthalpy is available to be converted to shaft work). The opposite is true for steam that is partly condensed (i.e. under the vapor dome with steam quality less than 1.0). Less shaft work can be delivered because condensed liquid decreases the steam's ability to turn a turbine because its specific volume is lower.

In Chapter 6, we also learned that ***isobars diverge*** as temperatures increase in the *superheated vapor* region of the $T - s$ diagram. This permits a greater enthalpy change (and greater shaft work out) for the same pressure drop through the turbine. Furthermore, turbine blades are subjected to lower vibration and reduced stress when the steam remains dry as it expands. Liquid water droplets can cause damage to turbine blades when they impact the blades with high relative velocity.

Economizing. An important device used for waste heat recovery in large boiler systems is the economizer. The economizer is a gas-to-liquid heat exchanger, installed in the combustion product path downstream of the convective superheater. It extracts heat from combustion gases that arrive with relatively high temperatures (>1000 °F) and transfers that heat to the pressurized feedwater, which then enters the boiler with only a few degrees of subcooling (small approach ΔT).

Air Preheating. Residual heat that remains in the exhaust gases after the economizer can still be extracted via an air preheater. Air preheaters are typically regenerative (porous, rotating disk introduced in Chapter 3) heat exchangers that borrow heat from the exhaust gas and pay it back it to the combustion air. Air preheating permits higher temperatures in the firebox, which can lead to higher cycle efficiencies, due to the reduction in waste heat.

Feedwater Heating. Another efficiency improvement technique used in large Rankine cycle installations is multistage steam expansion and interstage feedwater heating. The schematic in Figure 8.7 illustrates this method:

By splitting the pumping and expansion steps into two stages each and bleeding out a small quantity of medium-pressure steam to preheat the medium-pressure feedwater, three efficiency goals are accomplished: (i) less energy is required for combustion because the water is preheated, (ii) each of the two expansion steps has greater isentropic efficiency because the pressure ratios are smaller, and (iii) less heat is rejected in the condenser.

One important factor to prevent reverse flow is for $(p_6 = p_2) > p_3$. This layout improves cycle efficiency because heat input is reduced substantially, while the power output is reduced only by the percentage of working fluid displaced out of Stream 7. An end-of-chapter problem gives students an opportunity to analyze a feedwater heater upgrade to a boiler system.

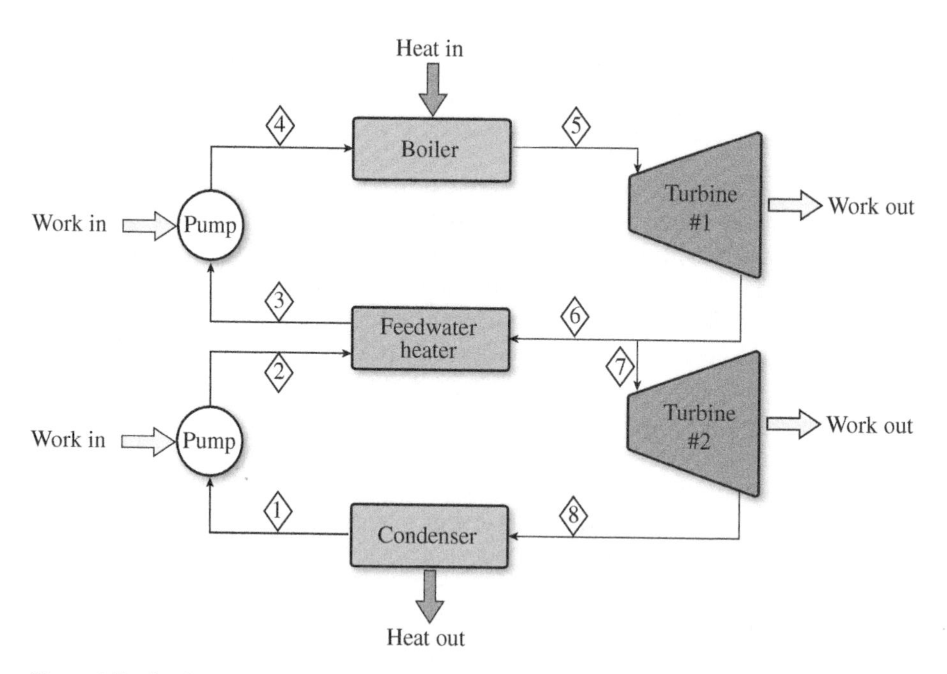

Figure 8.7 Feedwater heating in a two-stage Rankine cycle. *Source:* Martin Thermal Engineering Inc.

8.4 Controls and Safeguards

Boilers employ many of the same combustion safeguards as burners, as discussed in Chapter 7 of this textbook (e.g. flame sensor and interlock, redundant fuel safety shutoff valves with proof-of-closure, preignition purge with proof of minimum fresh air purge flow, proof of blower operation, and high- and low-gas pressure interlocks). Thermal system designers should review relevant boiler safety requirements to ensure the precise requirements are followed for the size and style of boiler being designed.

Irrespective of burner safety, the high pressures and temperatures on the waterside necessitate additional safeguards related to that working fluid. The following paragraphs address some of these additional safeguards.

Steam Drum Pressure Relief. By far, the most important safeguard for a boiler is the pressure relief valve installed at the top of the steam drum. Although the ordinary process controls are normally capable of preventing boiler vessels from reaching dangerous pressures, such controls sometimes fail, and the pressure relief valve is the last line of defense to prevent vessel rupture and BLEVE.

Pressure relief valves are spring-loaded, unidirectional, mechanical-only (i.e. not electromechanical) valves that open only when the internal pressure exceeds the value established by the spring adjustment setting. Once opened, the valve permits the escape of high-pressure steam (through a large pipe, to an approved location), which depressurizes the drum until a safe pressure is reached and the valve element closes again automatically.

Runaway heating can prompt multiple releases from pressure relief valves. If the boiler is operating unattended, and if other safeguards are not triggered to shut off the burner, the pressure relief valve could cycle opened and closed a number of times until the problem corrects itself, or an operator arrives to shut the system down and fix the problem.

Because pressure relief valves can fail shut, redundant installation is often recommended for large boiler vessels. Sizing of pressure relief valves is undertaken by a worst-case analysis, wherein the steam demand line is assumed completely shut off, and the firing rate of all burners is assumed to be maximized. Even residential hot water heaters (that do not boil water and have thermostatic controls set for ~140 °*F*) are equipped with pressure relief valves for these same reasons.

Pressure Control. Small boilers are frequently operated with a burner that fires at a single (design) heat release rate, and cycles on and off based on steam pressure. When *steam demand* is low (i.e. very little flow is going to the steam utilization equipment), the equilibrium pressure in the steam drum can remain high enough to permit the burner to remain off, until either (i) demand increases or (ii) the system loses enough heat that the saturation temperature drops, which causes the saturation pressure to drop too. Larger boilers may have a feedback loop that modulates (increases or decreases) the burner firing rate, based on the "error" between the pressure setpoint and the actual steam drum pressure ($\Delta p_{error} = p_{setpoint} - p_{actual}$, see also Chapter 15). If the error is large and positive, the burner firing rate increases; if the error is zero or negative, the burner shuts off completely.

Oxygen Control. Since both air and fuel are required for combustion, any firing rate adjustment should ensure that enough air is present to fully consume the fuel. This can be accomplished by setting the air flow rate at its maximum value and keeping it there while the fuel flow rate rises and falls to accommodate the changes in steam pressure. This technique is simple but wasteful at low firing rates. By reducing the fuel only, the combustion becomes leaner, and the excess oxygen in the exhaust rises. For most boilers, any amount of excess oxygen in the exhaust above the amount necessary to burn all the fuel is a waste of not only the oxygen, but also the 3.76 moles of nitrogen that come along for the ride. Boilers are typically designed to accommodate very high combustion temperatures, as long as the tubes and drums are not allowed to run dry. High combustion temperatures often are a factor in achieving high energy efficiency, due to the reduced waste of warm or hot exhaust flow, whenever excess oxygen is minimized. Thus, the strategy for maintaining low exhaust oxygen, and higher efficiency is to modulate the air flow in tandem with the fuel flow. If steam demand necessitates additional heat release from the burner, both fuel flow and air flow must increase, and to maintain exhaust oxygen constant, their relative flow rates much remain at approximately the same proportion (i.e. the same equivalence ratio). Achieving air and fuel balance can be accomplished by feedback controls using exhaust oxygen concentration as the measured process variable (with a setpoint of approximately 2–3 $\%\ vol\ O_2$) or by measuring fuel and air mass flow rates and maintaining a constant ratio of those two values. Air blower speed or air flow control damper position can be used to modulate the airflow to match a given fuel flow rate.

Water Level Control. Control of the burner firing rate and equivalence ratio (based on steam drum pressure and exhaust oxygen concentration) is not sufficient to maintain the functionality and safety of a boiler. Control of feedwater flow into the boiler is also vitally important. For feedwater flow control, the input variable is water level in the steam drum. Many small-to-medium boilers use a simple level switch (similar to a toilet tank float, but with electrical energize–de-energize contacts) to turn the feedwater pump on and off. If the level drops below the minimum value, the pump turns on, and if the level rises above the maximum value, the pump turns off. Larger and more sophisticated boilers may use a feedback loop with a continuous level element and a speed-controlled feedwater pump to keep the water level at precisely the desired elevation.

Superheat Temperature Control. Power boilers (i.e. boilers that generate steam to turn a turbine and generate electric power) are always operated such that the steam entering the turbine is superheated. Turbines spin at a rate of several thousand revolutions per second and the formation of liquid water droplets as the pressure and temperature decline can severely damage the blades. The level of superheat can be controlled by measuring the difference between superheater exit temperature and steam drum temperature. If the degrees of superheat ($\Delta T_{superheat}$) are too low, the steam drum pressure setpoint can be automatically increased to cause the firing rate to go up a bit, even though steam demand remains the same. This process is called "cascade control" and will be discussed further in Chapter 15.

Steam Backpressure Control. Some boilers with relatively large thermal inertia may incorporate steam backpressure regulators to ensure the steam drum pressure is not adversely affected by rapid changes in steam demand. The concept of backpressure regulation is to restrict outflow temporarily to the steam usage devices (e.g. air handler, autoclave) until the firing rate has had a chance to catch up and maintain steam drum pressure at a relatively constant level. Backpressure control regulators function much the same way as (forward) pressure regulators, as will be discussed further in Chapter 14.

Interlocks. For any of the control signals that are monitored and controlled, a high-level and/or low-level interlock can be implemented. For example, if the measured steam drum pressure exceeds the setpoint and approaches a dangerous value, an alarm can be triggered to warn the operator that action should be taken to avert the risk of a BLEVE. If the alarm sounds and the operator is unable to rectify the problem, a second trigger may be implemented that automatically shuts off all the burners and allows the unit to cool off. Similarly, interlocks for high and low superheat temperature, and for high and low water level can also be incorporated into the overall control system. Such interlocks can help avoid or limit damage and injury that might result from the underlying failures. Interlocks will also be discussed further in Chapter 15.

8.5 Blowdown and Water Treatment

The importance of water treatment in a boiler system cannot be overstated. Dissolved minerals (e.g. Ca^{2+}, Mg^{2+}, Fe^{3+}, Na^{+}, K^{+}, Zn^{2+}, Si^{4+}, and Se^{4+}) can accumulate inside the boiler and eventually precipitate (as solid oxides or salts) and adhere to heat exchange surfaces. These poorly conducting materials act as barriers to heat transfer (i.e. added resistance) between the water and the tube being heated, which causes the tube temperature to increase. If thick enough, the materials can drive tube temperatures so high, they weaken and fail – sometimes catastrophically.

Although industrial water sources are not regulated for water quality, EPA recommends drinking water should adhere to the secondary maximum contaminant level (SMCL) of 500 mg/l for total dissolved solids (TDS).

Clearly, a drinking water standard would not be enforceable for industrial boilers, but the contaminants are nonetheless problematic in such boilers. These concerns include scaling, corrosion, and staining of equipment (EPA 2021). Because most boiler operators rely on city water to make up for steam losses at various points in the cycle, TDS accumulation can become a serious problem. Water softeners only remove hardness minerals (i.e. Ca and Mg); they do not reduce TDS. Deionization equipment can remove TDS from water, but this option tends to be much more expensive than a relatively simple procedure called ***blowdown***. Figure 8.8 shows a schematic from a typical steam system, with the focus explicitly on the water, contaminants, and additives (i.e. fluid movers, burners, and tanks are ignored, and condensation is assumed to occur in conjunction with the demand).

Think Stop. In Figure 8.8, the red arrows are the four vectors that represent water entry into the cycle and water or steam exit out of the cycle. For simplicity, the chemical treatment vector is

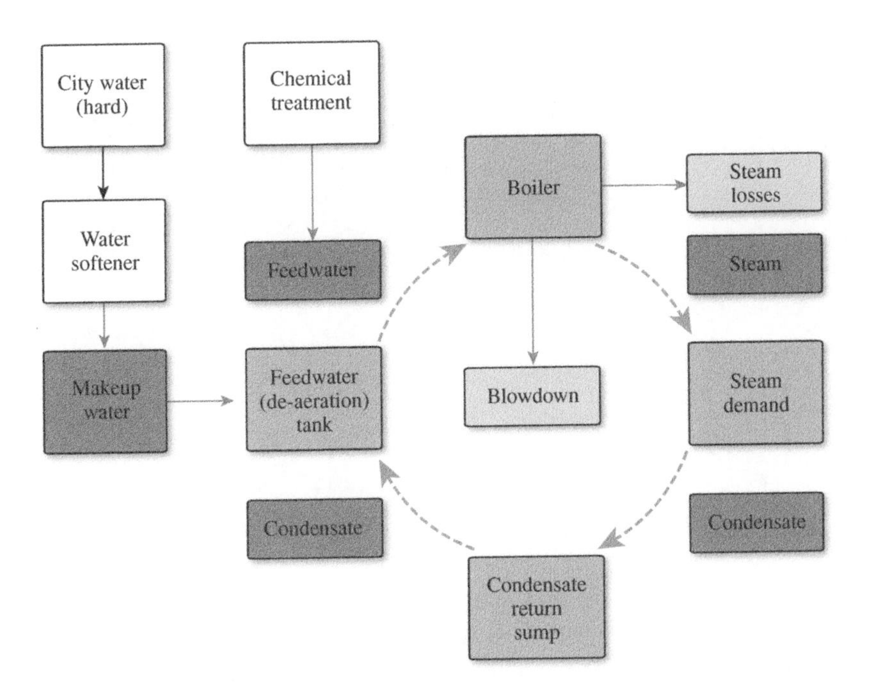

Figure 8.8 Typical boiler steam and condensate cycle, including water softening, chemical treatment, and blowdown. *Source:* Martin Thermal Engineering Inc.

ignored because the inflow rate of this additive is very small compared to the other flows, and its purpose is simply to encourage precipitated solids to agglomerate, so they can be removed from the system with the blowdown water.

The three vectors that remain are makeup water in, steam losses out, and blowdown out. The makeup water and blowdown water contain dissolved solids, but the steam is pure water vapor – no solids. When water boils, the dissolved solids remain in solution because their vapor pressures are tiny at the water saturation temperature. Because of this, boiling is an effective purification method (called distillation). When city water is vaporized and then condensed in a different vessel, essentially zero carryover of the dissolved solids occurs. Whenever the system is at steady state (no changes with time), the following mass conservation equations (Equations 8.2) must hold. Note that subscript w indicates ***water*** (liquid or vapor) and subscript s indicates ***solids*** (both dissolved and precipitated).

$$\dot{m}_{w,makeup} = \dot{m}_{w,steamloss} + \dot{m}_{w,blowdown} \tag{8.2a}$$

$$\dot{m}_{s,makeup} = \dot{m}_{s,blowdown} \tag{8.2b}$$

$$\dot{m}_{w,steamloss} = \dot{m}_{w,steam,demand} - \dot{m}_{w,condensate,return} \tag{8.2c}$$

It is also helpful to define a new term called number of cycles (see Equation 8.3), which is the average number of times a unit mass of water must flow around the circuit before its accompanying solids have all exited via blowdown. Or to put it differently, consider a 1.0 lb mass of city water that entered as makeup water at a specific point of time. Suppose the city water brought 500 mg of TDS with it. Depending on the blowdown rate, the water mass might travel as few as 2 or as many as 10 complete cycles, before all the TDS have exited with the blowdown. A smaller number of cycles means the TDS accumulation rate is lower, which helps keep its concentration low.

$$C = \frac{Y_{s,makeup}}{Y_{s,blowdown}} \doteq \frac{ppmw_{s,makeup}}{ppmw_{blowdown}} \tag{8.3}$$

Think Stop. Based on these equalities and definitions, the following formula (Equation 8.4) can be derived (see end-of-chapter problems).

$$BD = \frac{(S-R)}{(C-1)} \tag{8.4}$$

where

$$BD = \dot{m}_{w,blowdown} + \dot{m}_{s,blowdown} \left[\doteq \frac{lb_{w+s}}{h} \right]$$

$$S = \dot{m}_{w,steam,demand} \left[\doteq \frac{lb_w}{h} \right]$$

$$R = \dot{m}_{w,condensate,return} \left[\doteq \frac{lb_w}{h} \right]$$

Example 8.1

Table 8.2 presents six cases that illustrate the importance of steady-state blowdown to maintaining an acceptable TDS concentration inside the boiler. In all six cases, the steam demand (S) is identical and so is the concentration ($ppmw$) of TDS in the makeup water. The parameters being varied are condensate return flow $R \in \{900, 700, 500 lb_w/h\}$ and the steady blowdown flow

Table 8.2 Cases of boiler TDS versus blowdown rate for the schematic in Figure 8.8. Boiler TDS concentrations higher than 4500 *ppmw* (boxed) are unacceptable.

Attribute of steam cycle	Case 1	Case 2	Case 3	Case 4	Case 5	Case 6
S (steam demand, lb_w/h)	1000	1000	1000	1000	1000	1000
R (condensate return, lb_w/h)	900	700	500	900	700	500
BD (blowdown, lb_{w+s}/h)	10	10	10	50	50	50
M (makeup, lb_{w+s}/h)	110	310	510	150	350	550
C (*cycles*)	11	31	51	3	7	11
TDS in makeup (*ppmw*)	300	300	300	300	300	300
TDS in boiler (*ppmw*) (target = 3500 – 4500)	3300	[9300]	[15 300]	900	2100	3300
TDS inflow (lb_s/h)	0.033	0.093	0.153	0.045	0.105	0.165
TDS outflow (lb_s/h)	0.033	0.093	0.153	0.045	0.105	0.165

Source: Martin Thermal Engineering Inc.

$BD \in \{10, 50 lb_{w+s}/h\}$. Students should independently confirm mass is conserved and the computation of C given in the table is correct.

Answer. The most important finding in Table 8.2 is that for the low blowdown flow rate (10 lb/h or 1% of steam demand), the concentration of dissolved solids in the boiler reaches unacceptably high-values for moderate-to-high rates of steam loss, whereas for the higher blowdown flow rate (50 lb/h or 5% of steam demand), the boiler TDS is acceptable even for 50% steam loss.

8.6 Air Pollutant Reduction

Commercial and industrial boilers are responsible for a significant proportion of the pollutants emitted to the atmosphere each year. These pollutants are regulated by local, state, or federal governments and boiler users must install pollution control devices to reduce emissions to allowable levels. Regulated pollutants include carbon monoxide (CO), nitrogen oxides (NO_x), sulfur oxides (SO_x), and particulate matter (PM) (ash, soot). Table 8.3 (EPA 2019) provides an emission inventory by source type.

Boilers are also responsible for emitting large quantities of carbon dioxide, but those emissions are not addressed here because CO_2 removal processes have not been demonstrated at a commercial scale yet. Furthermore, the energy sources for society's future electricity demand may undergo a relatively complete transition to renewable resources (e.g. solar, wind, and geothermal) and away from polluting nonrenewable resources (e.g. coal, oil, and gas) years before any technology for CO_2 removal achieves feasibility.

Although not shown in the table, gaseous and liquid fuels generate more total mass of air pollutant emissions than the combustion of solid fuels because more of the nonsolid fuels are consumed annually. However, coal-fired boilers tend to emit more *types* of pollutants than any other combustion technology. Today, coal-burning, utility-scale boilers continue to emit nonzero amounts of all the pollutants addressed in the table, even though regulations require them be equipped with treatment devices that capture or destroy a large fraction of the total mass of

Table 8.3 U.S. emission inventory for 2019 (units are 1000 s of *tons/year*).

Source category	CO	NO_x	SO_x	VOC	PM10
Fuel comb – electric utility	588	996	1017	32	132
Fuel comb – industrial	812	1 032	377	112	246
Fuel comb – other	2 671	494	44	362	370
Chemical manufacturing	118	41	111	75	19
Metal processing	468	66	85	22	51
Petroleum production, refining	652	623	95	2493	27
Other industrial	447	321	146	346	767
Solvent utilization	2	1	0	2972	5
Storage, transport	7	5	1	697	36
Waste disposal, recycling	1 301	81	25	177	227
Highway vehicles	16 866	2 775	11	1496	253
Off-highway equipment	11 579	2 105	38	1190	135
Miscellaneous	28 676	409	218	6888	14 843

Source: Data from EPA (2019).

combustion-generated pollutants. Methods for reducing pollutant emissions from boilers that burn coal or natural gas are described below.

Carbon Monoxide. Carbon monoxide is a combustion by-product of all types of fuel burning equipment. If burners are designed, operated, and maintained properly, *CO* emissions will be low (well under 100 *ppmv*) and in such cases, the boiler would not be required to install post-treatment equipment to remove excess *CO*. However, when certain in-furnace methods for reducing NO_x are employed (see the next section), boiler operators may find it difficult to maintain *CO* emissions below allowable levels. Under such conditions, a boiler may need to be *de-rated* (operated at reduced firing rate *BTU*/*h*) to ensure all regulated pollutants remain under allowable annual emission levels.

Nitrogen Oxides. Utility boilers that burn coal must mitigate NO_x emissions caused by the small but consequential fraction of organo-nitrogen compounds in the coal solid matrix. During combustion, these compounds readily dissociate to form reactive nitrogen species such as *CN*, *NH*, or *N*, which then quickly combine with oxygen molecules (O_2) to form *NO* and a small amount of NO_2 (the sum $NO + NO_2 = NO_x$). To reduce NO_x emissions, these large combustion systems must employ one or more of the following NO_x reduction processes: (i) in-furnace methods (e.g. flue gas recirculation *FGR* and/or staged combustion) or (ii) post-furnace methods (e.g. selective catalytic reduction *SCR* or selective noncatalytic reduction *SNCR*).

FGR has been employed for more years than any other NO_x reduction strategy, but its effectiveness is limited. The necessary equipment is relatively simple and inexpensive: (i) two new ports – an extraction port downstream of the particulate collection device and an injection port upstream of the combustion air blower, along with an insulated duct to convey the flue gas back to the burner, (ii) a recirculation blower to boost the pressure of the flue gas so it can be blended with the combustion air, and (iii) a control valve to control the proportion of flue gas to fresh air. To avoid combustion instabilities, the *FGR* system should ensure the burner's oxygen concentration does not fall below 15 % *vol*.

Staged combustion is only slightly younger than *FGR*. It is characterized by (i) a fuel-rich zone, where fuel is combusted with insufficient air and *NO* is scavenged by hydrocarbon fragments, followed by (ii) a fuel-lean zone where *overfire* air is introduced to burn out the excess carbon monoxide and hydrocarbons left over from the first stage. Because staging is an in-furnace technique, it tends to be implemented as part of the original design and construction of the boiler, although retrofitting the boiler to achieve staging is also possible. Additional equipment includes: (i) extra ports to convey the *overfire* air into the furnace several feet downstream from the main combustion zone, and (ii) a separate *overfire* air blower and control damper to regulate the quantity of overfire air needed to accomplish the burnout. The term *overfire* arose because the firebox was typically oriented vertically, and the final air addition ports were literally "over" the fire. In coal-fired boilers, *overfire* air is alternatively referred to as *tertiary* air because *primary* and *secondary* air both enter the furnace at the main combustion elevation. *Primary* air pneumatically conveys the pulverized coal from the grinder to the furnace and it accounts for the smallest proportion of total combustion air (approximately 20%). *Secondary* air (approximately 50% of total air) is introduced at the same elevation as the coal and *primary* air, but through different ducts to improve combustion stability. *Tertiary* air (approximately 30% of the total) is then added well downstream of the burner elevation to effect the final burnout of fuel compounds.

Neither of the two in-furnace methods for NO_x reduction is sufficiently effective to meet today's regulatory NO_x emission limits, so post-boiler reduction techniques are typically required. The most common apparatus is selective catalytic reduction (***SCR***), which was patented by Engelhard Industries (Cohn et al. 1961). Selective noncatalytic reduction (***SNCR***) is less common than ***SCR*** because ***SNCR*** does not perform as well as ***SCR***, but it is also less costly to install because no catalyst is required. ***SNCR*** was patented by Exxon Research and Engineering (Lyon 1975).

SCR involves a large catalytic reactor typically installed in the exhaust gas stream, downstream of the economizer and upstream of the air preheater and other pollution control equipment. The optimum temperature is approximately $T \approx 250\ °C$ (~480 °*F*). Sufficient surface area is required to ensure *NO* molecules diffuse to the catalyst where *NO* combines with a reducing agent to form nitrogen N_2. See Equation (8.5) for the stoichiometry when ammonia (NH_3) is the reducing agent. In addition to the catalyst matrix, a grid of ammonia injectors is required far enough upstream to ensure good mixing into the exhaust gas before it reaches the catalyst.

$$2NO + 2NH_3 + \frac{1}{2}O_2 = 2N_2 + 3H_2O \tag{8.5}$$

Interestingly, a version of *SCR*, where urea $(NH_2)_2CO$ is the *NO* reducing agent, has had reasonable success in removing *NO* from the exhaust of heavy-duty diesel trucks. Diesel-powered trucks are required to carry diesel exhaust fluid (DEF) to effect the NO_x reduction required of their vehicles.

The removal efficiency of ***SNCR*** is lower than that of *SCR* for two reasons:

a) The ideal temperature for *SNCR* is $T \approx 1250\ K$ (~1800 °*F*), which occurs upstream of the economizer. The residence time available for contact between *NO* and NH_3 at this temperature is small and this limits the reaction's yield (completeness).
b) The ratio of NH_3 to *NO* tends to be higher than *SCR*. With the concurrently lower *NO* removal efficiency, the cost per unit of *NO* removed is higher.

So even though ***SNCR*** was once a promising technology (because it does not require expensive catalyst material or large equipment volumes), it has not achieved favorable reviews by regulators and is rarely chosen by coal-fired boiler operators.

Sulfur Oxides. The combustion of coal almost always involves the production of some sulfur oxides ($SO_2 + SO_3 = SO_x$). Coals mined in the western United States (e.g. Wyoming and Utah) tend to be lower in fuel-bound sulfur (~1% S by mass) and coals mined in the eastern United States (e.g. Illinois and Pennsylvania) tend to be higher in fuel-bound sulfur (~3 % S by mass). Martin et al. (1988) provided an early snapshot of the research that showed dry injection of sorbent (e.g. limestone) into a coal-fired furnace was far less effective at removing SO_x emissions than post-boiler removal of SO_x by ***scrubbing***. Scrubbing is shorthand for removal of pollutant gases from air by absorption into liquid water.

The most effective SO_x ***scrubbers*** are vertical counterflow vessels with spray nozzles near the top creating a heavy "rain" of alkaline water solution (i.e. sodium or calcium hydroxide dissolved in water at high pH) that falls downward under the influence of gravity and makes good contact with the SO_x-laden exhaust gas stream. Scrubbers not only absorb the sulfur oxides into the liquid water droplets, but they also facilitate a reaction between SO_x and alkaline hydroxides to form sulfates Na_2SO_4 or $CaSO_4$ that can be precipitated and filtered out of the scrubbing liquid and potentially recycled to become a chemical feedstock. See Equation (8.6) for the stoichiometry of SO_x removal with $NaOH$ as the aqueous alkaline agent.

$$SO_3 + 2NaOH = Na_2SO_4 + H_2O \tag{8.6}$$

Detailed design of scrubbers is outside the scope of this textbook. See EPA (2002) or Green (2008) for additional engineering guidance on scrubbers.

Particulate Matter. The contaminant whose mass fraction in coal is higher than any other is mineral matter. Because coal is an earth material, it inevitably contains mineral matter in the form of aluminosilicates (clays), oxides (silica, ferrite), sulfides (pyrite), and carbonates (calcite). None of these compounds are combustible, although some are reactive at high temperatures and contribute to gaseous pollutant emissions when coal is burned. The combustion by-product of these contaminants is called ***ash***, and it leaves the furnace in two distinct physical forms – bottom ash and fly ash.

Bottom ash is often referred to as slag, but technically, bottom ash is any mineral matter by-product of coal combustion that fails to leave the furnace with the combustion gases. Bottom ash can contain large particles that escaped the coal pulverizer and dropped into the ash pit at the bottom of the furnace under their own weight. Bottom ash can also contain agglomerations of fine particles that accumulated on boiler tubes and eventually fell off and landed in the ash pit. They can also be molten drippings of mineral matter that accumulated near burners and flowed into the ash pit where they solidified. Because bottom ash is not easily dispersed into air, it is considered a solid waste, rather than a potential air pollutant.

Fly ash is very fine PM that is carried out of the furnace with the combustion gases. Emission of PM into the air is regulated, so coal-fired boilers are required to install emission control equipment that removes a large majority (e.g. 99.9%) of the fly ash produced. Two types of PM removal devices typically meet regulatory requirements – baghouses and electrostatic precipitators (ESP).

A baghouse is essentially a large rectangular enclosure that contains a thousand or more fabric filter elements that resemble upside-down vacuum cleaner bags. The filtration process removes the fly ash from the air and occasionally the bags are cleaned by shaking, which causes the filter cake to fall into the hopper at the bottom of the enclosure. ESP are also housed in large enclosures, but they use high-voltage electrostatic fields to charge the particles and attract them to a series of grounded metal plates. Periodically, the plates are cleaned by rapping them with a mechanical hammer, which causes the adhered particulate mass to fall into the hopper at the bottom of the enclosure.

8.7 Organic Rankine Cycle

The organic Rankine cycle (ORC) is simply a Rankine cycle (see Figure 8.1) that uses organic working fluids (e.g. refrigerants, hydrocarbons) instead of water. ORC processes tend to be used in situations where the high-temperature heat source and low-temperature heat sink are both relatively close to local ambient temperature because many organic fluids boil and condense readily within that range. ORC systems are sometimes installed as "bottoming cycles," because they can generate incrementally more electric power from the low-temperature waste heat produced by larger power generation systems. The ORC process is conceptually identical to the water-based Rankine cycle, and Figures 8.1 and 8.2 can be viewed as analogous to the equipment and working fluid design of an ORC.

One innovative process that typically uses ORC is ocean thermal energy conversion (OTEC), where the relatively small temperature gradient present at different elevations below the ocean's surface drives a two-temperature power cycle. Heat addition from warmer surface water boils the working fluid and heat removal to the colder deep water (perhaps as much as 1000 m below the surface) recondenses it.

According to Barberis et al. (2019), small island nations that do not have reliable sources of fresh water may be able to rely on a combined OTEC/ORC/DESAL system to provide sufficient fresh water to supplement the natural but unpredictable rainfall supply. Pumping power tends to represent a significant parasitic loss for OTEC systems. Thermal insulation of pipes is vital to preserve the temperature difference that is the source of the thermodynamic power cycle.

ORC processes can also be used in a cycle where the boiler receives its heat from the sun. Combining a solar thermal collector (concentrating or flat, see Chapter 11) with an ORC cycle has the potential for rivaling the energy efficiency of solar photovoltaic collectors, but the ORC system is more complex. Components of a Solar ORC system may include the following:

- Organic working fluid
- Liquid pump
- Solar thermal collector/boiler
- Turbine with generator
- Finfan condenser
- Liquid receiver vessel (to help ensure pump is not pushing vapor)
- Interconnect piping, controls

ORC systems tend to work better at lower temperatures than SRC (steam Rankine cycle) systems because many light organic compounds have lower boiling points and lower freezing temperatures than water, and consequently they have substantially higher vapor pressures at low temperatures, which permit the system to be operated entirely at pressures greater than ambient. This is often a benefit from a maintenance perspective because leaks outward (from high internal pressure to low ambient pressure) are much easier to detect and fix than leaks inward (from ambient pressure to vacuum internal pressure). For this reason, working fluids that generate one atmosphere of saturation pressure at temperatures slightly higher than ambient (i.e. $p_{sat}(T_\infty) \geq 1\ atm$) are preferred. Reynolds (1979) provides a listing of substances ordered by saturation temperature at $T_{sat}(p_{sat} = 1.00\ atm)$ and several are reproduced here, along with selected other compounds.

Thermal designers should consider the breadth of the vapor dome at the ORC's condensing and evaporating temperatures as part of their assessment of organic compounds to select for their working fluid.

Table 8.4 Criteria for determining working fluid in an organic Rankine cycle.

Fluid	T_{sat} (1 atm) (K)	T_{crit} (K)	Fluid	T_{sat} (1 atm) (K)	T_{crit} (K)
R32	221.4	351.3	Ammonia	239.9	406.8
Propylene	225.4	364.9	Isobutane	261.3	409.1
Propane	231.3	369.8	Butane	272.7	424.0

Source: Reynolds (1979). Used with permission by Stanford University Mechanical Engineering Department.

While operating a boiler above the working fluid's supercritical temperature is not impossible, the effectiveness of the heat transfer will be reduced because the fluid's temperature will rise as it absorbs thermal energy from the heat source.

Appendix A provides thermodynamic properties over a range of temperatures and pressures for three of the fluids listed in Table 8.4: R32, propane, and ammonia. For other working fluids, see Linstrom and Mallard (2017–2021).

8.8 Boiler Failure Analysis

Boiler tubes fail for a variety of reasons (Combustion Engineering 1981), but the most common reasons are (i) short-term overheating and stress rupture, (ii) long-term overheating and creep rupture, (iii) vibration-induced fatigue, (iv) thermal-cycling-induced fatigue, (v) waterside corrosion, (vi) fireside corrosion, (vii) erosion, and (viii) manufacturing defects.

The following failure investigation summaries provide additional information for designers and operators to help them avoid boiler safety problems.

Inadequate Tube Inspection. A wood-waste boiler that was used to heat a low-temperature, lumber-drying kiln, experienced a pair of tube failures in rapid succession (one failure likely initiated the second because of the high-pressure steam jet impingement, see Figure 8.9 for an example of tube rupture). Operators heard the noises and saw a white mist escaping from the perimeter of the firebox inspection door, and they concluded that one or more boiler tubes had ruptured.

The operator immediately shut down the fuel conveyor, the fresh air blower, and the flow of feedwater to the boiler. Unfortunately, extinguishing the burning layer of wood waste at the bottom of the stoker was not a trivial matter. The combination of radiant heat from the smoldering sawdust pile and the lack of new feedwater entering the boiler led to overheating and weakening of the remaining tubes that did not rupture initially. As the water in the tubes boiled way, the tubes drooped one-by-one and created a tube tangle that looked like frozen spaghetti inside the firebox.

The tubes were examined by a metallurgist and substantial wall thinning was observed in the vicinity of the tube that ruptured first. While tube thinning is sometimes caused by overheating and stretching, the metallurgist in this case concluded it was caused by erosion from contact with high-velocity ash particles being carried by the hot combustion gases to the particulate removal device (baghouse) downstream of the boiler. If the boiler owner had been more diligent in performing tube inspections, the wall thinning could have been discovered, and the eroded tube(s) could have been replaced in plenty of time to avert the rupture and subsequent heat damage to the hundreds of other tubes in the boiler.

FGR Purge Damper Failure. We investigated a combustion explosion in a medium-sized boiler that was used for space heating (via air handlers) of several detached buildings in a commercial

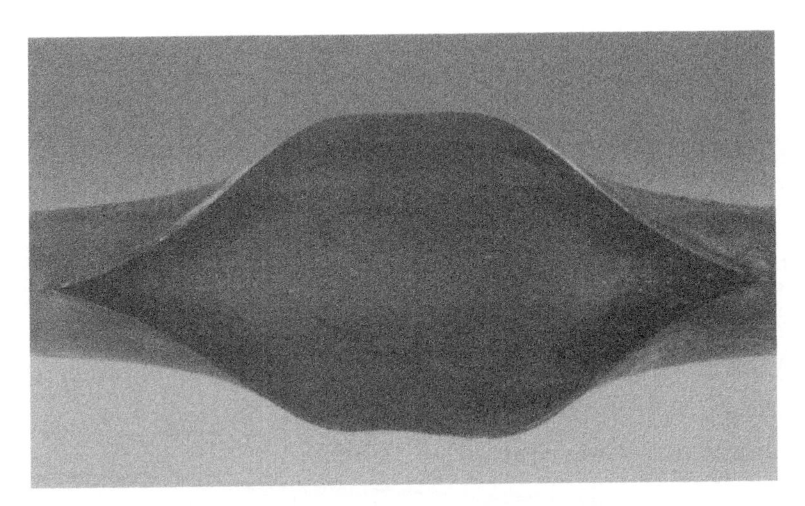

Figure 8.9 Example of a "fishmouth" longitudinal pipe rupture due to short-term overheating, which often occurs when pipes are under pressure but have insufficient water flow to cool them from inside. *Source:* Photo courtesy of The Babcock & Wilcox Company. See full citation: Babcock and Wilcox (2015, 2021).

campus of office and retail space. The investigation revealed a failure mode that has been addressed by applicable standards for some time, but perhaps with insufficient clarity. The failure mode occurs only in combustion chambers equipped with natural gas burners and FGR for control of NO_x emissions.

On smaller boilers equipped with FGR, an automatic binary damper is often installed at the upstream end of the FGR pipe. The damper is closed (no FGR flow) during startup and open (normal FGR flow rate) during normal boiler operation. However, if the FGR damper fails in the open position, the preignition purge flow will be a mixture of fresh air *and* residual fuel gas that was not ignited during the prior ignition attempt. Although the combustion safety codes do mandate that the source of purge gas must be "fresh air" and that "all passages" be prepurged before ignition, the possibility of FGR damper failure is easy to overlook.

Our recommended fix for this safety deficiency was for the purge timer to be set to a safer value (longer duration) whenever FGR is used for pollution control. Without FGR, the typical preignition purge timing is established to ensure at least four firebox volumes of fresh air enter and exit the system before ignition can be attempted. When an FGR damper fails in the open position, and the purge gas becomes a mixture of fresh air and recirculated (unburned), fuel/air mixture from prior ignition attempts, the timer should be extended to a new duration, based on the actual concentration of residual fuel contaminant that is recycled back to the burner. A "perfectly stirred reactor" model can be used to determine this new safety margin (Martin 2016) and establish the appropriate purge timer setting to reduce the risk of explosion resulting from an open damper failure.

Water Blowdown Failure. We investigated an incident where the convection surfaces of an unattended watertube boiler were destroyed by excessive heat from the burner a few months after the boiler was installed.

Generally, boiler tubes can survive contact with very high-temperature combustion gases, as long as water is flowing in the tubes, because the convection coefficient on the waterside is much higher than that on the fireside. A series resistance model concisely proves this point – the high value of h on the waterside causes tube temperature to be much closer to the water temperature because the resistance is lowest. However, when water ceases to flow, the tubes will fall victim to overheating by the combustion gases and oxidation (plus weakening) of the tubes will occur very quickly.

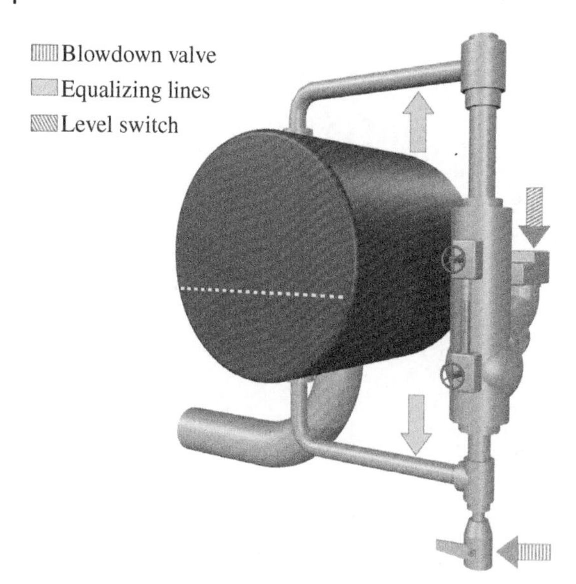

Figure 8.10 Steam drum of small watertube boiler, with associated level controls. The dotted line across the drum indicates normal water level. The diagonal hatch arrow indicates the external level control chamber. The vertical hatch arrow indicates the manual blowdown valve. The no hatch arrows indicate equalizing lines. *Source:* Martin Thermal Engineering Inc.

As discussed earlier in this chapter, most boilers have steam drums that are equipped with a water level controller and two low-level interlocks to ensure that the boiler never "runs dry" (or if it begins to, the burners are completely shut down to prevent damage). Based on the observed level of damage to the tubes, the subject boiler's safeguards must have failed to perform all three of these functions. Upon confirming this relationship, the investigation pivoted to the question of why the boiler ran dry and the burners kept operating.

Maintenance documentation established that the boiler operators were not performing blowdown frequently enough to prevent buildup of dissolved solids in the boiler. When the boiler was further examined, the external reservoir (i.e. the housing for the level control and low water cutoff LWCO#1 – indicated by a diagonal hatch arrow in Figure 8.10) was found full of water as a consequence of a hard, black solid material plugging the lower equalizing line that connected the steam drum to the level control chamber. Upon analysis, the material was found to be high in iron, with small amounts of calcium, copper, and chromium.

Witness marks also showed black and orange stains on the lowermost 7 in of LWCO#2 which was installed vertically into the top of steam drum (not shown in the figure). The mark indicated that a large volume of solids (probably sludge) was present at the bottom of the steam drum before the overheating occurred.

According to a mid-twentieth-century textbook *Steam-Boiler Operation, Maintenance, and Furnaces* (Andrews 1949):

> Sludges are the result of impurities in the boiler water. They consist of small particles of matter which do not readily attach themselves to the boiler metal... These sludges may settle out in the drums or headers in which water movement is slowed down. Blow-off valves have been provided for their removal. The general purpose of the chemical treatment of boiler water is to produce such sludges from unavoidable feedwater impurities and thereby prevent the impurities from depositing as scale. Settled-out sludges may block circulation through drums, headers, or supply tubes, and may pile up in drums to cause overheating... If such sludge deposits are found, the inspector should question the adequacy and the operating frequency of the blow-off equipment.

Thus, the operators' failure to blow down the drums frequently enough to prevent heavy buildup of sludge was singularly responsible for the following chain of events:

- Blockage of the lower equalizing line, which led to a false-positive indication (in the external level control chamber) of acceptable water level in the control chamber.
- Failure of the level control to energize the feedwater pump to bring fresh water into the boiler at a time when the boiler was desperately in need of cool feedwater.
- Failure of LWCO#1 to shut off the burners due to a false-high indicated water level in the external level control chamber.
- Failure of LWCO#2 to shut off the burners due to low a false-high indicated water level in the sensor's protection tube within the steam drum.
- Overheating and burning up of the water tubes after they ran dry.

Without investigating the matter thoroughly, one's inclination might be to blame the boiler failure on some unidentified latent defect associated with the manufacture or installation of a very new boiler (only a few months old). This demonstrates why a thorough investigation is so important to determine the root cause of a failure. Absent a determination of root cause (and correction of the problem), future failures of a similar nature are somewhat inevitable.

8.9 Homework Problems

8.1 Derive Equation (8.4), the expression for blowdown $BD = (S - R)/(C - 1)$. Show work.

8.2 Determine the missing mass flows, enthalpies, pressures, and temperatures for the stream table given below that is intended to represent the water side of an ideal Rankine cycle with superheat at state 3. Use Figure 8.1 to represent the schematic that accompanies this stream table. The pumping process (from state 1 to state 2) should be considered isentropic and the expansion process across the turbine (from state 3 to state 4) should also be considered isentropic. The quality at the turbine exit is unknown but assume State 4 to be slightly more dense than saturated steam (i.e. $\chi_4 < 1.00$). Let the quality at the pump inlet be $\chi_1 = 0.00$ (saturated liquid). For a water/steam mass flow rate $\dot{m}_{H_2O} = 4.10\ kg/s$, compute (a) net shaft power leaving the system after subtracting pumping power in from turbine power out and (b) cycle efficiency. Use the custom Excel functions for water in Appendix B.4 or interpolate from the property tables for water in Appendix A.4 to obtain values of thermodynamic properties (or use NIST Chemistry Webbook Fluid Properties for higher accuracy than interpolation).

Answer. $\dot{W}_{net} = \dot{W}_{turbine} - \dot{W}_{pump} = 3.003\ MW$; $\eta_{cycle} = 27.5\%$.

Property	State 1	State 2	State 3	State 4
Temperature (K)	373.14		630	
Pressure (kPa)	101.325	5000	5000	101.325
Density (kg/m^3)				
Enthalpy (kJ/kg)				
Entropy ($kJ/kg\ K$)				
Quality (%)	0.00	—	—	

8.3 Calculate the energy distribution (percent of total chemical enthalpy in fuel being burned) for a utility boiler (note – this is not an energy balance for the entire Rankine cycle, it is an energy balance around the boiler + superheater + economizer):

(a) Steam
(b) Losses to flue gases
(c) Losses to ash
(d) Other losses (e.g. walls)

Process Details. Fuel consumption is 600 ton/d of coal at STP. The lower heating value (LHV) of the coal is 12 500 BTU/lb_m. The air to coal ratio by mass is 19 : 1. Ash is generated at the rate of 55 ton/d and has an internal energy of $u_{ash} = 750\ BTU/lb_m$ when it exits the boiler enclosure. Subcooled liquid water ($\dot{m} = 350\ 000\ lb_m/h$) enters the boiler at $p = 1450\ psia$, 215 °F and leaves as superheated steam at 1450 $psia$, 1160 °F. Combustion air (assumed to be constant pressure throughout at $p = 14.696\ psia$) enters at $T_{air,\ in} = 80$ °F and exhaust gas exits at $T_{exh} = 395$ °F. Assume the exhaust gas thermodynamic properties are equivalent to those of air. Use property tables in Appendices A.1 and A.4, interpolating where necessary or use Linstrom and Mallard (2017–2021). Show work.

Answer. (a) 78.43%, (b) 13.17%, (c) 0.55%, and (d) 7.85 % .

8.4 Combustion gases at $p = 101.325\ kPa$ and $T_{econ,\ in} = 475$ °C enter a crossflow-type economizer and leave at 325 °C. High-pressure feedwater ($p_w = 10\ 000\ kPa$), with a mass flow rate of $\dot{m} = 0.5 \times 10^6\ kg/h$, enters the economizer at $T_{econ,\ in} = 205$ °C and leaves at $\Delta T_{econ,\ out,\ subcool} = -4.0$ °C. The overall heat transfer coefficient (which accounts for the convection and conduction resistances) is $U = 0.075\ kW/(m_2 \cdot K)$. Find the total length of heat exchanger tubing needed, if the tube outer diameter (OD) is 5.0 cm, and assuming the wall thickness is negligible. Do not use LMTD method for a crossflow heat exchanger (Equation 3.39) unless figures for correction factor F (not available in this textbook) are accessible. Assume the thermodynamic properties of combustion gases may be represented by the properties of air, Appendix A.1. For thermodynamic properties for water, interpolate from the values given in Appendix A.4 or use Linstrom and Mallard (2017–2021) or use the customized Excel functions for water in Appendix B.4. Show work.

Answer. $L_{tot} = 46.61\ km$.

8.5 For a feedwater heater (see Figure 8.7), determine the missing mass flows, enthalpies, pressures, and temperatures for the stream table given below. The working fluid is water. Assume the feedwater heater is adiabatic with respect to the environment. Use thermodynamic properties for water/steam in Appendix A.4 and interpolate where necessary, or use Linstrom and Mallard (2017–2021) or use the customized Excel functions for water in Appendix B.4.

Property	**State 2**	**State 3**	**State 6**
Temperature (K)	351.0		600.0
Pressure (kPa)	500	500	500
Spec volume (m^3/kg)	973.30		1.8242
Enthalpy (kJ/kg)			
Entropy ($kJ/kg\ K$)			
Mass flow rate (kg/s)	2.000	2.160	0.160
H_2O phase	Liquid	Liquid	Vapor

Cited References

Andrews, L.V.; (1949); *Steam-Boiler Operation, Maintenance, and Furnaces*; Scranton, PA: International Textbook Company.

ASME; (2021); *Richard Trevithick (Short Biography)*; American Society of Mechanical Engineers, Topics and Resources. https://www.asme.org/topics-resources/content/richard-trevithick.

Babcock & Wilcox; (1978); *Steam: Its Generation and Use* – 39 New York: The Babcock & Wilcox Company.

Babcock & Wilcox; (2015); *Steam: Its Generation and Use*, 42 Barberton, OH: The Babcock & Wilcox Company.

Babcock & Wilcox; (2021); The B&W Learning Center.; "*Finding the Root Cause of Boiler Tube Failures, General Failure Mechanisms, Short Term Overheat*"; https://www.babcock.com/en/resources/learning-center/finding-the-root-cause-of-boiler-tube-failure.

Barberis, S., Guigno, A., Sorzana, G., Lopes, M.F.P., and Traverso, A.; (2019); "Techno-economic analysis of multipurpose OTEC power plants"; *E3S Web Conf.*; **113**; 03021. From SUPEHR19 SUstainable PolyEnergy Generation and HaRvesting Volume 1. https://doi.org/10.1051/e3sconf/201911303021; https://www.e3s-conferences.org/articles/e3sconf/pdf/2019/39/e3sconf_supehr18_03021.pdf.

CA-DIR; (2021); *Cal/OSHA Pressure Vessels, Training Materials, 2016 PowerPoint, B&W Overview*. California Department of Industrial Relations. www.dir.ca.gov/dosh/pressure-vessels/documents/B&W-Overview.pdf

Cohn, J.G.E., Steele, D.R., and Anderson, B.C. (1961). Method of Selectively Removing Oxides of Nitrogen from Oxygen-Containing Gases. Assigned to Engelhard Industries, Inc. US Patent 2,975,025.

Collier, J.G.; (1981); *Convective Boiling and Condensation*, 2; New York: McGraw-Hill Book Company.

Combustion Engineering; (1981); *Combustion: Fossil Power Systems*, 3; Windsor, CT: Combustion Engineering, Inc..

EIA (2017–2021); "*Electricity Explained, Electricity in the United States*"; U. S. Department of Energy; Energy Information Administration. https://www.eia.gov/energyexplained/electricity/electricity-in-the-us.php

El-Wakil, M.M.; (2002); *Powerplant Technology*, 1 Fossil Fuel Steam Generators, See Chapter 3; New York: McGraw-Hill

EPA; (2002); *EPA Air Pollution Control Cost Manual*, 6. U.S. Environmental Protection Agency, Office of Air Quality Planning and Standards. PA/452/B-02-001

EPA; (2019); "*Criteria pollutants National Tier 1 for 1970–2019*"; Air Emission Inventories, Air Pollution Emission Trends Data, National Annual Emissions Trend. https://www.epa.gov/air-emissions-inventories/air-pollutant-emissions-trends-data.

EPA; (2021); "*Safe Drinking Water Act; Secondary Drinking Water Standards: Guidance for Nuisance Chemicals*"; United States Environmental Protection Agency; https://www.epa.gov/sdwa/secondary-drinking-water-standards-guidance-nuisance-chemicals.

Green, D.W., (2008); *Perry's Chemical Engineers' Handbook*, 8 Equipment for distillation, gas absorption, phase dispersion, and phase separation, see Chapter 14"; New York: McGraw-Hill.

Linstrom, P.J. and Mallard, W.G., (2017–2021); *NIST Chemistry WebBook, SRD 69*; "Thermophysical properties of fluid systems"; National Institute of Standards and Technology, U.S. Dept. of Commerce; http://webbook.nist.gov/chemistry/fluid

Lyon, R.K. (1975). Method for the reduction of the concentration of NO in combustion effluents using ammonia. Assigned to Exxon Research and Engineering Company. US Patent 3,900,554.

Martin, R.J. (2016). Blog: boiler purge causes explosion. http://www.martinthermal.com/2016/11/09/boiler-purge-explosion (accessed 1 May 2021).

Martin, R.J., Kelly, J.T., Ohmine, S., and Chu, E.K.; (1988); "Pilot-scale characterization of dry sorbent injection for SO_2 control in a low-NOx tangential system"; *ASME J. Eng. Power*; **110**; 111.

Reynolds, William C.; (1979); *Thermodynamic Properties in SI*; Stanford, CA: Stanford University Mechanical Engineering Department.

9

Combustion Turbines

A combustion turbine is more than a turbine; it is a power generation system that combines three major pieces of equipment together (compressor, combustor, and turbine) to form an open thermodynamic cycle in a single package. This product design strategy is contrasted with a steam turbine (i.e. Rankine cycle) system, where the liquid pump, boiler, and condenser are rarely, if ever, packaged together with the steam turbine into a compact, transportable system.

This chapter addresses design techniques for sizing and coordinating components in a combustion turbine system, but it does not address design details of the turbomachines themselves. For more information on that subject, refer to Korpela (2019) or Dixon and Hall (2013).

9.1 Turbomachinery

The two predominant styles used in turbomachines to convert shaft power into fluid power (and vice versa) are radial and axial (Figure 9.1):

As seen in the sketches, the radial style of compressor converts shaft power input into fluid pressure in two steps: (i) the flow is accelerated tangentially and radially by the impeller's shaft work, and (ii) the tangential flow is then decelerated to a low velocity by the (quasi-toroidal) housing, which converts the fluid's kinetic energy into mechanical energy (pressure). The converse steps are true for the turbine: (i) the flow is accelerated tangentially inside the housing by nozzles and (ii) the tangential flow is decelerated by doing work on the impeller blade that converts kinetic energy into shaft work.

In contrast, the axial style of turbomachine accelerates and decelerates the flow with multiple stages of rotors and stators, but the motion produced is only axial and tangential – not radial.

Turbomachines are also separated into four primary categories, based on size and the origin of the technology, as follows:

- Turbochargers. Small (1-50 kW_e), radial-style machines that are used for engine performance enhancement on autos, trucks, locomotives, earth movers, and marine diesels.
- Microturbines. Medium (25-500 kW_e), radial-style machines that are used for small-scale power generation, often with unconventional fuels.
- Aero-derivatives. Large (5-50 MW_e), axial-style machines, initially designed for turboprop and turbojet aircraft propulsion systems, and later adapted for stationary power generation.
- Frame (industrial, utility). Very large (1-350 MW_e), axial-style machines, designed specifically for power generation, especially load-peaking applications.

Thermal Systems Design: Fundamentals and Projects, Second Edition. Richard J. Martin.

Companion website: www.wiley.com\go\Martin\ThermalSystemsDesign2

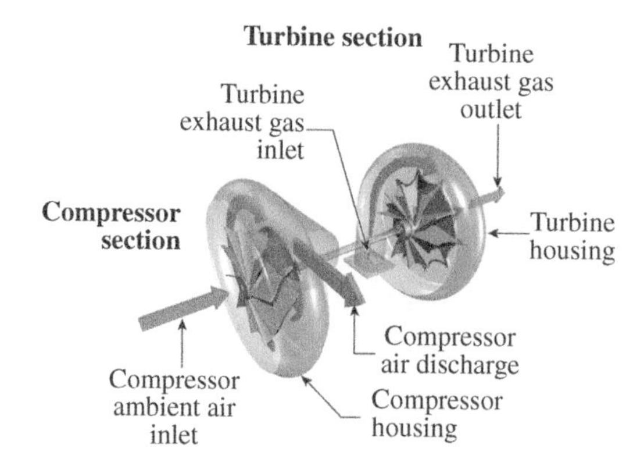

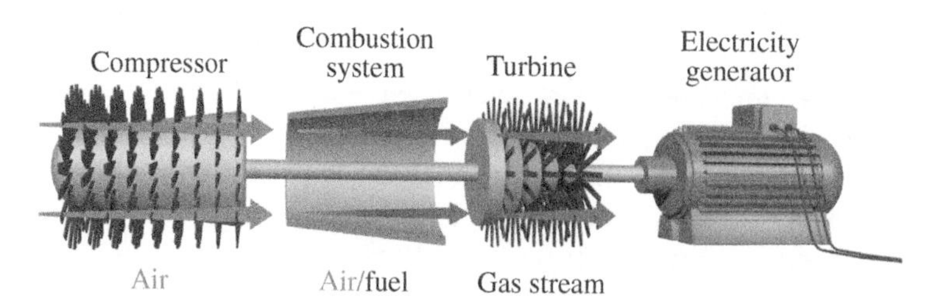

Figure 9.1 Turbomachinery: radial style (top) and axial style (bottom). *Source:* Martin Thermal Engineering Inc.

9.2 Brayton Cycle

The Brayton cycle is an open-loop power cycle involving a gas compression step, a heat addition step, and a gas expansion step. Figure 9.2 shows the classic (open) Brayton cycle, with air (including combustion products) as the working fluid.

Occasionally, a fourth step (cooling) is added, and the cycle becomes closed loop if the working fluid is of high value (e.g. helium). Unlike the Rankine cycle, the working fluid does not undergo phase change in the Brayton cycle. The open-loop cycle is also called an "air-breathing" cycle. As discussed below, ideal compressors and turbines undergo isentropic processes and ideal combustors undergo isobaric processes.

Combustion turbines, especially smaller ones, often operate with quite lean equivalence ratios ($\phi < 0.5$) and with exhaust oxygen levels around 15 % *vol*. Three design limitations are responsible for this level of air dilution in the exhaust: (i) the maximum turbine inlet temperature often must not exceed $T_3 \approx 2000\,^{\circ}F$, which puts an upper limit on the overall fuel–air ratio utilized; (ii) the air is preheated by the compression process reducing the need for fuel; and (iii) the air is often preheated by a recuperator, further reducing the need for fuel. With so much dilution air present, the combustion products become thermodynamically similar to air. For this reason, solving for the various states in a Brayton cycle is often performed with the assumption that properties of air can be substituted for the properties of dilute combustion products.

The cycle's inventor was George Brayton (ASME 2021) and his patent (Brayton 1872) was titled as "Improvements in Gas Engines." The invention, which contained separate cylinders and pistons for compression and power, looked like a strange internal combustion engine and it contained no

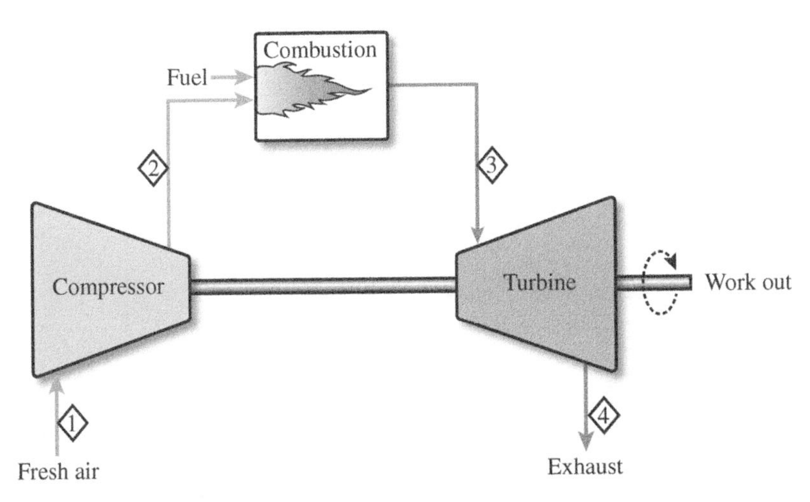

Figure 9.2 The (open) Brayton cycle. The closed form of the cycle would have a cooler between states 4 and 1, with none of the fluid going to exhaust, and no fuel or fresh air being admitted. *Source:* Martin Thermal Engineering Inc.

mention of turbomachinery. Today's gas turbine engines have an external combustion chamber that receives air from a compressor and delivers combustion products to the turbine, so it may seem like a stretch that Brayton should receive credit for the cycle that bears his name.

In fact, Brayton's main contributions were much more significant than they appear at first glance. Apparently, his perspective was to improve on limitations of the Otto cycle (*c. 1861*), not to compete with the Rankine cycle (*c. 1859*). The inventive content that differentiated Brayton's invention from the Otto cycle were: (i) employing separate devices for compression and power, (ii) delivering shaft power directly from the power piston to the compression piston, and (iii) employing a steady-flow burner to combust the fuel as it enters the power cylinder, instead of relying on a spark plug.

The efficiency of a Brayton cycle can be reported in two ways: (i) Carnot efficiency (dimensionless) or (ii) energy efficiency ratio (dimensioned):

$$\eta_{Brayton} = \frac{\dot{W}_{net,out}}{\dot{Q}_{fuel,in}} \left[\veeeq \frac{kW_{mech}}{kW_{chem}} = \langle dimensionless \rangle \right]$$

$$EER_{Brayton} = \frac{\dot{Q}_{fuel}}{\dot{W}_{elec}} \left[\veeeq \frac{(BTU/h)_{fuel}}{kW_e} \right]$$

As with any heat engine, the Carnot efficiency of a combustion turbine is governed by its maximum and minimum temperatures. However, with a Brayton cycle, the compressor power input is paid for out of the turbine power output, and the fuel consumed per unit of net shaft work goes down with higher pressure ratios, hence higher pressure ratios lead to higher cycle efficiencies.

The overall pressure ratio for a large, multistage compressor/turbine combination can reach high levels (e.g. $p_2/p_1 > 50$), but for microturbines and turbochargers, the overall pressure ratio is often much lower (e.g. $p_2/p_1 < 10$). Consequently, the cycle efficiency of many smaller combustion turbines is typically quite low compared to those of large frame and aero-derivative combustion turbines.

9.3 Polytropic Processes

Recalling Chapters 5 and 6, the following concepts are reiterated:

- ***Lines*** host the various ***states*** of the working fluid in a thermodynamic cycle.
- ***Equipment*** executes ***processes*** to change the working fluid's state as it traverses through the cycle.

Subject to that terminology distinction, a unique equation can be constructed (Equations 9.1) that embodies at least four unique ***processes***.

$$pv^n = constant \begin{bmatrix} isobaric \ n = 0 \\ isothermal \ n = 1 \\ isentropic \ n = k \\ isochoric \ n \to \infty \end{bmatrix} \tag{9.1a-d}$$

This formula, called the polytropic equation, is very useful for ideal gases, and it can be derived directly from the laws of mass, momentum, and energy conservation (with certain assumptions regarding irreversibilities and heat flows into or out of the control volume). The derivation is left to the student as a homework problem at the end of the chapter.

Figure 9.3 shows the four processes on a pressure versus volume graph. The value for n could theoretically be any positive real number between zero and infinity, but the four processes described are unique and worthy of attention by themselves.

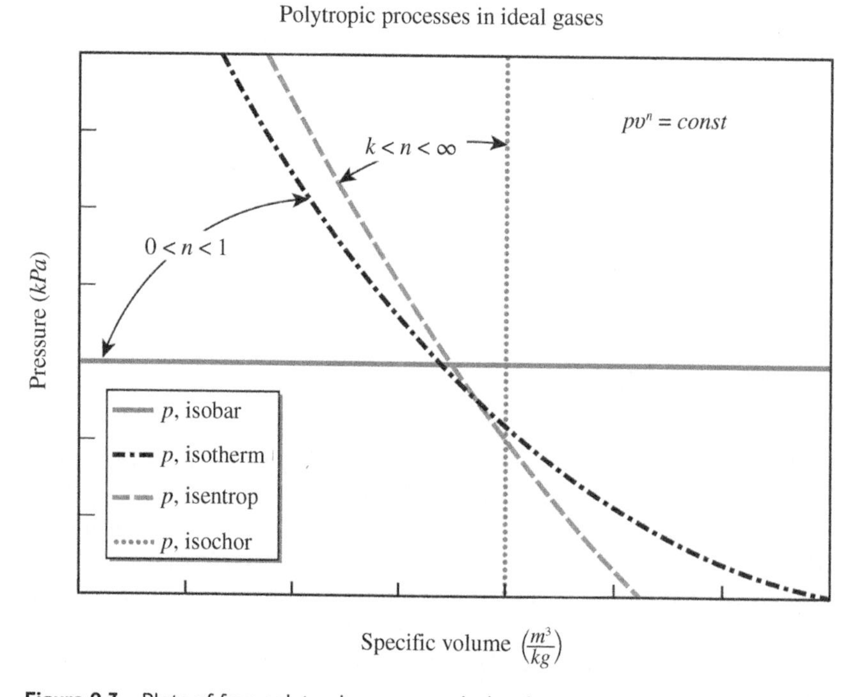

Figure 9.3 Plots of four polytropic processes: isobar (constant pressure), isotherm (constant temperature), isentrope (constant entropy), and isochor (constant specific volume). *Source:* Martin Thermal Engineering Inc.

The $n = 0$ and $n \to \infty$ cases should be self-explanatory. Ideal heat exchangers and ideal, isomolar reactors obey the isobaric process, because both have no friction and no shaft work. The idealized Otto cycle's heat addition and heat rejection steps obey the isochoric process because no shaft work occurs (no gas volume change) when the piston is momentarily stationary at its top dead center and bottom dead center positions.

The $n = 1$ case is relatively simple if the working fluid is an ideal gas. Recall the ideal gas relationship ($pv = RT$) for a specific gas with known molar mass $\hat{M}$, and consider also that when T is constant (i.e. for an *isothermal* process), the entire right side of the equation is a constant. Thus, an isothermal process involving an ideal gas ($pv = const$) is simply a polytropic process with =1 ($pv^1 = const$).

For the $n = k$ case, it was shown for an ideal gas in Section 6.3 that both ***isentropic*** compression and ***isentropic*** expansion processes obey the relationship (with the ratio of specific heats given as $k = c_p/c_v$):

$$pv^k = const \qquad (6.10)\ \{\text{IG, Ad, and Rev}\}$$

Returning to the general polytropic process for *ideal gases*, the following relationships (Equations 9.2) can be derived, all of which represent process characteristics for $n = 0$, $n = 1$, $n = k$, and $n \to \infty$ and other intermediate values for n:

$$p_1 v_1^n = p_2 v_2^n \qquad \frac{p_2}{p_1} = \left(\frac{v_2}{v_1}\right)^{-n} \qquad \boxed{\frac{v_2}{v_1} = \left(\frac{p_2}{p_1}\right)^{-\frac{1}{n}}} \qquad (9.2a)$$

$$\frac{p_2}{p_1} = \left(\frac{\left[\frac{RT_2}{p_2}\right]}{\left[\frac{RT_1}{p_1}\right]}\right)^{-n} = \left(\frac{T_2}{T_1}\right)^{-n}\left(\frac{p_2}{p_1}\right)^{n}$$

$$\therefore \left(\frac{T_2}{T_1}\right)^{-n} = \left(\frac{p_2}{p_1}\right)^{1-n} \qquad \boxed{\frac{T_2}{T_1} = \left(\frac{p_2}{p_1}\right)^{\frac{n-1}{n}}} \qquad (9.2b)$$

$$\left(\frac{v_2}{v_1}\right)^{-n} = \frac{p_2}{p_2} = \frac{\left[\frac{RT_2}{v_2}\right]}{\left[\frac{RT_1}{v_1}\right]} = \left(\frac{T_2}{T_1}\right)\left(\frac{v_2}{v_1}\right)^{-1} \quad \therefore \quad \boxed{\frac{T_2}{T_1} = \left(\frac{v_2}{v_1}\right)^{1-n}} \qquad (9.2c)$$

These relations are very helpful to determine the ending state of a (hypothetical) isentropic process, such as compression or expansion. When used together with the isentropic efficiency (addressed in the next section), the work output of *real* (not ideal) turbomachines can be predicted with confidence.

9.4 Isentropic Efficiency

If a compressor or turbine operates under adiabatic and reversible conditions, the gas will undergo an isentropic process between the inlet and outlet of the device. This condition represents the ideal performance for a device that converts shaft work to mechanical energy (of a gas) by increasing its pressure, or that converts mechanical energy of a gas to shaft work by reducing its pressure. Since real compressors and turbines operate under essentially adiabatic conditions (inconsequential heat loss to surroundings), the only factor that could cause them to deviate from isentropic behavior is irreversibility.

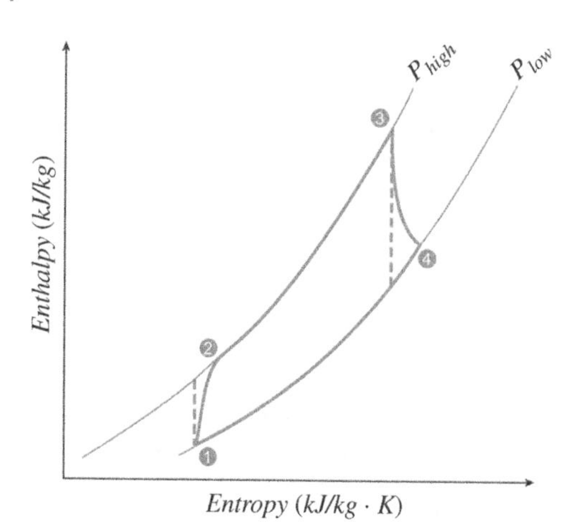

Figure 9.4 Ideal and real, closed-loop Brayton cycle plots of enthalpy versus entropy. State 1 is compressor inlet; state 2 is heater (or combustor) inlet; state 3 is turbine inlet; state 4 is cooler inlet (or exhaust if cycle is open). Unlabeled states 2*s* and 4*s* (connected by dashed lines) are ideal conditions that correspond to isentropic compressor and isentropic turbine, respectively. *Source:* Martin Thermal Engineering Inc.

As with all process equipment, the Second law compels real turbines and compressors to operate with some level of irreversibility. Engineers quantify this irreversibility in turbomachines with a parameter called the isentropic efficiency.

Think Stop. The isentropic efficiency applies to a piece of equipment, not to an entire system. As such, isentropic efficiency is very different from cycle efficiency. The isentropic efficiency is a measure of how reversibly the turbine or compressor is behaving, and the irreversibility always shows up as greater levels of entropy and enthalpy in the fluid exiting the device than the levels that would have been if the process had been fully reversible.

Figure 9.4 shows ideal (dashed) and real (solid) compressor and turbine processes in a Brayton cycle system. The states 2*s* and 4*s* represent the points in an ideal Brayton cycle that would correspond to isentropic compression or expansion and the states 2 and 4 represent real compressors and turbines that do not execute their processes reversibly.

The isentropic efficiency of a compressor and turbine are defined as shown in Equations (9.3). Irreversibilities in compressors and turbines can result from shaft friction, fluid friction, and irreversible accelerations and decelerations of the fluid as it is being compressed or expanded.

$$\boxed{\eta_c = \frac{W_{ideal}}{W_{actual}} = \frac{(h_{2s} - h_1)}{(h_2 - h_1)}} \quad \boxed{\eta_t = \frac{W_{actual}}{W_{ideal}} = \frac{(h_3 - h_4)}{(h_3 - h_{4s})}} \tag{9.3a-b}$$

As indicated previously and seen in Figure 9.4, the actual exit condition of the working fluid in each type of device has higher values for both entropy and enthalpy than the ideal exit condition. Not surprisingly, both conditions result in less favorable performance – more work into the compressor and less work out of the turbine. The production of entropy leads to a result where the fluid leaves with more enthalpy than the ideal case would have, and a penalty is assessed to the work.

Commercially available compressors and turbines have isentropic efficiencies in the 70–85% range.

Example 9.1

For a compressor-heater-turbine system with *T* and *p* given for all four states: (i) find the isentropic efficiencies of the compressor and turbine and (ii) compute the net shaft work transmitted out of the system per unit mass of the working fluid – air. Use property tables from chapter 19 and interpolate where necessary, or use *NIST Chemistry Webbook* (Linstrom and Mallard 2017–2021) for greatest accuracy.

The given conditions are:

$$p_1 = 101.33\ kPa,\quad T_1 = 300.00\ K \qquad p_2 = 500.0\ kPa,\quad T_2 = 500.00\ K$$

$$p_3 = 500.0\ kPa,\quad T_3 = 1000.0\ K \qquad p_4 = 101.33\ kPa,\quad T_4 = 725.00\ K$$

Answer. Look up all the actual enthalpies and entropies using the thermodynamic tables in Appendix A.1:

$$h_1 = 302.2245\frac{kJ}{kg}\ s_1 = 6.741\ 77\frac{kJ}{kgK}\ h_2 = 506.4852\frac{kJ}{kg}\ s_2 = 6.802\ 651\frac{kJ}{kgK}$$

$$h_3 = 1053.856\frac{kJ}{kg}\ s_3 = 7.556\ 94\ 8\frac{kJ}{kgK}\ h_4 = 745.5578\frac{kJ}{kg}\ s_4 = 7.657\ 391\frac{kJ}{kgK}$$

To determine η_s, find states 2*s* and 4*s*. To accomplish this, state 2*s* should be locked in by $p_{2s} = p_2$ and $s_{2s} = s_1$. Similarly, locking in state 4*s* requires $p_{4s} = p_4$ and $s_{4s} = s_3$. The new properties are found by identifying the temperatures and enthalpies that correspond to the known pressures and entropies:

$$p_{2s} = 500\ kPa\ \ T_{2s} = 471.54\ K\ \ p_{4s} = 101.33\ kPa\ \ T_{4s} = 660.74\ K$$

$$h_{2s} = 476.994\frac{kJ}{kg}\ s_{2s} = 6.741\ 77\frac{kJ}{kgK}\ h_{4s} = 676.147\frac{kJ}{kg}\ s_{4s} = 7.556\ 95\frac{kJ}{kgK}$$

The isentropic efficiencies of the two turbomachines are:

$$\eta_c = \frac{(h_{2s} - h_1)}{(h_2 - h_1)} = \frac{(476.99 - 302.22)}{(506.49 - 302.22)} = \boxed{85.6\%}$$

$$\eta_t = \frac{(h_3 - h_4)}{(h_3 - h_{4s})} = \frac{(1053.86 - 745.56)}{(1053.86 - 676.15)} = \boxed{81.6\%}$$

The net shaft work per unit mass exiting the system is $W_{net} = (h_3 - h_4) - (h_2 - h_1)$ $= 308.30 - 204.27 = \boxed{104.0\ kJ/kg}$.

Fuel Addition. For aero-derivative combustion turbine engines, the fuel is typically a liquid (e.g. jet fuel), and it is delivered to the combustion chamber by a high-pressure pump and a spray nozzle, where the droplet cloud ignites and burns in a continuous (standing) flame.

For gaseous fuels (e.g. natural gas), the system designer may be tempted to premix the fuel with the air and compress the premixture in a single stream to eliminate the need for a separate gas compressor and nozzle within the combustion chamber. While the case for simplicity and cost reduction are compelling, thermal system designers must not lose sight of the fact that fuel and air are typically not commingled until they just beyond the burner tip and within the flame envelope (as is the case with the nozzle-mix burner from Chapter 7).

Considerable safety issues arise when premixing fuel into air and conveying the mixture to a location where it is ignited and burns exothermically. Designers should avoid this option unless they can prove that under *all* circumstances, the accumulation of premixed air and fuel *cannot* undergo a deflagration that will damage or rupture the system's pressure containment envelope.

Conditions that may allow for this level of explosion prevention assurance include recuperative Brayton cycles (Section 9.7) and low-compression ratios. Whichever gaseous fuel is selected (e.g. methane or propane), the mixture's composition must remain comfortably below the fuel's lower flammability limit (LFL). Conventional wisdom for other combustion systems says that 25% of the fuel's LFL is a safe upper limit for fuel air premixtures. (In other words, if we suppose $LFL_{CH_4} \approx 5.0\,\%\,vol$, then the premixture should never exceed 25% of that concentration, or $\chi_{max\ CH_4\ \text{in air}} \approx 1.25\,\%\,vol$). Recuperation in a Brayton cycle may be sufficient to ensure the amount of fuel needed to reach the desired turbine inlet temperature that is below this level, but analysis is required to confirm.

However, even an upper limit of 25% LFL may not be sufficient to prevent damage to the engine under circumstances where the compression ratio is high and/or where the isentropic efficiency of the compressor is low. While a deflagration may not occur even if the non-isentropic compression process causes the fuel–air mixture to reach a temperature higher than the fuel's autoignition temperature (AIT), homogenous combustion of fuel may begin before the inlet mixture leaves the compressor, and the exothermicity of those premature combustion reactions could overheat and damage the compressor blades.

Hence, if a system designer wishes to admit a premixture of fuel and air into their compressor, they likely will be required to confirm that both of the following safety criteria are met: (i) ensure the fuel–air mixture never exceeds 25% of the fuel's LFL and (ii) ensure the temperature rise in the compressor falls comfortably below the fuel's AIT.

A further complication to this design analysis is that the temperature rise across the compressor (effectively a preheating of the fuel–air mixture) will serve to broaden its flammability limits. This means that the 25% safety margin should be applied to the temperature broadened LFL instead of the standard LFL determined at STP. NFPA 86 (2019) provides guidance about how to estimate the magnitude of LFL reduction that may occur when a fuel–air premixture exists at an elevated temperature. If conditions that could cause an explosion or premature combustion at any point in the system upstream of the combustor are even slightly probable, it is generally expected that separate compressors for the fuel and air are required.

9.5 Gas Property Relationships

From the prior analysis, it is apparent that any decline in its isentropic efficiency robs a turbomachine of the ability to transform a fluid's mechanical and thermal energy (i.e. pressure and temperature) into shaft work, or vice versa. This section presents a number of tools and concepts that may help students to analyze a real turbomachine or refine their intuition about the laws and properties that affect turbines and compressors.

Diverging Pressure Curves. The factor that is most responsible for the Brayton cycle's ability to generate net power from sequential compression, heating, and expansion processes is the divergence of the pressure lines as enthalpy and entropy are increased (Figure 9.4).

Since the isentropic work out of the turbine ($W_{t,s} = h_3 - h_{4s}$) is extracted from gas enthalpy at high temperature and the isentropic work into the compressor ($W_{c,\,s} = h_{2s} - h_1$) is delivered to gas at low temperature, the net shaft work output will be positive for a significant range of turbomachinery isentropic efficiencies that nonetheless fall short of $\eta_s = 100\%$. In general, the divergence of the pressure curves ensures that isentropic work terms follow this rule. Yes, one can imagine cases where the turbomachines' isentropic efficiencies (η_t, η_c) are so low that the net shaft work output is negative. However, with typical isentropic efficiencies and the employment of a high turbine inlet temperature plus a low compressor inlet temperature, combustion turbine systems are generally viable heat engines.

Density Recovery. Process inefficiencies often result in the transformation of highly useful mechanical energy into less useful thermal energy. For gases undergoing compression (e.g. mechanical shaft work in), any entropy-producing inefficiency results in ***reduced density*** of the outflowing gas compared to the isentropic case. The converse is not entirely intuitive, but it too makes sense upon examination. For gases undergoing expansion, any entropy-producing inefficiency *also* results in reduced density of the outflowing gas. Both of these results are obtained because lower gas density occurs concurrently with higher temperature and higher specific volume – the two hallmarks of increased entropy.

As a side note, it may also be slightly nonintuitive that higher temperature drives entropy higher, but if we recall the corollary to the Third law of thermodynamics – the entropy of a perfect crystal approaches zero as temperature approaches absolute zero – we recognize that larger volumes and larger temperatures both lead to greater disorder.

The concept of density recovery (i.e. what is the ratio of outlet gas density to inlet gas density) is another way of thinking about isentropic efficiency of turbomachinery. If we have another look at Equations (9.2), which were derived directly from the polytropic equation (Equation 6.10), we find that the process that maximizes density during compression of an ideal gas occurs when $n = 1$, the isothermal case. For an ideal gas in this case, the exponent of $pv^n = const$ is exactly 1.0, so 100% of the pressure rise shows up as density rise.

$$\frac{\rho_2}{\rho_1} = \left(\frac{p_2}{p_1}\right)^{\frac{1}{n}}$$

This occurs only when we are extracting (in the form of heat) exactly the same amount of energy as we have admitted (in the form of work). The $n = 1$ case actually results in a decrease in the entropy of the outflowing gas because of the heat removal (and the reduction in specific volume). When we increase pressure and density at constant temperature, the entropy goes down.

For the isentropic case (for ideal gases), we have $n = k$ and the density recovery is less than 100% but still reasonably high. If we go further and examine a case of *very poor* isentropic efficiency (i.e. any polytropic equation where $n \gg k$), very little of the pressure rise shows up as density rise. In the limit of $n \rightarrow \infty$, the density does not increase at all. This is because the incoming shaft work contributed just as much to the temperature increase as it did to the pressure increase.

Students should note that very large pressures will lead to nonideal gas effects, which must be evaluated on a case-by-case basis. In such cases, Appendix A.1 or the *NIST Chemistry Webbook* (Linstrom and Mallard 2017–2021) are acceptable resources and the ideal gas relations may be unacceptable.

9.6 Reheating, Intercooling

Larger combustion turbines often utilize intercooling between compression stages and reheating between turbine stages to increase net power output. The reheating process requires more fuel, which is a cost, but the benefit is particularly attractive if the expansion process is already being performed in two (or more) stages. By boosting the temperature between stages of expansion, the work output is increased significantly due to the divergence of the pressure lines at high temperatures.

Similarly, for intercooling, if compression is already being performed in two or more stages, passing the warm air through a cooler between compression stages permits a significant reduction in

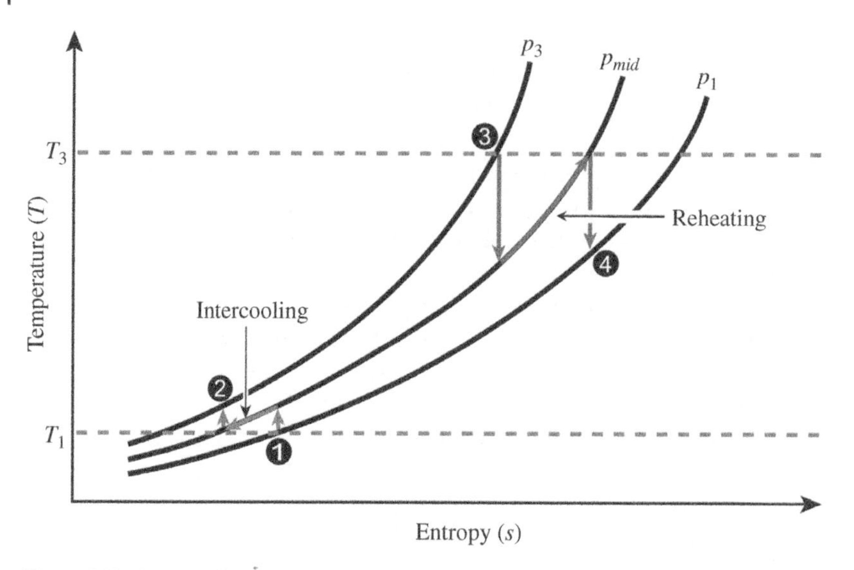

Figure 9.5 Intercooling and reheating to increase power output from multistage turbomachinery. Blue arrows show intercooling between a pair of compression stages to reduce compression power input. Red arrows show reheating between a pair of expansion stages to increase expansion power output. Compression and expansion are shown as isentropic (ideal). *Source:* Martin Thermal Engineering Inc.

power input for the latter stage(s). Again, this improvement is due to the convergence of the pressure lines at low temperatures. It takes less work to compress a cool gas than a warm one. Figure 9.5 shows the details of intercooling and reheating on a $T - s$ diagram.

Think Stop. Why would reheating be appropriate only between expansion stages and intercooling only be appropriate between compression stages? For reheating, greater enthalpy results in more power output, which is the primary goal of reheating. For intercooling, lower enthalpy change per compression stage results in lower overall power usage for compression. This also enhances the goal of producing more net power and/or making the cycle more efficient.

Think Stop. Is the cost of reheating worth the extra cost of hardware and fuel to reheat the gas after the first stage of expansion? The answer to this question is maybe. A combined application of reheating and intercooling actually causes the entire overall cycle efficiency to decline (see Problem 9.4 at the end of the chapter) because of all the extra fuel consumed in the reheating process. However, the ability to generate and sell more output power could have a significant positive economic benefit, with respect to the total investment in the equipment to do so.

See Table 9.1for a comparison of the related but opposite attributes of intercooling and reheating for combustion turbines.

9.7 Recuperation

As demonstrated in Section 7.5, recuperation delivers a benefit of fuel economy for thermal oxidation processes where the VOC-laden stream is highly diluted with air or an inert gas. The same benefit can be obtained by a combustion turbine under many circumstances. By preheating the air stream prior to combustion (and using the hot turbine exhaust gas to do so), the amount of fuel necessary to reach the desired turbine inlet temperature is reduced.

Table 9.1 Comparison of intercooling and reheating attributes for turbomachinery processes.

Feature	Intercooling	Reheating
Best-suited process	Compression Shaft work in	Expansion Shaft work out
Pressure line behavior	Converge for cooling Reduced work in	Diverge for heating Increased work out
Density optimization	High density fluids need less $p \cdot dv$ work to obtain a desired pressure rise	High temperatures deliver more $p \cdot dv$ work over a given pressure drop
Reversibility of shaft work to mechanical energy	Smaller pressure ratio (compression or expansion) may result in greater reversibility for each stage but does not result in greater overall cycle efficiency	

Source: Martin Thermal Engineering Inc.

Example 9.2

Assuming a recuperator could be added anywhere in the simple Brayton cycle (as long as the Second law does not thwart the heat recovery process because actual cycle temperatures will not permit heat to flow in the desired direction): (i) identify the most logical place from which to draw hot gas for the inlet of the hot side of the recuperator, and (ii) identify the most logical place from which to draw cool gas for the inlet side of the cold side the heat exchanger. See Figure 9.4 for guidance.

Answer. Based on Figure 9.4 and Example 9.1, there are only two places where "hot gas" could be extracted – upstream of the turbine ($T_3 = 1000\ K$) and downstream of the turbine ($T_4 = 734\ K$). Similarly, there are only two places where "cold gas" could be extracted – upstream of the compressor ($T_1 = 298\ K$) and downstream of the compressor ($T_2 = 465\ K$). Thankfully, for this example, both of the candidate "cold gas" temperatures are lower than both of the candidate "hot gas" temperatures, so the Second law will not thwart this example, no matter which points are chosen.

From a purely fuel-saving perspective, one might imagine that the best overall result would be to pick the highest hot temperature and the lowest cold temperature, because that would create the largest driving force and more heat would be recuperated, which would reduce fuel the most. Unfortunately, this logic is fallacious because it fails to account for the compressor work in and turbine work out.

In order to ensure the work into the compressor is as low as possible, the recuperator should refrain from preheating the combustion air until *after* it is pressurized. Similarly, in order to ensure the work out of the turbine is as high as possible, the recuperator should refrain from extracting heat from the combustion gases until *after* they are expanded.

The preferred recuperative combustion turbine cycle is therefore comparable to the schematic shown in Figure 9.6. Note that the recuperator must be built to withstand high-pressure air on the cold gas side with low-pressure exhaust products on the high-temperature side.

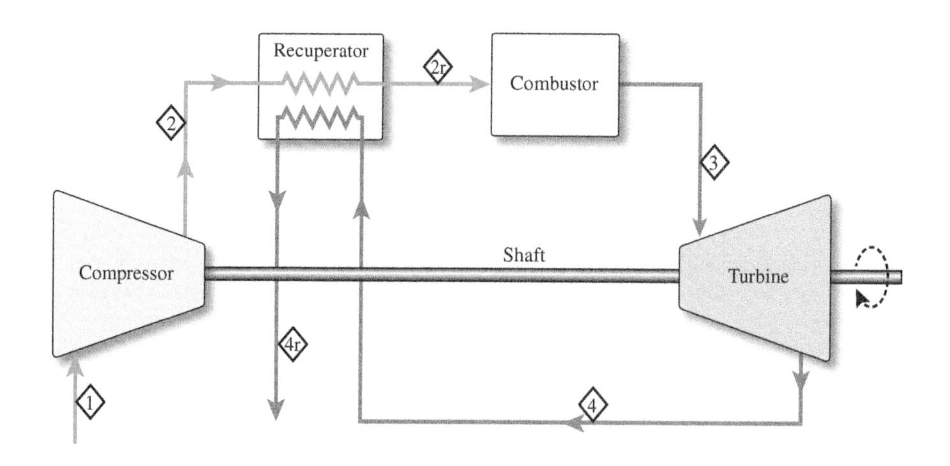

Figure 9.6 Recuperated combustion turbine cycle. The hot side of the heat exchanger is low-pressure exhaust gas (after exiting the turbine) and the cold side of the heat exchanger is high-pressure air (after exiting the compressor). *Source:* Martin Thermal Engineering Inc.

9.8 Homework Problems

9.1 Determine the missing properties in the empty boxes for the stream table given below. The working fluid is helium. The schematic for this stream table is shown in Figure 9.4 for an ideal, closed-loop, Brayton cycle (following dashed lines to unlabeled states 2*s* and 4*s* instead of solid lines to labeled states 2 and 4). Use the property tables for helium in Appendix A.6 to obtain values of thermodynamic properties and interpolate where necessary or use Linstrom and Mallard (2017–2021). Determine net work output per unit mass $W_{net} = ((h_3 - h_4) - (h_2 - h_1))[\doteq kJ/kg]$ and heat input per unit mass $Q_{in} = (h_3 - h_2)[\doteq kJ/kg]$.

Answer. $W_{net} = 1061.8\ kJ/kg$ and $Q_{in} = 1523.9\ kJ/kg$

Property	State 1	State 2s	State 3	State 4s
Temperature (K)	275.0		1200.0	
Pressure (kPa)	101.33	2000	2000	101.33
Spec Volume (m^3/kg)	5.6403			
Enthalpy (kJ/kg)	1433.6			
Entropy ($kJ/kg\ K$)	27.547	27.547		

9.2 Determine the missing enthalpies, entropies, pressures, and temperatures for the stream table given below. The schematic for this stream table is shown in for an open-loop Brayton cycle with a recuperator. Assume air and fuel enter the compressor at sea level *STP* conditions, and zero mass is added in the combustor, and also assume the working fluid properties are equal to those of air at all states. Ignore pressure losses in the heating and recuperation steps (i.e. $p_3 \approx p_2 \approx p_{2r}$ and $p_4 \approx p_{4r} \approx p_1$). Let the isentropic efficiencies of the compressor and turbine be $\eta_c = 80\%$ and $\eta_t = 70\%$. Use Appendix A.1 for air properties and interpolate where necessary or use Linstrom and Mallard (2017–2021) or use the customized Excel functions in Appendix B.3. Assuming recuperator effectiveness of $\varepsilon = 85\%$, compute the heat input per unit mass (i) without recuperation ($Q_{in} = h_3 - h_2$) and (ii) with recuperation ($Q_{in,\ r} = h_3 - h_{2r}$).

Property	State 1	State 2s	State 2	State 2r	State 3	State 4s	State 4	State 4r
Temperature (K)	298.15				1000.0			
Pressure (kPa)	101.33	500	500	500	500	101.33	101.33	101.33
Enthalpy (kJ/kg)								
Entropy ($kJ/kg\ K$)								

9.3 Estimate the isentropic efficiencies of the compressor and turbine for a combustion turbine system without recuperation that has a net cycle efficiency of $\eta_{cycle} = 24.0\%$, assuming $\eta_c \approx \eta_t$, $p_2/p_1 \approx p_3/p_4 = 5.0$, with state 1 conditions at STP and turbine inlet temperature of $T_3 = 1000\ K$. (a) Use polytropic relationships for air with constant specific heat assumptions $c_p = 1.07\ kJ/kgK$ and $k = 1.4$. Iterate on isentropic efficiencies to obtain desired cycle efficiency. (b) Compare results to those obtained using property tables for air in Appendix A.1, interpolating where necessary.

Answer. (a) $\eta_c = \eta_t = 0.870$, (b) $\eta_c = \eta_t = 0.877$

9.4 Consider a closed-loop, gaseous nitrogen Brayton cycle with one stage of intercooling between compressors and one stage of reheating between turbines. Assume the three pressure stages are $p_1 = p_6 = 200\ kPa$ and $p_2 = p_5 = 500\ kPa$ and $p_3 = p_4 = 1000\ kPa$, which implies that pressure losses in the heater, reheater, cooler, and intercooler can be considered negligible. State 1 conditions are $p = 200\ kPa$ and $T = 300.0\ K$ and the turbine inlet temperature is $T_4 = 1300\ K$. Use polytropic relationships for nitrogen with average specific heats assumed constant at $c_{p,\ avg} = 1.13\ kJ/kgK$ and $k_{avg} = 1.36$. Assign values of $\eta_c = \eta_t = 87\%$ for all stages of compression and expansion. Let the intercooler return the temperature to $T_{2i} = T_1 = 300\ K$, and let the reheater return the temperature to $T_{5r} = T_4 = 1300\ K$. (i) Compute net work output per unit mass $W_{net}[\doteq kJ/kg]$ and cycle efficiency η_{cycle} for the cycle described earlier. (ii) Compare values for net work output and cycle efficiency to a simple cycle with neither intercooling nor reheating. Discuss results.

Answer. (a) $W_{net} = 304.1\ kJ/kg$, $\eta_{cycle} = 24.0\%$ (b) $W_{net} = 236.4\ kJ/kg$, $\eta_{cycle} = 25.6\%$

9.5 An ideal gas, closed Brayton cycle uses helium (He) as a working fluid with a mass flow rate of $\dot{m} = 4.1\ lb_m/s$. The compressor and turbine isentropic efficiencies are $\eta_c = 0.77$ and $\eta_t = 0.79$, and their pressure ratios are $(p_2/p_1) = (p_3/p_4) = 3.75$. The compressor inlet and turbine inlet temperatures are $T_1 = 50\ ^\circ F$ and $T_3 = 1450\ ^\circ F$, respectively. A recuperator with effectiveness $\varepsilon = 0.85$ is installed to cross-exchange heat from the turbine exit to the heater inlet. (Note that a closed-loop Brayton cycle must have a cooler installed between the exhaust and the inlet to return the gas to its initial temperature at state 1). Calculate (a) the net power $\dot{W}_{net}[\doteq hp]$ and the cycle efficiency $\eta_{cycle} = \dot{W}_{net}/\dot{Q}_{heater}$ without recuperation; and (b) $\eta_{cycle,\ r}$ with recuperation. Use the polytropic relationship to determine temperatures T_{2s} and T_{4s}, and assume constant specific heats for Helium $c_p = 1.2403\ BTU/lb_mR$ and $c_v = 0.7442\ BTU/lb_mR$. (This is actually a very good assumption over a wide range of temperatures because Helium is a monatomic, noble gas. See tabulated specific heat data in Appendix A.6.)

Answer. (a) $\dot{W}_{net} = 1,139\ hp$; $\eta_{cycle} = 16.9\%$, (b) $\eta_{cycle,\ r} = 23.7\%$

Cited References

ASME; (2021); *George Brayton (Short Biography)*; American Society of Mechanical Engineers, Topics and Resources.; https://www.asme.org/topics-resources/content/george-brayton.

Brayton, G.B. (1872). Improvement in gas engines. *US Patent* **125**,166.

Dixon, S.L. and Hall, C.; (2013); *Fluid Mechanics and Thermodynamics of Turbomachinery*, **7**; Amsterdam: Elsevier B.V.

Korpela, S.A.; (2019); *Principles of Turbomachinery*, 2; Hoboken: Wiley.

Linstrom, P.J. and Mallard, W.G., Eds.; (2017–2021); *NIST Chemistry WebBook, SRD 69; Thermophysical Properties of Fluid Systems*; National Institute of Standards and Technology, U.S. Dept. of Commerce. http://webbook.nist.gov/chemistry/fluid

NFPA 86; (2019); *Standard for Ovens and Furnaces; Annex D: The Lower Limit of Flammability and the Autogenous Ignition Temperature of Certain Common Solvent Vapors Encountered in Ovens*; Quincy, MA: National Fire Protection Association.; https://nfpa.org/86.

Reynolds, W.C.; (1977); *Engineering Thermodynamics* 2; New York: McGraw-Hill

Reynolds, W.C.; (1979); *Thermodynamic Properties in SI*; Stanford University Mechanical Engineering Department.

10

Refrigeration and Heat Pumps

Broadly, the term ***heat pump*** carries a connotation of any process related to cooling or heating that utilizes a refrigerant fluid plus several mechanical components to move thermal energy uphill in the temperature terrain that is associated with a temperature altered (indoor) state versus an ambient temperature (outdoor) state.

Unlike a fluid pump that is a single piece of equipment, most ***heat pump*** systems require at least four pieces of equipment working together to accomplish the overall goal, which is to utilize electrical energy (or shaft power) to extract heat from a colder space and deposit it into a warmer space.

When the primary objective is to extract heat out of a cold (outdoor) environment and deposit it into a warmer, inhabited space, the system keeps the generic name of ***heat pump***. When the objective is to cool or dehumidify a warm, inhabited space or a space for cold storage of food or other heat-labile materials, the system is called an ***air conditioner*** or ***refrigerator***. The sections that follow include information about refrigeration cycles, measures of energy efficiency, functionality of the various system components, and examples of refrigeration system failures. For an excellent treatise on industrial refrigeration systems, see Stoecker (1998).

10.1 Vapor Refrigeration Cycle

The ***ideal*** vapor refrigeration cycle is nearly the converse of the ideal Rankine cycle. For comparison, the ideal Rankine cycle has two constant-pressure, phase-change processes and two constant-entropy, mechanical-shaft-work pressure-change processes. In contrast, the ***ideal*** vapor refrigeration cycle has two constant-pressure, phase-change processes and one constant-entropy, mechanical-shaft-work pressure-change process (and a second pressure-change process that involves no shaft work, so the enthalpy remains the same). See Figure 10.1 for a p - H diagram showing the ideal vapor refrigeration cycle.

While other authors may use a different convention for numbering states in a refrigeration cycle, this textbook adheres to a standard designation (that was also described in Chapters 8 and 9), where state 1 is always immediately upstream of the pressure-increasing device. For a vapor refrigeration cycle, this convention puts state 1 at the compressor inlet.

The four pieces of equipment that make up the ideal vapor refrigeration cycle are:

- From 1 to 2 is the isentropic ***compressor***, where the refrigerant goes from low-pressure saturated vapor to high-pressure superheated gas, along a line of constant entropy.

Thermal Systems Design: Fundamentals and Projects, Second Edition. Richard J. Martin.

Companion website: www.wiley.com\go\Martin\ThermalSystemsDesign2

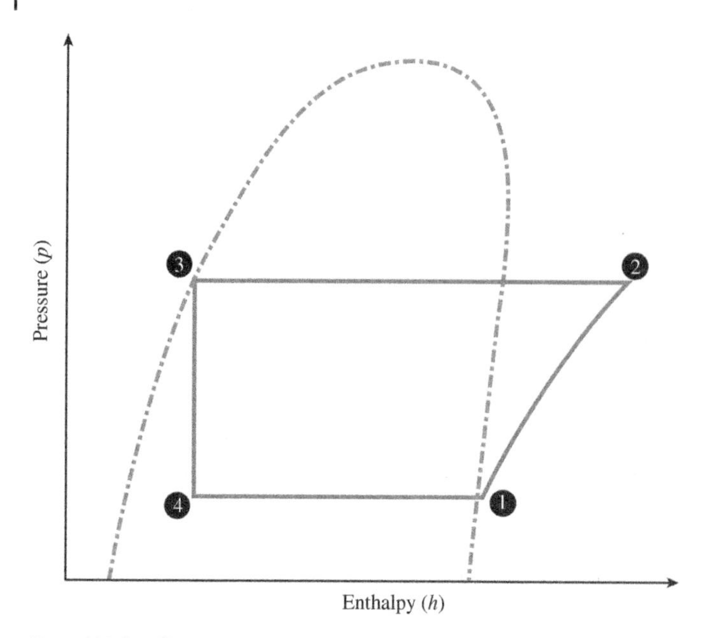

Figure 10.1 The ideal vapor refrigeration cycle is shown in solid line with the vapor dome of the working fluid shown dot-dash line. The unique features of this cycle are (i) isentropic compression from state 1 to state 2, (ii) both heat exchangers exiting precisely at their respective saturation conditions (states 3 and 1), and an isenthalpic expansion valve (from state 3 to state 4). *Source:* Martin Thermal Engineering Inc.

- From 2 to 3 is the isobaric ***condenser***, where the refrigerant rejects heat from high-temperature refrigerant to a slightly less warm environment at constant pressure.
- From 3 to 4 is the isenthalpic ***expansion valve***, where the refrigerant goes from high-pressure saturated liquid to a low-pressure saturated liquid–vapor mixture, along a line of constant enthalpy.
- From 4 to 1 is the isobaric ***evaporator***, where the refrigerant absorbs heat from a cold environment at constant pressure.

By contrast, the ***real*** vapor refrigeration cycle utilizes the same four pieces of equipment but differs from the ideal cycle in two aspects: (i) the real cycle succumbs to the Second law (which imposes a loss of efficiency in the compression step that leads to higher entropy and higher temperature at state 2 and greater work input for the same pressure rise) and (ii) the real cycle avoids p - T ambiguity at the heat exchanger exits (by deliberately crossing past the edges of the *vapor dome*). See Figure 10.2 for details of the real vapor refrigeration cycle.

In the real cycle, states 1 and 3 are intentionally relocated outside the saturated vapor and saturated liquid lines, respectively, for two reasons: (i) when these loci are at the very edge of the vapor dome, the control system has a difficult time determining the exact location of state 3 (because temperature and pressure sensors are unable to determine the condition where vapor quality equals exactly zero) and state 1 (because temperature and pressure sensors are unable to determine the condition where vapor quality equals exactly one); and (ii) safety margins are vital to ensure the gas entering the suction remains superheated and the liquid entering the expansion valve remains subcooled. As will be discussed in Section 10.10, liquid entering the compressor and vapor entering the expansion valve can lead to serious problems for system performance and longevity.

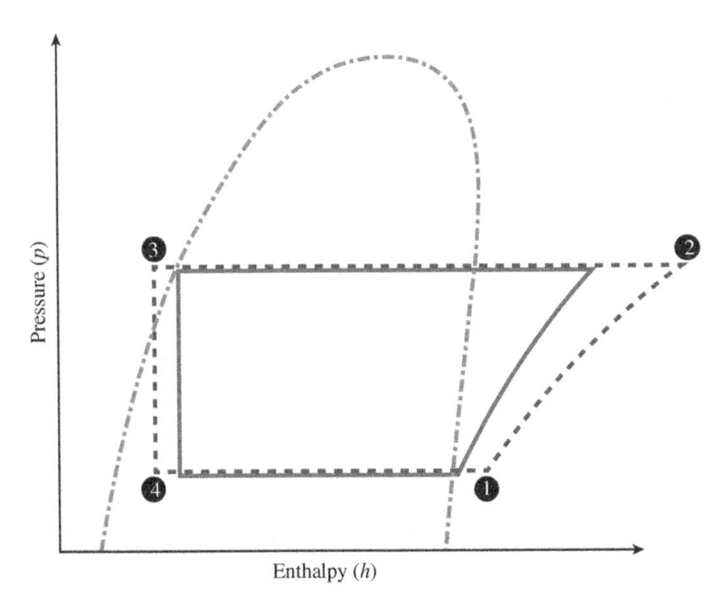

Figure 10.2 The real vapor refrigeration cycle is shown (dashed line) superimposed on the ideal vapor cycle (solid line), with the vapor dome shown in dot-dash line. The real cycle features are (i) non-isentropic compression from state 1 to state 2, (ii) subcooled conditions after the condenser at state 3, (iii) constant enthalpy expansion to state 4, and (iv) superheated conditions after the evaporator at state 1. *Source:* Martin Thermal Engineering Inc.

The performance of a ***real*** vapor refrigeration cycle depends on all of the following factors that the thermal engineer has control over:

- Pressure rise versus volumetric flow rate characteristic (i.e. performance curve) of the compressor for a given rotational speed.
- Rotational speed of the rotor shaft, as determined by motor speed.
- Heat-transfer area and overall heat-transfer coefficient (*UA* product) of the condenser.
- Degrees of subcooling, as determined by a small bypass expansion valve with temperature feedback.
- Thermostatic expansion valve (*TXV*) characteristics (i.e. flow resistance at wide open condition and pressure drop versus flow rate curve for different opening percentages).
- Heat-transfer area and overall heat-transfer coefficient (*UA* product) of the evaporator.
- Degrees of superheating as determined by the spring setting on the *TXV*. See also Chapter 14 for a deeper discussion of the *TXV*.
- Total internal volume associated with the hermetically sealed, refrigerant-wetted components of the system (including pipes, fittings, heat exchangers, compressor, valves, etc.).
- Mass of refrigerant added to the system after it has been evacuated of air and any other gas impurities.

Think Stop. Students may not have an intuitive awareness about the importance of the last two factors in the list given earlier, but correctly "charging" the system with refrigerant is a task that must not be overlooked. In one of two extreme conditions, it can be imagined that the greatest possible amount of refrigerant that could be added to a particular system with a given internal volume would be the mass that fills the system under saturated liquid conditions at ambient temperature.

By filling the system to this extent with liquid there is no spare volume available for refrigerant gas to occupy. The implication of this is that the compressor will be trying to compress an "incompressible" liquid, which can cause damage to mechanical components and/or overheating of the compressor's electric motor. The opposite extreme filling condition would be to add just enough refrigerant mass to fill the system volume with saturated vapor at ambient temperature. This situation compromises the ability of the evaporator to absorb heat from the cold space.

These two refrigerant charge quantities are obviously very different, and as one might imagine, the ability of the compressor, *TXV*, and heat exchangers to cause pressure changes and phase changes will depend on the refrigerant's charged density (charged mass divided by system volume). To determine the correct charge, a simulation should be performed that examines the hydrostatic forces in the installed system at steady state and provides operational safety margins.

10.2 Gas Refrigeration Cycle

The gas refrigeration cycle (see Figure 10.3) is very much the converse of the "closed" Brayton cycle located far to the right of the vapor dome on a T-s plot. Begin with air (or another ideal gas) at *STP* as the working fluid at initial state ①. Then, undergo a real (not isentropic) compression step to state ②, which will increase the temperature, pressure, and entropy. Follow that with an approximately

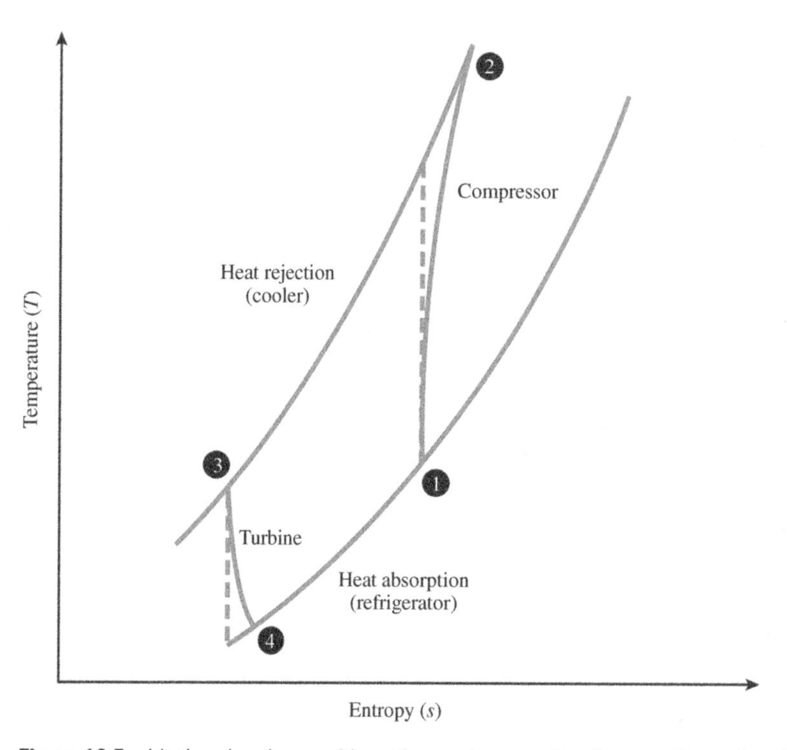

Figure 10.3 Ideal and real gas refrigeration cycles on a *T-s* diagram. Since the refrigerant never changes phase, it would be incorrect to use device names from the vapor refrigeration cycle (evaporator, condenser). Unfortunately, industry convention is confusing – the heat rejection device is called a "cooler" and the heat absorption device is called a "refrigerator," despite the apparent contradiction. *Source:* Martin Thermal Engineering Inc.

isobaric heat rejection step (akin to the condensing step in the vapor refrigeration cycle with ambient air as the heat sink) to reach state ③, which is higher in temperature than state ① and at higher pressure. Then, expand the gas using a turbine (with the shaft work transmitted to the compressor to reduce its required input work) to reach state ④, where the pressure is ambient, and the temperature is much colder than ambient. Finally, implement an isobaric heat absorption step (extracting heat from the desired cold space), ending back at state ①, with the working fluid at *STP*, where the cycle begins again.

From Figure 10.3, it is apparent that the work per unit mass going into the compressor is much higher than the work out of the turbine. As seen previously in Chapter 9, the divergence of the isobars is the primary reason for this consequence. In other words, the magnitude of the change in enthalpy (up or down) for a 5 : 1 pressure ratio change (compression or expansion) is much higher where the starting temperature and entropy are higher, and it is much lower when the starting temperature and entropy are lower.

Think Stop. A noteworthy observation about Figure 10.3 is that T_2 and T_4 are not constrained by Second law considerations to upper or lower temperature limits, but T_1 is constrained such that it cannot rise above a maximum value for the cold space and T_3 cannot fall below a minimum value for the warm space. Students are also encouraged to compare cycle energy efficiency values (see next section) for gas refrigeration versus vapor refrigeration.

10.3 Heat Pump Efficiency

Think Stop. *Multiple choice* – The efficiency of an ideal heat pump cycle (what you get divided by what you paid for it) is (i) always higher than an ideal heat engine cycle operating at the same high and low temperatures, (ii) always higher than an ideal refrigeration cycle operating at the same high and low temperatures, (iii) frequently greater than 1.0, or (iv) all of the above. The best answer is (iv), because the first two answers are completely correct, and the third answer is only slightly incorrect. The efficiency of a heat pump cycle is *always* greater than or equal to 1.0.

Recall the definition of efficiency (see Equation 6.28) for a two-temperature, Carnot heat engine, where the shaft work out is "what you get" and the heat flow in is "what you paid for."

$$\eta_{Carnot} = \frac{\overbrace{\dot{W}_{net,out}}^{\substack{What \\ you\ get}}}{\underbrace{\dot{Q}_{in}}_{\substack{What\ you \\ paid\ for\ it}}} \tag{6.28}$$

For the simplistic Carnot engine (where gases are ideal and specific heats are constant), the Carnot efficiency can be reduced to a quotient of temperatures (Equation 10.1):

$$\eta_{Carnot} \approx \frac{(T_h - T_c)}{T_h} \left[= 1 - \frac{T_c}{T_h} \right] \tag{10.1}$$

Clearly, the Carnot efficiency cannot exceed 1.0 and in fact can only approach 1.0 under extreme temperature conditions of very high T_h and very low T_c. As stated previously, students should ensure that temperatures are reported in absolute units (*K* or *R*) for a Carnot analysis.

In contrast to the engine, the heat pump's efficiency (Equation 10.2) is measured by the amount of heat transferred into the hot space divided by the amount of shaft work transferred into the system to effectuate the pumping of heat uphill on the temperature terrain. This efficiency measure is called the heating coefficient of performance (*HCOP*), and from the definition, it is easy to see *HCOP* can never drop below 1.0:

$$\eta_{HeatPump} = HCOP = \frac{\dot{Q}_{heating}\ [\doteq \text{kW}]}{\dot{W}_{in}\ [\doteq \text{kW}]} \approx \frac{T_h}{T_h - T_c} \tag{10.2}$$

The refrigeration coefficient of performance (*COP*; Equation 10.3) is related to the *HCOP* for the same cycle, with only the numerators being different:

$$\eta_{Refrigeration} = COP = \frac{\dot{Q}_{cooling}\ [\doteq \text{kW}]}{\dot{W}_{in}\ [\doteq \text{kW}]} \approx \frac{T_c}{T_h - T_c} = (HCOP - 1) \tag{10.3}$$

The relationship between *COP* and *HCOP* (i.e. that the difference between *HCOP* and *COP* for a given cycle is always 1) is a direct result of the First law. For simple heat pump cycles at steady state, only three energy flows cross the system boundary (i.e. heat enters the system at T_c, work enters the system by the shaft, and heat exits the system at T_h). This necessitates that the sum of the two energy inflows ($\dot{Q}_{cooling} + \dot{W}_{in}$) must equal the one energy outflow ($\dot{Q}_{heating}$).

Unfortunately, this relationship dooms the heat pump's efficiency to decrease as the difference between T_h and T_c increases. Thus, the most efficient heat pump will have a very small denominator and will pump a lot of heat up through a very small temperature difference. If the two heat reservoirs have temperatures that are far apart, the *HCOP* declines and eventually approaches 1.0. When *HCOP* is 1.0, the heat pump system could be replaced by an electric resistance heater, and the electrical cost of the heating would be identical.

The corollary to this condition is that whenever *HCOP* approaches one, *COP* approaches zero. Such a system is not accomplishing any cooling at all because all the work going into the compressor ends up as thermal energy (higher *T*), and none of it ends up as mechanical energy (higher *p*).

Two alternate measures of efficiency used by some system manufacturers are energy efficiency ratio (EER) and reciprocal efficiency kW_{elec}/ton_{refrig}. These are essentially the same as *COP*, but neither is nondimensional.

First, it is essential to introduce the concept of a ***ton*** of refrigeration and provide context for the new term. One ***ton*** of refrigeration (*TR*) was originally defined to be the rate of cooling that would freeze 2000 lb_m(=1.0 *ton*) of liquid water at 0 °*C* (32 °*F*) to ice at 0 °*C* (32 °*F*) in a time duration of $t = 24\ h$. One *TR* is approximately equal to 12 000 *BTU*/*h*, which is approximately equal to 3.5 kW_{th}. Additional refrigeration efficiencies are defined in Equations (10.4) and (10.5):

$$EER\left[\doteq \frac{BTU}{kWh}\right] = \frac{\dot{Q}_{cooling}\ [\doteq BTU/h]}{\dot{W}_{in}\ [\doteq kW_e]} = COP \cdot 3412 \tag{10.4}$$

$$\eta_{refr}^{-1}\left[\doteq \frac{kW_e}{TR}\right] = \frac{\dot{W}_{in}\ [\doteq kW_e]}{\dot{Q}_{cooling}\ [\doteq TR]} = \frac{3.5}{COP} \tag{10.5}$$

10.4 Sizing and Energy Usage

At first glance, a thermal engineer might think that sizing a heat pump system and estimating its electricity consumption over a typical year are two sides of the same coin. At a very high level, that may be true, but while the energy usage depends on system sizing, other factors such as seasonal variations in local climate play a more determinative role in annual energy usage.

Determining the system's size is like determining the horsepower of a vehicle's engine. If the vehicle's expected duty is to haul 50 *tons* of dirt up a 3-*mile* grade with an elevation change of 2000 *ft*, your engine choice will be a huge failure if it is the same size as one from a two-wheeled motorcycle on level ground. And by contrast, even though the motorcycle engine is tiny compared to the haul truck, the motorcycle might easily consume more gasoline in a year if it is used every day for hundreds of miles while the haul truck makes only one or two uphill trips per year.

Thus, it makes sense to ***size*** a heat pump system for the extreme temperature conditions (the coldest day of the year for heating and the hottest day of the year for cooling), but computing energy usage requires entirely different data. Thankfully, ASHRAE (2013) has computed average climate data for more than one thousand cities worldwide and published them for use by designers.

Two conceptual devices are important factors in these different design tasks: (i) ***heating dry bulb*** (99.6%) and ***cooling dry bulb*** (0.4%) temperatures (for sizing) and (ii) ***degree days*** (for annual energy consumption). Table 10.1 provides excerpts from ASHRAE (2013) Chapter 14 climatic design data for eight US airports. A glossary of acronyms is provided after the table.

The selection of the ***no-load*** outdoor temperature $T_{nl} \equiv 65\ °F$ is neither arbitrary nor skewed toward occupants whose preferred indoor temperature (summer and winter) is a bit toward the chilly side. The ***no-load*** condition (i.e., where neither a heating load nor a cooling load is required) does indeed occur at outdoor temperatures of approximately 65 °*F*. Colder outdoor temperatures do require occupants to expend energy for heating and warmer outdoor temperatures do require occupants to expend energy for cooling.

This phenomenon happens because whenever $T_{outdoor} = T_{nl} = 65\ °F$, the indoor temperature automatically becomes a more comfortable $T_{indoor} \approx 72.5\ °F$, as heat is being dispersed into the indoor environment by lights, electronics, appliances, and human body losses. Due to these interior heat sources, it is easy to see that if the outdoor temperature climbed to $T_{outdoor} = 72.5\ °F$, the indoor environment would become uncomfortably warm (i.e. it would reach approximately 80 °*F*), and the occupants would choose to expend energy for cooling.

Clearly, personal preference for indoor temperature varies widely among different individuals, and so does the presence of individual sources of energy dispersion. However, on average, the ***no-load*** outdoor temperature value of 65 °*F* has held up remarkably well for generic energy consumption analyses, and it has remained the basis for the degree-day method for more than 40 years.

For our simple sizing calculations (i.e. where the user chooses to keep the indoor control temperature at a constant value year-round), we elect to utilize the degree day method's ***no-load*** outdoor temperature as our indoor control temperature ($T_{indoor,\ control} = T_{outdoor,\ nl}$). To be clear, this is a shortcut. For more accurate sizing of HVAC equipment, thermal engineers should follow the methods given in chapters 14–19 of ASHRAE (2013, 2015), where the effects of fenestration, infiltration, human occupancy, lighting, appliances, and other heat loads are estimated explicitly for the habitable enclosure being examined.

One way to look at this issue is that the interior loads essentially represent a "free" source of energy for heating, thus diminishing the need to pay for extra heating in winter. However, in the summer season, they represent an energy "penalty" that the designer must provide extra cooling to accommodate.

By selecting $T_{indoor,\ control} = T_{outdoor,\ nl}$, we have created a shortcut whose accuracy is unknown and unexamined for individual cases, but nevertheless achieves a closer estimate for equipment sizing, by accepting the "free" thermal energy input as a "gift" and thereby reducing the size of the heating equipment for winter, and accepting the energy "penalty" as a cost and thereby enlarging the size of cooling equipment for summer. Two end-of-chapter problems are given for students

Table 10.1 Climatic design data for eight US airports. Units are degrees Celsius. Note 18.3 °C = 65 °F.

	Heating	Cooling		Evaporation		Dehumidification			Degree days	
IATA code	99.6% DB	0.4% DB	0.4% MCWB	0.4% WB	0.4% MCDB	0.4% DP	0.4% HR	0.4% MCDB	HDD 18.3	CDD 18.3
ANC	−22.9	21.9	15.0	15.8	20.5	13.6	9.7	17.0	5623	3
HNL	16.6	32.1	23.4	25.1	29.3	23.9	18.7	27.3	0	2599
IAH	−0.9	36.2	24.8	26.8	31.6	25.7	21.0	28.3	762	1699
JFK	−10.1	32.1	22.7	24.8	28.8	23.6	18.5	26.7	2961	547
MIA	8.7	33.2	25.3	26.8	30.5	25.8	21.2	28.6	70	2521
ORD	−18.6	33.0	23.5	25.4	31.0	23.7	19.0	28.7	3449	480
PHX	3.7	43.5	20.9	24.4	35.4	21.8	17.2	27.9	513	2570
SFO	3.9	28.2	17.1	18.6	25.3	16.2	11.5	20.1	1494	80

Acronym definitions follow in the bullet list

- IATA: International Air Transport Association (i.e. airport codes).
- Heating: Design case outdoor temperature for heating in winter.
  - 99.6% DB: The dry bulb temperature (°*C*) at which only 35 *h* per year are colder. (This is because a year with 365.25 days has 8766 *h* and 99.6% of that number is 8731 *h*.)
- Cooling: Design case outdoor temperature for cooling in summer.
  - 0.4% DB: The dry bulb temperature (°*C*) at which only 35 *h* per year are hotter.
  - 0.4% MCWB: The average wet bulb temperature (°*C*) that occurs coincidentally during the 35 *h* per year that meets the 0.4% DB condition.
- Evaporation:
  - 0.4% WB: The wet bulb temperature (°*C*) at which only 35 *h* per year are higher.
  - 0.4% MCDB: The average dry bulb temperature (°*C*) that occurs coincidentally during the 35 *h* per year that meets the 0.4% WB condition.
- Dehumidification:
  - 0.4% DP: The dew point temperature (°*C*) at which only 35 *h* per year are higher.
  - 0.4% HR: The corresponding humidity ratio (g_w/kg_{da}) that occurs coincidentally during the 35 *h* per year that meet the 0.4% DP condition.
  - 0.4% MCDB: The average dry bulb temperature (°*C*) that occurs coincidentally during the 35 *h* per year that meet the 0.4% DP condition.
- Degree Days:
  - HDD 18.3: The product of time (*d*) and temperature exceedance (°*C*) for all the hours where the outside temperature falls below 18.3 °*C* (65 °*F*).
  - CDD 18.3: The product of time (*d*) and temperature exceedance (°*C*) for all the hours where the outside temperature rises above 18.3 °C (65 °*F*).

Source: Data from ASHRAE (2013).

to undertake as a "practice run" for design projects that call for sizing and energy usage estimation for heating and/or cooling an indoor environment.

10.5 Refrigerants

Chapter 29 of ASHRAE (2013) contains a great deal of information on refrigerants, including thermodynamic property tables, Mollier diagrams, safety, global warming potential (GWP), and more. The unique characteristic of refrigerants is their ability to absorb heat from a (relatively) low-temperature space (typically by undergoing evaporation) and reject heat to a (relatively) high-temperature space (typically by undergoing condensation). Appendix A of this textbook provides thermodynamic properties for three exemplary refrigerants: $R32$, $R134a$, and $R717$; and a couple of atypical ones too: CO_2, and C_3H_8.

Think Stop. These steps are so fundamental to the concept of "heat pumping" that engineering students may underappreciate how a specific application (e.g. winter season home heating or commercial storage of frozen foods) may be addressed uniquely well by one refrigerant and exceptionally badly by another. Over limited ranges of temperatures, all refrigerants can accomplish their primary goal – absorb heat at low ambient temperature and reject heat at high ambient temperature.

Consider an ideal gas refrigeration cycle using helium as the working fluid. Such a cycle can be used to pump heat uphill for temperatures as low as $6\ K$ and also for very high temperatures ($\gg 1000\ K$). At the high end of the temperature range, refrigeration technologies are less practical, because the equipment (compressor, heat exchangers) may not survive well. At the low end of the temperature range, many refrigerants (other than helium) are incompressible liquids (or even solids), and the work of compression fails to provide a useful change in temperature of the working fluid.

Table 1 in chapter 29 of ASHRAE (2013) identifies chemical formula, molar mass, normal boiling point, and safety information for eight categories of ***single-component*** refrigerants (59 total), and the adjacent Table 2 of the same chapter provides a nearly identical set of information for refrigerant ***blends*** (79 total). The normal boiling point is simply the saturation temperature of the ***pure*** substance at standard pressure $p_{std} = 101.33\ kPa$. Refrigerant ***blends*** boil over a range of temperatures, so these fluids are classified by a combination of "bubble point" and "dew point" (see Chapter 6), which represents the lower and upper temperatures at which multiple phases can coexist, respectively.

The boiling point data is useful to place a refrigerant into an application category – substances that boil substantially below $25\ °C$ may be most useful in very low-temperature refrigeration/freezing applications, whereas substances that boil in the vicinity of STP may be most useful in air-conditioning (AC) or comfort heating applications. In either case, system designers should compare the actual work, heat, and cycle efficiency data for multiple refrigerants to determine which one is best for their application. Appendix A of this textbook provides thermodynamic properties for several refrigerant candidates.

Figure 10.4 shows a typical testing apparatus for air-conditioning/heat pump systems. The gauge on the left (connected to the left hose) measures both negative and positive gauge pressures, whereas the gauge on the right (connected to the right hose) measures positive pressures only.

When the system is operating in AC mode, the left hose is connected to an access port on the low-temperature vapor (evaporator exit; state ④) pipe, and the right hose is connected to the port on the high-temperature liquid (condenser exit; state ②) pipe. To reduce frictional losses (and save installation cost), lines containing refrigerant liquid are smaller diameters than lines containing refrigerant vapor. In dual-service (AC plus heat pump) systems, the small pipe and large pipe will swap identities each time the system switches modes. In other words, when the system is operating in heat pump mode, the left hose should be connected to the smaller pipe (which contains the moderately high temperature refrigerant liquid at the condenser exit; state ②) and the right hose should be connected to the larger pipe (which contains the high-temperature refrigerant vapor at the condenser inlet; state ①). The port connections in the refrigerant piping are typically equipped with check valves, which are sometimes colloquially known by the last name of the original inventor (Schrader 1893). The presence of the two check valves helps to prevent refrigerant loss when making or breaking test equipment connections.

The general categorization of refrigerants is as follows:

- Two-digit numbers (i.e. the first digit of a three-digit number is zero) are the methane series.
- Three-digit numbers beginning with numeral 1 are the ethane series.

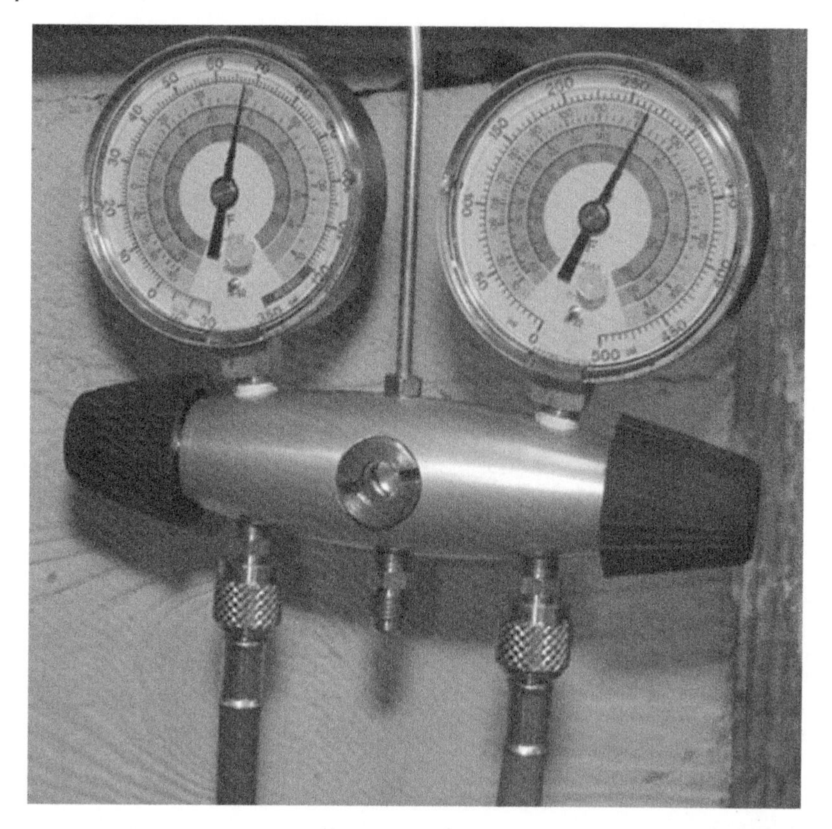

Figure 10.4 Refrigerant pressure testing assembly. *Source:* Martin Thermal Engineering Inc.

- Three-digit numbers beginning with numeral 2 are the propane series.
- Three-digit numbers beginning with numeral 4 are zeotrope blends.
- Three-digit numbers beginning with numeral 5 are azeotrope blends.
- Three-digit numbers beginning with numeral 6 are miscellaneous organic compounds.
- Three-digit numbers beginning with numeral 7 are inorganic compounds.
- Four-digit numbers are unsaturated organic compounds.

Think Stop. For inorganic compounds, identify which compound corresponds to each refrigerant number: (i) R704, (ii) R717, (iii) R718, (iv) R728, (v) R732, (vi) R744. It should be unnecessary to google the answer to this question. All the information necessary is available within each R-number.

Environmental Consequences. Chapter 29 of ASHRAE (2013) provides information (in tables 3 and 4) about the consequences of refrigerant releases to the environment. The tables report ozone depletion potential (ODP) relative to R11 (trichlorofluoromethane, CCl_3F), which is given a value of 1.0 and is deemed as the worst culprit for scavenging stratospheric ozone.

Also reported is GWP relative to carbon dioxide (CO_2), which is the benchmark for greenhouse effect metrics because its atmospheric concentration is by far the greatest. However, CO_2 is not the worst GWP molecule because its influence is dwarfed by larger molecules that better trap infrared radiation that would otherwise escape earth's atmosphere.

Many of the common refrigerants developed and used in the first half of the twentieth century are no longer sold in most countries due to their negative environmental consequences. The Montreal

Protocol, which phased out most CFCs (chlorofluorocarbons), has been a worldwide environmental success story, as the severity of ozone depletion events in each hemispheric spring has declined significantly since CFCs were banned from consumer products and HVAC systems.

Think Stop. For a vapor-compression cycle AC system with evaporator and condenser temperatures at $T_{evap} = 45\ °F$ and $T_{cond} = 86\ °F$, which one of the following refrigerants would be most suitable?

a) R50 (CH_4): $T_{bp} = -259\ °F$, $T_{crit} = -117\ °F$, $ODP = 0.0$, GWP = 21
b) R12 (CCl_2F_2): $T_{bp} = -22\ °F$, $T_1 = 92\ °F$, $p_1 = 108\ psia$, $p_1/p_4 = 1.9$, $ODP = 0.82$, GWP = 10 900
c) R32 (CF_2H_2): $T_{bp} = -61\ °F$, $T_1 = 116\ °F$, $p_1 = 280\ psia$, $p_1/p_4 = 1.9$, $ODP = 0.0$, GWP = 716
d) R717 (NH_3): $T_{bp} = -28\ °F$, $T_1 = 137\ °F$, $p_1 = 169\ psia$, $p_1/p_4 = 2.1$, $ODP = 0.0$, GWP < 1

10.6 Compressors

Much of the discussion about compressors in Chapter 9 can be recapped in this section without substantial change. Caution is urged to some degree because the compressor inlet (state 1 in Figure 10.2) is close enough to the vapor curve on the saturation dome, such that ideal gas behavior may not be valid at this state. If ideal gas behavior cannot be assumed at state 1, it may be necessary to employ the analysis method of computing enthalpy by interpolation from tables and matching entropies for the isentropic exit condition (also by interpolation) rather than simply utilizing the polytropic relation for the pressure rise and adjusting for irreversibilities with the compressor's isentropic efficiency. Once the isentropic exit state (2*s*) is determined by interpolation, the isentropic efficiency calculation should yield reliable results for the non-isentropic exit state (2).

When in doubt, students are urged to attempt more than one method of computing isentropic work for a given pressure ratio and to compare them to the tabulated values that represent real gases (and real mixtures of saturated vapor and liquid).

Types of Compressors. Refrigeration compressors tend to be the "positive-displacement" type. For a more thorough discussion of compressor and blower technologies, see Section 12.1. Specific compressor designs that are commonly used in refrigeration systems are:

- Reciprocating. The reciprocating compressor (see Figure 10.5) is the oldest design on this list and has been a workhorse for centuries. In fact, the first piston-cylinder vacuum pump was invented in the seventeenth century (Pumps and Systems 2021), and the first air compressor was patented in Great Britain just before the start of the nineteenth century (Train History 2021). The steam locomotive propulsion system can be thought of as an inverse technology to the reciprocating compressor.

  The purpose of that device was to use steam pressure from the boiler to move a piston back and forth inside a cylinder and with linkages (hinged moment arms) drive the wheels of a locomotive. The distinction of the reciprocating air compressor was that the shaft performed work on the air to increase its pressure. Commercial reciprocating compressors can achieve pressure ratios of several hundred and isentropic efficiencies of 70 – 90%.
- Scroll. A scroll compressor (see Figure 10.6) is a positive-displacement style of compressor that is used in small refrigerators and for air-conditioning of homes and small commercial buildings. Scroll compressors can generate pressure ratios of 10 or higher are available in sizes that deliver 3 – 30 kW_{th} of cooling capacity. They have fewer moving parts than reciprocating compressors; hence, their reliability tends to be greater.

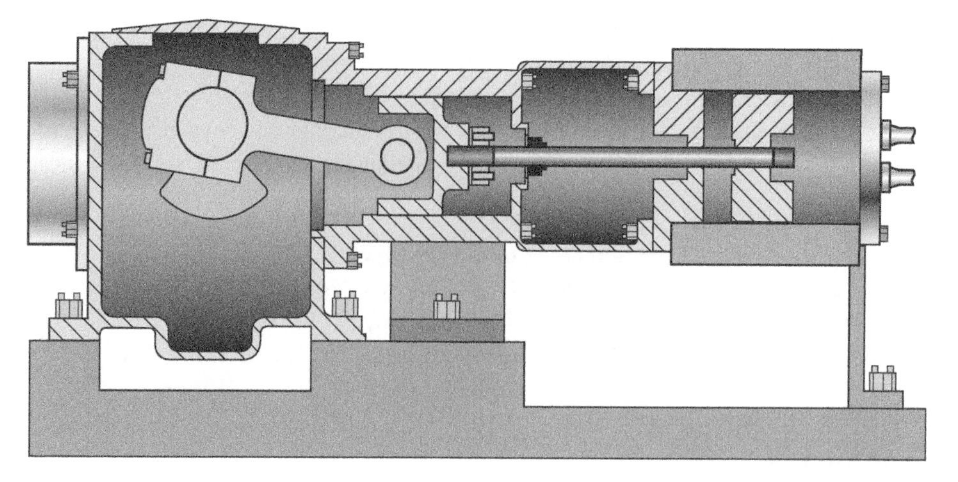

Figure 10.5 Reciprocating piston-type air compressor. *Source:* Martin Thermal Engineering.

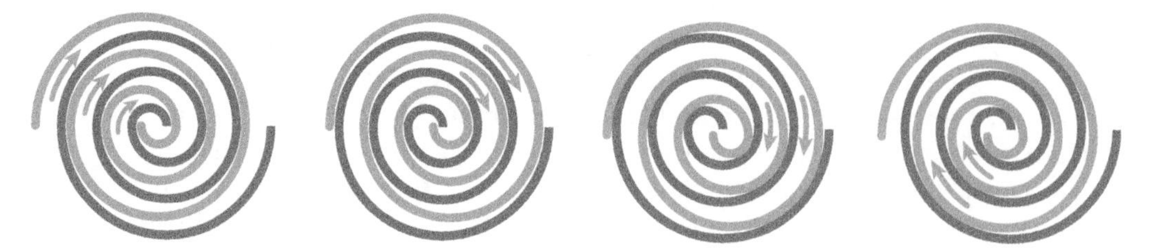

Figure 10.6 Four phases of a scroll compressor. As a mass of refrigerant vapor (arrows) is pushed in a circular motion between the stationary (blue) spiral and the moving (red) spiral, its volume decreases, which causes the pressure and temperature to rise. *Source:* Martin Thermal Engineering Inc.

- Centrifugal. The only refrigeration compressor on this list that is not "positive displacement" is the centrifugal style compressor (see Figure 10.7), which is analogous to centrifugal pumps and blowers that will be covered in Chapter 12. Briefly, the centrifugal compressor receives gas along the axial direction of the power shaft and expels it along the radial (and tangential) direction. One advantage of the centrifugal style is that lubricating oil is not needed because the clearances between the impeller blades and the housing wall are much larger than for the other styles, so metal-to-metal contact is avoided by geometry alone. Unfortunately, the greater clearance comes with a price: backflow. When the machine is operating, a positive pressure gradient is generated from the outlet side of the rotor to the inlet side (i.e. in the reverse direction of the flow). Because of the rotor's motion, backflow is a small percentage of the forward flow produced, but it does penalize the isentropic efficiency of the device.
- Rotary Screw. The rotary screw style of compressor (see Figure 10.8) is most often used for large, commercial refrigeration systems. Its precision machined surfaces, stout bearings, and low vibration qualify this device as both *high performance* and *high price*. Although the two screws are both rotating, the refrigerant that is being compressed is moving in a predominantly axial direction. The system requires oil lubrication to maintain its positive displacement characteristic. Rotary screw compressors are well suited to flow control by using motor speed control, e.g. using a variable speed drive. Customers often prefer screw compressors because of lower maintenance costs.

Figure 10.7 The centrifugal compressor receives low-pressure gas axially and spins it out radially at a high velocity, where it then decelerates with a concurrent increase in pressure. *Source:* Martin Thermal Engineering Inc.

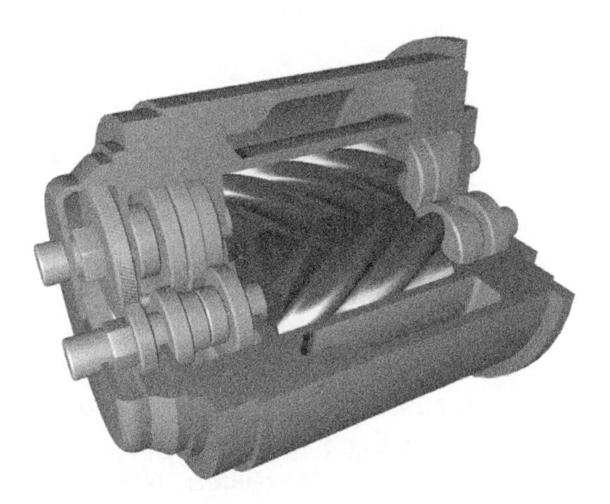

Figure 10.8 Rotary screw compressor with twin screws that rotate opposite to each other. *Source:* Martin Thermal Engineering Inc.

10.7 Air Handlers

Air handlers are simply enclosed heat exchangers that use convection to heat or cool the ambient air for comfort in an occupied space, typically relying on high- or low-temperature refrigerant that undergoes a phase change in response to the heat it rejects to the air or receives from the air. The terms air handling unit (AHU) or forced air unit (FAU) are typically applied to ***split*** heating and cooling systems (i.e. where the compressor and one coil are *outside* the conditioned space, and the other coil is *inside* the conditioned space). An exploded view of an air handler is shown in Figure 10.9.

Air handlers that operate in both heating and cooling modes may contain one or the other of these design features.

- A solo heat exchanger that circulates a single refrigerant and has the flexibility to act as a condenser (for winter heating) or an evaporator (for summer cooling).
- A pair of single-duty heat exchangers in series (different working fluids) for heating and cooling, respectively.

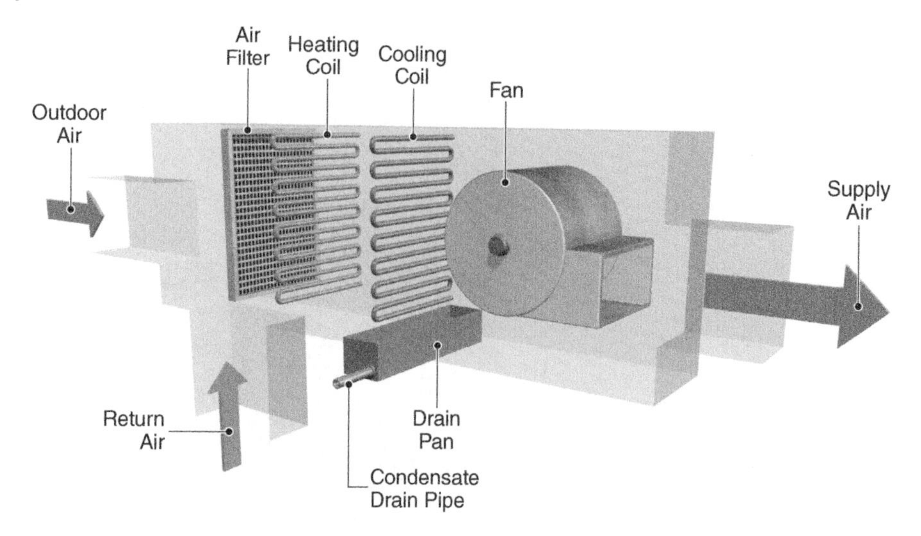

Figure 10.9 Exploded view of air handling unit (AHU) showing return air plenum, air filter, heating coil (working fluid hot water), cooling coil (working fluid refrigerant), and circulation fan. *Source:* Martin Thermal Engineering Inc.

In the single-duty, exchanger-pair arrangement, the cooling coil may contain refrigerant or a secondary heat-transfer fluid (e.g. glycol, brine) to extract heat from warm indoor air and the heating coil may contain hot water or steam from a boiler to add heat to cold indoor air. In some cases, the single-duty, exchanger-pair arrangement may comprise a gas-to-gas heat exchanger with burner to perform the heating function. Occasionally, an air handler will operate solely as air-conditioner (no heating functionality), or vice versa. Typically, the term "coil" refers to a long tube with bends or spirals that convey a heat-transfer liquid or a two-phase refrigerant. Gas-to-gas heat exchangers are usually not referred to as "coils."

Figure 10.10 shows a ***split-system*** schematic where the coil that would typically be designated "outdoor" is utilizing a separate loop of ground water to extract heat from or deposit heat to the earth instead of relying on outdoor air, which undergoes much greater temperature swings than groundwater.

Think Stop. Students should note the following unique features about Figure 10.10:

- The physical equipment does not move when the seasons change. The air handler remains fixed in one location indoors, and the geothermal unit can be located indoors or outdoors but also remains in a fixed location for all seasons. This is the essence of the ***split-system*** concept.
- A four-way switching valve is used to redirect the refrigerant flow such that the indoor coil becomes a condenser in winter and an evaporator in summer. The path of the four-way valve is unique in that its west and east ports always remain connected to the inlet (state ①) and outlet (state ②) of the compressor, while the north and south ports alternate between the following orientations:
  - *Winter*: North port directs refrigerant to the indoor coil, which is functioning as a condenser (at state ②) and south port receives refrigerant from the outdoor coil, which is functioning as an evaporator (at state ①).
  - *Summer*: North port receives refrigerant from the indoor coil, which is functioning as an evaporator (at state ①), and south port directs refrigerant to the outdoor coil, which is functioning as a condenser (at state ②).

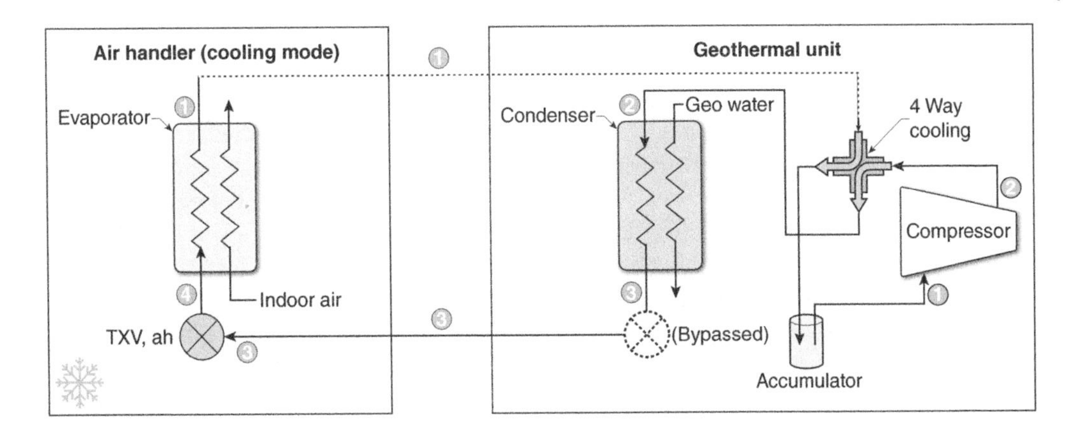

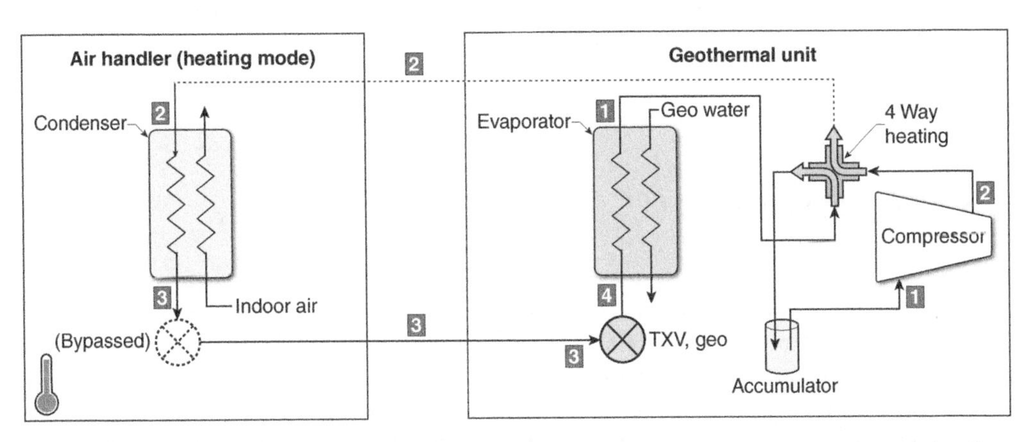

Figure 10.10 Split system HVAC system with geothermal fluid heat source/sink and separate air handler, shown with refrigerant path in both heating mode (winter season) and cooling mode (summer season). *Source:* Martin Thermal Engineering Inc.

- Two TXVs are provided, each having two pathways for refrigerant flow:
  - *Winter*: The outdoor *TXV* is active and the refrigerant is forced through the orifice that causes its pressure and temperature to drop. The indoor *TXV* is bypassed because a check valve opens to permit the refrigerant to flow around the orifice, thereby retaining essentially all of its pressure for later expansion in the outdoor *TXV*.
  - *Summer*: The indoor *TXV* is active, and the refrigerant is forced through the orifice that causes its pressure and temperature to drop. The outdoor *TXV* is bypassed because a check valve opens to permit the refrigerant to flow around the orifice, thereby retaining essentially all of its pressure, for later expansion in the indoor *TXV*.
- Two long refrigerant lines connect the air handler (attic or remote closet) to the ***geothermal unit*** (mechanical room). The liquid line has a small diameter (typically 3/8- to 1/2-*inch* copper for one to five tons of refrigeration capacity) and the vapor line has the larger diameter (typically 3/4- to 1 1/8-*inch* copper for the same capacity range).
- An ***accumulator*** may be provided to catch refrigerant droplets and help prevent liquid carryover into the compressor, which can damage the compressor blades or shaft. Since state 1 is nominally

superheated, any liquid refrigerant that temporarily accumulates is expected to evaporate and eventually enter the compressor as a gas.

- To ensure the refrigerant is sufficiently subcooled after the condenser, a number of strategies can be implemented, but the simplest is to install a ***receiver*** vessel in the condensed refrigerant line immediately after the condenser (not shown in Figure 10.10). The receiver is basically a liquid reservoir with liquid + vapor entering at the top and only liquid exiting at the bottom. By slowing down the velocity of the liquid refrigerant so that droplets fall to the bottom under the influence of gravity, the receiver helps to minimize the transport of vapor bubbles to the TXV and maximizes the transport of liquid refrigerant.
- Unfortunately, a ***receiver*** tends to de-subcool the liquid, raising its temperature back up to the saturation condition due to the condensation of refrigerant vapor present in the same vessel as the liquid. To return the desired level of subcooling to the liquid after it leaves the condenser, the liquid must undergo a pressure increase or lose additional heat. Either can be accomplished passively if at least one of two conditions is present: (i) if the TXV and evaporator are lower in elevation than the condenser or (ii) if the pipe carrying the liquid refrigerant from the receiver to the TXV is sufficiently long and the surroundings are sufficiently cool, the refrigerant can lose heat to ambient and improve its level of subcooling.

10.8 Refrigeration Control

When considering the control strategy for a heating/AC system, one sensed variable stands above all the others – the indoor temperature. This may cause one to ponder why so many HVAC systems are controlled by the simplest control method of all – thermostatic control (see Chapter 15 for further discussions of control strategies for thermal systems). When the indoor temperature gets too cold, the thermostat switches on the heating system. When it gets too hot, the thermostat switches on the AC system. When the temperature is just right, the thermostat switches off both heating and cooling systems entirely. It is a rather unsophisticated method, but it works well most of the time. The exception is when two or more occupants have incompatible psycho-thermal calibrations for their optimum body temperature (i.e. when one feels too hot at the same time the other feels too cold). Unfortunately, control thermostats can all too often become a battleground for occupants with opposing body thermostats.

The simplest AC systems have a fixed orifice (or fixed-length capillary line) that produces a constant pressure drop between states ③ and ④ while the compressor runs at a fixed speed and its high and low pressures never change. This is acceptable for small spaces as long as the design temperatures for the condenser and evaporator will reliably absorb heat from the cool space and reject heat to the warm space, respectively. In these systems, the cycling rate of the thermostat alone regulates the temperature of the indoor space and no *TXV* is present in the cycle.

The more complex refrigeration/heat pump systems also rely on internal refrigerant control devices to ensure the system functions effectively under changing ambient conditions. For such systems, it is not enough for the HVAC system to have automatic start-up and shutdown capabilities in response to changes in the indoor temperature. The HVAC system must also be able to adjust its refrigerant pressures to ensure that evaporation happens at a temperature that will absorb heat from the cool space, and that condensation happens at a temperature that will reject heat to the warm space. When the outdoor temperature climbs from 75 to 105 $°F$, the compressor must be able to respond with greater pressure rise to ensure the condenser pressure and temperature are high enough to reject heat to the higher outdoor temperature. The thermostat alone cannot accomplish

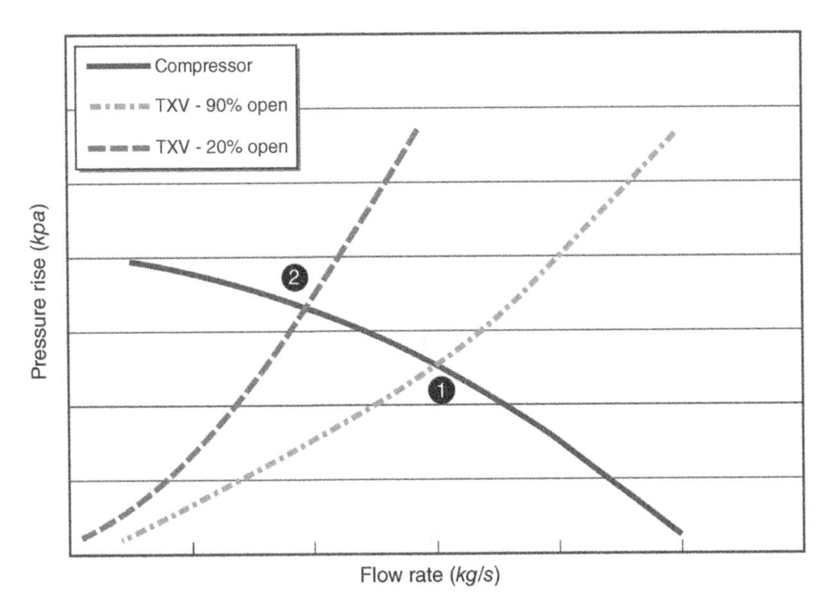

Figure 10.11 Generic pressure versus flow rate plots for a single compressor speed (solid curve) and two TXV opening percentages (dash-dot curve and dash curve). In this example, by closing the valve from 90% open to 20% open (TXV condition ① to TXV condition ②), the refrigerant flow will decrease by nearly half, and the compressor pressure rise will increase by approximately 50%. *Source:* Martin Thermal Engineering Inc.

this type of control – only the *TXV* can actively respond to such changes in outdoor conditions, while the compressor passively obeys its performance curve.

In Chapter 12, pump and blower performance are discussed in detail and in the context of many different systems. Students may want to skip ahead temporarily to review Section 12.7 before returning to this section. Students may also want to review Section 14.7, which discusses the functionality and mechanistic details of the *TXV*. The hot–cold temperature difference and working fluid's flow rate in a refrigeration system are governed by the interaction between compressor and *TXV*.

Examining Figure 10.11, we see two examples of the *TXV*'s performance (dot-dash curve or dashed curve), and both resemble a concave-up, half-parabola with a positive slope on a plot of pressure change (y-axis) versus flow rate (x-axis). The positive slope of the *TXV* lines displays the physics accurately. For any given valve position (e.g. **90** or **20**% as shown), a higher pressure differential across the valve delivers a higher refrigerant flow rate through the valve (all else equal).

In general terms, the ***compressor*** performance (solid curve) follows a concave-down, partial parabola with a negative slope. For a given compressor speed, any restriction in the fluid path that causes the flow to decrease simultaneously causes the compressor's pressure rise to increase, hence the negative slope. As a side note, most residential air conditioners have single-speed compressors. Variable speed compressors (examined more closely in Chapter 12) are generally found only in large commercial or industrial refrigeration applications.

Clearly, all the points on the ***TXV-20%*** curve are at lower flowrates (further left on the x-axis) than the ***TXV-90%*** curve. This observation is intuitively correct. When the *TXV* valve partially closes, the refrigerant flow is reduced, and when it opens, the refrigerant flow rate increases.

Another observation about Figure 10.11 is that the compressor responds passively to changes in the *TXV*'s valve position. If the system is assessed as the combination of two independent pieces of equipment (i.e. the compressor and the *TXV*), there can be only one operating point that corresponds to a unique percentage of *TXV* valve opening. That is the intersection between the compressor curve and the appropriate *TXV* curve (for that percent open). The consequences of closing the *TXV* valve from ***90%*** open (intersection ①) to ***20%*** open (intersection ②) are that the refrigerant flow rate decreases, and the pressure rise increases, which is exactly the anticipated "passive" response of the compressor to such an action by the *TXV*.

To ensure the refrigerant is sufficiently superheated after it leaves the evaporator, most *TXV* valves are equipped with a ***dual pressure feedback mechanism*** (see discussion in Section 14.7) that effectively maintains a constant level of superheat at the evaporator exit. Note that the evaporator ***pressure*** is not directly controlled. Only the degrees of superheat beyond the evaporator's saturation pressure are controlled. However, the evaporator temperature can be indirectly controlled by the *TXV*, as it forces the compressor to develop more or less pressure rise as a consequence of the throttling or unthrottling of the *TXV* as it responds to a change in superheat level. In practice, minimum superheat levels are of the order $\Delta T_{sat\text{ - }1} \approx 5\ °C$. Some manufacturers recommend $\Delta T_{sat\text{ - }1} \approx 20\ °F$ for an extra margin of safety.

A designer must be cautious if the system is intended to provide dual-purpose thermal management (e.g. heat pump duty in winter and AC duty in summer) – especially if a significant elevation difference exists between the indoor coil and the outdoor coil. The hydrostatics of heat pump system design is sometimes forgotten in the same way that the charging density (i.e. charged refrigerant mass divided by system interior volume) is taken for granted.

For example, consider a vertical separation distance of 10 *ft* from the floor of the first story to the floor of the second story. With this elevation change, a column of liquid refrigerant will lose or gain 5 - 7 *psi* (35 - 50 *kPa* or ⅓–½ *atm*) of pressure, depending on whether it is going uphill or downhill. When the elevation change is two or more stories, the situation is exacerbated even further.

10.9 Coil Defrost

For very low-temperature refrigeration systems (i.e. food or pharmaceutical freezers) or for heat pump systems that operate in deep freezing outdoor temperatures, atmospheric humidity may condense on the air side of the evaporator coil and accumulate in the form of frost. The immediate effect of frost is to reduce heat transfer by acting as a layer of thermal insulation between the warmer air and the cold tube exterior. If left unattended, frost can partially or fully obstruct flow across the air side of the evaporator and prevent nearly 100% of heat extraction from the cold space. In some circumstances, a large accumulation of frost could partly thaw, refreeze, accumulate more frost, and repeat the steps multiple times. This process often results in the formation of thick ice layer that can exert destructive forces on the refrigerant tubes and fins when the ice expands as its temperature drops.

Frost formation simply cannot be prevented whenever the air humidity level is higher than the ice–vapor saturation humidity at the coil temperature. This occurs frequently in freezers and sometimes even in nonfreezing refrigerator spaces. The proper maintenance practice is therefore to ensure incipient layers of frost never accumulate to the point where they are transformed into destructive ice. The best practice to prevent long-term frost accumulation is to periodically warm the evaporator (i.e. "defrost" it).

Four methods of defrosting are common for various types of refrigeration systems:

- Ambient thaw. When ambient temperatures are above 0 °*C*, the system can be shut off and left idle, during which time the ambient temperature slowly warms the frost, which melts and drips off into a pan and out of the system through a drain tube.

- When ambient temperatures are below 0 °C, consider the following options:
  - Electric heater. The system is shut off for a brief period (usually not to exceed 15 – 20 *min*), during which time electric heating elements in contact with the evaporator coil (and drain pan) are energized for a specified time period when the frost melts and flows away to a designated location.
  - Water spray. The system is shut off briefly, during which time warm water spray is directed at the frost where it initiates melting and liquid water flows to a designated location.
  - Hot gas defrost. With the system running, a switching valve is energized, directing hot gas from the compressor discharge to the evaporator coil (briefly converting the evaporator to a condenser) during which time the frost melts and flows away to a designated location.

Residential freezers are often marketed as "frost-free" which is a slight misnomer. While it is true that the freezer surfaces visible to the owner do not accumulate frost, the evaporator coil, which is hidden behind or below the freezer chamber, is actually working diligently to remove humidity from the air, trapping it as frost on the evaporator surfaces, and undergoing periodic defrost (typically via electric heaters) to melt and drain away the accumulated water.

On rare occasions, frost can be seen on visible surfaces of frost-free freezers, but it rarely lasts very long – unless the door seal is compromised, which allows very humid air to enter and deposit moisture on subfreezing surfaces persistently. The reason these rare frost-evident periods are typically so short is that the evaporator is very effective at trapping moisture in a non-visible location, and the air that is circulated into the freezer space is so dry that it very effectively causes surface frost to sublime, which sends water vapor to the evaporator, where it is trapped and rejected as liquid during defrosting.

10.10 Compressorless Refrigeration

The ammonia absorption refrigeration cycle was perhaps one of the most counterintuitive inventions of the twentieth century (Von Platen and Munters 1926) because it employed the use of heat as the only energy input to drive a refrigeration process. The original US patent, granted to Von Platen and Munters, claimed a type of refrigeration cycle that had did not need a mechanical compressor (or any other means of conveying shaft work into the flowing system). Instead of mechanical energy crossing the system boundary to effect the heat pumping phenomenon, it utilized a small heat source (along with some ingenious thermodynamic and fluid-dynamic design features involving three working fluids) to motivate the flow of heat out of a colder space and into a warmer ambient environment.

Today, coolers based on the Von Platen patent are widely used in recreational vehicles and hotel rooms as beverage coolers. In small spaces where sleeping and refrigerating take place in close proximity, the silent operation of a compressorless cooler makes it a more agreeable technology than compressor-based refrigerators.

A slightly different style of ammonia absorption refrigeration is used in industrial water chillers. This alternate technology uses a liquid pump and throttling valve to obtain suitable high and low refrigerant pressures, respectively but still includes an ammonia boiler and water absorber. By employing shaft work of this nature, the third working fluid is unnecessary.

The inventors, Von Platen and Munters were engineering students at the KTH Royal Institute of Technology in Sweden when they conceived the idea, which was quite a *multidisciplinary* excursion for the mechanical engineering program they had enrolled in. Today, a Swedish company (Dometic

2021) is a leading manufacturer of the Von Platen refrigeration technology. According to their website, Dometic prides itself in carrying on the tradition of their pioneering predecessors. The history page reveals that even Albert Einstein was impressed with the technology, calling it a "stroke of genius." Based on this device's unique ability to pump heat uphill in the temperature domain without a compressor (or any other moving parts), we believe the invention has earned the nickname *The Chemical Engineer's Refrigeration Cycle.*

Von Platen's three-fluid cycle is shown in Figure 10.12, and it relies on buoyancy and gravity forces to circulate the different fluids through the intricate piping arrangement without the large changes in fluid pressure you would find in a compressor-driven cycle.

The pure working fluids are ammonia (NH_3), water (H_2O), and hydrogen (H_2), with the water acting as a liquid solvent for the ammonia vapor, and the hydrogen acting as an evaporation aid for the ammonia liquid.

The cycle's main equipment units are the boiler $\left(\dot{Q}_{boil\ \langle into\ fluid\rangle}\right)$, condenser $\left(\dot{Q}_{cond\ \langle out\ of\ fluid\rangle}\right)$, evaporator $\left(\dot{Q}_{evap\ \langle into\ fluid\rangle}\right)$, and absorber $\left(\dot{Q}_{abs\ \langle out\ of\ fluid\rangle}\right)$. In addition, the system utilizes two counterflow heat exchangers, as well as a liquid–gas separator, and a liquid–gas mixer. If we ignore minor heat losses/gains from the remainder of the system, the First law ensures the four external heat transfers will be in balance at a steady state (since shaft work is absent):

$$\underbrace{\dot{Q}_{boil} + \dot{Q}_{evap}}_{heat\ inflows} = \underbrace{\dot{Q}_{cond} + \dot{Q}_{absorb}}_{heat\ outflows}$$

Three chemical processes in the cycle replace the traditional compressor and expansion valve:

a) Heat input (i.e. high temperature) in the boiler drives the desorption of ammonia gas out of the strong (highly concentrated) water–ammonia solution leaving behind a weak $NH_3 + H_2O$ solution and a gaseous stream with mostly NH_3 and a small fraction of H_2O.
b) Constant enthalpy temperature reduction in the mixer causes the working fluid to reach temperatures well below water's freezing point.
c) Heat removal (to reach ambient temperature) in the absorber drives the ammonia gas to redissolve into the weak solution, once again creating a strong $NH_3 + H_2O$ solution with no remaining ammonia gas.

Briefly, the process starts at the boiler and separator, where ammonia gas is produced, and weak ammonia–water solution is sent to the absorber. Next, the (moderately high temperature) gas that is largely ammonia is condensed to liquid by heat transfer to ambient air through the finned walls of the condenser. Then the rich ammonia liquid is sent to the evaporator where it receives heat from the chilled space and vaporizes at a low temperature.

But before it actually gets into the evaporator, the ambient liquid is tricked into reducing its temperature by forcing it to blend with hydrogen gas, which induces a small portion of the ammonia to evaporate without needing to reduce the stream's pressure through a valve (which would be impossible to achieve continuously without a compressor). Ingeniously, this is accomplished by blending a flow of hydrogen gas into the flowing ammonia-rich liquid. The H_2 gas bubbles constitute spaces where molecules of liquid ammonia can vaporize due to the initially zero ***partial pressure*** of NH_3 vapor in the H_2 gas stream. And this reduction in partial pressure can be accomplished without any change in the overall pressure of the two-phase, three-constituent mixture.

Referring to Figure 10.12, the strong ammonia solution enters the bottom of the boiler at a slightly preheated condition (state ①). The electric heating element increases the temperature of the

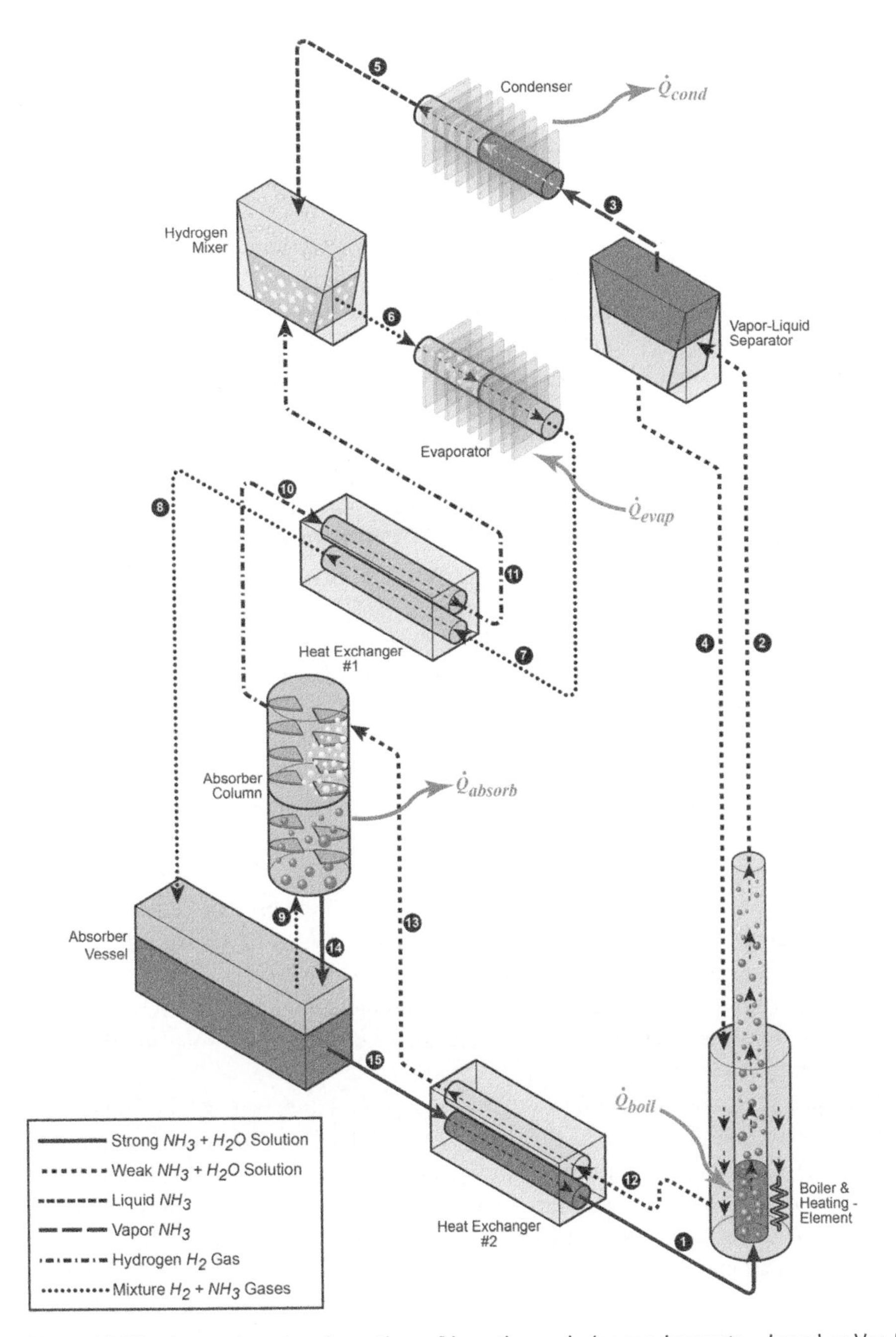

Figure 10.12 Ammonia water absorption refrigeration cycle (no moving parts – based on Von Platen patent). *Source:* Dometic (2021).

solution and ammonia vapor begins to desorb out of the solution. The boiler is sometimes called a "generator" because a large proportion of the water does not boil – ammonia vapor is desorbed or "generated" by raising the temperature of the strong solution. In hotel rooms, the heat source that drives the desorption of ammonia out of the strong solution is always an electric heating element, but in RVs, the heat source can be a small propane flame or an electric heating element, whichever is more plentiful.

A mixture of NH_3 vapor (with some H_2O vapor) plus droplets of weak ammonia solution exit the top of the boiler at state ②. Two separate streams then exit the separator – a high-temperature ammonia vapor (mixed with a small proportion of water vapor) at state ③ and a high-temperature liquid stream of weak ammonia solution at state ④. The weak solution ③ flows downward into an annular chamber around the boiler, where it remains hot.

The concentrated ammonia (+ water) vapor ④ is then cooled to ambient temperature in the externally finned condenser where it condenses to a liquid and is fed by gravity (state ⑤) into the hydrogen mixer where it combines with a flow of cooled hydrogen gas (state ⑪) to form a two-phase mixture (state ⑥). In the mixer, the hydrogen gas ⑪ creates a (nonliquid) head space where the partial pressure of ammonia vapor is initially zero. This causes some liquid ammonia to rapidly evaporate, causing the mixture temperature to drop to maintain the enthalpy flow rate constant in the adiabatic mixing device. The cold mixture of $H_{2,\ gas} + NH_{3,\ liq\ +\ vap} + H_2O_{liq}$ then flows into the evaporator (located in the refrigerator's cold space), and as heat flows into the very cold working fluid, more ammonia evaporates, just enough so the rate of heat input from the cold space equals the rate of heat absorbed by ammonia evaporation.

A mixture of $NH_3 + H_2$ (plus a small proportion of H_2O) flows out of the evaporator at state ⑦ and flows through heat exchanger HX1, which precools the incoming pure hydrogen stream (state ⑩) as it travels toward the inlet of the mixer (state ⑪). The $NH_3 + H_2$ mixture then exits HX1 and flows downward (state ⑧) into the absorber vessel, where it contacts the strong solution in the bottom of the vessel and creates a vapor shield of high concentration ammonia gas to buffer the strong $NH_3 + H_2O$ solution it contacts. The $NH_3 + H_2$ gas mixture leaves the absorber vessel and flows upward at state ⑨ through the absorption column, which induces very significant contact between the counterflowing fluids ($H_2 + NH_3$ gas flowing upward versus weak $NH_3 + H_2O$ solution flowing downward). This counterflow contact results in essentially complete removal of the NH_3 from the H_2 stream by the time it reaches state ⑩ while simultaneously creating the strong solution that eventually travels back to the boiler.

The fluid falling opposite to the upward flowing $NH_3 + H_2$ gas mixture is a downward flowing, precooled, weak ammonia solution (state ⑬) that enters at the top of the absorption column and flows downward while absorbing ammonia gas. This stream then exits the absorber column and flows into the absorber vessel as strong ammonia solution at state ⑭. The absorption column is externally finned to facilitate heat loss to the environment such that the column temperature remains close to ambient all along the contact path. The final step is through HX2, which extracts heat from the hot weak solution leaving the boiler's annulus (state ⑫) and delivers it to the cooler as strong ammonia solution (stream ⑮), permitting the cycle to begin again at state ①.

A homework problem is given at the end of this chapter which provides the mathematical counterpart to the written descriptions of the thermochemistry in the paragraphs below. The question requires students to compute flows, temperatures, and pressures for the example system described next. Conservation equations for some of the unit operations are given below, but some additional ***assumptions*** are necessary to render the problem tractable:

a) The separator is adiabatic; all of the liquid ($NH_3 + H_2O$) entering the separator at state ② returns to the annulus of the boiler at state ④, whereas all the vapor ($NH_3 + H_2O$) entering the separator flows toward the condenser at state ③.
b) The mixer is adiabatic; all of the hydrogen entering the mixer at state ⑪ combines with all the condensed liquid (largely NH_3 with residual H_2O) entering the mixer at state ⑤ and with the combined flow proceeding toward the evaporator at state ⑥.
c) Fluid flow is driven solely by hydrostatic forces (e.g. gravity and buoyancy) in the vertical liquid tubes.
d) In spite of the assumption above, *for thermodynamics purposes* – the fluid pressure is assumed constant at every point in the cycle.
e) For flow and pressure drop calculations, liquid densities may be assumed to be equivalent to water at the designated T and p. This is a reasonable assumption because the specific gravity of 36 %w ammonia–water solution is $\gamma = 0.88$ at room temperature. The gas densities may be found by averaging specific volumes from Appendices A.4 and A.12 at known T, p values and taking the reciprocal for density.
f) For tube sizing purposes, friction losses are assumed negligible in the unit operations (i.e. equipment) but not in the interconnect tubing.
g) Values for the inside diameter of different runs of tubing can be varied continuously (i.e. diameters need not be limited to standard size increments) to achieve a precise flow resistance (for known tubing lengths). This resistance will regulate the flow rates of gases and liquids from origins at higher hydrostatic pressures to destinations at lower hydrostatic pressures.
h) Ammonia concentrations in water for weak and strong solutions and the enthalpy changes associated with absorption/desorption for these solutions are available by employing VBA scripts (Appendices B.6 and B.7) or using the Tillner–Roth tables (Appendices A.13 and A.14).
i) Heat exchangers are designed as concentric tube pairs (with the stream whose inlet temperature is closest to ambient being conveyed in the outer annulus) and effectiveness is assumed constant irrespective of tube sizes.
j) The absorber column and absorber vessel may be considered as one device (i.e. it is not necessary to compute flows and properties at states ⑨ and ⑭ because neither crosses the control surface of the combined pieces of equipment); this treatment is also validated by considering that the strong liquid at ⑭ should have approximately the same properties as stream ⑮ and the gas mixture at state ⑨ should have approximately the same properties as stream ⑧.
k) The condenser and the absorber column/vessel combination are uninsulated and externally finned, and the outlet flows exiting from these devices (e.g. streams ⑤, ⑩, and ⑮) have exit temperatures slightly above ambient (e.g. $T_5 = T_{10} = T_{15} \approx 300\ K$).
l) Although not shown in the schematic, the finned evaporator is configured to cool without freezing, which implies the working fluid can enter the evaporator at a subfreezing temperature, but the fluid in closest contact with the fins and which leaves the evaporator at state ⑦ is slightly above the freezing point of water (i.e. $T_7 \approx 275\ K$).
m) For all nonadiabatic devices, the rate of heat flow (in or out) can be calculated by difference from the rates of enthalpy flows into and out of the device.

Properties of the pure substances (NH_3, H_2O) at their critical points are given in Table 10.2. Thermodynamic properties for pure water and pure ammonia (liquid and vapor) are available from the *NIST Chemistry Webbook* and are reproduced here as Appendices A.4 and A.12, respectively. Patek and Klomfar (1995) present a limited set of equations of state for saturated water–ammonia

Table 10.2 Properties of pure ammonia and pure water from Tillner–Roth and Friend (1998) and Keenan et al. (1969).

Property	Ammonia (NH_3)	Water (H_2O)
Critical temperature, $T_c(\doteq K)$	405.4	647.29
Critical pressure, $p_c(\doteq kPa)$	11 360	22 088
Critical density, $\rho_c(\doteq kg/m^3)$	225	317
Molar mass, $\hat{M}(\doteq kg/kmol)$	17.030 26	18.0153

Sources: Adapted from Tillner–Roth and Friend (1998) and Keenan et al. (1969).

mixtures over a moderate range of pressures, temperatures, and mole fractions of the constituents. (Density and entropy are not available in Patek, but enthalpy is.)

VBA scripts based on the Patek equations are provided in Appendix B.6. By copying and pasting the VBA scripts into Excel's Developer, customized Excel functions that compute $T(p,\hat{x})$, $T(p,\hat{y})$, $y(p,\hat{x})$, $h_{liq}(T,\hat{x})$, and $h_{vap}(T,\hat{y})$ become available to compute a wide range of saturated conditions for $NH_3 + H_2O$ mixtures of vapor and liquid. As the variable names affirm, the computed values of specific enthalpy h (without carat) are per kilogram, not per kilomole in Patek, even though the composition variables $(\hat{x},\hat{y})$ are per kilomole.

The parametric equations (for $T, h,$ and $\hat{y}$, as functions of $p,\hat{x}$) developed by Patek should be considered acceptable for academic analyses of the compressorless refrigeration cycle, but they are not sufficient to analyze systems where pressure varies due to turbomachinery shaft work. Even for single-pressure applications such as the Von Platen cycle, the accuracy of the Patek equations varies quite substantially in comparison to the more rigorous equations of state developed by Tillner–Roth and Friend (1998) and tabulated by pressure, temperature, and composition.

Until we obtain all the state equations employed by Tillner–Roth (which were published outside the United States and were not accessible at press time for this book) and develop corresponding VBA scripts for them, we recommend using the Patek equations (Appendix B.6) *exclusively* for academic analyses of the Von Platen cycle – not for designing actual thermal systems that utilize ammonia water mixtures. In our opinion, professional analyses of ammonia–water systems should not be carried out using Appendix B.6. Instead, Appendices A.13 and A.14 provide selected thermodynamic properties from the Tillner–Roth tabulations, and because they were published by the *Journal of Chemical and Physical Reference Data*, the values in the property tables are deemed higher quality than Patek's data.

However, the Tillner-Roth tables themselves suffer from a shortcoming because users can unknowingly introduce unacceptable interpolation errors due to the nonlinear nature of some of the property relationships. If users compare the results of the VBA scripts with the results of interpolating from the appendices, they will surely find subdomains of the tables that can cause interpolation errors – especially with the bubble point and dew point relations $x = f(p,y)$ and $y = f(p,x)$.

Think Stop. Students should refer to Chapter 6 to reinforce their understanding of the dew point and bubble point concepts. This is especially helpful when comparing property values from the saturation table (Appendix A.13) and the single-phase table (Appendix A.14). At temperatures lower than T_{bub} for a given p, the mixture is entirely subcooled liquid (a single-phase), and at temperatures higher than T_{dew} for a given p, the mixture is entirely superheated gas (also a single phase).

The system described here is based on four important design parameters given in Von Platen and Munters (1926) – composition at state ①, flow ratio out of the separator at states ③ and ④, mass ratio of hydrogen ⑪ to ammonia–water ⑤ in the mixer and charging pressure p_{tot}.

- The strong solution at state ① should comprise 36% ammonia (by mass) dissolved in water (i.e. $z_{①} \approx 0.36$).
- Mass flow of the weak solution should be at least 3× the mass flow of ammonia (plus minor water) gas leaving the separator (i.e. $\dot{m}_{④}/\dot{m}_{③} > 3$).
- The mass flow ratio of hydrogen gas $\dot{m}_{H_2⑪}$ to ammonia water liquid should be less than 40% (i.e. $\dot{m}_{H_2⑪}/\dot{m}_{NH_3\,+\,H_2O⑤} \leq 0.40$).
- The overall system pressure at ambient temperature should be $p \approx 1000\ kPa$ (including hydrogen). We suspect that this pressure is a lower limit and pressures up to 3× higher may be preferred.

To close out this description of the compressorless refrigeration cycle, we present control volume analyses of selected equipment units. Analysis of each unit in the sequence given here is likely to result in a tractable path to a solution.

Boiler. The boiler can be analyzed using the methodology given in Section 6.7. If the thermal power input to the boiler ($\dot{Q}_{boil,in}$) is initially known, that value can be used to scale *all* the individual component flow rates ($\hat{a}, \hat{b}, \hat{c}, \hat{d}, \hat{e}, \hat{f}, \dot{n}_{H_2,gas}$) based on the change in specific enthalpy from state ① to state ② at the very beginning of the analysis. If only the evaporator thermal duty ($\dot{Q}_{evap,in}$) is known, it may be preferable to use an arbitrary placeholder for the combined mole flow rates of ammonia and water entering the boiler (e.g. $\dot{n}_{①,comb} \triangleq 1.0\ kmol/s$) temporarily that is, until the evaporator is analyzed, at which time the value for $\dot{n}_{1,comb}$ can be scaled up or down so that the value determined for $\dot{Q}_{evap,in}$ matches the problem statement. Initially, the boiler inlet temperature may be assumed to be $T_{①} = 300\ K$, which is substantially below the bubble point for the 36% strong solution. When the entire cycle analysis is complete, this temperature should be modified to be consistent with the exit temperatures of both of the two streams leaving HX2. Iteration on $\dot{n}_{①,comb}\left(= \hat{e} + \hat{f}\right)$ entering the boiler may also be necessary at the end of the analysis, to match up all the heat flows to any pre-specified values.

Separator. Downstream of the boiler, a hot, two-component, two-phase stream of $NH_3 + H_2O$ enters the separator (assumed adiabatic), and two distinct streams (liquid, vapor) exit without any change in the combined enthalpy flow rate.

Think Stop. What is the consequence of adiabatic separation in terms of the temperatures of the three streams associated with this device (i.e. one inlet stream and two exit streams)? Is adiabatic separation a realistic assumption?

The liquid leaving the separator at state ④ is a weak ammonia–water solution, and the vapor leaving the separator at state ③ is mostly ammonia, but also contains a modest amount of water vapor, because of the high temperature at the boiler exit.

According to Von Platen and Munters (1926), the preferred mass flow rate of the weak solution ④ should be at least triple the mass flow rate of the concentrated ammonia vapor ③. For the given values of $\hat{z}_{①} = \hat{z}_{②} = 0.36$ and $p = 1000\ kPa$, values for T and $\hat{y}$ as functions of p and $\hat{x}$ entering the separator can be determined easily by using the Patek-sourced VBA custom functions given in Appendix B.6. We found a minimum value of $\hat{x}_{②} \approx 0.21$ that satisfies the Von Platen-preferred liquid to vapor mass ratio ($\dot{m}_{④}/\dot{m}_{③} > 3$). However, a more preferred value for the entire cycle is $\hat{x}_{②} = 0.25$. We verify this meets the preferred criterion as follows:

$$T_{px②}(1000, 0.25) = 388.08\ K; \hat{y}_{px②}(1000, 0.25) = 0.8572; \hat{X}_{②} = \frac{(\hat{z} - \hat{x})}{(\hat{y} - \hat{x})} = 0.1812$$

$$\hat{M}_{liq} = \hat{x} \cdot \hat{M}_{NH_3} + (1 - \hat{x}) \cdot \hat{M}_{H_2O} = 17.769; \quad \hat{M}_{vap} = \hat{y} \cdot \hat{M}_{NH_3} + (1 - \hat{y}) \cdot \hat{M}_{H_2O} = 17.171$$

$$\frac{\dot{n}_{liq④}}{\dot{n}_{vap③}} = \frac{(1 - \hat{X})}{\hat{X}} = 4.5202; \quad \frac{\dot{m}_{liq④}}{\dot{m}_{vap③}} = \frac{\hat{M}_{liq}}{\hat{M}_{vap}} \cdot \frac{\dot{n}_{liq}}{\dot{n}_{vap}} = \frac{17.769}{17.171} 4.5202 = 4.6776 > 3$$

The individual mole flow rates ($\triangleq kmol/s$) of NH_3 and H_2O leaving the two exit ports on the separator are determined from the values of $\hat{z}_{③}$ and $\hat{z}_{④}$ and the individual mole flow rates leaving the boiler. Their mass flow ratio is shown to be greater than 3, as the patent recommends.

Condenser. Two of our assumptions above were that flow in the interconnecting lines between equipment units was adiabatic and that pressure differences due to elevation and friction were inconsequential to thermodynamic property computation. This implies that stream ③ (concentrated ammonia vapor + minor water vapor) enters the condenser with $\hat{z}_{③} = \hat{y}_{②} = 0.8572$. We also know that the cooled stream exiting the condenser is approximately at ambient temperature ($T_{⑤} \approx 300\ K$), which we find is lower than the bubble point ($\hat{x}_{bub} = \hat{z}_{③}$) for $T_{px}(1000, 0.8572) \approx 303\ K$. This means the flow exiting the condenser is a subcooled liquid, which we will find is a convenient status for our analysis of the mixer. The heat flow rate out of the condenser to ambient can be computed by finding the decrease in enthalpy flow rate (kJ/s) from the inlet state $\dot{H}_{③}$ to the exit state $\dot{H}_{⑤}$.

Mixer. The Von Platen mixer is perhaps the most ingenious component in the entire Von Platen invention. This is the device whose parallel in conventional vapor refrigeration systems is the expansion valve. Both devices utilize pressure reduction to flash a cool liquid into a very cold vapor/liquid mixture. The valve relies on the compressor to generate the high-pressure stream that gets cooled and condensed before it expands across the valve. In contrast, the mixer needs no compressor to accomplish its mission of *pressure reduction*, because bulk pressure reduction is not needed and does not occur. The mixer's secret is that when hydrogen is isobarically mixed with the condensed ammonia–water liquid, it instantly reduces the ***partial pressure*** of ammonia in the vapor space and a portion of the ammonia quickly evaporates to ensure the two phases are in equilibrium.

Analyzing the mixer is significantly more complicated than analyzing the separator, but the concept is similar. Two streams enter and one combined stream exits adiabatically. Even though the First law mandates that the exit enthalpy flow rate is equal to the sum of the two inlet enthalpy flow rates (i.e. $\dot{H}_{⑥} = \dot{H}_{⑤} + \dot{H}_{⑪}$), the exit state is at a lower temperature because some of the liquid ammonia spontaneously expands to the vapor state (in the presence of the $H_{2,\ gas}$), which is an endothermic process. The mixer exit temperature can be found by iteration, while enforcing three conditions: (i) the First law is upheld by ensuring the ***enthalpy flow rates*** achieve balance ($\dot{H}_{out} = \dot{H}_{in}$), (ii) ***continuity*** is upheld by ensuring the molar flow rates of the three constituents ($NH_3 + H_2O + H_2$) achieve balance ($\dot{n}_{out,j} = \dot{n}_{in,j}$ for $j \in \{NH_3, H_2O, H_2\}$), and (iii) the ***$H_2$ mole fraction*** $\dot{n}_{H_2,gas⑥}/\dot{n}_{NH_3\ +\ H_2O\ +\ H_2,vap⑥}$ in the gas phase matches the ***H_2 partial pressure*** $p_{H_2⑥}/p_{tot⑥}$, which is a function of temperature (due to the vaporization of NH_3 and H_2O).

Thus, the complicating factor that enters the analysis here is that the phase distribution of the ammonia and water is dependent on both temperature and partial pressures of the hydrogen, ammonia, and water. This renders the analysis circular – temperature is determined by First law balance, but First law balance depends on the amount of NH_3 and H_2O vaporized, which depends on the equilibrium temperature. Thankfully, the algebra is stable and convergence to a solution is possible by iteration (assuming the initial guess for temperature is not wildly off the mark).

Students who are familiar with the many ways Excel can utilize iteration to find the solution to a set of simultaneous equations, should apply whichever method(s) they deem advantageous. We have developed a customized VBA function that solves for state ⑥ when given appropriate input data for states ⑤ and ⑪ of the Von Platen mixer. The solution technique, which can be employed manually in Excel without utilizing the VBA script provided, is described partly in Appendix B.7, and more fully in the solution key for one of Chapter 10's end-of-chapter problems.

Evaporator. The evaporator operates very much like the boiler, with the exception that the boiler exit temperature was determined from the ~3 : 1 liquid-to-vapor mass flow ratio designated in the Von Platen patent (and the author's subsequent optimizations), whereas the evaporator exit temperature is set by the desired temperature in the refrigerator's cold space (i.e. $T_{⑦} \approx 275\ K$). Consequently, new values of $\hat{x}$ and $\hat{y}$ must be determined for the evaporator exit state, because the heat input causes much of the liquid ammonia to vaporize out of the liquid state (along with a very small amount of liquid water).

First, we must recall that the custom VBA function $T_{px}(p, \hat{x})$ has one dependent variable (T) and two independent variables (p and $\hat{x}$). Because ammonia is vaporizing inside the evaporator, its partial pressure will increase, and the partial pressure of hydrogen will decrease to maintain the system's total pressure constant. But we must not forget that the value of p that must be used as the argument in the T_{px} function continues to be the partial pressure $p_{(non\text{-}H_2)}$, as defined earlier, not the total pressure p.

As with the mixer analysis, the evaporator exit conditions must undergo a double-iteration process to ensure the conservation laws are obeyed and the partial pressure of hydrogen is consistent with both $\hat{x}$ and the individual vapor molar flow rates.

Since the evaporator temperature is assigned, the specific enthalpy values only vary when $\hat{x}$ and $\hat{y}$ vary, as they must, due to the evaporation that is occurring. Using the same generic process (i.e. alternating between Calc A and Calc B) but with property values focused on the evaporator instead of the mixer will result in stable values for of $\hat{x}_{⑦}$ and $p_{H_2,gas⑦}$ that simultaneously satisfy all the physical and chemical requirements of the system.

Absorber. The absorber appears to be the most complex piece of equipment in the entire schematic. Indeed, it does have twice as many inflows and outflows as the boiler, condenser, and evaporator, and it is not even adiabatic.

However, solving for absorber properties is actually less complicated than the mixer or evaporator. This is because all inlet and outlet flow conditions are known (whenever we abide by the simplifying assumption that heat exchangers HX1 and HX2 are not present or have zero effectiveness).

- The flow condition at state ⑧ is the same as state ⑦
- The flow condition at state ⑬ is the same as state ⑫
- The flow condition at state ⑩ is the same as state ⑪
- The flow condition at state ⑮ is the same as state ①

These assumptions also permit us to avoid iteration to solve streams entering or exiting the absorber's control volume. All that remains is determining the rate of heat loss to the environment out of the absorber, by computing the decrease in the combined enthalpy flow rates exiting and entering the absorber's control volume:

$$\dot{Q}_{absorb,out} = \dot{H}_{⑩} + \dot{H}_{⑮} - \dot{H}_{⑧} - \dot{H}_{⑬}$$

Think Stop. While many of the calculations above do not necessitate the summing up of "overall" or "total" values for mole/mass/enthalpy flow rates, readers who are engaged in complex calculations should contemplate how their terminology might bring clarity (as opposed to confusion) among various sums that are needed.

For example, in the boiler, no hydrogen is present, so the flows at states ① and ② can be "totaled" in different ways. Using the nomenclature from Section 6.7, one might be tempted to designate molar flow rate $\hat{e}$ as "total ammonia" and molar flow rate $\hat{f}$ as "total water," but these labels can get confusing quickly. For example, what name would follow logically for the sum $\hat{e} + \hat{f}$? And if sums of flows in the hydrogen-free boiler are designated "total," how would sums of flows in the hydrogen-rich mixer be designated?

We suggest "combined" (with a guide word, such as "combined vapor" or "combined ammonia") might be a useful designator in cycle states where hydrogen is absent (to eliminate the connotation of totality). Similarly, "all-total" might be advantageous for cycle states where all three species (NH_3, H_2O, and H_2) are present. Other options are possible, depending on the designer's perspective and preferences.

10.11 Thermoelectric Coolers

Thermoelectric coolers are devices that utilize the Peltier effect to "pump" heat uphill on the temperature terrain, from a colder location to a warmer location. Peltier coolers consume electrical energy to effectuate this heat transfer.

The Peltier effect is the inverse of the Seebeck effect (which is the principle of operation for a thermocouple, to be further discussed in Chapter 14). The Seebeck effect is best illustrated by visualizing an electrical conductor (wire) that is positioned so that one end is in a cold space and the other end is in a warm space. With these boundary conditions, the Second law of thermodynamics establishes a driving force for the conductor's electrons to diffuse from the hot end (high energy electrons) to the cold end (low energy electrons) – even if they must overcome an existing (small) electromotive potential (voltage) to do so. The voltage generated by this temperature difference is a function of (i) the thermochemical properties of the material comprising the conductor and (ii) the temperature difference.

The electromotive potential for electron flow with temperature difference in a given material is called its Seebeck coefficient. Table 10.3 (Lasance 2006) provides examples of the Seebeck coefficient for selected metals. Platinum is arbitrarily designated to have a Seebeck coefficient equal to zero, but any two metals can be used to form a thermocouple, as long as their Seebeck coefficients are not so close to each other that the voltage generated per degree of temperature rise is difficult to distinguish from electrical noise.

A Peltier cooler can also be used as a thermoelectric generator. When operated as a cooler, a voltage is applied across the device, and as a result, a difference in temperature will build up between the electrodes. When operated as a generator, one electrode is heated to a temperature greater than the other electrode, causing a voltage difference to develop between the two sides (the Seebeck effect).

Table 10.3 Approximate Seebeck coefficients for selected metals.

Metal	Seebeck coefficient ($\frac{\mu V}{K}$)
Antimony (Sb)	+47
Nichrome (80Ni + 20Cr)	+25
Copper (Cu), Gold (Au)	+6.5
Lead (Pb)	+4.0
Aluminum (Al)	+3.5
Platinum (Pt)	0.0
Nickel (Ni)	−15
Constantan (55Cu + 45Ni)	−35
Bismuth (Bi)	−72

Source: Lasance (2006). Used with permission.

Thermoelectric coolers have a number of advantages compared to vapor-compression refrigerators, such as absence of moving parts and smaller size, but their primary disadvantages are high cost and lower energy efficiency.

10.12 Refrigeration System Failures

Compressor Failures. Although refrigeration systems can undergo any number of different failure modes, one of the most common is compressor failure. This observation raises an obvious question – how could it be that most "packaged" air conditioners (where all components are located in one enclosure, typically installed half inside and half outside a wall or window) and many "split" systems (installed partly inside and partly outside of the conditioned space) operate very well for years without compressor failures, whereas others (typically split units) somehow experience frequent compressor failures (perhaps as often as two or three in a calendar year).

While a compressor is not particularly fragile when the system is designed and maintained properly, it can be susceptible to rapid failure if exposed to certain conditions that are unforgiving (Springer 2016). An adage shared by many HVAC specialists is that if the cause of a compressor failure has not been determined and corrected, the replacement compressor will quickly fail too.

Slugging. One refrigeration system failure mode involves slugs of liquid refrigerant displacing refrigerant vapor in the suction of a positive displacement compressor (piston, scroll, or rotary screw). Positive displacement compressors are designed to reduce vapor volume progressively after the vapor enters the compression chamber, but they cannot reduce the volume of an incompressible liquid. Because the liquid slug displaces an equivalent volume of compression chamber space (permitting less vapor volume to enter the chamber), compressors can only tolerate slugging conditions where the liquid fraction of the intake fluid is low compared to the gas fraction.

For very high levels of intake liquid fraction, the mechanical forces acting on the compressor begin to exceed acceptable levels and damage occurs. Damage can include blown gaskets, stretched or fractured discharge bolts, broken or bent crankshafts, and locked-rotor current draw by the motor. Locked rotor current draw can be a factor of five higher than normal current draw at rated horsepower, and it can lead to overheating that can ignite nearby combustibles in extraordinary circumstances.

Slugging usually occurs at compressor start-up and may be caused by factors such as refrigerant overcharging (i.e. too much mass of refrigerant for the system's volume), improperly designed refrigerant piping (i.e. gravitational head not considered, line diameters too small, line lengths too long), or an oversized *TXV* orifice. The compressor may be spared damage if its design incorporates pressure relief means where liquid and/or vapor can escape out of the compression chamber(s) before pressures rise to the point of causing mechanical damage.

Slugging is especially problematic for split-system heat pumps where the condenser is located several floors above the *TXV*, evaporator and compressor. The long, vertical column of liquid refrigerant above the *TXV* increases the refrigerant pressure (and therefore its degree of subcooling), which increases both (i) the saturation temperature of the evaporating refrigerant, which diminishes the driving force for heat absorption from the low temperature heat source, and (ii) the enthalpy capacity of the evaporating refrigerant to a value beyond the design expectation (i.e. with minimal elevation change, and negligible, gravity-caused, extra subcooling). Collectively, these changes render it more difficult to achieve the necessary degree of superheat at the exit of the evaporator, which invites slugging.

Loss of Lubricant. Another compressor failure mode involves loss of lubricant, which can rapidly cause seizing of bearings and scoring of mating surfaces. Factors that may lead to lubricant loss include: washout by liquid refrigerant (also called crankcase flooding, which occurs during the off-cycle and leads to carryout of oil when the compressor is started); oil dilution by liquid refrigerant while running (due to insufficient superheat, which may indicate *TXV* failure or very low evaporator loading); insufficient lubricant charge (self-explanatory); and wrong lubricant (too viscous or not viscous enough).

Air Handler Refurbishment Defect. We investigated a major water failure in a residential AHU with an arrangement similar Figure 10.9. The subject AHU had separate coils – hot water from a shared hot water boiler was the working fluid for the air heating coil, and refrigerant was the working fluid for the air-cooling coil (evaporator). Figure 10.13 shows major cracks in three of the 180° copper elbows of the heating coil, which had been installed as a retrofit by an installation contractor when the original heating coil became plugged with calcium scale after many years of service. The

Figure 10.13 Three of four copper 180°-elbow fittings were split open at the U-bends due to excessive internal pressure. Approximately two months before the failure, this hot water coil had been installed immediately downstream of a refrigeration coil in the subject air handler, and when the water froze inside the tubes, the ice expanded. *Source:* Martin Thermal Engineering Inc.

residents discovered the problem when they heard what they thought was rainwater running down inside the walls.

In the original equipment, the cooling coil was located immediately downstream of the heating coil (along the air flow direction through the AHU), but due to space constraints, the installer swapped the coil sequence in the retrofit installation. The new placement had the evaporator coil in the upstream position and the heating coil in the downstream position. This amounted a defective installation, because of the properties and operating conditions of the two working fluids. Since the heating coil contained water, the installer should have ensured that the water tubing would never reach temperatures below the freezing point of water. Positioning the water coil downstream of an evaporator filled with refrigerant was risky and was irresponsible without performing a test.

Months after the installation, during the summer cooling season, when the evaporator was cooling the circulating air, much of the quiescent water in the heating coil froze to ice, which expanded and caused several of the water tubes to crack. However, since no hot water was flowing to the heating coil in summer, no one noticed the cracking. Later still, when the system was first used for winter heating, a hot water circulation valve was opened and a large volume of water flowed into the heating coil, much of which escaped out of the cracked copper elbows and caused major water damage to the attic space, ceiling, and walls of the residence.

Evaporator Defrost Failure. We investigated recurring defrost problems at an ultra-low-temperature, pharmaceutical warehouse freezer and discovered a system design defect that was the underlying cause of the failure. The warehouse evaporators (16 finfan units) were located near the ceiling of the cold space and they would accumulate frost as a normal consequence of humid air infiltration (i.e. whenever access doors were opened or through imperfect sealing at joints between the walls and floor). The compressor and condenser were located in a mechanical attic space above the warehouse, which consigned the evaporators to the lowest elevation in the refrigerant path – a red flag, but not conclusive proof of a defect. Periodically, the control system would transition to hot-gas-defrost mode to melt the frost and drain the accumulated water out of the system.

Initially, the failures manifested themselves as incomplete defrost along the bottom few rows of tubing for each evaporator (see Figure 10.14). The facility's solution was to defrost the exteriors of lower tubes manually, using spray bottles of warm water, but they were unable to regularly synchronize their personnel availability with the timing of the hot gas defrost. Eventually, the recurring freeze–thaw cycles turned frost into ice, and this led to crushing of some of the evaporator tubes, which further diminished their capacity for conveying refrigerant through the evaporator, whether in normal evaporation mode or in hot gas defrost mode.

The installation defect that caused the partial defrost failure was a simple fluid mechanical issue related to gravity and pressure. The evaporators were built with large diameter, vertical headers (on the left and right) with the inlet and outlet connections at the tops of the two headers. Smaller, horizontal evaporation tubes conveyed refrigerant from the inlet header to the outlet header, whether in evaporation mode or condensation mode (see Figure 10.14). When hot gas was first admitted to the frost-covered evaporator, it condensed and fell to the bottom under the influence of gravity. Since the inlet path was downward at the left and the exit path was upward at the right, the condensed refrigerant liquid was trapped at the bottom of the coil until the defrost cycle ended and the coil pressure declined back again to become an evaporator. This situation amounted to having a slug of unmoving liquid in the bottom rows, which prevented hot gas to enter that area, and caused the defrost to be ineffective there.

The problem could have been avoided if the system designer/installer had incorporated a drain at the bottom of the exit header, piped to an accumulator sized for the condensate volume anticipated for the programmed hot gas defrost duration. Judicious placement of switching valves could have

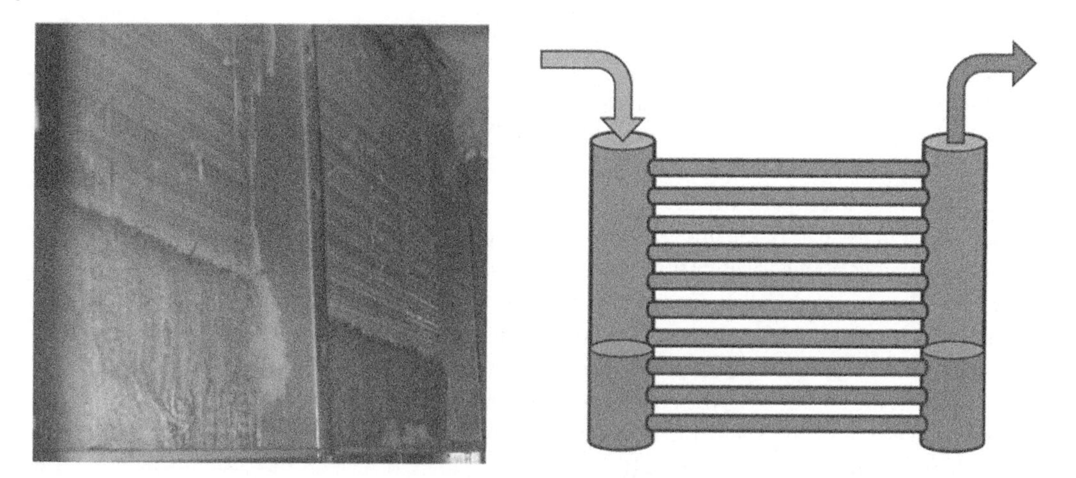

Figure 10.14 Defective evaporator defrost in low-temperature industrial freezer. Liquid refrigerant (i.e. condensate that formed when hot gas contacted ice-laden evaporator tubes) accumulated in the lower tube rows, obstructing additional hot gas from flowing through and continuing the defrosting, while hot, gaseous refrigerant continued to flow through the upper rows of tubes, effectively completing the defrost in this region. *Source:* Martin Thermal Engineering Inc.

allowed uncondensed gas to flow back to the attic during the defrost cycle and forced the collected condensate to boil off after defrosting, when the evaporator volume was pumped down to its more typical, low-pressure state.

10.13 Homework Problems

10.1 Among the refrigerants listed, select the best one for a refrigeration cycle that operates with a condenser temperature of $T_{sat,\ cond} = +30\ °C$ and an evaporator temperature of $T_{sat,\ evap} = -31.7\ °C$. For this question, ignore criteria such as *GWP* and *ODP* because those data are not provided in the list below. Boiling point temperatures are for 1.0 *atm* pressure.

(a) R50 (CH_4): $T_{bp} = -161.5\ °C$, $T_{crit} = -82.6$

(b) R32 (CH_2F_2): $T_{bp} = -51.7\ °C$, $T_{crit} = +78.1\ °C$

(c) R14 (CF_4): $T_{bp} = -121.8\ °C$, $T_{crit} = -45.6\ °C$

(d) R170 (C_2H_6): $T_{bp} = -88.6\ °C$, $T_{crit} = +32.2\ °C$

10.2 Consider an academic problem (i.e. not likely to occur is a practical design situation) where temperatures for a real (i.e. work is not isentropic), ideal gas refrigeration cycle are given at the turbomachinery exit states (T_2 and T_4), but temperatures at turbomachinery inlet states (T_1 and T_3) are unknowns. Derive two accurate engineering formulas, one for the compressor and one for the turbine, that will permit the inlet state temperatures to be computed using thermodynamic formulas and given values for isentropic efficiencies η_t and η_c.

10.3 An ideal vapor refrigeration cycle (conventional style, with isentropic compressor) that uses an ammonia–water refrigerant blend (36% NH_3 by mole) operates as a beverage cooler between an evaporator inlet temperature of $T_4 = T(p_4, \hat{x}_4) = +5.00°C$ and a condenser exit temperature of $T_3 = T_{bub}(p_3, \hat{x}_3) = +50.00°C$. Determine (a) the condenser pressure (kPa),

(b) the pressure in the evaporator (after the adiabatic *TXV*), (c) the molar quality of the refrigerant at state 4 ($\hat{X}_4$), and (d) the enthalpy of the refrigerant on both sides of the *TXV* (i.e. h_3, h_4). For thermodynamic properties of the $NH_3 + H_2O$ mixture, use custom Excel functions in Appendix B.6. Double iteration on p_4 and $\hat{x}_4$ to obtain the known values of T_4 and h_4 may be necessary.

Answer. (a) 308.02 kPa, (b) 2.86 kPa, (c) 66.92%, and (d) −13.67 kJ/kg.

10.4 Complete the analysis of the Von Platen mixer that was initiated in Section 10.10 and further described in Appendix B.7. Ensure the First law is upheld, such that the enthalpy flow rate at the mixer exit ($\dot{H}_{⑥}$) is equal to the sum of the enthalpy flow rates at the two mixer inlet ports ($\dot{H}_{⑤} + \dot{H}_{⑪}$). This can be accomplished by finding the exit temperature $T_{⑥}$ and the exit partial pressures of ammonia vapor, water vapor, and hydrogen ($\chi_{NH_3⑥}, \chi_{H_2O⑥}, \chi_{H_2⑥}$) in the gas phase that also accomplishes conservation of all species ($\dot{n}_{out,j} = \dot{n}_{in,j}$ for $j \in \{NH_3, H_2O, H_2\}$) in the two-phase mixture. Manually perform the iterations necessary to reach the final goal and compare the manual result to the results obtained by utilizing the customized Excel functions that are created from the VBA scripts in Appendix B.7.

10.5 Determine the compressor size (i.e. horsepower) that meets the following design condition for a recreational vehicle: RV outside dimensions =35 ft × 8 ft × 10 ft; RV insulation $k = 0.3165\ BTU \cdot in./(h \cdot ft^2 \cdot °F)$, thickness $\Delta x = 3.8$ in; Air exchange rate $ACH = 4\ h^{-1}$; Location: Phoenix; $p_\infty = 0.9735\ bar$; $T_{indoor,\ conditioned} = 65\ °F$; $\phi_{indoor,\ conditioned} \leq 60\%$; $\phi_{outdoor,\ 0.4\ \%\ DB} = 20\%$; Assume system efficiencies $HCOP_{heating} = 4.0$, $COP_{cooling} = 3.0$.

10.6 For the system and location of Problem 10.4, estimate the annual electrical energy consumption (kWh/y) for (a) heating and (b) cooling. Determine the annual dollar cost if the price of electricity is fixed at $\$0.15/kWh$.

10.7 Determine the missing mass flows, enthalpies, pressures, and temperatures for the stream table given below. The working fluid is R32. The schematic for this stream table is shown in Figure 10.1 for an ideal, vapor-refrigeration cycle. Assume pressure remains constant through the evaporator and condenser and compression is isentropic. Use Appendix A.10 for thermodynamic properties and interpolate where necessary or use Linstrom and Mallard (2017–2021).

Property	State 1	State 2	State 3	State 4
Temperature (K)			344.33	279.77
Pressure (MPa)				
Spec. volume (m^3/kg)				
Enthalpy (kJ/kg)				
Entropy ($kJ/kg \cdot °C$)				
Quality (%)	1.00		0.00	

10.8 A vapor refrigeration cycle utilizes a compressor which has isentropic efficiency of $\eta_c = 80\%$. The refrigerant is R134a, with mass flow rate $\dot{m} = 0.12\ kg/s$. The compressor inlet condition is $T_{sat} = 246.79\ K,\ \chi = 100\%$, and the condenser exit condition is $T_{sat} = 312.54\ K,\ \chi = 0\%$. Determine (a) the COP, (b) thermal cooling power (*TR*), and (c) mass flow of geothermal water through the condenser having a temperature rise of $\Delta T_w = +20\,°C$ (assuming the specific heat of liquid water is constant at $c_{p,w} = 4.18\ kJ/kgK$. Use Appendix A.11 for thermodynamic properties of R134a and interpolate where necessary or use Linstrom and Mallard (2017–2021).

Answer. (a) $COP = 3.90$ (b) $\dot{Q}_{evap} = 6.12\ TR$ (c) $0.291\ kg/s$.

10.9 A small, cryogenic, ideal gas refrigeration cycle (see Figure 10.3) utilizes a compressor and a turbine with isentropic efficiencies of $\eta_c = 91\%$ and $\eta_t = 89\%$, respectively. The working fluid is helium, and the low-temperature heat removal power is $\dot{Q}_{4-1} = 550\ W$. Assume the low-temperature cooling coil and high-temperature heating coil operate at constant pressures of $p_1 = 0.50\ MPa$ and $p_2 = 10.00\ MPa$, respectively. The minimum and maximum temperatures are $T_4 = -160\,°C$ and $T_2 = 280\,°C$, respectively. Determine the missing enthalpies, entropies, pressures, and temperatures in the stream table given below. Assume the specific heats of helium are constant at $c_p = 5.19$ and $c_v = 3.11$ and use the polytropic relations (Equations 9.1c, b) to determine temperature changes across the turbine and compressor. Use Equation (1.14) to compute missing entropy values, based on known values. Calculate (a) the best-case (i.e. lowest) average temperature of the cryogenic space, (b) the net input power required, and (c) the COP for the cycle.

Answer. (b) $\dot{W}_{shaft,net-in} = 2.74\ kW$ (c) $COP = 0.20$

Property	State 1	State 2	State 3	State 4
Temperature (°C)		280.0		−160.0
Pressure (MPa)	0.50	10.0	10.0	0.50
Spec volume (m^3/kg)				
Enthalpy (kJ/kg)	815.58			
Entropy ($kJ/kg \cdot °C$)	21.279			
Helium mass. flow (kg/s)				

10.10 A compressorless recreational vehicle refrigerator is powered by a small propane burner that causes ammonia to desorb out of the strong solution in the boiler, per Figure 10.12. Assume 90% of the thermal energy from the propane flame enters the strong solution (state 1) to drive the desorption. For simplicity, assume $\varepsilon_{HX1} = \varepsilon_{HX2} = 0\%$, so that all the heat exchanger's exit states (i.e. states 8, 11, 13, and 1) have the same property and flow values as their corresponding inlet states (i.e. states 7, 10, 12, and 15). Ignore states 9 and 14 but do perform mole and enthalpy balances around the vessel/absorber combination, with $\dot{n}_{8i+13i} = \dot{n}_{10i+15i}$, where $i \in \{NH_3, H_2O, H_2\}$ and $\dot{H}_{8+13} = \dot{H}_{10+15}$. The design cooling load for the refrigerated space is $\dot{Q}_{evap} = 0.125\ kW_{th}$. Fill in the missing entries in the following heat and material balance table. Compute (a) combustion heat release $\dot{Q}_{boil}[\triangleq kW_{th}]$ from the propane burner, (b) heat rejected to ambient from the condenser $\dot{Q}_{cond}$ and from the absorber $\dot{Q}_{absorb}$, and (c) coefficient of performance $\left(COP = \dot{Q}_{evap}/\dot{Q}_{boil}\right)$.

Answer. (a) 0.861 kW_{th}. See the following stream table.

Property	**States 1, 15**	**State 2**	**States 4, 12, 13**	**State 3**	**State 5**	**State 6**	**States 7, 8**	**States 10, 11**
Temperature, T (K)	300				300		275	300
Pressure, p (kPa)	1000	1000	1000	1000	1000	1000	1000	1000
Hydrogen partial pressure, p_{H_2} (kPa)	0	0	0	0	0			1000
Mole ratio ammonia liquid, $\hat{x}$ $(kmol_{a,\,f}/kmol_{a\,+\,w,\,f})$		0.25						
Mole ratio ammonia vapor, $\hat{y}$ $(kmol_{a.\,g}/kmol_{a\,+\,w,\,g})$								
Mole ratio ammonia combined, $\hat{z}$ $(kmol_{a.\,comb}/kmol_{a\,+\,w,\,comb})$	0.36							
Molar vapor quality, $\hat{X}$ $(kmol_{a\,+\,w,\,g}/kmol_{a\,+\,w,\,comb})$	Subcooled liquid		0.00	1.00				
Liquid phase enthalpy, $\hat{h}_f$ $(kJ/kmol)$								
Vapor phase enthalpy, $\hat{h}_g$ $(kJ/kmol)$								
Hydrogen enthalpy, $\hat{h}_{H2}$ $(kJ/kmol)$								
Mole flow ammonia liquid, $\hat{a}$ $\left(kmol_{NH_3,f}/s\right)$								
Mole flow water liquid, $\hat{b}$ $\left(kmol_{H_2O,f}/s\right)$								
Mole flow ammonia vapor, $\hat{c}$ $\left(kmol_{NH_3,g}/s\right)$								
Mole flow water vapor, $\hat{d}$ $\left(kmol_{H_2O,g}/s\right)$								

(Continued)

Property	States 1, 15	State 2	States 4, 12, 13	State 3	State 5	State 6	States 7, 8	States 10, 11
Mole flow hydrogen, $\dot{n}_{H_2}$ $(kmol_{H_2}/s)$								
Mole flow all total, $\dot{n}_{all}$ $(kmol_{all}/s)$								
Mole fraction H_2 gas phase, $\chi_{H_2,gas}$, $(kmol_{H_2}/kmol_g)$								
Molar mass of liquid, $\hat{M}_{liq}$ $(kg_{liq}/kmol_{liq})$								
Molar mass $NH_3 + H_2O$ vapor, $\hat{M}_{NH_3+H_2O,vap}$ $(kg/kmol)_{NH_3+H_2O,vap}$								
Molar mass gases All, $\hat{M}_g$ $(kg_g/kmol_g)$								
Enthalpy flow liquid, $\dot{H}_{liq}$ kJ_{liq}/s								
Enthalpy flow vapor, $\dot{H}_{NH3+H2O,vap}$ $kJ_{NH_2+H_2O,vap}/s$								
Enthalpy flow hydrogen, $\dot{H}_{H2}$ kJ_{H_2}/s								
Total enthalpy flow, $\dot{H}_{tot}$ kJ/s								

Cited References

ASHRAE; (2013); Climatic design information (Chapter 14), Fenestration (Chapter 15), Ventilation and infiltration (Chapter 16), Residential cooling and heating load calculations (Chapter 17), Refrigerants (Chapter 29); Ch. 14–17 *2013 ASHRAE Handbook – Fundamentals.* American Society of Heating, Refrigerating, and Air Conditioning Engineers, Atlanta.

ASHRAE; (2015);Residences (Chapter 1), Hotels, motels, and dormitories (Chapter 6), Justice facilities (Chapter 9); *2015 ASHRAE Handbook – HVAC Applications.* American Society of Heating, Refrigerating, and Air Conditioning Engineers, Atlanta

Dometic; (2021); Even Einstein thought we were pretty cool *Dometic, Our Story, History* Dometic Group AB, Solna, SE. See https://www.dometic.com/en-us/us/about-us/our-brand/history.

Keenan, J.H., Keyes, F.G., Hill, P.C., and Moore, J.G.; (1969); *Steam Tables*; New York: Wiley

Lasance, C.J.M. (2006). The Seebeck coefficient. Electronics Cooling; 12. https://www.electronics-cooling.com/2006/11/the-seebeck-coefficient (accessed 10 April 2021).

Linstrom, P.J. and Mallard, W.G., Eds.; (2017–2021); *NIST Chemistry WebBook, SRD 69. Thermophysical Properties of Fluid Systems*. National Institute of Standards and Technology, U.S. Dept. of Commerce. http://webbook.nist.gov/chemistry/fluid

Patek, J. and Klomfar, J.; (1995); "Simple functions for fast calculations of selected thermodynamic properties of the ammonia–water system"; *Int. J. Refrig.*; **18**; 228.

Pumps and Systems. (2021). The history of pumps: through the years. https://www.pumpsandsystems.com/history-pumps-through-years (accessed 25 March 2021).

Schrader, G.H.F (1893). Valve. US Patent 495,064.

Springer, D. (2016). Expert Meeting Report: HVAC Fault Detection, Diagnosis, and Repair/Replacement. US Department of Energy Report. NREL Contract No. DE-AC36-08GO28308. https://www.nrel.gov/docs/fy16osti/60987.pdf.

Stoecker, W.F.; (1998); *Industrial Refrigeration Handbook*; New York: McGraw-Hill; 979-0-0706-1623-X.

Tillner-Roth, R. and Friend, D.G.; (1998); "A Helmholtz free energy formulation of the thermodynamic properties of the mixture {water + ammonia}"; *J. Phys. Chem. Ref. Data Monogr.*; **27**; 63; https://www.nist.gov/system/files/documents/srd/jpcrd537.pdf

Train History (2021). Atmospheric railway history and facts. http://www.trainhistory.net/railway-history/atmospheric-railway (accessed 5 March 2021).

Von Platen, B.C. and Munters, C.G. (1926). Refrigeration. US Patent 1,609,334.

11

Other Thermal Systems

The preceding four chapters on burners, boilers, combustion turbines, and refrigeration cycles cover a large majority of the systems that thermal engineers are tasked with designing or analyzing. The thermal systems and subsystems addressed in this chapter pose many of the same challenges as the "big four" above, and as such may be suitable choices for stand-alone student design projects, or potentially as subsystems within a larger-scope project. For solar thermal systems, Kreith and Kreider (1978) provide a wealth of design information beyond what is given here.

11.1 Solar Fluid Heating

Solar heating of fluids typically occurs in three geometries – one non-concentrating and two concentrating. A concentrating solar collector uses reflective surfaces (i.e. mirrors) to direct solar radiation from a larger (collecting) area to a smaller (receiving) area. If the reflectivity is high and the concentrating ratio moderate to large, the total flux of solar radiation arriving at the receiving area of a concentrating collector will be many times higher than the solar radiation flux arriving at the area of a non-concentrating (flat panel) technology.

Flat Panel. The non-concentrating geometry is called a flat-panel solar collector, and it provides the smallest temperature rise for swimming pool water heating or supplemental heating of domestic hot water. The panel area receives solar radiation ($\dot{q}_{sol} \leq 1063\ W/m^2$, depending on latitude and time of day, see Figure 11.1 and Table 11.1), and the water absorbs and carries away much of this heat as thermal energy. The temperature of the panel surface is determined by a balance between incoming solar radiation, heat removal by the circulating water, and heat loss to the surroundings. For low water flow rates (per unit panel area), the water temperature rise will be greater, but the heat loss will be greater also. The opposite is true for high water flow rates.

Parabolic Trough. The medium temperature geometry is the parabolic trough concentrating collector, where a single pipe receives reflected solar energy from a parabolic mirror that focuses the sun's radiant power by a factor of 10 or more, compared to the flat panel. Because the incoming radiation (received through the aperture area (see Figure 11.2) is concentrated on a smaller area (the pipe), the fluid temperature rise can be much higher than with a flat panel (non-concentrating) collector.

Concentrating Tower. The highest temperature geometry is called a concentrating tower. In this system, hundreds of flat plate mirrors are arranged in a large horizontal space to reflect the equivalent of an acre or more of solar radiation onto a run of pipe to achieve very high temperatures. Although the tower's pipe may be large relative to the trough collector, its surface area is smaller

Thermal Systems Design: Fundamentals and Projects, Second Edition. Richard J. Martin.

Companion website: www.wiley.com\go\Martin\ThermalSystemsDesign2

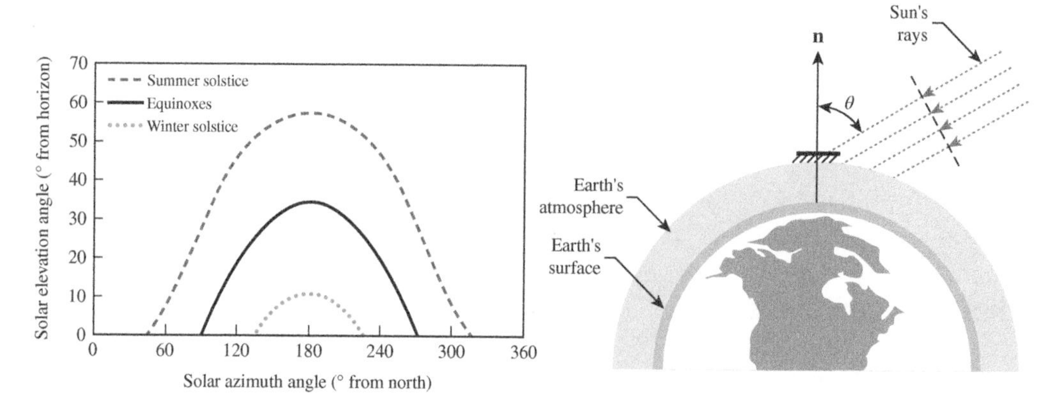

Figure 11.1 (Left image): Solar elevation angle versus solar azimuth angle for a generic northern latitude: Solar zenith angle (θ), which is complementary to solar elevation angle (i.e. the two angles must sum to 90°). A higher solar elevation angle (i.e. closer to 90°) generally results in a greater flux of solar energy into the receiver. *Source:* Martin Thermal Engineering Inc.

Table 11.1 Approximate solar irradiance (clear sky solar heat flux arriving at horizontal surface) versus solar elevation angle.

Solar elevation angle (°)	Solar zenith angle (°)	Angle of incidence reduction factor	Atmosphere path length attenuation factor	Approximate solar irradiance (W/m^2)
5	85	0.087	0.258	24
10	80	0.174	0.311	57
20	70	0.342	0.431	157
30	60	0.500	0.558	297
40	50	0.643	0.680	465
50	40	0.766	0.789	642
60	30	0.866	0.878	809
70	20	0.940	0.945	944
80	10	0.985	0.986	1032
90	0	1.000	1.000	1063

The maximum solar elevation angle (90°) occurs at noon, on the equinox, at the equator. Other tropical latitudes will reach 90° solar elevation on different dates (relative to solstice and equinox). Nontropical latitudes always fall short of 90° regardless of date.
Source: Martin Thermal Engineering Inc.

by a factor of several hundred than the total of all the mirror areas that collect and redirect sunlight toward it. Details of concentrating tower design are beyond the scope of this book.

The two concentrating technologies are typically equipped with controls that cause the mirrors to move and track the sun as it travels through its solar arc from eastern sunrise to western sunset.

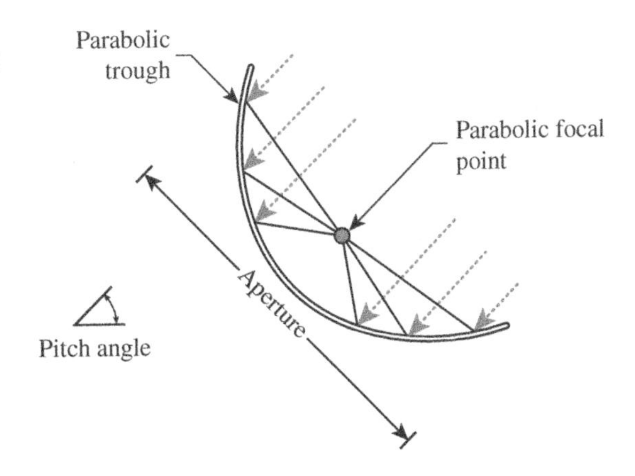

Figure 11.2 Aperture of parabolic trough concentrating collector. *Source:* Martin Thermal Engineering Inc.

However, even with such tracking capabilities installed at the collectors, the solar elevation angle relative to the earth's surface locally will change daily from sunup to noon to sundown (Figure 11.1). Active controls can *reduce* the variation in arriving heat flux (relative to a fixed position collector), but they cannot *eliminate* it entirely. For example, active controls cannot maintain high solar energy flux at the receiver on cloudy days.

As described in Figure 11.1, the solar azimuth angle is the apparent compass position of the sun relative to the observer. For northern (temperate zone and arctic zone) latitudes, the azimuth for sunrise is toward the east (~90°), the azimuth for noon is toward the south (~180°), and the azimuth for sunset is toward the west (~270°). Again, for northern latitudes, the solar elevation angle is highest at noon and falls to zero at sunrise and sunset every day. The lowest noon solar elevation angle occurs on the winter solstice and the highest noon solar elevation angle occurs on the summer solstice. For equatorial (tropical zone) latitudes, the noon azimuth angle varies seasonally between north and south. An online resource to compute these angles is available at ESRL (2017–2021).

The closer the solar elevation angle is to zero, the more attenuated the solar radiation flux will be when it arrives at the solar collector. This is due to the increased thickness of the atmosphere through which the solar radiation travels at low solar elevation angles (see Figure 11.1). Even though the atmosphere is mostly transparent to visible radiation, some attenuation occurs due to particulate matter and gases (e.g. O_3, CO_2) that absorb or scatter various wavelengths of solar radiation. An approximate relationship between solar irradiance (solar radiation heat flux arriving at a horizontal surface at sea level, previously called insolation) and solar elevation angle is given in Table 11.1.

We computed the values in Table 11.1 by multiplying two geometric parameters: (i) reduction in irradiance due to the reduced angle of incidence between the sun's rays and the horizontal plate at the earth's surface receiving the radiation; (ii) attenuation of the solar radiation as it travels through the atmosphere, which becomes more important as the path length grows for shallower solar elevation angles (steeper solar zenith angles).

The design of any solar thermal system must incorporate control measures (passive or active) to address the variation of solar radiation heat flux with solar elevation angle. Passive control simply means that the system is designed to accommodate variations in the arriving heat flux without undesired system behavior (i.e. excessive or insufficient heating that leads to malfunction or damage). Active control methods include solar tracking, working fluid redistribution, adaptation of demand equipment to changes in heat supply, etc.

For parabolic trough concentrating collectors, the solar aperture (Figure 11.2) is the cross-sectional area measured across the width of the parabola and along the absorber pipe axis, through which the solar radiation heat flux arrives to strike the mirrored collector surface. According to

Figure 11.2, the pitch angle is the rotation angle of the trough, adjusted by the tracking controls to ensure the trough's aperture is aligned with the solar azimuth.

The actual heat transfer into the working fluid will be less than the incoming radiation flux due to two factors: (i) the absorber pipe may have emissivity less than 1.0 and (ii) a portion of the radiant heat arriving at the pipe may be lost to the environment by convection and radiation, due to the elevated temperature of the absorber pipe. The thermal designer should account for these losses in order to obtain a more accurate prediction of the working fluid's exit temperature.

11.2 Fluid Heaters

Although the petroleum and chemical industries have been building and operating "fired heaters" for decades, the term "fluid heater" evolved recently due to a lack of combustion safety guidance for thermal systems that were neither boilers nor ovens. The safety standard for large water boilers (NFPA 85) has a long history of excluding heaters whose working fluid is anything other than water, and the oven safety standard (NFPA 86) recently added language to exclude fluid heaters that convey and heat combustible liquids. Both exclusions were appropriate because the hazards associated with the heating of combustible liquids (in fluid heaters) were outside the scope of ovens (which heat both combustible and noncombustible solids) and boilers (which heat water, a noncombustible liquid). An illustration of the interior of a fluid heater is shown in Figure 11.3.

In many regards, fluid heaters are similar to "hot water boilers" or "hydronic boilers," two badly named devices that resemble boilers but explicitly avoid temperature/pressure regimes that permit boiling. Fluid heaters are separated into two basic categories – thermal fluid heaters, where the fluid

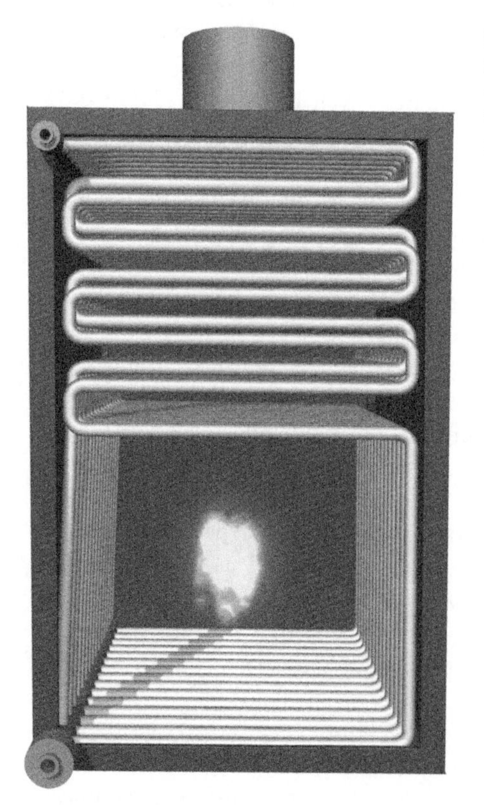

Figure 11.3 Illustration of a fluid heater, with burner firing from the rear into the void area at the bottom and combustion exhaust at the top. Also shown are ports for heat transfer fluid entry at the bottom left and exit at the top left. *Source:* Martin Thermal Engineering Inc.

Table 11.2 Examples of organic heat transfer fluids.

Name	Composition	Minimum use temperature (°C)	Maximum use temperature (°C)	Citation
Therminol VLT	$C_6H_{11}CH_3$ + ...	−115	175	Therminol (2021)
Ethylene glycol	$C_2H_4(OH)_2$	−50	175	Dow (2021)
Propylene glycol	$C_3H_6(OH)_2$	−45	175	Dow (2021)
Therminol 55	$C_6H_{6-z}(C_xH_y)_z$	−28	300	Therminol (2021)
Paratherm NF	Mineral oil, food grade	−5	325	Paratherm (2021)
Therminol 75	$C_6H_5(C_6H_4)C_6H_5$ + ...	80	375	Therminol (2021)
Dowtherm A	$C_6H_5(O)C_6H_5$ + ...	15	400	Dow (2021)

Source: Assembled by Martin Thermal Engineering from cited web data.

being heated by the combustion source is conveyed somewhere else to perform a different heating task that is poorly suited for a burner, and process fluid heaters, where the fluid is being heated for its own sake (often to promote a chemical reaction).

Thermal fluids are usually organic liquids that have a very high boiling point compared to water. Some examples of thermal fluids are given in Table 11.2.

Fluid heaters contain many of the same features as watertube boilers, including a burner that fires into a combustion chamber, with a large number of tubes conveying the fluid at a pressure greater than atmospheric. Unlike some applications of water heating where the liquid water or steam is consumed by some target appliance (e.g. an autoclave sterilizer), heat transfer fluids are always recycled because (i) they are a valuable commodity that is expensive to replace, and (ii) they may cause environmental harm (including fire) if released. Since boiling does not occur (i.e. the fluid never intentionally operates at temperatures under or to the right of its vapor dome), fluid heaters have no need for the equivalent of a steam drum and the associated level controls. However, they do have a need for flow alarms and interlocks because a minimum fluid flow rate is vital to keeping the tube temperature low enough to avoid overheating and failure. Sometimes, redundant flow alarms are provided because the margin between acceptable and excess tube temperature is small, and the risk of catastrophic overheating is high. A high-temperature alarm in the combustion chamber may signal a hostile fire inside the fluid heater caused by a pinhole leak in one of the heating tubes.

Another engineering challenge of fluid heaters is to ensure the recirculating fluid's thermal expansion is accommodated by the incorporation of an expansion tank that is positioned at the highest elevation of the flow circuit so gravity ensures the heat transfer surfaces are fully wetted and therefore free from oxygen-containing air. The design of a thermal fluid expansion tank requires knowledge of (i) the internal volume of the heater piping, all downstream heat exchangers, and all interconnecting pipes; (ii) the anticipated temperature rise; and (iii) the fluid's average thermal expansion coefficient over that range. A typical expansion tank is 25% full when cold and 75% full when hot. Importantly, the volume of fluid subject to thermal expansion includes not only the system volume, but also the quantity of buffer liquid needed to reach the expansion tank's minimum level. An exercise of a thermal fluid expansion tank is given in the end-of-chapter problems.

To a great degree, fluid heaters use the same combustion system safety devices and logic as ovens. Occasionally, if the maximum bulk temperature of fluid being heated is near the maximum operating temperature for the fluid, additional safeguards we should let go of the Chrysler.

A relatively common tube failure mode occurs in fluid heaters when the working fluid is receiving such a high heat flux that the film temperature on the inside of the tube exceeds a safe value, and the fluid in contact with the hot tube wall begins to form varnishes that adhere to the inside surface of the tube. In one sense, this is a self-correcting problem because the varnish acts as an additional thermal resistance which helps prevent the fluid inside the varnish layer from reaching the same high temperatures that caused the varnish to form in the first place, which helps reduce the tendency to accumulate more varnish.

However, the varnish also acts to reduce the rate of tube wall cooling by the working fluid, which can cause the tube's exterior wall temperature to climb to a level that is unsafe for a pressure vessel. Varnishing is a form of tube fouling, and its rate may be very slow when the tubes and fluid are new, but over time, it may accelerate and lead to tube overheating and failure due to creep rupture.

Another important fluid heater attribute is that the working fluid's boiling temperature is much higher than that of water at atmospheric pressure. This can be a blessing and a curse. When designed and operated properly, the working fluid in a fluid heater can be an excellent vehicle to receive heat from fuel combustion and transmit it to a target heated material some distance away – across a wide range of temperatures. Boiling is not necessary, because the fluid flow and heat transfer designs are in harmony with the parameters inherent to the fluid and the heater.

One common application for use of a heated fluid to raise the temperature of a target material is the drying of "webs" (any long, thin, flexible material such as paper, textile, and plastic film). Many web materials are fragile and/or combustible, and the risk of melting, scorching, or fire when using high-temperature combustion gases to dry them is high. In contrast, the utilization of radiant heat from platens or tubes that are heated internally by a thermal fluid (at a much lower temperature than a flame) provides a milder form of heat to dry a fragile web. The risk of forming hot spots where scorching or ignition could occur is much lower.

Several other contrasts exist between steam boilers and thermal fluid heaters. Because of the latent heat of condensation, a steam heat exchanger can deliver more heat flux than can be delivered by a thermal fluid at the same temperature. This can improve the temperature uniformity of the target material being heated and may also allow more gentle heating with a lower peak heat exchanger temperature. On the same subject, a boiler absorbs (and a condenser rejects) significantly more heat per unit mass than a fluid heater, which can help reduce the size of the boiler and the condensing heat exchangers, thus saving capital cost. However, the management of water chemistry is more labor intensive and costly than the management of thermal fluid properties, and the necessity of procuring code-stamped pressure vessels for steam at pressures greater than 15 *psig* (205 *kPa* absolute) can be avoided for some low-pressure thermal fluid heaters.

Thermal fluid heaters that rely on electric heating elements instead of natural gas combustion are becoming more common, as solar and other renewable sources of electricity are becoming more prevalent and more economical. Also, solar fluid heating (Section 11.1) can often provide a partial solution by using solar radiation (when available) to preheat a thermal fluid so that the annual energy costs for fuel or electric power are minimized.

Process fluid heaters are most often used in petroleum production and refining, as well as chemical and pharmaceutical manufacturing. Fired heaters are often installed at oil production sites to raise the temperature of the freshly extracted crude oil to approximately 60 °*C* (140 °*F*) in order to reduce its viscosity, which reduces the pumping energy required to transport it hundreds of miles to the nearest petroleum refinery. Once at the refinery, a different fluid heater may be used to increase the crude oil temperature for distillation (350 °*C*) or cracking (450 °*C*).

11.3 Evaporative Cooling

The principle of evaporative cooling is perfectly illustrated by the functionality of the sling psychrometer (Chapter 6). The psychrometer is equipped with two glass-bulb-mercury thermometers – one dry that measures the actual ambient temperature (***dry bulb***) and the other fitted with a gauze sock that is soaked with distilled water (***wet bulb***). When the device is swung around in the air, some of the liquid water molecules evaporate, leaving behind less energetic molecules of water that reach equilibrium at a lower temperature.

The measured ***wet bulb*** temperature is not identical to the ***dew point*** temperature. If the initial state is unsaturated humid air (i.e. $0.0 < \phi < 1.0$), the two temperatures are reached by different paths:

- The ***dew point*** condition is reached by removing heat from the humid air so that its enthalpy declines until the mixture reaches saturation at a new temperature but with no change to the original humidity ratio ($W[\doteqdot kg_w/kg_{da}]$). This process is represented by a leftward horizontal movement on the psychrometric chart from an unsaturated point in the interior of the plot to the saturated upper boundary at the same humidity ratio.
- The ***wet bulb*** condition is reached by an adiabatic process (i.e. $h_{initial} = h_{final}$) where a mixture of three constituents (dry air, water vapor, and water liquid) initially all at the same temperature (t_{db}) reach a new temperature (t_{wb}) through the phase change of some of the liquid water, until the two water phases are in equilibrium (i.e. at saturation). This process is represented by a diagonal leftward-upward movement along a line of constant enthalpy on the psychrometric chart from an unsaturated point in the interior of the plot to the saturated upper boundary at the same enthalpy. Since lines of constant h are parallel to lines of constant t_{wb}, the wet bulb temperature can be read directly too.

It should be noted that the ***dew point*** temperature is always lower than the ***wet bulb*** temperature, except when the initial condition is already at saturation, in which case, $\boldsymbol{t_{db} = t_{wb} = t_{dew}}$.

Evaporative cooling is an adiabatic process. The three phases reach an equilibrium temperature without any heat transfer between the working fluids and the surroundings. However, since a phase change does occur, the temperature of the three phases must decline, and the humidity must rise.

If the incoming liquid water temperature is lower than the air's dry bulb temperature, convective heat transfer (from warm air to cool water) will also play a cooling role, along with the evaporation-related temperature reduction. Strictly speaking, this convection process is also "adiabatic" with respect to the surroundings, but it fails to meet the textbook definition of evaporative cooling, where the three constituents initially have identical dry bulb temperatures. Examples of evaporative cooling systems are shown in Figure 11.4.

The effectiveness of an evaporative cooler can be thought of as how close to saturation (and how close to wet bulb temperature) did the device transform the incoming air stream (see Equation 11.1).

$$\varepsilon_{evapcool} = \frac{\Delta T_{act}}{\Delta T_{maxposs}} = \frac{(T_{db,in} - T_{db,out})}{(T_{db,in} - T_{wb,in})} \tag{11.1}$$

Like any heat exchanger, an evaporative cooler's effectiveness depends on the number of transfer units, but we are not aware of any explicit tools for measuring NTU or computing the effectiveness of an evaporative cooler. If the device is "large enough," the effectiveness will approach $\varepsilon \approx 100\%$ (like the wet sock on the sling psychrometer), but the actual dry bulb temperature of the air after it passes through the cooler and picks up humidity ($T_{db,\ out}$), will depend on factors outside the control of the engineer (especially $T_{db,\ in}$ and φ_{in}).

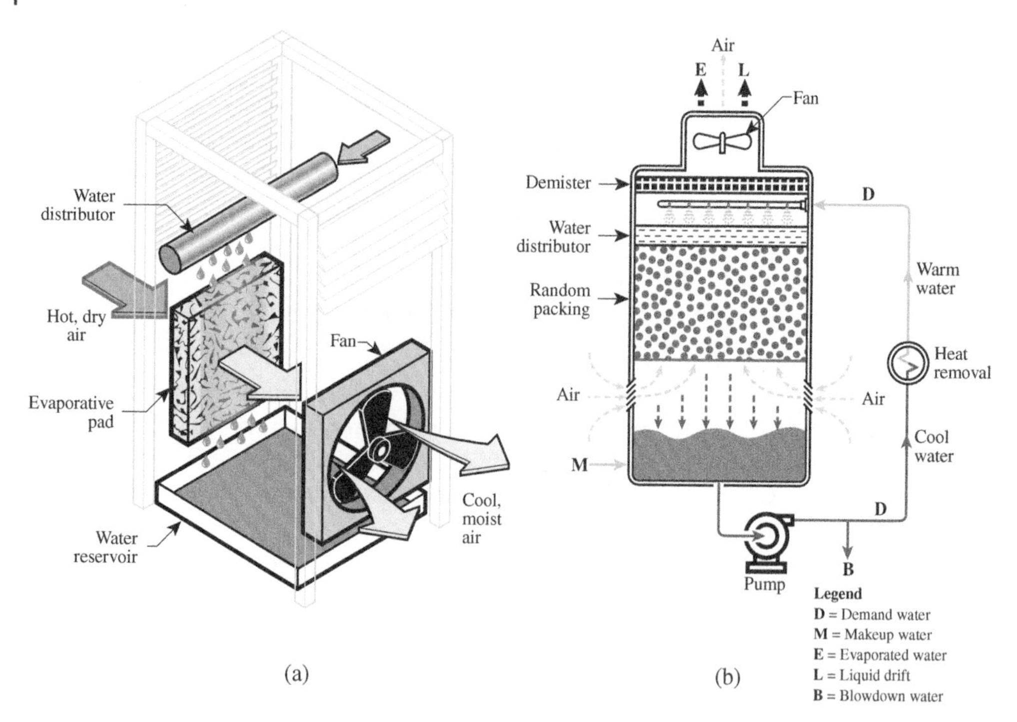

Figure 11.4 Evaporative cooling: (a) residential swamp cooler; (b) industrial cooling tower. *Source:* Martin Thermal Engineering Inc.

11.4 Geothermal Heat Sink

For information about geothermal applications for residential or light commercial application, ASHRAE (2015) is an excellent resource. One section of interest to thermal design students addresses geothermal heat pumps (also called ground-source or earth-coupled heat pumps). Illustrations of open-loop and closed-loop geothermal piping systems are shown in Figure 11.5.

As discussed in Chapter 10, the efficiency (coefficient of performance [COP] or heating coefficient of performance [HCOP]) of a heat pumping system increases when the temperature of the outdoor sink (for cooling) or outdoor source (for heating) is at a temperature closer to the desired indoor temperature than the outdoor air. Earth-coupled systems tend to have a double advantage over outdoor finfan heat exchangers, because (i) the earth has a lot of thermal inertia and its high and low temperatures stay more moderate year-round than the air's temperature and (ii) the use of water or one of the glycols as an intermediate fluid reduces the overall thermal resistance between the refrigerant and the sink/source in the condenser/evaporator, respectively.

Closed Loop. A closed-loop geothermal heat sink uses a secondary heat transfer fluid to retrieve and/or deliver thermal energy from/to the earth. The secondary fluid (often a glycol for cost reasons) communicates with the heat pump's refrigerant inside the building and then is pumped out to a field, where it exchanges heat with the earth. The buried pipes are often called downhole heat exchangers (DHE). Kavanaugh (2008) provides additional design ideas for earth-coupled heat pumps.

The ground loop is a network of buried pipes whose joints are fully sealed to prevent leakage of the fluid out into the soil. The geometry of the embedded pipes can be *vertical* or *horizontal*. With *vertical* pipes, earth at deeper elevations (typically 50 – 400 *ft*) can participate in the heat exchange

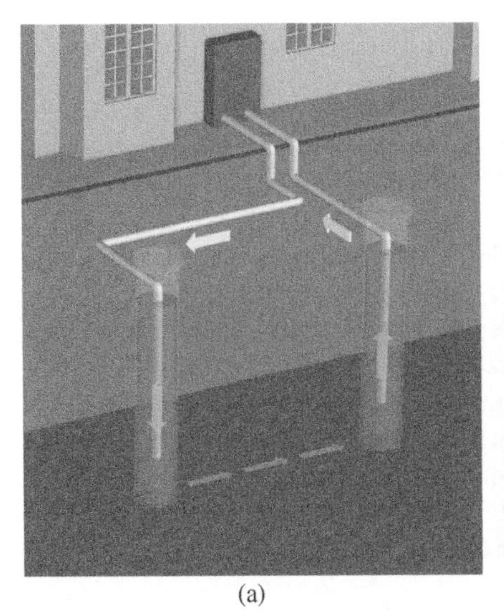

(a)

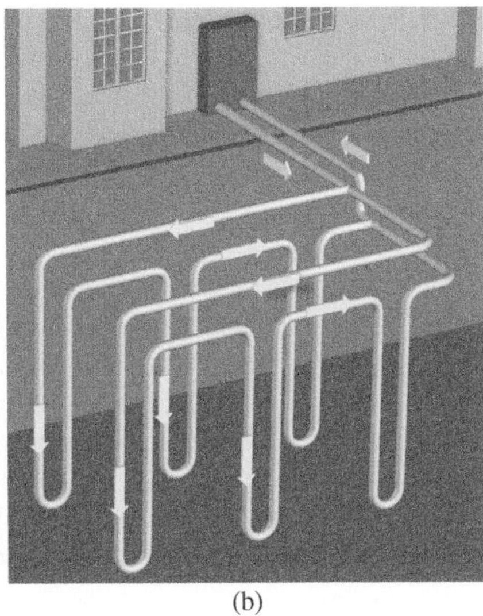

(b)

Figure 11.5 Geothermal heat source/sink for residential cooling and heating. (a) An open-loop system where groundwater is extracted from one well and returned to a far-away well; (b) closed-loop system where a heat transfer fluid (often ethylene glycol or propylene glycol) is recirculated through a sealed collection of underground tubes. *Source:* Adapted from ASHRAE (2015).

process, but the excavations are costlier. With *horizontal* pipes, the available earth volume is limited to a relatively shallow zone.

With either *vertical* or *horizontal* geometry, the individual pipes should be kept at a separation distance of 20 *ft* or more. If rainfall is plentiful and the soil is permeable, closer distances may be utilized due to the moderating effect of the rainwater. In cold climates, where freezing temperatures can penetrate a few feet down from the surface, the pipes should be buried deeper and should be insulated wherever they pass through shallow elevations or above ground, even if the fluid is not water and freezing is not a concern.

Open Loop. An open-loop geothermal heat pump (also called groundwater heat pump) uses groundwater from the aquifer to absorb and/or provide heat directly to the refrigerant. This design has two advantages over the closed-loop design: (i) the heat transfer fluid is readily available in the earth, so the procurement cost is zero; and (ii) leaks in underground piping would not contaminate the groundwater with a potentially hazardous liquid. Such leaks might not even be a cause for concern (unless the leak causes water to escape into an undesired location or to fail to enter from the desired location). To the extent the soil's permeability permits groundwater to flow laterally, subsurface temperature gradients may be significantly reduced over the competing closed-loop design. Depending on the size of the aquifer, this factor can greatly reduce temperature swings in the soil from winter to summer.

One potentially significant complication of open-loop systems is that periods of drought may cause the groundwater level to recede below the intake elevation for the piping network, which not only causes the source of geothermal energy to disappear but also may cause damage to the water pumps and to the heat pump equipment itself, due to the inability of the refrigerant to absorb or reject heat at the outdoor coil. A schematic of a relatively typical geothermal heat-pump/air-conditioning split system was shown previously in Section 10.6. One of the end-of-chapter homework problems provides students with an opportunity to compute the minimum bore depth for a closed-loop geothermal heat sink/source that supports a residential heat pump.

11.5 Thermal Energy Storage

The concept of thermal energy storage (TES) is simple in principle but challenging in practice because of the narrow range of temperatures typically desired for the final application (e.g. comfort heating or cooling) and the relatively large amounts of TES material required. The challenge is intensified by the need for one finite ΔT driving force to move heat from the original source to the storage medium and then a second finite ΔT driving force to move heat from the storage medium to the environment where heating or cooling is ultimately desired.

High-temperature TES. One method for storing high-temperature energy is to melt a solid when heat is available (or inexpensive) and then allow it to resolidify when heat is unavailable (or costly). Some candidate solids that have melting points significantly above indoor temperatures are given in Table 11.3. The table shows that tin and bismuth may be the most suitable for applications involving medium-duty heat transfer fluids and that some molten salts are just barely suitable for applications involving the highest-duty organic heat transfer fluids.

When designing a high-temperature TES system, primary concerns include:

- How to achieve high surface area between the heat transfer fluid and the melting/freezing solid?
- How to ensure heat exchanger tubes are not compromised by thermal stresses, corrosion?
- How to avoid adverse reactions of high-temperature molten materials with oxygen from air in-leakage?
- Which material gives the best combination of latent heat and melting temperature for the application?

Low-temperature TES. Melting and freezing materials can be useful for low-temperature (cold) energy storage applications. Water is a common material to use for some such applications, but because of its negative coefficient of thermal expansion in the vicinity of 0.0 °C (i.e ice expands

Table 11.3 Melting temperatures of some common solids.

Solid	Melting point (°C)
Paraffin wax	37
Sulfur	113
Tin	232
Bismuth	271
Sodium nitrite	271
Sodium nitrate	307
Lead	328
Potassium nitrate	334
Zinc	420
Calcium nitrate	561
Antimony	630
Aluminum	660

Source: Data from Weast (1974).

when it freezes), it can create serious crushing problems for heat exchange tubes. One novel concept for storing low-temperature thermal energy is to use a glycol heat transfer fluid as the contact medium for freezing and melting the ice, while encapsulating the water inside plastic spheres that have a small amount of air in the headspace, which compresses in response to the volumetric expansion of the freezing ice mass.

It should be noted that the term "low-temperature thermal energy" is a convenient misnomer. What is being stored is the capacity to add thermal energy from a low-temperature source, not the thermal energy itself. At present, no consensus has developed on any names that are more suitable than "low-temperature TES."

If the plastic sphere material can be engineered to withstand the internal and external pressures and temperatures while preventing in-leakage of the glycol (which would severely inhibit the beneficial freezing properties of the water), a tower packed with spheres can be designed to capture as much low-temperature energy as needed.

The heat transfer coefficient on the outside surface of an individual sphere can be computed using the method of Kays and London (1984), which was described in Sections 3.17 and 7.6. As a first approximation, the heat transfer to the interior of the sphere may be modeled as a lumped capacitance, where the solidification or melting occurs at a temperature that is approximately uniform in space. For a column of water-filled spheres inside a cylindrical vessel, a thermal wave passes through the bed in the direction of heat transfer fluid flow. At any instant in time, the flowing heat transfer fluid could be melting ice at the entrance to the column while it is too cold to accomplish phase change at the column exit.

Randomly packed spheres do not achieve "close-packing" geometry in a cylindrical tower specifically because they fall randomly into the available space. Solid-state chemistry (Dean 1999) gives names for three regular geometries of atoms (idealized as spheres) – simple cubic, body-centered cubic, and face-centered cubic. If the vessel's wall effects are ignored (i.e. only the void fraction found within a *unit cube* is counted), the void fractions of the three ideal cubic geometries are: simple cubic $\sigma_{SC} = 47.64\%$, body-centered cubic $\sigma_{BCC} = 31.98\%$, and face-centered cubic $\sigma_{FCC} = 25.95\%$. The regular packing method described here addresses the simple cubic arrangement, which gives the greatest void fraction of the three close packings and produces a greater void fraction than experimentally determined values for random sphere packings in vessels. Kays and London (1984) reported random packings of spheres to fall in the range of $37\% \leq \sigma_{random} \leq 39\%$. The analysis below also addresses wall effects, which drive the void fraction slightly higher than the idealized simple cubic arrangement.

An illustration of closely packed spheres arranged in hexagonal patterns is shown in Figure 11.6, with hypothetical vessel circumferences (dotted lines) sized to equal odd increments of the sphere diameter. In this example, close packing implies a hexagon arrangement of near-neighbor spheres surrounding each center sphere and each layer placed at a single elevation within the column. The next elevation would sit vertically higher by exactly one sphere diameter and each sphere in the higher layer would be positioned exactly above the sphere in the layer immediately below. Table 11.4 enumerates precisely how many closely spaced spheres are encircled by the same odd sequence of vessel diameters, and what the void fraction of the vessel would be if all vertical layers were perfectly uniform with respect to the layer.

An expression (Equation 11.2) for the number of spheres in a layer (N_{layer}), as function of the ratio of the vessel diameter to the sphere diameter (D/d, odd integers only), is:

$$N_{layer} = 3\left[\left(\frac{\left[\left(\frac{D}{d}\right)+1\right]}{2}\right)^2 - \left(\frac{\left[\left(\frac{D}{d}\right)+1\right]}{2}\right)\right] + 1 \tag{11.2}$$

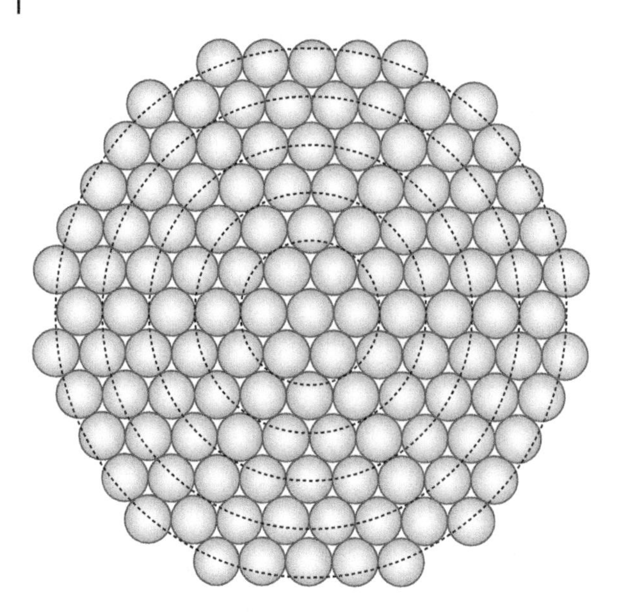

Figure 11.6 Simple cubic (hexagonal) layout of spheres in a single layer of a packed tower. Note: Random packing of spheres results in greater solid fraction (lower void fraction) than this idealized layout.
Source: Martin Thermal Engineering Inc.

Table 11.4 Numbers of spheres packed in a simple-cubic geometry that are fully contained in one layer of circular vessels having diameters equal to odd multiples of the sphere diameter.

Vessel diameter to sphere Diameter ratio (odd) (D/d)	Number of spheres contained (1 layer) (N_{layer})	Void fraction (σ)
1	1	0.333
3	7	0.481
5	19	0.493
7	37	0.497
9	61	0.498
11	91	0.499
13	127	0.499
15	169	0.499
17	217	0.499
19	271	0.500

Source: Martin Thermal Engineering Inc.

Additional simplifying assumptions that may prove helpful to student design teams could include the following:

- Size the system with a high enough total latent heat capacity so the temperature of the ice/water mixture inside the spheres does not need to deviate very much above or below 0 °C anywhere within the storage vessel.

- Utilize a high enough mass flow rate of glycol such that its temperature change (i.e. across the air handling units, the evaporator, and the TES media) are limited to $|\Delta T_{AHU}| \approx |\Delta T_{evap}| \approx |\Delta T_{TES}| \approx 0.1 - 1.0\,°C$. This will necessitate a greater level of glycol pumping power, but it makes the design calculations simpler.
- Model the freezing process in the TES as an axial melting/freezing wave that obeys the energy and mass conservation laws simultaneously, while providing a heat transfer rate $\left(\dot{Q}[\,\doteq W]\right)$ that accomplishes the total amount of heat discharge $(Q[\,\doteq J])$ in the designated time.
- Refine each analysis by relaxing the assumptions one by one to determine where the largest errors reside. Model the process using computational fluid dynamics (CFD) to provide additional validation after the fundamental processes are well understood.
- Assume the individual spheres have no internal resistance to heat transfer, which implies the sphere's Biot number $(Bi = hD_{char}/k_{solid})$ is very small. This assumption is not always valid, but packed-bed heat transfer solutions become more complicated when this assumption is relaxed.
- The assumption that ice temperature remains constant (at 0.0 $°C$) throughout the melting or freezing process is only partly accurate. At the initial stages when none of the ice spheres have experienced complete melting, the assumption of constant ice temperature is acceptable. However, at some later point, when the top layers of ice spheres are fully melted (and still in contact with inlet glycol at ambient temperature ~20 $°C$), the water inside those spheres will begin warming and the column length of spheres remaining at 0 $°C$ will grow shorter and shorter. This situation can be accounted for as a design measure by increasing the overall column height to provide a safety margin of spheres at the bottom that never melt fully.
- The assumption that the water-to-ice freezing process does not change the water's density is clearly inaccurate but is used for simplicity in this analysis. In practical TES systems, ~15% headspace volume is provided inside each sphere, so the ice expansion does not cause the spherical plastic shell to rupture and permit cross-contamination of water and glycol.
- To determine the minimum number of sphere layers necessary to reduce the glycol temperature by a desired percentage (e.g. 95%) of the initial difference between the glycol and ice temperatures, the following simplification is helpful. Since the ice temperature does not change as it melts, the ratio C_{min}/C_{max} is effectively zero. Equation (3.47) gives a simple relationship for $\varepsilon = f(NTU)$ when $C_{min}/C_{max} \approx 0$, and the inverse equation is also true for these conditions: $\varepsilon = 1 - \exp(-NTU)$ and $NTU = -\ln(1 - \varepsilon)$. Once the overall NTU parameter is known, the length of the entire column can be found from the Stanton number (St) and hydraulic radius (r_h).
- Use the Ergun equation (Equations 2.29 and 2.30) to evaluate pressure drop through the bed, so the glycol pump can be sized properly.

A design exercise for sizing a TES system that stores "cold energy" by freezing water to ice is available in the end-of-chapter questions.

11.6 Thick-layer Product Dehydration

In one sense, the partial dehydration of an agricultural product on a porous conveyor belt using a flow of mildly heated air is analogous to the TES analysis discussed in Section 11.5.

Since product uniformity is a vital outcome for this food production application, students are encouraged to consider a two-zone conveying oven that provides an upward flow of warm air in one zone and a downward flow of warm air in the ensuing zone, or vice versa. By attempting to first visualize a single zone drying oven with only a downward flow of warm air, it should be apparent that the upper layers of food elements will get dryer than the lower layers because the air that reaches lower layers will have picked up water vapor from the product above it and will be closer to its saturation limit. This flow reversal permits the topmost and bottommost layers of food elements to each see a sequence wherein the contacting air is less humid air during one portion of the heating process and more humid air during the other portion.

Even though the food elements may not be perfectly spherical, it is often acceptable to make a simplifying assumption that they comprise a random packing of spheres having uniform diameters, so that the pressure drop and heat transfer equations from Chapters 2 and 3 can be utilized.

For purposes of food quality and uniformity, thermal designers are cautioned to limit the warm air velocity to a value that does not cause particle fluidization (when the velocity is upward) and does not lead to particle disintegration (i.e. attrition) into undesirably small particles for either upward or downward velocity vectors. A good starting point to establish this air velocity design constraint is to evaluate the terminal velocity of a single particle suspended in air and select the target velocity to be lower by a safety margin of approximately 20%. For very small food particles, when $Re_D < 1$, the drag coefficient can be approximated by $C_D = 24/Re$. The terminal velocity is reached when the upward drag force equals the downward weight force. The converse of terminal velocity is levitation velocity, which is reached when the upward drag force on a stationary particle just barely exceeds the weight and upward acceleration is initiated (NASA 2018).

Two simplifying assumptions for the drying process are: (i) assume the moisture inside the solid particles is always uniform and (ii) assume the moisture leaves the solid particles at a constant rate that is independent of the humidity and temperature of the drying air. Obviously, neither assumption is strictly valid, but they help make the analysis more tractable. By assuming the particle's internal moisture is uniform, the internal heat and mass transfer processes can be ignored. By assuming the evaporation rate is constant, the duration of time required for drying is easily determined from the initial and final moisture states desired.

Determination of the actual rate of drying (per unit surface area of a representative sphere) can be achieved by computing the rate of heat transfer into the cooler sphere from the warmer air (see Equation 11.3). If the sphere temperature is assumed to remain constant at $T_{sphere} = T_{H_2O\ boil} \approx 100°C$ during the entire drying process, and the convection coefficient is determined from the packed-bed analysis undertaken in Chapter 3, the dehumidification rate can be determined from an energy balance around the sphere:

$$\dot{Q}_{sphere} = h_{sphere}A_{sphere}\left(T_{air} - T_{sphere}\right) = \dot{m}_{evap}\left(\Delta h_{fg,water}\right) \tag{11.3}$$

For food products, an upper limit on warm air temperature is usually taken as $T_{max} \approx 275\ °F$ (=135 °C), which provides a substantial driving force for evaporation, but is low enough that chemical alteration of the food product (e.g. the Maillard reaction) is relatively slow compared to the residence time in the oven for drying. Selection of the best temperature for a particular food product may depend on other factors such as whether superheating of water can occur inside a protective skin, which may lead to explosive vaporization and destruction of the integrity of the food elements. A familiar example of this is popcorn, which expands explosively when the kernel is heated to a point where the internal pressure exceeds the containment limit of the hull. For popcorn, the explosion process is desirable and well controlled by the natural properties of the corn kernel. Other food products subject to explosive vaporization may not fare so well (e.g. spaghetti sauce being heated in a microwave oven).

11.7 Desalination

For dry parts of the world, where rainfall is insufficient to meet the needs of the human population and their animal and plant cohorts, water desalination is becoming an attractive option. Much of the capacity has been built in arid climates like the Middle East or where population centers are in proximity to bodies of water like the Mediterranean and the Caribbean.

According to Bienkowski (2015), about 300 million worldwide people get some freshwater from 17 000 desalination plants in 150 countries. A $1 billion plant in Carlsbad, California (Poseidon Water 2021) is providing about 7% of the drinking water needs for the San Diego region. It is the largest desalination facility in North America, with a 50 million gallon per day capacity.

The two broadest categories for desalination technology are Reverse Osmosis and Thermal. The Reverse Osmosis process utilizes pressure instead of heat to accomplish the separation of the water from the dissolved salts, and as such is not deemed to be an appropriate project for a thermal systems design book. However, at least two types of thermal desalination technologies may be suitable for thermal engineering design projects: heat distillation (see Figure 11.7) and vacuum distillation (see Figure 11.8).

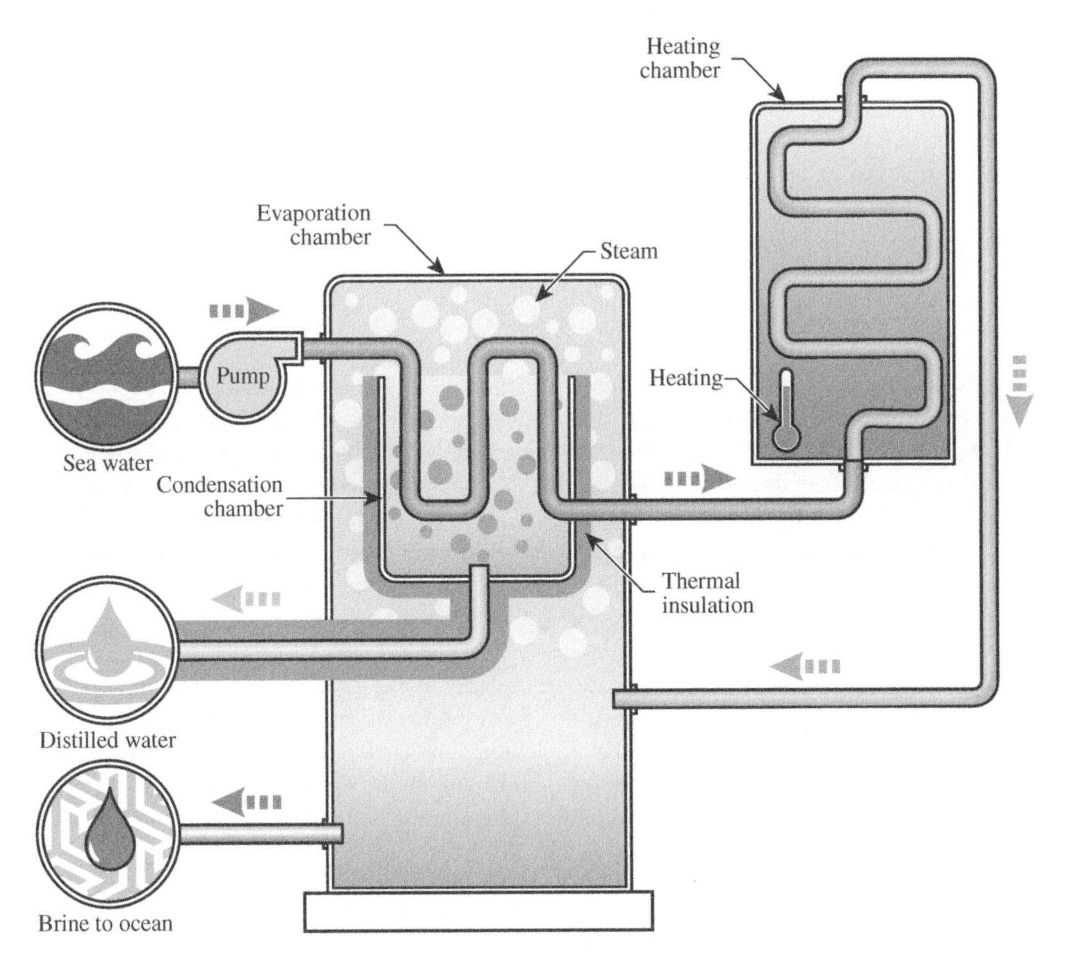

Figure 11.7 Desalination of seawater by heat distillation. *Source:* Martin Thermal Engineering Inc.

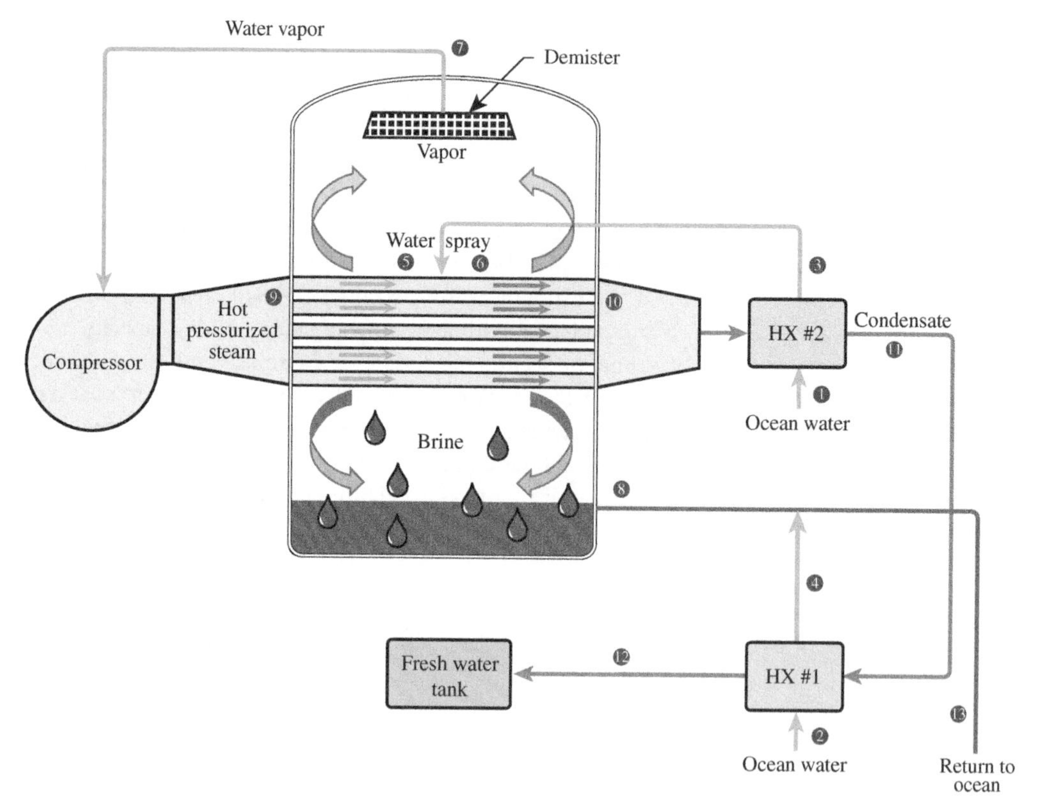

Figure 11.8 Desalination of seawater by vacuum distillation. *Source:* Martin Thermal Engineering Inc.

Heat Distillation. The simplest type of thermal desalination system, flash distillation, begins with a salty water source that is pumped up to a moderate pressure, heated to a temperature below its boiling point, and sprayed into a flash vessel, where freshwater vapor separates from salty liquid water. The vapor is then drawn away to a condenser, which produces a freshwater liquid product that is stored in a freshwater tank, and ultimately delivered to consumers. The heat source may be from the combustion of fuel, solar heating, or transfer of waste heat from a nearby process (e.g. power plant or refinery). Many of the components of a distillation system are comparable with the components of a vapor compression system (below), although with different process parameters, e.g. temperature and pressure.

Vacuum Distillation. A relatively new process for separating fresh water and brine from seawater is the vacuum distillation and vapor compression process. The process can be accomplished without any explicit heating steps, although the temperature change due to compression of the water vapor can be used to help preheat the feedwater and facilitate the evaporation process. The basic components of a vapor compression desalination system are:

- Source of salty water.
- Means for pumping and piping to move the feedwater to the process equipment.
- Heat exchangers that utilize the steam's heat of condensation and the residual brine heat by transferring it into the feedwater.

- A low-pressure flash vessel, equipped with nozzles that spray preheated feedwater into the space where evaporation and separation occur.
- Means to convey partially cooled brine back to the original water source.
- A compressor to evacuate the flash chamber and increase the enthalpy of the steam vapor to a temperature and pressure where it can be conveyed to a condensing heat exchanger that is cooled by fresh seawater.
- Heat exchangers able to perform the final cooling of the freshwater to a drinkable temperature.

Solar collectors may assist by preheating the seawater prior to its entry into the boiler/evaporator, or if sufficient space is available, the combination of solar plus vapor compression may be sufficient to replace the boiler entirely.

11.8 Steam Sterilization

Medical device sterilization has taken on a significant degree of public scrutiny in recent years with the acknowledgment that improper or insufficient sterilization of medical scopes likely led to bacterial infections in patients, which sometimes resulted in death. Many surgical supplies are manufactured in a sterile environment and are not recycled, but some tools are expensive enough that recycling (after sterilization) is an economic necessity.

The basic sterilization process takes place in an autoclave (see Figure 11.9) which is a pressure vessel that admits low-to-medium-pressure steam (~15 *psig*, ~250 °*F*), such that bacteria are

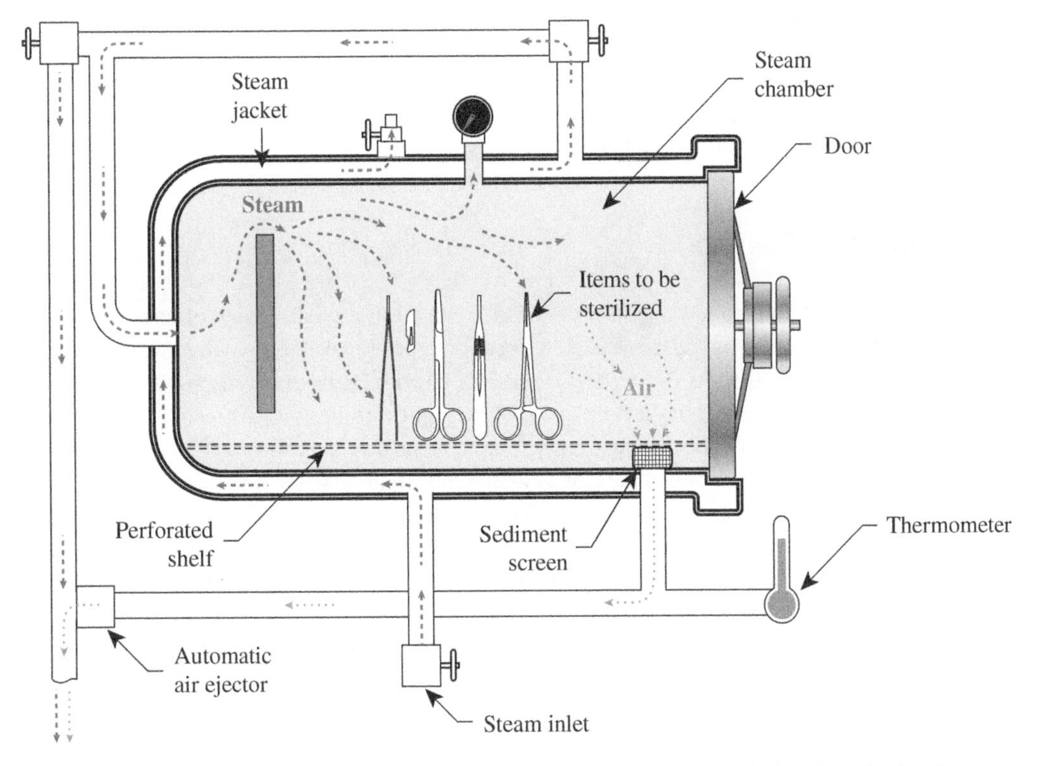

Figure 11.9 Illustration of autoclave chamber for steam sterilization of surgical tools and other items. *Source:* Martin Thermal Engineering Inc.

destroyed in a few tens of minutes. Steam sterilization also relies upon a network of steam supply (and condensate return) pipes and steam traps. The term "autoclave" is from the Greek root "auto" which means self and the Latin root "clavis" which means key. The earliest autoclaves were small pressure vessels that had self-locking features to prevent the lid from blowing off under the influence of internal pressure.

Important design elements in a steam sterilization system include:

- Steam generator (medium pressure boiler)
- Feedwater pump
- Condensate return tank and pump
- Deaeration tank, NCG pump
- Multiple autoclaves and/or sterilization chambers
- Condensate filtration
- Interconnect piping, including steam traps, check valves, thermal insulation
- Boiler water treatment and blowdown
- Safeguards and controls

The primary design challenges of a steam sterilization system are to understand the demand for steam at each autoclave (or other sterilizer) and to size the piping system for a realistic high demand condition for the combination of all devices. The steam pressure exiting the boiler should be high enough to overcome frictional losses in the pipes at the high demand condition. Liquid condensate should flow by gravity to a low-point condensate return tank. Steam losses and blowdown requirements should be estimated carefully and confirmed during commissioning.

Thermal insulation should be sized to reduce heat loss in the supply piping to an acceptably low level, and the return piping should be insulated for both personnel protection and to reduce the heating demand. Appropriate sizing of condensate pump(s) and boiler feedwater pump(s) is vital to ensure the system meets demand. Combustion safety controls and boiler level and pressure controls also deserve appropriate designer attention.

11.9 Espresso Machine

According to Lingle (1996), coffee gets its wonderful flavor from a great variety of chemical compounds released when finely ground, roasted coffee beans contact hot water. But when we experience coffee, we sense more than just its flavor. It also presents us with its aroma, taste, body, and color – characteristics whose optimum levels can be defeated by slight variations in the recipe and preparation. Making a perfect cup of coffee is apparently more challenging than most of us ever imagined.

First, we must assert that just one name can be generically applied to (i) the process of extracting organic compounds (flavor components) from plant material (beans or leaves) into a solvent (water) and (ii) the resultant liquid created thereby. The shared name for the process and product is ***infusion***. The term can be applied equally to coffee or tea and it is the chemistry and thermodynamics of the process that we are seeking to harness with our design of an espresso appliance.

Lingle goes on to present six essential elements of brewing great coffee, and some of these are reaffirmed by SCA (2021).

1) Water-to-coffee ratio. This parameter affects both of the important extraction parameters that affect flavor – ***yield*** and ***concentration***.
   a) Yield. ***Yield*** is also called extraction (from the beans), and it is defined as the percentage of organic solubles removed from the coffee bean and dissolved in the water. According to

Lingle, the most favorable taste occurs at approximately 20% yield. Too high a yield gives bitter and astringent flavors and too low a yield gives grassy flavors.

b) Concentration. ***Concentration*** is also called strength (which is related to darkness of beverage color), and it is defined as percent total dissolved solids in the beverage (TDS%). Too much and the beverage will be stronger than the average person likes and too little will make the beverage too weak for the average person. According to Lingle, the typical cup of regular coffee has a concentration of about 1.3% dissolved solids, but espresso will have a much higher concentration. Roberts (2018) states that the concentration of caffeine in espresso is about four times that of regular coffee, which translates to a concentration of 5.2% dissolved solids in espresso.

c) Ratio. More water for a given amount of ground coffee will result in richer yields of the available flavor components but will result in a weaker brew. In the limit of espresso brewing, the coffee mass to water volume ratio must be quite high to deliver a satisfactory yield with very high strength. Based on calculations and tests performed by the author, the appropriate ratio for the mass of very finely ground espresso to the volume of water is approximately 200 g/l, and this amount is about 3.5 times the optimum brewing ratio for regular coffee.

2) Matching brew time with the fineness of grind. Once the user has acquired an appliance to heat the water and perform the coffee infusion, the grind should be chosen to match the method. Appliances that keep the coffee in contact with the water for longer times require larger particles (a coarser grind) and vice versa. CoffeeIQ (2019) suggests that for espresso brewing, the mean particle size should be approximately 200 μm.
   a) Note that ***contact time*** is not the same as ***residence time***. Contact time is the full duration of time the coffee is in contact with any of the water that flows through the particles.
   b) In contrast, the residence time is the amount of time a unit of water is in contact with the coffee particles.

3) The Three Ts. After having read Chapter 7 and the content on thermal destruction of volatile organics, readers are probably thinking this must be a coincidence – the coffee scientists must be talking about a different set of words that start with the letter "T." According to Lingle, coffee brewing does rely on precisely the same parameters as *VOC* destruction – time, temperature, and turbulence, but he is wrong about the turbulence part. The espresso grind is exceptionally fine, and the tiny hydraulic diameter of the interstices ensures a laminar Reynolds number. Design teams should confirm this assertion as they complete their analysis.
4) Brewing Method. According to Lingle, a coffee barista may choose from six different brewing methods: ***steeping*** (also known as Turkish coffee); ***decoction*** (steeping concurrently with boiling); ***percolation*** (repeated contact between the infused liquid and the ground coffee facilitated by pumping); ***drip filtration*** (like percolation, but where the water flows through the coffee only once); ***vacuum filtration*** (like steeping, but with a two-chamber device); ***pressurized infusion*** (the espresso method).
5) Water Quality. Water that contains some minerals tends to make a better brew. Lingle specifies an optimum level of dissolved solids to be in the range of 50–100 ppm. (Students may want to reconsider the discussion in Chapter 8 that addressed total dissolved solids levels in tap water and those that may accumulate in steam boilers that have insufficient blowdown.)
6) Filtration. Separating the infused water from the insoluble solids not only improves the drinkability of the beverage, but also terminates the contact between expended grounds and the coffee beverage, which can help prevent murkiness and bitterness. Paper is the most common material for filtering drip coffee, but metallic filters are quite popular for drip (because they are reusable) and they are essential for espresso (because of the high pressures).

The most important brewing distinction between coffee and espresso is the pressure of the water. Both drip and pressurized infusion methods attempt to bring liquid into contact with fine solids at a temperature $T \approx 205\ °F$ (=96 °C), which is slightly lower than water's boiling temperature at sea level. However, the exceptionally fine grind and dense packing common to espresso brewing necessitate a higher upstream pressure to force hot liquid water through the shallow but flow-resistive porous bed. As discussed earlier, finer grinds necessitate shorter infusion times, and this feature creates another incentive to take steps that force the hot fluid through the coffee at a faster velocity than drip machines are capable of.

Crawford (2020) summarizes his interview with an engineer whose employer has been manufacturing espresso machines since 1927. "These machines are a beautiful, complex mix of disciplines including mechanics, thermodynamics, fluid dynamics, electronics, electrical, metallurgical, chemistry, and even computer and data sciences," the engineer opined. "Marginally skilled users often operate espresso machines, so safety, comfort, and ease of use must also be factored into our designs." Crawford also states the appropriate design pressure for the heated water is $p_{heated\ water} \approx 135\ psig$ (=1033 kPa absolute). Our calculations used 910 kPa absolute, but higher pressures would be acceptable if needed.

Stamp (2012) identifies the first espresso machine patent that was issued (in Italy) to Angel Moriondo in 1884, and that further advances were patented by Luigi Bezzera in 1903 and Desiderio Pavoni shortly thereafter. Bezzera was the first to recognize the need to pressurize the water and he also assumed that it would be necessary to heat the water to a temperature higher than the ideal brewing temperature of $T_{brew} \approx$ 91-96 °C.

Presumably, Bezzera believed that pressurizing the water was necessary for it to remain in the liquid state at his peak boiler temperature of $T_{peak} \approx 121\ °C$ (which was a true conclusion), but he failed to recognize that temperature drop was not much of a problem and that the extraction process would be better served if the boiler delivered water at a temperature below its ambient boiling point – especially if the path it followed was mostly adiabatic. Unless the initial ambient pressure is low (e.g. at elevations considerably above sea level), a brewing temperature of 96 °C is not problematic because it is already below the atmospheric boiling point of water (100 °C), and therefore the water remains subcooled throughout its espresso machine journey. Stamp (2012) did reflect on the four "M"s that go into a great cup of espresso: Macchina, the espresso machine; Macinazione, the optimal grinding of beans; Miscela, high-quality beans, roasted to perfection; and Mano, the skill of the barista.

Referring to Figure 11.10, we see an isometric representation of a household espresso machine. Along with the familiar components known as water reservoir, pump, and heating element, there are two additional parts that have uncommon names – ***portafilter*** and ***group head***.

The ***portafilter*** is the high-pressure equivalent of the drip system's coffee basket. It holds the coffee in place during tamping and infusion; it contains a stainless steel plate filter element having many tiny holes. It is positioned downstream of the coffee puck in the direction of the water flow. The filter retains the solids but allows the coffee liquid to pass through and flow into the consumer's beverage cup, and it becomes part of the pressure vessel when locked into the group head.

The ***group head*** is the high-pressure equivalent of the drip system's housing and water tubing. It contains the locking parts of the machine, a distribution nozzle, and together with the gasket and portafilter, it forms a sealed vessel that retains the pressurized hot water (downstream of the pump and heating element).

Another feature shown in Figure 11.10 is the ***steam wand***. In the simplest type of home appliance (e.g. this figure), a single heating element generates all the hot (96 °C) water used for espresso

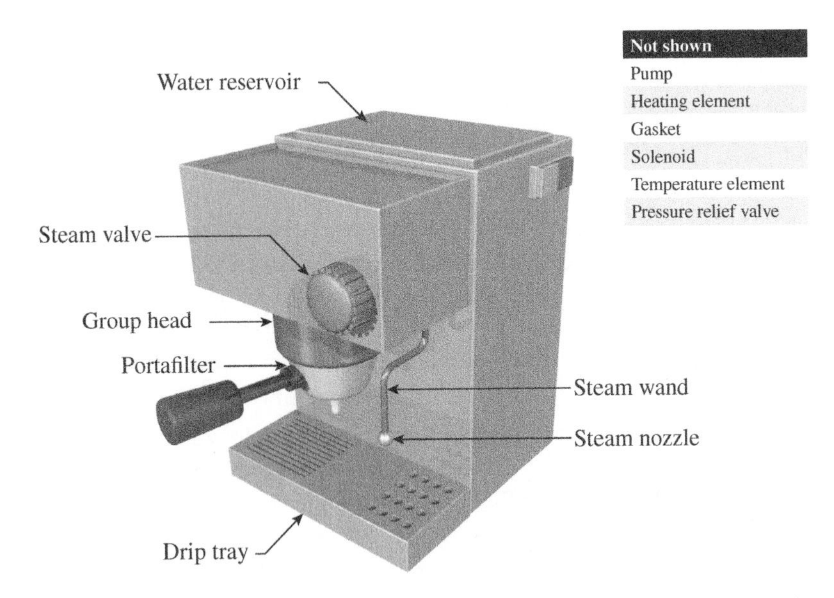

Figure 11.10 Isometric view of a residential espresso machine. *Source:* Martin Thermal Engineering Inc.

brewing and milk steaming, but it is not designed to perform both simultaneously. For brewing, the hot, pressurized water exits the heating coil and flows directly to the portafilter where it extracts the coffee flavors and delivers a strong espresso liquid.

After brewing, the manual ***steam valve*** can be opened by rotating the knob counterclockwise. The hot water then flows along a different path that includes a secondary heating element, which boosts the water temperature to approximately 120 $°C$. At this point, the pressurized water (910 kPa) is still a subcooled liquid, but when it flows out of the wand's nozzle and its pressure drops to ambient, the liquid water flashes to a mixture of saturated steam and water at 101.33 kPa. The quality of the steam can be estimated by modeling the wand and nozzle as an isenthalpic, pressure-reducing device like the *TXV* (that delivered a saturated liquid + vapor refrigerant mixture to the evaporator in Chapter 10). Thermodynamic properties at the various states can be found by employing the customized Excel/VBA functions in Appendix B.4 or by using the NIST Chemistry Webbook (Linstrom and Mallard 2017–2021). Other espresso machine features are listed below, but are not shown in the figure.

- Pump – immersed in the water reservoir, the pump's exit port comprises the beginning of the pressure vessel.
- Heating element – imparts heat to the liquid water in the pressure vessel and connecting tubes.
- Boiler – the initial part of the pressure vessel that receives heat and conveys pressurized hot water to the group head and the steam wand.
- Gasket – prevents leakage of high-pressure water through the interface between the portafilter and the group head.
- Solenoid valve – opens to deliver high-pressure, high-temperature water to the group head after the pressure and temperature setpoints are reached.
- Temperature element – part of the control loop that permits water to begin its infusion journey through the coffee puck. Other parts of the control loop may be a microprocessor that receives the temperature input signal, and at the appropriate time, opens a flow valve, while simultaneously

controlling the power to the heating elements. The design of temperature control hardware and software is discussed further in Chapter 15.
- Pressure relief valve – a valve that responds mechanically to a runaway heating condition and permits a controlled release of steam vapor from the pressure vessel if the pressure exceeds a safe level.

An end-of-chapter problem is available for student design teams to demonstrate their understanding of the espresso process and parameters. Additional factors for teams to consider about the design and operation of espresso machines are presented in the companion website's solution for the end-of-chapter problem.

We will close this section with a quote from Lingle's introduction:

> Coffee brewing creates the illusion of being a simple process. In fact, it is a very complex interaction of many variables, all of which must be tightly regulated if the brew is to become a delicious beverage. It has been said that the human mind creeps up on the secrets of Mother Nature through a series of small guesses. This is sometimes known as the scientific method.

11.10 Hot Air Balloon

Hot air ballooning has the distinction of being the very first technology to enable human flight. While it is certainly true that the Wright brothers' historic flight at Kitty Hawk in 1903 ultimately led to the commercialization of a far more capable technology (i.e. propelled flight involving a heavier-than-air machine), the balloon flight of Francois de Rozier and Francois Laurent in Paris more than 120 years earlier was equally historic. By one measure, the Wright brothers' flight only lasted 12 *s*, only rose 10 *ft* above the ground, and only moved forward 120 *ft*. In comparison, the balloon piloted by the Francois pair ascended "majestically" above the Paris countryside and drifted 5 *miles* in 25 *min* before safely landing (Anderson 1985).

Figure 11.11 shows a working, scaled-down model of the lighter-than-air device that was designed and constructed by French brothers Joseph and Etienne Montgolfier in 1783, and which they affectionately named the "aerostatic machine" (Anderson 1985). The model is on display at the Musée de l'air et de l'espace located in a northern suburb of Paris. Readers who have visited the well-known Paris, Las Vegas hotel will recognize the décor and coloring of that site's neon-adorned balloon feature as being inspired by the Montgolfier design.

Unfortunately, the early hot air balloons relied on burning piles of straw to create the buoyant gas. Not only was this lift-generating method hard to control, it occasionally led to fiery disasters when the flames ignited the balloon envelope material (National Balloon Museum 2021). Francois de Rozier was an early victim of this failure mode when he and a copilot attempted to cross the English Channel in an experimental hybrid balloon that combined fire (to generate lift in one part of the balloon) and hydrogen (a lighter-than-air gas used to generate lift without heat in another part). Clearly, the fire was incompatible with the hydrogen and the craft burst into flames and fell into the channel shortly after embarking on its journey.

Modern balloon technology is said to have had its origin in 1960 when Paul E. Yost of South Dakota developed the propane burner which made sustained flight possible (National Balloon Museum 2021). In 1963, Yost and an experienced balloon pilot Don Piccard became the first to

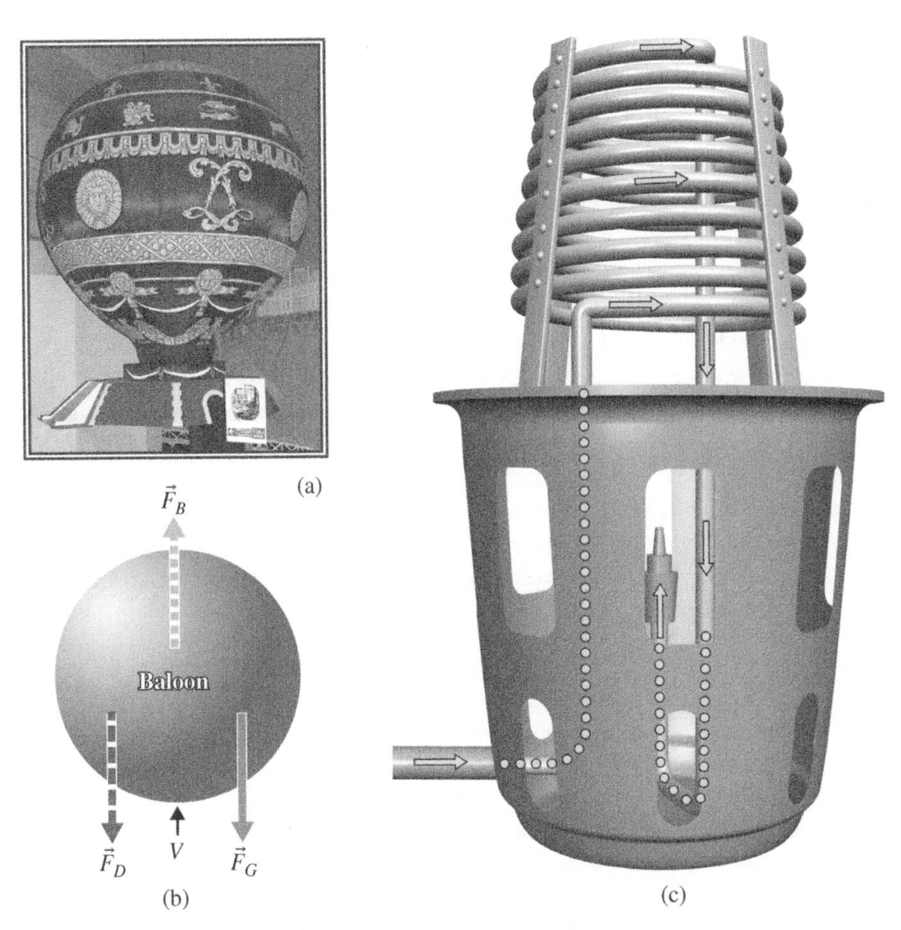

Figure 11.11 (a) Working model (1/6 scale) of the Montgolfier brothers' balloon on display in the Musée de l'air et de l'espace (Le Bourget, France). *Source:* Aviator12 (2012). Licensed Under CC BY-SA 3.0. (b) Free-body diagram of a spherical balloon (assumes basket and skirt are counted with $\vec{F}_G$ for the force balance). $\vec{F}_G$ is gravity (downward), $\vec{F}_D$ is drag (downward when body's velocity is upward), and $\vec{F}_B$ is buoyancy (upward). Wind and draft are assumed to be zero. (c) Burner with liquid propane evaporator coil is shown without control valves. *Source:* Martin Thermal Engineering Inc.

fly a hot air balloon across the English Channel thereby proving the practicality of hot air ballooning.

Balloon Technology. Hot air balloons used for recreational passenger flights come in many sizes but nearly all are equipped with the same technical features.

- Basket. The payload carrying component sized to hold the pilot, passengers, fuel, and ballast.
- Burner(s). Usually redundant because burners are the only devices that are capable of filling, launching, and controlling the altitude of the balloon. If one burner fails, a backup could be a lifesaver. For each main burner, there is a separate pilot burner.
- Burner valves. Spring-loaded valves equipped with a lever arm that are manually operated by the pilot and send the design (i.e. maximum) flow rate of fuel to the main burner.
- Cord. Rope that is accessible to the pilot and when pulled, causes the deflation port to open.

- Deflation Port. A mechanically operated "valve" (actually more of a fabric plug) located at the top of the balloon that, when actuated, permits hot air to escape, thereby reducing the net buoyancy force. This device is sometimes called a parachute valve because of the shape of the fabric plug.
- Envelope. The strong, lightweight, temperature-resistant fabric that holds the hot air to create the buoyancy force.
- Load Tapes. Straps or cables made of steel or Kevlar that run vertically up the sides of the envelope to distribute the basket load more evenly up and over the height of the balloon (and therefore without the associated stress concentrations at the mouth due to grommets or other single-point attachment means).
- Mouth. The opening at the bottom of the envelope where burner gases and fresh air enter. The basket is attached to the envelope's load tapes by a network of short suspension cables.
- Propane Cylinders. Redundant, pressurized cylinders sized to hold enough fuel (including a safety reserve) to keep the balloon aloft for a designated time period. The payload weight and the balloon volume also play a role in determining the quantity of stored fuel required. Fuel cylinders are equipped with an internal dip tube so that liquid propane is delivered to burner control valves. See NFPA 58 (2020) for more safety information about portable propane cylinders.
- Skirt or Scoop. A fireproof extension of the envelope below the mouth to protect the burner from extinguishment by horizontal winds.

Operation. Training by a certified instructor is mandatory for pilots, and knowledge of the balloon's technology (and the underlying physics) is a required element of the training. Space will not permit an extensive review of either topic in its entirety, but readers are encouraged to obtain a copy of the *Balloon Flying Handbook* (FAA 2008) for more information. Explicit regulations for balloon airworthiness and balloon operation can be found in FAA (2011).

FAA (2008) asserts the following key motor skills are a requirement for those learning to pilot a balloon:

- Coordination – the ability to take physical action in the proper sequence.
- Timing – the application of muscle coordination at the correct time.
- Control – the ability to interpret and predict the balloon's reaction to a stimulus based on visual cues and instrument readings.
- Awareness – the ability to sense instantly variations of altitude, speed, and direction.

To control the upward or downward motion of a hot air balloon, the operator only has two active tools – (i) to promote upward velocity, pull a cord to open a small propane valve (which fires the burner) for a few seconds and (ii) to promote downward velocity, pull a cord to open the deflation port (which allows warm air to escape and be replaced by cooler ambient air) for a few seconds.

The handbook also enumerates four sources of risk associated with piloting a balloon: pilot competency, airship worthiness, environment, and operation/mission.

Design and Theory. Legend has it that Archimedes discovered buoyancy more than 22 centuries ago when he immersed the king's crown in his bathtub to determine whether the craftsman had made it out of 100% gold or a mixture of base and precious metals that was much less dense. Because the crown was a very irregular shape, it was impossible to determine its volume simply by examining its dimensions. Archimedes reasoned that if he immersed the crown into a full container of water and collected the overflow, he would have the ability to precisely estimate the crown's density and then compare it to standard gold shapes whose volumes were easily calculated. While he was conducting the density examination, he recognized the buoyancy effect that he later became famous for.

The physical principle he identified states that the buoyancy force acting on an object immersed in a fluid is equal to the weight of the fluid displaced by the object. By this principle, a balloon will float neutrally (i.e. with neither upward nor downward acceleration) when it has displaced just enough air to equal its original weight (including envelope, enclosed hot air, basket, passengers, etc.). Effectively, there is only one item on the list that the balloon pilot can control in flight – the weight of the air enclosed in the envelope which contains a known volume of gas. The pilot can change this weight by changing its temperature, and therefore its density. Thus, for a balloon to lift a weighty payload, the balloon must be exceptionally large to ensure that the cold/hot air weight difference covers the weight of the payload too.

Figure 11.2 shows a sketch of the forces that may act on a balloon when it is rising at a constant velocity.

The force balance on the free-body diagram of Figure 11.2 is:

$$\vec{F}_B - \vec{F}_G - \vec{F}_D = 0$$

where

$$\vec{F}_{Buoyancy} = \frac{1}{6}\pi D^3 \rho_\infty \frac{g}{g_c}$$

$$\vec{F}_{Gravity} = \left(M_{balloon,nonair} + \frac{1}{6}\pi D^3 \rho \right) \frac{g}{g_c}$$

$$\vec{F}_{Drag} = \frac{1}{2} C_D \rho_\infty V^2 \pi \frac{D^2}{4}$$

An end-of-chapter problem provides an opportunity for students to evaluate the balloon's forces for a case of constant velocity (no acceleration) and the matching burner firing rate to achieve that velocity.

11.11 Homework Problems

11.1 Select a satisfactory heat transfer fluid from Table 11.2 for a hot press that is used to form oriented-strand board (OSB) from wood chips and adhesive. The thermal fluid increases press temperature to $T_{high} = 215\ °C$, and the pipe is exposed to outdoor temperatures, which routinely fall to $T_{low} = -15\ °C$ in winter.

11.2 Determine the required volume of an expansion tank for a fluid heater system having an internal volume $V_{system,no\_tank} = 9.0\ m^3$, a thermal fluid with coefficient of thermal expansion $CTE = 0.001\ mm^3/(mm^3 \cdot\ °C)$, and a temperature rise of $\Delta T = 200\ °C$.

11.3 Select one or more satisfactory phase-change solid(s) from Table 11.3 for a TES system that is charged with heat during the day from a solar-thermal system having propylene glycol for its circulating thermal fluid. The stored heat is discharged at night to an organic Rankine cycle system having a working fluid of R134a, which operates with a boiler pressure $p_3 \approx p_2 = 4000\ kPa$ and a condenser pressure $p_1 \approx p_4 = 200\ kPa$.

11.4 Discuss the pros and cons of open-loop versus closed-loop geothermal heat sinks.

11.5 Compute the exiting air temperature (dry bulb) for an evaporative cooler with effectiveness of $\varepsilon = 90\%$ and inlet air conditions of $t_{db} = 33\ °C$ and $\phi = 25\%$. Assume liquid water enters the cooler at the same temperature as $t_{db,\ air}$. Use custom Excel functions from Appendix B.5 or use the psychrometric tables in Chapter 1 of ASHRAE (2013).

Answer. $T_{cooler,\ out} = 20.37\ °C$.

11.6 Consider a TES system comprised of a cylindrical tower filled with hollow, thin-walled, plastic spheres, each containing liquid water (85% of interior volume for expansion), whose purpose is to store "cold energy" by freezing the water to ice. A heat transfer fluid, propylene glycol, flows down through the packed tower to freeze or thaw the spheres.

(a) For a cooling load (i.e. discharging "cold energy" from storage bed to occupied space) $\dot{Q}_{load} = 1.0\ TR$ and a discharge duration of $t_{discharge} = 8.0\ h$, compute the minimum storage capacity of the TES system $\left(Q_{capacity,tower}[\ \doteq kJ]\right)$ and the mass of ice required to achieve that capacity by melting from solid at $0\ °C$ to liquid at $0\ °C$.

(b) For the cold energy capacity determined in part (a), determine the height of the vessel and the total number of spheres in the vessel, for the following parameters: $D_{sphere} = 0.05\ m$; $D_{vessel} = 0.75\ m$; $\sigma \approx 0.50$ if the layout of spheres on each vertical layer is hexagonal, per Figure 11.6 and Table 11.4.

(c) For the same vessel and sphere diameters given in part (b), determine the vessel height and the total number of spheres for the following parameters: $T_{glycol,\ in} = 2.1\ °C$; $T_{glycol,\ out} = 0.1\ °C$; $C_{max(ice)} = 35\,170\ W/°C$. Cengel and Ghajar (2015) present a set of inverse equations for the determination of NTU as a function of effectiveness. The equation of interest is the one for a counterflow heat exchanger: $NTU = \dfrac{1}{\left(\dfrac{C_{min}}{C_{max}} - 1\right)} \ln\left(\dfrac{\varepsilon - 1}{\varepsilon \dfrac{C_{min}}{C_{max}} - 1}\right)$

Use Equations (3.48)–(3.50) to find h for a packed bed of spheres and use Figure 11.6 and Equation (11.2) as the assumed packing geometry and porous media properties.

(d) Which height and number of spheres (b) or (c) is the best design choice for this system? Why?

Answers. (a) 101 290 kJ; (b) 1.650 m; (c) 10.54 m, 35 659 *spheres*.

11.7 Calculate (a) the rate of drying $\dot{m}_{w,removal}[\ \doteq lb_w/s]$ and (b) the recommended belt speed $V_{belt}[\ \doteq ft/s]$ for a spherical food element in a packed bed passing through a tunnel oven having properties below. Assume particle temperature is constant, $T_{food} = 212\ °F$. $L_{belt} = 50\ ft$

- $W_{belt} = 45\ in.$
- $\dot{m}_{air} = 0.12\ lb/s$
- $D_{sphere} = 0.38\ in.$
- $\sigma = 0.50$
- $T_{air} = 275°F$
- $\rho_{food} = 70.0\ lb/ft^3$
- $Y_{w,init} = 0.18\ lb_w/lb_{tot}$
- $Y_{w,fin} = 0.05\ lb_w/lb_{tot}$

Answer. (a) $\dot{m}_{w,removal,sphere} = 1.50 \times 10^{-8}\ lb/s$;
(b) $V_{belt} = 0.232\ ft/min$

11.8 For the desalination process depicted in Figure 11.8, fill in the blank entries (empty boxes) in the heat and material balance table below. For both heat exchangers (HX1 and HX2), (i) identify which fluid has the minimum value of $\dot{m}c_p$ and (ii) compute the heat transfer effectiveness. Use properties of water (Appendix A.4) for both seawater and brine.

Property	State 1	State 2	State 3	State 4	State 10	State 11	State 12
Temperature (K)	275.0	275.0			373.13	373.13	315.0
Pressure (kPa)	500	500			101.33	101.33	101.33
Enthalpy (kJ/kg)	8.263 8	8.263 8					
Quality (%)	—	—	—	—	100.0	0.0	—
Water mass flow (kg/s)	0.01	0.002			0.001 250		
Salt mass flow (kg/s)	0.000 15	0.000 03			0.000 0	0.000 0	0.000 0

Answer. (a) HX1: $(\dot{m}c_p)_{11-12}$; HX2: $(\dot{m}c_p)_{1-3}$, (b) HX1 $\varepsilon = 0.59$; HX2 $\varepsilon = 0.68$.

11.9 For an espresso machine that meets the requirements below, determine (a) the coffee puck thickness, (b) the residence time of the water as it passes through the porous puck, and (c) the total brewing time. Use Appendix B.4 for saturated water properties and use the methods of Section 12.8 to estimate subcooled water properties from the saturated values. For the porous media pressure drop, use the equations in Section 2.14. TDS means total dissolved solids (as a percent of the brewed infusion). Assume the pump can push hot water through the espresso puck or out of the steam wand, but not both.

- $m_{fine\ ground\ coffee} = 14.10\ g$
- $m_{espresso\ liquid\ final} = 56.65\ g$
- $Y_{strength} = 5\% TDS\ in\ brew$
- $Y_{extraction} = 20\% coffee\ extracted$
- $R_{brew} = 210\ g_{coffee\ grind}/L_{hot\ water}$
- $R_{dampness} = 0.95\ g_{damp,w}/g_{solid\ remains}$
- $p_{hot\ water} = 910\ kPa_{abs}$
- $T_{hot\ water} = 96°C$
- $\sigma_{puck} = 0.26$
- $D_{portafilter} = 50.8\ mm$
- $D_{coffee\ particle} = 180\ \mu m$

Answer. (a) 11.75 mm; (b) 1.91 s; (c) 20.72 s

11.10 Estimate the balloon diameter and steady burner heat release rate necessary to lift a hot air balloon at an upward velocity of $V = 5.0\ m/s$, assuming the elevation is sea level, and the local air temperature and pressure are those given for the standard atmosphere. Assume the balloon (and basket) comprises a perfect sphere and let the total mass of the envelope plus payload (including people and fuel but excluding the hot air inside the envelope) be $M_{balloon,\ non-air} = 500\ kg$. Assume still air (i.e. $V_{wind,\ horizontal} = V_{draft,\ vertical} = 0$) and the hot air temperature is uniform within the envelope and must not exceed $T_{hotair} = 110\ °C$ (for adequate longevity of the envelope material).

Cited References

Anderson, J.D.; (1985); *Introduction to Flight*, 2; New York: McGraw-Hill.

ASHRAE (2013). *ASHRAE Handbook – Fundamentals*; Chapter 1. Psychrometrics; p. 1.1; Atlanta: American Society of Heating, Refrigerating, and Air Conditioning Engineering.

ASHRAE; (2015); “Geothermal energy (Chapter 34)”; *2015 ASHRAE Handbook – Heating, Ventilating, and Air-Conditioning Applications*; Atlanta: American Society of Heating, Refrigerating, and Air Conditioning Engineers.

Aviator12 (2012). Model Montgolfier balloon (Le Bourget). https://commons.wikimedia.org/w/index.php?curid=25718465 (accessed 30 March 2021).

Bienkowski, B.; (2015); “*Desalination Is an Expensive Energy Hog, but Improvements Are on the Way*”; Cambridge, MA: PRX Public Radio.; https://www.pri.org/stories/2015-05-15/desalination-expensive-energy-hog-improvements-are-way

Cengel, Y.A. and Ghajar, A.J.; (2015); *Heat and Mass Transfer: Fundamentals Applications* 5; New York: McGrawHill.

CoffeeIQ (2019). Tamaño de Partículas y Extracción, Colombia. https://www.coffeeiq.co/molienda-tamano-de-particulas-y-extraccion (accessed 20 February 2021).

Crawford, M. (2020). *How Espresso Machines Work: The Engineering Inside*; American Society of Mechanical Engineers, Topics and Resources. https://www.asme.org/topics-resources/content/how-espresso-machines-work-the-engineering-inside

Dean, J.A.; (1999); *Lange's Handbook of Chemistry* 15; New York: McGraw-Hill.

Dow (2021). Lubricants, heat transfer and deicing fluids: enabling formulation success. https://www.dow.com/en-us/product-technology/pt-lubricants.html (accessed 25 April 2021).

ESRL; (2017–2021); “*Global Monitoring Laboratory – NOAA Solar Calculator*”; Earth System Research Laboratories; National Oceanic and Atmospheric Administration; https://www.esrl.noaa.gov/gmd/grad/solcalc.

Federal Aviation Administration; (2008) *Balloon Flying Handbook*; Oklahoma City, OK: US Dept of Transportation, Federal Aviation Administration, Flight Standards Service; FAA-H-8083-11A.

FAA (2011). 49 Code of Federal Regulations; Aeronautics and Space, Part 31. Airworthiness standards: manned free balloons. https://www.govinfo.gov/app/details/CFR-2011-title14-vol1/CFR-2011-title14-vol1-part31 (accessed 5 April 2021).

Kavanaugh, S.; (2008); “A 12-step method for closed-loop ground-source heat-pump design”; *ASHRAE Transactions*; **114**; (2); 328, Report Number SL-08-032.

Kays, W.M and London, A.L.; (1984); *Compact Heat Exchangers*3; New York: McGraw-Hill.

Kreith, F., Kreider, J.F.; (1978); *Principles of Solar Engineering*; New York: McGraw-Hill.

Lingle, T.R.; (1996); *The Basics of Brewing Coffee*; Santa Ana: Specialty Coffee Association of America.

Linstrom, P.J. and Mallard, W.G. (2017–2021); *NIST Chemistry WebBook, SRD 69. Thermophysical Properties of Fluid Systems.* National Institute of Standards and Technology, U.S. Dept. of Commerce. http://webbook.nist.gov/chemistry/fluid

NASA; (2018); *Terminal Velocity: Gravity and Drag*. Glenn Research Center, National Aeronautics and Space Administration. https://www.grc.nasa.gov/www/k-12/airplane/termv.html.

National Balloon Museum (2021). History of ballooning. https://www.nationalballoonmuseum.com/about/history-of-ballooning (accessed 30 March 2021).

NFPA 58; (2020); *Liquefied Petroleum Gas Code*; Quincy, MA: National Fire Protection Association.

Paratherm (2021). Paratherm heat transfer fluids. https://www.paratherm.com (accessed 5 April 2021).

Poseidon Water (2021). The Claude 'bud' Lewis Carlsbad desalination plant. https://www.poseidonwater.com/carlsbad-desal-plant.html (accessed 25 April 2021).

Roberts, C. (2018). Is There More Caffeine in Espresso Than in Coffee? Consumer Reports (Coffee). https://www.consumerreports.org/coffee/is-there-more-caffeine-in-espresso-than-in-coffee.

SCA (2021). *Brewing Standards*. Santa Ana: Specialty Coffee Association. https://sca.coffee/research/coffee-standards (link to "Golden Cup Standard" at bottom of page).

Stamp, J. (2012). The long history of the espresso machine. Smithsonian. https://www.smithsonianmag.com/arts-culture/the-long-history-of-the-espresso-machine-126012814 (accessed 25 February 2021).

Therminol (2021). Heat transfer fluids by Eastman. https://www.therminol.com (accessed 5 April 2021).

Weast, R. C.; (1974); *CRC Handbook of Chemistry and Physics*55; Cleveland: CRC Press.

12

Pipe and Fluid Mover Analysis

Selection of pumps and blowers is one of the last design steps taken in a thermal system design project; selection of refractory, thermal insulation, and somc control elements can often be performed later, along with completion of the project documentation. Part of the reason for this lateness in the process is that fluid mover selection simply cannot be accomplished until the heat and material balance is completed, the major equipment is sized, and the piping network is laid out. The equipment and pipes define the areas where pressure drop is occurring, and without knowledge of their characteristics, it is impossible to know the necessary performance requirements of pumps and blowers. Even then, the system designer typically performs an economic balancing act to coordinate the selection of fluid movers that create pressure gains and the sizing of ducts and pipes that greatly influence pressure losses.

The objective of this chapter is to present various techniques for sizing fluid movers and the conveying means to circulate the working fluid around a thermal system. The first topics address basic categories of pipes and fluid movers, rules they obey, and design guidelines. The balance of the chapter addresses special design problems, such as pump and blower curves, the incompressible liquid assumption, piping networks, friction in long pipes, and the chimney effect.

Because pumps and blowers are essentially commodities (i.e. technical innovation in the field is infrequent), academic treatises covering these devices are either very old or relegated to standards documents (or both), as can be seen in the citation list at the end of this chapter. This fact does not negate the inherent challenge and complexity the thermal engineer will face when evaluating different options for fluid movers and piping systems. The engineering knowledge needed to properly select fluid movers and pipes may not be changing fast, but its span is very broad and its successful implementation requires a deep and encyclopedic knowledge of the options available and their capabilities and limitations.

12.1 Fluid Mover Categories

Fluid movers can be categorized according to several different characteristics: fluid phase, flow/pressure, and pumping mechanism.

Categorization by fluid phase:

- Gas movers – fans, blowers, and compressors
- Liquid movers – pumps

Thermal Systems Design: Fundamentals and Projects, Second Edition. Richard J. Martin.

Companion website: www.wiley.com\go\Martin\ThermalSystemsDesign2

Categorization by flow and pressure:

- Pressure devices – moderate-to-high-pressure devices, with low-to-moderate volume flow rates
- Flow devices – low-pressure devices, with low-to-high volume flow rates

Categorization by mechanism:

- Positive displacement – piston, peristaltic, roots, rotary screw, and scroll. (These motivate a constant volumetric inflow rate because no backward leakage occurs, but they should not be considered constant mass flow devices because they can cause inlet density to vary.)
- Velocity recovery – centrifugal, radial, axial, and ejector. (These accelerate the fluid [often via shaft work], then decelerate it to convert kinetic energy into mechanical energy.)

Fans. Fans are devices that impart axial motion to air via a set of airfoil blades rotating around a common shaft. Pressures developed are very low ($\Delta p \ll 1.0$ *inch WC*) but volume flows can be very high.

Blowers. The most common air moving device in thermal fluid systems is the blower. Blowers have a centrifugal impeller that receives inlet air along the center axis, accelerates and propels the air radially and tangentially toward the outer wall of the blower housing, which decelerates it again to recover the kinetic energy as mechanical energy (i.e. pressure). Pressure rise tends to be low ($\Delta p < 30$ *inch WC*) and volume flow rates may be low to high.

The terms fan and blower are often used interchangeably, but strictly speaking, a blower's motion inducement is centrifugal and a fan's motion inducement is axial.

Compressors. Compressors are gas pressurization devices. Their design may be centrifugal, axial, or positive displacement, with centrifugal and axial designs falling into the velocity recovery style. Pressure ratios are much higher ($2 < p_{exit}/p_{inlet} < 50$) than other air moving devices. Scroll compressors, rotary screw compressors, and Roots blowers are positive displacement devices that rely on contact between pairs of rotating lobes that move gas from inlet to exit, while increasing the gas density along the path. Piston compressors are similar to gasoline and diesel engines in that the piston moves up and down, and gas is admitted at low pressure when the intake valve opens and rejected at high pressure when the exhaust valve opens.

Pumps. Pumps tend to be used for liquids exclusively and they tend to be equipped with centrifugal impellers that work according to the velocity recovery principle. Positive displacement pumps for liquids also exist and they deliver a relatively steady flow rate at a wide range of pressure rise values. Deadheading a positive displacement liquid pump (i.e. blocking the exit path entirely) can cause damage to pump, its motor, gaskets, and even pipe integrity because the liquid is virtually incompressible, and one part of the system must become the weakest link.

Ejectors. Ejectors are fluid moving devices that have no moving parts. Ejectors can operate on gases or liquids, and the principle of operation is to utilize a high-pressure stream to create fluid motion in a venturi cavity that establishes a vacuum condition which draws the other fluid out of a connected passage. Eductors are a type of ejector that is designed to deliver a pressurized fluid, as opposed to motivating the flow of a low-pressure fluid.

Additional Fluid Mover Terminology.

- The inlet nozzle on a fluid mover is called the intake or suction port and the outlet nozzle is called the exit or discharge port.
- Suction Head, Suction Lift. The term suction here does not mean vacuum conditions; it means the location of the respective "head" or "lift" (namely, in the suction pipe, immediately upstream of the pump's inlet). Suction head refers to the situation where a reservoir feeding the pump is

higher in elevation than the centerline of the pump's inlet port, and the gauge pressure at the inlet is positive. Suction lift refers to the situation where a reservoir feeding the pump is lower than the inlet centerline, and the gauge pressure at the inlet is negative.

- Affinity Laws. To be covered in Section 12.7, these are the laws that relate impeller speed and diameter to volume flow rate, pressure rise, and mechanical horsepower associated with a blower or pump.
- Tip Leakage. Centrifugal blowers and pumps are not designed for contact between the impeller tip and the housing. Centrifugal devices have no mechanical seals (i.e. no piston rings). When the blower is operating and developing downstream pressure, the fluid wants to flow back upstream through the finite gap between the impeller tip and the housing wall, to a location where the pressure is lower. Fluid always wants to flow from high pressure to low pressure, even when an impeller is motivating the fluid to flow forward. Because the gap is small, the rate of reverse flow is only a small fraction of the forward flow but is referred to as tip leakage. It is an element of inefficient (i.e. entropy generating) behavior.

Procurement of blowers and pumps requires knowledge of these terms and how to interpret the content given in catalogs and specification sheets. For example, one blower specification sheet presents the following data:

- Inlet diameter $D_{in} = 9.0\ inch\ (OD)$
- Outlet area $A_{c,\ out} = 0.429\ ft^2$
- Impeller diameter $D_{wheel} = 14\ inch$
- Cubic feet per minute $500 \leq \dot{V}_{inlet} \leq 2500\ acfm$
- Outlet velocity $1166 \leq V_{avg} \leq 5828\ fpm$
- Pressure rise $2 \leq \Delta p_{rise} \leq 22\ inch\ WC$
- Rotational speed (Ω, rpm varies with Δp_{rise} and $\dot{V}_{inlet}$)
- Power ($\dot{W}_{shaft}$, bhp varies with Δp_{rise} and $\dot{V}_{inlet}$)

The data in the table could be used to produce a series of fan curves, with each curve having a different value of rotational speed. At least one blower vendor (IAP 2021) provides an app that generates printable fan curves for each model and rotational speed they manufacture.

12.2 Conveying Means Categories

Means for conveying fluids (e.g. pipes, ducts) through thermal systems are almost never open (e.g. a three-sided channel for liquid water) because the working fluids are often *valuable*, or they must not be released to the environment. Loss of working fluids can even be costly when the working fluid is (i) water, if it has undergone treatment processes that help reduce corrosion and scaling or (ii) air, if it has undergone intermediate processing such as pressurization or heating.

By the same token, the conveying means in thermal systems are typically able to withstand moderate to high pressure, with the major exception being heating, ventilating, and air conditioning (HVAC) indoor air recirculation systems that use nonrigid ductwork.

Among these moderate ($p > 1.0\ psig$) to high ($p > 15.0\ psig$) pressure conveying means, some of the terms will be familiar to students. Their typical usage is provided here.

- Black pipe – Wrought steel used for moderate-to-high-pressure air, natural gas, or treated water ($T < 600\ °F$). See ASTM A53 (2020).

- Galvanized pipe – Steel with a galvanic coating (zinc) to reduce corrosion in low-temperature service ($T < 350\ °F$). See ASTM A53 (2020).
- Stainless steel pipe – Chrome-nickel steel used for moderate temperature ($T > 600\ °F$) or moderate corrosion resistance. Like black and galvanized pipe, this pipe is manufactured by many different entities. See ASME B36.19 (2018).
- Alloy steel pipe – High nickel steel with other alloys used for highly corrosive streams, or very high-temperature service ($T > 1200\ °F$). See Haynes (2020), Rolled Alloys (2021), and Special Metals (2008).
- Copper pipe – Class K, L, or M pipe used predominantly for drinking water or refrigerant service at relatively high pressures ($p > 100\ psig$) and very low temperatures ($T > 5\ K$). See CDA (2020).
- Fiber-reinforced plastic (FRP) is a highly corrosion-resistant material that is generally used for liquids at or near ambient temperature and moderate pressure. See Engineering Toolbox (2001).
- Tubes made of many of the same materials as the pipes cited above are available commercially. (Exceptions are FRP and galvanized, which are only available in pipe, not in tube.) See Tube-Web (2021).

Most piping is sold by its ***nominal*** inside diameter (ID), whereas most tubing is sold by its actual outside diameter (OD). ***Nominal*** ID means that the actual ID varies with wall thickness, but it is approximately equal to the nominal value for at least one of the thicknesses offered.

Even though piping is labeled according to a nominal ID, the product is sold with a consistent OD for *all* wall thicknesses. Copper pipe is an exception, in that it may be specified either way (ID or OD). Engineers should specify both ID and OD along with ***nominal*** size if any doubt exists as to the actual product being procured. ASME (2020a, 2020b, and 2020c) publishes three important standards (B31.1, B31.3, and B31.9) for specifying and using different categories of pipes and fittings for various temperatures, pressures, and levels of fluid aggressiveness.

Pipe is sold in a range of wall thicknesses, referred to as the ***schedule***. All the nominal pipe sizes (diameters) are available in one or more of the schedules from 5 to 160. The most common schedules are 40 and 80.

The numerical value of a pipe's schedule is interpreted as a multiplier for the ratio of the internal fluid pressure divided by the allowable hoop stress in the pipe wall (see Equation 12.1). For example, if the allowable hoop stress for a particular pipe material and outside diameter is $\sigma_{circumferential,\ max} = 6000\ psi$, the maximum internal pressure for ***schedule***-40 pipe would be determined from:

$$Schedule = \frac{1000 \cdot p_{internal,\ max}}{\sigma_{circumferential,\ max}} \tag{12.1}$$

$$\text{Example} : p_{max} = \frac{40 \cdot 6000}{1000} = 240\ psig$$

12.3 Leak Prevention

Fluid moving and conveying systems are occasionally assembled with hermetic seals throughout the entire flow circuit (e.g. refrigerant compressors and pipes), but most flowing systems are designed for disassembly and reassembly when repairs are needed. Piping joints must be sealed to prevent or minimize fluid leakage during normal operation. Rotating shafts that connect electric motor shafts to pump or blower impellers must pass through a leak prevention mechanism that

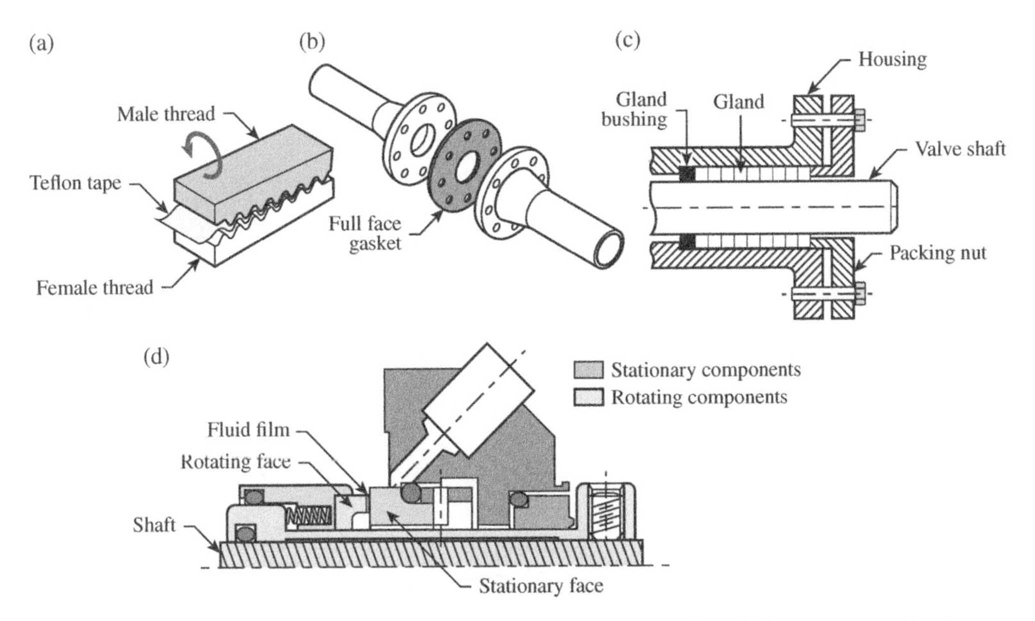

Figure 12.1 Four types of fluid sealing technologies. (a) Pipe-thread sealant; (b) flange gasket; (c) valve steam packing; and (d) pump-shaft mechanical seal. *Source:* Martin Thermal Engineering Inc.

permits the shaft to turn but minimizes the escape of fluid outside the pump housing. These leak prevention materials are categorized as follows and illustrated in Figure 12.1.

- Sealants – viscous material applied between threaded joints to prevent fluid leakage out of pipe joints.
- Gaskets – compressible material clamped between two stationary surfaces (e.g. flanges) to obstruct fluid passage through adjoining surfaces.
- Packings – compressible material compressed axially to exert pressure radially around a shaft that rotates slowly or periodically (e.g. valve shaft), to prevent fluid leakage.
- Mechanical seals – segments of rigid and compressible material arranged with a spring and surrounding a shaft such that fluid pressure and spring pressure exert forces that oppose fluid leakage even when the shaft is rotating at high speed; can either be end face seals or radial "lip" seals.

12.4 Pressure Rise and Drop

Students should become familiar with the terms ***rise*** and ***drop*** in various engineering contexts. Pumps and blowers cause pressure to rise, whereas valves, fittings, and lengths of pipe cause pressure to drop. When defined as the exit pressure minus the inlet pressure, the variable associated with pressure change (Δp) is reported as a positive number for fluid movers (e.g. $\Delta p_{pump} = (p_{exit} - p_{inlet}) > 0$) and negative for pipes, valves, and fittings (e.g. $\Delta p_{friction} = (p_{exit} - p_{inlet}) < 0$), respectively.

The pressure rises and drops in a fluid circuit are analogous to the voltage rises and drops in a direct current (DC) electrical circuit, with the battery replacing the pump and the resistor replacing

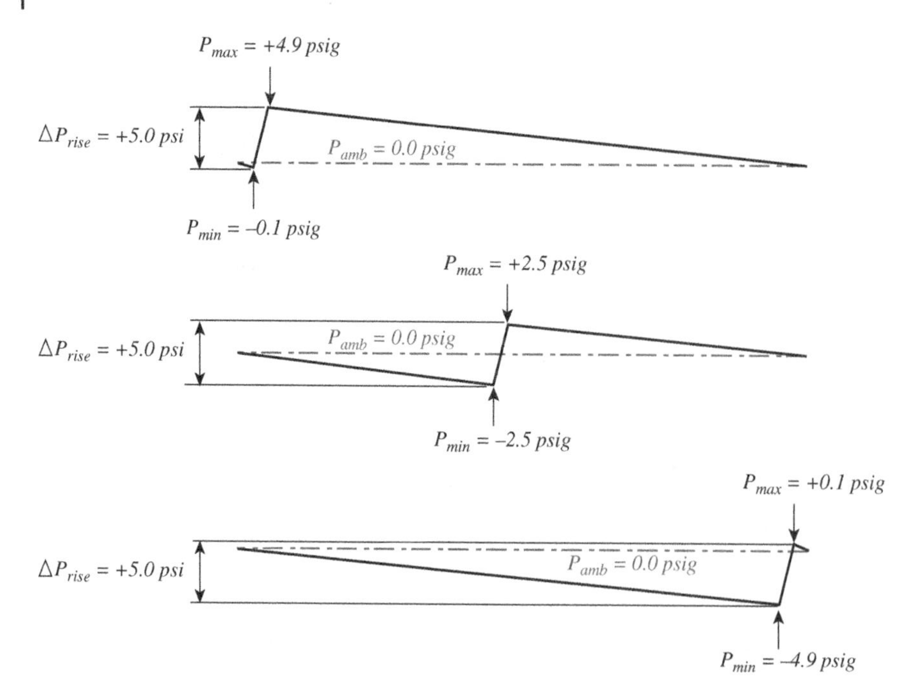

Figure 12.2 Three examples of pressure rise and pressure drop for pumps located near the inlet, middle, and exit of a long pipe. *Source:* Martin Thermal Engineering Inc.

the pipe. Heat conduction through thermally conducting materials can also be addressed with the term temperature drop.

Think Stop. Consider a pipe filled with a working fluid but having no pump or blower to motivate any flow. If the working fluid's density differs from that of the ambient air, it will move out of the pipe under the influence of gravity, but otherwise, the fluid will remain stationary. Thus, most thermodynamic cycles require a fluid mover to generate mechanical energy (i.e. gauge pressure) that will be dissipated by friction as the fluid moves through the pipes, fittings, and equipment that comprise the cycle. The exceptions to this rule are processes where the fluid moves under the influence of the thermosiphon effect (e.g. heat pipes and absorption refrigerators), which can operate without a mechanical fluid mover that creates a pressure rise.

For a straight run of pipe with an inlet and exit (i.e. not a cycle), the fluid mover can be located near the pipe entrance, near the exit, or anywhere in the middle. The location of the fluid mover determines the minimum and maximum pressures that the fluid will experience along its flow path. See Figure 12.2 for three plots of fluid pressure versus distance where the friction-related pressure drop is $\Delta p_{friction} = -5.0\ psi$ for a particular fluid, flow rate, and pipe specification (length and diameter).

12.5 Electricity Analogy for Flow

In Chapter 3, the electric circuit analogy was seen to apply nicely for conductive and convective heat transfer. The concept was also shown to be useful for radiative heat transfer if the concepts of surface and view resistances were introduced, along with the concepts of blackbody emissive power and radiosity as driving force potentials for the radiative heat flow.

A similar analogy exists for fluid flow, but as with the radiative model, some minor adjustments must be made so the model will comply with the physics. To recap the discussion in Chapter 3, we recall that Equation (12.2) provides the basic relationship for the flow of electricity in a power circuit:

$$\overbrace{I}^{\substack{\text{Current}\\\text{flow}}} = \frac{\overbrace{\Delta V}^{\substack{\text{Voltage}\\\text{drop}}}}{\underbrace{R_{elec}}_{\substack{\text{Electrical}\\\text{resistance}}}} \tag{12.2}$$

The analogous terms for fluid flow are:

- I ~ $\dot{V}^2$ Volume flow rate squared
- ΔV ~ Δp Pressure drop
- R_{elec} ~ $R_{\substack{fluid\\flow}}$ Fluid flow resistance

The first important difference between the fluid flow circuit and the electric circuit is that the electric current flow is raised to the first power, but the fluid volume flow is squared. This distinction arises because pressure drop is related to velocity pressure, which varies as the square of the velocity.

The analogous equations for the fluid flow circuit are:

$$\overset{\substack{\text{Volume}\\\text{flow}\\\text{squared}}}{\dot{V}^2} = \frac{\overbrace{\Delta p}^{\substack{\text{Pressure}\\\text{drop}}}}{\underbrace{R_{fluid}}_{\substack{\text{Fluid flow}\\\text{resistance}}}} \tag{12.3a}$$

$$\dot{V} = \frac{\dot{m}}{\rho} = A_c V_{avg} \left[\stackrel{\vee}{=} \frac{m^3}{s} \right] \tag{12.3b}$$

$$R_{fluid} = \frac{8\rho f L}{\pi^2 D^5 g_c} \left[\stackrel{\vee}{=} \frac{Pa \cdot s^2}{m^6} = \frac{kg}{m^7} \right] \tag{12.3c}$$

The end-of-chapter problems provide an exercise for students to show that the expression for R_{fluid} is correctly derived from the expression for friction-caused pressure drop in the modified Bernoulli equation. When checking the variables in the resistance term to see if they make sense physically, it is apparent that increases in density, friction factor, and pipe length cause the flow resistance to increase (as expected), and an increase in diameter causes the resistance to decrease dramatically (also as expected).

Finally, the electric circuit and the fluid flow circuit must be compared in the context of resistances in series or in parallel. The series relationships (Equations 12.4) are identical, but the parallel relationships (Equations 12.5) are not.

Specifically, for ***series*** resistors:

- Pressure drops add

$$\Delta p_{01} + \Delta p_{12} = \Delta p_{02} \tag{12.4a}$$

- Resistances add

$$R_1 + R_2 = R_{tot} \tag{12.4b}$$

- Volume flow conserved

$$\dot{V}_1^2 = \dot{V}_2^2 = \dot{V}_{tot}^2 \tag{12.4c}$$

- Circuit relationship

$$\dot{V}_{tot}^2 = \frac{\Delta p_{02}}{R_{tot}} \tag{12.4d}$$

And for ***parallel*** resistors:

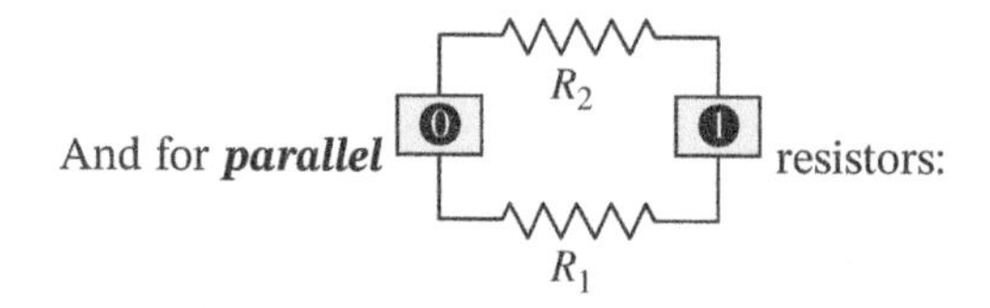

- Pressure drops conserved

$$\Delta p_1 = \Delta p_2 = \Delta p_{01} \tag{12.5a}$$

- Resistance (reciprocal) adds differently

$$\frac{1}{R_1} + \frac{1}{½\sqrt{R_1 R_2}} + \frac{1}{R_2} = \frac{1}{R_{tot}} \tag{12.5b}$$

- Volume flow squared adds differently

$$\dot{V}_1^2 + 2\dot{V}_1\dot{V}_2 + \dot{V}_2^2 = \dot{V}_{tot}^2 \tag{12.5c}$$

- Circuit relationship

$$\Delta p_{01} = R_{tot}\dot{V}_{tot}^2 \tag{12.5d}$$

While the fluid flow analogy to the electric circuit is not exact, it is similar enough that engineers may use it effectively as long as the correct formulas are applied.

12.6 Piping Network Rules

The rules for fluid flow in piping networks with multiple pipe lengths and multiple connections (or tees) are similar to the rules given in the prior section because the electricity analogy holds for multiple pipes just as it does for multiple wires conducting electricity. Here are a few of the extended rules (maintaining the same assumptions of steady, incompressible flow throughout) for multiple (three or more) pipe segments:

- Pipe segments in series:
  - Volume flow rate is identical in each segment $\left(\dot{V}_1 = \dot{V}_2 = \dot{V}_3 = \cdots = \dot{V}_n\right)$.
  - Total pressure drop is the sum of all the individual pressure drops ($\Delta p_{tot,\ series} = \Delta p_1 + \Delta p_2 + \Delta p_3 + \cdots + \Delta p_n$).
- Pipe segments in parallel:
  - Volume flow rates sum to obtain total flow rate $\left(\dot{V}_{tot,parallel} = \dot{V}_1 + \dot{V}_2 + \dot{V}_3 + \cdots + \dot{V}_n\right)$.
  - Total pressure drop is identical to individual pressure drop in each parallel segment ($\Delta p_{tot} = \Delta p_1 = \Delta p_2 = \Delta p_3 = \cdots = \Delta p_n$).
- Pipe segments connecting reservoirs together at a single junction:
  - If all flows are labeled positive when they flow *into* the junction, the flow rates must sum to zero $\left(\dot{V}_{tot,junction} = \dot{V}_1 + \dot{V}_2 + \dot{V}_3 + \cdots + \dot{V}_n = 0\right)$.
  - The pressure at the common junction is identical, regardless of which flow path is taken to arrive there ($p_{junction} = uniform$).
  - If the upper surface of all reservoirs is assumed to have the same ambient pressure, the pressure drop in each segment can be determined by enforcing the first and second reservoir rules $\left(\dot{V}_{tot,junction} = 0, p_{junction} = uniform\right)$.
  - Accounting for frictional pressure drop and gravitational effects on pressure, a system of n equations (each with the volume flow rate unknown) can be solved for the individual flow rates that achieve the common value of p_J. The process is iterative – guess a value for $p_{junction}$, compute the respective pressure drops, and solve for flow velocities in each flow path from Equation (12.6):

$$\Delta p_{i\text{-}junction} = \rho \frac{V_i^2}{2g_c} f \left(\frac{L}{D}\right)_{i\text{-}junction} + \rho \frac{g}{g_c} \Delta z_{i\text{-}junction} \tag{12.6}$$

  - Continue by evaluating whether the total volume flow rate into the common junction is zero and iteratively adjust the value for $p_{junction}$ until the flows into the junction are in balance.

See also White (2016) for more information about multiple pipe networks.

12.7 Blower and System Curves

Approach to Cost Minimization. Proper sizing of a fluid mover (blower or pump) and a piping network invariably requires an iterative approach to obtain a design that meets performance objectives and is economical. Although cost optimization is beyond the scope of this textbook, Figure 12.3 shows a generic curve of the amortized cost of the initial capital investment versus the recurring operating costs to run the equipment. If smaller pipes are selected, the amortized initial cost is reduced, but the pressure drop to circulate the fluid is increased, requiring additional pump work and electricity cost. The combined lifetime cost as a function of pipe diameter looks like Figure 12.3, and the optimum diameter is selected by minimizing the combined lifetime cost curve.

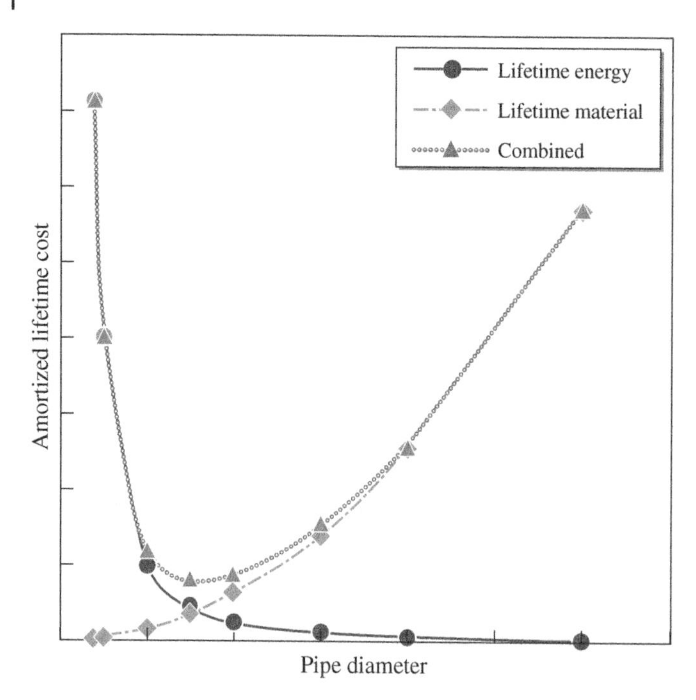

Figure 12.3 Generic lifetime cost curves for energy, material, and combined. *Source:* Martin Thermal Engineering Inc.

Since the piping design process is iterative, it is advisable to begin by selecting a pipe diameter based on a recommended (generic) fluid velocity. We suggest the following initial choices (see Equations (12.7)):

$$V_{liquids,generic} \approx 400\ \frac{ft}{min} \approx 2\ \frac{m}{s} \tag{12.7a}$$

$$V_{gases,generic} \approx 2000\ \frac{ft}{min} \approx 10\ \frac{m}{s} \tag{12.7b}$$

The above values will often favor lower energy costs, and consequently, they may result in a capital cost that outweighs the energy cost over the system's lifetime.

Affinity Laws. Pump and blower affinity laws are the basic rules that relate rotational speed of the impeller to the performance of the device. These laws (see Equations 12.8 and 12.9) are generic in nature – they do not predict actual performance of a specific fluid mover. Rather, they predict relative performance $\left(\dot{V}_{inlet}, \Delta p\right)$ as a function of relative changes in impeller size (D) or rotational speed (ω). The affinity laws are:

For a given impeller size (D):

$$\dot{V}_{inlet} \sim \omega; \quad \Delta p \sim \omega^2; \quad \dot{W}_{shaft} \sim \omega^3 \tag{12.8}$$

For a given rotational speed (ω):

$$\dot{V}_{inlet} \sim D; \quad \Delta p \sim D^2; \quad \dot{W}_{shaft} \sim D^3 \tag{12.9}$$

Performance Map. Centrifugal pumps and blowers are designed to function according to a performance map for a given fluid at a given temperature and pressure. The performance parameters plotted on the map are flow rate (x-axis) and pressure rise (y-axis), with a family of parametric curves associated with varying impeller rotational speeds for the given impeller and housing. Contours of mechanical efficiency ($\eta = \dot{E}_{fluid}/\dot{W}_{shaft}$) are sometimes included on the map (see also Section 12.8).

Blower Curve versus System Curve. The ***blower curve*** (note that this methodology also applies to pumps and compressors) is a single curve on the performance map relating inlet volumetric flow rate (on the x-axis) to pressure rise (on the y-axis) for one size housing and impeller and one shaft rotational speed. Because inlet density affects the performance of a fluid mover, manufacturers usually plot their blower curves under an assumed inlet temperature and pressure (often unique to that manufacturer rather than *STP*), and they provide a correction formula to estimate differences when inlet density varies.

By comparison, the ***system curve*** is a single plot that represents the flow resistance associated with all the pipe lengths, fittings, valves, and pieces of equipment through which the working fluid passes on its way through the system. The system is usually defined by a "baseline system curve" that reflects its nominal condition (i.e. all valves fully open, all filters and strainers clean, no pipe fouling, and all equipment operating normally). The baseline system curve is a single plot of pressure drop (y-axis) versus flow rate (x-axis). Although blower curves typically present flow rate as actual (or inlet) cubic feet per minute (*acfm* or *icfm*), the system curve is more commonly represented with mass flow rate or mole flow rate as the dependent variable. If plotted together, both curves (***system*** and ***blower***) should be adjusted to the same inlet conditions and the same flow rate units.

The operation point for the system and blower under nominal conditions is the point where the ***blower (fan) curve*** intersects the ***system curve*** (Figure 12.4). Indeed, whether the system designer

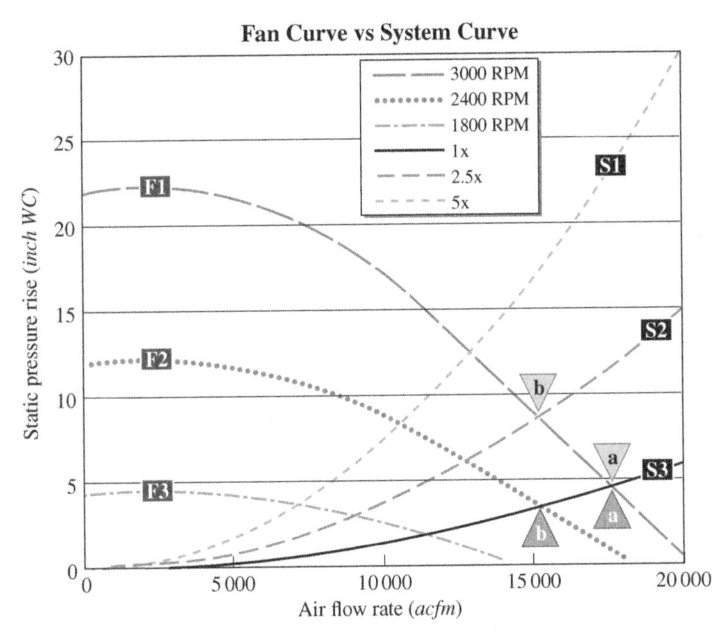

Figure 12.4 Fan curves and system curves. Upward-pointing triangles (a) and (b) indicate response of reduced fan speed to a demand for lower flow rate; downward-pointing triangles (a) and (b) indicate response of increased damper restriction to the same demand for lower flow rate. The initial and final flows are identical, but the fan speed change results in a much lower pressure rise across the fan and pressure drop across the system. *Source:* Martin Thermal Engineering Inc.

chooses to plot blower and system curves to predict the anticipated operating condition for a thermal system, the blower and the system will find the appropriate curve-curve intersection and operate there without any designer involvement until something changes (with either the system or the blower or both).

Given that only one "baseline system curve" and only one "fixed speed blower curve" exist for a given installation, the baseline system has virtually no flexibility to operate at any condition other than the intersection point. To increase flexibility, the designer can add components that alter either the system curve (e.g. control valves) or the blower curve (e.g. rotational speed controller). Figure 12.4 shows the effect of such changes on the intersection point.

In Figure 12.4, the three ***blower curves*** represent different rotational speeds with the same impeller and housing. The affinity laws govern the curve locations, with higher pressure rises associated with higher speeds. A fixed-speed blower can operate anywhere along its ***blower curve***, which has the appearance of a downward-facing half-parabola. (We must emphasize that the manufacturer's empirically determined ***blower curve*** is a far better choice for a system flow analysis than utilizing a generic parabola as a representative blower curve, and, therefore, the curves illustrated in Figure 12.4 should be considered very rough.)

The three ***system curves*** represent different total resistances associated with all the pipes, fittings, valves, and equipment combined. Each ***system curve*** individually reflects the Darcy–Weisbach relationship that pressure drop increases as the square of fluid velocity; hence, the system curves are legitimately upward-facing half-parabolas. This parabolic nature is correct as long as secondary effects (e.g. nonlinear pressure changes related to condensation, evaporation, and sonic flow) are ignored.

Figure 12.4 shows the initial condition as being identical for ***fan*** and ***system***:

- Initial. Downward triangle ***a*** on fan curve **F-1** at the same *x*, *y* location as upward triangle ***a*** on system curve **S-3**. Approximate conditions are $\dot{V}_1 \approx 17\,500$ *acfm* and $\Delta p_1 \approx 4.5$ *inch WC*.

And two different final conditions as:

- Final, same fan curve. Downward triangle ***b*** on the same fan curve **F-1**, where it intersects with the new system curve **S-2**, at approximate conditions of $\dot{V}_1 \approx 15\,500$ *acfm* and $\Delta p_1 \approx 8.5$ *inch WC*.
- Final, same system curve. Upward triangle ***b*** on the same system curve **S-3**, where it intersects with the new blower curve **F-2**, at approximate conditions of $\dot{V}_1 \approx 15\,500$ *acfm* and $\Delta p_1 \approx 3.5$ *inch WC*.

What is noteworthy about the two final points is that they end at the same volume flow rate (as intended), but the ***b*** point (at the intersection of **F-1** and **S-2**) is at a much higher pressure rise than the ***b*** point (at the intersection of **S-3** and **F-2**). The change that occurred along the original fan curve **F-1** involved only a change from one system curve to another, and the change that occurred on the original system curve **S-3** involved only a change from one fan curve to another.

Think Stop. From Figure 12.4, students should determine (i) which change involved closing a valve to reduce the flow rate, (ii) which change involved slowing the fan speed to reduce the flow rate, and (iii) which of the two changes is likely to be more energy efficient.

Thermal system designers are cautioned again that inlet gas density has a significant effect on the performance of a blower, and most blower manufacturers provide equations to correct for density changes caused by variations of inlet temperature and pressure. Inlet density affects mass flow, pressure rise, and horsepower to first order, and inlet volume flow to second order.

The phenomenon can be understood by examining Figure 12.4 again. First, consider the system curves **S-1** and **S-2**. The effect of volume flow rate (and velocity) on pressure drop is seen in Equation (2.10), the Darcy–Weisbach equation:

$$\Delta p_{friction} = f\frac{L}{D}\frac{\rho V^2}{2} \tag{2.10}$$

The pressure drop goes up with the square of increasing velocity, but the pressure drop also declines linearly with declining density. The velocity change is reflected by moving upward along a single system curve (e.g. **S-2**) but the density change is reflected by moving to a new, higher curve (i.e. from **S-2** to **S-1**) that reflects greater resistance for the same volumetric flow rate.

Next, consider the blower curves. The physical effect on blower performance associated with changing gas density is more like a system curve effect than a blower effect, but the underlying phenomenon is slightly different. A centrifugal blower generates pressure by first accelerating the fluid with the impeller, and then decelerating it at the blower exit to recover the kinetic energy as mechanical energy (i.e. stagnation pressure). As was seen with the Pitot Probe analyses in Chapter 2, the "dynamic pressure" portion of the stagnation pressure is $\rho V^2/2$.

This means that when a gas flow is decelerating from a high velocity to a low velocity in the exit of a blower, *a higher density gas will cause a greater increase in pressure than a lower density gas* will cause. Although not consistent with the legend in Figure 12.4, the effect of increasing blower inlet density is analogous to moving to a new higher blower curve (i.e. from **F-2** to **F-1**).

12.8 Pump and Blower Work

When specifying a fluid mover, not only must the pump or blower be sized correctly for the ranges of mass flow rates and pressure rises desired, but an electric motor must also be chosen that can deliver the necessary torque and horsepower. Motor efficiency (mechanical power delivered to the shaft divided by electrical power consumed) tends to be highest when the motor is fully loaded (i.e. operating at or slightly below its faceplate rating). Application of a "service factor" (i.e. additional margin of safety) is required by some users for motors installed at their facilities. A typical value for service factor is 1.15. This implies that the motor's rated horsepower should be 15% higher than the maximum mechanical power required by the fluid mover.

Liquid Pumps. The shaft power required by a pump to impart mechanical energy (i.e. pressure) to a subcooled liquid is $\dot{W}_{pump} = \dot{m}(h_2 - h_1)$. If we had a reliable way to determine the enthalpy of the inlet and outlet states, the problem would be straightforward. However, readers will note that the customized VBA functions for water (Appendix B.4) are not very helpful in the liquid domain (to the left of the vapor dome) because the equations of state that work well for the saturated and superheated regions are not reliable for subcooled liquids. The equations for enthalpy and entropy are acceptable for subcooled liquids – if the density is known, but unfortunately that is a serious problem in the subcooled region. The problem is that the density is usually an unknown, and you must iterate on ρ to obtain a match for the known values of p and T. That outcome is explicitly impossible with these VBA functions because the equation of state for $p(T, \rho)$ is totally unacceptable to the left of the dome. Attempting to find a value for `H2O_p(T, ρ)` for densities in the subcooled liquid region will produce a value of −1 in Excel.

To find the enthalpy for a subcooled liquid, even if the subcooled density is unknown, either of two approximate methods can be utilized, and the discrepancy between them is small. Two

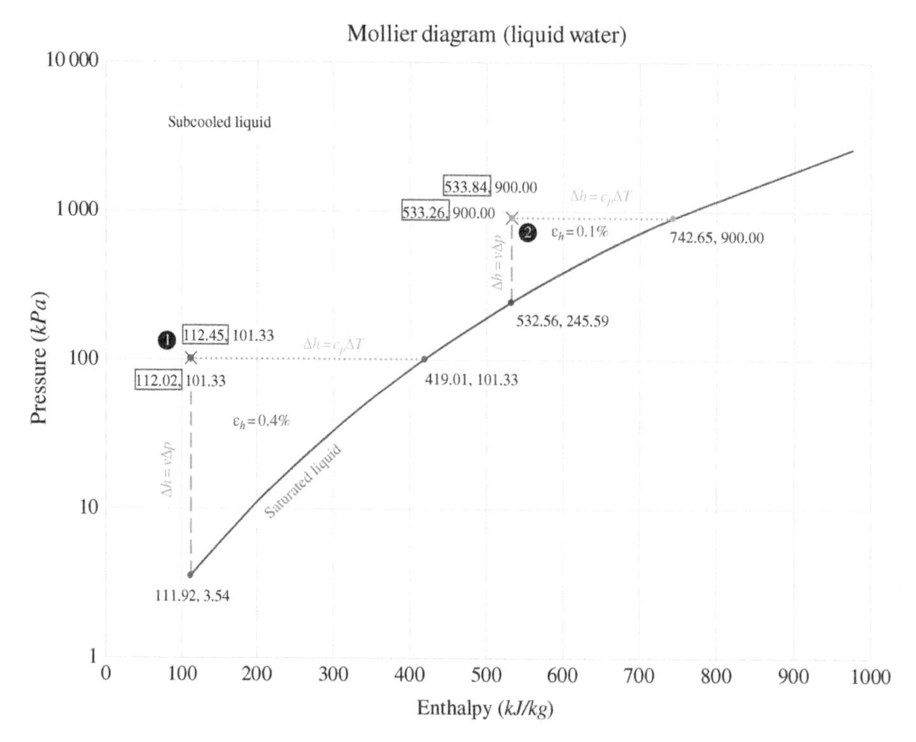

Figure 12.5 Determining enthalpy of subcooled liquids by two methods – enthalpy change along an isotherm or enthalpy change along an isobar. *Source:* Martin Thermal Engineering Inc.

examples are shown in Figure 12.5, and their agreement is very good for low pressures ($p < 2000\ kPa$), and still acceptable for high pressures. Real gas effects and temperature variations of specific heat near the critical point are the causes of the discrepancies.

Both methods utilize a Mollier diagram, and each starts at a different saturated liquid state (which are accurately represented by the VBA functions). One of the methods follows a line of constant temperature and the other follows a line of constant pressure. The formulas are given in Equations (12.10) and the method is illustrated in Figure 12.5.

$$\Delta h_{isotherm} \approx v\Delta p \qquad (12.10a)\{L\}$$

$$\Delta h_{isobar} \approx c_p \Delta T \qquad (12.10b)\{L\}$$

The isotherm method Equation (12.10a) starts at the saturated liquid curve where the temperature is equal to the known subcooled value ($T_❶ = 300\ K$). The saturated starting point is identified by the (h, p) data pair (111.92, 3.54) at the lower left of Figure 12.5. Using the VBA functions, the saturated density is obtained by iterating on density until quality is exactly zero for the known temperature $\mathbf{X}(T,\rho) = 0.0$. Then, along an isotherm, we can approximate enthalpy change as follows:

$$\Delta h = \Delta u + \Delta(pv) = \Delta u + p\Delta v + v\Delta p \approx v\Delta p\ \{T, v\ \text{const}\}$$

The approximate term on the far right is not unreasonable because (i) the liquid's temperature does not rise very much with pressure, which implies that $\Delta u \approx 0$, and (ii) the liquid's density is assumed to be unaffected by pressure changes, so $p\Delta v \approx 0$. In Figure 12.5, we arrive at the desired (h, p) point by following the *nearly* vertical dashed line upward from the saturation point. The isotherm is *nearly* vertical because enthalpy lines are precisely vertical on the Mollier diagram, and the temperature lines bend slightly forward to larger enthalpy values when pressure is rising. With this method, we obtain $h_{❶T}(T, p) = 112.02\ kJ/kg$.

For the isobar method Equation (12.10b), we again start at the saturated liquid curve, but in this case, we begin at the point where the pressure is equal to the known subcooled p value ($p_{❶} = 101.33\ kPa$). The saturated starting point is identified by the (h, p) data pair (419.01, 101.33) near the center of Figure 12.5.

From this starting point, we use Equation (12.10b) to determine the change of enthalpy from the higher temperature at the saturation line to the lower temperature in the subcooled region. In Figure 12.5, we arrive at the desired (h, p) point by following the horizontal dotted line leftward from the saturation point. Because specific heat varies with temperature, it is best to use the average of the initial (saturated) and final (subcooled) temperatures to find c_p from Appendix A.4. Using this method, we obtain $h_{❶p}(T, p) = 112.45\ kJ/kg$. The error in enthalpy between the two methods is approximately 0.4%. A second case (point ❷) is shown in the upper right area of Figure 12.5 and students are encouraged to use the VBA functions and Equation (12.10) to confirm the values shown.

Blowers. Because air and other gases are not incompressible *fluids* (regardless of whether they can participate in incompressible *flows* when Mach numbers are low), the specific volume for a gas must *not* be assumed constant when it passes through a blower or compressor. In fact, the isothermal assumption is also poor for gas compression even if the pressure ratio is as little as $p_2/p_1 \approx 1.1$. As an alternative to the relationship above for a constant volume liquid, the work performed by a blower on an ideal gas may be estimated by assuming constant entropy. This relationship was discussed previously in Chapter 9 with the polytropic equation for compressors and turbines. Irrespective of the above assumptions or non-assumptions, the fundamental energy relation (Equation 12.11) is the same for blowers as for pumps (as long as they are assumed adiabatic with respect to the environment):

$$\dot{W}_{blower} = \dot{m}\Delta h \tag{12.11}$$

The rotational speed of the shaft plays a very important role in determining the performance of a pump or blower. Many blower catalogs will report motor *RPM* values that permit a specific blower (impeller and housing) to deliver a certain flow rate at a certain pressure rise. In the prior section, it was seen that use of a variable speed drive to enable blowers and pumps to meet different flow and pressure conditions can save electricity and broaden system performance, as compared to the use of dampers or control valves.

All fluid movers (pumps, blowers, and compressors) are subject to energy inefficiencies just as compressors and turbines are. Although the design of the impeller (curved/flat, forward/backward), the flow path (centrifugal, axial), and the pressure ratio all affect the device's efficiency, a particular piece of equipment, operating at a particular rotational speed will have a particular point of maximum efficiency – essentially its "sweet spot." This point corresponds to a single operating condition $(\dot{V}, \Delta p)$ in the flow versus pressure-rise domain. All other operating conditions will have lower efficiency, and many pump curves provide efficiency contours to assist the system designer with the tasks of predicting electricity consumption and sizing electric motors. Figure 12.6 shows a set of generic pump curves for a single fluid operating at a single rotational speed, where the solid curves represent the pump curves for different impeller diameters and dashed curves represent efficiency contours.

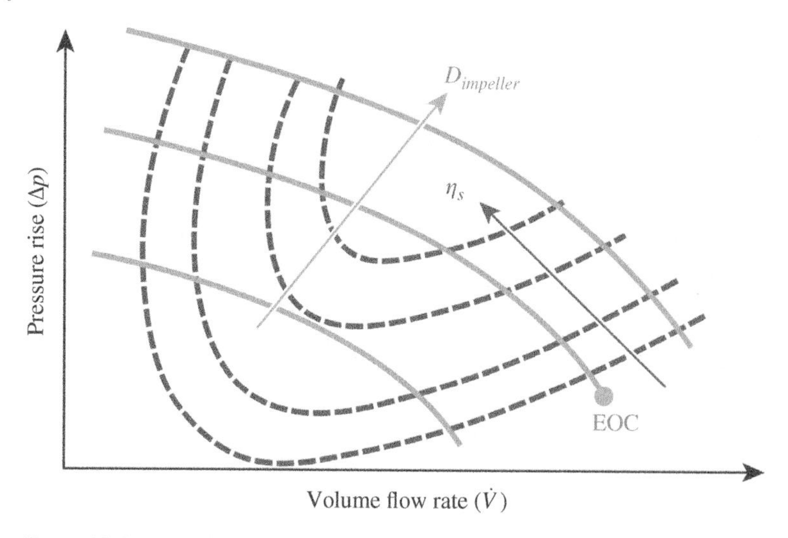

Figure 12.6 Generic pump curve for a single rotation speed, showing $\Delta p(\doteq kPa)$ on the vertical axis and $\dot{V}(\doteq m^3/s)$ on the horizontal axis, with solid lines indicating performance curves for different impeller diameters and dashed lines indicating isentropic efficiency contours over the entire domain of possible operating conditions. The point EOC (end of curve) signifies a conventional choice of the condition by which to size a motor for the given pump, speed, and impeller diameter. *Source:* Martin Thermal Engineering Inc.

Think Stop. Is the efficiency of a pump or blower as shown in Figure 12.6 the same as its isentropic efficiency?

Answer: The pump efficiency is a measure of the mechanical power delivered to the fluid ($\dot{W}_{mech} \approx \dot{V}\Delta p$ for an incompressible fluid) in the numerator, divided by the shaft power provided by the electric motor ($\dot{W}_{shaft} = \tau \bullet \omega$) in the denominator.

Frequently, these efficiency values are in the range of $50\ \% < \eta < 80\%$, which implies that up to 50% of the motor's electric power is "lost." The First law necessitates that the "lost" energy must end up somewhere, and the primary recipient is the fluid's thermal energy, as reflected by a temperature increase. So, ***yes***, blower and pump inefficiency is precisely what occurs due to irreversibilities (entropy-producing factors) in a compressor or turbine, where the transfer of mechanical energy between the fluid and the impeller is never 100% efficient, and the inefficiency results in temperature, enthalpy, and entropy increases for the discharged fluid, relative to the predicted levels for an isentropic process (see Section 9.4).

Another design consideration worth noting is that the inefficiency of the pump must be accounted for when sizing the electric motor. (Although motors also incur an efficiency cost for the conversion of electric power to shaft power, motor inefficiencies are typically much smaller than pump inefficiencies, so they are ignored here.) Since the pump's operating domain cannot be characterized by a single value for isentropic efficiency, the motor's size should be selected based on a conservative estimate for efficiency, based on the anticipated operating condition that exhibits the lowest efficiency ($\dot{W}_{shaft} = \dot{V}\Delta p/\eta_s$). For a fixed-speed pump with a fixed-diameter impeller, this is point often chosen (somewhat arbitrarily) as the ***end of curve*** of a representative impeller line (see point ***EOC*** in Figure 12.6).

12.9 Compressibility in Long Pipes

The thermodynamics of compressible flows (i.e. flows having moderate-to-high Mach numbers, $0.3 < M < 4.0$) are mostly beyond the scope of this textbook, except for a brief discussion in this section on long pipes. The study of compressible flow in nozzles (which may be of interest to student teams designing the steam wand for an espresso machine, see Section 11.9) requires more development than this book can sufficiently address. For readers wishing a broader coverage of compressible flow in nozzles and elsewhere, please consider John and Keith (2006), Korpela (2019), or White (2016). Korpela is particularly noteworthy for its analysis of flow choking when liquid water flashes to steam during its acceleration through a nozzle.

Long Pipes. The effect of friction on an ideal gas flowing in a very long pipe is addressed relatively easily if the flow is assumed to be isothermal and the exit Mach number remains low ($M < 0.3$). The assumption here is that the flow is steady, the pipe's cross section is constant, and the conditions are isothermal (i.e. the fluid exchanges heat through the tube wall with the environment and the thermal energy associated with the frictional dissipation is all transferred out to the environment, so the fluid's temperature remains constant). The steady-state continuity equation requires that $\dot{m}(x)$ is constant for all values of x, which implies, $\rho A_c V = const$ must hold. This implies that as pressure and density decline, velocity must increase. If the pipe is long and the pressure drop is big enough, the fluid's local velocity can approach the speed of sound for the gas at the local ambient temperature. This can lead to choked flow (i.e. a maximum flow is reached that is far less than the design flow rate that was anticipated if pressure drop had been better controlled).

One way to diminish the likelihood of sonic Mach number at the exit is to install compressor stations along the pipeline at strategic locations to boost pressure and ensure the gas density never declines into the regime where Mach number is high.

Example 12.1

Consider a pipe transporting methane gas $\left(\hat{M}_{CH_4} = 16.04\ kg/kmol\right)$ to an exit location where the desired pressure is $p_{exit} = 1.00\ atm$.

a) What is the highest mass flow rate possible in a smooth pipe with $D_i = 12.0\ inch$ that will ensure the Mach number never exceeds $M = 0.3$. Assume the flow is isothermal ($T = 298.15\ K$) and the critical velocity (i.e. where $M = 0.3$) is $V_{max} = 125.4\ m/s$.
b) Assuming a single gas compressor is located at the upstream end of the pipe, and the pipe length is $L = 1.00\ km$, estimate the pressure rise necessary to ensure the mass flow computed above can be delivered at the desired exit conditions without reaching compressible flow conditions (i.e. $M > 0.3$).

Answer.

a) The density of the gas at the exit condition is determined from the ideal gas law:

$$\rho_{exit} = \frac{p_{exit}\hat{M}}{\hat{R}T} = 0.6557\ \frac{kg}{m^3}$$

For the given value of V_{max}, the maximum mass flow is:

$$\dot{m}_{max} = \rho_{exit} A_c V_{max} = 6.00\ \frac{kg}{s}$$

b) First, it must be recognized that the Reynolds number does not change along the pipe because the flow is isothermal, and the ideal gas law is assumed to apply. From continuity, $\dot{m} = \boxed{\rho A_c V = const}$, which implies $\rho V = const$, since the cross-sectional area is unchanging. Also, it was discussed in Chapter 1that dynamic viscosity (μ) does not change with pressure, so that ***Reynolds number*** (and ***friction factor***) are *constant* along the length of the pipe.

However, the velocity will increase as the pressure and density decline in the direction from entrance to exit. If an initial pressure at the pumping station inlet is assumed, the average density and average velocity can be computed, and the pressure drop can be obtained from the Darcy–Weisbach equation (Equation 2.10). Finally, the pipe's initial pressure downstream of the inlet compressor can be determined by iteration, where the initial pressure is varied until p_{exit} reaches the desired value (1.00 *atm*).

Initially, guessing $p_{compr\ exit} = 200\ kPa$ gives the following intermediate results:

$$p_{init} = 200\ kPa; \quad \rho_{init} = 1.294\ \frac{kg}{m^3}$$

$$\rho_{avg} = 0.975\ \frac{kg}{m^3}; \quad V_{avg} = 84.34\ \frac{m}{s}; \quad Re = 2.278 \times 10^6\ (turbulent)$$

$$f_{smooth} = 0.003\ 86; \quad f\frac{L}{D} = 12.65; \quad \left(\rho\frac{V^2}{2}\right)_{avg} = 3.468\ kPa$$

$$\Delta p = -43.858\ kPa; \quad p_{exit} = 156.142\ kPa \qquad \text{too high!}$$

Upon iteration to obtain a small residual error, the approximate initial pressure of $p_{compr\ exit} =$ 153.24 *kPa* is computed. This value provides the required mass flow rate at the required exit pressure, given the pressure drop along the kilometer-long pipe.

Interestingly, for this problem, accuracy is not improved by discretizing the overall pipe length into smaller increments and evaluating average density, velocity, and pressure drop locally for each segment. The reason for this is that for isothermal flow, the pressure decrease and velocity increase are both linear with distance along the pipe. This behavior leads to a situation where the arithmetic mean and the logarithmic mean are equal, so utilization of the pipe length average for density everywhere gives the same result as discretization followed by integration.

12.10 Chimney Effect

The old expression "heat rises" is a common cliché for the physical process called buoyancy-driven flow. However, a closer examination reveals that the phrase is a misnomer that might be misunderstood to mean that thermal energy (heat) moves in opposition to gravity. This is clearly not true of heat conduction in solids, which are fully immune from any type of material flow (buoyancy-driven or otherwise).

Nevertheless, the phenomenon called *chimney effect*, which has been readily observable throughout the long technology history of human beings (Butler 2013), can be a difficult concept for engineering students to fully grasp. Because the system is flowing, the momentum equation is written first to guide the design team's understanding of the fluid *dynamics*. Because the fluid is buoyant, concepts from fluid *statics* should be applied to inform students about the forces involved. Finally, because the flow is *internal*, students must ensure the Moody friction factor is applied correctly to account for the shear forces associated with the no-slip condition at the wall. A potentially confounding bit of data is the fact that the mean velocity is unknown because the flow is not driven by a fluid mover, but rather by the buoyancy forces inherent to the lower density air.

For simplicity, the chimney is modeled as a vertical, cylindrical tube (length $= L$, inside diameter $= D$) that is open to the atmosphere at both ends. A heat source uniformly imparts thermal energy to the air entering at the inlet plane, raising its temperature from T_∞ to T_{hot}.

Consider a free-body diagram of an upward-moving fluid element configured as a cross-sectional slice of the interior of a cylindrical chimney, as seen in Figure 12.7. Make the following assumptions to simplify the analysis: steady state, steady flow, one-dimensional flow, adiabatic chimney walls, isothermal and isochoric conditions hold separately for both inside and outside the chimney, and fully developed turbulent flow all along the chimney length.

From the free-body diagram, the net pressure force is seen to be positive in the upward direction $dF_{pressure} = (p_z - p_{z+dz})A_c$. This is balanced by the two downward forces, weight $dF_{gravity} = -\rho g A_c dz$ and shear force at the wall $dF_{shear} = -\tau_w A_w$. Rewriting this force balance in the form of the momentum equation, Equation (12.12) is obtained. We note that the steady-state and constant temperature and density assumptions permit us to eliminate the convective and unsteady terms from the momentum equation, so we are left with a force balance.

$$\underbrace{-\frac{dp}{dz}\frac{\pi D^2}{4}dz}_{\substack{pressure\\force}} = \underbrace{\rho\frac{g}{g_c}\frac{\pi D^2}{4}dz}_{\substack{gravity\\force}} + \underbrace{\tau_w \pi D dz}_{\substack{shear\\force}} \tag{12.12}$$

Think Stop. A pair of related questions – How do we know the flow is turbulent? Which flow parameter is the relevant one for chimneys – Grashof number or Reynolds number?

Clearly, the turbulent flow assumption should be checked for validity, but most chimneys have large enough diameters that the Reynolds number will be in the turbulent range for moderately large temperature differences. The fully developed assumption renders Grashof number irrelevant because it applies to external flow. Once velocity is determined, Reynolds number can be computed to confirm the validity of the turbulent assumption (and to refine the friction factor estimate).

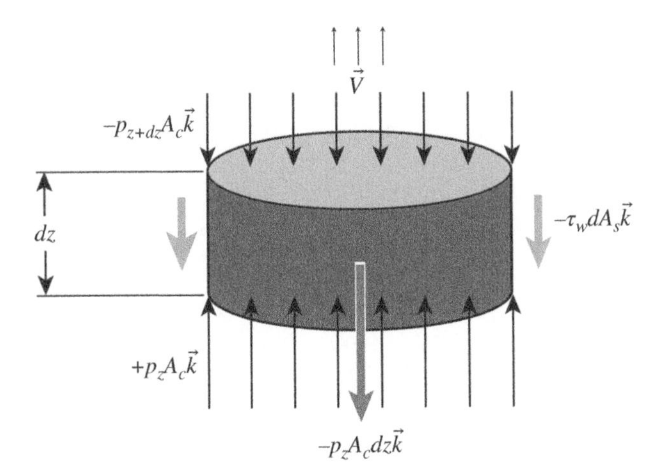

Figure 12.7 Infinitesimal slice of fluid in a cylindrically symmetric chimney to illustrate free-body diagram. *Source:* Martin Thermal Engineering Inc.

Because the air in the chimney is open to atmospheric pressure at both the ends, the pressure boundary conditions (see Equations 12.13) are:

$$p_{z=0} = p_{\infty\, z=0} \equiv p_{local\ barometric} \tag{12.13a}$$

$$p_{z=L} = p_{\infty\, z=L} = \left(p_{\infty\, z=0} - \rho_\infty \frac{g}{g_c} L\right) \tag{12.13b}$$

The local value of barometric pressure is assigned to the pressure at the bottom of the chimney, and gradient of pressure going upward vertically (positive z-direction) is constant for all z (i.e. the pressure is linear with z) both inside and outside the chimney, per Equation (2.14):

$$\left.\frac{dp}{dz}\right]_{inside} = \left.\frac{dp}{dz}\right]_{outside} = -\rho_\infty \frac{g}{g_c} \tag{12.14}$$

Think Stop. Equation (12.14) may seem counterintuitive because the air inside the chimney has a lower density than ρ_∞, but it must indeed hold for the given boundary conditions and the one-dimensional flow assumption inside the chimney that holds concurrently with the static condition outside the chimney. Equation (12.14) also assumes the air densities inside and outside the chimney are (separately) constant with elevation, which is not strictly correct but is valid approximately for most practical buoyancy problems. The Boussinesq approximation asserts that density variations are negligible in the inertial (velocity) terms of the momentum equation, but important in the gravity terms. The practical implication is that although ρ and ρ_∞ are different inside and outside the chimney, respectively, variations in density with pressure, temperature, and velocity inside the chimney are ignored.

We can now combine two of the terms to simplify the momentum equation, integrate over dz from 0 to L, and divide both sides by the chimney volume, as seen in Equations 12.15–12.17:

$$\underbrace{-\left(-\rho_\infty \frac{g}{g_c}\right)\frac{\pi D^2}{4} dz}_{\substack{pressure\\ force}} = \underbrace{\rho \frac{g}{g_c}\frac{\pi D^2}{4} dz}_{\substack{gravity\\ force}} + \underbrace{\tau_w \pi D dz}_{\substack{shear\\ force}} \tag{12.15}$$

$$\underbrace{(\rho_\infty - \rho)\frac{g}{g_c}\frac{\pi D^2 L}{4}}_{\substack{overall\ buoyancy\\ force}} = \underbrace{\tau_w \pi D L}_{\substack{overall\ friction\\ force}} \tag{12.16}$$

$$\underbrace{(\rho_\infty - \rho)\frac{g}{g_c}}_{\substack{buoyancy\\ pressure\\ gradient}} = \underbrace{\frac{4\tau_w}{D}}_{\substack{friction\\ pressure\\ gradient}} \tag{12.17}$$

Because the pressure drop in a pipe is caused by wall shear stress, the shear term on the right can be replaced with Equation (2.15), the Darcy–Weisbach expression (for friction-based pressure drop per unit length) to obtain Equation (12.18):

$$\underbrace{(\rho_\infty - \rho)\frac{g}{g_c}}_{\substack{\text{buoyancy} \\ \text{pressure} \\ \text{gradient}}} = \underbrace{\frac{f}{D}\frac{\rho V^2}{2g_c}}_{\substack{\text{friction} \\ \text{pressure} \\ \text{gradient}}} \tag{12.18}$$

All the terms are known, except chimney velocity and friction factor, which are related to each other through the Reynolds number and the wall's relative roughness. The equation above can be solved for velocity (see Equation 12.19):

$$V = \sqrt{\frac{2gD}{f}\frac{(\rho_\infty - \rho)}{\rho}} \tag{12.19}$$

If the friction factor is guessed and an iteration is performed using Equation 2.19 (the Chen equation), an acceptable margin of error is eventually obtained for friction factor and average chimney velocity.

$$f = \left(-2\log_{10}\left(\frac{\varepsilon/D}{3.7065} - \frac{5.0452}{Re_D}\cdot \log_{10}\left(\frac{(\varepsilon/D)^{1.1098}}{2.8257} + \frac{5.8506}{Re_D^{0.8981}}\right)\right)\right)^{-2} \tag{2.19}$$

Thermal design engineers who are responsible for ensuring chimney performance and maintenance should consider reviewing NFPA 211 (2019), which provides additional design and safety information about chimneys.

12.11 Homework Problems

12.1 Beginning with the Darcy–Weisbach equation, prove that the resistance term in the fluid flow circuit (Equation 12.3) is $R_{fluid} = (8\rho f L)/(\pi^2 D^5 g_c)$, and the flow-pressure-resistance relationship given in Equations (12.2) and (12.3) is correct.

12.2 A vessel containing dry steam at $p = 4.5\ bar,\ g$ has three holes drilled through a sidewall, with diameters $D_1 = 1.0\ mm$, $D_2 = 2.5\ mm$, and $D_3 = 12.5\ mm$. One at a time, a student measures the steam leakage rate through the three holes and reports the amounts to be $\dot{m}_1 = 4.1\ kg/h, \dot{m}_2 = 26.0\ kg/h,$ and $\dot{m}_3 = 568.1\ kg/h$. Which one of the flow rate values was reported erroneously by the student? Show your work.

Answer. 568.1 is the erroneous value.

12.3 The pressure drop (per unit length) of liquid water at $T = 68\ °F$ flowing in a 1 *inch*, *sch* – 40 NPS pipe is $\Delta p/L = 0.008\ psi/ft$. Multiple choice: From the set of values given below, select the pressure drop per unit length that is likely to be closest to that of liquid water flowing in a 1/2-*inch*, *sch* – 40 NPS pipe with the same relative roughness and same mass flow rate, assuming the flow is fully turbulent. (a) 0.016 *psi/ft*, (b) 0.111 *psi/ft*, (c) 0.198 *psi/ft*, and (d) 0.306 *psi/ft*.

12.4 A flow of methane gas ($\dot{V}_{stp} = 31\ 000\ Nm^3/h$) is to be delivered from a well to a distribution facility using a compressor located at the well-head. The gas must arrive at the distribution facility with $p_{exit} = 200\ kPa,\ abs$ through a long pipeline with inside diameter $D = 0.3\ m$, length $L = 2.25\ km$, and absolute roughness $\varepsilon = 0.3\ mm$. The gas leaves the compressor at $T = 293\ K$ and the flow is isothermal throughout the long pipe run. The dynamic viscosity of the methane may be assumed constant at $\mu = 1.1 \times 10^{-5}\ kg/m \bullet s$. What pressure must be developed at the compressor discharge to accomplish this objective? Show your work.

Answer. $\approx 429\ kPa$ (absolute)

12.5 A gas-fired fluid heater is intended to operate in a remote area where electricity is limited. Thus, the air flow to the burners must be supplied by natural-draft means (i.e. with benefit of a long, vertical chimney), rather than forced-draft blowers. Assuming the chimney is adiabatic (no heat loss to surroundings), determine whether dimensions $D_i = 0.5\ m$, $L = 30.0\ m$, and $\varepsilon = 0.2\ mm$ will provide enough draft to exhaust $\dot{V}_{stp} = 6000\ Nm^3/h$ when the outside temperature is $T_o = 25\ °C$ and the chimney temperature is $T_i = 200\ °C$. Assume the molar mass for the exhaust gas is $\hat{M} = 29.0\ kg/kmol$, and use Appendix A.1 for dynamic viscosity of air at the chimney temperature. Show your work.

Answer. Yes. With no other flow resistances, the draft-induced flow rate could be greater than 6000 Nm^3/h.

12.6 Which of the following blowers, operating at a fixed speed $\omega = 1800\ rpm$, is best suited for a system that requires a maximum flow rate of $\dot{V}_{stp} = 4000\ scfm$ of air and will be responsible for a pressure drop of $\Delta p = 14.0\ inch\ WC$ at the maximum flow rate and nominal inlet conditions of STP. Provide backup data for your answer.

(a) $Model\ 294;\ D_{impeller} = 29.6\ inch;\ \Delta p_{max-rise}(3798\ scfm) = 16.0\ inch WC;$
$\dot{W}_{\Delta p, max} = 16.6 bhp;\ \dot{V}_{stp}(2.0\ inch\ WC) = 14\ 488\ scfm$

(b) $Model\ 224;\ D_{impeller} = 22.6\ inch;\ \Delta p_{max-rise}(1821\ scfm) = 16.0\ inch\ WC;$
$\dot{W}_{\Delta p, max} = 16.6\ bhp;\ \dot{V}_{stp}(2.0\ inch\ WC) = 6320\ scfm$

(c) $Model\ 334;\ D_{impeller} = 33.0\ inch;\ \Delta p_{max-rise}(10\ 328\ scfm) = 16.0\ inch\ WC;$
$\dot{W}_{\Delta p, max} = 45.4\ bhp;\ \dot{V}_{stp}(2.0\ inch\ WC) = 19\ 227\ scfm$

(d) $Model\ 264;\ D_{impeller} = 26.1\ inch;\ \Delta p_{max-rise}(1804\ scfm) = 16.0\ inch\ WC;$
$\dot{W}_{\Delta p, max} = 13.3\ bhp;\ \dot{V}_{stp}(2.0\ inch\ WC) = 1804\ scfm$

Answer. (a)

12.7 Consider a blower with a given impeller diameter, whose empirically determined fan curve (at $\omega = 1200\ rpm$) is given by $\boxed{\Delta p_{rise} = -5.0 \times 10^{-7}\dot{V}^2 + 2.0 \times 10^{-3}\dot{V} + 20}$, with inlet conditions at STP, and with units $\dot{V} \doteq acfm$, and $\Delta p_{rise} \doteq inch\ WC$. Also, consider a process cycle (i.e. system curve), whose overall resistance to air flow (including all straight pipe runs and all minor losses with valves and dampers in the fully open position) is $\boxed{3.0 \times 10^{-7}\ inch\ WC/acfm^2}$.

(a) Find the pressure and flow rate at the intersection of the fan curve and system curve with all dampers wide open.

(b) Find the new pressure and flow rate if the overall resistance changes to 5.0×10^{-7} *inch WC/acfm*^{2} when a damper is closed to 70% of full open.
(c) Assuming the blower's isentropic efficiency is constant at $\eta_s = 55\%$, compute the shaft power required $\dot{W}_{shaft}[\doteqdot bhp]$ for both the initial case (a) and the secondary case (b).

Answer. (a) 6404 *acfm*, 12.30 *inch WC*; (b) no answer provided; and (c) initial case $\dot{W}_{shaft} = 22.1\ bhp$, secondary case not provided.

12.8 Consider a water pump with a given impeller diameter and a fixed operating speed, having the following optimum or extreme properties: $T_{max} = 140\ °F$, $\Delta p_{max} = 200\ psi$, $\dot{V}_{max} = 35\ gpm$, and $\eta_{max} = 75\%$. Determine whether each of the following is likely true or likely false:
(a) True or False. Volume flow rate at maximum efficiency is less than volume flow rate at maximum pressure rise.
(b) True or False. Efficiency at maximum volume flow rate is equal to maximum efficiency.
(c) True or False. Minimum shaft power occurs at maximum pressure rise.
(d) True or False. Maximum temperature is likely to be exceeded for all operating conditions if inlet temperature is $T_{inlet} = 135\ °F$.

Cited References

ASME B31.1 (2020a). *ASME B31.1 Power Piping*; New York: American Society of Mechanical Engineers; https://www.asme.org/codes-standards/find-codes-standards/b31-1-power-piping (accessed 25 April 2021).

ASME B31.3; (2020b); *ASME B31.3 Process Piping*; New York: American Society of Mechanical Engineers; https://www.asme.org/codes-standards/find-codes-standards/b31-3-process-piping (accessed 25 April 2021).

ASME B31.9; (2020c); *ASME B31.9 Building Services Piping*; New York: American Society of Mechanical Engineers; https://www.asme.org/codes-standards/find-codes-standards/b31-9-building-services-piping (accessed 25 April 2021).

ASME B36.19M (2018); *ASME B36.19M Stainless Steel Pipe*; New York: American Society of Mechanical Engineers; https://www.asme.org/products/codes-standards/b3619m-2004-stainless-steel-pipe (accessed 25 April 2021).

ASTM A53; (2020); *ASTM A53 Standard Specification for Pipe, Steel, Black and Hot-Dipped, Zinc-Coated, Welded and Seamless*, West Conshohocken, PA: American Society of Testing and Materials; https://www.astm.org/Standards/A53.htm (accessed 20 December 2020).

Butler, O.R. (2013). Smoke gets in your eye: the development of the house chimney. The Ultimate History Project. http://ultimatehistoryproject.com/chimneys.html (accessed 20 April 2021).

CDA; (2020); *Copper Tube Handbook: Industry Standard Guide for the Design and Installation of Copper Piping Systems*; CDA Publication A4015-14/20; MacLean, VA: Copper Development Association; https://www.copper.org/publications/pub_list/pdf/copper_tube_handbook.pdf (accessed 15 January 2018).

Engineering Toolbox (2001). Fiberglass pipes: common standards – Commonly used standards for fiberglass pipes and their applications. http://www.engineeringtoolbox.com/fiberglass-pipes-standards-d_789.html (accessed 5 April 2021).

Haynes; (2020); *Hastelloy C276 Alloy Principal Features*; Kokomo, IN: Haynes International; http://haynesintl.com/docs/default-source/pdfs/new-alloy-brochures/corrosion-resistant-alloys/brochures/c-276.pdf?sfvrsn=2 (accessed 20 April 2021).

IAP; (2021); *Fan Pro*; Phillips, WI: Industrial Air Products; https://www.iapfan.com/fanprosoftware.html (accessed 20 April 2021).

John, J.E.A. and Keith, T.; (2006); *Gas Dynamics*, 3; Boston: Pearson (Allyn Bacon).

Korpela, S.A.; (2019); *Principles of Turbomachinery*, 2; Hoboken: Wiley.

NFPA 211; (2019); *Standard for Chimneys, Fireplaces, Vents, and Solid-Fuel Burning Appliances*; Quincy, MA: National Fire Protection Association.

Rolled Alloys; (2021); *RA 253MA*; Temperance, MI: Rolled Alloys Inc.; https://www.rolledalloys.com/alloys/stainless-steels/ra-253-ma/en (accessed 10 March 2018).

Special Metals; (2008); *Inconel Alloy 600*; New Hartford, NY: Special Metals Inc.; http://www.specialmetals.com/assets/smc/documents/alloys/inconel/inconel-alloy-600.pdf (accessed 15 March 2018).

TubeWeb; (2021); *TubeWeb: Technical Tube Information and Tube Specification*; Dayton, OH: Production Tube Cutting, Inc.; http://www.tubeweb.com (accessed 25 April 2021).

White, F.M.; (2016); *Fluid Mechanics*, 8; Multiple-pipe systems; New York: McGraw-Hill.

13

Thermal Protection

This chapter covers a diverse set of thermal engineering problems that designers often face after they have resolved the basic thermodynamic, fluid flow, and heat transfer issues that govern the functionality of their thermal system. Nevertheless, these peripheral issues can be both challenging and indisputably vital. They involve protecting people and materials from the detrimental effects of high and low temperatures.

The protection of important structural, mechanical, electrical, or biological substances from high or low temperatures requires one of two approaches: design structural, mechanical, or electrical components with materials and/or shapes that can withstand the high- or low-temperature environment they are subjected to or interpose one or more layers of protecting materials between the hot or cold environment and the materials needing protection. The first category invites solutions based on the management of thermal expansion or thermal shock, and the second category invites solutions comprised of insulating and refractory materials.

The terms ***refractory*** and ***insulation*** are sometimes used interchangeably, but their meanings are more complementary than redundant. ***Refractory*** materials are materials that can withstand high temperatures without breaking down. The adjective "high" is relative here. In some contexts, ***refractory*** could indicate a material's ability to survive conditions of just a few hundred degrees Fahrenheit, and in others, it may designate a material's tolerance of 2000 °*C*.

Conversely, ***insulation*** materials have low thermal conductivity and help prevent heat transfer in or out of a device or system. Here too, the adjective "low" must be viewed in the proper context. Some ***refractory*** materials have relatively poor ***insulation*** properties (e.g. they only reduce temperature by a few hundred degrees Fahrenheit over a 12 – *inch* thickness), but that temperature drop might turn out to be vital to protect the next layer of ***insulation*** that is a better insulator but could not withstand the aggressive chemical and thermal atmosphere that is well tolerated by the innermost lining of ***refractory***.

This chapter introduces students to some of the varieties of high- and low-temperature materials that they may want to consider specifying for a thermal system and some design tools for eliminating problems associated with thermal expansion and thermal shock.

13.1 Refractory Ceramics

Ceramic refractory materials can be loosely characterized as any nonmetallic, inorganic solid (typically comprised of various oxides) that retains most of its strength (without creep) at furnace temperatures. Most ceramics are suitable for oxidizing environments (e.g. presence of substantial levels of O_2),

Thermal Systems Design: Fundamentals and Projects, Second Edition. Richard J. Martin.

Companion website: www.wiley.com\go\Martin\ThermalSystemsDesign2

whereas only certain specialized ceramics are recommended for high temperature reducing environments (e.g. atmospheres with substantial amount of H_2, CO, or other reducing compounds). Ceramic materials may have moderately high thermal conductivities, but they are valued for their ability to withstand high temperature, corrosive, and/or erosive environments. The process of making cement in a high-temperature rotary kiln is one where the kiln's innermost refractory material (high-density firebrick comprised of MgO, Al_2O_3, ZrO_2, and other oxides) is a poor insulator but survives 6–12 months at 2400 °F while in continuous moving contact with highly corrosive, partially molten cement feed materials (CaO, Al_2O_3, SiO_2) and highly erosive solid cement clinker.

Refractory ceramics may be applied to the inside of a furnace in one of the following physical forms, depending on the nature of the process and the shape of the furnace.

- Firebrick. Firebrick may be of high density and moderately thermally conductive (called "dense firebrick") or of low density and a good insulator (called "insulating firebrick"). Brick shapes may be rectangular (for flat furnace walls) or truncated wedges (also called "arch shapes" for cylindrical walls or arched ceilings). Mortar is often a necessary adhesive to retain flat walls in place, but cylindrical walls are often installed without any need for mortar because of the interlocking nature of the geometric shapes.
- Castable. Castable refractory is the preferred term for refractory concrete. Castable materials are so named because the powdered raw material is mixed with water into a thick slurry (or thin paste) that can be poured or "cast" into a form, where it pre-solidifies due to cementitious bonding between the particles. The *green* (i.e. unfired) material later forms fused bonds during the high temperature curing schedule. Like firebrick, castable refractories can be "insulating" or "dense."
- Shotcrete. Shotcrete refractory (also called "gunite" or "gunnable" refractory) is a slurry applied pneumatically (with an air gun) to a wall, which is later fired to develop strength and high-temperature resistance. The properties of shotcrete after firing are like the properties of castable refractory after curing.
- Ceramic fiber. Ceramic fiber is spun from molten ceramics having specialized mixtures of constituents to form fibrous blankets, mats, and blocks that withstand high temperature and are very good insulators against heat loss. The final products are "nonwoven," which means the fibers are not spun into long threads of well-controlled diameter and then woven into fabrics on a loom. Rather, the fibers are formed and layered while still in a partly molten state so that substantial void space remains in the interstices between the fibers after they solidify. Ceramic fiber blankets and mats are generally flexible, but excessive flexing can cause disintegration of the fibers into dust, which causes the blanket to lose mechanical integrity. Ceramic fiber blocks are often chosen to line the inside walls of heat treating furnaces. The blocks are larger and lighter than firebricks, but their staggered arrangement is similar to overlapped bricks. They are anchored to the inside of the steel shell with anchoring mechanisms. Because of their high void fraction, ceramic fiber refractory products usually have very low density and very low thermal conductivity.
- Plastic. The term "refractory plastic" is frequently misunderstood as being a super-high-temperature organic polymer (e.g. polyethylene, polystyrene, or polycarbonate), but any such material would not be suitable at all for furnace duty. In the context of high-temperature refractory, the term *plastic* implies that the *green* (unfired) state of the material is plastic (i.e. not elastic) and can be manipulated into cavities or formed into shapes (e.g. noses). After firing, the hardened materials bond strongly to most substrates they contact, and they perform like a monolithic castable refractory.

13.2 Refractory Metals

Refractory metals are never used for insulation but rather may be used for heat exchanger tubes, thermocouple protection wells, or other applications where high thermal conductivity and strength at high temperature is desired. Refractory metals often contain high concentrations of nickel and/or chromium and may be ferrous or nonferrous.

Metals such as tungsten are reserved for high-temperature environments that are devoid of oxygen (e.g. incandescent light bulbs filled with inert gas or vacuum tubes), but under those conditions, they often retain strength at exceptionally high temperatures.

Inconel (approximate composition: 70% nickel, 20% chromium, and 10% iron) is a favored metal for thermocouple wells and sheaths (Cleveland Electric 2016) because of its good oxidation resistance and high thermal conductivity. The material designated as RA-253MA (20% chromium, 10% nickel, 2% silicon, and 65% iron) is often selected for use in high-temperature (gas-to-gas) heat exchangers (Rolled Alloys 2011). Both materials may be used at 2000 °F for limited amounts of time.

One refractory metal designated Haynes 214 (Haynes 2021) was developed for corrosion resistance and moderate strength retention at relatively high temperatures. The secret ingredient that helps Alloy 214 withstand corrosion from HCl at furnace temperatures is a small amount of metallic Al, which, when exposed to oxygen at high temperatures, readily forms a durable coating of Al_2O_3, which protects the underlying metal from attack by corrosive gases.

Common stainless steels (e.g. $SS304$, $SS316$) are not considered highly refractory but may be useful for some applications where the service temperature is less than 1200 °F. $SS304$ is sometimes referred to 18/8 stainless and $SS316$ is sometimes referred to 16/10 stainless – for their relative proportions of chromium and nickel, respectively.

Carbon steel and nonferrous metals (e.g. copper, brass, and aluminum) lose strength and/or oxidize at temperatures below 800 °F and should not be classified as refractory metals. These materials are only suitable for use at low to medium temperatures (e.g. ovens, dryers, and pipes).

13.3 Thermal Insulation

Thermal insulation is mostly used to minimize the flow of heat into or out of a temperature-controlled space. Refrigerators rely on thermal insulation to keep materials (e.g. food) and living things in a cool environment. Conversely, ovens rely on thermal insulation to achieve high temperatures that are needed to accomplish a processing objective (e.g. cooking, drying, and curing).

However, in addition to the simple goal of enabling an enclosure's temperature to be increased or decreased relative to ambient, thermal insulation is also tailored so the energy consumption (fuel or electricity) required to attain the desired temperature difference is minimized. The tailoring usually involves encapsulating tiny pockets of air to act as immovable obstacles that thwart convection currents which can greatly augment heat flow. The tailoring sometimes also involves arraying opaque solid fibers to act as radiation shields.

While stationary air is an excellent insulator against thermal conduction (only very low-pressure gases and vacuum conditions are more effective), air movement (i.e. air susceptible to natural convection currents) amplifies the molecular conduction of heat and makes the air into a poor thermal insulator.

The vacuum flask was invented by James Dewar in 1892 (RIGB 2021). It comprised inner and outer glass shells with an evacuated annulus. He improved the devices thermal insulating capability by applying a thin coat of highly reflective silver to the inner and outer surfaces of the annulus to

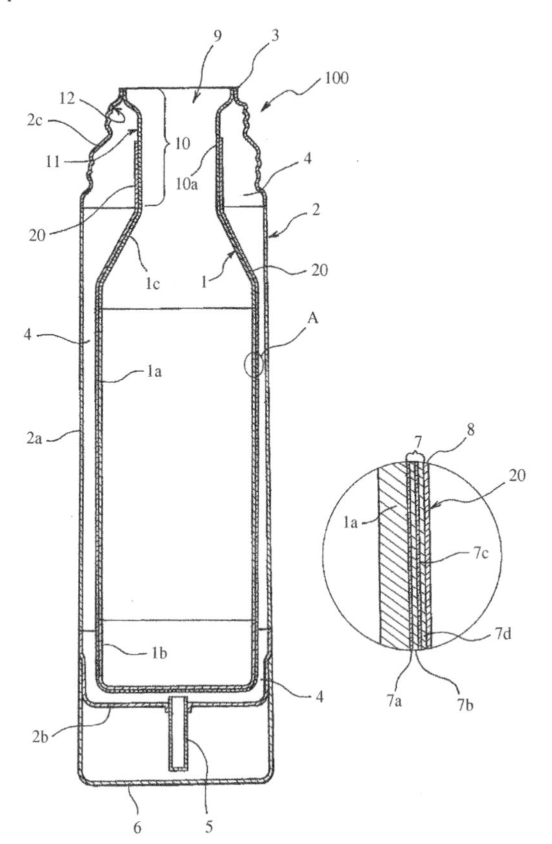

Figure 13.1 Patent artwork for stainless steel vacuum flask. *Source:* Komeda and Fujiyama (1984).

reduce radiative heat transfer. The invention was first made commercially in 1904 by German company Thermos.

Think Stop. A popular brand of insulated drink containers employs a unique geometry (similar to Figure 13.1) wherein a double-wall, stainless steel enclosure with an evacuated annulus (location 4 on the figure) greatly reduces the heat loss from a hot or cold beverage held in the central space (see location 9 on the figure).

We have found ice cubes persisting for more than 24 hours in the bottom third of such a container in summer, whereas full containers rarely maintain solid ice more than 3–4 hours. Develop a hypothesis why the liquid in a full container gains or loses heat much faster than that in a container that is only half full. *Hint*: In the figure, the inner and outer stainless steel shells are welded together at location 3.

When considering an insulating material for very high temperatures, most products are not capable of surviving furnace temperatures for any length of time. Ceramic fiber-based insulation has the best survivability in furnaces.

Examples of low-to-high temperature thermal insulation used for thermal systems are given below. The cited temperature ranges are from Engineering Toolbox (2005).

- Polystyrene (−60 to 165 *°F*). Also called Styrofoam, the expanded or foam configurations offer very good insulating properties with closed cells, lightweight, and low cost. Polystyrene is unsuitable for high-temperature applications but may be ideal for insulating low-temperature pipes or processes. It tends to be brittle, so durability may be a concern in an industrial environment. It is also combustible, which may trigger fire protection concerns in a large facility.
- Polyurethane foam (−350 to 250 *°F*). Polyurethane foam (PUF) is a widely used, durable, and organic polymer that has become nearly ubiquitous for refrigerant pipe insulation. PUF pipe insulation sleeves are hollow cylinders with a longitudinal slit that can be easily wrapped around straight runs of pipe. They are sold in 36 inch lengths for very low cost. PUF is characterized by its lightweight, cellular structure, and flexibility. The product is combustible, but more importantly, its isocyanate binders can release toxic cyano-compounds when subject to high temperatures (as in a fire).
- Cellular glass (−450 to 900 *°F*). Referred to as a "closed cell" material because the pores are not interconnected (i.e. the material is impermeable to fluid flow). Cellular glass is rigid but can be formed in any shape and applied with adhesive to cover pipes, cylinders, and flat walls.
- Fiberglass batts (−20 to 1000 *°F*). Lightweight, spun-glass fibers comprising batts or blanket much like ceramic fiber refractory only less durable and less refractory. Contains organic binders that render it unsuitable for high furnace temperatures.
- Calcium silicate block (0 to 1200 *°F*). Comprised of calcium silicate cement with a fibrous additive for strength, "Cal-Sil" is often sold in clamshell form (pairs of semicylindrical lengths) that is secured with aluminum sheathing and band clamps around pipe runs and fittings. The product is soft enough to be cut with a reciprocating saw. Calcium silicate block was originally sold with asbestos as the strengthening fiber, but glass or ceramic fiber was substituted into Cal-Sil products approximately 50 years ago, and the asbestos-containing version is no longer available commercially.
- Mineral wool (32 to 1400 *°F*). Also called "rock wool" or "slag wool," mineral wool fibers have the property of having high thermal conductivity along the fiber axis but when assembled into batts, conduction from fiber to fiber is very low because the contact points are very small in area and spaced far apart. The air present in the open cells between the nonwoven fibers remains effectively motionless (even under the influence of buoyancy forces) because of the no-slip condition and the presence of multiple fiber surfaces on all sides (which comprises a tiny hydraulic diameter).
- Ceramic fiber (<2200 *°F*). In addition to the forms discussed in the refractory section earlier, ceramic fiber insulation is sold in very thin (e.g. ¼ *inch*) sheets called "paper" and are used for backup insulation behind brick or castable refractories to protect the steel shell in the case of refractory failure.

Thermal insulation is also used for fireproofing. Because fires are transient events, the goal of fireproofing insulation is to help retain the integrity of the encapsulated material long enough so that it can survive the fire's lifetime without catastrophic damage. Structural steel beams in tall buildings are often required to be treated with a thick layer of a fibrous, adhesive coating so the underlying members do not weaken and fail during short fires that are extinguished quickly.

Noncombustible firewalls and fire doors in buildings are designed to have a 1-hour or 2-hour survivability rating to help prevent the spread of fire that would occur rapidly if the materials were made of wood or other combustible materials. Fire safes are insulated to temporarily protect valuables and important papers in the event of a fire.

13.4 Radiative-Convective Insulation Systems

A common cause of structure fires is the positioning of a hot exhaust pipe too close to combustible construction materials (e.g. wood joists, rafters, and beams). This situation occurs routinely when a combustion exhaust duct must pass through a ceiling and roof to be released outdoors. The sources of the combustion gases are typically heating or cooking appliances. Building codes call for ample clearance between the hot flue pipes and the combustibles, but space constraints sometimes make it impossible to provide the clearance required by code.

Often, a valid solution for such circumstances are pre-engineered, double-wall exhaust vents, also known as type B gas vents (Cote 2008). Such vents are lightweight, factory built, and can be snapped together to provide concentric cylindrical paths for (i) the hot exhaust gases (inner volume) and (ii) a buffer stream of cool air (annular outer volume). Type B vent pipes are suitable for use in tight spaces because its installation is often acceptable even when the clearance to combustibles is as small as $d \geq 1$ *inch*.

The inner tube conveys the hot combustion gases away from the fuel-burning appliance by utilizing the chimney effect (Section 12.10). The outer annulus conveys a flow of cooler ambient air that is also driven by buoyancy forces, which are created by the air's contact with the hot wall that separates the inner tube from the annulus. The inner gases, which are the hottest, and potentially contain combustion generated pollutants CO and NO, are shielded from human contact and inhalation by two steel pipe walls that are separated by a buffer layer of warm, moving, ambient air.

The inner pipe is typically aluminum (high reflectivity, low thermal mass), and the outer pipe is typically galvanized steel (higher strength). Neither of these materials is suitable for very high temperatures ($T_{limit} \approx 400\ °F$), but type B vent pipes are only intended for use with a draft hood (truncated cone) whose purpose is to blend in a substantial amount of ambient air along with the combustion appliance exhaust to produce a vented mixture that is hot enough to produce a chimney effect and prevent condensation of combustion moisture but not hot enough to damage the materials of construction. An end-of-chapter problem addresses some of these issues.

13.5 Skin Contact Burns

One of the important purposes of thermal insulation is to protect workers and consumers from skin contact injury (either burns or freezing). Two standards (ASTM C1055 2014; ASTM C1057 2017) provide data to assess the time-temperature threshold for first-degree burns (where epidermal tissue is reddened and painful, but the damage is reversible) and a separate threshold for second-degree burns (where 100% of the epidermal tissue has been heated to the point of necrosis and blistering usually occurs, but no dermal tissue is permanently affected). Burn injuries of the third degree (not plotted in C1055) involve significant necrosis of dermal tissue and permanent scarring or the need for skin grafts. A summary of points on the two plots is given in Table 13.1.

Although the US Occupational Safety and Health Administration (OSHA 2020) does not explicitly regulate hot surfaces by designating a specific temperature that requires workplace safeguards, California's counterpart, Cal-OSHA has promulgated such regulations for industrial workers (Cal-OSHA 2011) and the applicable temperature threshold is $T_{regulated} \geq 60\ °C\ (=140\ °F)$.

Table 13.1 Approximate times and temperatures for first- and second-degree burn injuries.

Time (*s*) at epidermis/dermis interface	Approximate temperature (°*C*) for first-degree burn threshold	Approximate temperature (°*C*) for second-degree burn threshold
10 000	44	45
1 000	46	48
100	50	52
10	55	58
1	65	70

Source: Adapted from ASTM C1055 (2014).

US OSHA indirectly regulates hot surfaces by requiring employers to abide by the so-called General Duty Clause (US Code 1970), which obligates them to "...provide a workplace that is free from recognized hazards that are causing or are likely to cause death or serious harm." The General Duty Clause was an integral part of the Occupational Safety and Health Act of 1970, and the language above appears in Section 5(a)(1) of the Act. According to federal OSHA regulations (29 CFR, 2020), an employer who complies with the standards set forth in the CFR regulations is deemed to be in compliance with the General Duty Clause.

Pipes or other exposed surfaces having an external surface temperature of 140 °*F* (60 °*C*) or higher and located within 7 *ft* measured vertically from floor or working level or within 15 *inches* measured horizontally from stairways, ramps, or fixed ladders shall be covered with a thermal insulating material or otherwise guarded against contact. This order does not apply to operations where the nature of the work or the size of the parts makes guarding or insulating impractical.

The designation of 60 °*C* as a surface temperature that requires insulation, guarding, or personal protective equipment (PPE) is a recognition by Cal-OSHA that workers may accidentally contact a hot surface for a few seconds, and that second-degree burns are unlikely unless the contact time exceeds five *seconds*. A person with normal reflexes and good mobility will, in most cases, be able to withdraw a limb or digit from a hot surface within one to two seconds of contact. Disabled persons may have more difficulty avoiding longer contact times with hot surfaces than workers, so most residential water heaters sold since the 1990s have a more limited range of temperature adjustability – typically from 110 to 130 °*F*, with the default setting at 120 °*F*.

13.6 Protection Against Thermal Expansion

One of the most pernicious problems faced by thermal engineers is thermal expansion. Most materials expand when their temperature increases and if the expansion is improperly constrained or if a differential thermal expansion occurs between two adjacent materials, the parts can exert great forces and create severe damage. Unlike an explosion or a mechanical fracture, the rate of damage caused by thermal expansion is usually slow enough that the naked eye does not perceive what is happening in real time.

Thermal expansion damage occurs when a material's tendency to grow when heated finds itself opposed to an immovable surface that obstructs that growth.

Mathematically, the thermal expansion of a material is typically computed from Equation (13.1), where ΔL is the object's change in length, L is the object's original (unexpanded) length, α is the material's coefficient of thermal expansion, and ΔT is the temperature change to which the item is subjected.

As shown, the coefficient α typically has units of inverse temperature, but the more appropriate way of thinking about this is that the thermal environment has introduced a certain level of longitudinal strain ($\Delta L/L$) per unit temperature. Clearly, whether the coefficient is expressed in units of reciprocal Celsius or reciprocal Fahrenheit must not be ignored because the units matter to the results.

$$\frac{\Delta L}{L} = \alpha \Delta T \qquad \text{where, } \alpha \doteq {}^{\circ}C^{-1}\left[= \frac{m}{m \cdot {}^{\circ}C}\right] \tag{13.1}$$

The immovable object portion of the damage scenario can be thought of in terms of the relationship between strain and stress of an elastic deformation, as seen in Equation (13.2), where $\epsilon(\doteq m/m)$ is compressive strain, $\sigma(\doteq kPa)$ is axial stress, and $E(\doteq kPa)$ is the material's modulus of elasticity. If the stress exceeds the material's yield point, plastic deformation occurs, and the member is irreparably damaged.

$$\epsilon = \frac{\Delta L}{L} = \frac{\sigma}{E} \tag{13.2}$$

In some cases, thermal expansion can cause elastic buckling to occur, and the deformation is reversible. Compelling time-lapse videos of thermal expansion of pipes can be seen at Walraven (2020). The video also succinctly describes the proper way of managing thermal expansion so it does not cause irreparable damage: direct the expansion to a known location where forces will be predicably elastic. This is accomplished, they argue, by immobilizing the pipe at critical points and by supporting the weight of long pipe runs with sliding devices that prevent sagging but do not constrain axial movement due to thermal expansion.

Anchoring is often accomplished by selecting a pipe elbow to be the immovable fixed point and welding a heavy plate onto the exterior wall of the elbow that can be bolted to a concrete pad or footing. This anchoring technique can be applied equally well to vertical or horizontal pipe runs. Although anchoring pipe with a friction fit can work in some instances, a friction-held anchor is less reliable than a welded one.

Pipe hangers and pipe rollers are often used as sliding supports that oppose gravity but not axial motion. Spring loaded hangers are sometimes needed for supporting very weighty piping or equipment that is expected to undergo vertical growth due to thermal expansion.

The designer must visualize all the motions that can occur and account for them with appropriate strain accommodating devices. Expansion joints are most often used to accommodate axial motion, but they are sometimes useful where pipe shear, angular deflection (bending), or torsion is problematic. Figure 13.2 shows examples of pipe expansion joints, supports, and anchors that are intended to accommodate thermal growth.

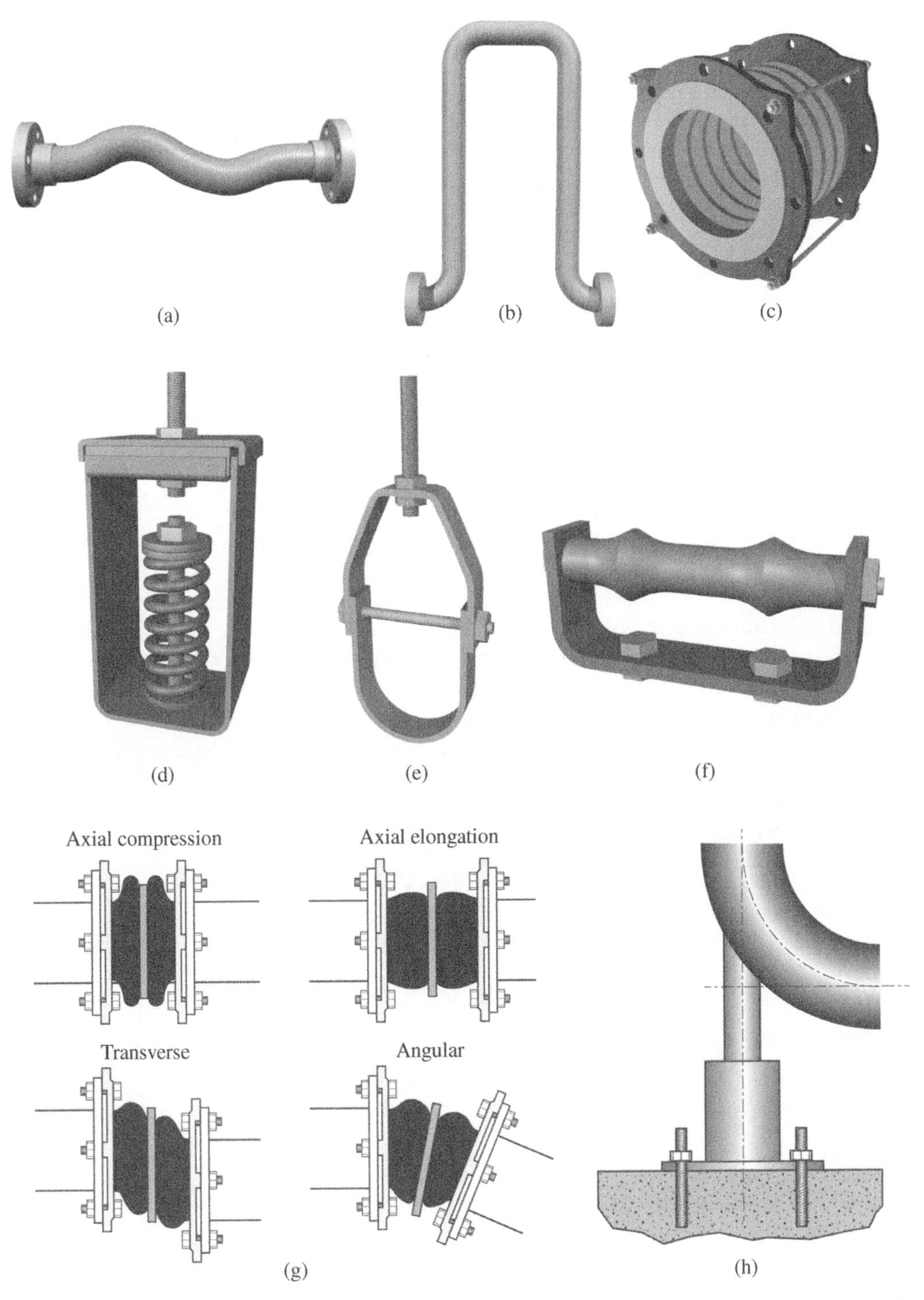

Figure 13.2 Items that help prevent equipment and pipe damage due to thermal expansion: (a) corrugated stainless steel flex line with protective steel braid; (b) upside down expansion loop; (c) bellows-style expansion joint; (d) spring hanger (for very heavy loads that grow vertically); (e) clevis-style pipe hanger; (f) pipe roller; (g) elastomer expansion spool; and (h) rigid pipe anchor. *Source:* Martin Thermal Engineering Inc.

13.7 Protection Against Thermal Shock

Thermal shock damage is related to thermal expansion damage, but with two important differences: (i) thermal shock failures are always brittle, whereas thermal expansion failures are usually ductile, and (ii) thermal shock damage occurs due to temperature gradients in space, whereas thermal expansion damage occurs due to temperature changes over time.

Thermal shock damage usually occurs when a substance (e.g. a component of a heated device) sees high temperatures followed very quickly by low temperatures (or vice versa). Oddly enough, this statement does not contradict the statement above where thermal shock damage was asserted to be linked with large temperature gradients, as opposed to temperature transients. It is true that for a thermally shock-labile material, quenching (i.e. rapid cooling) can trigger a thermal shock related failure, but the actual failure mechanism occurs because different depths of the material are shrinking at different rates (due to low thermal conductivity), and the layers of molecules essentially tear themselves apart because of the internal forces developed due to the differential strain associated with the large temperature gradients. An interesting video of thermal shock causing surface fragmentation of soda-lime glass can be found at Kyocera (2017).

Materials that are resistant to thermal shock tend to be substances with (i) high toughness (a hybrid property that incorporates strength and ductility) and (ii) high thermal conductivity (because it helps reduce temperature gradients in the material).

The classic case of thermal shock occurs when a ceramic or glass material has been heated slowly to a high temperature (e.g. red hot, $T > 1200\ °F$) and it is then immersed in a cool fluid (liquids are more damaging than gases, but both can create quenching at a sufficiently high rate to induce thermal shock). The ceramic or glass has expanded a moderate amount due to the temperature rise, and then upon quenching, because its thermal conductivity is so low, a steep temperature gradient occurs radially, with the outer layers being cooler and wanting to shrink more rapidly than the inner layers. Internal pressure (radial compressive stress) builds up and the exterior layers fail in a brittle fracture.

Examples of thermal shock failure includes:

- Subzero ice cubes cracking when immersed quickly in warm water.
- Splashing cold water onto the glass envelope of a hot incandescent light bulb.
- Pouring water on a red-hot cast-iron stove.

Examples of materials/items susceptible to thermal shock includes:

- Cast-iron cookware
- Some epoxies
- Soda-lime glass
- Alumina
- Porcelain

Examples of materials/items resistant to thermal shock includes:

- Silicon nitride
- Silicon carbide (carborundum)
- Pyrex glass (borosilicate glass)
- Quartz (fused crystalline silica)
- Graphite-reinforced carbon fiber (space shuttle nose material)
- Thermoplastic polymers
- Metals

Li et al. (2014) evaluated the thermal shock resistance of a wide range of ceramic materials and ranked them according to a factor they called thermal shock resistance index (TRSI). For comparison, fused silica (quartz) earned a rating of 96 and high-purity alumina earned a rating of 19.

13.8 Homework Problems

13.1 Conduct a web search to identify at least one insulating material that has a maximum use temperature not lower than $T_{max} \geq 500\ °C$ and a maximum thermal conductivity no greater than $k_{max}(T_{max}) \leq 0.1\ W/m\ °C$.

13.2 Conduct a web search to identify at least one refractory brick that has a maximum density no greater than $\rho \leq 1000\ kg/m^3$, a cold crushing strength not less than $\sigma \geq 2.0\ MPa$, a maximum use temperature not less than $T_{max} \geq 1400\ °C$, and a thermal conductivity no greater than $k \leq 0.5\ W/m\ °C$. Identify the recommended mortar for the brick you identified and the recommended mortar thickness.

13.3 The interior of a rotary cement kiln is one of the most extreme environments for refractory materials. Temperatures must be very high ($T_{max} \approx 2400\ °F$) to convert the raw materials (clay, lime, and sand) into cement, the molten cement is chemically reactive with practically all refractory materials, and the rotating kiln environment causes rapid mechanical erosion of the interior lining by chunks of hardened cement clinker. Cement kilns typically need to fully replace their refractory brick linings at six-month interval.

After performing internet research, evaluate which one of each pair of properties given below is more likely to provide a longer-lasting refractory lining in a cement kiln:

(a) High vs. low density
(b) Medium vs. low thermal conductivity
(c) High vs. low porosity
(d) High alumina (Al_2O_3) vs. high silica (SiO_2)
(e) High calcium (Ca) vs. high sodium (Na)

13.4 Identify an alternative, low-temperature thermal ($T_{cold} \approx -100\ °C$) insulation that could be competitive with PUF in the sense that a cumulative assessment of its properties (low thermal conductivity, low density, and high strength) is equivalent or better than PUF.

13.5 Consider a 12 *ft* tall, 4 *inch* diameter, single-wall (i.e. not type B), uninsulated, cylindrical, and vertical flue pipe (chimney) that passively conveys exhaust gas from a pool heater that burns natural gas at a firing rate of 400 000 BTU/h and has a thermal efficiency of 80%. Assume that 100% of the ($\varphi = 1$) exhaust gas is entrained into the flue but the amount of dilution air must be calculated from the laws of fluid mechanics in the chimney. Assume negligible radiative and convective heat loss from the exterior of the flue pipe (only a good assumption for high oven burner release rates and short chimneys). Use properties of commercial steel for roughness of the interior flue pipe surface. Assume the combustion products have the properties of air (i.e. properties can be obtained from Appendix A.1). State whether the flue pipe temperature is low enough that thermal insulation is unnecessary.

Cited References

ASTM C1055; (2014); *Standard Guide for Heated System Surface Conditions that Produce Contact Burn Injuries*. West Conshohocken, PA: American Society for Testing and Materials.

ASTM C1057; (2017); "*Standard Practice for Determination of Skin Contact Temperature from Heated Surfaces Using a Mathematical Model and Thermesthesiometer*"; West Conshohocken, PA: American Society for Testing and Materials.

Cal-OSHA (2011). California Code of Regulations Title 8, Subchapter 7, "General Industry Safety Orders", Group 2 "Safe Practices and Personal Protection", Article 7 "Miscellaneous Safe Practices", Section 3308 "Hot Pipes and Hot Surfaces"; 8 CCR 3308. www.dir.ca.gov/title8/3308.html (accessed 1 March 2021).

Cleveland Electric (2016). "Tube Specifications and Applications"; Cleveland Electric Laboratories, Resources, Downloads, Thermocouple Literature, Thermocouple Protection Tube Documents. http://www.clevelandelectriclabs.com/wp-content/uploads/2016/02/Tube-Specifications-and-Applications-pg.-13.pdf (accessed 10 March 2021).

Cote, A.E. (2008). Section 10: building services. *NFPA Fire Protection Handbook* 20; Quincy, MA: National Fire Protection Association.

Engineering Toolbox (2005). Insulation materials – temperature ranges. https://www.engineeringtoolbox.com/insulation-temperatures-d_922.html (accessed 30 December 2020).

Haynes; (2021); *Haynes® 214 Alloy Brochure*. Haynes International, Inc., Alloys, Technical Library; https://haynesintl.com/docs/default-source/pdfs/new-alloy-brochures/high-temperature-alloys/brochures/214-brochure.pdf (accessed 25 February 2021).

Komeda, M.K. and Fujiyama, M.N. (1984). Stainless steel thermos bottle. US Patent 4,427,123 (issued 24 January 1984). Assigned to Zojirushi Vacuum Bottle Company, Ltd.

Kyocera; (2017); "*Fine Ceramics Characteristics – Thermal Shock Resistance*"; Kyocera Corporation; https://www.youtube.com/watch?v=ENIiNOsLiQU&t=3s (accessed 25 April 2021)

Li, K., Wang, D., Chen, H., and Guo, L.; (2014); *J. Adv. Ceramics*; **3**; 250; "Normalized evaluation of thermal shock resistance for ceramic materials"; https://link.springer.com/content/pdf/10.1007/s40145-014-0118-9.pdf

Occupational Safety and Health Standards (1910). s 5 (A); https://www.osha.gov/laws-regs/regulations/standardnumber/1910/1910.5

Occupational Safety and Health Act (1970). s 5(a) (1) https://www.osha.gov/laws-regs/oshact/section_5

RIGB; (2021); "*James Dewar's Vacuum Flask*"; Royal Institution Great Britain; https://www.rigb.org/our-history/iconic-objects/iconic-objects-list/dewar-flask (accessed 24 April 2021).

Rolled Alloys (2011). RA 253 MA vs 304, 316; Rolled Alloys, Inc.; Alloys, Stainless Steels, RA 253 MA, Related Literature; Bulletin No. 141USe 09/12; https://www.rolledalloys.com/shared-content/technical-resources/alloy-comparison/RA-253-MA-V-S-304-316_AC_US_EN.pdf.

Walraven (2020). Thermal Expansion and Contraction of Pipes. Walraven USA. https://www.youtube.com/watch?v=eGR77WUARKc (accessed 25 April 2021).

14

Piping and Instrumentation Diagrams

The ***piping and instrumentation diagram (P&ID)*** is perhaps the most widely used drawing in a project's design package.

- It identifies all process components, including equipment, lines, instruments, valves, and off-page connectors.
- It illustrates the basic control philosophy showing sensors, transmitters, controllers, and final control elements.
- It forms the basis for bills of materials and other lists.
- It is used by the process hazard assessment (PHA) team to evaluate the safety of the process and whether additional safeguards are needed before construction begins.

In essence, the process flow diagram (PFD) provides the road map for what, where, and how the process fluids are being manipulated; the P&ID provides the parts list and the basic methodology for the hardware and software that will control and facilitate the desired fluid manipulation. In addition to guidance on preparing the P&ID, this chapter also focuses on various types of instruments, valves, and actuators. Further information resources that cover P&ID drawings can be found in the citations at the end of Chapter 5.

In its broadest context, ***instruments*** are categorized into ***primary elements*** (i.e. those that sense a property of the fluid such as temperature and pressure) and ***final elements*** (i.e. those that cross the system boundary and interact with the fluid to cause a change, such as reduced flow or increased speed or power). Broadly, therefore, ***valves*** must be considered instruments, even though instruments are colloquially thought of as sensing devices only. This broad viewpoint is corroborated by the exclusion of the term ***valve*** and inclusion of only the word ***instrument*** in the drawings title – P&ID.

14.1 Design Packages

A design package is the culmination of the process design work carried out by a thermal engineering team. Typically, several parts of the design package are completed at the proposal stage and submitted to the potential customer as part of the engineer's quotation of scope and price.

After the engineering team wins the project, they complete the remaining elements of the design package and submit the completed set of drawings and other documents to the customer for final approval before procurement and fabrication/construction begins.

If significant differences exist between the proposal drawings and the final drawings, the engineering team must communicate those to the client – with a memorandum (or ***engineering change***

Thermal Systems Design: Fundamentals and Projects, Second Edition. Richard J. Martin.

Companion website: www.wiley.com\go\Martin\ThermalSystemsDesign2

request) and as ***revisions*** noted appropriately in the drawings themselves. Additional information about the growing field of ***change management*** within the various disciplines of engineering, construction, and operations is given in Chapter 17. A list of drawing and document types that may be typical for a thermal system is given in Table 14.1.

Table 14.1 Candidate drawings and documents for a thermal system.

Category	Drawing name	Stage needed
Process	Process flow diagram	Proposal
Process	P&ID standards, legends	Approval
Process	Piping and instrumentation diagram	Approval
Architectural	Site plans	Approval
Architectural	General arrangement – plans, elevations, sections, details	Approval
Mechanical	Equipment assembly (exploded views)	Approval
Mechanical	Component details	Approval
Electrical	One-line control diagram	Approval
Control	Control system description	Approval
Mechanical	Data sheets	Procurement
Fabrication	Subcomponent (e.g. Weldment) details	Construction
Structural	Structure assembly, details	Construction
Construction	Foundation plans, details	Construction
Construction	Piping isometrics	Construction
Electrical	Power schematic	Construction
Electrical	Wiring termination drawing	Construction
Control	Ladder logic	Commissioning
Process	Thermal load calculations	Turnover
Structural	Structural load calculations	Turnover
Control	Lists of alarms, interlocks, loops (with recommended setpoints, parameters)	Turnover
General	Bills of materials	Turnover
General	Recommended spare parts list	Turnover
General	Operation and maintenance manual	Turnover

Source: Martin Thermal Engineering Inc.

14.2 Temperature Sensors

Temperature sensors can be divided into categories of indicators and elements. Temperature indicators include thermometers and dial indicators, where the temperature reading is directly readable, and the sensor visibly moves or changes its appearance in response to temperature changes. Temperature elements change electrically in response to temperature changes, but they must be connected to an electronic display device for the response to be discernable as a temperature reading. Temperature elements include thermocouples and resistance temperature detector (RTD). Examples are shown in Figure 14.1.

Thermocouples. The principle behind ***thermocouple*** performance is the Seebeck effect, which was discussed previously in Section 10.9. A thermocouple is manufactured by electrically joining (usually by welding) a pair of dissimilar wires at the junction or measurement end (to be placed in the zone of temperature to be sensed, often called the "hot junction") and connected to opposing terminals of a digital readout at the reading end (to remain in a space that is held at a reference temperature, typically 0 °C, and often called the "reference junction"). An electromotive force (voltage) is generated between the dissimilar wires due to the temperature difference between the wire ends at the hot junction and the wire ends at the reference junction. The voltage developed is correlated to the temperature difference by a calibration curve for the dissimilar metal pair. See Figure 14.2 for three examples of thermocouple voltage differentials.

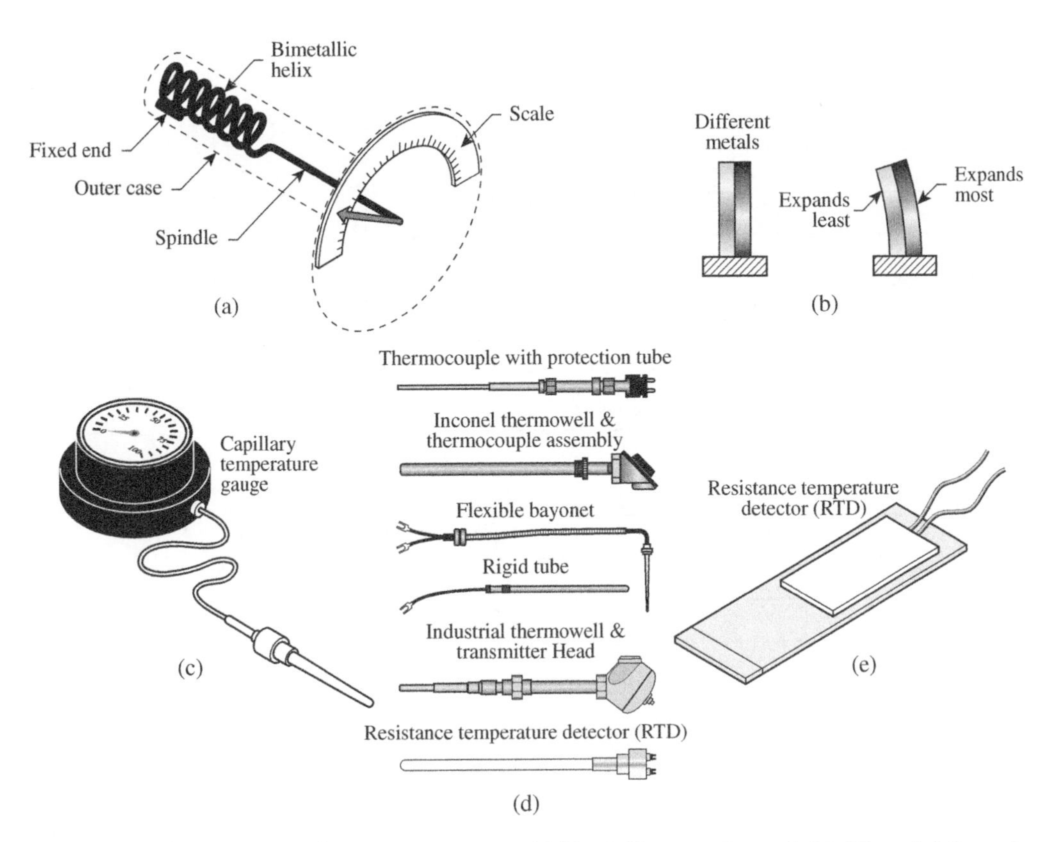

Figure 14.1 Different types of temperature gauges. (a) Bimetallic expanding coil; (b) differential thermal expansion; (c) thermocouple assemblies (various); (d) bulb and capillary with expanding liquid; and (e) resistance temperature detector. *Source:* Martin Thermal Engineering Inc.

Resistance Temperature Detectors. The principal behind the ***RTD*** is that a conducting wire's electrical resistance changes with temperature. When an applied current is passed through the RTD, a voltage drop occurs and can be measured. The voltage drop is correlated to the resistance and can be compared to a calibration curve to assess temperature.

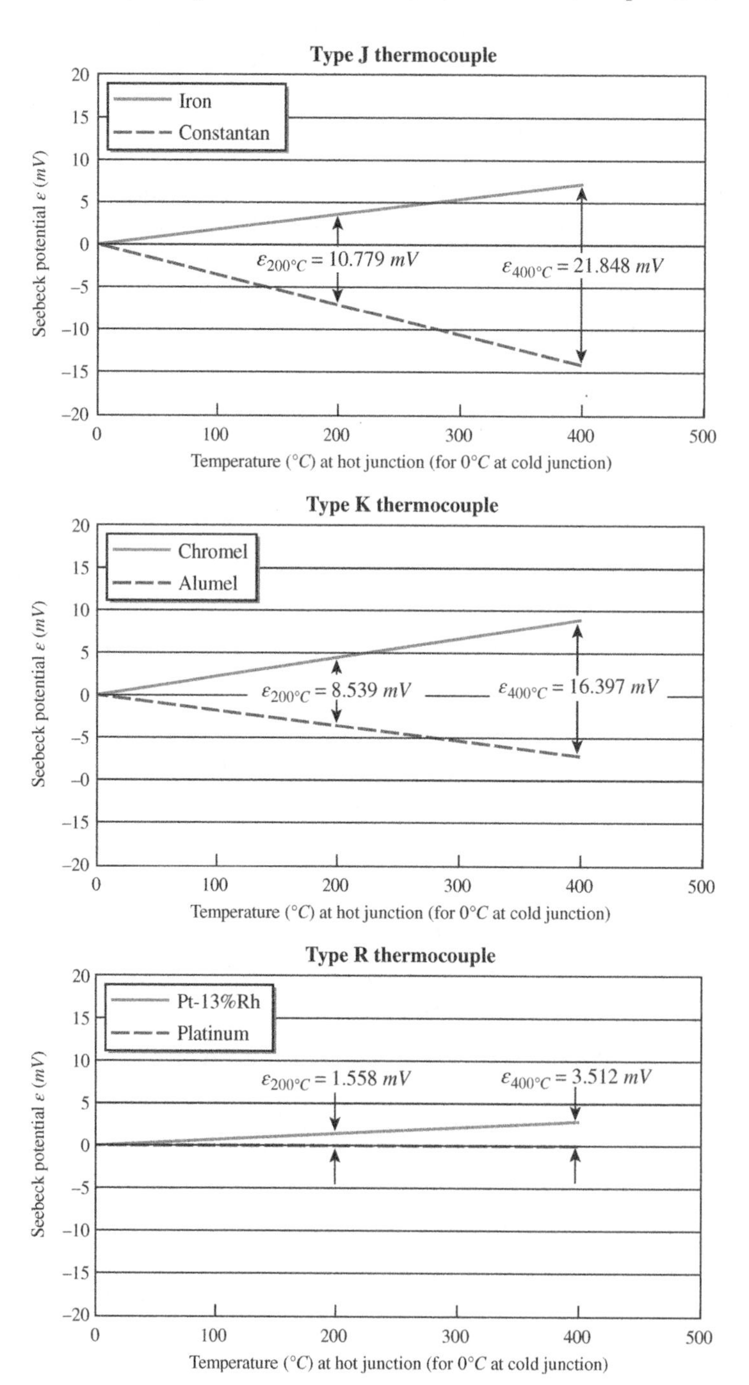

Figure 14.2 Seebeck potential vs temperature at hot junction for three thermocouple calibrations (types J, K, and R). *Source:* Martin Thermal Engineering Inc.

Thermometers. ***Thermometers*** were the first temperature-measuring devices invented and are simply a glass capillary tube with a liquid (usually mercury) that expands by a known fractional amount per incremental temperature change ($\Delta cm^3/[cm^3 \cdot °C]$). The mercury expansion can be seen in the capillary tube and the location of the top of the mercury column can be correlated to the temperature of the liquid mercury in the bulb. Markings are usually added to the capillary tube to simplify the direct readout process.

Indicating thermometers are usually of two types – capillary or bimetallic. The capillary-style temperature gauge contains a bulb with an expanding fluid and a capillary tube connected to the dial mechanism. As the liquid expands in response to the temperature change at the bulb, the liquid pushes against a diaphragm that is connected to a gear mechanism that turns a dial. The bimetallic style relies on the thermal expansion difference between two dissimilar metals. As one metal expands more for a given temperature change, an offset occurs, which causes a dial to move.

14.3 Pressure Sensors

Pressure sensors can be categorized by their principle of operation – gravity, elastic deformation, electrical property response, gas property response, and luminescent coating. See Figure 14.3 for examples.

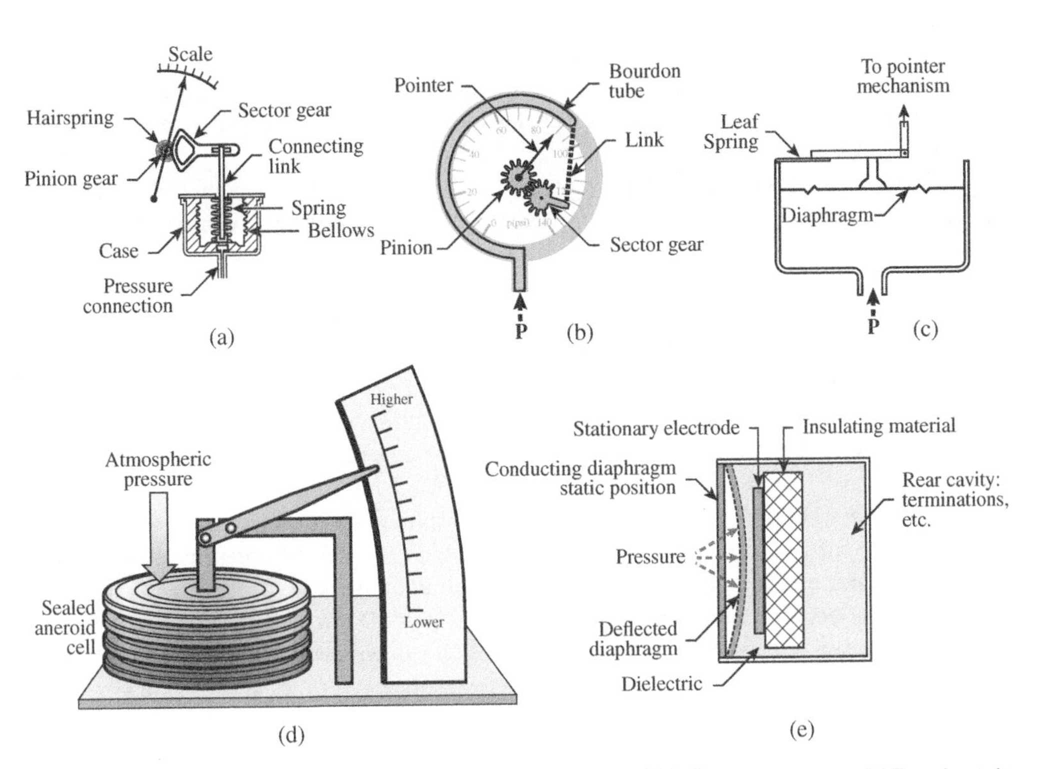

Figure 14.3 Examples of various types of pressure instruments: (a) bellows-type gauge; (b) Bourdon tube gauge; (c) diaphragm gauge; (d) aneroid barometer for ambient pressure; and (e) dielectric capacitance gauge. *Source:* Martin Thermal Engineering Inc.

Gravity Sensors. The most well-known and versatile pressure instrument is a ***manometer***, which is a pair of connected columns containing a liquid that can be partially displaced (up or down) by a pressure difference at the upper surface of the two columns. A ***barometer*** is essentially a manometer with one of the legs closed at the top and with an evacuated space above the measuring liquid. A ***deadweight piston*** is essentially a calibration device, used to compare a gauge response to a manometer. In it, fluid pressure is varied by an adjusting screw that varies the height of a liquid column, whose position can be directly compared to the response of the gauge-under-test.

Elastic Deformation. A ***Bourdon tube*** gauge relies on the principle that a flattened tube tends to straighten when pressurized internally. The strain is magnified by forming the tube into a helical shape. A ***diaphragm*** gauge uses a flexible membrane to separate regions of different pressure. It relies on a reproducible deflection of the membrane with changing pressure, and converts this motion to a dial rotation using gears and linkages. A ***bellows*** instrument can be used to measure ambient or relative pressure, based on the principle that the device expands longitudinally based on the internal charge of a gas, relative to the external pressure.

Electrical Property Response. A ***dielectric capacitance gap*** instrument measures pressure by the compression of an elastic dielectric material, whose capacitance is proportional to its thickness, which depends on the applied pressure. A ***strain gauge*** is a device where an elastic diaphragm is embedded with a strain gauge, whose electrical resistance changes when it is stretched. Similarly, the diaphragm may be comprised of a thin sheet of metal that stretches elastically and changes resistance without requiring separate materials for the strain gauge and the diaphragm. A ***piezoresistive strain gauge*** is a thin sheet of silicon whose electrical resistivity is sensitive to pressure. Digital manometers may use any of these principles to convert a pressure impulse to an electric signal.

Gas Property Response. A ***Knudsen gauge*** relies on the principle of molecular impact for accurate *vacuum* measurements. An ionization gauge creates a stream of electrons that collide with gas molecules in a partially evacuated space to create ions that generate a current that depends on the number of ions per unit volume.

Luminescent Coatings. Certain coatings respond to pressure with variation in emitted light that is inversely proportional to the surface pressure. These ***luminescent coatings*** are often used to illustrate relative variations in pressure in wind tunnels or low-Reynolds number flows.

Think Stop. If fluid mechanics was a fair subject for the television show Jeopardy, a question under the category "Fluid Scientists" might go like this... *Your Final Jeopardy answer is:* "Because the pump makers told him that water could not rise more than 10 m under the influence of suction." *Contestant number one, your answer please* – "Why did Archimedes invent the water screw pump?" *Contestant number two* – "Why did Torricelli invent the mercury barometer?" *Contestant number three* – "Why did Poiseuille investigate the laminar flow characteristics of the circulatory system?"

Contestant number two is the winner. Torricelli's chief invention was the mercury barometer. The name comes from the Greek words for "weight" and "gauge." The device represented the balance between the weights of (i) a column of air and (ii) a column of mercury.

To wrap up the Jeopardy question, a favorite question for PhD qualifying exams is, "A homeowner wants to pump water uphill from a pond located 50 *ft* downhill from her residence. Where should she install the pump – at the inlet to the pipe (pressure end) or the exit from the pipe (suction end)?"

The pump makers who told Torricelli that 10 *m* was the maximum water height possible under suction were correct, and the location of the pump at the suction end simply would not work, because 50 *ft* of elevation change is greater than 10 *m* (33 *ft*).

14.4 Flow Sensors

The number of operating principles that have been applied to the measurement of flow is almost as great as the number applied to the measurement of pressure. Several different flow measurement methods are described in this section. See Figure 14.4 for illustrations.

Flow measurement by an ***orifice plate*** equipped with a pressure differential gauge is the most common and cost-effective means of quantifying flow rate. The technique requires knowledge of upstream fluid density to compute a mass flow rate from the throat velocity determined by the pressure differential and the orifice flow area. Flow measurement by a ***venturi tube*** is nearly identical to that of an orifice plate, with the primary differences being the cost of the venturi meter (much higher than the orifice) and the (undesirable) permanent pressure loss of the orifice plate (much higher than the venturi).

A properly sized and maintained ***turbine meter*** (also called vane-type flow meter) provides a very good measure of volumetric flow rate if the bearings are in good shape and if the fluid slip around the edges of the vanes is minimal. The meter is typically equipped with a magnet on one of the vane blades and a pulse pickup to count the rotational speed of the device.

The ***rotameter*** (also called variable area flowmeter) is a device that relies on a balance between the (downward) weight of the float and the (upward) drag force of the fluid on the exterior of the float body. As the flow rate increases, the float rises higher in the tube until the velocity at the narrowest point creates a drag force that barely lifts the float. This ensures the local fluid velocities are approximately equal at every float position, and the flow rate is quantified by a scale on the side of the tube.

A ***Pitot-averaging flow meter*** performs a mechanical averaging of the velocity pressures at multiple locations around the cross-section of the pipe. This method is preferred over relying on a

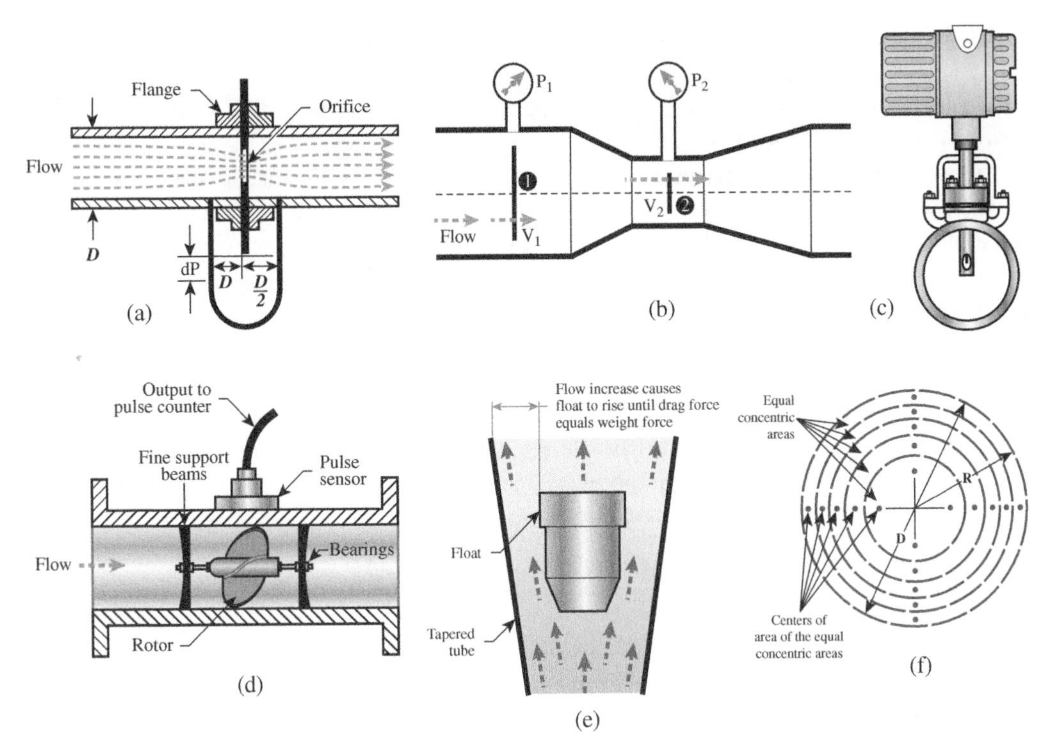

Figure 14.4 Examples of various types of flow instruments: (a) orifice plate with manometer; (b) venturi flow tube with pressure gauges; (c) thermal anemometer; (d) turbine volume flow meter; (e) rotameter; and (f) averaging velocity meter. *Source:* Martin Thermal Engineering Inc.

single-point Pitot-probe velocity measurement, especially where upstream or downstream fittings may be too close to ensure the velocity profile is axisymmetric.

The ***thermal anemometer flow meter*** is a device that utilizes an electrically heated probe to correlate heat loss to the fluid with flow rate past the probe. The most common arrangement is for the device to maintain a constant probe temperature by supplying a varying rate of ohmic heating to the probe as the fluid flow rate varies up or down. The rate of heat addition (i.e. the electric power delivered, which is expressed as $\dot{Q}_{elec} = I^2R$) is directly related to the mass flow rate. If the device is calibrated for the fluid being measured, it can produce an output in units of kg/s.

The technologies that are automatically suitable to be used as sensors in a feedback control loop are the turbine meter and the thermal anemometer. The devices that rely on a measurement of pressure differential (i.e. orifice, venturi, and Pitot-averaging) may be rendered suitable for electronic control if a digital pressure transducer is tasked with quantifying the Δp. The rotameter is almost never equipped for feedback control, and as such should be reserved for flow rate indication-only applications.

Orifice Calculations. The ***orifice*** is the flow meter of choice for many applications. However, the proper interpretation of orifice pressure differentials must address all of the many factors (e.g. diameter of orifice [d], diameter of pipe [D], upstream density [ρ_1], Reynolds number [Re_d], and pressure-tap location [flange, vena contracta, $-D + ½D$]) that affect the calibrated flow result. The different pressure-tap locations represent varying degrees of apparatus convenience versus scientific purity. The vena contracta location is the location of the narrowest streamline convergence in the entire flow field. This is where the flow area is the smallest and where the velocity is the greatest – a scientifically strong location for the downstream tap, but not a very practical one because the location varies with flow rate. By contrast, flange taps are customarily provided with predrilled holes in the left and right flange bodies, and as such are the most practical choice because no additional drilling through the pipe is necessary. The detrimental aspect of flange taps is that they generate the lowest pressure drop per unit mass flow rate among the three options. Therefore, the $-D + ½D$ tap choice is a popular compromise between the other two.

The underlying physics of the orifice is attributable to the Bernoulli effect (i.e. mechanical energy in the form of pressure is converted to kinetic energy in the form of velocity), but the actual geometry of the application may introduce factors that complicate the simple Bernoulli model. Because the measured pressure drop is not necessarily equal to the permanent pressure loss, friction is not the appropriate model, and the modified Bernoulli equation (Equation 2.28) is not particularly helpful. Nevertheless, when applied according to the well-established rules for orifice meters, a dimensionless flow coefficient $C\ (= \dot{V}_{actual}/\dot{V}_{theoretical})$ provides the appropriate correlation between the measured pressure change at the installed tap locations and the upstream and downstream velocities, where the reversible and irreversible phenomena of the flow, including approach, discharge, and acceleration effects are lumped together in the empirically derived coefficient C.

Applying this empirical result to the basic Bernoulli equation, with a substitution of volume flow rate divided by cross-sectional area for the velocity term, leads to Equation (14.1) (which assumes a horizontal pipe and steady, single-phase, incompressible flow):

$$(p_1 - p_2) = \frac{1}{C^2}\left(\frac{1}{2g_c}\right)\left[\frac{\rho\dot{V}^2}{A_{c,2}^2} - \frac{\rho\dot{V}^2}{A_{c,1}^2}\right] \quad \text{where } C \text{ varies with } d, D, \text{and } Re_D$$

$$\Delta p = \frac{1}{C^2}\frac{\rho\dot{V}^2}{2g_c}\frac{1}{\left(\frac{\pi d^2}{4}\right)^2}\left[1-\beta^4\right] \quad \text{where } \beta = \frac{d}{D}$$

$$\boxed{\dot{V} = \frac{C}{\sqrt{1-\beta^4}}\left(\frac{\pi d^2}{4}\right)\sqrt{\frac{2g_c\Delta p}{\rho}}} \tag{14.1}$$

The flow coefficient (C) varies with the Reynolds number (Re_D, based on pipe diameter) and with the diameter ratio β. For large values of Re_D and/or small values of β, a value of $C \approx 0.61$ may be assumed, irrespective of the specific pipe and orifice diameters. Crane (1988) provides some useful plots for flow coefficient as a function of Reynolds number.

When greater accuracy is needed (or at the opposite ends of the ranges for β and Re_D), Bean (1971) provides different sets of equations to compute C, depending on the location of the pressure taps. For Equations (14.2), upstream pressure should be measured one pipe diameter upstream ($-D$) of the leading face of the orifice plate and the downstream pressure should be measured one-half pipe diameter downstream ($+½D$) of the orifice's leading face.

A more universally applicable solution for the discharge coefficient C_d (with pressure taps at $-D$ and $+½D$) can be obtained from Equations (14.2), where all parameters are dimensionless except pipe inside diameter D, which must have dimensions of actual (rather than nominal) ***inches***.

$$C = K\sqrt{1-\beta^4} \qquad K = K_0 + b\lambda \qquad \lambda = \frac{1000}{\sqrt{Re_D}} \tag{14.2}$$

$$b = \left(0.0002 + \frac{0.0011}{D}\right) + \left(0.0038 + \frac{0.0004}{D}\right)\left(\beta^2 + (16.5 + 5D)\beta^{16}\right) \quad \{D \doteq inch\}$$

$$K_0 = \left(0.6014 - 0.013\,25D^{-1/4}\right) + \left(0.3760 + 0.072\,57D^{-1/4}\right)\left(\frac{0.000\,25}{(D^2\beta^2 + 0.0025D)} + \beta^4 + 1.5\beta^{16}\right)$$

$$\{D \doteq inch\}$$

If the Mach number at the throat exceeds $M \geq 0.3$, the acceleration is probably too large to permit the flow to be considered incompressible. In such cases, Equations (14.2) must be modified to handle the effects. Students should refer to Bean (1971) for additional details about compressibility effects in orifices.

14.5 Level Sensors

Like most other instruments, level sensors can function according to a number of different operating principles. The most basic division of level sensor types is between point-level sensors (which are usually operated as level switches) and continuous-level sensors (which provide analog signals that correspond to actual liquid level in millimeters or inches). Floats can be used to elicit responses 3that can be used for either point-level or continuous-level sensing. Electrical conductivity, acoustic reflection, and electrical capacitance can also be applied for either type of level sensing task. A variety of level sensors are summarized in Figure 14.5 and explained further in the text that follows.

Float Switch. A ***float switch*** operates in a manner very similar to that of a float valve inside a toilet tank. When rising liquid level reaches the float, it begins to rise and when the lever-arm angle passes a design threshold, a microswitch closes (or opens) in the body of the switch. The closed or open circuit condition can be utilized by a digital controller to take some action such as turning a feedwater pump on or off.

Floating Magnet and Reed Switch. The principle of operation for a ***floating magnet*** device is that a cylindrical magnet is assembled into the float and the assembly floats up and down along the axis of a smaller diameter cylindrical probe. When the magnet is at a design threshold level, it attracts a magnetic reed internal to the smaller probe. The motion of the reed makes or breaks contact with a nearby internal electrode, thereby opening or closing an electric circuit. Multiple reeds can be

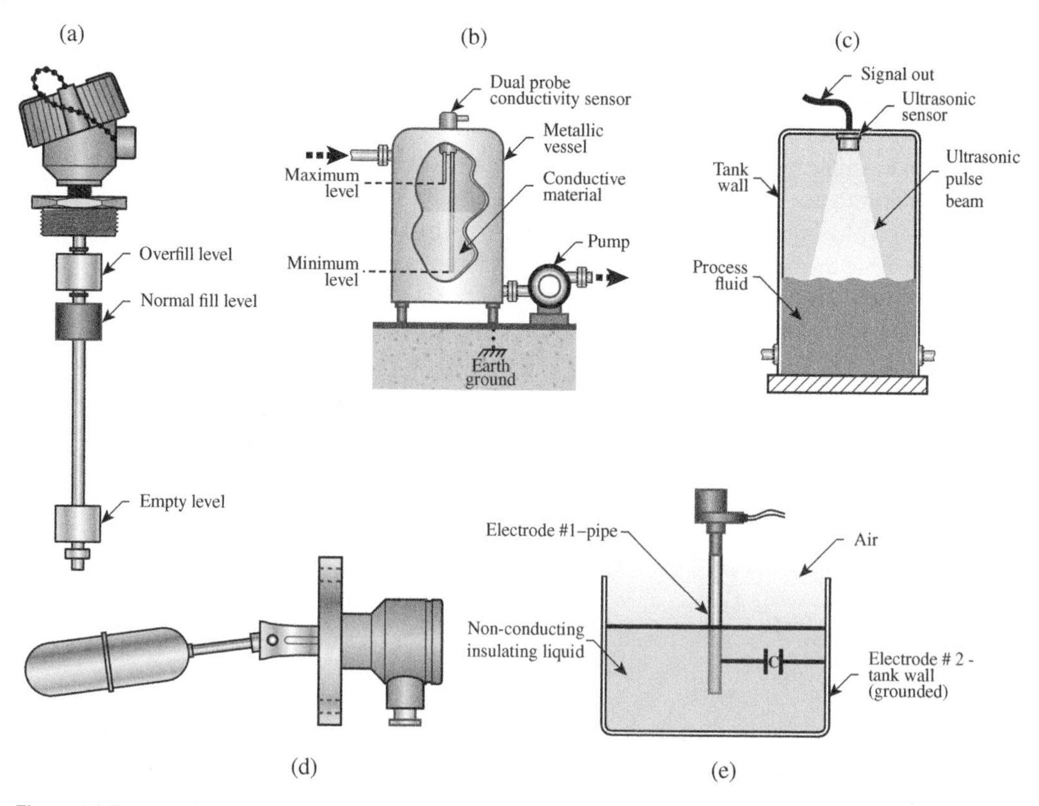

Figure 14.5 Level sensors and switches. (a) Floating magnet with reed switch; (b) electrical conductivity level switch; (c) ultrasonic level sensor; (d) float level switch; and (e) electrical capacitance level sensor. *Source:* Martin Thermal Engineering Inc.

associated with one or more magnets to accomplish a piecewise-continuous sort of digital-level sensing and/or switching.

Electrical Conductivity Level Switch. When the liquid whose level is to be detected is electrically conductive, a simple ***electrical conductivity*** probe can be used to detect level at a point. If the conducting liquid is entirely below the probe, the circuit is open, but whenever the liquid is in contact with the probe, the circuit is closed. Multiple probes can be installed for multiple-level determination. Protection circuitry to safeguard against electric overcurrent and excessive storage of electrical charge for the overall system (tank, liquid, and sensors) is mandatory for this type of device.

Acoustic Reflection Sensor. The ***ultrasonic*** level sensor is a continuous device that incorporates an acoustic wave generator and a reflected acoustic wave sensor to detect and quantify the level of liquids or granular solids in a vessel. The principle of operation relies on the known speed of sound for the headspace fluid (typically air or nitrogen) and a determination of the time-of-flight for the outgoing and returning waves. The wave generator and sensor are both located near the uppermost surface of the vessel, and the time-of-flight depends on the distance between the upper surface of the stored material and the generator/sensor assembly.

Electrical Capacitance Sensor. An ***electrical capacitance*** level sensor is another continuous device that is based on the principle that the electrical permittivity of liquids and solids is greater than that of air. Electrical permittivity ($\varepsilon \stackrel{\vee}{=} F/m$, or farads per meter) is a measure of electron flux through a nonconducting medium as a function of the charge accumulated on unit-area-sized electrode plates separated by a unit length of the medium. Materials with higher permittivity pass more electron flux than materials with lower permittivity. The permittivity of a material is inversely related to its relative dielectric strength. A perfect vacuum is defined to have a permittivity of 1.0 and the relative permittivity of air is just six-hundredths of a percent greater than 1.0. The electrical capacitance probe is often installed vertically in the vessel and extends from its top to just above its bottom. When the vessel is empty, the capacitance of the probe-vessel system is low, because the dielectric separating the two electrodes is air. When the liquid or solid level rises to cover a fraction of the probe length, the capacitance increases, and the increase is directly correlated to level height.

14.6 Exhaust Gas Analyzers

Most industrial combustion processes (and some non-combustion processes) require the exhaust/effluent to be monitored for pollutant flow rates exiting to the environment. Waterborne pollutants (e.g. oil and/or hazardous substances) are monitored in the aqueous phase and airborne pollutants (e.g. "criteria pollutants" and "hazardous pollutants") are monitored in the gas phase.

For airborne pollutants, a continuous emissions monitoring system (CEMS) is frequently required by regulatory authorities to quantify the appropriate process variables and transmit the data directly to the agency. The reporting frequency is typically four times per hour.

The components in a CEMS package usually include: (i) a stack-installed sampling port where a small stream of exhaust gas is continuously extracted from the process exhaust; (ii) a heated, stainless steel sampling line to transport the sampled gas stream to the sampling interface; (iii) a gas conditioning system (i.e. sampling interface), which contains a suction/booster pump, a particulate filter, a moisture removal device, and a flow splitter; (iv) a rack containing one or more species analyzers, each receiving a discrete flow of sampled and conditioned gas from the sampling interface; (v) a data-acquisition system that records and averages the analysis data; (vi) a data modem that periodically transfers the averaged data to the regulatory authority; and sometimes (vii) an opacity meter installed directly in the stack that quantifies particulate matter via the attenuation of light from an installed source, across the stack width.

Chemical species analyzers are often based on optical principles, but some utilize other fundamental analysis principles such as electrochemistry or high-temperature ionization. The types of analyzers that are often installed in a CEMS package are summarized in Figure 14.6 and explained further in the text that follows.

Photon Absorption. ***Spectrophotometry*** relies on the principle of photon flux attenuation (absorption plus scattering) to quantify the concentration of a specific molecule in a gas sample. A light source of a particular wavelength (as selected with a monochromator) is passed through the sample and the attenuation of the light intensity is directly correlated with the number of molecules per volume in the sample cell. Spectrophotometry can be very accurate, as long as the wavelength selected is not absorbed by other species in the gas mixture – which can seriously confound the measurement of the species of interest. Carbon monoxide and carbon dioxide are typically measured by spectrophotometry using infrared wavelengths. A variation of this technique is known as Fourier transform infrared (FTIR) spectroscopy. FTIR is less susceptible to confounding by other molecules, because absorption across a wide range of IR wavelengths is performed, and the absorption pattern is compared to the known absorption spectrum of the molecule of interest, through a

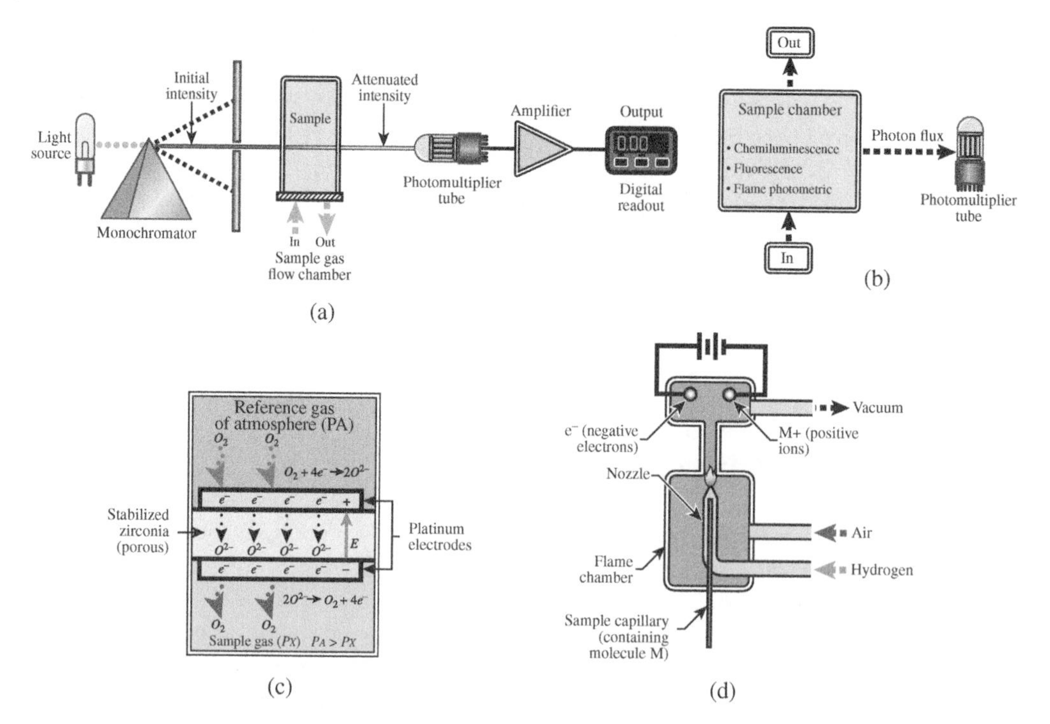

Figure 14.6 Analyzer types. (a) Photon absorption (also called spectrophotometry); (b) photon emission (includes chemiluminescence, fluorescence, and flame photon emission); (c) electrochemical (micro-fuel cell); and (d) flame ionization. *Source:* Martin Thermal Engineering Inc.

Fourier transform mathematical technique, so extraneous absorption peaks produced by nontarget molecules are filtered out of the results.

Photon Emission. Three categories of photon emission analyzers are used to detect and quantify different species in process exhaust gas – chemiluminescence, fluorescence, and flame photometric.

The first category, ***chemiluminescence***, relies on a chemical reaction in the sample chamber that results in the emission of a photon as the bonds are rearranged and product compounds are formed from reactant compounds. Most NO_x analyzers are based on the chemiluminescence principle, where the sample gas is exposed to a small flow of ozone (O_3), which reacts with NO to form NO_2 and gives off a photon of a certain wavelength.

The next category, ***fluorescence***, relies on a molecule's propensity to absorb a photon at one wavelength, and emit a photon of a different wavelength. An analyzer based on fluorescence must have a light source of the desired wavelength to trigger the emission of the fluorescent photon, and the photomultiplier tube is typically placed at right angles to the source of the fluorescence-triggering photons. Sulfur dioxide (SO_2) is often measured by the fluorescence principle.

The last category, ***flame photometric***, is a form of chemiluminescence wherein the molecule of interest is heated to a very high temperature in a hydrogen flame, and the decomposition process results in the emission of photons of a characteristic wavelength. The flux of photons detected at the photomultiplier tube is proportional to the concentration of the target species. Reduced sulfur compounds (e.g. H_2S) and certain organophosphorus compounds are quantified using flame photometric analyzers.

Electrochemical. Analyzers that are based on ***electrochemical*** principles are essentially micro-fuel cells. Oxygen concentration is frequently quantified using an electrochemical sensor, with the "stabilized zirconia" sensor illustrated in Figure 14.6c being one of the most common. The fuel cell analysis technology is based on a principle wherein at least two chemical reactions occur at two electrodes separated by an electrolyte. In the subject case, the electrolyte is an Yttria-stabilized zirconia, which facilitates the transport of O^{2-} ions from the cathode to the anode. At the platinum cathode, oxygen molecules from an adjacent reference gas atmosphere (typically pure O_2) react with electrons transported by a conductor from the anode, forming a pair of O^{2-} ions. The oxygen ions then travel through the electrolyte, where they reach the platinum cathode, give up their excess electrons, and reform O_2 molecules that diffuse into the adjacent sample gas atmosphere. The electrochemical potential developed across the two electrodes is a function of the difference in oxygen molecule concentrations at the reference gas electrode and the sample gas electrode.

Flame Ionization. The ***flame ionization*** principle of gas analysis is similar to the flame photometric principle, but the measured flux is electrons, rather than photons. Flame ionization detectors (FID) are most commonly used to measure hydrocarbons in a sample gas stream. Like the flame photometric detector, the FID relies on a very high-temperature hydrogen air flame to dissociate species that are present in the sample gas. The high temperatures cause electrons (negative charge) to be stripped away from some of the molecules, forming ions (positive charge). The concentration of ions is proportional to the concentration of hydrocarbon molecules that entered the flame chamber. The ion concentration is quantified by measuring current (μA) that flows between a pair of electrodes located near the flame.

14.7 Combustion Safety Instruments

The technology of combustion safety has evolved greatly in the past 50 years, most notably in instruments that detect conditions that represent an explosion risk and trigger a safety shutdown. Among these devices are flame sensors that ensure fuel flow to a burner is permitted only when flame is present; lower flammability limit (LFL; see also Section 4.14) analyzers that ensure flammable gas does not accumulate in a low-temperature oven; various sensors that detect the failure of fuel shut-off valves to close properly, which might permit fuel gas to leak into an inactive furnace. Many of these safety accomplishments have been initiated or reduced to practice by volunteer committees who write consensus safety codes and standards for the National Fire Protection Association (NFPA) and other similar organizations. The primary documents that capture these rules are NFPA 85 Boiler and Combustion Systems Code, NFPA 86 Standard for Ovens and Furnaces, and NFPA 87 Standard for Fluid Heaters. The following paragraphs provide some background details on the instruments themselves. Readers are urged to obtain copies of the NFPA documents for information on their implementation, maintenance, and use.

Flame Sensors. One of the most vital components in a combustion safety system is the flame sensor. The simplest flame sensors are ***temperature*** based – the fuel valve closes if the sensor ***temperature*** is too low. These are typically used on low-cost, tank-style domestic hot water heaters. The sensor permits the fuel gas valve to remain open when its temperature is high, or it initiates valve closure if the temperature drops too low. In domestic water heaters especially, no external source of power is needed. Valve opening is accomplished with the relatively low voltage produced by a thermopile and valve closure occurs by mechanical spring whenever the voltage output drops too low. A thermopile incorporates several two-junction thermocouples connected in series (where alternating junctions reside in the hot zone and the ambient zone). This increases the magnitude of the voltage output.

Another category of ***temperature***-based safety sensor is 1400 ° F bypass interlock, which senses furnace temperature rather than flame temperature. This device presupposes that in a furnace with multiple burners, loss of flame at one burner would not create a dangerous condition if the entire furnace is above the autoignition temperature of the fuel gas, since unburned gas accumulation is not possible if the gas combusts immediately upon entry into the furnace chamber.

After temperature, the next most common flame sensors are ***optical***, which work by sensing photons emitted by the hot flame gases. ***Optical*** flame sensors are divided into three types: infrared, ultraviolet, and multispectral. Infrared flame sensors were the earliest available for industrial combustion systems, but they soon fell out of favor when it was discovered that the sensors were fooled into providing a false-positive indication of flame if they could "see" hot (glowing) refractory lining the furnace walls. Ultraviolet sensors overcame this limitation because hot refractory did not produce a high enough flux of ultraviolet photons (as compared to the flame itself) to trigger a false-positive response. More recently, multispectral flame sensors that respond to light at multiple wavelengths have offered a more robust characterization of the flame's ***optical*** signature. By looking at changes in light emission at different wavelengths, these sensors can provide early warning information to the operator that the flame's stability and/or chemistry is changing – perhaps seconds or minutes prior to actual flame failure.

The third principle of flame detection is gas molecule ***ionization***. In a high-temperature flame, some of the gas molecules lose an electron because of high-speed molecular collisions. Although the electrons quickly recombine with the ***ionized*** molecules, the population of ions and electrons in the flame volume at any one instant makes the gas a relatively good conductor of electricity. By applying a voltage between the burner and a sensing electrode and sensing the current (typically μA microamperes), the loss of flame can be reliably detected when the current falls too low. These devices are often called "flame rods."

LFL Analyzers. LFL analyzers have been employed in drying and curing ovens where flammable solvent vapors might accumulate in the oven during the drying process. While ovens are invariably designed to limit the concentration of flammable solvent vapors therein, accidents involving loss of exhaust outflow or loss of fresh air inflow (collectively loss of ventilation airflow) can lead to an unsafe condition. The purpose of these analyzers is to warn the user when the solvent concentration in the chamber exceeds 25% LFL and shut down the heating system before it reaches 50% LFL. The shutdown safety margin is large by design because the signals generated by these instruments historically lagged the actual oven concentration by several tens of seconds.

To a lesser degree, LFL analyzers have also been used to confirm that the preignition, fresh air purge of an oven, furnace, boiler, or heater has been successfully accomplished and the burner light-off is safe to proceed, because flammable gases are not present at dangerous concentrations.

The operating principles of LFL analyzers are nearly identical to those of hydrocarbon analyzers discussed in Section 14.6. However, unlike exhaust gas analyzers, which do not require rapid responsiveness, LFL analyzers that respond quickly and accurately are the most valued. Principles of operation for LFL analyzers include ***catalytic combustion*** (Tozier et al. 1985), ***flame ionization*** (Wills 1944), and ***flame temperature analysis*** (Schaeffer et al. 2010).

Valve Leakage Prevention. The explosion safeguard devices discussed earlier in the chapter are effective if and only if their interlock signal effectively accomplishes a cessation of fuel flowing into the oven or furnace. This outcome can be thwarted if an automatic fuel shutoff valve fails to close due to sticking or blockage of the flow-controlling element (e.g. gate, ball, globe, and wafer).

To counteract this type of failure, NFPA safety committees typically require that automatic fuel shutoff valves are "proved closed" prior to restarting a burner. A proved closed condition can be accomplished with a ***proof-of-closure*** switch or a ***valve proving system***. The proof-of-closure switch is a proximity switch that detects whether the valve stem has traveled to the "fully closed" position. The valve proving system detects pressure anomalies across one or more automatic shutoff valves and trips an interlock if the anomaly indicates a dangerous leak. Pros and cons associated with these devices are discussed by Carlisle (2002).

14.8 Valves and Actuators

Valves can be sorted into two basic categories – flow valves and specialty valves. Under the flow valve heading are subcategories binary (on–off) valves and modulating (variable flow rate) valves. Specialty valves (e.g. check valves, proof-of-closure valves, cam/poppet valves, and rotary valves) have unique features that go beyond the binary or modulating functionality of the basic flow valves.

Actuators impart axial or rotational motion to a shaft that enters the valve body and modifies the position of the valve control element. Actuators can be powered manually (by hand) or by electrical, pneumatic, or hydraulic sources. Manual actuators are typically provided with the valve body, but mechanical actuators tend to be procured separately and affixed to the valve body upon field installation. Some actuators employ electrical or other forms of energy to open the valve, and when that energy source is removed, a spring will mechanically close the valve. These are often referred to as ***energize-open/spring-close***, or similar nomenclature. The inverse configuration (***energize-close/ spring-open***) is also utilized for certain applications (e.g. where a power failure could introduce a hazardous condition if the valve remained closed).

See Figure 14.7 for illustrations of several valve types and discussions of common actuators therefor.

Binary Flow Valves. ***Binary valves*** are used to start and stop flow but are notoriously bad for variable flow control. Specific valve types under this category include:

- Gate valve – a valve with a flat gate that travels up and down. In theory, gate valves should be able to provide acceptable modulation control, but their downside is that the flow is not linear with position, with perhaps four-fifths of the flow variation occurring in the first 25% of upward gate movement (from closed to open), and the remaining 75% of gate movement only accomplishes one-fifth of the flow modulation.
- Plug valve – a quarter-turn valve with a cylindrical plug having a slot that can be rotated to permit or prevent flow.
- Ball valve – another quarter-turn valve (like a plug valve), but one that uses a spherical rotating element and a cylindrical flow passage.
- Poppet valve – a disk-shaped flow obstruction attached to a stem that can lift the disk out of the way (to permit flow) or cover the sealing surface (to prevent flow).

Modulating Flow Valves. ***Modulating valves*** are used to maintain precise control of flow rate over a range of flows that typically covers a 10 : 1 turndown (or more). Valve types in this category include:

- Butterfly valve – Typically used for large diameter pipes where high flow rates are to be controlled with reasonable precision, and the fluid being controlled is a gas. The butterfly valve does not adhere to a linear relationship between controller output (or angle, 0 – 90°) and flow rate, but is nevertheless, one of the most frequently selected valves in process service.

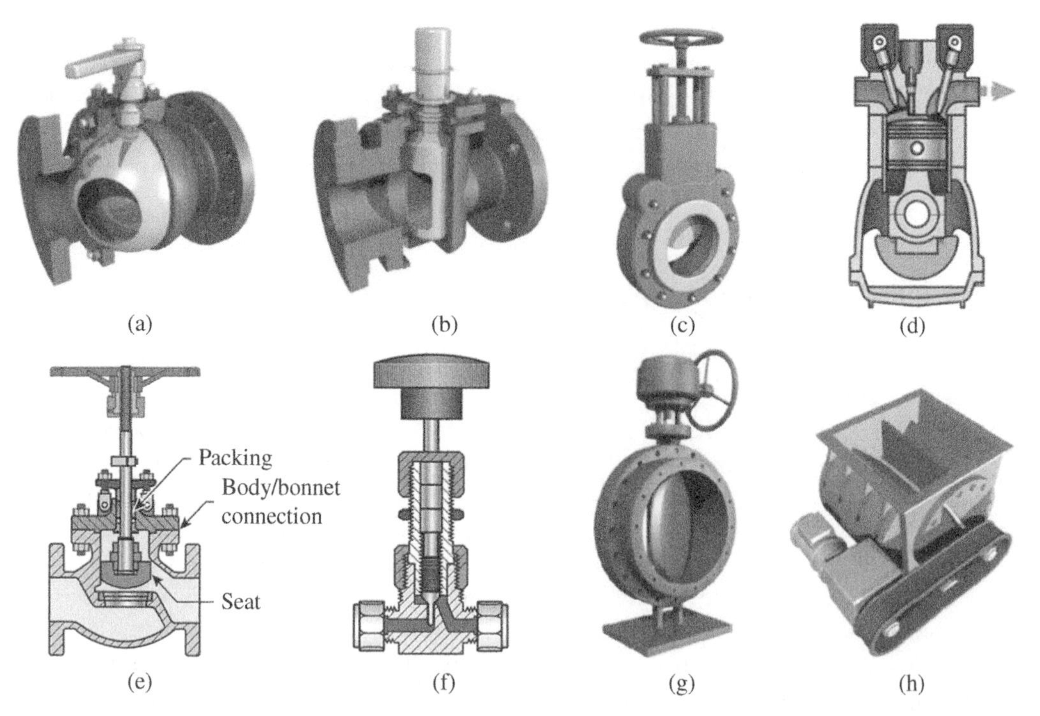

Figure 14.7 Examples of various types of flow valves. The top row shows on–off valves, from left to right: (a) ball valve with quarter-turn hand actuator; (b) plug valve; (c) gate valve with handwheel actuator; and (d) poppet valve for cylinder intake or exit, with cam actuators. The bottom row shows modulating valves, from left to right: (e) globe valve with handwheel actuator; (f) needle valve with manual knob actuator; (g) butterfly damper; and (h) rotary valve for particulate solids with motor and belt-drive actuator. Not shown are hydraulic, pneumatic, and solenoid actuators. *Source:* Martin Thermal Engineering Inc.

- Globe valve – A globe valve is typically used for manual control of flow at moderate flow rates, and more often for liquids than for gases. Globe valves have knobs and shafts that turn several revolutions to control the liquid flow rate.
- Needle valve – A needle valve is a manual valve used for control of very low flows of (predominately) gases.
- Rotary valve – A rotary valve is used for metering pulverized solids into a thermal system for processing and/or combustion. The solids' flow rate is controlled by the speed of the vanes, which look like a sideways revolving door or a steamboat paddle wheel.

Specialty Valves. ***Check valves*** are either spring-loaded or gravity closed. They are intended to add very little resistance to the flow path when flow is moving in the approved direction but to provide a strong seal that prevents essentially all flow in the reverse direction. Often the gravity-based designs include a feature that incorporates downstream pressure to improve the sealing quality whenever reverse flow is possible due to pressure variations. A cutaway of a check valve is shown in Figure 14.8.

The ***pressure relief valve*** (PRV) is a mechanical safety device that is installed on pressure vessels to release a small amount of fluid when other control measures fail to maintain a safe internal pressure. Since the valve is entirely mechanical (components include an inlet nozzle, outlet nozzle, and spring-loaded shaft with moveable sealing surface, stationary sealing surface), it can function even during instances when power is interrupted. A cutaway of a PRV is shown in Figure 14.9.

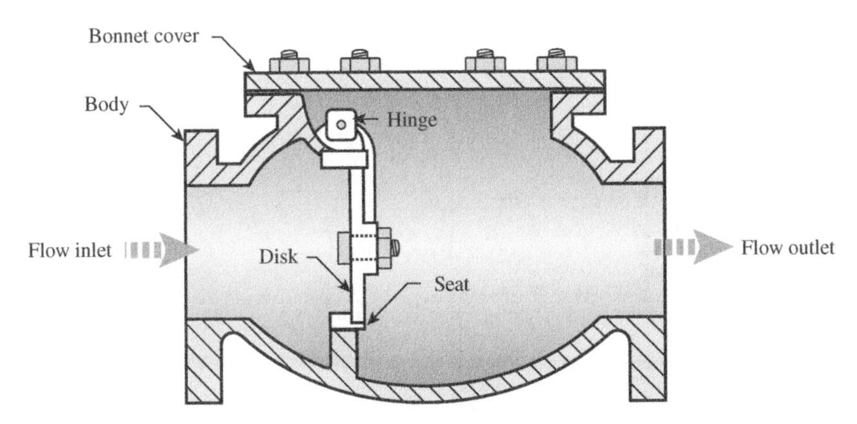

Figure 14.8 Cutaway of swing check valve. *Source:* Martin Thermal Engineering Inc.

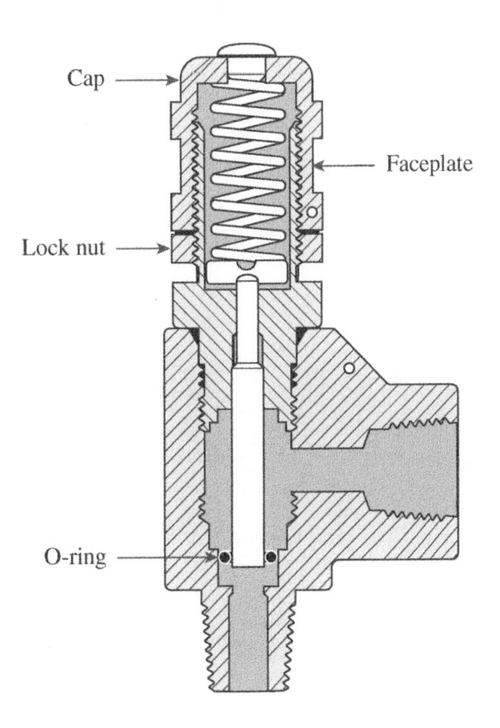

Figure 14.9 Cutaway of pressure relief valve. *Source:* Martin Thermal Engineering Inc.

Pressure regulators are devices that control the downstream pressure of high-pressure sources (e.g. compressed gas cylinders). Pressure regulators contain a diaphragm that is loaded on the underside by the fluid's exit pressure and loaded on the topside by a spring. When the two forces are in equilibrium, the poppet body remains stationary at whatever degree of opening maintains the balance between pressure and flow. A cutaway of a pressure regulator is shown in Figure 14.10, with the dark fluid representing the high pressure (inlet) side, and the light fluid representing the lower pressure (outlet) side.

The ***thermostatic expansion valve*** (TXV) was described briefly in Chapter 10, and the detailed mechanism is addressed here (see Figure 14.11). The valve functions automatically and entirely by mechanical forces. The inset figure at the lower right of the figure shows the opposition of two

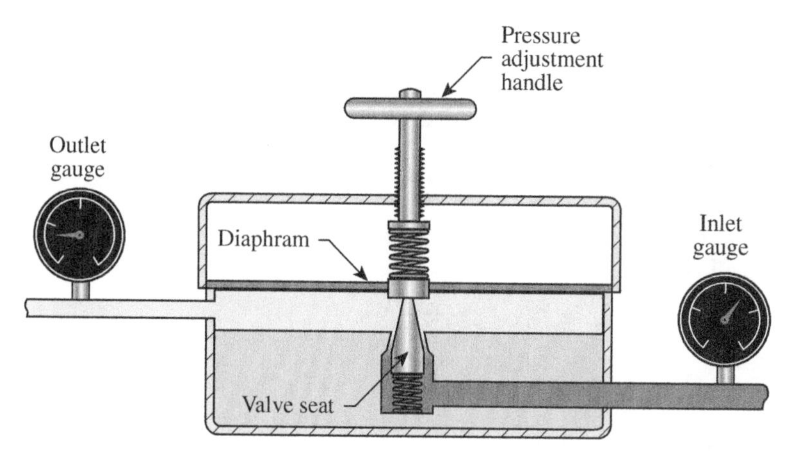

Figure 14.10 Schematic of a pressure regulator. *Source:* Martin Thermal Engineering Inc.

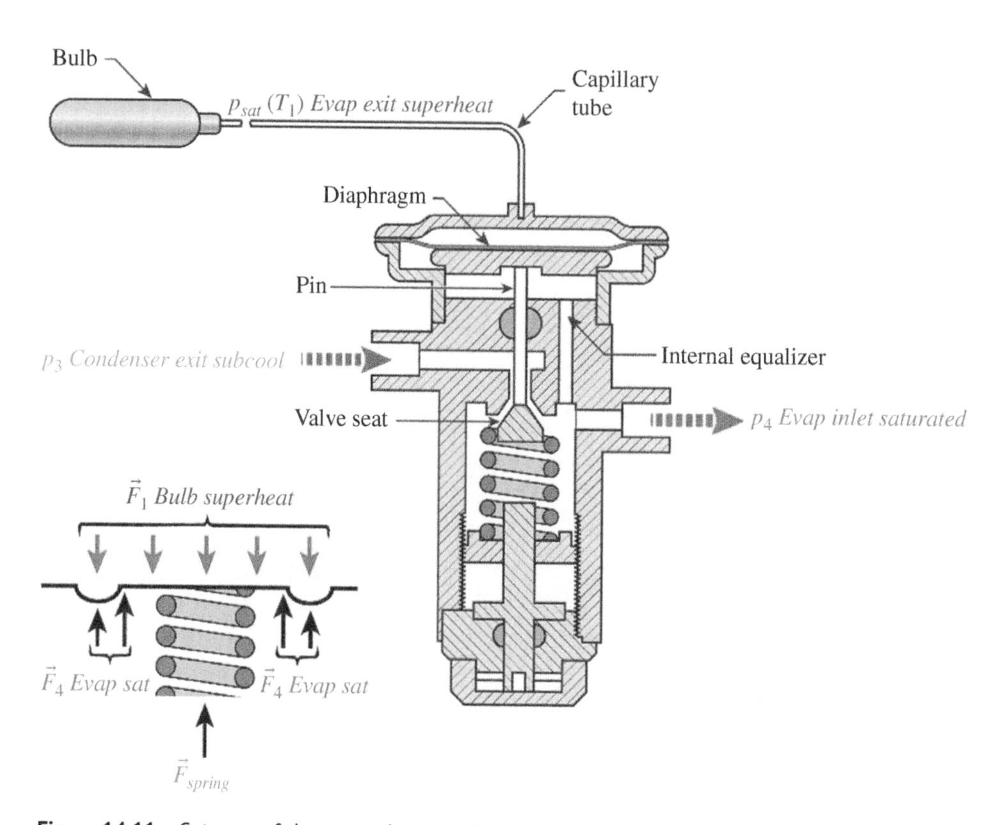

Figure 14.11 Cutaway of thermostatic expansion valve. *Source:* Martin Thermal Engineering Inc.

upward forces (F_{spring} and F_{sat}) and one downward force ($F_{bulb\ superheat}$). When the three forces are balanced, the pin+seat are stationary and the flow rate past the seat remains steady.

Think Stop. Students should note that the downward pressure on the top surface of the diaphragm is generated by the *temperature* of the bulb, which is in contact with the refrigerant just downstream of the evaporator exit (T_1). The fluid inside the bulb and capillary tube is the same as the refrigerant in the system, so its pressure–temperature relationship will match the pressure–temperature relationship of the fluid passing through the valve. The refrigerant passing through the valve delivers a pressure force ($F_{underside} = p_4 A_{diaphragm}$) to the underside of the diaphragm, and the difference between those two fluid forces is applied by the spring. If the spring is adjustable, its setting represents the "degrees of superheat" that the valve is attempting to control.

The first step is to imagine a situation with no superheat (i.e. the pressure in the bulb is equal to the pressure of the fluid entering the evaporator, or $p_1 = p_4$). Because the two fluid forces are equal, the spring force will find its neutral position forcing the valve seat, pin, and diaphragm upward, which chokes off more of the refrigerant flow until the compressor suction draws down the evaporator pressure enough to create a consistent force differential.

During the normal operation, if the temperature at the bulb declines (i.e. due to insufficient superheating of the refrigerant by the warm air passing over outside of the evaporator coil), the diaphragm will move upward a small distance and the pin travel will close the cross-sectional area at the valve seat by a small amount. This action restricts the refrigerant flow and permits greater warming of the refrigerant, which increases the refrigerant's superheat back to the desired level.

Conversely, if the temperature at the bulb increases (i.e. due to a heavier load of warm air flowing across the outside of the coil), the diaphragm will move downward causing the pin and valve seat to open the cross-sectional area by a small amount. This action increases refrigerant flow, which extracts more heat flow out of the ambient air while reducing the refrigerant's superheat back to setpoint value.

14.9 ISA Tag Glossary

Many engineers maintain a P&ID legend page as the first page of their P&ID drawings. The legend page usually contains their tag nomenclature as well as samples of the icons and lines they customarily use.

ISA 5.1 (2009) is the standard resource for instrument nomenclature and symbols, and it is usually preferable for a thermal engineer to use the ISA standard labeling methodology rather than to invent nomenclature that conflicts with ISA. Where the ISA standard is silent (i.e. blank cells in the identification table), an engineer's unique nomenclature may be added without trepidation. Table 14.2 provides highlights of the ISA standard.

Instrument tags usually have two or three letters followed by two or three unique numbers, for example, PI-101 or TSH-503. The first letter is always the initiating process variable that defines the instrument or valve, and the second letter is always the output function. The third letter, a modifier, is optional and usually reserved for switches and alarms.

Some examples of common instrument and valve tags are listed as follows:

- ***FE*** Flow element – an orifice plate, an averaging pitot device, or similar device that responds to the process variable "flow" by creating a pneumatic or other signal that is directly correlated to the flow rate. Often, temperature, pressure, and molar mass of the fluid must be known independently to compute the fluid density which then permits volumetric flow rate to be determined mathematically, through Bernoulli's equation and the ideal gas law.
- ***FSL*** Flow switch low – a switch that responds to the process variable "flow" and closes or opens an electric circuit when the flow drops below the setpoint.
- ***FV*** Flow valve – a valve that responds automatically (via a control loop) to changes in "flow" as measured by a flow element, attempting to maintain flow rate as close as possible to the setpoint value.

Table 14.2 Examples of instrument tag nomenclature for P&ID.

First letter	Initiating variable	Second letter	Output function	Third letter	Modifier
A	Analysis	A	Alarm	C	Closed
B	Burner flame	C	Controller	H	High
F	Flow rate	E	Element	L	Low
H	Manual (hand)	I	Indicator	O	Open
L	Level	S	Switch		
M	Humidity (moisture)	T	Transmitter		
P	Pressure	U	Multifunction		
T	Temperature	V	Valve		
Z	Position	Y	Relay		

Source: ISA 5.1 (2009).

- ***HV*** Hand valve – a valve that is manually opened and closed by a (human) operator.
- ***LAH*** Level alarm high – an alarm that is triggered automatically (via control hardware) when the liquid level in a tank exceeds a "first high" threshold. (The alarm is a warning that permits the operator to consider making adjustments that could bring the system back into the proper range prior to the point where a shutdown interlock is triggered.)
- ***PI*** Pressure indicator – an instrument that is responsive to pressure and produces a visual readout that can be monitored and or logged by an operator but is generally unsuitable for utilization as part of a control loop, because the readout cannot be converted to an electronic signal (i.e. 4 – 20 mA).
- ***PSLL*** Pressure switch low low – an interlock that is triggered automatically (via control hardware) when the pressure falls below a "second low" threshold and sends the system into safety shutdown mode.
- ***TE*** Temperature element – a thermocouple or thermistor that generates an electric signal that is responsive to temperature. The signal often is not strong enough to transmit itself long distances without a power boost.
- ***TT*** Temperature transmitter – a powered device that sees the millivolt signal from a thermocouple and sends out a 4 – 20 mA signal to the process controller that is directly correlated to the temperature calibration of the thermocouple. Transmitters are nearly always installed hand-in-hand with elements.
- ***TY*** Temperature relay – a simple analog-input, digital-output device that receives one or more temperature signals and sends out one or more on–off (or open–closed) control instructions to motors, valve actuators, etc.
- ***ZSC*** Position switch closed – an on–off switch that responds to valve position, essentially proving the valve has reached its fully closed position, to enable purge or some other function requiring a permissive that the valve is indeed closed.

Other tags with different letter combinations should be self-explanatory.

Example 14.1

Suppose a heating system has a burner whose temperature is controlled by modulating the fuel flow rate with a butterfly valve while keeping air flow constant. Suppose the unique identifying number for this control loop is 104. Identify the initiating variable for the control loop and give the four tag numbers for the four components in the control loop:

Answer. Temperature is the initiating variable, and the four tag numbers for the four components in the control loop are *TE*104 → *TT*104 → *TC*104 → *TV*104.

14.10 P&ID Techniques

Creating a useful, beautiful, and accurate P&ID is a process that requires a substantial amount of checking and revising, repeatedly. We have found that students who are enrolled in a thermal systems design class likely will not have sufficient exposure to "good P&ID techniques" in one or two lectures to master the process at a high level. Furthermore, while it may be possible for a single chapter of a textbook such as this to be successful at outlining the overarching principles for creating a good P&ID, it could never be successful at identifying each of the myriad of poor P&ID practices and serious errors that novices will undoubtedly explore and attempt. For some such errors, see Table 14.3.

In light of this, we have found that students can benefit greatly from a lecture, followed by a homework assignment, followed by detailed instructor feedback on the homework. It is recommended that the instructor alone assumes responsibility for grading the DRAFT P&ID submissions by the students – a grader or teaching assistant (TA) rarely has enough experience to point out all the erroneous content and provide examples of appropriate alternatives.

Table 14.3 Common P&ID errors.

Common P&ID errors
Stream tags do not belong on a P&ID and neither does a stream table. They clutter up the drawing and do not fit with the overall purpose.
Failing to keep loop numbers constant for all four devices in a loop (sensor element, transmitter, controller, and final element).
Wrong type of blower for the application (positive displacement blowers can generate dangerously high pressures and should not be used unless absolutely needed).
Shell and tube heat exchanger used with hot gas on the inside of the tubes and water on the outside – possibility of thermal expansion problems.
Lines going in and out of the wrong ports on compressors and turbines.
Two lines going into a piece of equipment and none coming out, or vice versa.
Using TC instead of TY for a thermostat, which gives on–off control signals only.
Connecting two sensed variables into a single loop controller (preferred is TE-TT-TC-TV, for example). For a computational algorithm that looks at multiple inputs to govern multiple outlets, use a multifunction device (e.g. TU).

(Continued)

Table 14.3 (Continued)

Common P&ID errors
Adding a blower or compressor for natural gas when the source pressure is probably high enough for most burners.
Improper use of a four-way valve when a pair of three-way valves is more appropriate.
Mismatch of actuator with the type of valve (e.g. solenoid and butterfly).
For line specs, pipe material or diameter not consistent with process fluid and flow rates.
Placing two (modulating) control valves into a single line. For example, if fuel gas is being modulated by TV101 in response to TC101 and TE101, but the same line is also equipped with PV102 that is looped together with PC102 and PE102, the two valves may try to fight each other if the pressure (PE102) is too low and the temperature (TE101) is too high.
Creating a feedback loop with a temperature sensor, element, controller, and valve, but the controller and valve are labeled *F* for flow rather than *T* for temperature. While it is intuitive to think of all valves as modulating "flow" and therefore deserving of the first letter *F*, the correct labeling procedure is that all instruments, controllers, and valves in a single loop should have the same first letter and same loop number, but each individual component has a different second letter.

Source: Martin Thermal Engineering Inc.

14.11 Homework Problems

It is recommended that students utilize a commercially available computer-aided design (CAD) software package to produce these schematics. Some commercially available software tools are cited in the reference list at the end of Chapter 5. Instructors may deem manually drawn schematics to be acceptable, but the learning outcomes will be stronger with a CAD software package.

14.1 Referring to HW Problem 5.1, draw a DRAFT piping and instrumentation diagram for a ***lumber kiln*** that is used to dry batches of green lumber. Add valves, instruments, and control loops for this process, but do not include stream tags or a stream table.

14.2 Referring to HW Problem 5.2, draw a DRAFT piping and instrumentation diagram for a ***landfill gas thermal oxidizer***. Add valves, instruments, and control loops for this process, but do not include stream tags or a stream table.

14.3 Referring to HW Problem 5.3, draw a DRAFT piping and instrumentation diagram for a ***split-system, heat-pump*** that provides forced-air heating in the winter season and forced-air cooling in the summer season to a two-story (i.e. two-zone) multi-student ***residence hall***. Add valves, instruments, and control loops for this process, but do not include stream tags or a stream table.

14.4 Referring to HW Problem 5.4, draw a DRAFT piping and instrumentation diagram for a natural gas-fired, ***swimming pool water heating and cooling*** system. Add valves, instruments, and control loops for this process, but do not include stream tags or a stream table.

14.5 Referring to HW Problem 5.5, draw a DRAFT piping and instrumentation diagram for a ***refrigeration system*** capable of freezing and maintaining an appropriate subfreezing temperature of an ***ice rink in an indoor arena***. Add valves, instruments, and control loops for this process, but do not include stream tags or a stream table.

14.6 Referring to HW Problem 5.6, draw a DRAFT piping and instrumentation diagram for a gas-fired, evaporative-distillation-style ***water desalination process***. Add valves, instruments, and control loops for this process, but do not include stream tags or a stream table.

14.7 Referring to HW Problem 5.7, draw a DRAFT piping and instrumentation diagram for a continuous ***grain-drying oven***, including a horizontal, air-permeable conveyor belt with two heating zones. Add valves, instruments, and control loops for this process, but do not include stream tags or a stream table.

14.8 Referring to HW Problem 5.8, draw a DRAFT piping and instrumentation diagram for a ***split air-conditioning system*** for a three-zone (i.e. three-story), 24-inmate ***county jail building***. Add valves, instruments, and control loops for this process, but do not include stream tags or a stream table.

14.9 Referring to HW Problem 5.9, draw a DRAFT piping and instrumentation diagram for a ***hot air balloon*** including pressurized LP gas cylinder, burner, balloon envelope, exhaust valve, and passenger basket for a payload of 12 adults. Add valves, instruments, and control loops for this process, but do not include stream tags or a stream table.

14.10 Referring to HW Problem 5.10, draw a DRAFT piping and instrumentation diagram for an ***exothermic gas generator for a heat-treating furnace***. Add valves, instruments, and control loops for this process, but do not include stream tags or a stream table.

14.11 Referring to HW Problem 5.11, draw a DRAFT piping and instrumentation diagram for a ***tenter-frame fabric drying oven*** that removes moisture from a moving web. Add valves, instruments, and control loops for this process, but do not include stream tags or a stream table.

14.12 Referring to HW Problem 5.12, draw a DRAFT piping and instrumentation diagram for a ***commercial espresso machine***. Add valves, instruments, and control loops for this process, but do not include stream tags or a stream table.

14.13 Referring to HW Problem 5.13, draw a DRAFT piping and instrumentation diagram for a ***chemical engineer's refrigerator***. Add valves, instruments, and control loops for this process, but do not include stream tags or a stream table.

14.14 Referring to HW Problem 5.14, draw a DRAFT piping and instrumentation diagram for a ***passenger aircraft biohazard destruction system*** that conveys droplets and biohazard particles into a heated, metal foam while permitting air to pass through. Add valves, instruments, and control loops for this process, but do not include stream tags or a stream table.

Cited References

Bean, H.S., Ed.; (1971); *Fluid Meters, Their Theory and Application* – **6**; New York: American Society of Mechanical Engineers; https://www.asme.org/publications-submissions/books/find-book/fluid-meters-theory-application-sixth-edition (accessed 25 March 2021).

Carlisle, K. (2002). 10 Tips on valve-proving systems for industrial heating. Process Heating (1 August 2002). https://www.process-heating.com/articles/88401-tips-on-valve-proving-systems-for-industrial-heating (accessed 30 March 2021).

Crane; (1988); *Flow of Fluids Through Valves, Fittings, and Pipe*; Technical Paper #410; Joliet, IL: Crane Company.

ISA 5.1; (2009); *Instrumentation Symbols and Identification*; Research Triangle Park: International Society of Automation; https://www.isa.org/products/ansi-isa-5-1-2009-instrumentation-symbols-and-iden.

Schaeffer, C.G., Schaeffer, M.J., and Patel, P. (2010). Gas analyzer for measuring the flammability of mixtures of combustible gases and oxygen. U.S. Patent 7,704,748 (issued 27 April 2010); Assigned to Control Instruments Corporation.

Tozier, J.E., Anouchi, A., and Critchlow, R. (1985). Constant temperature catalytic gas detection instrument. U.S. Patent 4,541,988 (issued 17 September 1985); Assigned to Bacharach Instrument Company.

Wills, W.P. (1944). Control apparatus. U.S. Patent 2,352,143 (issued 20 June 1944); Assigned to the Brown Instrument Company.

15

Control of Thermal Systems

The field of thermal system control can be divided into three main areas: process control, safety control, and sequencing control. Switches, relays, and discrete (microprocessor) controllers are suitable for very simple thermal systems, but many systems require hardware and software with greater capability and complexity, such as programmable logic controllers, ladder logic, and distributed control networks. The treatment of this topic in this chapter is necessarily brief and should be supplemented by additional in-depth study for students wishing to specialize in control methodology and control system architecture. Readers are referred to Seborg et al. (2016) or Dunn (2005) for more thorough treatments of the subject.

15.1 Control Nomenclature

The primary purpose of process control is to ensure all the equipment comprising the thermal system is responding to changes in the environment or process streams while maintaining appropriate parameters (e.g. p, T, and $\dot{m}$) within the ranges anticipated by the system designer.

A ***control loop*** is the combination of components (hardware) and instructions (software) that work together to accomplish the desired control objective. The parameter (e.g. temperature) whose control is sought for a given control loop is called the ***process variable*** (*PV*). The desired value for the process variable is called the ***setpoint*** (*SP*). The difference between the process variable and the setpoint is called the ***error*** (ϵ).

In a home, comfort control is aimed at one variable – indoor temperature and the strategy is often to employ ***thermostatic control***. (Indoor humidity may also be managed concurrently with temperature.) Industrial process control is usually thought of as continuous or ***feedback control***, wherein a process variable *PV* is monitored and instructions are given to a final control element (e.g. valve stem position actuator) to minimize the error ϵ between the *PV* and the *SP* by making small changes to the partially open valve position.

For simplicity, this chapter focuses on the temperature control, but other process variables (e.g. pressure, mass flow, and level) may be appropriate or necessary to implement for a properly functioning thermal system.

15.2 Thermostatic Control

Thermostatic control (with subcategories ***on–off***, ***hysteresis***, and ***deadband***) is by far the most common comfort control strategy. Thermostatic control is simple to design, implement, and maintain.

Thermal Systems Design: Fundamentals and Projects, Second Edition. Richard J. Martin.

Companion website: www.wiley.com\go\Martin\ThermalSystemsDesign2

Characteristically, only one environmental variable (temperature, T) is measured, and just one or two thermal functions (heating and/or cooling equipment) are instructed to turn on or off based on the status of the environmental variable relative to the thermostatic controller setpoint(s). If control of a process variable is sought for multiple physical locations, the control should be set up in ***zones***, where each zone has its own sensor and actuator.

Suppose that during the heating season, the outflow of indoor heat (due to losses through walls and cold air infiltration) is steady (e.g. $\dot{Q}_{loss} = -10\ kW$), and that the heating unit is capable of a heat input rate that is significantly higher (e.g. $\dot{Q}_{gain} = +25\ kW$), the net heat input rate will vary between two different values when the heater is on and when it is off:

$$\dot{Q}_{net,in,htr:\boldsymbol{on}} = \dot{Q}_{gain} + \dot{Q}_{loss} = +25\ kW - 10\ kW = +15\ kW$$

$$\dot{Q}_{net,in,htr:\boldsymbol{off}} = \dot{Q}_{gain} + \dot{Q}_{loss} = +0\ kW - 10\ kW = -10\ kW$$

Thus, the slope of the temperature–time curve will vary from negative to positive, depending on whether the heating unit is off or on. See Figure 15.1 for an example of the time–temperature profile for the heated space described earlier.

The amplitude of the sawtooth waveform for temperature is dependent on (i) the values of the net heat input rates during the off-cycle and the on-cycle, (ii) the thermal mass of the room and sensor plus other nonthermal delays (e.g. electric motor start-up, blower rotation speed, and combustion safeguard timers), and (iii) the time duration the heating unit is on versus the time duration it is off.

On–Off. Pure on–off control is neither desirable nor fully achievable in practice, but it does represent a simplified control strategy that is worthy of study. In the theoretical limit, on–off control is achieved with a massless thermal system (i.e. no storage of thermal energy in the air, furnishings, or walls, and no thermal energy storage in the temperature sensor). For this extreme case (assuming only active heating is required, no active cooling), only the nonthermal time delays (electrical, mechanical, and other) affect the sawtooth wave profile, making its on–off frequency very short and its amplitude very small, which implies the root-mean-square (RMS) error (ϵ_{rms}) for temperature is tiny.

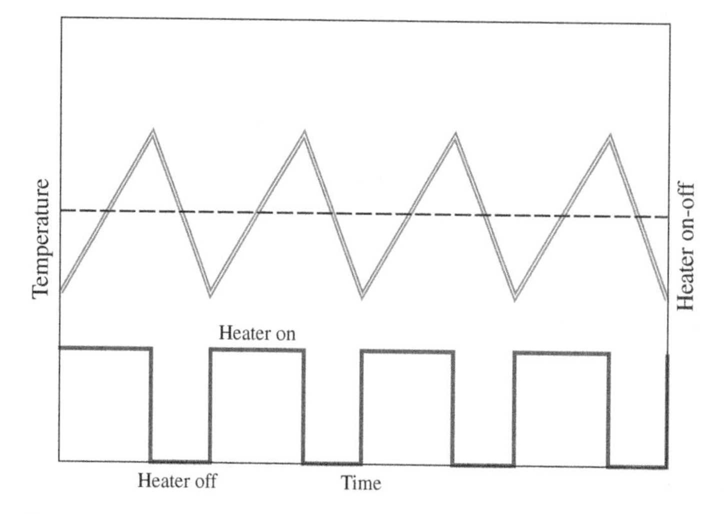

Figure 15.1 Thermostatic control trend history. On–off history (single line square wave) versus temperature profile (double line sawtooth wave with different slopes for heater on and heater off). The dashed line is the temperature setpoint. *Source:* Martin Thermal Engineering Inc.

RMS error is found by recording the error at a prescribed frequency (e.g. every second or every 10 seconds) for a duration that involves a substantial number (10 or more is preferred) of changes in state of the heater. Like the sawtooth of Figure 15.1, some of the error measurements will be negative and some will be positive. To obtain an estimate of the magnitude of the error (without regard for the sign of the error), the individual error values are squared, summed, and the sum square rooted. If we stopped here, the result would be called root sum square (RSS) because the number of error measurements has not been factored out. To obtain RMS, we must take the sum of squares value and divide it by the number of reported error values that were summed – then apply the square root.

Think Stop. It is probably obvious that the squaring of the errors is essential regardless of whether the RSS or RMS is sought. Due to the thermostatic ***on–off*** response, the process variable will spend approximately equal lengths of time above and below the setpoint. This behavior will nearly always cause the sum and average of all the errors to be very close to zero, regardless of whether the swings are large or small. Equation (15.1) shows the proper way to compute the amplitude of the swings is to compute the RMS of the error at some regular time interval:

$$\epsilon_{rms} = \sqrt{\frac{1}{n}\sum_{i}^{n}(T_{PV} - T_{SP})_i^2} \tag{15.1}$$

where ϵ_{rms} is the root-mean-square error, i is the index for temperature measurements recorded, n is the total number of measurements, T_{pv} is measured process variable temperature, and T_{sp} is the setpoint temperature. The sequence of the letters in the acronyms is a mnemonic for how to arrange the operators in Equation (15.1). *Root* is first, which means it is the operator we encounter first in the expression (but it is actually the last operation performed). *Mean* is second, which signifies the averaging is done inside the square root, but outside the summation sign. *Squared* is third in the displayed sequence, but it ends up being the first operation performed after the recording of all the errors.

Due to their simplicity, ***on–off*** systems have essentially no ***hysteresis*** – the temperature setpoint triggers an immediate change of state (off-to-on or on-to-off) as soon as the sensor's temperature value passes the setpoint (either going up or going down). The dashed line in Figure 15.1 indicates the setpoint, and the maximum to minimum amplitude variation may appear large, but for true ***on–off*** control the amplitudes are quite small in magnitude because the nonthermal time delays are typically very short.

Think Stop. Rapid on–off transitions may cause early failure of some types of electromechanical equipment (e.g. solenoid valves, motor contactors). ***On–off*** is rarely the best control strategy for a thermal system.

Hysteresis. ***Hysteresis*** control is nearly the same as ***on–off*** control, except that an additional delay associated with triggering the on/off switching is added to the system by the control hardware/software. Essentially, hysteresis is a control outcome whereby the system triggers on at a different temperature than it triggers off. On average, occupants would experience temperatures that deviated $\pm\epsilon_{rms}$ from the setpoint.

One way to achieve hysteresis in a thermal control system is if the thermal sensor is relatively massive. The mass of the temperature sensor causes a time lag between when the atmosphere's temperature reaches the setpoint and when the sensor reaches it. An outcome of hysteresis is that the temperature amplitude (high–low temperature cycle difference) is larger than on–off control without hysteresis. This may help alleviate early hardware failures because hysteresis reduces the cycling frequency. However, with too much hysteresis, comfort may be sacrificed – occupants may feel unwanted high temperatures before the sensor shuts off the heating unit, or occupants may feel unwanted low temperatures before the sensor turns the heating unit back on again.

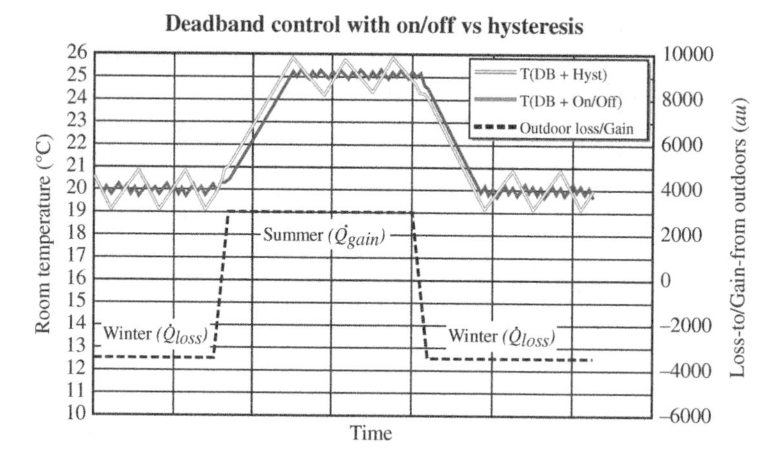

Figure 15.2 Deadband control with either on/off or hysteresis setpoint interpretation. The winter setpoint is $T_{heat,\ SP}$ = 20 °C and the summer setpoint is $T_{cool,\ SP}$ = 25 °C, which creates a deadband between the two temperatures, where neither heating nor cooling occurs. The hysteresis values that affect the setpoint interpretation are δT_{rising} = + 0.6 °C and $\delta T_{falling}$ = − 0.7 °C. Together these incremental setpoint broadeners cause the frequency of appliance start-up and shutdown to be significantly reduced compared to on–off interpretation. *Source:* Martin Thermal Engineering Inc.

Deadband. Deadband control is sometimes confused with hysteresis control. Specifically, deadband control is only applicable for systems that have both active heating and active cooling capabilities. The so-called deadband is the temperature range where neither active cooling nor active heating is switched on. When the temperature exceeds the maximum setpoint, the cooling unit is turned on. When the temperature falls below the minimum setpoint, the heating unit is turned on. Deadband thermostat controllers are designed to ensure the temperature difference between the minimum and maximum setpoints accomplishes two goals: (i) heating and cooling functions are never operating simultaneously and (ii) occupant comfort is not sacrificed at the upper or lower bounds of the deadband.

A common deadband value for mechanical thermostat controllers is $\Delta T_{deadband} = 5\ °C = 9\ °F$. For example, suppose the minimum setpoint is $T_{min,\ SP} = 20\ °C$ (68 °F), and the maximum setpoint is $T_{max,\ SP} = 25\ °C$ (77 °F). The heating unit would be turned on when the room sensor falls below $T_{min,\ SP}$, the cooling unit would be turned on when the room sensor rises above $T_{max,\ SP}$, and neither unit would be on when the room sensor is in the "deadband" between the two setpoints. Many thermostatic controllers sold since the 1980s are designed such that the user is unable to change the deadband designation to be smaller than 5 °C, but larger deadband values are available for the user to choose.

The phenomena of hysteresis and deadband control are shown in Figure 15.2.

15.3 PID Control

Proportional, integral, derivative (*PID*) is the most common type of ***feedback control*** used for process control today. Unlike ***thermostatic*** control, the heating (or cooling) system is *always on* for *PID* control. *PID* control is more costly than thermostatic control, but it is also more capable of meeting quality and productivity goals.

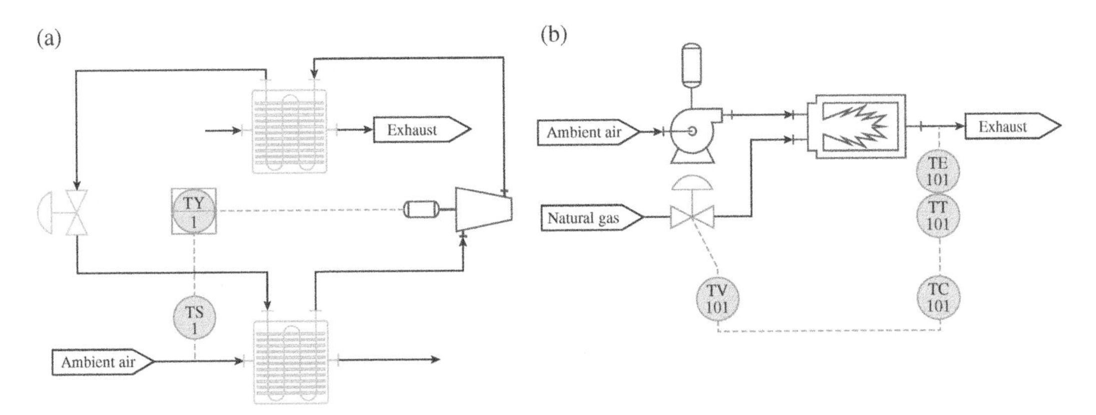

Figure 15.3 (a) Thermostatic control configuration and (b) PID control loop. *Source:* The author has adapted certain images/illustrations from AutoCAD® P&ID 2016 software and used them with the permission of Autodesk, Inc. Autodesk and AutoCAD are registered trademarks of Autodesk, Inc., in the United States and other countries.

In addition to the terminology introduced for thermostatic control (i.e. ***Process Variable*** [*PV*], ***Setpoint*** [*SP*], and ***Error*** [ϵ]), a fourth variable ***controller output*** (*CO*) is necessary to govern the status of the final control element, as it modulates between its minimum (0 % *CO*) and maximum (100 % *CO*) settings.

Figure 15.3 contains a thermostatic control schematic and a *PID* control schematic. For thermostatic control (Figure 15.3a), two functions are required – a temperature switch (*TS*) based on ambient air temperature, and a temperature relay (*TY*) that switches motors on and off based on the status of the temperature switch. In a home heating system, the *TS* is the "thermostat" that is located centrally and instructs the heater controller (the *TY* that is located at the furnace) when it is time to turn on the blower, open the gas valve, and energize the flame igniter.

For PID control (Figure 15.3b), four components are included in the loop – a sensor (*TE*), a transmitter (*TT*), a controller (*TC*), and a final control element (*TV*). In an industrial oven, the *TE* could be a thermocouple inserted through the wall of the oven with the hot junction at a representative place in the oven's atmosphere. The *TT* would be affixed to the head of the thermocouple protection tube (just outside the exterior skin of the furnace but only a foot or so away from the thermocouple junction). The *TC*, which receives an electric signal from the *TT*, could be situated in a local control panel (where the oven operator could view the digital display to observe the temperature). And the *TV*, which receives a different signal from the *TC*, would be positioned in the gas line to regulate the flow of gas in response to the *TC*'s PID control algorithm to maintain the oven's curing temperature as close as possible to the setpoint.

Typically, in a PID loop, the input signal received by the *TC* from the *TT* is an electric current (i) in the range $4\ mA \leq \boldsymbol{i_{PV}} \leq 20\ mA$ that is directly correlated to the value of the *PV* (e.g. temperature) in engineering units (e.g. $°C$). If the sensor element is nonlinear (e.g. an orifice plate that produces a differential pressure signal that is proportional to volumetric flow rate *squared*), a linearization calculation can be carried out either in the *TC* or in the *TT* that converts the raw element process signal (e.g. Δp for an orifice plate) to an electronic signal that can be understood by the *TC*.

Separately, the *TC* sends an output signal *CO* electronically (again, typically $4\ mA \leq \boldsymbol{i_{CO}} \leq 20\ mA$) to the actuator that responds by varying the position, angle, or rotational speed of the final control element *TV*. The correlation for this signal is that 4 mA represents the 0% position of the final control element and 20 mA represents the 100% position of the final control element. Occasionally, an actuator will require an additional device that responds like an inverse transmitter (i.e. electric current-to-pneumatic pressure $[i - \text{to} - p]$ signal generator). The need for this extra "signal translator" depends on the input and output specifications of the actuator and valve.

The expression for how a PID control device mathematically responds to variations in the process variable is given in Equation (15.2), with the terms defined in the following list.

$$CO_{act} = CO_{SP} + \underbrace{K_p\epsilon}_{proportional} + \underbrace{K_i\int\epsilon\cdot dt}_{integral} + \underbrace{K_d\frac{d\varepsilon}{dt}}_{derivative} \tag{15.2}$$

- CO_{act}: Controller output (signal actually transmitted to the TV that varies from 0 to 100%)
- $CO_{SP}(nom)$: Controller output (nominal), also called offset (a predetermined controller output value associated on a one-to-one basis with the prevailing SP value, when conditions are nominal, i.e. when error is zero)
- PV: Process variable (engineering units, e.g. °C, K, $psig$, bar_{abs}, and lb_m/h)
- K_p: Proportional control coefficient
- K_i: Integral control coefficient
- K_d: Derivative control coefficient
- t: Time
- ϵ: Error, $\epsilon = (SP - PV)$ computed in units of PV

Think Stop. From a common arithmetic perspective, this definition of error may seem counter intuitive. In noncontrol applications, we intuitively expect a "positive error" to correspond to the current PV (actual) value being larger than the current SP (desired) value, but in control applications, the opposite is true. This convention is established to permit the control coefficients (K_p, K_i, and K_d) to always be expressed with nonnegative values.

For example, consider a linear flow valve, which is a valve whose flow coefficient (C_v) is directly proportional to the valve's actuator position (0–100%). In other words, for identical upstream conditions (i.e. T, p, and *composition*) the actual flow rate delivered for an actuator position of 50% is exactly one-half of the actual flow rate delivered for an actuator position of 100%.

Suppose this linear flow valve (FV) is part of a feedback control loop where output flow rate is measured with a flow instrument (FE/FT) that sends its process variable (PV) signal to a flow controller (FC) that sends its controller output signal (CO) to the flow valve (FV) in response to the difference between the setpoint (SP) and process variable (PV). If the upstream conditions do not change, the actual controller output signal (CO_{act}) should be identical to the nominal controller output signal (CO_{SP}) because the error ($\epsilon = SP - PV$) would be zero.

Now, suppose the flow valve is subjected to a decrease in upstream fluid pressure, all else equal. This will cause the actual flow rate through the valve to drop and will cause the error ($\epsilon = SP - PV$) to become positive (which will begin to drive the CO_{act} to a value greater than the CO_{SP}). This action is the result of the feedback control loop's goal of maintaining a constant flow rate, even under varying upstream conditions.

From a practical perspective, whenever the three error-related terms in Equation (15.2) sum up to a positive value, the actual controller output value (CO_{act}) will be larger than the nominal controller output value (CO_{SP}).

To recap, consider a temperature valve (TV101) on a natural gas line, whose actuator position (0–100% open) varies based on instructions sent from a temperature controller (TC101), whose mission is to maintain the reading of a furnace temperature instrument (*TE*101) constant at the setpoint value (e.g. $T_{SP} = 1000\ °F$).

Now, suppose a cold object is admitted to the furnace, and the temperature (*TE*101) signal declines to $T_{act} = 900\ °F$. This causes the error to become positive, because $T_{SP} > T_{act}$. Based on the proportional and integral ($K_p\epsilon + K_i \int \epsilon \cdot dt$) signals alone, this positive error would cause the gas valve (*TV*101) to open up a little more (higher CO_{act}), thus adding more fuel to the burner, increasing the rate of heat release in the furnace, and ultimately causing the furnace temperature to rise.

The same is true of the derivative coefficient, K_d, but the logic is less straightforward. Consider a situation where the derivative ($d\varepsilon/dt$) is positive and large. This could occur when a negative error ($SP - PV < 0$) is becoming less negative so rapidly that the *PV* is likely to climb past the *SP* with such momentum that a large overshoot occurs. In such a case, the derivative term ($K_d \cdot d\varepsilon/dt$) would be large and positive, and it would act to reduce the actual controller output (CO_{act}) to a value that is (temporarily) less than the nominal controller output (CO_{SP}).

To summarize, ***proportional action responds moment-by-moment*** and moves the final control element to a more appropriate position to reduce error; ***integral action responds slowly to past errors*** and tunes the final control elements position to eliminate small biases that accumulate over time; ***derivative action responds quickly to future errors*** by anticipating overshooting and negating temporarily excessive P or I response.

15.4 Safety Controls and Interlocks

Safety control is a form of digital control that helps prevent dangerous excursions of process variables beyond designated safe minimum and maximum thresholds. The two primary forms of safety controls are ***shutdown interlocks*** and ***permissive interlocks***. An ***interlock*** is a control feature that creates a mutual dependency between two states – "interlocking" them like puzzle pieces that fit together. A good example of an engineering interlock is a food processor that receives large chunks of food (e.g. cheese) and shreds them with a perforated blade that spins at a high rate. To prevent people from sticking their fingers into the area with the spinning blade, the processor has an electric switch built into the bowl-lid assembly that (i) prevents the blade motor from being energized when the lid is not securely in place (***permissive***) and (ii) shuts off power to the motor if the lid is rotated to an unsecure position (***shutdown***).

Shutdown Interlocks. Shutdown interlocks in thermal systems function to bring an operating system to a ***safe state*** if one or more process variables exceed their design limits. The term ***safe state*** generally implies that certain sources of energy (e.g. electricity, fuel flow, raw material feed, and sometimes air flow) are shut down or "de-energized." Examples of shutdown interlocks include: *TSHH* (e.g. high-high combustion temperature), *PSLL* (e.g. low-low gas pressure), and *LSHH* (e.g. high-high tank level). If the *PV* is simply high (*H*) or low (*L*) (i.e. not *HH* or *LL*), the system designer usually considers that threshold as a warning that warrants an ***alarm***, rather than a ***shutdown interlock*** response. A critical temperature sensor might have a *TAH* with a setpoint of 950 °*F* and a *TSHH* with a setpoint of 1050 °*F*. Entering the ***alarm high*** condition ostensibly gives the operator a bit of time to manually intervene to halt the temperature runaway before it reaches the dangerous level that necessitates the automatic actuation of the ***high-high shutdown interlock***.

Permissive Interlocks. Permissive interlocks function to prevent a system from initiating operation if one or more process variables exceed their design limits. For example, the burner start-up sequence is not permitted to begin until all designated process variables are in their "safe range." Often, the physical safeguards (often switches that change state when a process variable exceeds a designated limit) that function as shutdown interlocks are also called upon to function as permissive interlocks. If a process variable's high or low state creates enough concern to initiate a safety shutdown, it also should prevent the initiation of a normal start-up.

Shutdown and permissive interlocks are either ***digital*** or ***pseudodigital*** by nature. A ***digital*** safeguard is a local device (i.e. one mounted on the process equipment) that responds to input from a primary element (e.g. thermocouple or pressure transducer) and actuates a binary switch (*open circuit, closed circuit*) whenever the process variable exceeds the designated threshold. If the actuation performed is the *opening* of a switch (i.e. to de-energize a control circuit, or to switch a control bit to zero), the switch is said to be ***normally closed*** (NC). If the actuation performed is the closing of a switch (i.e. to energize a control circuit or switch a control bit to one), the switch is said to be ***normally open*** (NO).

By contrast, a ***pseudodigital*** safeguard is executed by software coded into the process control hardware. The safety logic instructions look at all the relevant analog process variables (e.g. furnace temperature, gas line pressure, tank level, and flame sensor current) and depending on whether the values are within a safe range or outside that range, digital logic is executed that results in a safe outcome. The term ***pseudodigital*** implies that the energize/de-energize action is not hardwired to the actual process sensor, but rather, is a logical outcome of the analog inputs coming from the process that are translated to digital outputs by the software, in a location remote from the process sensors.

The traditional method for implementing safety controls was to employ ***digital*** safeguards in a "daisy-chain" format, wherein the individual switches are wired in series, and the safety control circuit is de-energized until/unless all the individual safety switches are in the closed position. In this way, permissive interlocking is accomplished by de-energizing the ***start-up*** function until *all* process variables are within their safe range, and shutdown interlocking is accomplished by de-energizing the ***run*** function as soon as *any* individual process variable migrates outside its safe range. This method is still quite popular for smaller, less complicated processes and systems.

Recently, with the advent of the safety instrumented systems (SIS) philosophy of safety control (IEC 61511 2016), larger and more complex processes tend to undergo a risk-based analysis to assess how safeguards should be implemented. The SIS approach gives the safety design team more flexibility with their design choices, but often precludes the implementation of a simple "daisy-chain" hardware solution, because such systems typically cannot perform *self-checking* and other *watchdog* tasks that are required from SIS installations. Safety standards and the SIS approach are discussed further in Section 16.4.

15.5 Sequencing Control

Sequencing control is another form of digital control that serves to automate transitions between different modes of operation for a thermal system. A good example of this is a conveyor oven, the operation of which would normally include all the following stages:

- Stage 1 – Power-up. The control system is energized and begins checking out its own safety functions. The status of permissive interlocks is also addressed during this stage. The operator may need to manually reset certain alarms or switches to allow the system to exit this stage and enter the next stage.

- Stage 2 – Preignition purge. When all the power-up permissives are satisfied, the sequencing control system can be advanced (automatically or manually) to the preignition purge step. Purge is a safety requirement for all industrial combustion systems, and the action is to provide four or five volumes of fresh air flow into and out of the furnace plus interconnected ducts and equipment to ensure any residual fuel gas or flammable vapor is removed before igniters are energized. Additional permissives such as valve position switches and blower speed thresholds may be employed at the beginning of this stage, prior to initiation of the purge timer.
- Stage 3 – Trial for ignition. Upon successful timeout of the purge timer, the burner ignition system is ready to be (manually or automatically) initiated. At this stage, the igniter is energized, and the pilot/main fuel shutoff valves are opened. The control system begins to look for an acceptable flame sensor signal, and if verified, the fuel control valve(s) will remain open, permitting the burner to heat up the system. If no flame is sensed within the "trial-for-ignition" period (typically 15 seconds or less), the system enters safety shutdown mode, which closes the fuel shutoff valves and waits for approval to enter purge mode again.
- Stage 4 – Warm-up. Regardless of whether the system is started from a cold or hot condition, the warm-up stage follows immediately after ignition. Typically, this stage involves burner operation only, with no product entering the oven via conveyor belt. When the appropriate temperature sensors reach a threshold value, the system indicates its readiness for the next stage.
- Stage 5 – Run. Processing of the feed material occurs when the oven is up to temperature, and all permissives and interlocks are in a "safe" condition. Feedback temperature control (and other feedback control loops such as humidity) may be enabled to maintain processing conditions where desired.
- Stage 6 – Shutdown. If any interlock is tripped, or if the operator manually causes the system to enter shutdown mode, the system closes fuel valves and reverts to a safe state. Whether blowers and conveyors remain energized or are also de-energized along with the fuel shutoff valves can vary from process to process, based on the individual safety concerns with respect to fire, explosion, or thermal runaway.

Other thermal equipment may go through different sequences, but the principle of time sequencing is common to many control systems.

15.6 Ladder Logic

Ladder logic was originally a technique for organizing and documenting a collection of hardwired relays that performed logical tasks in an automated manufacturing process. A relay is an electrical device, typically utilizing an electromagnet, that is activated by a current or signal in one circuit and automatically opens or closes a switch in a different circuit. For example, if a manual switch upstream of the relay's input is closed, energizing the relay's input circuit will cause the relay's output switch to close (via the electromagnet), which provides continuity and power to a different circuit.

Just as binary logic utilizes "1s" and "0s" to accomplish tasks such as "on and off" or "open and close," ladder logic performs binary tasks by "energizing and de-energizing" various output devices, including other relays.

The "ladder" is represented by two vertical lines (labeled *L*1 and *L*2), where *L*1 represents line voltage (e.g. 120 *VAC* or 24 *VDC*) and *L*2 represents ground (0 *VAC* or 0 *VDC*). Horizontal lines (called "rungs" of the ladder) are nominally drawn from a node on *L*1 to a parallel node on *L*2 with at least one input device and exactly one output device placed on the rung.

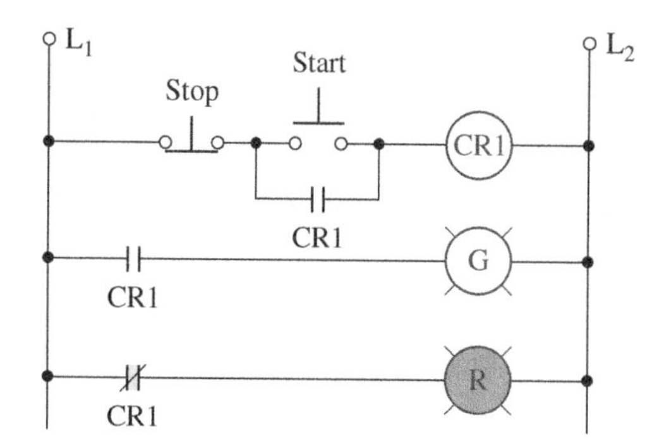

Figure 15.4 Simple example of ladder logic diagram. The red light in the third rung is shown as energized. This means the normally closed CR1 switch in the third rung is permitting current to flow (yellow highlight). This also means that CR1 is in its normal (de-energized) state. Having CR1 in its normal state also means the NO switch in CR1 is open, so the NO symbols shown in both the first and second rungs must be open with no current passing through them. *Source:* Martin Thermal Engineering Inc.

The ***input device(s)*** are switches that may be manually actuated (e.g. *HS* – a hand switch) or automatically actuated by a process instrument (e.g. *TSH* – temperature switch high). The ***output device*** can be a motor (energized or de-energized), an alarm (flashing/sounding or unlit/silent), an indicating light (green or red), a solenoid valve (open or closed), or another relay (energized or de-energized).

A switch or relay can be wired in the *NO* or *NC* configuration. When wired as *NO*, if the input state is de-energized (i.e. normal), the output state will be open circuit. When wired as *NC*, if the input state is de-energized (i.e. normal), the output state will be closed circuit.

Symbolically, on a ladder logic diagram, a ***NO*** switch is shown as a pair of contacts with ***no continuity*** between the two sides and a ***NC*** switch is shown as a pair of contacts with a ***diagonal line creating continuity*** between the two sides.

Relays are usually labeled *CR*, for control relay, with a unique device number, such as *CR*01 or *CR*02. When a de-energized *CR* becomes energized, all switches associated with that relay move to the opposing condition – NO becomes energized closed and NC becomes energized open. Thus, energizing a *CR* causes the continuity of the *CR's* input device symbol to be the opposite of how it is shown on the ladder rung.

For example, see Figure 15.4, which illustrates a very simple circuit having:

- On the top rung, there are three input devices and one output device:
  - First input is a momentary ***stop*** switch.
  - Second input is a momentary ***start*** switch.
  - Third input is the NO output terminals for *CR*1. It is in parallel with the momentary ***start*** switch.
  - First output device on the top rung is: *CR*1 (yes, this is the same *CR*1 that is acting as an input device in the top rung), which can be energized or de-energized depending on the input devices before it.
- On the second rung, there is one input and one output:
  - First (and only) input is the NO output for *CR*1. This appears to be a duplicate of the NO output for *CR*1 in the top rung. This is implemented with multiple NO contacts in parallel and different wire pairs connected to each pair of NO contacts.
  - First output is a green lamp that is de-energized.
- On the third and bottom rung, there is one input and one output:
  - First input is the *NC* output contacts for *CR*1.
  - First output is a red lamp that is energized.

By inspection, *CR*1 is ***de-energized*** whenever the *Start* button is in its normal state (i.e. not depressed) *and* either (i) the ***Stop*** button is being depressed or (ii) the *NO* output of *CR*1 is in its normal state (i.e. ***open circuit***).

Since occurrence (ii) above happens whenever *CR*1 is de-energized, the first rung is designed to "remember" whatever button was pushed last:

- If the ***Start*** button is pushed, the circuit's continuity causes *CR*1 to become energized, which flips both the *CR*1 *NO* output switches (in the first two rungs) to become ***closed***, which then creates a continuous path to maintain *CR*1 in the ***energized*** state.
- Similarly, when the ***Stop*** button is pushed, the circuit's lack of continuity causes *CR*1 to become de-energized, which flips the *CR*1 *NO* output switch back to ***open***, and which then creates a discontinuous path to maintain *CR*1 in the ***de-energized*** state.

The other two rungs of the ladder logic diagram show how the control relay's *NO* and *NC* output switches can be used to alternately light red or green panel lights, depending on whether *CR*1 is energized or de-energized. Since the input switch on the second rung is *CR*1's *NO* output switch, it will create a continuous path to energize the ***green*** lamp only when *CR*1 is ***energized***. Similarly, because the input switch on the third rung is *CR*1's NC output switch, it will create a continuous path to energize the ***red*** lamp only when *CR*1 is ***de-energized***.

15.7 Homework Problems

15.1 Give process examples that illustrate proportional action, integral action, and derivative action.

15.2 Give an example of where cascade control may be useful.

15.3 For the figure below, draw a sketch of the integral signal for the process variable shown.

15.4 For the figure below, draw a sketch of the derivative signal for the process variable shown.

15.5 Among (i) thermostatic control, (ii) electronic control, and (iii) pneumatic control, identify which control practice is best supported by which control component below.
(a) Bimetallic switch.
(b) Mercury switch.
(c) Level switch.

15.6 Give two examples of permissive interlocks that would be appropriate safeguards for energizing fuel gas shutoff valves to open and admit gas flow to a burner.

15.7 Give two examples of shutdown interlocks that would be appropriate safeguards for de-energizing a pump to shut down and stop filling a water tank.

15.8 Create a ladder logic diagram that performs the following tasks upon the press of a momentary start button: (a) opens a pilot gas solenoid valve, (b) energizes a high-voltage, spark igniter, (c) starts a 15-second timer, and (d) de-energizes both the solenoid valve and igniter when both of the following occur: (1) the timer has run out, and (2) a flame sensor detects the absence of flame.

Cited References

Dunn, W.; (2005); *Fundamentals of Industrial Instrumentation and Process Control*; New York: McGraw-Hill.

IEC 61511; (2016); *Functional Safety – Safety Instrumented Systems for the Process Industry Sector*; Geneva: International Electrotechnical Commission; Part 1: Framework, definitions, system, hardware and application programming requirements, https://webstore.iec.ch/publication/24241; Part 2: Guidelines for the application of IEC 61511, https://webstore.iec.ch/publication/25510; Part 3: Guidance for the determination of the required safety integrity levels, https://webstore.iec.ch/publication/25480.

Seborg, D.E., Edgar, T.F., Mellichamp, D.A., Doyle III, F.J.; (2016); *Process Dynamics and Control*, 4; Hoboken: Wiley.

16

Process Safety

Many sources of process safety information are available for students to review. Two important sources cited in this chapter are the National Safety Council (NSC 2016) and the International Society of Automation (ISA 2021).

The most important goals of this chapter are to help students understand why safety is the reciprocal of risk and to appreciate how these seemingly nontechnical concepts can be quantified, standardized, and achieved more reliably in the built environment.

16.1 Safety Terminology

The terms ***safe*** and ***safety*** are widely misunderstood by many people. We believe that learning the fundamentals of safety terminology is more compelling for thermal engineers than for any other engineering discipline, because the level of risk associated with thermal systems (explosion, fire, carbon monoxide inhalation, and skin burns) is arguably greater than that associated with electricity (electrocution, arc flash burns), mechanisms (rotating machinery dismemberment/injury), or buildings (falls, engulfment, and structural collapse).

Unfortunately, online sources of information such as Merriam-Webster, Dictionary.com, and Wikipedia are not particularly helpful on this issue. Students who believe they have a sufficient understanding of safety terminology based on the common vernacular may feel that the discussion here is excessively precise or more complex than engineers need to care about, but we disagree wholeheartedly with that perspective.

We have studied, provided consultation on, and testified about risk and safety issues for more than two decades, and we wish to strongly encourage students to fully comprehend and adopt these concepts, unless or until some better paradigm is developed. For additional evidence that the methods promoted here are appropriate and necessary, readers should consider participating in a process hazard analysis (PHA) team (described in Section 16.5).

Risk and ***safety*** are essentially reciprocals of each other, and to a large extent, both are assessed uniquely by each human being. The threshold level of risk that causes a person to deem an item or activity ***unsafe*** is in the eye of the beholder. For example, some people think mountain climbing and bungee jumping are well within their own personal ***safety*** envelope, while others think those activities are much too ***risky***, and therefore ***unsafe***.

The actual ***risk*** level of owning some item or performing some activity can be measured with relative objectivity by the ***likelihood*** and ***severity*** of the ***negative consequence*** that might occur,

Thermal Systems Design: Fundamentals and Projects, Second Edition. Richard J. Martin.

Companion website: www.wiley.com\go\Martin\ThermalSystemsDesign2

but that mathematical outcome may have little bearing on an individual's emotional placement of the dividing line that separates ***safety*** from ***danger*** in their mind.

A device or activity is ***hazardous*** if using it or participating in it carries the possibility of a bad outcome or negative consequence. Thus, a ***hazard*** is nothing more than the potential for something bad to happen. Again, with strong emphasis, the elements of likelihood and severity are reserved for the term ***risk***, and acknowledgement of the existence of a ***hazard*** provides no definitive information about whether the item or activity is ***dangerous*** or ***safe***. The terms ***danger*** and ***hazard*** are not synonymous and should not be used interchangeably.

For example, nearly everyone will agree that the likelihood of cutting yourself with a carving knife at Thanksgiving is greater than zero. But when the ***likelihood*** of an accidental skin cut is combined with the ***severity*** of having to compress the wound for a few minutes and wear a bandage for a couple of days, nearly everyone concludes that carving knives are ***safe*** because their benefit outweighs the risk of accidental harm. Even if the negative outcome is assumed to be so ***severe*** that a digit is lost, the ***likelihood*** of that scenario occurring is even more remote than the surface cut, so that most people still consider the knife to be ***safe***.

The sharp edge of the knife is its inherent ***hazard***. Nothing can be done to eliminate the injury ***hazard*** from the knife because the knife ceases to be an effective tool for cutting meat if its sharpness is taken away. It is impossible for a knife to be both ***useful*** and ***hazard***-free.

However, because the ***risk*** of serious injury is low, it is very much possible for knives to be simultaneously ***hazardous*** and ***safe***. (Obviously, we are considering accidental injury here. The level of ***danger*** one faces in a room with a knife-wielding attacker is quite high, but that higher likelihood of injury is fully dependent upon the distinction that the injury causation mode would be intentional versus accidental.)

Many components of thermal systems are ***hazardous***. They can cause injury or property damage if not designed, built, used, or maintained correctly. A pump shaft can break, and the rapidly spinning impeller can disintegrate, fracture the housing, and cause injury or damage. An oven can accumulate combustible baking crumbs/debris that can catch fire. However, each of these components can accurately be considered ***safe*** (or *not* ***dangerous***) if made *and* used with safety in mind.

We have observed instances in litigation proceedings, where a science or an engineering expert has egregiously (perhaps willfully) obfuscated the distinction between ***hazard*** and ***risk***, to the detriment of jurisprudence. An expert witness fails to perform their ethical duty if they fail to inform the trial participants that ***hazards*** themselves are *not* a problem when the humans who build and use the hazardous device do so in a ***safe*** manner. One of the primary duties of an expert witness is to teach juries (and attorneys and judges) about their field of expertise. When an expert uses the word ***hazard*** dozens of times on the witness stand and fails to differentiate it from the term ***danger***, they are not being informative – they are likely trying to make something appear more ***dangerous*** than it really is.

The opposite misuse of safety terminology can also occur. Excessive proclamations of a device's beneficial features (perhaps even the valid enumeration of many inherent or engineered safety features) are not tantamount to an affirmation of its overall safety. Decades ago, a certain model of automobile was designed and built with all the mandatory safety devices and features applicable at the time, and it was consequently deemed ***safe*** by the manufacturer and regulators. However, once in the hands of actual drivers, a recurring pattern emerged where gasoline-fueled flash fires would ensue shortly after a rear-end collision.

After some time, it was found that these fires were the result of a design flaw in the placement and attachment of the fuel tank with respect to the vehicle frame. The manufacturer had followed the standard safety practices published at the time, but they nevertheless failed to foresee that their design was ***unsafe*** in rear-end collisions, despite the vehicle's numerous other safety features. After this series of unfortunate incidents, the applicable safety rules (NHTSA 1997) for the quantity of fuel permissible to be released during/after a test collision were updated to include rear-end collisions above a certain velocity threshold.

16.2 Safety Hierarchy

The National Safety Council (NSC) publishes its Accident Prevention Manual (Hagan et al. 2020) every four years, and one of its most well-known features is the hierarchy of controls to reduce risk in the workplace. The intent of the hierarchy is for businesses to implement the highest level of hazard control possible because each successively lower level provides less protection. The hierarchy is summarized as follows, with each level being preceded by the phrase:

When possible,

1) Eliminate hazards in the design.
2) Substitute lower-risk materials.
3) Incorporate safeguards.
4) Provide warning systems.
5) Apply administrative controls.
6) Use personal protective equipment.

Sometimes, the hierarchy is shortened to three levels – eliminate, guard, and warn, but the full list is more instructive and provides more options for safety. The NSC suggests that the first three tend to be more effective because they are focused on the hazard rather than the worker. For illustrative purposes, following are some examples of the six levels of the hierarchy, using examples from industry, consumer products, and environmental protection.

Eliminate. The pressure vessels that comprised early steam boilers and steam locomotives were assembled with rivets, which experienced high rates of failure and caused many catastrophic vessel ruptures. Eventually, boiler designers were able to eliminate the rivets entirely by improving methods for rolling sheet steel into a complete circle that was seam welded along one longitudinal joint.

Substitute. The health hazard associated with inhaling lead emissions from the combustion of gasoline was reduced greatly when air quality regulations forced the automobile and petroleum industries to develop substitutes for tetraethyl lead, which was subsequently phased out over a period of years.

Guard. The very large circular saws that are used in lumber mills to cut logs into long boards are not good candidates for elimination or substitution. Since logs will not cut themselves into boards, a tool with sharp teeth is required. Unfortunately, a tool that is capable of cutting tree limbs is also capable of cutting human limbs. To reduce the rate of accidental dismemberment, mechanical guards exist at lumber mills that prevent the sawing area from being accessed by workers whenever the power is on. Guards and safeguards are also known as ***engineering controls***.

Warn. Mechanical conveyors are capable of causing serious injury if workers reach into the machine and a digit or limb is caught by the moving belt or chain. Even though guards may be employed to prevent workers reaching inside a conveyor enclosure under normal production conditions, occasionally, the conveyor requires maintenance or testing with guards intentionally removed. If the guard can and must be removed for some special circumstance, warnings affixed to the entry portal can remind workers to use extra caution to avoid entanglement and dismemberment.

Establish Procedures. For some hazards, procedural protections (also called ***administrative controls***) are helpful to reduce the frequency or severity of accidents that cause workplace injury or property damage. Examples of administrative controls include worker training to understand and react effectively to process upsets; workplace rules such as "never leave the cooking appliance unattended"; and scheduling risky tasks exclusively in daylight hours to prevent accidents related to reduced visibility. Administrative controls can be effective as a backup safeguard if an engineering control (e.g. machine guard or process interlock) fails to perform its primary function.

Use Personal Protective Equipment (PPE). Assignment of PPE to workers is essentially an acknowledgment that certain workplace risks are either too frequent or too consequential to permit workers to perform tasks without wearing the protective gear. Examples of PPE are respirators to reduce inhalation risks, footwear to reduce crush and penetration risks, hardhats to mitigate being struck by falling items, and visibility vests to reduce pedestrian-vehicle collisions.

The safety hierarchy is not a panacea for safety because accidents continue to happen, even when all parties (manufacturer, employer, and employee) do their best to live up to the ideals in the safety hierarchy. In our experience, missing or defective safeguards or interlocks are a contributing cause to approximately half of all fires and explosions, and operator error, inadequate maintenance, and failure to follow procedures are a contributing cause to over half of such incidents.

In cases where hazard elimination, substitution, and guarding are impossible or impractical, warnings, procedures, and PPE end up being the best options available. Unfortunately, these tools will never be 100% effective at preventing accidents since they rely on human awareness, comprehension, and decision making for their ultimate effectiveness.

16.3 Safeguards and Warnings

For thermal systems, especially where combustible fuels or feed materials are present, applicable safety standards may call for numerous safeguards in the form of process interlocks. As discussed in Chapters 7, 14, and 15, an interlock is a digital safeguard that protects a process from operating unsafely. Interlocks are triggered whenever a process variable measurement is outside its safe operating range. A list of selected interlocks for combustion systems installed in ovens and furnaces was given in Table 7.1. Other combustion equipment (e.g. boilers) may require additional or different interlocks (e.g. water pressure or flow). Heating systems that do not rely on burners (e.g. solar concentrating receivers or electric heating elements) may have yet other safeguards.

The National Electrical Manufacturers Association is the custodian of ANSI Z535 (2017), which is a series of six documents that provide guidance about warnings for consumer products and industrial facilities. Their purpose is to enlist the human sciences (e.g. visibility/audibility, neuroscience, perception/cognition, and psychology) into the development of tools that can enhance accident prevention measures that rely on optimum human responses to be effective.

The six documents cover the following areas:

- Safety colors.
- Environmental and facility safety signs.
- Criteria for safety symbols.

- Product safety signs and labels.
- Safety tags and barricade tapes for temporary hazards.
- Safety information in product manuals, instructions, and other collateral materials.

National Fire Protection Association (NFPA) publishes documents on fire alarms NFPA 72 (2019), and the human factors associated with egress and evacuation in emergencies (NFPA 101 2021), which address visual and audible warnings to promote life safety.

16.4 History of Safety Standards

Among the earliest of acknowledgments of workplace hazards were the late nineteenth-century "boiler and machinery" insurance policies that were first offered commercially in that era. The exceptional efforts of The Hartford Steam Boiler Inspection and Insurance Company (HSB 2015), which was founded in 1866 stand out for the demonstrated reduction in accidents they helped foster. The company's name proclaims its unique linkage of inspection (promoted by engineers) and insurance (promoted by financiers) – each link being seen as necessary to support the other.

The HSB engineers understood that the analysis of boiler failures was critical to the goal of reducing boiler explosions. Likewise, the HSB financiers understood they could afford to offer discounts on insurance premiums to policyholders who agreed to implement safeguards and permit their facilities to be inspected. Armed with the scientific knowledge of how to make boilers safer, the policyholders would endure fewer boiler explosions, and the payouts by the insurance company would be lower than their competitors.

In the ensuing century, safety requirements evolved from multiple sets of mandates prescribed by numerous insurance carriers, and they became "standards" or "codes" developed by bodies of knowledgeable individuals having different backgrounds and perspectives. These safety organizations included the American Society of Mechanical Engineers and the National Fire Protection Association, among others, and their primary missions were to help reduce the occurrence of accidents that caused property damage and personal injury by promulgating their standards at all levels – federal, state, and municipal.

The codes were called "consensus" because the membership of each safety committee was intentionally balanced among competing interests – equipment manufacturers, equipment users, equipment installers, equipment insurers, special experts, and firefighting authorities. The committees were deemed balanced as long as no single interest group comprised more than one-third of the committee membership. As a practical matter, this rule typically produced committees with four or more or interests represented, and the documents they produced benefitted from the latest knowledge as well as an approach that considered all sides of the safety equation.

Toward the end of the twentieth century and beyond, many standards began to acknowledge that ***prescriptive*** requirements may not provide the highest level of safety for some systems and facilities. When adopted, ***prescriptive*** rules are mandatory (and therefore inflexible) and they usually describe the minimum level of safety, rather than an optimal level of safety for a particular facility.

The alternative approach, ***performance-based*** rules, was based on statistics and placed a heavy emphasis on redundancy of safeguards and controls, when feasible. Generally, larger companies are more suited for ***performance-based*** rules, whereas smaller companies prefer the ***prescriptive*** approach, because they can comply without having a team of engineers analyzing statistics and adapting technologies and procedures in response to real data.

Even though the ***prescriptive*** rules are theoretically inferior to the ***performance-based*** approach (especially for very large facilities like chemical plants and petroleum refineries), a decision to follow the ***prescriptive*** approach ensures a much higher level of safety than ignoring the standards altogether. In a sense, the ***prescriptive*** rules provide the minimum required level of safety, and the ***performance-based*** approach is expected to produce safety results that exceed the minimum level because they are customized to the risks associated with the actual equipment and facility.

Many standards development organizations (SDO) are revising their documents to permit users to develop their own ***performance*** standards that may exclude one or more of the document's nominal safeguards, as long as the overall level of safety can be demonstrated (via a quantitative risk assessment) to be higher with the ***performance-based*** approach than it would be by simply following the ***prescriptive*** requirements.

Recently, the American and European versions of the standard for safety instrumented systems were merged into a single document (IEC 61511 2018). The previous American designation was ISA 84.00.01-2004, now obsolete for developing and implementing a ***performance-based*** safety approach for a large facility or a complex process.

Think Stop. Students who are interested in further study of the safety instrumented systems approach may want to (i) investigate subjects such as safety integrity level (SIL) for sensors, actuators, and controllers and (ii) review the differences in structure and procedure of a PHA versus a semiquantitative risk analysis (SQRA).

16.5 Process Hazard Analysis

The United States Occupational Safety and Health Administration (OSHA) is a division of the U.S. Department of Labor, and its mission is "...to assure safe and healthful working conditions for working men and women by setting and enforcing standards and by providing training, outreach, education and assistance."

One of OSHA's areas of emphasis is "Process Safety Management of Highly Hazardous Chemicals," also known as PSM (OSHA 1992). The standard primarily applies to chemical manufacturing and related industries, but it also affects the manufacturing and distribution of fuels and electricity, as well as large wastewater treatment and sanitation. Within the PSM regulation is OSHA's requirement for when and how a PHA should be performed.

The hallmark of a PHA is that it is a systematic approach for identifying and analyzing process hazards and assigning controls (engineering, administrative, or both) to reduce the risks associated with those hazards in a particular facility. For PSM-covered facilities, PHA results should be reviewed and updated every five years, or sooner if the process undergoes major changes. Several PHA methods are cited by the regulation (e.g. What-If, HazOp, and Fault-Tree), but only the HazOp (hazards and operability analysis) will be covered here.

Team. Many formulas exist for populating the different skill areas on a HazOp team. The HazOp team should be large enough that all safety and operability stakeholders are represented, but not so large that the group gets bogged down or distracted from their primary purpose. The team must have at least two individuals and a common number of participants is 4–5 (with 8–10 participants being relatively common for large, complex processes). The participants on the HazOp team should represent the following areas of expertise: facilitator, scribe, owner's representative, operations representative, and installer's representative.

Often, it is beneficial to retain a third-party HazOp facilitator. The facilitator may also act as the scribe, but it is more common for the scribe duties to be assigned to a different participant, frequently an employee of the process owner. The following additional skills or perspectives may be valuable to invite stakeholders and subject matter experts as participants: safety representative, maintenance representative, equipment manufacturer's representative, fire/explosion protection specialist, and other process specialists (e.g. chemists, materials scientists, and instrumentation/programming specialist), as well as the local fire marshal.

Terminology. The following terms and abbreviations are used in the HazOp analysis:

- *HazOp*: Hazards and operability analysis – the systematic process of analyzing a piping and instrumentation diagram (P&ID) for the risks associated with different failure modes.
- *Hazard*: A condition or circumstance that could lead or contribute to an unplanned or undesirable event.
- *Node*: Location on a process flow diagram (PFD) or P&ID where process parameters are investigated for deviations from the design intent.
- *GW*: Guide words – simple words to stimulate brainstorming failure scenarios (e.g. "more," "less," "none," or "late").
- *Deviation*: How the process could fail to meet design intent ("high pressure," "low temperature," and/or "reverse flow").
- *Causes*: Why do failures happen? ("Plant lost power," "valve fails open," and/or "regulator diaphragm breaks.") Note – the abbreviation *NCC* stands for ***no credible cause***.
- *Consequences*: What the negative result is ("gas explosion," "ignition of combustibles," and/or "hazardous material release"). Note – the abbreviation *NSC* stands for ***no significant consequence***.
- *Safeguards*: Protections (engineering controls) already in place to prevent incident or mitigate outcome. Note – safeguards do not include administrative controls.
- *S*: Severity – How bad? ("Serious," "high," "medium," "low," or "none.") A factor from 1 to 5, with 5 being the worst (e.g. single or multiple fatalities, massive destruction of property)
- *L*: Likelihood – How likely? ("High," "moderate," or "low.") A factor from 1 to 5, with 5 being the worst (e.g. more than 10 causative deviations per year, more than 1000 defective failures per million manufactured)
- *R*: Risk – a mathematical combination of severity and likelihood (on a scale from 1 to 10, with 10 being most dangerous).
- *Recommendations*: Follow-up tasks and task owner(s). These may include engineering or administrative controls. The HazOp is not complete until the recommendations have all been addressed.

Samples of HazOp results are contained in Table 16.1.

Process safety is an important consideration for all thermal systems, and designers should resolutely and methodically address the causes and consequences of technology failures and human errors to help prevent accidents that injure workers, cause property damage, or harm the environment.

Think Stop. Give examples of how a PHA might fail to prevent an accident. (This question is given as a homework problem at the end of the chapter.)

Table 16.1 Samples of HazOp report content for a process that utilizes a compressed toxic gas.

Node	GW	Dev	Cause	Consequence	Current safeguard	*S*	*L*	*R*	Recommend	Own
1	More	High T	TE failure	NSC	Safety shutdown	1	3	3	None	—
	More	High T	Fire	Equip damage	None	4	2	7	Install CO_2 suppression	RJM
2	Less	Low p	Gas utility loss	Flameout	PSL + safety shutdown	2	1	2	None	—
	More	High p	NCC	Flame impingement	PSH + safety shutdown	3	1	3	None	—
2	Also	Glycol + water	Heat exchanger leak	Contamination	None	3	2	6	Redesign exchanger	RJM
3	Reverse	Backflow	Pump failure	Water released to storm drains	None	1	2	2	None	—

Source: Martin Thermal Engineering Inc.

16.6 Homework Problems

16.1 Give two examples that illustrate the difference between hazard and risk.

16.2 Give two examples that illustrate how an administrative control might be inferior to an engineering control.

16.3 Perform a web search and provide an example of (a) a pair of similar safety instruments with different safety integrity levels (SILs) and (b) a pair of similar safety valves with different SIL. Each comparison should contain manufacturer name and model number for both devices. At least one of the comparisons should include a SIL-3 capable device.

16.4 Perform a guide word analysis on the third node of the simple process identified in Table 16.1. Use at least three guide words for your analysis.

16.5 (a) Give an example of an unacceptable PHA procedure that might contribute to the causation of an accident, and (b) give an example of how an acceptable PHA might nevertheless fail to prevent an accident.

Cited References

ANSI Z535 (2017). *Set of Six Standards for the Design, Application, and Use of Safety Signs, Colors, and Symbols*. National Electrical Manufacturers Association. https://www.nema.org/standards/z535/ansi-z535-brief-description-of-all-six-standards-and-safety-color-chart (accessed 15 December 2020).

Hagan, P., Montgomery, J.F., and O'Reilly, J.T.; (2020); *Accident Prevention Manual for Business and Industry*, 14. Itasca, IL: National Safety Council.

HSB (2015). *Hartford Steam Boiler: 150 Years of Innovation – An Enduring Mission*. https://vimeo.com/132970055 (accessed 15 March 2021).

IEC 61511; (2018); *Functional Safety – Safety Instrumented Systems for the Process Industry Sector*; Geneva: International Electrotechnical Commission; Part 1: Framework, definitions, system, hardware and application programming requirements https://webstore.iec.ch/publication/24241, Part 2: Guidelines for the application of IEC 61511 https://webstore.iec.ch/publication/25510, and Part 3: Guidance for the determination of the required safety integrity levels https://webstore.iec.ch/publication/25480.

ISA (2021). *ISA's History*. International Society of Automation. https://www.isa.org/about-isa/history-of-isa (accessed 10 March 2021).

NFPA 101 (2021); *Life Safety Code*; Quincy, MA: National Fire Protection Association; http://www.nfpa.org/101.

NFPA 72 (2019); *National Fire Alarm and Signaling Code*; Quincy, MA: National Fire Protection Association; http://www.nfpa.org/72.

NHTSA (1997). FMVSS 301 fuel system integrity. *49 CFR 571.301. Federal Motor Vehicle Safety Standards*, National Highway Traffic Safety Administration. https://www.govinfo.gov/content/pkg/CFR-1997-title49-vol5/pdf/CFR-1997-title49-vol5-sec571-301.pdf.

NSC (2016); "*National Safety Council: Leading the Way to a Safer World*"; Itasca, IL: National Safety Council; https://www.youtube.com/watch?v=tpHHgvHzC4Q

OSHA (1992). *Process Safety Management of Highly Hazardous Chemicals*. 29 CFR 1910.119; see https://www.gpo.gov/fdsys/pkg/CFR-2017-title29-vol5/pdf/CFR-2017-title29-vol5-sec1910-119.pdf.

17

Process Quality Methods

Manufacturing quality is a unique field of study with significant breadth, depth, and history. The field's origin is often traced to Eli Whitney, who first recognized the value of having ***interchangeable parts*** for the musket and cotton gin mechanisms he manufactured.

A classic textbook, ***Juran's Quality Handbook*** was first published in 1951 is now in its seventh edition (DeFeo 2010), with over 1000 pages of theory and practice focused on achieving and maintaining product quality. Among its more than 40 chapters, titles include: *How to Think About Quality, The Quality Planning Process, The Quality Control Process, Process Management, Total Quality Management, Training for Quality, Process Industries,* and *Quality and Society*. Another resource that provides substantial detail about management of change (MOC) is Center for Chemical Process Safety (CCPS 2008). A new offering called ***Planning and Executing Credible Experiments*** by Moffat and Henk (2021) has an encyclopedic feel like Juran but with a far greater emphasis on the irreplaceable value of well-planned and well-executed experiments to achieve *credibility* in research, production, safety/risk, and most of all the reputation of the individual or organization.

This chapter addresses the application of quality techniques to thermal processes and includes an example of a statistics based, process quality methodology for a combustion technology.

17.1 Quality Terminology

Two related but distinct definitions of ***quality*** are addressed in the literature:

1) Product features that meet the needs of customers and thereby provide satisfaction.
2) Fitness for use; freedom from deficiencies.

Think Stop. Question – In an organization, who is most responsible for quality? Which of these answers is the most accurate?

- Product development team
- Engineering design team
- Procurement team
- Manufacturing team
- Incoming parts quality control (QC) team
- Sales team
- Outgoing product quality assurance (QA) team
- Service team
- Senior management team

Thermal Systems Design: Fundamentals and Projects, Second Edition. Richard J. Martin.

Companion website: www.wiley.com\go\Martin\ThermalSystemsDesign2

Of course, the best answer is *all* of the above. A chain is only as strong as its weakest link, and breakdowns in any of these areas can be a recipe for disaster. However, since "all of the above" was not offered as a possible right answer, we would choose the senior management team as the entity most responsible for quality in any organization. If senior management has no interest in quality, the organization as a whole will not be motivated to make quality a priority, regardless of how many individual employees perform their work in an excellent manner.

The standard for industrial quality has been ***six-sigma*** since the middle of the twentieth century. Bhargav (2021) gives a succinct history of the method, which traces its origins to 1809 when Friedrich Gauss developed the formula for the normal distribution. The ***six-sigma*** concept originally held that manufacturing processes should be employed in a way that the mean value of each critical dimension in a mass manufactured part would be very close to the specified dimension, and the standard deviation for each critical dimension in the manufactured part would be one-sixth of the difference between the upper tolerance limit and the lower tolerance limit.

For example, if a critical dimension on a part is given as 5.00 *mm* on a design drawing, and the tolerance band is given as ±0.05 *mm*, then the mean value for a population N of the manufactured parts should be 5.00 *mm* and the standard deviation for the population should be 0.0167 *mm* (=[5.05 − 4.95]/6). Thus, parts that measure less than 4.95 *mm* or greater than 5.05 *mm* would be rejected, but more importantly if the rejection rates were found to be exceeding the proportions predicted by the normal distribution (i.e. 1350 *ppm* on the low side and another 1350 *ppm* on the high side), then the process was considered out of control, and an investigation would be undertaken to determine the cause of the unacceptable deviations. While rejecting nonconforming parts is certainly a better practice than fraudulently trying to sell them as conforming, if the rejection rate is too high, profitability can suffer.

For many chemical processes, where negative and positive deviations are often not equally detrimental, the six-sigma method assumes that the mean of the measurement is not at the nominal value, but rather it is displaced by 1.5-sigmas toward either the upper or lower limit – whichever gives the more benign production outcome.

Consequently, if a chemical process operator is achieving six-sigma quality in the manner described above, the Gaussian tail on the side of the mean away from the displacement will be irrelevant, and the Gaussian tail on the side of the mean corresponding to the displacement will be the value computed for 4.5-sigmas or 3.4 *ppm*. Thus, a process that is adhering to six-sigma guidelines will have no more than 3 out of a million process deviations that are unacceptable.

17.2 Advanced Statistical Methods for Quality in Thermal Processes

To illustrate how statistical methods for quality can be applied to thermal systems, a thermal oxidation (TO) process is examined. We start with the approach from Martin (2019) that is based on the first-order kinetics methodology of Cooper et al. (1982) as introduced in Chapter 7 without any detail. The problem is framed by defining the ***governing equation*** (Equation 17.1) for each regulated volatile organic compounds (VOC) entering the thermal oxidizer.

$$\boxed{\frac{\chi_{VOC,exit}}{\chi_{VOC,inlet}} = e^{-k\tau}} \tag{17.1}$$

where,

- χ_{VOC} $[\doteq kmol_{VOC}/kmol_{total}]$ is the ***mole fraction*** of the *VOC* (in the inlet or outlet stream).
- k $[\doteq s^{-1}]$ is the thermal destruction ***rate coefficient*** and is a function of *VOC* and *TO* properties.
- $\tau[\doteq s]$ is the ***residence time*** inside the thermal oxidizer at the average temperature.

The ***rate coefficient,*** *k*, is a function of the molecular collision rate, the mole fraction of oxygen present, and the probability of reaction between *VOC* and O_2 at each collision. To oversimplify the chemistry, when a *VOC* molecule collides with an oxygen molecule, the probability of reaction depends on the geometry and collision angle of the molecules (referred to as the steric factor) and on the relative kinetic energy (ΔKE) of the collision (versus the activation energy barrier that must exceed for the $VOC + O_2$ reaction to proceed). The kinetic theory of gases relates average molecular kinetic energy to the gas temperature and the individual molecules obey a Gaussian-like distribution of kinetic energies that tails off in both directions away from the mean.

Cooper and Alley (2014) illustrate how to compute the value of *k* for a given *VOC* and temperature, and we loosely follow their example in the following steps:

1) The rate coefficient *k* is computed from the ***pre-exponential coefficient*** and the ***activation energy*** for the given *VOC*, as shown in Equation (17.2).

$$\boxed{k[\doteq s^{-1}] = A \cdot \exp\left(\frac{-E}{\hat{R}T}\right)} \tag{17.2}$$

2) The ***pre-exponential coefficient*** *A* is computed from Equation (17.3):

$$\boxed{A[\doteq s^{-1}] = \frac{Z' S \chi_{O_2} p}{\hat{R}}} \tag{17.3}$$

3) The ***collision rate factor*** Z' as given by Cooper and Alley is computed from Equation (17.4), where $\hat{M}_{VOC}[\doteq kg/kmol]$. The equation appears to be in good agreement with the underlying experiments reproduced in the book.

$$\boxed{Z'[\doteq s^{-1}] = \begin{cases} 3.22 \times 10^9 \cdot \hat{M}_{VOC} + 4.392 \times 10^{10} & alkanes \\ 3.03 \times 10^9 \cdot \hat{M}_{VOC} + 3.418 \times 10^{10} & alkenes \\ 3.92 \times 10^9 \cdot \hat{M}_{VOC} - 7.620 \times 10^{10} & aromatics \end{cases}} \tag{17.4}$$

4) The ***steric factor*** *S* is computed from Equation (17.5), where both the numerator and the denominator share consistent units, e.g. $\hat{M}_{VOC}[\doteq kg/kmol]$ and *S* is a dimensionless parameter less than 1. Cooper and Alley assert that the steric factor is extremely difficult to compute theoretically, but their simplification (where methane was chosen as the reference molecule) gave "reasonably consistent" results when compared to the theoretical data.

$$\boxed{S = \frac{16}{\hat{M}_{VOC}}} \tag{17.5}$$

5) For good results, the ***oxygen mole fraction*** $\chi_{O_2}[\doteq kmol_{O_2}/kmol_{total}]$ should be relatively high ... typically in the range $0.12 \leq \chi_{O_2} \leq 0.21$. A default value of $\chi_{O_2} \approx 0.15$ is often used for thermal

oxidizers where oxygen is not continuously measured. However, users are urged to analyze gas samples periodically to ensure the O_2 value is not drifting below its effectiveness threshold.

6) The ***pressure*** is frequently taken as $p \approx 101.325\ kPa$ for thermal oxidizers operating at elevations near sea level. For higher elevations, p should be adjusted according to the equations for the U.S. standard atmosphere (Equation 6.18a).
7) The ***ideal gas constant*** $\widehat{R} = 8.314\,462\ kJ/(kmol \cdot K)$ is selected to have consistent units with all the other factors in the pre-exponential coefficient.
8) The ***activation energy*** E, as reported by Cooper and Alley, is computed from Equation (17.6), where molar mass of the VOC is in the same units as given in Equation (17.5):

$$\boxed{E[\stackrel{\vee}{=} kJ/kmol] - 0.040\,444 \cdot \widehat{M}_{VOC} + 193.01} \tag{17.6}$$

One advantage of this approach to estimating system performance is its analytical nature – once the VOC has been identified, all the equations can be entered into a spreadsheet (or other computational tool) and the results can be computed automatically, without iteration. This advantage also permits a discretized solution for a thermal oxidizer vessel that has a non-isothermal temperature profile. Destruction can be computed one axial slice at a time, and variations in local temperature are incorporated into changes in both the volumetric flow rate (and local residence time) as well as the local destruction efficiency.

The disadvantages of this approach are mostly related to its limited validation domain – alkane, alkene, and aromatic hydrocarbons. Users who attempt to apply this method to oxygenated or halogenated hydrocarbons should be very careful to revalidate their results based on test results from actual equipment or complex chemical kinetic models, perhaps even ones that incorporate computational fluid dynamics.

An end-of-chapter problem is given to illustrate the method of using statistical mathematics to optimize the thermal oxidizer performance for destruction efficiency and energy savings. We will develop the first part of the statistics here for handy reference, but much of the content provided in the solution key (see companion website) is also found in Martin (2019).

We presume the air quality agency automatically receives a telemetry report every 15 minutes showing the average exit (i.e. exhaust stack) VOC concentration for the prior time interval. We are also told that the processing facility wants to ensure the operating parameters are set to values that will yield no more than one VOC exceedance per year, since any exceedance of the permitted exit VOC concentration will result in a notice of violation and require payment of a fine.

If the thermal oxidizer's temperature control loop has a setpoint of $T = 935.0\ °K$ (i.e. right at the nominal value) and if all the process variables (including temperature) are assumed to vary randomly around their nominal values, the thermal oxidizer will meet the minimum required destruction level (i.e. 99.0% destruction efficiency [DE]) slightly more than 50% of the time! This means that the air pollution control system will likely be in violation of the facility's air quality management district (AQMD) operating permit nearly half (i.e. approximately 17 500) of the reporting periods each year – a *totally unacceptable* result.

Assuming each of the four process variables obeys a normal distribution subject to the given means and standard deviations, determine how many 15-minute VOC exceedances per year are likely. If the result with the current parameters is greater than one, determine a new value for the temperature setpoint that is required to achieve exactly one exceedance per year.

First, we compute the number of 15-minute intervals that exist in a year.

$$N = 4 \cdot 24 \cdot 365.25 = 35\,064\ intervals$$

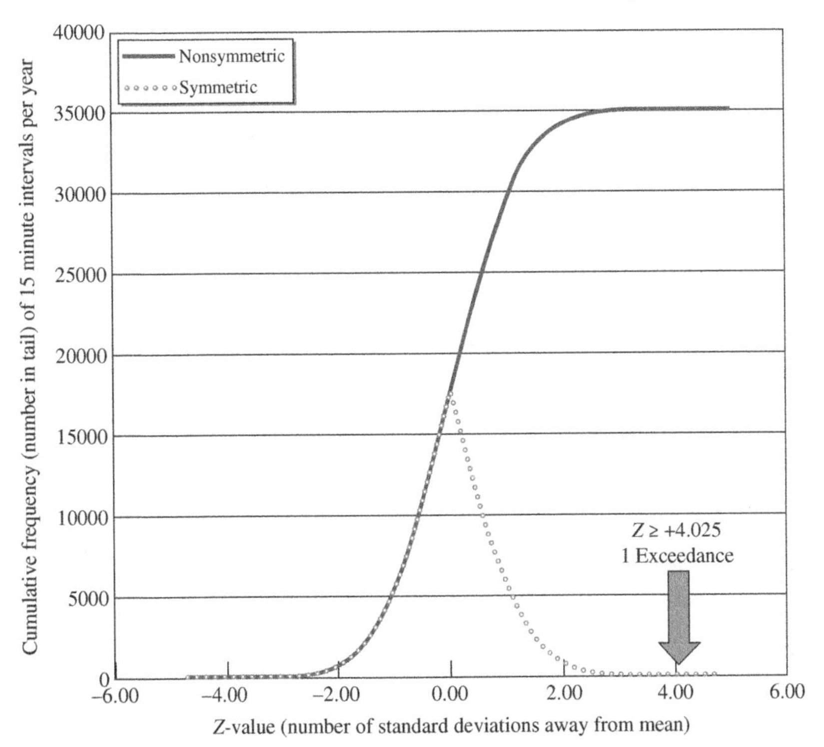

Figure 17.1 Plot of symmetric cumulative normal distribution for number of 15-minute $[VOC]_{exit}$ exceedances in 1 year versus Z-value ($Z = (x - \mu)/\sigma$). Z represents the number of standard deviations (σ) greater than the mean (μ), which should also be the targeted value. Downward arrow indicates Z-value where the random process parameter variations are collectively so unfavorable that the $[VOC]_{exit}$ concentration exceeds the maximum value allowed per the permit granted to the user by the air pollution control agency. As shown here, this exceptionally high $[VOC]_{exit}$ result occurs at a Z-value of +4.025. *Source:* Martin (2019).

Next, we find the Z-value that corresponds to a cumulative frequency of one exceedance out of N. In Microsoft Excel, the cumulative frequency as a function of Z-value is obtained by invoking the built-in function `NORM.DIST(Z,0,1,TRUE)`. This distribution is plotted in Figure 17.1.

The user must operate the *TO* at a temperature significantly higher than the nominal value to ensure that *VOC* emissions only exceed the permitted value when the combination of all process variable values deviate in a way that gives a *VOC* exit concentration that is in the right tail of Figure 17.1 at a Z-value of approximately +4.025 standard deviations above the mean.

$$f = \frac{1}{35\,064} = \left[\frac{1}{\sqrt{2\pi}} \int_{Z}^{+\infty} \exp\left(-\frac{z^2}{2}\right) dz\right] \rightarrow \boldsymbol{Z} = 4.0247$$

In order to find the standard deviation of the dependent variable ($\chi_{VOC,\ exit}$) in the ***governing equation***, we must first determine the effect of varying each of the independent variables by their standard deviations. This is done by partial differentiation of the ***governing equation*** with respect to each of the independent variables followed by multiplying the resulting derivative by that variable's known standard deviation (Equation (17.7)) and combining the individual deviations in a statistically appropriate manner (Equation (17.8)). The analysis is given below:

$$\underbrace{(\delta Y)_i}_{\substack{\textit{finite deviation of dependent variable } Y \\ \textit{with respect to deviation of} \\ \textit{independent variable } X_i}} \approx \underbrace{\frac{\partial Y}{\partial X_i}}_{\substack{\textit{partial derivative} \\ \textit{of } Y \textit{ with } X_i}} \bullet \underbrace{\delta X_i}_{\substack{\textit{finite deviation} \\ \textit{of } X_i}} \tag{17.7}$$

The combined effect of deviations in all the individual process variables is treated as a root sum square (RSS), because of the statistical implausibility that all the deviations would simultaneously line up to be at their extreme values and thereby induce the maximum effect on the dependent variable.

The RSS approach is akin to summing individual variances (i.e. deviations squared) and taking the square root of the sum to determine the overall standard deviation.

$$\overbrace{\delta Y}^{\substack{\textit{combined deviation of } Y \\ \textit{from all } X_i \textit{ deviations}}} \approx \sqrt{\sum_{i=1}^{\overbrace{N_{X_i}}^{\textit{number of independent variables}}} (\delta Y)_i^2} \tag{17.8}$$

The remainder of the derivation is available in the solution key for the relevant end-of-chapter problem.

17.3 Management of Change for Quality, Stewardship, and Safety

MOC is a strategy whereby an organization analyzes, plans, and implements changes in a methodical way (rather than a reactionary manner), ostensibly for the benefit of all interested parties. MOC may be applied to personnel, equipment, procedures, or outcomes. Change management is deemed to be essential to maintaining a ***culture of quality*** in an organization, particularly one that focuses on the values, goals, practices, and processes that help deliver quality consistently. Thomas (2018) provides an organizational behavior perspective of why change implementation sometimes fails due to *sacred cows* and *entrenched stakeholders*.

Quality. Just as ***product quality*** implies a certain level of reproducibility associated with what is delivered to customers, an organization's QA efforts generally require it to thoughtfully organize its people and processes. The process of ***change*** is not exempt from the goal of reproducibility, nor the requirement for intentionality.

When quality is part of the culture at an organization, customer satisfaction is a routine by-product. Nevertheless, the MOC process may benefit explicitly from elements such as quantification and trend analysis of customer fulfilment, to help discern problems and correct them before the firm's reputation is damaged by a series of low-quality outcomes.

From a thermal systems design perspective, MOC may focus on changes involving thermodynamic calculations, process drawings, process operating conditions, feed material identity, process variable tolerances, control loop tuning, operator procedures, and/or QA procedures. To the extent any of these inputs is changed, the MOC process should take responsibility for documenting the changes and validating the results.

A change management task might be as simple as entry of maintenance data into a computer database to retain accurate records of when belts and gaskets are replaced, or it might be as complicated as scrapping an old "enclosed-flare" style thermal oxidizer and building a new regenerative thermal oxidizer to achieve better environmental performance at lower operating cost.

Stewardship. In the environmental realm, ***corporate stewardship*** is one hallmark of quality, and may help deliver side benefits of good reputation and consumer satisfaction to an organization.

Environmental stewardship is comprised of three primary factors: (i) responsible use of natural resources; (ii) protection of the environment (land, water, and air) by effective management of wastes, effluents, and emissions; and (iii) sustainable practices that mitigate or eliminate negative consequences to local and global ecosystems.

The concept of ***cradle-to-grave*** responsibility for raw materials and wastes was developed as part of the federal Resource Conservation and Recovery Act of 1976 (RCRA), which tasked the US Environmental Protection Agency (EPA) with regulating the treatment, disposal, and recycling of hazardous wastes; conserving valuable natural resources; and delivering relevant information to citizens so they may be part of the decision-making processes in their communities. The hazardous waste program, under RCRA Subtitle C, established a system for controlling hazardous waste from the time it is generated until its ultimate disposal. For example, every *generator* of waste mercury lamps (e.g. a construction contractor who manages demolition of old structures) is responsible for the discarded lamp at the time it becomes a waste (upon removal from the structure) to the time of its ultimate recycling or disposal.

Product stewardship, a term coined shortly after the turn of the millennium, extends the concept of ***cradle-to-grave*** responsibility to manufacturers – not simply waste generators. The concept of product stewardship involves sustainable use of natural resources and foresight about how consumers use and ultimately dispose of products. The so-called ***life cycle analysis*** (LCA) is a hallmark of product stewardship. Here, the term "life cycle" implies that a thorough assessment of the environmental impact of a product requires an analysis of raw material provision, the manufacturing process, distribution, use, and disposal caused by the mere existence of the product.

Safety. In addition to the motivation of maintaining high quality, safety also can be enhanced by well-implemented MOC processes. Referring again to Occupational Safety and Health Administration (OSHA)'s Process Safety Management (PSM) standard, an important aspect of PSM is that senior managers at process companies covered by the standard are responsible for both (i) the development of the safety program and (ii) the maintenance of that program. A vital component of safety program maintenance is the proper MOCs to the facility and its processes.

According to the PSM regulation, OSHA requires companies to thoroughly evaluate anticipated process changes to fully assess their impact on employee safety and health before they implement them. With one exception (for maintenance activities qualifying as a *replacement in kind*), modifications to a PSM-regulated process must begin as written procedures to safely carry out MOC. Types of changes for which the regulations apply include (i) different process chemicals, (ii) new technologies, (iii) upgraded equipment, (iv) changes to infrastructure or utilities, and (v) updated operating procedures. And conversely, if one of the above-listed changes (i) through (v) is being evaluated, the potential impact of that change on the other four should also be evaluated concurrently.

The MOC plan must ensure that the following points are addressed prior to the implementation of any change:

- Technical analysis of the proposed change.
- Impact of the change on employee safety and health.
- Necessary modifications to operating procedures.
- Detailed schedule to implement the change.
- Authorization requirements for each phase of the proposed change.

Contract employees (including operators, maintenance technicians, and safety personnel) whose tasks may be affected by a change must be informed and trained prior to start-up. If necessary, process safety information, such as safety data sheets, must be updated accordingly.

17.4 Homework Problems

17.1 Contemplate (i.e. research) the implementation of an LCA for ceramic fiber insulation installed in a high-temperature furnace used for heat-treating steel parts. List at least five specific aspects of environmental or personnel impact factors that may be associated with the production, use, and/or disposal of the material.

17.2 Which of the following approaches is a more effective way to evaluate consequences to process quality associated with the change from one feedstock chemical to another (e.g. substituting $n\text{-}C_4H_{10}$ for $i\text{-}C_4H_{10}$)? Discuss your response.

(a) Acquire a small batch of the new chemical and run one or two limited-duration production campaigns (not longer than one shift each) with existing operating personnel and equipment, so the controllability of the process and the quality of the final product can be compared to the known characteristics of the old product.

(b) Build a pilot-scale facility and permit engineering staff to operate the pilot process with both old and new feedstock chemicals, under varying conditions, to develop a "governing equation" for the process, which can then be used to predict quality in the full-scale process.

17.3 True or False. A process change involving a new supplier for a highly hazardous chemical need not adhere to OSHA's "management of change" requirements, but it should be subjected to a quality assessment to evaluate new chemical's purity variance and the identity and concentration of contaminants. Discuss your reasoning.

17.4 Write the governing equation for a TO process that destroys toluene, based on the methodology in Section 17.2.

17.5 Consider a hypothetical thermal oxidizer with internal volume $\mathcal{V} = 15.57\ m^3$, which is guaranteed to achieve 99.0% destruction efficiency of benzene C_6H_6. Suppose the *TO* is designed to obey the governing equation below and is subjected to the following operating restrictions mandated by the Air Quality Management District.

- $T \geq 963.7\ °K$
- $\dot{\mathcal{V}}_{STP} \leq 4.956\ m^3_{STP}/s$
- $\chi_{O_2,stack} \geq 14.5\%vol$
- $\chi_{C_6H_6,stack} \leq 5.00\ ppmv$ The governing equation for this case is:

$$\frac{\chi_{VOC,stack}}{\chi_{VOC,inlet}} = e^{-k\tau} = \exp\left(-5.7413 \times 10^{11} \cdot \chi_{O_2stack} \cdot e^{\frac{-22\,834}{T}} \cdot \frac{4643.6}{\dot{\mathcal{V}}_{STP} \cdot T}\right)$$

where,

$$\hat{R} = 8.314\,462\ kJ/(kmol \cdot K)$$

$$\hat{M}_{VOC} = 78.11\ kg/kmol$$

$$p = p_{std} = 101.325\ kPa$$

$$\tau = \frac{\mathcal{V}}{\dot{\mathcal{V}}_{STP}\frac{T}{T_{std}}} = \frac{4643.6}{\dot{\mathcal{V}}_{STP} \cdot T}\ s$$

$$k = A \cdot \exp\left(\frac{-E}{\hat{R}T}\right) s^{-1} \quad \textit{from Equation (17.2)}$$

$$A = \left(5.7413 \times 10^{11} \cdot \chi_{O_2,stack}\right) s^{-1} \quad \textit{from Equation (17.3)}$$

$$Z' = 2.2999 \times 10^{11}\ s^{-1} \quad \textit{from Equation (17.4)}$$

$$S = 0.204\,839 \quad \textit{from Equation (17.5)}$$

$$E = 189\,851\ kJ/mol \quad \textit{from Equation (17.6)}$$

The following list provides nominal (design) values for the four independent process variables and their measured standard deviations over time. In practice, these values are determined by the user from historical data-logging records for the TO process. In this example, the standard deviations are uniformly taken to be 1% of the nominal design parameters for simplicity.

- $T \pm \sigma_T \rightarrow 935.0 \pm 9.350°K$
- $\dot{\mathcal{V}}_{STP} \pm \sigma_{\dot{\mathcal{V}}} \rightarrow 2.500 \pm 0.0250\ m^3_{STP} \cdot s^{-1}$
- $\chi_{O_2,stack} \pm \sigma_{\chi_{O_2}} \rightarrow 0.1650 \pm 0.001\,65\ kmol_{O_2}/kmol_{total}$
- $\chi_{C_6H_6,inlet} \pm \sigma_{\chi_{O2}} \rightarrow 0.000\,400 \pm 0.000\,004\ kmol_{C_6H_6}/kmol_{total}$

Notably, the nominal inlet mole fraction of benzene is 400 *ppmv*. If we plug the nominal values for the four independent variables into the governing equation, we obtain a nominal result for the dependent variable and the destruction efficiency:

$$\chi_{C_6H_6,stack,nominal} = 3.778\ ppmv$$

$$DE \approx \left(1 - \frac{3.778\ ppmv}{400\ ppmv}\right) = 99.056\%$$

Find the temperature setpoint that will result (statistically) in exactly one exceedance per year of the C_6H_6 emission limit.

Cited References

Bhargav, R.; (2021). *History and Evolution of Six Sigma*. Simplilearn Solutions. https://www.simplilearn.com/history-and-evolution-of-six-sigma-article

CCPS; (2008); *Guidelines for the Management of Change for Process Safety* ; Hoboken: Center for Chemical Process Safety, Wiley.

Cooper, C.D. & Alley, F.C.; (2014); *Air Pollution Control: A Design Approach*, 4; Prospect Heights, IL: Waveland Press; 352–356.

Cooper, C.D., Alley, F.C. and Overcamp, T.J.; (1982); "Hydrocarbon vapor incineration kinetics"; *Environ. Prog.*; **1**; 129.

DeFeo, J.; (2010); *Juran's Quality Handbook – The Complete Guide to Performance Excellence*, 6. McGraw-Hill

Martin, R.J. (2019). Advanced quality methods for thermal oxidizer operation. *Paper #71EA-0175; 11th U.S. National Combustion Meeting at Pasadena, CA* (24–27 March 2019). Pittsburgh, PA: The Combustion Institute.

Moffat, R.J. and Henk, R.W.; (2021); *Planning and Executing Credible Experiments: A Guidebook for Engineering, Science, Industrial Processes, Agriculture, and Business*; Hoboken: Wiley.

Thomas, S.; (2018); "Why change efforts fail and what to do about it"; *Inspectioneer. J.*; (July/August 2018); https://inspectioneering.com/journal/2018-08-29/7880/why-change-efforts-fail-and-what-to-do-about-it.

18

Procurement, Operation, and Maintenance

Unlike a fireworks extravaganza, the finale for this textbook does not contain the most beautiful, the most unusual, or the most remarkable content for readers to devour. Nevertheless, procurement, operation, and maintenance are vitally important subjects for any product or service provider, and students should become motivated to learn certain basics as part of their overall commitment to engineering excellence. Hopefully, the material covered in this chapter will be informative and useful, even if it is not sizzling with heat and light.

18.1 Engineering Design Deliverable

A discussion of the types of drawings and documents that might be included in an engineering design package was initially provided in Chapter 14, and a list was included there as Table 14.1.

The precise content of an engineering deliverable transmitted from engineer to client may be subject to negotiations between the parties. For example, a manufacturer may want to protect certain trade secret information about their designs. They could attempt to protect these details in several ways: (i) exclude certain design details from drawings and intentionally distort shapes and sizes, acknowledging the omissions and inaccuracies in the title block or notes as part of their intellectual property protection program; (ii) exclude certain drawings from the engineering design package; (iii) obtain copyright protection for control logic software; and (iv) password protect control logic software, so that users may only change settings and may not alter fundamental control instructions.

Notwithstanding the above, users who purchase a system from a manufacturer or from a designer/fabricator may want to obtain as much information as possible about a design, because they intend to operate the equipment for many years and the designer/manufacturer cannot guarantee they will be in business that long to provide technical support. In such cases, the design deliverable may end up being "the whole enchilada," as identified in Table 14.1. From the design engineer's perspective, accuracy and completeness should be the hallmark of the design package. The designer's goal should be that the information they leave behind them is sufficient for (i) a manufacturing engineer to use as a basis to build a duplicate system and (ii) a process engineer to use as a basis for operating that system.

18.2 Engineering Data Sheets

Engineering data sheets are procurement documents that contain far more information than is possible (or advisable) to include on a drawing or a bill of materials. The purpose of an engineering

Thermal Systems Design: Fundamentals and Projects, Second Edition. Richard J. Martin.

Companion website: www.wiley.com\go\Martin\ThermalSystemsDesign2

data sheet is to summarize the important specifications for a component that is part of an assembly or system, so a purchasing agent can obtain quotes and search for competing products at a better price. The engineer completes as much of the information as possible, and turns it over to procurement staff, who obtain product quotes from various sources. If the data sheet contains enough information for the purchasing department to eliminate noncompliant bidders, the engineer need not play any further role in the final issuance of the purchase order.

Furthermore, even though the engineer may be responsible for recommending a particular supplier as part of their purchase requisition, many companies do not authorize engineers to actually place orders with any suppliers – only the purchasing department is authorized to do that.

If an engineer has a favorite supplier or manufacturer for a specific part (e.g. ball valve, pressure transmitter, or programmable logic controller [PLC]) but the purchasing agent finds a highly regarded substitute at a lower price, the engineer must either accept the substitute, or provide valid data that demonstrate why the part sold by the initially requested vendor is uniquely better than the alternate part found by the buyer.

In some cases, additional data such as mean time between failures (MTBF) or safety integrity level (SIL) may play a role in the ultimate decision, and engineering substantiations containing such data may be persuasive to both purchasing and management.

Figure 18.1 provides an example of an equipment data sheet (for a blower) that hopefully represents useful information to be transferred from engineer to buyer. Figure 18.2 provides an example of an instrument data sheet (for a set of orifice flanges and plates). As seen in the examples, the data are artfully arranged onto a single sheet of information that accompanies the purchase requisition, which is generated by the engineer and transmitted to purchasing.

18.3 Construction and Commissioning

We have long considered the construction phase of a project to be like an extended Christmas holiday. Each day, a new "toy" arrives on the loading dock, and the project engineer usually has the responsibility for ensuring it is the right item, with the right tag and the right documentation. When something goes wrong, it means Christmas is over (temporarily) and Boxing Day has arrived (when the unwanted packages are returned to the store).

The project engineer also watches the construction crew assemble and build the structures and must be prepared to make decisions if miscalculations or misalignments are discovered. Young engineers will discover the utility of a "drift pin" when a pipe flange does not quite align with the nozzle flange on a vessel. But even drift pins cannot rescue an assembly where the tolerances were not managed properly at the design stage (see Section 17.2). Other quality assurance tasks to be performed by the project engineer include: verification of gasket placement or thread sealant at pipe joints, verification of drainage slope and direction for pipes that may convey condensate, confirmation that debris and other obstructions have been cleared out of pipe interiors, confirmation that wire marker numbers and wire termination points are consistent with electrical drawings, verification of bolt/nut/washer grade and size at all bolted joints, etc.

After construction is complete, the equipment must be commissioned. This step can take days to weeks depending on the complexity of the project. Stages in commissioning may include:

- Commissioning plan: Prior to commencement of commissioning, the engineer should prepare a punch list of commissioning tasks, and worksheets for each task that provide acceptable ranges

Engineer Logo Here		BLOWER			SHEET 1 OF 1		
Project:	No.	BY	DATE	REVISION	SPEC		REV
Customer:	A				CONTRACT	DATE	
Job No.:						P.R.	
Plant:					BY	CHK'D	APPR.
Location:							

GENERAL	1	Tag Number	
	2	Service	
	3	Rated Volume Flow (acfm)	
	4	Rated Volume Flow STP (scfm)	
	5	Rated Mass Flow (lb/h)	
	6	Installed Elevation (ft above sea level)	
INLET CONDITIONS	7	Inlet Temperature (°F)	
	8	Inlet Pressure (psig)	
	9	Inlet Molecular Weight (lb/lbmol)	
	10	Inlet Relative Humidity (%)	
	11	Inlet Specific Gravity (Dry Air = 1.0)	
PERFORMANCE	12	Nominal Pressure Rise (inch WC)	
	13	Nominal Rotational Speed (rpm)	
	14	Maximum Rotational Speed (rpm)	
	15	Maximum Inlet Flow (acfm)	
	16	Nominal Power (hp)	
CONSTRUCT MATERIALS	17	Case	
	18	Bearings	
	19	Belt or Direct Drive	
	20	Impeller	
	21	Shaft	
DIMENSIONS	22	Floor Space (ft^2)	
	23	Skid Dimensions (inch × inch)	
	24	Anchor Bolts (number, dia × depth)	
	25	Installed Weight (lb)	
	26	Inlet Connection (NPS, Flange Rating)	
	27	Outlet Connection (NPS, Flange Rating)	
MFG	28	Recommended Manufacturer	
	29	Vendor Name, Location, Phone	
	30	Model Name, Number	
Notes:	1	Supply stainless steel tags engraved with tag number	
	2		

Figure 18.1 Example of a blower data sheet (used for communicating specifications from engineering to procurement). *Source:* Martin Thermal Engineering Inc.

for the measurements taken at each item being commissioned during that task. A commissioning worksheet (checklist) for a blower might include:

- Rotational direction and speed of a motor shaft driving a blower.
- Rotational speed of the blower shaft, if connected to the motor by belt drive.
- Outside and inside pulley diameters.
- Impeller rotation direction.
- Presence of belt guards.
- Presence of drain valve.
- Electrical disconnect with lock-out–tag-out capability located within view of blower.
- Measurements of inlet- and outlet-plane temperature, static pressure, and centerline velocity for both low- and high-inlet resistance (i.e. damper position).
- Measurements of voltage and current to motor during low- and high-resistance flow testing, etc.

Engineer Logo Here		Restriction Orfices					Sheet 1 of 1		
Project:		No.	By	Date	Revision	Spec.			Rev.
Customer:						Contract			
Job No.:						Req.			
Plant:						By	Checked	Approved	
Location:									

Orifice Plates				Orifice Flanges			
1	Concentric ☐	Other:		8	Type:	Weld Neck ☐	Slip On ☐
2	ISA Standard ☐	Other:				Threaded ☐	
3	Bore: Maximum Rate ☐	Nearest 1/8" ☐		9	Material:	☐	Other:
4	Material: ☐	Other:		10	Flanges Included ☐		By Others ☐
	☐			11	Flange Rating:	See Note 3	
5	Ring Material and Type:						
6	Manufacturer:						
7	Model Number:						

Section	No.	Item				
Fluid Data	12	Tag No.				
	13	Service				
	14	Line Number				
	15	Fluid				
	16	Fluid State				
	17	Maximum Flow				
	18	Normal Flow				
	19	Pressure				
	20	Temperature				
	21	Specific Gravity at Base				
	22	Operating Specif. Gravity				
	23	Supercomp. Factor				
	24	Mol. Weight | Cp / Cv				
	25	Operating Viscosity				
	26	Quality % or Deg. Superheat				
	26	Base Press. | Base Temp.				
Plate and Flange	27	Differential Range				
	28	Beta = d/D				
	29	Orifice Bore Diameter				
	30	Line I.D.				
	31	Flange Rating				
	32	Vent or Drain Hole				
	33	Plate Thickness				
	34	Plate Material				

Notes:

1 Supply stainless steel tags engraved with tag numbers.

2 Orifice will be installed between PN10 pipe flanges, with gaskets.

3 All flange specifications shall comply with ASME B-16.36.

Figure 18.2 Example of an orifice/flange data sheet (used for communicating specifications from engineering to procurement). *Source:* Martin Thermal Engineering Inc.

- Pre-commissioning validation: During construction, just prior to commissioning, the following steps are advised:
  - Ensure tags on equipment, instruments, valves, and lines are in place and accurate.
  - Photograph all pieces of equipment and instruments, in the context of their installation.
  - Review instruction manuals for purchased items to ensure test plans and installed equipment are consistent with and comprehensive of all relevant operating functions.
- Component testing: Immediately prior to testing of the entire system, component testing is recommended:
  - Power up each instrument and piece of equipment individually and validate basic functionality.
  - Observe equipment at upper and lower ends of turndown range.
  - Perform capacity tests on pumps, blowers, burners, heaters, and chillers.
  - Validate alarm and interlock functionality by driving process variables beyond alarm and shutdown limits or introducing other process anomalies.
- System testing. The next-to-last stage of commissioning is to operate the thermal system in a thorough manner to attempt to expose problems before the system is turned over to the customer. The following steps are recommended:
  - Operate the system through all process stages, including power up, purge, ignition, warm up, normal operation, and shutdown.
  - Test functionality of automatic control loops by introducing step changes to process variables such as flow or temperature and observing corrective measures taken automatically by controllers.
  - Where appropriate, tune PID (proportional, integral, derivative) loops to optimize control response to process variable changes.
- Warranty testing. If a performance guarantee was offered by the system provider, the customer will want proof that the equipment has met its performance targets before they will sign off on the commissioning. Or, if any performance anomalies develop during the warranty period, a series of recommissioning tests may be necessary to validate any warranty claims made by the new owner.
  - For example, suppose a thermal oxidizer is purchased with a minimum performance guarantee of 99.9% destruction efficiency for a certain inlet mixture, when the chamber is at a certain operating temperature.
  - Also, suppose the system passed its initial performance test.
  - Now, suppose that months later, by occasional spot checking, the purchaser believes the destruction efficiency is no longer meeting the guaranteed level.
  - In order to bring a successful claim against the manufacturer, the user will have to demonstrate that all the conditions of the warranty are being met, and the system performance somehow deteriorated unexpectedly all by itself.
  - A full (or nearly full) recommissioning effort may need to be undertaken to determine whether the user's actions did or did not cause or contribute in a major way to the reduced performance.

18.4 Inspection, Maintenance, and Training

Each of the three fire safety standards for industrial heating equipment, NFPA 85 (2011) for boilers, NFPA 86 (2019) for ovens, and NFPA 87 (2018) for fluid heaters, contains requirements for system inspection and maintenance. Obviously, the overall responsibility for system maintenance belongs to the equipment user after the manufacturer has turned over the system to the user when commissioning is complete. In some instances, the user may contract the maintenance service out to an installer/maintainer, who then establishes schedules for different maintenance tasks, in coordination with the operating team.

Preventative Maintenance. It is an unavoidable fact that some facility managers argue (correctly) that a maintenance strategy of waiting until equipment and components fail irreversibly before they are replaced is the best way to maximize equipment life and minimize replacement cost. While true in a narrow sense, this approach ignores other, non-equipment costs (e.g. lost production, consequential damage due to a catastrophic failure of the item not replaced preventatively) that may far exceed the benefit of the extended equipment life obtained.

For example, savings obtained by performing preventative maintenance (PM) according to a predetermined schedule (including "slightly premature" component replacement) can reduce or eliminate process upsets, which at best necessitates the rejection of some amount of product for failure to meet quality specifications (during the upset) and at worst, causes property damage or personal injury, including long downtime periods without any production. According to Jackson and Döttboom (2014), poor scheduling of PM tasks can result in maintenance consuming 40% of the operating budget of a large processing facility. Conceivably, management's repeated refusal to perform maintenance preventatively could be considered willful, and if the poorly maintained component ultimately fails in a way that involves loss of life, the company may be forced to pay a steep price to resolve valid lawsuits.

In addition to periodic maintenance of lubrication systems, drive belts, and air filters (and other maintenance activities that have parallels in automotive maintenance), industrial heating systems also require regular inspection and testing of safety instruments and interlocks, such as temperature switches, pressure switches, flame safeguards, and more. Furthermore, boilers require testing and maintenance of water chemistry and fluid heaters require regular sampling and analysis of thermal fluids to look for signs of thermal decomposition or contamination. An example of a maintenance checklist recommended for certain types of ovens and furnaces is given in Table 18.1. Intervals may be daily, weekly, monthly, or annually, depending on the type of system and the manufacturer's recommendations.

Spare Parts. One of the most cost-effective maintenance programs is to keep a supply of spare parts available on site for all thermal systems. Although the goal of preventative maintenance is to preclude unanticipated downtime, there are no guarantees that all the components will function properly for their anticipated life, and sometimes, at the worst possible time, a necessary part breaks down. When spare parts are on hand, such breakdowns can be dealt with rapidly, and downtime can be as short as a couple of hours or less. Furthermore, the older the system, the higher the likelihood of component failures without warning.

As common sense would confirm, the following benefits may result from an effective spare parts program:

- Reduced downtime for repairs.
- Reduced labor for repair tasks.
- Immediate availability of items that are back ordered or discontinued.
- Reduced energy losses (due to removal of worn-out thermal insulation).
- Reduced process variability and increased product quality (due to replacement of unreliable burners, pumps, etc.).
- Less downtime equates to higher production rates and greater revenue.

Table 18.1 Maintenance tasks recommended for classes A and B ovens/furnaces.

Recommended maintenance for class A and class B ovens	
Inspect flame sensor for cleanliness	Check operation of modulating valve
Inspect thermocouple wires for short/open circuit, anomalous readings	Inspect and test orifice plates and differential pressure gauges
Check setting and function of high-temperature limit switches; check setpoint calibration against known device	Inspect ignition cables and transformers; replace cables showing any sign of insulation deterioration
Check setting and function of high- and low-pressure limit switches; check setpoint calibration against known device	Inspect and maintain burners and pilots; clean away obstructions; replace damaged pieces
Test visual/audible alarms for function, level	Lubricate plug valves
Verify spark gap on igniters	Test purge timer and trial-for-ignition timer
Check electric heating elements for cracks, contamination, and distortion	Check ductwork for integrity (leaks) and cleanliness
Test safety shutoff valves for leakage rate	Test shutdown sequence for all interlocks
Test manual fuel valves for signs of leakage	Check wiring terminals for tightness
Inspect all switches for signs of arcing	Replace filters and clean strainers
Inspect radiant tubes for signs of deterioration	Inspect air, water, fuel, and impulse piping for leaks

Source: Reproduced with permission of NFPA from NFPA 86 (2019). © 2018, National Fire Protection Association. For a full copy of NFPA 86, please visit www.nfpa.org.

Training. Staff who operate thermal systems should be trained before being initially authorized to operate equipment and should receive follow-up training at least annually to maintain such authorization. The training should be based on the written instructions provided by the manufacturer or installer as part of the commissioning process, and should cover startup, mode transition, normal operation, shutdown (normal and emergency), and lockout procedures. Training should always emphasize the potential danger of bypassing interlocks rather than addressing the underlying cause, and examples of failure to heed such warnings should be given. The training program should include a written and/or practical competency exam.

Records of training, inspection, and maintenance should be retained for one year or until the next training, whichever is longer.

18.5 Operation and Maintenance Manual

The equipment manufacturer should assemble and transfer an operation and maintenance (O&M) manual to the owner at or before the time of system turnover. The purpose of the O&M manual is fivefold:

- Reconfirm the design basis, including process flow diagram (PFD), P&ID, and control system description.
- Provide instructions for the operation and maintenance staff.

- Identify the potential hazards associated with heating equipment and provide safety instructions and warnings to emphasize the negative consequences that may follow if safety procedures are not followed.
- Provide maintenance guidelines, including a list of recommended spare parts that the user should keep on hand to minimize downtime in the event of unanticipated failure or breakdown.
- Provide manufacturer's instructions for subcomponents that were integrated into the overall system.

Warnings that may be deemed appropriate for a thermal system's O&M manual include:

- NEVER jumper or bypass any safety interlock.
- NEVER operate the system at a temperature higher (lower) than ___ or a flow rate higher (lower) than ___.
- ALWAYS ensure the purge cycle is being performed with fresh air; avoid indeterminate purge gases that may contain residual flammable gases.
- NEVER rely exclusively on automatic valves to prevent leakage of fuel during prolonged shutdowns. Close manual valves too.
- NEVER enter an enclosed space without proper safety training and appropriate lock-out–tag-out procedures in place.

The following chapter arrangement may provide a useful guide for the manufacturer to follow when assembling their O&M manual.

Chapter 1 – Introduction and Scope
- **1.1** Scope and Purpose of O&M Manual
- **1.2** Description of System
- **1.3** Major Components and their Function
- **1.4** Sequence of Operation
- **1.5** Maintenance Tasks and Frequency

Chapter 2 – Bill of Materials
- **2.1** Process Equipment, Valves, and Instruments
- **2.2** Electrical Components

Chapter 3 – Drawings
- **3.1** Process Drawings (PFD, P&ID)
- **3.2** Mechanical Drawings (Site Plan, General Arrangement, Equipment Sections, and Elevations)
- **3.3** Electrical Schematics (Power, Control, and Wiring Termination)

Chapter 4 – PLC Ladder Logic Program

Chapter 5 – Vendor Literature (Subcomponents)
- **5.1** Control Panel
- **5.2** PLC and HMI
- **5.3** Field Instrumentation and Valves
- **5.4** Ceramic Packing, Insulation, and Refractory
- **5.5** Blowers, Compressors, Motors, and variable frequency drive (VFD)
- **5.6** Burner and Combustion System

The O&M manual is a vital document that should be assembled with as high a degree of workmanship as the system itself. If the instructions in the O&M manual provide inaccurate or unclear warnings, a user will someday perform a task erroneously and the author of the manual may be considered responsible for inadequate warnings or improper instructions.

18.6 Homework Problems

18.1 Give two examples of equipment and three examples of instruments that may be recommended for keeping on site as "spare parts."

18.2 Give an additional example of a warning (beyond those given in Section 18.5) that may be helpful and appropriate to be included in an operation and maintenance manual, to maintain safety in combustion systems.

18.3 Give an example of an important combustion safety topic that may not be well known to operators of thermal systems, and therefore, may be a helpful subject to cover in an operator training curriculum.

18.4 For each commissioning objective below, provide a recommendation to the system's commissioning engineer of techniques that may be useful to confirm the commissioning objective has been met satisfactorily.

(a) Confirm all pumps and blowers are properly wired to their designated electric power source.

(b) Confirm all temperature interlocks trip to the appropriate "safe state" when the process variable reaches the setpoint value.

(c) Confirm all installed instruments are equipped with stamped metal tags that bear the correct instrument tag number.

(d) Confirm PID tuning parameters for all control loops provide acceptable control response for that loop's control objective.

Cited References

Jackson, M. and Dröttboom, M.; (2014); How effective is preventative maintenance in saving money?; *Process Worldwide*; Vogel Communications.; https://www.process-worldwide.com/how-effective-is-preventative-maintenance-in-saving-money-a-415850.

NFPA 85; (2011); *Boiler and Combustion Systems Hazards Code*; See section 4.4, maintenance, inspection, training, and safety; Quincy, MA: National Fire Protection Association.

NFPA 86 (2019); *Standard for Ovens and Furnaces*; See section 7.4 inspection, testing, and maintenance; Quincy, MA: National Fire Protection Association.

NFPA 87 (2018); *Standard for Fluid Heaters*; See section 7.5 inspection, testing, and maintenance; Quincy, MA: National Fire Protection Association.

Appendix A

Property Tables

The property tables that follow were obtained from the sources cited at the beginning of each table, with full citations given at the end of this appendix.

Unless otherwise noted, the units for the various properties are:

- $p \doteq kPa$
- $\mathcal{V} \doteq m^3$
- $\hat{M} \doteq kg.kmol^{-1}$
- $v \doteq m^3.kg^{-1}$
- $\rho \doteq kg.m^{-3}$
- $u \doteq kJ.kg^{-1}$
- $h \doteq kJ.kg^{-1}$
- $s \doteq kJ.kg^{-1}.K^{-1}$
- $c \doteq kJ.kg^{-1}.K^{-1}$
- $\mu \doteq kg.m^{-1}.s^{-1}$
- $k \doteq W.m^{-1}.K^{-1}$

A.1 Properties of Air (Gas, 3.76 N_2 + 1.00 O_2, by Mole)

The enthalpy of formation of air is $\Delta\hat{h}_f^0 = 0.0\ kJ/kmol.$ The molar mass of air is $\hat{M} = 28.85\ kg/kmol.$ These properties were derived from the oxygen and nitrogen data sources by Linstrom and Mallard (2017–2021).

T (K)	p (kPa)	v (m^3/kg)	u (kJ/kg)	h (kJ/kg)	s (J/g·K)	c_v (J/g*K)	c_p (J/g*K)	μ (Pa*s)	k (W/m*K)	Air phase
275	**101.33**	0.781 698	197.6982	276.9092	6.653 656	0.722 516	1.012 388	1.729E−05	0.024 270	vapor
298.15	**101.33**	0.847 709	214.4522	300.3551	6.735 526	0.723 407	1.012 950	1.843E−05	0.025 915	vapor
300	**101.33**	0.852 980	215.7949	302.2239	6.741 763	0.723 495	1.013 087	1.852E−05	0.026 044	vapor
325	**101.33**	0.924 243	233.9149	327.5672	6.822 926	0.724 900	1.014 194	1.970E−05	0.027 776	vapor
350	**101.33**	0.995 468	252.0682	352.9438	6.898 145	0.726 770	1.015 914	2.085E−05	0.029 472	vapor
375	**101.33**	1.066 650	270.2771	378.3683	6.968 275	0.729 140	1.018 144	2.196E−05	0.031 139	vapor

(*Continued*)

Thermal Systems Design: Fundamentals and Projects, Second Edition. Richard J. Martin.

Companion website: www.wiley.com\go\Martin\ThermalSystemsDesign2

T (K)	p (kPa)	υ (m^3/kg)	u (kJ/kg)	h (kJ/kg)	s (J/g·K)	c_v (J/g*K)	c_p (J/g*K)	μ (Pa*s)	k (W/m*K)	Air phase
400	**101.33**	1.137 911	288.5540	403.8532	7.034 067	0.732 012	1.020 921	2.304E−05	0.032 784	vapor
425	**101.33**	1.209 091	306.9036	429.4160	7.096 045	0.735 380	1.024 130	2.409E−05	0.034 411	vapor
450	**101.33**	1.280 247	325.3410	455.0668	7.154 764	0.739 223	1.027 965	2.511E−05	0.036 024	vapor
475	**101.33**	1.351 427	343.8788	480.8179	7.210 422	0.743 511	1.032 162	2.611E−05	0.037 625	vapor
500	**101.33**	1.422 506	362.5299	506.6746	7.263 474	0.748 193	1.036 771	2.709E−05	0.039 216	vapor
525	**101.33**	1.493 686	381.3066	532.6569	7.314 165	0.753 236	1.041 766	2.804E−05	0.040 799	vapor
550	**101.33**	1.564 842	400.2042	558.7679	7.362 779	0.758 569	1.047 051	2.897E−05	0.042 376	vapor
575	**101.33**	1.635 945	419.2382	585.0152	7.409 411	0.764 153	1.052 613	2.989E−05	0.043 945	vapor
600	**101.33**	1.707 101	438.4185	611.4011	7.454 388	0.769 938	1.058 370	3.078E−05	0.045 506	vapor
625	**101.33**	1.778 281	457.7505	637.9363	7.497 713	0.775 861	1.064 289	3.166E−05	0.047 061	vapor
650	**101.33**	1.849 361	477.2217	664.6156	7.539 507	0.781 890	1.070 316	3.253E−05	0.048 609	vapor
675	**101.33**	1.920 517	496.8500	691.4494	7.580 078	0.787 976	1.076 349	3.337E−05	0.050 147	vapor
700	**101.33**	1.991 620	516.6252	718.4357	7.619 318	0.794 093	1.082 465	3.421E−05	0.051 677	vapor
725	**101.33**	2.062 776	536.5581	745.5765	7.657 382	0.800 189	1.088 535	3.503E−05	0.053 198	vapor
750	**101.33**	2.133 856	556.6357	772.8651	7.694 347	0.806 253	1.094 612	3.584E−05	0.054 709	vapor
775	**101.33**	2.205 012	576.8710	800.3060	7.730 411	0.812 254	1.100 542	3.664E−05	0.056 210	vapor
800	**101.33**	2.276 092	597.2539	827.8922	7.765 421	0.818 169	1.106 502	3.742E−05	0.057 699	vapor
825	**101.33**	2.347 195	617.7821	855.6261	7.799 555	0.823 985	1.112 316	3.820E−05	0.059 177	vapor
850	**101.33**	2.418 351	638.4557	883.5053	7.832 866	0.829 691	1.118 006	3.896E−05	0.060 643	vapor
875	**101.33**	2.489 431	659.2723	911.5252	7.865 300	0.835 265	1.123 550	3.971E−05	0.062 097	vapor
900	**101.33**	2.560 587	680.2219	939.6804	7.897 034	0.840 710	1.128 970	4.046E−05	0.063 539	vapor
925	**101.33**	2.631 667	701.3046	967.9687	7.928 068	0.846 013	1.134 267	4.120E−05	0.064 969	vapor
950	**101.33**	2.702 770	722.5202	996.3823	7.958 402	0.851 174	1.139 418	4.192E−05	0.066 386	vapor
975	**101.33**	2.773 926	743.8665	1024.9396	7.988 036	0.856 194	1.144 468	4.264E−05	0.067 790	vapor
1000	**101.33**	2.845 005	765.3335	1053.5969	8.017 093	0.861 062	1.149 319	4.336E−05	0.069 182	vapor
275	**200**	0.395 842	197.4782	276.6445	6.456 928	0.722 819	1.014 363	1.731E−05	0.024 301	vapor
300	**200**	0.432 053	215.5972	302.0069	6.545 135	0.723 735	1.014 671	1.854E−05	0.026 073	vapor
325	**200**	0.468 226	233.7349	327.3802	6.626 398	0.725 100	1.015 561	1.972E−05	0.027 803	vapor
350	**200**	0.504 377	251.9082	352.7868	6.701 717	0.726 940	1.017 005	2.086E−05	0.029 497	vapor
375	**200**	0.540 509	270.1295	378.2337	6.771 923	0.729 280	1.019 044	2.197E−05	0.031 163	vapor
400	**200**	0.576 630	288.4164	403.7439	6.837 815	0.732 139	1.021 714	2.305E−05	0.032 806	vapor
425	**200**	0.612 731	306.7759	429.3267	6.899 817	0.735 488	1.024 900	2.410E−05	0.034 432	vapor
450	**200**	0.648 825	325.2234	454.9898	6.958 536	0.739 323	1.028 563	2.512E−05	0.036 043	vapor
475	**200**	0.684 908	343.7711	480.7509	7.014 271	0.743 601	1.032 667	2.612E−05	0.037 643	vapor
500	**200**	0.720 982	362.4322	506.6276	7.067 345	0.748 281	1.037 260	2.710E−05	0.039 234	vapor
525	**200**	0.757 053	381.2089	532.6199	7.118 013	0.753 306	1.042 243	2.805E−05	0.040 817	vapor
550	**200**	0.793 116	400.1142	558.7409	7.166 651	0.758 636	1.047 442	2.898E−05	0.042 391	vapor
575	**200**	0.829 180	419.1582	584.9982	7.213 359	0.764 220	1.052 994	2.990E−05	0.043 960	vapor
600	**200**	0.865 241	438.3485	611.3941	7.258 260	0.769 996	1.058 747	3.079E−05	0.045 522	vapor
625	**200**	0.901 294	457.6728	637.9317	7.301 584	0.775 919	1.064 589	3.167E−05	0.047 076	vapor

T (K)	p (kPa)	υ (m^3/kg)	u (kJ/kg)	h (kJ/kg)	s (J/g·K)	c_v (J/g*K)	c_p (J/g*K)	μ (Pa*s)	k (W/m*K)	Air phase
650	**200**	0.937 340	477.1541	664.6262	7.343 455	0.781 945	1.070 516	3.253E−05	0.048 622	vapor
675	**200**	0.973 430	496.7823	691.4601	7.384 026	0.788 031	1.076 625	3.338E−05	0.050 160	vapor
700	**200**	1.009 458	516.5652	718.4487	7.423 267	0.794 139	1.082 665	3.422E−05	0.051 690	vapor
725	**200**	1.045 486	536.5005	745.5972	7.461 331	0.800 237	1.088 735	3.504E−05	0.053 210	vapor
750	**200**	1.081 514	556.5857	772.8881	7.498 395	0.806 298	1.094 812	3.584E−05	0.054 721	vapor
775	**200**	1.117 540	576.8210	800.3366	7.534 382	0.812 292	1.100 742	3.664E−05	0.056 220	vapor
800	**200**	1.153 568	597.2039	827.9252	7.569 393	0.818 206	1.106 602	3.743E−05	0.057 709	vapor
825	**200**	1.189 673	617.7321	855.6667	7.603 504	0.824 022	1.112 416	3.820E−05	0.059 188	vapor
850	**200**	1.225 701	638.4134	883.5460	7.636 814	0.829 719	1.118 106	3.896E−05	0.060 653	vapor
875	**200**	1.261 729	659.2223	911.5682	7.669 348	0.835 300	1.123 650	3.972E−05	0.062 107	vapor
900	**200**	1.297 758	680.1819	939.7311	7.701 082	0.840 746	1.129 147	4.047E−05	0.063 549	vapor
925	**200**	1.333 786	701.2646	968.0217	7.732 093	0.846 046	1.134 444	4.120E−05	0.064 978	vapor
950	**200**	1.369 814	722.4802	996.4637	7.762 374	0.851 210	1.139 518	4.193E−05	0.066 395	vapor
975	**200**	1.405 819	743.8265	1025.0233	7.792 084	0.856 219	1.144 568	4.265E−05	0.067 799	vapor
1000	**200**	1.441 847	765.2958	1053.6805	7.821 142	0.861 088	1.149 419	4.336E−05	0.069 190	vapor
275	**500**	0.158 088	196.7852	275.8305	6.190 350	0.723 744	1.020 321	1.737E−05	0.024 402	vapor
300	**500**	0.172 685	214.9842	301.3306	6.279 057	0.724 472	1.019 445	1.859E−05	0.026 166	vapor
325	**500**	0.187 244	233.1919	326.8139	6.360 697	0.725 706	1.019 528	1.976E−05	0.027 888	vapor
350	**500**	0.201 783	251.4152	352.3081	6.436 216	0.727 455	1.020 365	2.091E−05	0.029 576	vapor
375	**500**	0.216 299	269.6841	377.8327	6.506 722	0.729 723	1.021 907	2.201E−05	0.031 236	vapor
400	**500**	0.230 798	288.0110	403.4052	6.572 738	0.732 517	1.024 102	2.309E−05	0.032 875	vapor
425	**500**	0.245 287	306.4005	429.0481	6.634 893	0.735 833	1.026 984	2.414E−05	0.034 497	vapor
450	**500**	0.259 766	324.8780	454.7612	6.693 711	0.739 626	1.030 356	2.516E−05	0.036 105	vapor
475	**500**	0.274 235	343.4511	480.5699	6.749 493	0.743 874	1.034 335	2.615E−05	0.037 701	vapor
500	**500**	0.288 702	362.1345	506.4843	6.802 644	0.748 524	1.038 734	2.713E−05	0.039 289	vapor
525	**500**	0.303 161	380.9312	532.5066	6.853 412	0.753 536	1.043 529	2.808E−05	0.040 869	vapor
550	**500**	0.317 610	399.8566	558.6576	6.902 150	0.758 849	1.048 621	2.901E−05	0.042 442	vapor
575	**500**	0.332 059	418.9129	584.9449	6.948 858	0.764 411	1.053 994	2.992E−05	0.044 007	vapor
600	**500**	0.346 498	438.1109	611.3608	6.993 836	0.770 176	1.059 647	3.081E−05	0.045 567	vapor
625	**500**	0.360 936	457.4528	637.9260	7.037 183	0.776 090	1.065 466	3.169E−05	0.047 119	vapor
650	**500**	0.375 375	476.9441	664.6329	7.079 131	0.782 098	1.071 316	3.255E−05	0.048 664	vapor
675	**500**	0.389 807	496.5923	691.4945	7.119 648	0.788 174	1.077 325	3.340E−05	0.050 200	vapor
700	**500**	0.404 243	516.3852	718.5030	7.158 966	0.794 279	1.083 365	3.424E−05	0.051 729	vapor
725	**500**	0.418 675	536.3228	745.6616	7.197 030	0.800 365	1.089 335	3.506E−05	0.053 247	vapor
750	**500**	0.433 103	556.4180	772.9725	7.234 094	0.806 419	1.095 312	3.586E−05	0.054 757	vapor
775	**500**	0.447 525	576.6610	800.4233	7.270 081	0.812 410	1.101 242	3.666E−05	0.056 255	vapor
800	**500**	0.461 951	597.0539	828.0296	7.305 169	0.818 322	1.107 102	3.744E−05	0.057 743	vapor
825	**500**	0.476 380	617.5921	855.7811	7.339 303	0.824 131	1.112 916	3.822E−05	0.059 220	vapor
850	**500**	0.490 802	638.2734	883.6780	7.372 613	0.829 827	1.118 583	3.898E−05	0.060 685	vapor

(*Continued*)

T (K)	p (kPa)	v (m^3/kg)	u (kJ/kg)	h (kJ/kg)	s (J/g·K)	c_v (J/g*K)	c_p (J/g*K)	μ (Pa*s)	k (W/m*K)	Air phase
875	**500**	0.505 220	659.1000	911.7102	7.405 124	0.835 401	1.124 050	3.973E−05	0.062 138	vapor
900	**500**	0.519 639	680.0596	939.8755	7.436 858	0.840 836	1.129 447	4.048E−05	0.063 579	vapor
925	**500**	0.534 068	701.1522	968.1814	7.467 892	0.846 136	1.134 744	4.122E−05	0.065 008	vapor
950	**500**	0.548 487	722.3702	996.6357	7.498 249	0.851 290	1.139 894	4.194E−05	0.066 424	vapor
975	**500**	0.562 906	743.7242	1025.1976	7.527 883	0.856 308	1.144 868	4.266E−05	0.067 827	vapor
1000	**500**	0.577 318	765.1935	1053.8549	7.556 941	0.861 168	1.149 719	4.337E−05	0.069 217	vapor
275	**1 000**	0.078 846	195.6445	274.4903	5.986 438	0.725 274	1.030 288	1.747E−05	0.024 587	vapor
300	**1 000**	0.086 237	213.9735	300.2103	6.075 945	0.725 698	1.027 496	1.868E−05	0.026 335	vapor
325	**1 000**	0.093 594	232.2812	325.8736	6.158 108	0.726 711	1.026 098	1.984E−05	0.028 043	vapor
350	**1 000**	0.100 925	250.5999	351.5255	6.234 204	0.728 298	1.025 842	2.098E−05	0.029 719	vapor
375	**1 000**	0.108 238	268.9411	377.1778	6.305 010	0.730 449	1.026 577	2.208E−05	0.031 369	vapor
400	**1 000**	0.115 532	287.3280	402.8603	6.371 303	0.733 158	1.028 158	2.315E−05	0.032 999	vapor
425	**1 000**	0.122 816	305.7752	428.5931	6.433 681	0.736 394	1.030 458	2.419E−05	0.034 613	vapor
450	**1 000**	0.130 091	324.3027	454.3915	6.492 623	0.740 127	1.033 423	2.521E−05	0.036 213	vapor
475	**1 000**	0.137 354	342.9181	480.2726	6.548 658	0.744 323	1.037 011	2.620E−05	0.037 804	vapor
500	**1 000**	0.144 610	361.6369	506.2470	6.601 933	0.748 932	1.041 111	2.717E−05	0.039 386	vapor
525	**1 000**	0.151 863	380.4636	532.3270	6.652 801	0.753 907	1.045 617	2.812E−05	0.040 961	vapor
550	**1 000**	0.159 107	399.4212	558.5280	6.701 539	0.759 195	1.050 505	2.905E−05	0.042 528	vapor
575	**1 000**	0.166 352	418.5052	584.8553	6.748 346	0.764 729	1.055 760	2.996E−05	0.044 091	vapor
600	**1 000**	0.173 586	437.7332	611.3189	6.793 424	0.770 472	1.061 200	3.085E−05	0.045 646	vapor
625	**1 000**	0.180 821	457.0951	637.9164	6.836 848	0.776 363	1.066 843	3.173E−05	0.047 195	vapor
650	**1 000**	0.188 055	476.6064	664.6610	6.878 819	0.782 360	1.072 669	3.259E−05	0.048 736	vapor
675	**1 000**	0.195 280	496.2646	691.5526	6.919 413	0.788 418	1.078 502	3.344E−05	0.050 270	vapor
700	**1 000**	0.202 514	516.0776	718.5911	6.958 731	0.794 505	1.084 442	3.427E−05	0.051 795	vapor
725	**1 000**	0.209 739	536.0405	745.7720	6.996 895	0.800 589	1.090 388	3.509E−05	0.053 312	vapor
750	**1 000**	0.216 962	556.1457	773.1029	7.033 959	0.806 625	1.096 288	3.589E−05	0.054 819	vapor
775	**1 000**	0.224 180	576.4010	800.5891	7.070 046	0.812 611	1.102 119	3.669E−05	0.056 315	vapor
800	**1 000**	0.231 404	596.8115	828.2153	7.105 057	0.818 508	1.107 955	3.747E−05	0.057 801	vapor
825	**1 000**	0.238 619	617.3598	855.9792	7.139 268	0.824 304	1.113 616	3.825E−05	0.059 276	vapor
850	**1 000**	0.245 843	638.0534	883.8937	7.172 578	0.830 001	1.119 259	3.901E−05	0.060 739	vapor
875	**1 000**	0.253 058	658.8877	911.9460	7.205 112	0.835 562	1.124 726	3.976E−05	0.062 190	vapor
900	**1 000**	0.260 272	679.8496	940.1289	7.236 846	0.840 998	1.130 123	4.050E−05	0.063 630	vapor
925	**1 000**	0.267 487	700.9522	968.4471	7.267 880	0.846 290	1.135 321	4.124E−05	0.065 057	vapor
950	**1 000**	0.274 703	722.1855	996.8961	7.298 214	0.851 442	1.140 394	4.197E−05	0.066 471	vapor
975	**1 000**	0.281 916	743.5418	1025.4604	7.327 948	0.856 444	1.145 368	4.269E−05	0.067 873	vapor
1000	**1 000**	0.289 132	765.0288	1054.1223	7.357 006	0.861 310	1.150 195	4.340E−05	0.069 263	vapor
275	**2 000**	0.039 239	193.3431	271.8250	5.778 450	0.728 286	1.050 586	1.767E−05	0.025 019	vapor
300	**2 000**	0.043 026	211.9475	297.9950	5.869 557	0.728 103	1.043 668	1.886E−05	0.026 722	vapor
325	**2 000**	0.046 779	230.4752	324.0337	5.952 897	0.728 692	1.039 344	2.001E−05	0.028 394	vapor
350	**2 000**	0.050 506	248.9662	349.9779	6.029 816	0.729 967	1.036 881	2.113E−05	0.030 039	vapor

T (K)	p (kPa)	v (m^3/kg)	u (kJ/kg)	h (kJ/kg)	s (J/g·K)	c_v (J/g*K)	c_p (J/g*K)	μ (Pa*s)	k (W/m*K)	Air phase
375	**2 000**	0.054 214	267.4628	375.8878	6.101 322	0.731 885	1.035 916	2.222E−05	0.031 664	vapor
400	**2 000**	0.057 906	285.9720	401.7880	6.168 214	0.734 411	1.036 114	2.328E−05	0.033 272	vapor
425	**2 000**	0.061 585	304.5345	427.7009	6.230 993	0.737 497	1.037 397	2.431E−05	0.034 865	vapor
450	**2 000**	0.065 254	323.1520	453.6593	6.290 388	0.741 116	1.039 469	2.532E−05	0.036 449	vapor
475	**2 000**	0.068 915	341.8474	479.6857	6.346 646	0.745 220	1.042 355	2.631E−05	0.038 026	vapor
500	**2 000**	0.072 569	360.6462	505.7824	6.400 198	0.749 749	1.045 857	2.727E−05	0.039 595	vapor
525	**2 000**	0.076 217	379.5406	531.9801	6.451 289	0.754 659	1.049 861	2.822E−05	0.041 158	vapor
550	**2 000**	0.079 861	398.5559	558.2787	6.500 227	0.759 877	1.054 344	2.914E−05	0.042 716	vapor
575	**2 000**	0.083 500	417.6999	584.6960	6.547 235	0.765 369	1.059 141	3.004E−05	0.044 268	vapor
600	**2 000**	0.087 135	436.9678	611.2396	6.592 412	0.771 062	1.064 277	3.093E−05	0.045 815	vapor
625	**2 000**	0.090 767	456.3798	637.9149	6.635 937	0.776 913	1.069 696	3.181E−05	0.047 355	vapor
650	**2 000**	0.094 396	475.9287	664.7271	6.678 007	0.782 879	1.075 222	3.266E−05	0.048 891	vapor
675	**2 000**	0.098 026	495.6270	691.6710	6.718 702	0.788 905	1.080 879	3.351E−05	0.050 417	vapor
700	**2 000**	0.101 649	515.4676	718.7672	6.758 119	0.794 971	1.086 619	3.433E−05	0.051 937	vapor
725	**2 000**	0.105 271	535.4628	746.0058	6.796 360	0.801 016	1.092 365	3.515E−05	0.053 447	vapor
750	**2 000**	0.108 894	555.6057	773.3843	6.833 524	0.807 040	1.098 065	3.596E−05	0.054 949	vapor
775	**2 000**	0.112 507	575.8886	800.9129	6.869 611	0.812 999	1.103 872	3.675E−05	0.056 441	vapor
800	**2 000**	0.116 130	596.3192	828.5767	6.904 722	0.818 881	1.109 532	3.753E−05	0.057 923	vapor
825	**2 000**	0.119 743	616.8951	856.3806	6.938 933	0.824 669	1.115 069	3.830E−05	0.059 394	vapor
850	**2 000**	0.123 363	637.6087	884.3275	6.972 343	0.830 336	1.120 636	3.906E−05	0.060 853	vapor
875	**2 000**	0.126 976	658.4630	912.4174	7.004 877	0.835 889	1.126 003	3.981E−05	0.062 301	vapor
900	**2 000**	0.130 589	679.4526	940.6303	7.036 688	0.841 310	1.131 300	4.055E−05	0.063 737	vapor
925	**2 000**	0.134 202	700.5652	968.9763	7.067 745	0.846 593	1.136 421	4.129E−05	0.065 161	vapor
950	**2 000**	0.137 814	721.8185	997.4169	7.098 079	0.851 729	1.141 471	4.201E−05	0.066 572	vapor
975	**2 000**	0.141 427	743.1949	1026.0625	7.127 813	0.856 726	1.146 345	4.273E−05	0.067 971	vapor
1000	**2 000**	0.145 038	764.6918	1054.7291	7.156 871	0.861 577	1.151 095	4.344E−05	0.069 357	vapor
275	**5 000**	0.015 525	186.4448	264.0694	5.490 079	0.736 909	1.112 836	1.836E−05	0.026 678	vapor
300	**5 000**	0.017 145	205.8908	291.6134	5.585 956	0.734 996	1.092 292	1.946E−05	0.028 181	vapor
325	**5 000**	0.018 729	225.0908	318.7413	5.672 819	0.734 382	1.078 564	2.054E−05	0.029 695	vapor
350	**5 000**	0.020 289	244.1318	345.5803	5.752 338	0.734 789	1.069 277	2.161E−05	0.031 213	vapor
375	**5 000**	0.021 830	263.0807	372.2325	5.825 944	0.736 043	1.063 142	2.266E−05	0.032 731	vapor
400	**5 000**	0.023 356	281.9753	398.7604	5.894 436	0.738 050	1.059 386	2.368E−05	0.034 251	vapor
425	**5 000**	0.024 871	300.8602	425.2132	5.958 592	0.740 727	1.057 456	2.469E−05	0.035 769	vapor
450	**5 000**	0.026 375	319.7653	451.6393	6.019 010	0.744 008	1.057 000	2.567E−05	0.037 287	vapor
475	**5 000**	0.027 872	338.7161	478.0757	6.076 169	0.747 837	1.057 699	2.663E−05	0.038 805	vapor
500	**5 000**	0.029 362	357.7272	504.5378	6.130 443	0.752 138	1.059 485	2.758E−05	0.040 324	vapor
525	**5 000**	0.030 846	376.8269	531.0554	6.182 188	0.756 845	1.061 994	2.850E−05	0.041 842	vapor
550	**5 000**	0.032 326	396.0099	557.6418	6.231 626	0.761 899	1.065 272	2.941E−05	0.043 360	vapor
575	**5 000**	0.033 802	415.3115	584.3267	6.279 134	0.767 244	1.069 024	3.030E−05	0.044 877	vapor
600	**5 000**	0.035 275	434.7272	611.0980	6.324 711	0.772 807	1.073 207	3.118E−05	0.046 392	vapor

(Continued)

T (K)	p (kPa)	υ (m^3/kg)	u (kJ/kg)	h (kJ/kg)	s (J/g·K)	c_v (J/g*K)	c_p (J/g*K)	μ (Pa*s)	k (W/m*K)	Air phase
625	**5 000**	0.036 744	454.2668	637.9886	6.368 612	0.778 545	1.077 826	3.204E−05	0.047 904	vapor
650	**5 000**	0.038 211	473.9410	664.9931	6.410 983	0.784 407	1.082 653	3.289E−05	0.049 412	vapor
675	**5 000**	0.039 675	493.7470	692.1224	6.451 877	0.790 351	1.087 686	3.372E−05	0.050 916	vapor
700	**5 000**	0.041 137	513.6876	719.3762	6.491 595	0.796 326	1.092 849	3.454E−05	0.052 413	vapor
725	**5 000**	0.042 598	533.7781	746.7625	6.530 036	0.802 309	1.098 095	3.535E−05	0.053 904	vapor
750	**5 000**	0.044 057	554.0011	774.2810	6.567 323	0.808 261	1.103 472	3.614E−05	0.055 386	vapor
775	**5 000**	0.045 514	574.3640	801.9372	6.603 587	0.814 169	1.108 802	3.693E−05	0.056 862	vapor
800	**5 000**	0.046 970	594.8699	829.7211	6.638 898	0.819 992	1.114 086	3.770E−05	0.058 327	vapor
825	**5 000**	0.048 425	615.5158	857.6403	6.673 208	0.825 728	1.119 399	3.847E−05	0.059 783	vapor
850	**5 000**	0.049 879	636.2947	885.6872	6.706 719	0.831 356	1.124 589	3.922E−05	0.061 229	vapor
875	**5 000**	0.051 331	657.2090	913.8671	6.739 430	0.836 860	1.129 733	3.997E−05	0.062 663	vapor
900	**5 000**	0.052 784	678.2563	942.1777	6.771 287	0.842 243	1.134 754	4.071E−05	0.064 088	vapor
925	**5 000**	0.054 235	699.4343	970.6037	6.802 498	0.847 486	1.139 751	4.144E−05	0.065 500	vapor
950	**5 000**	0.055 686	720.7352	999.1372	6.832 932	0.852 597	1.144 524	4.216E−05	0.066 901	vapor
975	**5 000**	0.057 136	742.1516	1027.7992	6.862 689	0.857 559	1.149 198	4.287E−05	0.068 290	vapor
1000	**5 000**	0.058 585	763.6962	1056.6355	6.891 846	0.862 380	1.153 849	4.358E−05	0.069 666	vapor
275	**10 000**	0.007 715	175.1099	252.2653	5.252 268	0.749 673	1.213 766	1.980E−05	0.030 100	vapor
300	**10 000**	0.008 596	196.0572	282.0178	5.355 825	0.745 308	1.169 525	2.067E−05	0.031 152	vapor
325	**10 000**	0.009 447	216.3975	310.8647	5.448 251	0.742 996	1.140 006	2.159E−05	0.032 322	vapor
350	**10 000**	0.010 275	236.3471	339.0940	5.531 940	0.742 149	1.119 556	2.253E−05	0.033 572	vapor
375	**10 000**	0.011 085	256.0407	366.8948	5.608 616	0.742 456	1.105 178	2.348E−05	0.034 870	vapor
400	**10 000**	0.011 883	275.5623	394.3890	5.679 655	0.743 710	1.095 110	2.443E−05	0.036 204	vapor
425	**10 000**	0.012 670	294.9795	421.6765	5.745 834	0.745 786	1.088 312	2.537E−05	0.037 564	vapor
450	**10 000**	0.013 448	314.3370	448.8296	5.807 875	0.748 573	1.083 860	2.630E−05	0.038 947	vapor
475	**10 000**	0.014 220	333.6877	475.8884	5.866 410	0.751 980	1.081 386	2.722E−05	0.040 348	vapor
500	**10 000**	0.014 985	353.0541	502.9051	5.921 785	0.755 931	1.080 441	2.812E−05	0.041 763	vapor
525	**10 000**	0.015 746	372.4639	529.9204	5.974 553	0.760 338	1.080 780	2.901E−05	0.043 191	vapor
550	**10 000**	0.016 502	391.9345	556.9567	6.024 868	0.765 134	1.082 049	2.989E−05	0.044 628	vapor
575	**10 000**	0.017 255	411.4839	584.0317	6.072 999	0.770 247	1.084 201	3.075E−05	0.046 072	vapor
600	**10 000**	0.018 005	431.1272	611.1706	6.119 176	0.775 620	1.087 037	3.161E−05	0.047 521	vapor
625	**10 000**	0.018 751	450.8798	638.3835	6.163 654	0.781 184	1.090 356	3.245E−05	0.048 974	vapor
650	**10 000**	0.019 496	470.7364	665.6934	6.206 448	0.786 886	1.094 106	3.327E−05	0.050 428	vapor
675	**10 000**	0.020 238	490.7177	693.0927	6.247 819	0.792 687	1.098 216	3.409E−05	0.051 882	vapor
700	**10 000**	0.020 978	510.8236	720.6042	6.287 860	0.798 540	1.102 555	3.489E−05	0.053 336	vapor
725	**10 000**	0.021 717	531.0618	748.2281	6.326 601	0.804 404	1.107 125	3.569E−05	0.054 784	vapor
750	**10 000**	0.022 454	551.4201	775.9643	6.364 241	0.810 255	1.111 725	3.647E−05	0.056 230	vapor
775	**10 000**	0.023 190	571.9183	803.8105	6.400 752	0.816 061	1.116 532	3.724E−05	0.057 670	vapor
800	**10 000**	0.023 925	592.5443	831.7821	6.436 316	0.821 806	1.121 292	3.801E−05	0.059 102	vapor
825	**10 000**	0.024 658	613.2955	859.8790	6.470 850	0.827 462	1.126 053	3.876E−05	0.060 528	vapor
850	**10 000**	0.025 391	634.1821	888.0935	6.504 561	0.833 010	1.130 820	3.950E−05	0.061 946	vapor

T (K)	p (kPa)	υ (m^3/kg)	u (kJ/kg)	h (kJ/kg)	s (J/g·K)	c_v (J/g*K)	c_p (J/g*K)	μ (Pa*s)	k (W/m*K)	Air phase
875	**10 000**	0.026 123	655.1917	916.4211	6.537 371	0.838 452	1.135 563	4.024E−05	0.063 354	vapor
900	**10 000**	0.026 854	676.3344	944.8717	6.569 482	0.843 768	1.140 284	4.097E−05	0.064 754	vapor
925	**10 000**	0.027 584	697.5923	973.4353	6.600 739	0.848 960	1.144 881	4.169E−05	0.066 144	vapor
950	**10 000**	0.028 313	718.9786	1002.0736	6.631 373	0.854 016	1.149 378	4.240E−05	0.067 523	vapor
975	**10 000**	0.029 042	740.4803	1030.9145	6.661 307	0.858 928	1.153 828	4.311E−05	0.068 892	vapor
1000	**10 000**	0.029 771	762.0926	1059.7765	6.690 541	0.863 699	1.158 102	4.381E−05	0.070 250	vapor
275	**20 000**	0.003 997	154.7088	234.6585	4.991 661	0.768 813	1.359 279	2.381E−05	0.037 918	vapor
300	**20 000**	0.004 471	178.2504	267.6646	5.106 565	0.761 683	1.285 789	2.389E−05	0.037 824	vapor
325	**20 000**	0.004 926	200.6066	299.1356	5.207 374	0.757 196	1.234 931	2.427E−05	0.038 168	vapor
350	**20 000**	0.005 368	222.1707	329.5314	5.297 436	0.754 621	1.198 723	2.483E−05	0.038 805	vapor
375	**20 000**	0.005 798	243.1879	359.1538	5.379 222	0.753 538	1.172 357	2.549E−05	0.039 614	vapor
400	**20 000**	0.006 219	263.8221	388.2039	5.454 224	0.753 669	1.152 974	2.621E−05	0.040 545	vapor
425	**20 000**	0.006 633	284.1903	416.8433	5.523 696	0.754 797	1.138 587	2.698E−05	0.041 567	vapor
450	**20 000**	0.007 040	304.3694	445.1744	5.588 484	0.756 794	1.128 162	2.776E−05	0.042 659	vapor
475	**20 000**	0.007 442	324.4395	473.2764	5.649 266	0.759 526	1.120 631	2.855E−05	0.043 807	vapor
500	**20 000**	0.007 839	344.4329	501.2248	5.706 587	0.762 891	1.115 554	2.935E−05	0.045 001	vapor
525	**20 000**	0.008 233	364.4049	529.0641	5.760 955	0.766 793	1.112 299	3.016E−05	0.046 231	vapor
550	**20 000**	0.008 623	384.3833	556.8497	5.812 616	0.771 140	1.110 552	3.096E−05	0.047 492	vapor
575	**20 000**	0.009 011	404.3903	584.6093	5.861 947	0.775 863	1.110 111	3.176E−05	0.048 778	vapor
600	**20 000**	0.009 396	424.4512	612.3653	5.909 201	0.780 882	1.110 677	3.255E−05	0.050 085	vapor
625	**20 000**	0.009 778	444.5739	640.1482	5.954 626	0.786 142	1.111 973	3.334E−05	0.051 408	vapor
650	**20 000**	0.010 160	464.7835	667.9704	5.998 273	0.791 568	1.114 053	3.412E−05	0.052 743	vapor
675	**20 000**	0.010 538	485.0824	695.8520	6.040 291	0.797 117	1.116 562	3.489E−05	0.054 089	vapor
700	**20 000**	0.010 916	505.4837	723.8035	6.081 008	0.802 740	1.119 479	3.566E−05	0.055 441	vapor
725	**20 000**	0.011 293	525.9873	751.8298	6.120 349	0.808 404	1.122 825	3.642E−05	0.056 799	vapor
750	**20 000**	0.011 667	546.6009	779.9436	6.158 413	0.814 065	1.126 379	3.717E−05	0.058 159	vapor
775	**20 000**	0.012 041	567.3321	808.1475	6.195 477	0.819 696	1.130 109	3.791E−05	0.059 520	vapor
800	**20 000**	0.012 413	588.1734	836.4491	6.231 388	0.825 273	1.133 969	3.865E−05	0.060 880	vapor
825	**20 000**	0.012 786	609.1377	864.8460	6.266 375	0.830 784	1.137 983	3.938E−05	0.062 237	vapor
850	**20 000**	0.013 156	630.2096	893.3482	6.300 409	0.836 202	1.141 973	4.010E−05	0.063 591	vapor
875	**20 000**	0.013 527	651.4046	921.9458	6.333 520	0.841 518	1.146 017	4.082E−05	0.064 940	vapor
900	**20 000**	0.013 897	672.7125	950.6464	6.365 854	0.846 719	1.150 060	4.152E−05	0.066 283	vapor
925	**20 000**	0.014 265	694.1358	979.4400	6.397 465	0.851 796	1.154 134	4.223E−05	0.067 620	vapor
950	**20 000**	0.014 634	715.6751	1008.3903	6.428 299	0.856 754	1.158 108	4.292E−05	0.068 951	vapor
975	**20 000**	0.015 002	737.3198	1037.3569	6.458 456	0.861 571	1.162 005	4.361E−05	0.070 273	vapor
1000	**20 000**	0.015 369	759.0675	1066.4165	6.487 890	0.866 256	1.165 932	4.430E−05	0.071 587	vapor

Source: Linstrom and Mallard (2017–2021).

A.2 Properties of Nitrogen (Gas)

The enthalpy of formation of nitrogen is $\Delta \hat{h}_f^0 = 0.0\ kJ/kmol$. The molar mass of nitrogen is $\hat{M} = 28.012\ 86\ kg/kmol$. These properties were derived from Linstrom and Mallard (2017–2021).

T (K)	p (kPa)	v (m^3/kg)	u (kJ/kg)	h (kJ/kg)	s (J/g*K)	c_v (J/g*K)	c_p (J/g*K)	μ (Pa*s)	k (W/m*K)	N_2 phase
275	**101.33**	0.805 15	203.57	285.16	6.7511	0.742 9	1.0414	1.67E−05	0.024 159	Vapor
298.15	**101.33**	0.873 13	220.79	309.27	6.8353	0.743 14	352.07	1.781E−05	0.025 735	vapor
300	**101.33**	0.878 56	222.17	311.19	6.8417	0.743 17	1.0414	1.79E−05	0.025 858	vapor
325	**101.33**	0.951 95	240.77	337.23	6.9251	0.743 69	1.0416	1.90E−05	0.027 505	vapor
350	**101.33**	1.025 3	259.38	363.28	7.0023	0.744 53	1.0423	2.01E−05	0.029 107	vapor
375	**101.33**	1.098 6	278.02	389.35	7.0742	0.745 77	1.0434	2.12E−05	0.030 671	vapor
400	**101.33**	1.172	296.7	415.45	7.1416	0.747 46	1.045	2.22E−05	0.032 205	vapor
500	**101.33**	1.465 1	371.99	520.45	7.3759	0.759 21	1.0564	2.61E−05	0.038 143	vapor
600	**101.33**	1.758 2	448.82	626.98	7.5701	0.778 07	1.0751	2.96E−05	0.043 917	vapor
700	**101.33**	2.051 2	527.77	735.62	7.7375	0.801 1	1.0981	3.29E−05	0.049 605	vapor
800	**101.33**	2.344 2	609.1	846.64	7.8857	0.825 35	1.1223	3.59E−05	0.055 197	vapor
900	**101.33**	2.637 2	692.83	960.05	8.0192	0.848 84	1.1457	3.88E−05	0.060 666	vapor
1000	**101.33**	2.930 1	778.82	1075.7	8.1411	0.870 52	1.1674	4.16E−05	0.065 991	vapor
1100	**101.33**	3.223	866.86	1193.5	8.2533	0.889 99	1.1868	4.43E−05	0.071 16	vapor
1200	**101.33**	3.516	956.74	1313	8.3573	0.907 2	1.2040	4.68E−05	0.076 173	vapor
1300	**101.33**	3.808 9	1048.2	1434.2	8.4543	0.922 31	1.2191	4.93E−05	0.081 038	vapor
1400	**101.33**	4.101 8	1141.1	1556.8	8.5451	0.935 52	1.2323	5.17E−05	0.085 763	vapor
1500	**101.33**	4.394 7	1235.3	1680.6	8.6306	0.947 09	1.2439	5.41E−05	0.090 361	vapor
1600	**101.33**	4.687 7	1330.5	1805.5	8.7112	0.957 23	1.254	5.64E−05	0.094 843	vapor
1700	**101.33**	4.980 6	1426.7	1931.4	8.7875	0.966 15	1.2630	5.86E−05	0.099 222	vapor
1800	**101.33**	5.273 5	1523.7	2058.1	8.8599	0.974 03	1.2708	6.08E−05	0.103 51	vapor
1900	**101.33**	5.566 4	1621.5	2185.5	8.9288	0.981 02	1.2778	6.30E−05	0.107 71	vapor
2000	**101.33**	5.859 3	1719.9	2313.6	8.9945	0.987 25	1.2841	6.52E−05	0.111 84	vapor
275	**200**	0.407 76	203.35	284.9	6.5485	0.743 2	1.0434	1.67E−05	0.024 192	vapor
300	**200**	0.445 05	221.97	310.98	6.6392	0.743 41	1.043	1.79E−05	0.025 889	vapor
325	**200**	0.482 3	240.59	337.05	6.7227	0.743 89	1.043	1.90E−05	0.027 534	vapor
350	**200**	0.519 53	259.22	363.13	6.8000	0.744 7	1.0434	2.01E−05	0.029 134	vapor
375	**200**	0.556 74	277.87	389.22	6.872	0.745 91	1.0443	2.12E−05	0.030 697	vapor
400	**200**	0.593 94	296.56	415.35	6.9395	0.747 59	1.0458	2.22E−05	0.032 229	vapor
500	**200**	0.742 6	371.89	520.41	7.1739	0.759 3	1.0569	2.61E−05	0.038 162	vapor
600	**200**	0.891 17	448.75	626.98	7.3681	0.778 13	1.0755	2.96E−05	0.043 934	vapor

T (K)	p (kPa)	v (m^3/kg)	u (kJ/kg)	h (kJ/kg)	s (J/g*K)	c_v (J/g*K)	c_p (J/g*K)	μ (Pa*s)	k (W/m*K)	N_2 phase
700	**200**	1.039 7	527.71	735.64	7.5356	0.801 15	1.0983	3.29E−05	0.049 619	vapor
800	**200**	1.188 1	609.05	846.68	7.6838	0.825 39	1.1224	3.59E−05	0.055 209	vapor
900	**200**	1.336 6	692.79	960.11	7.8174	0.848 88	1.1459	3.88E−05	0.060 677	vapor
1000	**200**	1.485	778.78	1075.8	7.9393	0.870 55	1.1675	4.16E−05	0.066 00	vapor
1100	**200**	1.633 5	866.84	1193.5	8.0515	0.890 01	1.1869	4.43E−05	0.071 169	vapor
1200	**200**	1.781 9	956.72	1313.1	8.1555	0.907 22	1.2041	4.68E−05	0.076 182	vapor
1300	**200**	1.930 3	1048.2	1434.3	8.2525	0.922 33	1.2192	4.93E−05	0.081 045	vapor
1400	**200**	2.078 7	1141.1	1556.9	8.3433	0.935 54	1.2324	5.17E−05	0.085 77	vapor
1500	**200**	2.227 1	1235.3	1680.7	8.4287	0.947 11	1.2439	5.41E−05	0.090 368	vapor
1600	**200**	2.375 6	1330.5	1805.6	8.5094	0.957 25	1.2541	5.64E−05	0.094 849	vapor
1700	**200**	2.524	1426.7	1931.5	8.5857	0.966 17	1.263	5.86E−05	0.099 227	vapor
1800	**200**	2.672 4	1523.7	2058.2	8.6581	0.974 04	1.2709	6.08E−05	0.103 51	vapor
1900	**200**	2.820 8	1621.5	2185.6	8.727	0.981 03	1.2778	6.30E−05	0.107 71	vapor
2000	**200**	2.969 2	1719.9	2313.7	8.7927	0.987 26	1.2841	6.52E−05	0.111 84	vapor
275	**500**	0.162 9	202.65	284.1	6.274	0.744 13	1.0494	1.68E−05	0.024 299	vapor
300	**500**	0.177 93	221.35	310.32	6.3652	0.744 15	1.0478	1.80E−05	0.025 988	vapor
325	**500**	0.192 92	240.04	336.5	6.4491	0.744 5	1.047	1.91E−05	0.027 626	vapor
350	**500**	0.207 89	258.72	362.67	6.5266	0.745 22	1.0468	2.02E−05	0.029 22	vapor
375	**500**	0.222 84	277.42	388.84	6.5989	0.746 36	1.0472	2.12E−05	0.030 777	vapor
400	**500**	0.237 77	296.15	415.03	6.6665	0.747 97	1.0482	2.23E−05	0.032 305	vapor
500	**500**	0.297 4	371.59	520.29	6.9013	0.759 55	1.0584	2.61E−05	0.038 223	vapor
600	**500**	0.356 92	448.51	626.97	7.0958	0.778 32	1.0764	2.96E−05	0.043 984	vapor
700	**500**	0.416 39	527.53	735.72	7.2634	0.801 3	1.0990	3.29E−05	0.049 662	vapor
800	**500**	0.475 82	608.90	846.81	7.4117	0.825 52	1.1229	3.60E−05	0.055 247	vapor
900	**500**	0.535 23	692.67	960.28	7.5453	0.848 98	1.1462	3.89E−05	0.060 711	vapor
1000	**500**	0.594 63	778.68	1076	7.6672	0.870 64	1.1678	4.16E−05	0.066 031	vapor
1100	**500**	0.654 02	866.75	1193.8	7.7794	0.890 09	1.1871	4.43E−05	0.071 196	vapor
1200	**500**	0.713 41	956.65	1313.4	7.8835	0.907 3	1.2043	4.68E−05	0.076 207	vapor
1300	**500**	0.772 78	1048.2	1434.6	7.9805	0.922 39	1.2193	4.93E−05	0.081 069	vapor
1400	**500**	0.832 16	1141.1	1557.2	8.0713	0.935 6	1.2325	5.17E−05	0.085 792	vapor
1500	**500**	0.891 53	1235.2	1681	8.1568	0.947 16	1.2441	5.41E−05	0.090 388	vapor
1600	**500**	0.950 9	1330.5	1805.9	8.2374	0.957 3	1.2542	5.64E−05	0.094 868	vapor
1700	**500**	1.010 3	1426.7	1931.8	8.3137	0.966 21	1.2631	5.86E−05	0.099 245	vapor
1800	**500**	1.069 6	1523.7	2058.5	8.3861	0.974 09	1.2709	6.08E−05	0.103 53	vapor
1900	**500**	1.129	1621.5	2186	8.455	0.981 07	1.2779	6.30E−05	0.107 73	vapor
2000	**500**	1.188 4	1719.9	2314.1	8.5207	0.987 3	1.2841	6.52E−05	0.111 86	vapor
275	**1 000**	0.081 291	201.5	282.79	6.0641	0.745 66	1.0594	1.69E−05	0.024 495	vapor
300	**1 000**	0.088 899	220.33	309.23	6.1561	0.745 38	1.0559	1.80E−05	0.026 169	vapor
325	**1 000**	0.096 472	239.12	335.59	6.2405	0.745 51	1.0536	1.92E−05	0.027 793	vapor

(Continued)

T (K)	p (kPa)	v (m^3/kg)	u (kJ/kg)	h (kJ/kg)	s (J/g*K)	c_v (J/g*K)	c_p (J/g*K)	μ (Pa*s)	k (W/m*K)	N_2 phase
350	**1 000**	0.104 02	257.9	361.92	6.3186	0.746 07	1.0523	2.02E−05	0.029 374	vapor
375	**1 000**	0.111 55	276.67	388.22	6.3912	0.747 09	1.0519	2.13E−05	0.030 921	vapor
400	**1 000**	0.119 06	295.46	414.52	6.4591	0.748 62	1.0523	2.23E−05	0.032 439	vapor
500	**1 000**	0.149	371.09	520.09	6.6946	0.759 97	1.0608	2.62E−05	0.038 329	vapor
600	**1 000**	0.178 84	448.13	626.97	6.8894	0.778 63	1.078	2.97E−05	0.044 072	vapor
700	**1 000**	0.208 63	527.22	735.85	7.0572	0.801 54	1.1001	3.29E−05	0.049 736	vapor
800	**1 000**	0.238 38	608.66	847.04	7.2056	0.825 72	1.1238	3.60E−05	0.055 312	vapor
900	**1 000**	0.268 11	692.46	960.58	7.3393	0.849 16	1.1469	3.89E−05	0.060 768	vapor
1000	**1 000**	0.297 83	778.52	1076.3	7.4613	0.870 8	1.1683	4.16E−05	0.066 082	vapor
1100	**1 000**	0.327 54	866.62	1194.2	7.5736	0.890 23	1.1875	4.43E−05	0.071 242	vapor
1200	**1 000**	0.357 24	956.54	1313.8	7.6776	0.907 42	1.2046	4.68E−05	0.076 249	vapor
1300	**1 000**	0.386 94	1048.1	1435	7.7747	0.922 5	1.2196	4.93E−05	0.081 107	vapor
1400	**1 000**	0.416 63	1141	1557.6	7.8655	0.935 71	1.2327	5.17E−05	0.085 828	vapor
1500	**1 000**	0.446 32	1235.2	1681.5	7.951	0.947 26	1.2442	5.41E−05	0.090 421	vapor
1600	**1 000**	0.476 01	1330.4	1806.4	8.0316	0.957 39	1.2543	5.64E−05	0.094 9	vapor
1700	**1 000**	0.505 69	1426.6	1932.3	8.1079	0.966 29	1.2632	5.86E−05	0.099 274	vapor
1800	**1 000**	0.535 38	1523.7	2059	8.1804	0.974 16	1.271	6.09E−05	0.103 56	vapor
1900	**1 000**	0.565 06	1621.4	2186.5	8.2493	0.981 14	1.278	6.30E−05	0.107 76	vapor
2000	**1 000**	0.594 74	1719.9	2314.6	8.315	0.987 36	1.2842	6.52E−05	0.111 88	vapor
275	**2 000**	0.040 503	199.18	280.19	5.8501	0.748 66	1.0797	1.71E−05	0.024 952	vapor
300	**2 000**	0.044 398	218.29	307.08	5.9437	0.747 79	1.0721	1.82E−05	0.026 581	vapor
325	**2 000**	0.048 259	237.3	333.82	6.0293	0.747 5	1.0669	1.93E−05	0.028 169	vapor
350	**2 000**	0.052 095	256.25	360.44	6.1082	0.747 75	1.0634	2.04E−05	0.029 719	vapor
375	**2 000**	0.055 911	275.18	387	6.1815	0.748 54	1.0613	2.14E−05	0.031 239	vapor
400	**2 000**	0.059 711	294.09	413.52	6.25	0.749 89	1.0603	2.24E−05	0.032 735	vapor
500	**2 000**	0.074 807	370.09	519.7	6.4869	0.760 81	1.0656	2.63E−05	0.038 558	vapor
600	**2 000**	0.089 804	447.36	626.97	6.6824	0.779 25	1.0811	2.98E−05	0.044 258	vapor
700	**2 000**	0.104 75	526.61	736.11	6.8506	0.802 04	1.1023	3.30E−05	0.049 893	vapor
800	**2 000**	0.119 66	608.17	847.49	6.9993	0.826 13	1.1254	3.60E−05	0.055 447	vapor
900	**2 000**	0.134 55	692.07	961.17	7.1332	0.849 51	1.1481	3.89E−05	0.060 887	vapor
1000	**2 000**	0.149 43	778.19	1077	7.2552	0.871 1	1.1692	4.17E−05	0.066 188	vapor
1100	**2 000**	0.164 3	866.35	1194.9	7.3676	0.890 5	1.1883	4.43E−05	0.071 338	vapor
1200	**2 000**	0.179 16	956.32	1314.6	7.4717	0.907 66	1.2052	4.69E−05	0.076 336	vapor
1300	**2 000**	0.194 02	1047.9	1435.9	7.5688	0.922 73	1.2201	4.94E−05	0.081 188	vapor
1400	**2 000**	0.208 87	1140.9	1558.6	7.6597	0.935 91	1.2331	5.18E−05	0.085 902	vapor
1500	**2 000**	0.223 72	1235.1	1682.5	7.7452	0.947 44	1.2446	5.41E−05	0.090 49	vapor
1600	**2 000**	0.238 57	1330.3	1807.5	7.8258	0.957 56	1.2546	5.64E−05	0.094 964	vapor
1700	**2 000**	0.253 41	1426.5	1933.4	7.9022	0.966 46	1.2634	5.87E−05	0.099 335	vapor
1800	**2 000**	0.268 26	1523.6	2060.1	7.9746	0.974 31	1.2712	6.09E−05	0.103 61	vapor

T (K)	p (kPa)	v (m^3/kg)	u (kJ/kg)	h (kJ/kg)	s (J/g*K)	c_v (J/g*K)	c_p (J/g*K)	μ (Pa*s)	k (W/m*K)	N_2 phase
1900	**2 000**	0.283 1	1621.4	2187.6	8.0435	0.981 28	1.2782	6.31E−05	0.107 81	vapor
2000	**2 000**	0.297 94	1719.8	2315.7	8.1092	0.987 5	1.2844	6.52E−05	0.111 93	vapor
275	**5 000**	0.016 087	192.27	272.7	5.5539	0.757 19	1.1411	1.77E−05	0.026 677	Supercrit
300	**5 000**	0.017 748	212.21	300.95	5.6522	0.754 66	1.1203	1.88E−05	0.028 114	supercrit
325	**5 000**	0.019 375	231.89	328.77	5.7413	0.753 2	1.1059	1.98E−05	0.029 546	supercrit
350	**5 000**	0.020 978	251.39	356.28	5.8228	0.752 6	1.0957	2.08E−05	0.030 969	supercrit
375	**5 000**	0.022 562	270.77	383.58	5.8982	0.752 74	1.0885	2.19E−05	0.032 381	supercrit
400	**5 000**	0.024 131	290.07	410.73	5.9683	0.753 58	1.0836	2.28E−05	0.033 786	supercrit
500	**5 000**	0.030 31	367.15	518.7	6.2092	0.763 28	1.0793	2.66E−05	0.039 35	supercrit
600	**5 000**	0.0363 95	445.11	627.08	6.4068	0.781 09	1.0901	3.00E−05	0.044 89	supercrit
700	**5 000**	0.042 429	524.83	736.98	6.5762	0.803 5	1.1086	3.32E−05	0.050 418	supercrit
800	**5 000**	0.048 434	606.73	848.9	6.7256	0.827 35	1.13	3.62E−05	0.055 895	supercrit
900	**5 000**	0.054 419	690.89	962.99	6.8599	0.850 55	1.1516	3.91E−05	0.061 277	supercrit
1000	**5 000**	0.060 392	777.22	1079.2	6.9823	0.872 01	1.172	4.18E−05	0.066 533	supercrit
1100	**5 000**	0.066 355	865.55	1197.3	7.0949	0.891 31	1.1905	4.45E−05	0.071 647	supercrit
1200	**5 000**	0.072 312	955.65	1317.2	7.1992	0.908 39	1.207	4.70E−05	0.076 616	supercrit
1300	**5 000**	0.078 264	1047.3	1438.7	7.2964	0.923 39	1.2216	4.95E−05	0.081 444	supercrit
1400	**5 000**	0.084 212	1140.4	1561.5	7.3874	0.936 51	1.2344	5.19E−05	0.086 138	supercrit
1500	**5 000**	0.090 157	1234.7	1685.5	7.473	0.948	1.2456	5.42E−05	0.090 708	supercrit
1600	**5 000**	0.096 1	1330	1810.5	7.5537	0.958 07	1.2555	5.65E−05	0.095 167	supercrit
1700	**5 000**	0.102 04	1426.3	1936.5	7.6301	0.966 94	1.2642	5.87E−05	0.099 525	supercrit
1800	**5 000**	0.107 98	1523.4	2063.3	7.7026	0.974 76	1.2719	6.10E−05	0.103 79	supercrit
1825	**5 000**	0.109 47	1547.8	2095.2	7.7201	0.976 58	1.2737	6.15E−05	0.104 85	supercrit
1850	**5 000**	0.110 95	1572.3	2127	7.7375	0.978 34	1.2754	6.20E−05	0.105 9	supercrit
1875	**5 000**	0.112 43	1596.8	2158.9	7.7546	0.980 05	1.2771	6.26E−05	0.106 94	supercrit
1900	**5 000**	0.113 92	1621.3	2190.9	7.7715	0.981 71	1.2787	6.31E−05	0.107 98	supercrit
2000	**5 000**	0.119 86	1719.8	2319.1	7.8373	0.987 89	1.2848	6.53E−05	0.112 09	supercrit
275	**10 000**	0.008 052	181.04	261.56	5.3108	0.769 66	1.2373	1.91E−05	0.030 107	supercrit
300	**10 000**	0.008 951	202.43	291.93	5.4165	0.764 86	1.1949	1.99E−05	0.031 138	supercrit
325	**10 000**	0.009 82	223.22	321.42	5.511	0.761 8	1.1658	2.08E−05	0.032 252	supercrit
350	**10 000**	0.010 668	243.61	350.29	5.5966	0.76	1.145	2.17E−05	0.033 417	supercrit
375	**10 000**	0.011 499	263.73	378.72	5.675	0.759 23	1.1299	2.27E−05	0.034 616	supercrit
400	**10 000**	0.012 317	283.65	406.82	5.7476	0.759 34	1.1189	2.36E−05	0.035 838	supercrit
500	**10 000**	0.015 504	362.47	517.51	5.9946	0.767 23	1.1002	2.71E−05	0.040 888	supercrit
600	**10 000**	0.018 609	441.51	627.6	6.1953	0.784 09	1.104	3.04E−05	0.046 11	supercrit
700	**10 000**	0.021 668	521.98	738.66	6.3665	0.805 91	1.1184	3.36E−05	0.051 423	supercrit
800	**10 000**	0.024 7	604.43	851.42	6.5171	0.829 36	1.1373	3.65E−05	0.056 745	supercrit

(Continued)

T (K)	p (kPa)	v (m^3/kg)	u (kJ/kg)	h (kJ/kg)	s (J/g*K)	c_v (J/g*K)	c_p (J/g*K)	μ (Pa*s)	k (W/m*K)	N_2 phase
900	**10 000**	0.027 714	689.01	966.15	6.6522	0.852 27	1.1572	3.94E−05	0.062 012	supercrit
1000	**10 000**	0.030 716	775.67	1082.8	6.7751	0.873 52	1.1763	4.21E−05	0.067 179	supercrit
1100	**10 000**	0.033 71	864.26	1201.4	6.888	0.892 65	1.194	4.47E−05	0.072 223	supercrit
1200	**10 000**	0.036 697	954.59	1321.6	6.9926	0.909 6	1.2098	4.72E−05	0.077 135	supercrit
1300	**10 000**	0.039 68	1046.5	1443.3	7.09	0.924 48	1.2239	4.97E−05	0.081 915	supercrit
1400	**10 000**	0.042 66	1139.7	1566.3	7.1812	0.937 51	1.2363	5.20E−05	0.086 569	supercrit
1500	**10 000**	0.045 636	1234.1	1690.5	7.2669	0.948 92	1.2472	5.44E−05	0.091 106	supercrit
1600	**10 000**	0.048 611	1329.6	1815.7	7.3477	0.958 93	1.2568	5.67E−05	0.095 536	supercrit
1700	**10 000**	0.051 584	1426	1941.8	7.4241	0.967 73	1.2653	5.89E−05	0.099 869	supercrit
1800	**10 000**	0.054 555	1523.2	2068.7	7.4967	0.975 51	1.2729	6.11E−05	0.104 11	supercrit
1900	**10 000**	0.057 525	1621.1	2196.4	7.5657	0.982 41	1.2796	6.33E−05	0.108 28	supercrit
2000	**10 000**	0.060 495	1719.7	2324.6	7.6315	0.988 55	1.2856	6.54E−05	0.112 38	supercrit
275	**20 000**	0.004 227	161.24	245.78	5.047	0.788 43	1.3679	2.27E−05	0.037 616	supercrit
300	**20 000**	0.004 705	185.01	279.11	5.163	0.781 13	1.3021	2.30E−05	0.037 647	supercrit
325	**20 000**	0.005 168	207.69	311.04	5.2653	0.776 06	1.2549	2.34E−05	0.038 034	supercrit
350	**20 000**	0.005 617	229.62	341.96	5.3569	0.772 64	1.2203	2.40E−05	0.038 638	supercrit
375	**20 000**	0.006 056	251.01	372.13	5.4402	0.770 55	1.1944	2.46E−05	0.039 386	supercrit
400	**20 000**	0.006 486	272.01	401.73	5.5166	0.769 59	1.1749	2.53E−05	0.040 236	supercrit
500	**20 000**	0.008 146	353.9	516.82	5.7736	0.774 59	1.1348	2.84E−05	0.044 244	supercrit
600	**20 000**	0.009 743	434.89	629.75	5.9795	0.789 79	1.1275	3.14E−05	0.048 811	supercrit
700	**20 000**	0.011 305	516.71	742.81	6.1538	0.810 56	1.1353	3.44E−05	0.053 669	supercrit
800	**20 000**	0.012 844	600.15	857.04	6.3063	0.833 27	1.15	3.72E−05	0.058 659	supercrit
900	**20 000**	0.014 369	685.5	972.88	6.4427	0.855 65	1.167	4.00E−05	0.063 671	supercrit
1000	**20 000**	0.015 883	772.78	1090.4	6.5666	0.876 49	1.1842	4.26E−05	0.068 639	supercrit
1100	**20 000**	0.017 39	861.87	1209.7	6.6802	0.895 3	1.2003	4.52E−05	0.073 523	supercrit
1200	**20 000**	0.018 892	952.62	1330.5	6.7853	0.911 98	1.215	4.76E−05	0.078 304	supercrit
1300	**20 000**	0.020 389	1044.8	1452.6	6.8831	0.926 65	1.2282	5.00E−05	0.082 976	supercrit
1400	**20 000**	0.021 884	1138.4	1576	6.9745	0.939 5	1.2399	5.24E−05	0.087 539	supercrit
1500	**20 000**	0.023 375	1233.1	1700.6	7.0604	0.950 75	1.2503	5.47E−05	0.091 998	supercrit
1600	**20 000**	0.024 865	1328.8	1826.1	7.1414	0.960 62	1.2594	5.70E−05	0.096 361	supercrit
1700	**20 000**	0.026 354	1425.3	1952.4	7.218	0.969 31	1.2675	5.92E−05	0.100 64	supercrit
1800	**20 000**	0.027 841	1522.7	2079.5	7.2907	0.976 99	1.2748	6.14E−05	0.104 83	supercrit
1875	**20 000**	0.028 956	1596.2	2175.3	7.3428	0.982 16	1.2797	6.30E−05	0.107 93	supercrit

Source: Linstrom and Mallard (2017–2021).

A.3 Properties of Oxygen (Gas)

The enthalpy of formation of oxygen is $\Delta\hat{h}_f^0 = 0.0\ kJ/kmol$. The molar mass of oxygen is $\hat{M} = 31.998\ 06\ kg/kmol$. These properties were derived from Linstrom and Mallard (2017–2021).

T (K)	p (kPa)	υ (m^3/kg)	u (kJ/kg)	h (kJ/kg)	s (J/g*K)	c_v (J/g*K)	c_p (J/g*K)	μ (Pa*s)	k (W/m*K)	O_2 phase
275	**101.33**	0.704 5	178.37	249.75	6.3329	0.655 42	0.916 89	1.92E−05	0.024 637	vapor
298.15	**101.33**	0.764 03	193.59	271.01	6.4071	0.658 45	0.919 63	2.046E−05	0.026 508	vapor
300	**101.33**	0.768 78	194.81	272.71	6.4128	0.658 73	0.919 89	2.06E−05	0.026 657	vapor
325	**101.33**	0.833 04	211.35	295.76	6.4866	0.663 05	0.923 98	2.19E−05	0.028 667	vapor
350	**101.33**	0.897 27	228	318.92	6.5553	0.668 31	0.929 06	2.32E−05	0.030 672	vapor
375	**101.33**	0.961 48	244.79	342.22	6.6196	0.674 4	0.935 01	2.45E−05	0.032 679	vapor
400	**101.33**	1.025 7	261.74	365.68	6.6801	0.681 16	0.941 66	2.57E−05	0.034 689	vapor
500	**101.33**	1.282 3	331.39	461.33	6.8934	0.711 93	0.972 16	3.04E−05	0.042 75	vapor
600	**101.33**	1.538 9	404.18	560.12	7.0735	0.743 17	1.003 3	3.47E−05	0.050 738	vapor
700	**101.33**	1.795 5	479.94	661.87	7.2303	0.771 03	1.031	3.86E−05	0.058 499	vapor
800	**101.33**	2.051 9	558.26	766.18	7.3695	0.794 53	1.054 5	4.23E−05	0.065 933	vapor
900	**101.33**	2.308 4	638.72	872.63	7.4949	0.813 95	1.073 9	4.58E−05	0.072 997	vapor
1000	**101.33**	2.564 9	720.94	980.84	7.6089	0.829 93	1.089 8	4.91E−05	0.079 685	vapor
275	**200**	0.356 61	178.15	249.47	6.1555	0.655 73	0.918 78	1.92E−05	0.024 661	vapor
300	**200**	0.389 27	194.62	272.47	6.2355	0.658 97	0.921 42	2.06E−05	0.026 679	vapor
325	**200**	0.421 9	211.17	295.55	6.3094	0.663 25	0.925 24	2.19E−05	0.028 687	vapor
350	**200**	0.454 5	227.84	318.74	6.3782	0.668 48	0.930 12	2.32E−05	0.030 691	vapor
375	**200**	0.487 08	244.65	342.07	6.4425	0.674 54	0.935 91	2.45E−05	0.032 696	vapor
400	**200**	0.519 65	261.61	365.54	6.5031	0.681 28	0.942 43	2.58E−05	0.034 704	vapor
500	**200**	0.649 82	331.3	461.26	6.7166	0.712 01	0.972 61	3.04E−05	0.042 762	vapor
600	**200**	0.779 89	404.11	560.09	6.8967	0.743 22	1.003 6	3.47E−05	0.050 748	vapor
700	**200**	0.909 91	479.88	661.86	7.0535	0.771 06	1.031 2	3.86E−05	0.058 507	vapor
800	**200**	1.039 9	558.21	766.19	7.1928	0.794 56	1.054 6	4.23E−05	0.065 939	vapor
900	**200**	1.169 9	638.68	872.65	7.3182	0.813 97	1.074	4.58E−05	0.073 003	vapor
1000	**200**	1.299 8	720.91	980.87	7.4322	0.829 94	1.089 9	4.92E−05	0.079 69	vapor
275	**500**	0.142 25	177.48	248.61	5.915	0.656 64	0.924 6	1.93E−05	0.024 741	vapor
300	**500**	0.155 42	194.03	271.74	5.9955	0.659 7	0.926 11	2.07E−05	0.026 75	vapor
325	**500**	0.168 56	210.65	294.93	6.0697	0.663 84	0.929 1	2.20E−05	0.028 75	vapor
350	**500**	0.181 68	227.37	318.2	6.1387	0.668 98	0.933 35	2.33E−05	0.030 749	vapor
375	**500**	0.194 77	244.22	341.6	6.2033	0.674 96	0.938 65	2.46E−05	0.032 749	vapor
400	**500**	0.207 85	261.22	365.14	6.2641	0.681 65	0.944 78	2.58E−05	0.034 753	vapor
500	**500**	0.260 07	331.01	461.04	6.4779	0.712 23	0.974	3.05E−05	0.042 799	vapor
600	**500**	0.312 19	403.88	559.98	6.6582	0.743 37	1.004 5	3.47E−05	0.050 777	vapor

(Continued)

T (K)	p (kPa)	υ (m^3/kg)	u (kJ/kg)	h (kJ/kg)	s (J/g*K)	c_v (J/g*K)	c_p (J/g*K)	μ (Pa*s)	k (W/m*K)	O_2 phase
700	**500**	0.364 26	479.7	661.83	6.8152	0.771 17	1.031 9	3.86E−05	0.058 531	vapor
800	**500**	0.4163	558.06	766.21	6.9545	0.794 63	1.055 1	4.23E−05	0.065 96	vapor
900	**500**	0.468 32	638.55	872.71	7.0799	0.814 03	1.074 3	4.58E−05	0.073 021	vapor
1000	**500**	0.520 33	720.8	980.96	7.194	0.829 99	1.090 2	4.92E−05	0.079 706	vapor
275	**1 000**	0.070 798	176.37	247.17	5.7308	0.658 17	0.934 46	1.94E−05	0.024 888	vapor
300	**1 000**	0.077 476	193.05	270.52	5.8121	0.660 91	0.934	2.08E−05	0.026 88	vapor
325	**1 000**	0.084 12	209.77	293.89	5.8869	0.664 83	0.935 57	2.21E−05	0.028 867	vapor
350	**1 000**	0.090 739	226.57	317.31	5.9564	0.669 8	0.938 75	2.34E−05	0.030 854	vapor
375	**1 000**	0.097 338	243.5	340.83	6.0213	0.675 67	0.943 22	2.47E−05	0.032 845	vapor
400	**1 000**	0.103 92	260.56	364.48	6.0823	0.682 26	0.948 69	2.59E−05	0.034 841	vapor
500	**1 000**	0.130 16	330.52	460.68	6.2969	0.712 6	0.976 3	3.05E−05	0.042 864	vapor
600	**1 000**	0.156 29	403.51	559.8	6.4775	0.743 62	1.005 9	3.47E−05	0.050 829	vapor
700	**1 000**	0.182 38	479.4	661.78	6.6346	0.771 35	1.032 9	3.87E−05	0.058 574	vapor
800	**1 000**	0.208 44	557.81	766.25	6.7741	0.794 77	1.055 8	4.24E−05	0.065 996	vapor
900	**1 000**	0.234 47	638.34	872.81	6.8996	0.814 13	1.074 9	4.59E−05	0.073 052	vapor
1000	**1 000**	0.260 5	720.62	981.12	7.0137	0.830 07	1.090 6	4.92E−05	0.079 733	vapor
275	**2 000**	0.035 08	174.13	244.29	5.5426	0.661 22	0.954 75	1.97E−05	0.025 239	vapor
300	**2 000**	0.038 51	191.07	268.09	5.6255	0.663 3	0.950 08	2.10E−05	0.027 184	vapor
325	**2 000**	0.041 906	208.01	291.82	5.7014	0.666 78	0.948 64	2.23E−05	0.029 135	vapor
350	**2 000**	0.045 277	224.99	315.54	5.7718	0.671 43	0.949 59	2.36E−05	0.031 093	vapor
375	**2 000**	0.048 628	242.06	339.31	5.8374	0.677 06	0.952 36	2.48E−05	0.033 061	vapor
400	**2 000**	0.051 963	259.25	363.17	5.899	0.683 46	0.956 5	2.60E−05	0.035 038	vapor
500	**2 000**	0.065 203	329.56	459.97	6.1148	0.713 34	0.980 87	3.06E−05	0.043 008	vapor
600	**2 000**	0.078 348	402.76	559.46	6.2962	0.744 11	1.008 9	3.48E−05	0.050 941	vapor
700	**2 000**	0.091 442	478.79	661.68	6.4537	0.771 7	1.035	3.87E−05	0.058 665	vapor
800	**2 000**	0.104 51	557.31	766.32	6.5934	0.795 02	1.057 3	4.24E−05	0.066 072	vapor
900	**2 000**	0.117 55	637.92	873.02	6.719	0.814 32	1.076	4.59E−05	0.073 118	vapor
1000	**2 000**	0.130 58	720.26	981.42	6.8332	0.830 23	1.091 5	4.92E−05	0.079 79	vapor
275	**5 000**	0.013 677	167.27	235.66	5.28	0.670 15	1.019 8	2.05E−05	0.026 681	vapor
300	**5 000**	0.015 159	185.09	260.88	5.3679	0.670 27	1.000 1	2.17E−05	0.028 402	vapor
325	**5 000**	0.016 604	202.71	285.73	5.4474	0.672 44	0.988 58	2.29E−05	0.030 184	vapor
350	**5 000**	0.018 023	220.24	310.36	5.5204	0.676 16	0.982 3	2.41E−05	0.032 018	vapor
375	**5 000**	0.019 422	237.77	334.88	5.5881	0.681 08	0.979 67	2.53E−05	0.033 885	vapor
400	**5 000**	0.020 806	255.33	359.36	5.6513	0.686 93	0.979 68	2.65E−05	0.035 78	vapor
500	**5 000**	0.026 243	326.71	457.92	5.8712	0.715 46	0.994 26	3.09E−05	0.043 53	vapor
600	**5 000**	0.031 588	400.55	558.49	6.0545	0.745 54	1.017 6	3.51E−05	0.051 337	vapor
700	**5 000**	0.036 884	477.01	661.43	6.2131	0.772 71	1.041	3.89E−05	0.058 981	vapor
800	**5 000**	0.042 151	555.83	766.59	6.3535	0.795 77	1.061 7	4.26E−05	0.066 333	vapor
900	**5 000**	0.047 4	636.67	873.67	6.4796	0.814 9	1.079 3	4.60E−05	0.073 339	vapor

T (K)	p (kPa)	v (m³/kg)	u (kJ/kg)	h (kJ/kg)	s (J/g*K)	c_v (J/g*K)	c_p (J/g*K)	μ (Pa*s)	k (W/m*K)	O_2 phase
1000	**5 000**	0.052 637	719.18	982.36	6.5941	0.830 68	1.094 1	4.93E−05	0.079 981	vapor
275	**20 000**	0.003 242	133.21	198.05	4.8095	0.704 24	1.330 9	2.73E−05	0.038 914	supercrit
300	**20 000**	0.003 699	156	229.99	4.9208	0.697 67	1.232 1	2.69E−05	0.038 406	supercrit
325	**20 000**	0.004 133	177.29	259.95	5.0167	0.695 1	1.169 2	2.71E−05	0.038 61	supercrit
350	**20 000**	0.004 548	197.65	288.62	5.1017	0.695 31	1.127 7	2.76E−05	0.039 355	supercrit
375	**20 000**	0.004 95	217.44	316.44	5.1785	0.697 54	1.099 8	2.83E−05	0.040 364	supercrit
400	**20 000**	0.005 341	236.87	343.68	5.2489	0.701 26	1.080 8	2.91E−05	0.041 563	supercrit
500	**20 000**	0.006 831	313.27	449.89	5.486	0.724 38	1.052 2	3.26E−05	0.047 492	supercrit
600	**20 000**	0.008 252	390.09	555.14	5.6778	0.751 56	1.055 3	3.63E−05	0.054 277	supercrit
700	**20 000**	0.009 636	468.53	661.24	5.8414	0.777	1.067 4	3.99E−05	0.061 275	supercrit
800	**20 000**	0.010 996	548.75	768.67	5.9848	0.798 95	1.081 2	4.34E−05	0.068 19	supercrit
900	**20 000**	0.012 342	630.62	877.46	6.1129	0.817 32	1.094 3	4.67E−05	0.074 883	supercrit
1000	**20 000**	0.013 677	713.93	987.47	6.2288	0.83 257	1.105 8	4.99E−05	0.081 292	supercrit

Source: Linstrom and Mallard (2017–2021).

A.4 Properties of Water (Liquid–Vapor)

The enthalpy of formation of water is $\Delta\hat{h}_f^0 = -241\,826\ kJ/kmol$. The molar mass of water is $\hat{M} = 18.014\,71\ kg/kmol$. These properties were derived from Linstrom and Mallard (2017–2021).

T (K)	p (kPa)	v (m³/kg)	u (kJ/kg)	h (kJ/kg)	s (J/g*K)	c_v (J/g*K)	c_p (J/g*K)	μ (Pa*s)	k (W/m*K)	H_2O phase
275	**2.0**	0.001	7.759	7.761	0.028 31	4.2133	4.2139	0.001 682	0.564 53	liquid
290.64	**2.0**	0.001 001	73.426	73.428	0.260 56	4.1661	4.1863	0.001 066	0.593 9	liquid
290.64	**2.0**	66.987	2398.9	2532.9	8.722 6	1.4336	1.903	9.66E−06	0.018 07	vapor
300	**2.0**	69.16	2412.3	2550.6	8.782 6	1.4193	1.8858	9.92E−06	0.018 629	vapor
325	**2.0**	74.952	2447.7	2597.6	8.933 2	1.417	1.8811	1.07E−05	0.020 272	vapor
350	**2.0**	80.733	2483.2	2644.7	9.072 7	1.4226	1.8857	1.15E−05	0.022 096	vapor
375	**2.0**	86.51	2518.9	2691.9	9.203 1	1.4311	1.8936	1.24E−05	0.024 067	vapor
400	**2.0**	92.284	2554.8	2739.4	9.325 6	1.4416	1.9038	1.33E−05	0.026 165	vapor
500	**2.0**	115.37	2701.5	2932.3	9.755 8	1.4941	1.9559	1.73E−05	0.035 565	vapor
600	**2.0**	138.45	2853.9	3130.8	10.118	1.5549	2.0165	2.14E−05	0.046 233	vapor
700	**2.0**	161.53	3012.7	3335.7	10.433	1.6198	2.0814	2.56E−05	0.057 881	vapor
800	**2.0**	184.6	3178	3547.2	10.716	1.6879	2.1494	2.97E−05	0.070 321	vapor
900	**2.0**	207.68	3350.3	3765.7	10.973	1.7581	2.2196	3.37E−05	0.083 409	vapor
1000	**2.0**	230.76	3529.7	3991.2	11.21	1.8292	2.2907	3.76E−05	0.097 034	vapor
1100	**2.0**	253.83	3716.1	4223.8	11.432	1.8996	2.3611	4.14E−05	0.111 1	vapor

(Continued)

T (K)	p (kPa)	v (m^3/kg)	u (kJ/kg)	h (kJ/kg)	s (J/g*K)	c_v (J/g*K)	c_p (J/g*K)	μ (Pa*s)	k (W/m*K)	H_2O phase
1200	**2.0**	276.91	3909.5	4463.4	11.64	1.968	2.4295	4.51E−05	0.125 55	vapor
1275	**2.0**	294.22	4059	4647.4	11.789	2.0173	2.4788	4.77E−05	0.136 58	vapor
275	**5.0**	0.001	7.7591	7.7641	0.028 31	4.2132	4.2139	0.001 682	0.564 53	liquid
300	**5.0**	0.001 004	112.56	112.57	0.393 09	4.1305	4.1809	0.000 854	0.610 28	liquid
306.02	**5.0**	0.001 005	137.74	137.75	0.476 2	4.1053	4.1797	0.000 751	0.620 02	liquid
306.02	**5.0**	28.185	2419.8	2560.7	8.393 8	1.448	1.9217	1.01E−05	0.019 086	vapor
325	**5.0**	29.953	2447.1	2596.9	8.508 3	1.4275	1.8957	1.07E−05	0.020 334	vapor
350	**5.0**	32.274	2482.8	2644.2	8.648 7	1.4279	1.8934	1.15E−05	0.022 141	vapor
375	**5.0**	34.589	2518.6	2691.6	8.779 5	1.4342	1.8982	1.24E−05	0.024 101	vapor
400	**5.0**	36.902	2554.6	2739.1	8.902 2	1.4435	1.9067	1.33E−05	0.026 19	vapor
500	**5.0**	46.142	2701.4	2932.2	9.332 7	1.4945	1.9566	1.73E−05	0.035 574	vapor
600	**5.0**	55.377	2853.9	3130.8	9.694 7	1.5551	2.0169	2.14E−05	0.046 237	vapor
700	**5.0**	64.609	3012.6	3335.7	10.01	1.6199	2.0816	2.56E−05	0.057 884	vapor
800	**5.0**	73.84	3178	3547.2	10.293	1.6879	2.1495	2.97E−05	0.070 323	vapor
900	**5.0**	83.072	3350.3	3765.7	10.55	1.7581	2.2197	3.37E−05	0.083 41	vapor
1000	**5.0**	92.302	3529.7	3991.2	10.788	1.8292	2.2908	3.76E−05	0.097 035	vapor
1100	**5.0**	101.53	3716.1	4223.8	11.009	1.8996	2.3611	4.14E−05	0.111 11	vapor
1200	**5.0**	110.76	3909.5	4463.3	11.218	1.968	2.4295	4.51E−05	0.125 55	vapor
1275	**5.0**	117.69	4059	4647.4	11.366	2.0173	2.4788	4.77E−05	0.136 58	vapor
275	**10**	0.001	7.7591	7.7691	0.028 31	4.2132	4.2139	0.001 682	0.564 54	liquid
300	**10**	0.001 004	112.56	112.57	0.393 09	4.1305	4.1809	0.000 854	0.610 28	liquid
318.96	**10**	0.001 01	191.8	191.81	0.649 2	4.0466	4.1805	0.000 588	0.638 38	liquid
318.96	**10**	14.67	2437.2	2583.9	8.148 8	1.4615	1.94	1.05E−05	0.020 037	vapor
325	**10**	14.954	2446	2595.5	8.185 1	1.4488	1.9243	1.07E−05	0.020 438	vapor
350	**10**	16.12	2482.1	2643.3	8.326 8	1.4371	1.9065	1.15E−05	0.022 217	vapor
375	**10**	17.282	2518.2	2691	8.458 3	1.4394	1.906	1.24E−05	0.024 157	vapor
400	**10**	18.441	2554.3	2738.7	8.581 4	1.4466	1.9116	1.33E−05	0.026 232	vapor
500	**10**	23.066	2701.3	2932	9.012 5	1.4952	1.9579	1.73E−05	0.035 589	vapor
600	**10**	27.686	2853.8	3130.7	9.374 7	1.5553	2.0174	2.14E−05	0.046 244	vapor
700	**10**	32.303	3012.6	3335.6	9.690 4	1.62	2.0819	2.56E−05	0.057 888	vapor
800	**10**	36.919	3178	3547.2	9.972 8	1.688	2.1497	2.97E−05	0.070 326	vapor
900	**10**	41.535	3350.3	3765.6	10.23	1.7582	2.2198	3.37E−05	0.083 413	vapor
1000	**10**	46.151	3529.6	3991.2	10.468	1.8292	2.2908	3.76E−05	0.097 038	vapor
1100	**10**	50.766	3716.1	4223.8	10.689	1.8996	2.3612	4.14E−05	0.111 11	vapor
1200	**10**	55.382	3909.5	4463.3	10.898	1.968	2.4296	4.51E−05	0.125 55	vapor
1275	**10**	58.843	4059	4647.4	11.046	2.0173	2.4789	4.77E−05	0.136 59	vapor
275	**20**	0.001	7.7592	7.7792	0.028 31	4.2132	4.2139	0.001 682	0.564 54	liquid
300	**20**	0.001 004	112.56	112.58	0.393 08	4.1304	4.1809	0.000 854	0.610 29	liquid
325	**20**	0.001 013	217.06	217.08	0.727 68	4.0174	4.1821	0.000 53	0.645 71	liquid

T (K)	p (kPa)	v (m^3/kg)	u (kJ/kg)	h (kJ/kg)	s (J/g*K)	c_v (J/g*K)	c_p (J/g*K)	μ (Pa*s)	k (W/m*K)	H_2O phase
333.21	**20**	0.001 017	251.4	251.42	0.832 02	3.9764	4.1852	0.000 466	0.654 4	liquid
333.21	**20**	7.648	2456	2608.9	7.907 2	1.479	1.9649	1.09E−05	0.021 192	vapor
350	**20**	8.043 7	2480.7	2641.6	8.002 9	1.4562	1.934	1.15E−05	0.022 37	vapor
375	**20**	8.628 7	2517.2	2689.8	8.135 8	1.45	1.9218	1.24E−05	0.024 27	vapor
400	**20**	9.210 9	2553.6	2737.8	8.259 8	1.453	1.9216	1.33E−05	0.026 317	vapor
500	**20**	11.529	2701	2931.6	8.692 1	1.4967	1.9604	1.73E−05	0.035 619	vapor
600	**20**	13.84	2853.7	3130.5	9.054 5	1.5558	2.0184	2.14E−05	0.046 257	vapor
700	**20**	16.15	3012.5	3335.5	9.370 4	1.6203	2.0824	2.56E−05	0.057 897	vapor
800	**20**	18.458	3177.9	3547.1	9.652 8	1.6881	2.15	2.97E−05	0.070 332	vapor
900	**20**	20.767	3350.2	3765.5	9.910 1	1.7582	2.22	3.37E−05	0.083 419	vapor
1000	**20**	23.075	3529.6	3991.1	10.148	1.8293	2.291	3.76E−05	0.097 043	vapor
1100	**20**	25.383	3716.1	4223.7	10.369	1.8997	2.3613	4.14E−05	0.111 11	vapor
1200	**20**	27.691	3909.5	4463.3	10.578	1.968	2.4296	4.51E−05	0.125 55	vapor
1275	**20**	29.421	4058.9	4647.4	10.727	2.0173	2.4789	4.77E−05	0.136 59	vapor
275	**50**	0.001	7.7595	7.8095	0.028 311	4.213	4.2137	0.001 682	0.564 56	liquid
300	**50**	0.001 004	112.56	112.61	0.393 08	4.1303	4.1808	0.000 854	0.610 3	liquid
325	**50**	0.001 013	217.06	217.11	0.727 66	4.0173	4.182	0.000 53	0.645 73	liquid
350	**50**	0.001 027	321.75	321.8	1.038	3.8895	4.1946	0.000 369	0.668 01	liquid
354.47	**50**	0.001 03	340.49	340.54	1.091 2	3.866	4.1979	0.000 349	0.670 77	liquid
354.47	**50**	3.24	2483.2	2645.2	7.593	1.5136	2.0157	1.16E−05	0.023 14	vapor
375	**50**	3.436 3	2514.3	2686.1	7.705 1	1.4833	1.9719	1.24E−05	0.024 613	vapor
400	**50**	3.672 5	2551.5	2735.1	7.831 6	1.4729	1.9525	1.33E−05	0.026 573	vapor
500	**50**	4.605 7	2700.2	2930.5	8.267 5	1.5009	1.9681	1.73E−05	0.035 709	vapor
600	**50**	5.532 7	2853.2	3129.8	8.630 8	1.5574	2.0215	2.14E−05	0.046 298	vapor
700	**50**	6.457 7	3012.2	3335.1	8.947 1	1.621	2.084	2.56E−05	0.057 922	vapor
800	**50**	7.381 9	3177.7	3546.8	9.229 7	1.6885	2.1509	2.97E−05	0.070 352	vapor
900	**50**	8.305 6	3350.1	3765.3	9.487	1.7585	2.2206	3.37E−05	0.083 437	vapor
1000	**50**	9.229 2	3529.5	3990.9	9.724 7	1.8294	2.2914	3.76E−05	0.097 059	vapor
1100	**50**	10.153	3716	4223.6	9.946 3	1.8998	2.3616	4.14E−05	0.111 13	vapor
1200	**50**	11.076	3909.4	4463.2	10.155	1.9681	2.4299	4.51E−05	0.125 56	vapor
1275	**50**	11.768	4058.9	4647.3	10.304	2.0174	2.4791	4.77E−05	0.136 6	vapor
275	**101.33**	0.001	7.76	7.8613	0.028 313	4.2128	4.2135	0.001 682	0.564 59	liquid
298.15	**101.33**	0.001 0030	104.82	104.92	0.367 20	4.1376	4.1813	0.000 890	0.607 19	liquid
300	**101.33**	0.001 004	112.55	112.65	0.393 06	4.1302	4.1806	0.000 854	0.610 32	liquid
325	**101.33**	0.001 013	217.05	217.15	0.727 64	4.0171	4.1819	0.000 53	0.645 75	liquid
350	**101.33**	0.001 027	321.74	321.84	1.038	3.8894	4.1945	0.000 369	0.668 03	liquid
373.13	**101.33**	0.001 043	418.96	419.06	1.306 9	3.7683	4.2156	0.000 282	0.679 08	liquid
373.13	**101.33**	1.673 1	2506	2675.5	7.354 4	1.5558	2.0799	1.23E−05	0.025 094	vapor
375	**101.33**	1.682 2	2509	2679.4	7.364 8	1.5499	2.0716	1.23E−05	0.025 219	vapor

(Continued)

T (K)	p (kPa)	v (m^3/kg)	u (kJ/kg)	h (kJ/kg)	s (J/g*K)	c_v (J/g*K)	c_p (J/g*K)	μ (Pa*s)	k (W/m*K)	H_2O phase
400	**101.33**	1.801 9	2547.7	2730.3	7.496 2	1.5092	2.0093	1.33E−05	0.027 02	vapor
500	**101.33**	2.267 8	2698.7	2928.5	7.938 6	1.5084	1.9816	1.73E−05	0.035 865	vapor
600	**101.33**	2.727 2	2852.4	3128.7	8.303 5	1.56	2.0269	2.14E−05	0.046 368	vapor
700	**101.33**	3.184 6	3011.6	3334.3	8.620 3	1.6223	2.0868	2.56E−05	0.057 965	vapor
800	**101.33**	3.641 2	3177.3	3546.3	8.903 2	1.6892	2.1526	2.97E−05	0.070 385	vapor
900	**101.33**	4.097 5	3349.8	3765	9.160 7	1.7589	2.2217	3.37E−05	0.083 466	vapor
1000	**101.33**	4.553 4	3529.3	3990.7	9.398 4	1.8297	2.2921	3.76E−05	0.097 086	vapor
1100	**101.33**	5.009 3	3715.8	4223.4	9.620 2	1.9	2.3621	4.14E−05	0.111 15	vapor
1200	**101.33**	5.465	3909.2	4463	9.828 7	1.9682	2.4303	4.51E−05	0.125 58	vapor
1275	**101.33**	5.806 8	4058.7	4647.1	9.977 5	2.0175	2.4794	4.77E−05	0.136 61	vapor
275	**200**	0.001	7.761	7.961	0.028 316	4.2123	4.213	0.001 682	0.564 64	liquid
300	**200**	0.001 003	112.55	112.75	0.393 03	4.1298	4.1804	0.000 854	0.610 37	liquid
325	**200**	0.001 013	217.04	217.24	0.727 59	4.0169	4.1817	0.000 53	0.645 8	liquid
350	**200**	0.001 027	321.71	321.92	1.037 9	3.8892	4.1943	0.000 369	0.668 08	liquid
375	**200**	0.001 045	426.83	427.04	1.328	3.7584	4.2176	0.000 276	0.679 71	liquid
393.36	**200**	0.001 061	504.49	504.7	1.530 2	3.6652	4.2439	0.000 232	0.683 21	liquid
393.36	**200**	0.885 68	2529.1	2706.2	7.126 9	1.6184	2.1782	1.30E−05	0.027 493	vapor
400	**200**	0.902 47	2540.1	2720.6	7.163	1.5916	2.1386	1.32E−05	0.027 915	vapor
500	**200**	1.144 2	2695.9	2924.7	7.619	1.5232	2.0085	1.72E−05	0.036 167	vapor
600	**200**	1.379	2850.8	3126.6	7.987 1	1.5652	2.0374	2.14E−05	0.046 504	vapor
700	**200**	1.611 7	3010.6	3333	8.305 1	1.6247	2.0921	2.56E−05	0.058 048	vapor
800	**200**	1.843 6	3176.6	3545.3	8.588 5	1.6905	2.1557	2.97E−05	0.070 45	vapor
900	**200**	2.075 1	3349.2	3764.3	8.846 3	1.7597	2.2237	3.37E−05	0.083 524	vapor
1000	**200**	2.306 4	3528.8	3990.1	9.084 2	1.8303	2.2935	3.76E−05	0.097 139	vapor
1100	**200**	2.537 5	3715.4	4223	9.306 1	1.9003	2.3631	4.14E−05	0.111 19	vapor
1200	**200**	2.768 6	3909	4462.7	9.514 6	1.9685	2.431	4.51E−05	0.125 61	vapor
1275	**200**	2.941 8	4058.5	4646.9	9.663 5	2.0177	2.4801	4.77E−05	0.136 63	vapor
275	**500**	0.001	7.7638	8.2638	0.028 327	4.211	4.2116	0.001 681	0.564 8	liquid
300	**500**	0.001 003	112.52	113.02	0.392 95	4.1289	4.1795	0.000 854	0.610 5	liquid
325	**500**	0.001 013	216.99	217.5	0.727 45	4.0161	4.181	0.000 53	0.645 94	liquid
350	**500**	0.001 027	321.65	322.16	1.037 7	3.8886	4.1936	0.000 369	0.668 23	liquid
375	**500**	0.001 045	426.74	427.26	1.327 7	3.758	4.2169	0.000 276	0.679 88	liquid
400	**500**	0.001 067	532.59	533.13	1.601	3.6321	4.2548	0.000 219	0.683 79	liquid
424.98	**500**	0.001 093	639.54	640.09	1.860 4	3.515	4.312	0.000 18	0.681 72	liquid
424.98	**500**	0.374 81	2560.7	2748.1	6.820 7	1.7593	2.4103	1.41E−05	0.031 87	vapor
450	**500**	0.401 4	2604.8	2805.5	6.951 9	1.6352	2.2159	1.51E−05	0.033 294	vapor
500	**500**	0.451 76	2686.8	2912.7	7.177 9	1.5726	2.0984	1.72E−05	0.037 114	vapor
600	**500**	0.548 2	2846	3120.1	7.556 1	1.5813	2.0703	2.14E−05	0.046 921	vapor
700	**500**	0.642 52	3007.5	3328.8	7.877 7	1.6321	2.1085	2.56E−05	0.058 302	vapor
800	**500**	0.735 99	3174.4	3542.4	8.162 9	1.6946	2.1653	2.97E−05	0.070 647	vapor

T (K)	p (kPa)	υ (m^3/kg)	u (kJ/kg)	h (kJ/kg)	s (J/g*K)	c_v (J/g*K)	c_p (J/g*K)	μ (Pa*s)	k (W/m*K)	H_2O phase
900	**500**	0.829 04	3347.6	3762.1	8.421 6	1.7622	2.2299	3.37E−05	0.083 701	vapor
1000	**500**	0.921 86	3527.5	3988.5	8.66	1.8319	2.2977	3.76E−05	0.097 3	vapor
1100	**500**	1.014 5	3714.4	4221.7	8.882 2	1.9015	2.3662	4.14E−05	0.111 33	vapor
1200	**500**	1.107 1	3908.1	4461.7	9.091	1.9694	2.4333	4.51E−05	0.125 72	vapor
1275	**500**	1.176 5	4057.7	4646	9.24	2.0184	2.4819	4.77E−05	0.136 7	vapor
275	**1 000**	0.001	7.7685	8.7682	0.028 343	4.2087	4.2093	0.001 68	0.565 07	liquid
300	**1 000**	0.001 003	112.48	113.48	0.392 81	4.1272	4.1781	0.000 854	0.610 73	liquid
325	**1 000**	0.001 013	216.91	217.93	0.727 21	4.0149	4.1798	0.000 53	0.646 17	liquid
350	**1 000**	0.001 027	321.53	322.56	1.037 4	3.8876	4.1925	0.000 369	0.668 48	liquid
375	**1 000**	0.001 045	426.59	427.64	1.327 4	3.7572	4.2158	0.000 277	0.680 15	liquid
400	**1 000**	0.001 066	532.4	533.47	1.600 5	3.6315	4.2535	0.000 219	0.684 1	liquid
453.03	**1 000**	0.001 127	761.39	762.52	2.138 1	3.3954	4.4045	0.000 15	0.673 37	liquid
453.03	**1 000**	0.194 36	2582.7	2777.1	6.585	1.9271	2.7114	1.50E−05	0.036 427	vapor
500	**1 000**	0.220 64	2670.6	2891.2	6.825	1.6699	2.2795	1.71E−05	0.038 799	vapor
600	**1 000**	0.271 22	2837.7	3109	7.222 4	1.6098	2.1292	2.13E−05	0.047 636	vapor
700	**1 000**	0.319 43	3002.3	3321.7	7.550 4	1.6447	2.1368	2.56E−05	0.058 735	vapor
800	**1 000**	0.366 77	3170.7	3537.5	7.838 4	1.7014	2.1816	2.97E−05	0.070 983	vapor
900	**1 000**	0.413 67	3344.8	3758.5	8.098 6	1.7663	2.2402	3.37E−05	0.084	vapor
1000	**1 000**	0.460 34	3525.4	3985.7	8.338	1.8346	2.3048	3.76E−05	0.097 573	vapor
1100	**1 000**	0.506 86	3712.7	4219.5	8.560 8	1.9034	2.3713	4.14E−05	0.111 57	vapor
1200	**1 000**	0.553 29	3906.7	4460	8.769 9	1.9708	2.4371	4.51E−05	0.125 89	vapor
1275	**1 000**	0.588 07	4056.5	4644.5	8.919 1	2.0196	2.485	4.78E−05	0.136 83	vapor
275	**2 000**	0.000 999	7.7776	9.7758	0.028 373	4.2042	4.2046	0.001 678	0.565 6	liquid
300	**2 000**	0.001 003	112.4	114.4	0.392 54	4.1239	4.1753	0.000 853	0.611 18	liquid
325	**2 000**	0.001 012	216.76	218.78	0.726 73	4.0124	4.1776	0.000 531	0.646 64	liquid
350	**2 000**	0.001 026	321.31	323.36	1.036 7	3.8857	4.1903	0.000 369	0.668 99	liquid
375	**2 000**	0.001 044	426.3	428.39	1.326 6	3.7557	4.2135	0.000 277	0.680 7	liquid
400	**2 000**	0.001 066	532.02	534.16	1.599 6	3.6303	4.2508	0.000 219	0.684 71	liquid
450	**2 000**	0.001 122	747.47	749.71	2.107 2	3.4069	4.3881	0.000 153	0.675 41	liquid
485.53	**2 000**	0.001 177	906.14	908.5	2.446 8	3.2737	4.5655	0.000 126	0.655 32	liquid
485.53	**2 000**	0.099 585	2599.1	2798.3	6.339	2.1585	3.191	1.61E−05	0.042 572	vapor
500	**2 000**	0.104 41	2632.6	2841.4	6.426 5	1.9462	2.8185	1.68E−05	0.042 679	vapor
600	**2 000**	0.132 62	2820.4	3085.6	6.873 2	1.6733	2.2629	2.12E−05	0.049 154	vapor
700	**2 000**	0.157 86	2991.6	3307.4	7.215 1	1.671	2.1965	2.55E−05	0.059 637	vapor
800	**2 000**	0.182 15	3163.3	3527.6	7.509 2	1.7153	2.215	2.97E−05	0.071 676	vapor
900	**2 000**	0.205 99	3339.3	3751.3	7.772 6	1.7747	2.2613	3.38E−05	0.084 618	vapor
1000	**2 000**	0.229 58	3521.1	3980.2	8.013 7	1.8401	2.3191	3.77E−05	0.098 136	vapor
1100	**2 000**	0.253 03	3709.2	4215.2	8.237 7	1.9072	2.3815	4.15E−05	0.112 05	vapor
1200	**2 000**	0.276 38	3903.8	4456.5	8.447 6	1.9735	2.4447	4.51E−05	0.126 27	vapor

(Continued)

T (K)	p (kPa)	v (m^3/kg)	u (kJ/kg)	h (kJ/kg)	s ($J/g*K$)	c_v ($J/g*K$)	c_p ($J/g*K$)	μ ($Pa*s$)	k ($W/m*K$)	H_2O phase
1275	**2 000**	0.293 85	4054	4641.7	8.597 2	2.0218	2.4913	4.78E−05	0.137 09	vapor
275	**5 000**	0.000 998	7.802	12.79	0.028 443	4.1908	4.191	0.001 672	0.567 21	liquid
300	**5 000**	0.001 001	112.15	117.16	0.391 7	4.1142	4.167	0.000 853	0.612 54	liquid
325	**5 000**	0.001 011	216.3	221.35	0.725 31	4.005	4.1708	0.000 531	0.648 04	liquid
350	**5 000**	0.001 025	320.65	325.77	1.034 8	3.8799	4.1839	0.000 37	0.670 5	liquid
375	**5 000**	0.001 042	425.42	430.63	1.324 2	3.7512	4.2066	0.000 278	0.682 36	liquid
400	**5 000**	0.001 064	530.9	536.22	1.596 8	3.6268	4.243	0.000 22	0.686 54	liquid
425	**5 000**	0.001 09	637.48	642.93	1.855 5	3.5107	4.2974	0.000 181	0.684 78	liquid
450	**5 000**	0.001 12	745.68	751.28	2.103 2	3.4046	4.3755	0.000 154	0.677 71	liquid
475	**5 000**	0.001 156	856.19	861.97	2.342 6	3.309	4.4856	0.000 134	0.665 19	liquid
500	**5 000**	0.001 2	969.96	975.95	2.576 4	3.224	4.6429	0.000 118	0.646 52	liquid
537.09	**5 000**	0.001 286	1148.2	1154.6	2.921	3.1204	5.0368	0.000 1	0.604 15	liquid
537.09	**5 000**	0.039 446	2597	2794.2	5.973 7	2.5922	4.438	1.80E−05	0.055 203	vapor
550	**5 000**	0.041 752	2637.7	2846.5	6.069 8	2.2995	3.7464	1.87E−05	0.054 223	vapor
600	**5 000**	0.049 036	2760.7	3005.9	6.347 8	1.9212	2.8334	2.11E−05	0.054 653	vapor
700	**5 000**	0.060 818	2957.8	3261.9	6.743 3	1.7584	2.4045	2.55E−05	0.062 68	vapor
800	**5 000**	0.071 343	3140.4	3497.1	7.057 5	1.7587	2.3229	2.98E−05	0.073 95	vapor
900	**5 000**	0.081 362	3322.4	3729.2	7.330 9	1.8002	2.3271	3.39E−05	0.086 626	vapor
1000	**5 000**	0.091 12	3508	3963.6	7.577 8	1.8566	2.363	3.78E−05	0.099 971	vapor
1100	**5 000**	0.100 73	3698.7	4202.3	7.805 2	1.9186	2.4127	4.16E−05	0.113 64	vapor
1200	**5 000**	0.110 23	3895.1	4446.3	8.017 5	1.9819	2.4678	4.53E−05	0.127 51	vapor
1275	**5 000**	0.117 32	4046.4	4633	8.168 4	2.0286	2.5102	4.79E−05	0.138 01	vapor
275	**10 000**	0.000 995	7.8335	17.785	0.028 491	4.1692	4.1692	0.001 662	0.569 89	liquid
300	**10 000**	0.000 999	111.74	121.73	0.390 29	4.0984	4.1536	0.000 852	0.614 81	liquid
325	**10 000**	0.001 009	215.55	225.63	0.722 94	3.9928	4.1599	0.000 532	0.650 37	liquid
350	**10 000**	0.001 022	319.56	329.79	1.031 7	3.8705	4.1734	0.000 371	0.673 01	liquid
375	**10 000**	0.001 04	423.98	434.38	1.320 3	3.7438	4.1955	0.000 279	0.685 1	liquid
400	**10 000**	0.001 061	529.06	539.67	1.592 1	3.621	4.2302	0.000 221	0.689 57	liquid
500	**10 000**	0.001 193	965.25	977.18	2.566 9	3.2211	4.6022	0.000 12	0.651 64	liquid
575	**10 000**	0.001 406	1339.8	1353.9	3.267 2	3.0476	5.7427	8.57E−05	0.546 85	liquid
584.15	**10 000**	0.001 453	1393.5	1408.1	3.360 6	3.0438	6.1237	8.18E−05	0.526 83	liquid
584.15	**10 000**	0.018 03	2545.2	2725.5	5.616	3.1065	7.1408	2.03E−05	0.076 543	vapor
600	**10 000**	0.020 091	2619.1	2820	5.775 6	2.6239	5.1365	2.10E−05	0.071 11	vapor
700	**10 000**	0.028 285	2894.5	3177.4	6.330 5	1.9338	2.8741	2.57E−05	0.069 301	vapor
800	**10 000**	0.034 356	3100.2	3443.7	6.686 7	1.8367	2.5313	3.01E−05	0.078 476	vapor
900	**10 000**	0.039 804	3293.5	3691.6	6.978 7	1.8439	2.4458	3.42E−05	0.090 516	vapor
1000	**10 000**	0.044 963	3485.8	3935.5	7.235 7	1.8843	2.4397	3.81E−05	0.103 5	vapor
1100	**10 000**	0.049 959	3681	4180.6	7.469 3	1.9377	2.4661	4.19E−05	0.116 73	vapor
1200	**10 000**	0.054 854	3880.6	4429.2	7.685 5	1.9957	2.507	4.55E−05	0.13	vapor

T (K)	p (kPa)	υ (m^3/kg)	u (kJ/kg)	h (kJ/kg)	s (J/g*K)	c_v (J/g*K)	c_p (J/g*K)	μ (Pa*s)	k (W/m*K)	H_2O phase
1275	**10 000**	0.058 48	4033.7	4618.5	7.838 6	2.0398	2.542	4.81E−05	0.139 97	vapor
275	**20 000**	0.000 99	7.8644	27.672	0.028 343	4.1282	4.1288	0.001 643	0.575 24	liquid
300	**20 000**	0.000 995	110.94	130.84	0.387 41	4.068	4.1282	0.000 851	0.619 36	liquid
325	**20 000**	0.001 004	214.08	234.17	0.718 24	3.9693	4.1389	0.000 534	0.655 05	liquid
350	**20 000**	0.001 018	317.45	337.81	1.025 5	3.852	4.1533	0.000 374	0.678 02	liquid
375	**20 000**	0.001 035	421.19	441.89	1.312 7	3.7294	4.1742	0.000 282	0.690 57	liquid
400	**20 000**	0.001 056	525.5	546.62	1.583	3.6096	4.206	0.000 224	0.695 61	liquid
500	**20 000**	0.001 181	956.44	980.07	2.548 9	3.2155	4.5308	0.000 122	0.661 52	liquid
600	**20 000**	0.001 481	1456.8	1486.4	3.468 2	3.0038	6.1168	7.98E−05	0.518 25	liquid
625	**20 000**	0.001 687	1627.6	1661.3	3.753 5	3.0609	8.455	6.83E−05	0.458 14	liquid
638.9	**20 000**	0.002 04	1786.4	1827.2	4.015 6	3.5181	22.997	5.62E−05	0.415 51	liquid
638.9	**20 000**	0.005 865	2295	2412.3	4.931 4	4.2431	45.55	2.75E−05	0.234 14	vapor
650	**20 000**	0.007 906	2467	2625.1	5.262 2	3.249	11.643	2.58E−05	0.135 92	vapor
700	**20 000**	0.011 577	2730.2	2961.8	5.763 9	2.4016	4.689	2.69E−05	0.094 691	vapor
800	**20 000**	0.015 774	3010.9	3326.4	6.253	2.011	3.0783	3.10E−05	0.091 106	vapor
900	**20 000**	0.019 006	3232.7	3612.8	6.590 7	1.9351	2.7193	3.50E−05	0.100 5	vapor
1000	**20 000**	0.021 884	3440.3	3878	6.870 2	1.9405	2.6058	3.89E−05	0.112 36	vapor
1100	**20 000**	0.024 581	3645.1	4136.7	7.116 8	1.9757	2.578	4.26E−05	0.124 48	vapor
1200	**20 000**	0.027 171	3851.4	4394.8	7.341 4	2.0231	2.5875	4.61E−05	0.136 43	vapor
1275	**20 000**	0.029 066	4008.2	4589.5	7.498 8	2.0619	2.6068	4.87E−05	0.145 25	vapor

Source: Linstrom and Mallard (2017–2021).

A.5 Properties of Carbon Dioxide (Liquid–Vapor)

The enthalpy of formation of carbon dioxide is $\Delta\hat{h}_f^0 = -393\ 522\ kJ/kmol$. The molar mass of carbon dioxide is $\hat{M} = 44.007\ 66\ kg/kmol$. These properties were derived from Linstrom and Mallard (2017–2021).

T (K)	p (kPa)	υ (m^3/kg)	u (kJ/kg)	h (kJ/kg)	s (J/g*K)	c_v (J/g*K)	c_p (J/g*K)	μ (Pa*s)	k (W/m*K)	CO_2 phase
275	**10**	5.192	435.5	487.42	3.109	0.630 69	0.820 17	1.38E−05	0.014 773	vapor
300	**10**	5.665	451.6	508.25	3.1815	0.657 16	0.846 51	1.50E−05	0.016 752	vapor
325	**10**	6.137 7	468.35	529.73	3.2503	0.682 4	0.871 65	1.62E−05	0.018 801	vapor
350	**10**	6.610 4	485.72	551.82	3.3158	0.706 32	0.895 51	1.74E−05	0.020 891	vapor
375	**10**	7.083	503.67	574.5	3.3783	0.728 95	0.918 09	1.86E−05	0.023	vapor
400	**10**	7.555 6	522.16	597.72	3.4383	0.750 34	0.939 45	1.97E−05	0.025 113	vapor
500	**10**	9.445 5	601.08	695.53	3.6562	0.825 25	1.014 3	2.40E−05	0.033 467	vapor

(Continued)

T (K)	p (kPa)	υ (m^3/kg)	u (kJ/kg)	h (kJ/kg)	s (J/g*K)	c_v (J/g*K)	c_p (J/g*K)	μ (Pa*s)	k (W/m*K)	CO_2 phase
600	**10**	11.335	686.77	800.12	3.8467	0.886 54	1.075 5	2.80E−05	0.041 535	vapor
700	**10**	13.225	778.05	910.3	4.0165	0.937 43	1.126 4	3.17E−05	0.049 279	vapor
800	**10**	15.114	873.98	1025.1	4.1697	0.979 91	1.168 9	3.51E−05	0.056 695	vapor
900	**10**	17.003	973.8	1143.8	4.3095	1.015 4	1.204 4	3.83E−05	0.063 786	vapor
1000	**10**	18.893	1076.9	1265.8	4.438	1.045 2	1.234 1	4.13E−05	0.070 56	vapor
1100	**10**	20.782	1182.7	1390.5	4.5568	1.070 2	1.259 1	4.41E−05	0.077 031	vapor
275	**20**	2.594 3	435.42	487.31	2.9778	0.631 04	0.821 08	1.38E−05	0.014 777	vapor
300	**20**	2.831 1	451.54	508.16	3.0504	0.657 4	0.847 17	1.50E−05	0.016 756	vapor
325	**20**	3.067 7	468.3	529.65	3.1192	0.682 56	0.872 15	1.62E−05	0.018 805	vapor
350	**20**	3.304 3	485.67	551.76	3.1847	0.706 44	0.895 9	1.74E−05	0.020 894	vapor
375	**20**	3.540 7	503.63	574.44	3.2473	0.729 04	0.918 4	1.86E−05	0.023 003	vapor
400	**20**	3.777 1	522.13	597.67	3.3072	0.750 42	0.939 7	1.97E−05	0.025 117	vapor
500	**20**	4.722 4	601.06	695.5	3.5252	0.825 28	1.014 4	2.40E−05	0.033 47	vapor
600	**20**	5.667 4	686.75	800.1	3.7158	0.886 56	1.075 6	2.80E−05	0.041 537	vapor
700	**20**	6.612 3	778.03	910.28	3.8855	0.937 44	1.126 5	3.17E−05	0.049 281	vapor
800	**20**	7.557 1	873.97	1025.1	4.0388	0.979 92	1.168 9	3.51E−05	0.056 697	vapor
900	**20**	8.501 9	973.79	1143.8	4.1786	1.015 4	1.204 4	3.83E−05	0.063 787	vapor
1000	**20**	9.446 6	1076.9	1265.8	4.307	1.045 2	1.234 1	4.13E−05	0.070 561	vapor
1100	**20**	10.391	1182.7	1390.5	4.4259	1.070 2	1.259 1	4.41E−05	0.077 032	vapor
275	**50**	1.035 7	435.19	486.97	2.8038	0.632 1	0.823 84	1.38E−05	0.014 792	vapor
300	**50**	1.130 8	451.35	507.89	2.8766	0.658 12	0.849 17	1.50E−05	0.016 769	vapor
325	**50**	1.225 7	468.14	529.42	2.9456	0.683 07	0.873 66	1.62E−05	0.018 817	vapor
350	**50**	1.320 5	485.53	551.56	3.0112	0.706 81	0.897 07	1.74E−05	0.020 906	vapor
375	**50**	1.415 3	503.5	574.27	3.0738	0.729 32	0.919 33	1.86E−05	0.023 014	vapor
400	**50**	1.510 0	522.02	597.52	3.1338	0.750 63	0.940 46	1.97E−05	0.025 126	vapor
500	**50**	1.888 5	600.98	695.41	3.352	0.825 37	1.014 8	2.40E−05	0.033 478	vapor
600	**50**	2.266 8	686.7	800.04	3.5425	0.886 6	1.075 8	2.80E−05	0.041 544	vapor
700	**50**	2.644 9	777.99	910.23	3.7123	0.937 47	1.126 6	3.17E−05	0.049 286	vapor
800	**50**	3.022 9	873.93	1025.1	3.8656	0.979 94	1.169	3.51E−05	0.056 702	vapor
900	**50**	3.400 9	973.76	1143.8	4.0054	1.015 5	1.204 5	3.83E−05	0.063 792	vapor
1000	**50**	3.778 9	1076.8	1265.8	4.1339	1.045 2	1.234 2	4.13E−05	0.070 565	vapor
1100	**50.**	4.156 8	1182.7	1390.5	4.2527	1.070 2	1.259 2	4.41E−05	0.077 036	vapor
275	**101.33**	0.509 34	434.79	486.4	2.6689	0.633 92	0.828 62	1.38E−05	0.014 817	vapor
298.15	**101.33**	0.553 08	449.80	505.84	2.7368	0.657 49	0.850 85	1.493E−05	0.016 643	vapor
300	**101.33**	0.556 57	451.02	507.42	2.7421	0.659 35	0.852 62	1.50E−05	0.016 792	vapor
325	**101.33**	0.603 65	467.86	529.03	2.8113	0.683 94	0.876 25	1.62E−05	0.018 838	vapor
350	**101.33**	0.650 63	485.29	551.22	2.877	0.707 45	0.899 08	1.74E−05	0.020 925	vapor
375	**101.33**	0.697 55	503.29	573.98	2.9398	0.729 8	0.920 94	1.86E−05	0.023 032	vapor
400	**101.33**	0.744 41	521.83	597.26	2.9999	0.751	0.941 77	1.97E−05	0.025 143	vapor
500	**101.33**	0.931 52	600.85	695.24	3.2183	0.825 52	1.015 5	2.40E−05	0.033 491	vapor

T (K)	p (kPa)	v (m^3/kg)	u (kJ/kg)	h (kJ/kg)	s (J/g*K)	c_v (J/g*K)	c_p (J/g*K)	μ (Pa*s)	k (W/m*K)	CO_2 phase
600	**101.33**	1.118 4	686.6	799.92	3.4089	0.886 68	1.076 2	2.80E−05	0.041 555	vapor
700	**101.33**	1.305 1	777.91	910.16	3.5788	0.937 52	1.126 9	3.17E−05	0.049 296	vapor
800	**101.33**	1.491 7	873.87	1025	3.7321	0.979 97	1.169 2	3.51E−05	0.05 671	vapor
900	**101.33**	1.678 3	973.71	1143.8	3.8719	1.015 5	1.204 6	3.83E−05	0.063 799	vapor
1000	**101.33**	1.864 8	1076.8	1265.8	4.0004	1.045 2	1.234 3	4.13E−05	0.070 572	vapor
1100	**101.33**	2.051 3	1182.6	1390.5	4.1193	1.070 2	1.259 3	4.41E−05	0.077 042	vapor
275	**200**	0.256 37	434.02	485.29	2.5377	0.637 46	0.838 03	1.38E−05	0.014 866	vapor
300	**200**	0.280 61	450.39	506.51	2.6115	0.661 74	0.859 38	1.50E−05	0.016 836	vapor
325	**200**	0.304 7	467.33	528.27	2.6811	0.685 62	0.881 31	1.62E−05	0.018 878	vapor
350	**200**	0.328 69	484.83	550.57	2.7473	0.708 67	0.903	1.74E−05	0.020 962	vapor
375	**200**	0.352 61	502.89	573.41	2.8103	0.730 72	0.924 05	1.86E−05	0.023 066	vapor
400	**200**	0.376 48	521.47	596.77	2.8706	0.751 7	0.944 29	1.97E−05	0.025 176	vapor
500	**200**	0.471 61	600.61	694.93	3.0893	0.825 82	1.016 7	2.40E−05	0.033 517	vapor
600	**200**	0.566 47	686.41	799.71	3.2802	0.886 84	1.077	2.80E−05	0.041 576	vapor
700	**200**	0.661 19	777.77	910	3.4501	0.937 61	1.127 4	3.17E−05	0.049 314	vapor
800	**200**	0.755 83	873.75	1024.9	3.6035	0.980 03	1.169 6	3.51E−05	0.056 726	vapor
900	**200**	0.850 43	973.6	1143.7	3.7433	1.015 5	1.204 9	3.83E−05	0.063 814	vapor
1000	**200**	0.944 99	1076.7	1265.7	3.8719	1.045 2	1.234 5	4.13E−05	0.070 584	vapor
1100	**200**	1.039 5	1182.5	1390.4	3.9907	1.070 2	1.259 4	4.41E−05	0.077 054	vapor
275	**500**	0.100 46	431.61	481.84	2.3557	0.648 62	0.868 69	1.38E−05	0.015 026	vapor
300	**500**	0.110 55	448.42	503.7	2.4318	0.669 16	0.880 96	1.51E−05	0.016 979	vapor
325	**500**	0.120 49	465.68	525.92	2.5029	0.690 81	0.897 26	1.63E−05	0.019 008	vapor
350	**500**	0.130 32	483.41	548.57	2.5701	0.712 44	0.915 23	1.74E−05	0.021 081	vapor
375	**500**	0.140 07	501.65	571.68	2.6339	0.733 53	0.933 7	1.86E−05	0.023 176	vapor
400	**500**	0.149 77	520.37	595.26	2.6947	0.753 86	0.952 09	1.97E−05	0.025 278	vapor
500	**500**	0.188 23	599.86	693.97	2.9147	0.826 72	1.020 7	2.40E−05	0.033 597	vapor
600	**500**	0.226 41	685.85	799.06	3.1061	0.887 31	1.079 4	2.80E−05	0.041 642	vapor
700	**500**	0.264 45	777.32	909.54	3.2763	0.937 89	1.128 9	3.17E−05	0.049 371	vapor
800	**500**	0.302 42	873.38	1024.6	3.4299	0.980 22	1.170 7	3.51E−05	0.056 776	vapor
900	**500**	0.340 34	973.29	1143.5	3.5699	1.015 7	1.205 7	3.83E−05	0.063 857	vapor
1000	**500**	0.378 22	1076.4	1265.5	3.6985	1.045 3	1.235 2	4.13E−05	0.070 624	vapor
1100	**500**	0.416 09	1182.3	1390.3	3.8174	1.070 3	1.26	4.41E−05	0.077 089	vapor
275	**1000**	0.048 403	427.36	475.76	2.209	0.668 84	0.928 21	1.39E−05	0.015 341	vapor
300	**1000**	0.053 823	445.01	498.84	2.2894	0.682 17	0.920 89	1.51E−05	0.017 248	vapor
325	**1000**	0.059 06	462.85	521.91	2.3632	0.699 74	0.925 94	1.63E−05	0.019 245	vapor
350	**1000**	0.064 18	481	545.18	2.4322	0.718 85	0.936 81	1.75E−05	0.021 295	vapor
375	**1000**	0.069 218	499.55	568.77	2.4973	0.738 3	0.950 52	1.86E−05	0.023 371	vapor
400	**1000**	0.074 198	518.52	592.72	2.5591	0.757 51	0.965 55	1.98E−05	0.025 458	vapor
500	**1000**	0.093 773	598.61	692.38	2.7813	0.828 23	1.027 3	2.41E−05	0.033 735	vapor

(Continued)

T (K)	p (kPa)	v (m^3/kg)	u (kJ/kg)	h (kJ/kg)	s (J/g*K)	c_v (J/g*K)	c_p (J/g*K)	μ (Pa*s)	k (W/m*K)	CO_2 phase
600	**1000**	0.113 06	684.91	797.97	2.9736	0.888 08	1.083 3	2.80E−05	0.041 756	vapor
700	**1000**	0.132 21	776.57	908.78	3.1443	0.938 35	1.131 5	3.17E−05	0.049 467	vapor
800	**1000**	0.151 28	872.76	1024	3.2982	0.980 52	1.172 5	3.51E−05	0.056 859	vapor
900	**1000**	0.170 31	972.77	1143.1	3.4383	1.015 9	1.207 1	3.83E−05	0.063 931	vapor
1000	**1000**	0.189 3	1076	1265.3	3.5671	1.045 5	1.236 2	4.13E−05	0.070 69	vapor
1100	**1000**	0.208 27	1181.9	1390.2	3.6861	1.070 4	1.260 8	4.41E−05	0.077 15	vapor
275	**2000**	0.022 178	417.68	462.04	2.0415	0.718 81	1.098 7	1.41E−05	0.016 241	vapor
300	**2000**	0.025 368	437.62	488.36	2.1332	0.711 16	1.020 6	1.53E−05	0.017 934	vapor
325	**2000**	0.028 302	456.86	513.47	2.2136	0.718 79	0.992 76	1.64E−05	0.019 815	vapor
350	**2000**	0.031 09	475.98	538.16	2.2868	0.732 25	0.985 06	1.76E−05	0.021 789	vapor
375	**2000**	0.033 783	495.23	562.8	2.3548	0.748 12	0.987 07	1.87E−05	0.023 812	vapor
400	**2000**	0.036 411	514.74	587.56	2.4187	0.764 94	0.994 23	1.99E−05	0.025 856	vapor
500	**2000**	0.046 551	596.09	689.19	2.6453	0.831 26	1.040 9	2.41E−05	0.034 03	vapor
600	**2000**	0.056 391	683.04	795.82	2.8396	0.889 64	1.091 2	2.81E−05	0.041 994	vapor
700	**2000**	0.066 092	775.08	907.27	3.0113	0.939 28	1.136 7	3.18E−05	0.049 667	vapor
800	**2000**	0.075 717	871.53	1023	3.1657	0.981 14	1.176 2	3.52E−05	0.057 032	vapor
900	**2000**	0.085 295	971.73	1142.3	3.3062	1.016 3	1.209 9	3.83E−05	0.064 083	vapor
1000	**2000**	0.094 842	1075.1	1264.8	3.4352	1.045 9	1.238 4	4.13E−05	0.070 826	vapor
1100	**2000**	0.104 37	1181.1	1389.8	3.5544	1.070 7	1.262 5	4.41E−05	0.077 272	vapor
275	**5000**	0.001 077	197.67	203.06	1.0052	0.937 65	2.473 5	9.97E−05	0.110 8	liquid
287.43	**5000**	0.001 209	231.82	237.87	1.1289	0.983 14	3.357 2	7.56E−05	0.092 76	liquid
287.43	**5000**	0.006 383	385.74	417.66	1.7544	0.994 65	3.114 2	1.68E−05	0.027 323	vapor
300	**5000**	0.007 788	407.01	445.95	1.8509	0.852 58	1.802 5	1.67E−05	0.023 185	vapor
325	**5000**	0.009 659	435.33	483.63	1.9718	0.789 51	1.328 9	1.74E−05	0.022 826	vapor
350	**5000**	0.011 158	459.02	514.81	2.0643	0.777 39	1.187 5	1.83E−05	0.024 02	vapor
375	**5000**	0.012 494	481.17	543.64	2.1438	0.779 74	1.125 9	1.93E−05	0.025 629	vapor
400	**5000**	0.013 736	502.7	571.37	2.2155	0.788 24	1.096 5	2.04E−05	0.027 41	vapor
500	**5000**	0.018 24	588.42	679.62	2.457	0.840 35	1.084	2.45E−05	0.035 058	vapor
600	**5000**	0.022 411	677.42	789.47	2.6572	0.894 25	1.115 3	2.83E−05	0.042 792	vapor
700	**5000**	0.026 439	770.65	902.85	2.8319	0.942 02	1.152 3	3.19E−05	0.050 324	vapor
800	**5000**	0.030 391	867.89	1019.8	2.9881	0.982 96	1.187 1	3.53E−05	0.057 591	vapor
900	**5000**	0.034 298	968.64	1140.1	3.1298	1.017 6	1.217 9	3.85E−05	0.064 571	vapor
1000	**5000**	0.038 174	1072.4	1263.3	3.2595	1.046 9	1.244 6	4.14E−05	0.071 258	vapor
1100	**5000**	0.042 03	1178.8	1388.9	3.3792	1.071 5	1.267 4	4.42E−05	0.077 661	vapor

Source: Linstrom and Mallard (2017–2021).

A.6 Properties of Helium (Gas)

The enthalpy of formation of helium is $\Delta\hat{h}_f^0 = 0.0\ kJ/kmol$. The molar mass of helium is $\hat{M} = 4.002\ 602\ kg/kmol$. These properties were derived from Linstrom and Mallard (2017–2021).

T (K)	*p* (kPa)	*v* (m³/kg)	*u* (kJ/kg)	*h* (kJ/kg)	*s* (J/g*K)	c_v (J/g*K)	c_p (J/g*K)	*μ* (Pa*s)	*k* (W/m*K)	He phase
275	**101.33**	5.640 3	862.1	1433.6	27.547	3.116	5.193	1.88E−05	0.146 88	vapor
298.15	**101.33**	6.114 8	934.24	1553.9	27.967	3.1160	5.1930	1.985E−05	0.155 31	vapor
300	**101.33**	6.152 7	940	1563.5	27.999	3.116	5.193	1.99E−05	0.155 97	vapor
325	**101.33**	6.665 2	1017.9	1693.3	28.415	3.116	5.193	2.11E−05	0.164 85	vapor
350	**101.33**	7.177 7	1095.8	1823.1	28.8	3.116	5.193	2.22E−05	0.173 53	vapor
375	**101.33**	7.690 1	1173.7	1952.9	29.158	3.116	5.193	2.32E−05	0.182 03	vapor
400	**101.33**	8.202 6	1251.6	2082.8	29.493	3.116	5.193	2.43E−05	0.190 37	vapor
500	**101.33**	10.252	1563.2	2602.1	30.652	3.1159	5.193	2.84E−05	0.222 3	vapor
600	**101.33**	12.302	1874.8	3121.4	31.599	3.1159	5.193	3.22E−05	0.252 4	vapor
700	**101.33**	14.352	2186.3	3640.7	32.399	3.1159	5.193	3.59E−05	0.281 05	vapor
800	**101.33**	16.402	2497.9	4160	33.093	3.1159	5.193	3.94E−05	0.308 52	vapor
900	**101.33**	18.452	2809.5	4679.3	33.704	3.1159	5.193	4.28E−05	0.334 99	vapor
1000	**101.33**	20.502	3121.1	5198.6	34.251	3.1159	5.193	4.62E−05	0.360 6	vapor
1100	**101.33**	22.552	3432.7	5717.9	34.746	3.1159	5.193	4.94E−05	0.385 47	vapor
1200	**101.33**	24.602	3744.3	6237.2	35.198	3.1159	5.193	5.25E−05	0.409 68	vapor
1300	**101.33**	26.652	4055.8	6756.5	35.614	3.1159	5.193	5.56E−05	0.433 3	vapor
1400	**101.33**	28.702	4367.4	7275.8	35.999	3.1159	5.193	5.86E−05	0.456 39	vapor
1500	**101.33**	30.752	4679	7795.1	36.357	3.1159	5.193	6.15E−05	0.479	vapor
275	**200.**	2.859 1	862.14	1434	26.135	3.1163	5.1929	1.88E−05	0.146 95	vapor
300	**200**	3.118 7	940.04	1563.8	26.587	3.1162	5.1929	1.99E−05	0.156 05	vapor
325	**200**	3.378 3	1017.9	1693.6	27.003	3.1162	5.1929	2.11E−05	0.164 93	vapor
350	**200**	3.637 9	1095.8	1823.4	27.387	3.1161	5.1929	2.22E−05	0.173 61	vapor
375	**200**	3.897 6	1173.7	1953.2	27.746	3.1161	5.1929	2.32E−05	0.182 11	vapor
400	**200**	4.157 2	1251.6	2083.1	28.081	3.1161	5.1929	2.43E−05	0.190 44	vapor
500	**200**	5.195 7	1563.2	2602.4	29.24	3.116	5.1929	2.84E−05	0.222 37	vapor
600	**200**	6.234 2	1874.8	3121.6	30.186	3.116	5.1929	3.22E−05	0.252 47	vapor
700	**200**	7.272 8	2186.4	3640.9	30.987	3.1159	5.1929	3.59E−05	0.281 12	vapor
800	**200**	8.311 4	2498	4160.2	31.68	3.1159	5.193	3.94E−05	0.308 58	vapor
900	**200**	9.349 9	2809.5	4679.5	32.292	3.1159	5.193	4.28E−05	0.335 05	vapor
1000	**200**	10.388	3121.1	5198.8	32.839	3.1159	5.193	4.62E−05	0.360 66	vapor
1100	**200**	11.427	3432.7	5718.1	33.334	3.1159	5.193	4.94E−05	0.385 53	vapor
1200	**200**	12.466	3744.3	6237.4	33.786	3.1159	5.193	5.25E−05	0.409 74	vapor
1300	**200**	13.504	4055.9	6756.7	34.201	3.1159	5.193	5.56E−05	0.433 36	vapor

(Continued)

T (K)	p (kPa)	υ (m^3/kg)	u (kJ/kg)	h (kJ/kg)	s (J/g*K)	c_v (J/g*K)	c_p (J/g*K)	μ (Pa*s)	k (W/m*K)	He phase
1400	**200**	14.543	4367.5	7276	34.586	3.1159	5.193	5.86E−05	0.456 45	vapor
1500	**200**	15.581	4679	7795.3	34.945	3.1159	5.193	6.15E−05	0.479 05	vapor
275	**500**	1.145 4	862.25	1434.9	24.232	3.1169	5.1928	1.88E−05	0.147 18	supercrit
300	**500**	1.249 2	940.16	1564.7	24.684	3.1167	5.1927	1.99E−05	0.156 28	supercrit
325	**500**	1.353	1018.1	1694.6	25.1	3.1166	5.1927	2.11E−05	0.165 15	supercrit
350	**500**	1.456 8	1096	1824.4	25.484	3.1166	5.1926	2.22E−05	0.173 83	supercrit
375	**500**	1.560 7	1173.9	1954.2	25.843	3.1165	5.1926	2.32E−05	0.182 33	supercrit
400	**500**	1.664 5	1251.8	2084	26.178	3.1164	5.1926	2.43E−05	0.190 66	supercrit
500	**500**	2.079 9	1563.3	2603.3	27.336	3.1163	5.1926	2.84E−05	0.222 58	supercrit
600	**500**	2.495 2	1874.9	3122.5	28.283	3.1162	5.1927	3.22E−05	0.252 67	supercrit
700	**500**	2.910 6	2186.5	3641.8	29.084	3.1161	5.1927	3.59E−05	0.281 32	supercrit
800	**500**	3.326	2498.1	4161.1	29.777	3.116	5.1928	3.94E−05	0.308 78	supercrit
900	**500**	3.741 4	2809.7	4680.4	30.389	3.116	5.1928	4.29E−05	0.335 24	supercrit
1000	**500**	4.156 8	3121.2	5199.6	30.936	3.116	5.1928	4.62E−05	0.360 85	supercrit
1100	**500**	4.572 2	3432.8	5718.9	31.431	3.116	5.1929	4.94E−05	0.385 71	supercrit
1200	**500**	4.987 6	3744.4	6238.2	31.883	3.116	5.1929	5.25E−05	0.409 92	supercrit
1300	**500**	5.403	4056	6757.5	32.298	3.1159	5.1929	5.56E−05	0.433 53	supercrit
1400	**500**	5.818 5	4367.6	7276.8	32.683	3.1159	5.1929	5.86E−05	0.456 61	supercrit
1500	**500**	6.233 9	4679.1	7796.1	33.041	3.1159	5.1929	6.15E−05	0.479 22	supercrit
275	**1 000**	0.574 12	862.43	1436.6	22.793	3.1179	5.1924	1.88E−05	0.147 56	supercrit
300	**1 000**	0.626 02	940.35	1566.4	23.245	3.1176	5.1923	2.00E−05	0.156 65	supercrit
325	**1 000**	0.677 92	1018.3	1696.2	23.66	3.1174	5.1922	2.11E−05	0.165 51	supercrit
350	**1 000**	0.729 82	1096.2	1826	24.045	3.1173	5.1922	2.22E−05	0.174 19	supercrit
375	**1 000**	0.781 72	1174.1	1955.8	24.403	3.1171	5.1922	2.33E−05	0.182 68	supercrit
400	**1 000**	0.833 62	1252	2085.6	24.738	3.117	5.1922	2.43E−05	0.191 01	supercrit
500	**1 000**	1.041 3	1563.6	2604.8	25.897	3.1167	5.1922	2.84E−05	0.222 92	supercrit
600	**1 000**	1.248 9	1875.1	3124	26.844	3.1165	5.1923	3.22E−05	0.253	supercrit
700	**1 000**	1.456 6	2186.7	3643.3	27.644	3.1163	5.1924	3.59E−05	0.281 63	supercrit
800	**1 000**	1.664 2	2498.3	4162.5	28.337	3.1163	5.1925	3.94E−05	0.309 08	supercrit
900	**1 000**	1.871 9	2809.9	4681.8	28.949	3.1162	5.1925	4.29E−05	0.335 54	supercrit
1000	**1 000**	2.079 6	3121.4	5201	29.496	3.1161	5.1926	4.62E−05	0.361 14	supercrit
1100	**1 000**	2.287 3	3433	5720.3	29.991	3.1161	5.1926	4.94E−05	0.386	supercrit
1200	**1 000**	2.495	3744.6	6239.5	30.443	3.1161	5.1927	5.25E−05	0.410 2	supercrit
1300	**1 000**	2.702 7	4056.2	6758.8	30.859	3.116	5.1927	5.56E−05	0.433 81	supercrit
1400	**1 000**	2.910 3	4367.7	7278.1	31.243	3.116	5.1927	5.86E−05	0.456 89	supercrit
1500	**1 000**	3.118	4679.3	7797.4	31.602	3.116	5.1928	6.16E−05	0.479 48	supercrit
275	**2 000**	0.288 5	862.8	1439.8	21.354	3.1199	5.1919	1.89E−05	0.148 28	supercrit
300	**2 000**	0.314 43	940.72	1569.6	21.806	3.1194	5.1916	2.00E−05	0.157 35	supercrit
325	**2 000**	0.340 37	1018.6	1699.4	22.222	3.119	5.1914	2.11E−05	0.166 21	supercrit
350	**2 000**	0.366 3	1096.6	1829.2	22.606	3.1187	5.1914	2.22E−05	0.174 87	supercrit

T (K)	p (kPa)	v (m^3/kg)	u (kJ/kg)	h (kJ/kg)	s (J/g*K)	c_v (J/g*K)	c_p (J/g*K)	μ (Pa*s)	k (W/m*K)	He phase
375	**2 000**	0.392 24	1174.5	1958.9	22.965	3.1184	5.1913	2.33E−05	0.183 35	supercrit
400	**2 000**	0.418 18	1252.4	2088.7	23.3	3.1182	5.1913	2.43E−05	0.191 67	supercrit
500	**2 000**	0.521 95	1564	2607.9	24.458	3.1175	5.1914	2.84E−05	0.223 55	supercrit
600	**2 000**	0.625 73	1875.5	3127	25.405	3.1171	5.1916	3.22E−05	0.253 61	supercrit
700	**2 000**	0.729 53	2187.1	3646.2	26.205	3.1169	5.1917	3.59E−05	0.282 23	supercrit
800	**2 000**	0.833 34	2498.7	4165.4	26.898	3.1167	5.1919	3.95E−05	0.309 67	supercrit
900	**2 000**	0.937 15	2810.3	4684.6	27.51	3.1165	5.192	4.29E−05	0.336 11	supercrit
1000	**2 000**	1.041	3121.8	5203.8	28.057	3.1164	5.1921	4.62E−05	0.361 7	supercrit
1100	**2 000**	1.144 8	3433.4	5723	28.552	3.1164	5.1922	4.94E−05	0.386 55	supercrit
1200	**2 000**	1.248 6	3745	6242.2	29.003	3.1163	5.1923	5.25E−05	0.410 73	supercrit
1300	**2 000**	1.352 5	4056.5	6761.4	29.419	3.1162	5.1924	5.56E−05	0.434 33	supercrit
1400	**2 000**	1.456 3	4368.1	7280.7	29.804	3.1162	5.1924	5.86E−05	0.457 4	supercrit
1500	**2 000**	1.560 1	4679.7	7799.9	30.162	3.1162	5.1925	6.16E−05	0.479 99	supercrit
275	**5 000**	0.117 14	863.85	1449.5	19.455	3.1258	5.1904	1.90E−05	0.150 31	supercrit
300	**5 000**	0.127 49	941.83	1579.3	19.907	3.1246	5.1897	2.01E−05	0.159 31	supercrit
325	**5 000**	0.137 85	1019.8	1709	20.322	3.1237	5.1892	2.12E−05	0.168 11	supercrit
350	**5 000**	0.148 2	1097.7	1838.7	20.707	3.1229	5.189	2.23E−05	0.176 72	supercrit
375	**5 000**	0.158 56	1175.7	1968.5	21.065	3.1222	5.1888	2.34E−05	0.185 16	supercrit
400	**5 000**	0.168 92	1253.6	2098.2	21.399	3.1216	5.1888	2.44E−05	0.193 45	supercrit
500	**5 000**	0.210 37	1565.2	2617.1	22.557	3.12	5.189	2.85E−05	0.225 23	supercrit
600	**5 000**	0.251 84	1876.8	3136	23.503	3.119	5.1894	3.23E−05	0.255 23	supercrit
700	**5 000**	0.293 32	2188.3	3654.9	24.303	3.1184	5.1898	3.60E−05	0.283 8	supercrit
800	**5 000**	0.334 81	2499.9	4173.9	24.996	3.1179	5.1901	3.95E−05	0.311 2	supercrit
900	**5 000**	0.376 31	2811.4	4693	25.608	3.1176	5.1904	4.29E−05	0.337 61	supercrit
1000	**5 000**	0.417 81	3123	5212	26.155	3.1173	5.1907	4.62E−05	0.363 17	supercrit
1100	**5 000**	0.459 32	3434.5	5731.1	26.649	3.1171	5.1909	4.94E−05	0.387 99	supercrit
1200	**5 000**	0.500 83	3746.1	6250.2	27.101	3.117	5.1911	5.26E−05	0.412 16	supercrit
1300	**5 000**	0.542 34	4057.6	6769.3	27.516	3.1168	5.1913	5.56E−05	0.435 73	supercrit
1400	**5 000**	0.583 86	4369.1	7288.5	27.901	3.1167	5.1914	5.86E−05	0.458 78	supercrit
1500	**5 000**	0.625 38	4680.7	7807.6	28.259	3.1167	5.1916	6.16E−05	0.481 35	supercrit
275	**10 000**	0.060 02	865.52	1465.7	18.021	3.1353	5.1887	1.92E−05	0.153 42	supercrit
300	**10 000**	0.065 183	943.58	1595.4	18.473	3.133	5.1871	2.03E−05	0.162 26	supercrit
325	**10 000**	0.070 347	1021.6	1725.1	18.888	3.1312	5.1861	2.14E−05	0.170 93	supercrit
350	**10 000**	0.075 512	1099.6	1854.7	19.272	3.1296	5.1854	2.24E−05	0.179 43	supercrit
375	**10 000**	0.080 678	1177.6	1984.3	19.63	3.1283	5.1851	2.35E−05	0.187 78	supercrit
400	**10 000**	0.085 845	1255.5	2114	19.965	3.1272	5.1849	2.45E−05	0.195 99	supercrit
500	**10 000**	0.106 53	1567.2	2632.5	21.122	3.124	5.185	2.86E−05	0.227 55	supercrit
600	**10 000**	0.127 22	1878.8	3151	22.067	3.1221	5.1857	3.24E−05	0.257 42	supercrit
700	**10 000**	0.147 93	2190.3	3669.6	22.866	3.1208	5.1865	3.60E−05	0.285 91	supercrit
800	**10 000**	0.168 65	2501.8	4188.3	23.559	3.12	5.1872	3.96E−05	0.313 25	supercrit

(Continued)

T (K)	p (kPa)	v (m^3/kg)	u (kJ/kg)	h (kJ/kg)	s (J/g*K)	c_v (J/g*K)	c_p (J/g*K)	μ (Pa*s)	k (W/m*K)	He phase
900	**10 000**	0.189 37	2813.3	4707	24.17	3.1193	5.1878	4.30E−05	0.339 62	supercrit
1000	**10 000**	0.210 1	3124.9	5225.8	24.717	3.1188	5.1883	4.63E−05	0.365 15	supercrit
1100	**10 000**	0.230 83	3436.4	5744.7	25.211	3.1184	5.1888	4.95E−05	0.389 94	supercrit
1200	**10 000**	0.251 57	3747.9	6263.6	25.663	3.1181	5.1891	5.26E−05	0.414 08	supercrit
1300	**10 000**	0.272 32	4059.4	6782.5	26.078	3.1178	5.1895	5.57E−05	0.437 63	supercrit
1400	**10 000**	0.293 06	4370.9	7301.5	26.463	3.1176	5.1898	5.87E−05	0.460 66	supercrit
1500	**10 000**	0.313 81	4682.4	7820.5	26.821	3.1175	5.19	6.16E−05	0.483 21	supercrit
275	**20 000**	0.031 463	868.57	1497.8	16.593	3.1529	5.188	1.96E−05	0.159 14	supercrit
300	**20 000**	0.034 033	946.82	1627.5	17.044	3.1487	5.1843	2.05E−05	0.167 62	supercrit
325	**20 000**	0.036 603	1025	1757.1	17.459	3.1452	5.1819	2.16E−05	0.175 97	supercrit
350	**20 000**	0.039 174	1103.1	1886.6	17.843	3.1424	5.1802	2.27E−05	0.184 2	supercrit
375	**20 000**	0.041 746	1181.2	2016.1	18.2	3.1399	5.1792	2.37E−05	0.192 32	supercrit
400	**20 000**	0.044 318	1259.2	2145.5	18.535	3.1378	5.1785	2.48E−05	0.200 32	supercrit
500	**20 000**	0.054 616	1571	2663.3	19.69	3.1318	5.178	2.88E−05	0.231 29	supercrit
600	**20 000**	0.064 926	1882.7	3181.2	20.634	3.1281	5.1789	3.25E−05	0.260 79	supercrit
700	**20 000**	0.075 248	2194.2	3699.1	21.433	3.1257	5.1802	3.62E−05	0.289 05	supercrit
800	**20 000**	0.085 578	2505.7	4217.2	22.124	3.1239	5.1814	3.97E−05	0.316 25	supercrit
900	**20 000**	0.095 916	2817.1	4735.4	22.735	3.1227	5.1825	4.31E−05	0.342 52	supercrit
1000	**20 000**	0.106 26	3128.6	5253.7	23.281	3.1217	5.1835	4.64E−05	0.367 98	supercrit
1100	**20 000**	0.116 61	3440	5772.1	23.775	3.1209	5.1844	4.96E−05	0.392 72	supercrit
1200	**20 000**	0.126 96	3751.4	6290.6	24.226	3.1203	5.1852	5.27E−05	0.416 83	supercrit
1300	**20 000**	0.137 31	4062.9	6809.2	24.641	3.1198	5.1858	5.58E−05	0.440 36	supercrit
1400	**20 000**	0.147 67	4374.3	7327.8	25.025	3.1194	5.1864	5.87E−05	0.463 37	supercrit
1500	**20 000**	0.158 03	4685.8	7846.4	25.383	3.1191	5.1869	6.17E−05	0.485 9	supercrit

Source: Linstrom and Mallard (2017–2021).

A.7 Properties of Methane (Gas)

The enthalpy of formation of methane is $\Delta\hat{h}_f^0 = -74\,873.1\ kJ/kmol$. The molar mass of methane is $\hat{M} = 16.040\,96\ kg/kmol$. These properties were derived from Linstrom and Mallard (2017–2021).

T (K)	p (kPa)	v (m^3/kg)	u (kJ/kg)	h (kJ/kg)	s (J/g*K)	c_v (J/g*K)	c_p (J/g*K)	μ (Pa*s)	k (W/m*K)	CH_4 phase
275	**101.33**	1.403 3	716.66	858.86	6.4961	1.6602	2.1843	1.04E−05	0.031 171	vapor
298.15	**101.33**	1.522 3	755.70	909.95	6.6744	1.7086	2.2317	1.119E−05	0.034 296	vapor
300	**101.33**	1.531 8	758.87	914.08	6.6883	1.7129	2.2359	1.12E−05	0.034 553	vapor

T (K)	p (kPa)	υ (m^3/kg)	u (kJ/kg)	h (kJ/kg)	s (J/ g*K)	c_v (J/ g*K)	c_p (J/ g*K)	μ (Pa*s)	k (W/m*K)	CH_4 phase
325	**101.33**	1.660 2	802.52	970.74	6.8696	1.7763	2.2985	1.20E−05	0.038 137	vapor
350	**101.33**	1.788 5	847.86	1029.1	7.0425	1.8487	2.3703	1.28E−05	0.041 93	vapor
375	**101.33**	1.916 7	895.1	1089.3	7.2087	1.9283	2.4494	1.35E−05	0.045 931	vapor
400	**101.33**	2.044 9	944.39	1151.6	7.3695	2.0134	2.534	1.43E−05	0.050 127	vapor
425	**101.33**	2.173	995.86	1216.1	7.5258	2.1023	2.6226	1.50E−05	0.054 505	vapor
450	**101.33**	2.301 2	1049.6	1282.8	7.6782	2.1939	2.7139	1.57E−05	0.059 048	vapor
475	**101.33**	2.429 2	1105.6	1351.8	7.8275	2.287	2.8068	1.63E−05	0.063 74	vapor
500	**101.33**	2.557 3	1164	1423.1	7.9738	2.3809	2.9006	1.70E−05	0.068 564	vapor
525	**101.33**	2.685 3	1224.7	1496.8	8.1176	2.4749	2.9945	1.76E−05	0.073 507	vapor
550	**101.33**	2.813 3	1287.7	1572.8	8.2591	2.5687	3.0881	1.82E−05	0.078 555	vapor
575	**101.33**	2.941 3	1353.1	1651.2	8.3984	2.6617	3.181	1.88E−05	0.083 697	vapor
600	**101.33**	3.069 3	1420.9	1731.9	8.5357	2.7539	3.2731	1.94E−05	0.088 921	vapor
625	**101.33**	3.197 3	1490.9	1814.8	8.6712	2.8448	3.364	2.00E−05	0.094 22	vapor
275	**200**	0.709 34	715.88	857.74	6.1408	1.6615	2.1914	1.04E−05	0.031 243	vapor
300	**200**	0.774 78	758.18	913.13	6.3336	1.7138	2.2415	1.13E−05	0.034 618	vapor
325	**200**	0.840 09	801.9	969.92	6.5153	1.777	2.3031	1.20E−05	0.038 196	vapor
350	**200**	0.905 32	847.3	1028.4	6.6886	1.8493	2.3741	1.28E−05	0.041 985	vapor
375	**200**	0.970 48	894.59	1088.7	6.855	1.9288	2.4526	1.36E−05	0.045 982	vapor
400	**200**	1.035 6	943.92	1151	7.0159	2.0138	2.5368	1.43E−05	0.050 175	vapor
425	**200**	1.100 6	995.43	1215.6	7.1724	2.1027	2.625	1.50E−05	0.054 55	vapor
450	**200**	1.165 7	1049.2	1282.3	7.325	2.1941	2.716	1.57E−05	0.059 09	vapor
475	**200**	1.230 7	1105.2	1351.4	7.4743	2.2872	2.8087	1.63E−05	0.063 779	vapor
500	**200**	1.295 6	1163.6	1422.8	7.6207	2.3811	2.9022	1.70E−05	0.068 601	vapor
525	**200**	1.360 6	1224.4	1496.5	7.7646	2.4751	2.9959	1.76E−05	0.073 542	vapor
550	**200**	1.425 5	1287.4	1572.5	7.9061	2.5688	3.0893	1.82E−05	0.078 588	vapor
575	**200**	1.490 4	1352.9	1650.9	8.0455	2.6619	3.1822	1.88E−05	0.083 729	vapor
600	**200**	1.555 3	1420.6	1731.6	8.1829	2.754	3.2741	1.94E−05	0.088 952	vapor
625	**200**	1.620 2	1490.6	1814.6	8.3184	2.8449	3.3649	2.00E−05	0.094 249	vapor
275	**500**	0.281 77	713.46	854.35	5.6571	1.6654	2.2134	1.05E−05	0.031 462	vapor
300	**500**	0.308 34	756.05	910.22	5.8516	1.7168	2.259	1.13E−05	0.034 815	vapor
325	**500**	0.334 79	800.01	967.4	6.0346	1.7793	2.3172	1.21E−05	0.038 376	vapor
350	**500**	0.361 15	845.6	1026.2	6.2088	1.8512	2.3858	1.28E−05	0.042 151	vapor
375	**500**	0.387 43	893.04	1086.8	6.376	1.9303	2.4624	1.36E−05	0.046 135	vapor
400	**500**	0.413 66	942.51	1149.3	6.5375	2.015	2.5451	1.43E−05	0.050 317	vapor
425	**500**	0.439 85	994.13	1214.1	6.6944	2.1037	2.6322	1.50E−05	0.054 683	vapor
450	**500**	0.466	1048	1281	6.8474	2.195	2.7223	1.57E−05	0.059 216	vapor
475	**500**	0.492 12	1104.1	1350.2	6.997	2.288	2.8141	1.64E−05	0.063 898	vapor
500	**500**	0.518 22	1162.6	1421.7	7.1437	2.3817	2.907	1.70E−05	0.068 714	vapor
525	**500**	0.544 3	1223.4	1495.5	7.2878	2.4757	3.0002	1.76E−05	0.073 649	vapor

(*Continued*)

T (K)	p (kPa)	v (m^3/kg)	u (kJ/kg)	h (kJ/kg)	s (J/g*K)	c_v (J/g*K)	c_p (J/g*K)	μ (Pa*s)	k (W/m*K)	CH_4 phase
550	**500**	0.570 36	1286.5	1571.7	7.4296	2.5693	3.0932	1.83E−05	0.078 69	vapor
575	**500**	0.596 41	1352	1650.2	7.5691	2.6623	3.1856	1.89E−05	0.083 826	vapor
600	**500**	0.622 44	1419.8	1731	7.7066	2.7544	3.2772	1.95E−05	0.089 045	vapor
625	**500**	0.648 46	1489.8	1814	7.8422	2.8453	3.3677	2.00E−05	0.094 338	vapor
275	**1 000**	0.139 24	709.38	848.63	5.2831	1.6719	2.2518	1.06E−05	0.031 838	vapor
300	**1 000**	0.152 87	752.48	905.35	5.4805	1.7217	2.2889	1.14E−05	0.035 152	vapor
325	**1 000**	0.166 37	796.84	963.2	5.6657	1.7832	2.3413	1.21E−05	0.038 682	vapor
350	**1 000**	0.179 77	842.75	1022.5	5.8415	1.8542	2.4056	1.29E−05	0.042 431	vapor
375	**1 000**	0.193 1	890.46	1083.6	6.0099	1.9328	2.479	1.36E−05	0.046 394	vapor
400	**1 000**	0.206 37	940.15	1146.5	6.1724	2.0171	2.5592	1.44E−05	0.050 558	vapor
425	**1 000**	0.2196	991.95	1211.6	6.3301	2.1054	2.6443	1.51E−05	0.054 908	vapor
450	**1 000**	0.232 79	1046	1278.8	6.4837	2.1965	2.7327	1.57E−05	0.059 427	vapor
475	**1 000**	0.245 96	1102.2	1348.2	6.6339	2.2892	2.8233	1.64E−05	0.064 097	vapor
500	**1 000**	0.259 1	1160.8	1419.9	6.781	2.3828	2.9151	1.70E−05	0.068 902	vapor
525	**1 000**	0.272 22	1221.7	1494	6.9255	2.4766	3.0074	1.77E−05	0.073 828	vapor
550	**1 000**	0.285 32	1285	1570.3	7.0675	2.5701	3.0996	1.83E−05	0.078 86	vapor
575	**1 000**	0.298 41	1350.5	1648.9	7.2073	2.663	3.1914	1.89E−05	0.083 988	vapor
600	**1 000**	0.311 49	1418.4	1729.9	7.3451	2.755	3.2824	1.95E−05	0.089 2	vapor
625	**1 000**	0.324 55	1488.5	1813.1	7.4809	2.8458	3.3725	2.01E−05	0.094 487	vapor
275	**2 000**	0.067 986	701.01	836.98	4.8934	1.6854	2.3349	1.08E−05	0.032 654	vapor
300	**2 000**	0.075 15	745.22	895.52	5.0971	1.7317	2.3522	1.15E−05	0.035 871	vapor
325	**2 000**	0.082 176	790.42	954.77	5.2868	1.7909	2.3913	1.23E−05	0.039 326	vapor
350	**2 000**	0.089 102	837.01	1015.2	5.466	1.8603	2.4461	1.30E−05	0.043 016	vapor
375	**2 000**	0.095 953	885.27	1077.2	5.637	1.9378	2.5126	1.38E−05	0.046 931	vapor
400	**2 000**	0.102 75	935.42	1140.9	5.8015	2.0212	2.5875	1.45E−05	0.051 055	vapor
425	**2 000**	0.109 5	987.61	1206.6	5.9607	2.1089	2.6685	1.52E−05	0.055 371	vapor
450	**2 000**	0.116 21	1042	1274.4	6.1157	2.1994	2.7537	1.58E−05	0.059 86	vapor
475	**2 000**	0.122 89	1098.5	1344.3	6.2669	2.2917	2.8416	1.65E−05	0.064 504	vapor
500	**2 000**	0.129 55	1157.4	1416.5	6.4149	2.385	2.9312	1.71E−05	0.069 286	vapor
525	**2 000**	0.136 19	1218.5	1490.9	6.5601	2.4785	3.0216	1.77E−05	0.074 192	vapor
550	**2 000**	0.142 82	1281.9	1567.6	6.7028	2.5718	3.1123	1.84E−05	0.079 206	vapor
575	**2 000**	0.149 43	1347.6	1646.5	6.8431	2.6644	3.2028	1.90E−05	0.084 317	vapor
600	**2 000**	0.156 02	1415.6	1727.7	6.9813	2.7562	3.2927	1.95E−05	0.089 514	vapor
625	**2 000**	0.162 61	1485.9	1811.1	7.1176	2.8469	3.3818	2.01E−05	0.094 788	vapor
275	**5 000**	0.025 27	674.07	800.42	4.3211	1.7274	2.644	1.15E−05	0.035 784	supercrit
300	**5 000**	0.028 595	722.46	865.43	4.5475	1.7616	2.5703	1.22E−05	0.038 48	supercrit
325	**5 000**	0.031 749	770.67	929.42	4.7523	1.8134	2.5554	1.29E−05	0.041 582	supercrit
350	**5 000**	0.034 791	819.54	993.49	4.9423	1.878	2.5752	1.35E−05	0.045 013	supercrit
375	**5 000**	0.037 753	869.6	1058.4	5.1213	1.9521	2.6172	1.42E−05	0.048 728	supercrit
400	**5 000**	0.040 654	921.21	1124.5	5.2919	2.033	2.6743	1.49E−05	0.052 693	supercrit

T (K)	p (kPa)	υ (m^3/kg)	u (kJ/kg)	h (kJ/kg)	s (J/ g*K)	c_v (J/ g*K)	c_p (J/ g*K)	μ (Pa*s)	k (W/m*K)	CH_4 phase
425	**5 000**	0.043 51	974.61	1192.2	5.456	2.1187	2.7418	1.55E−05	0.056 878	supercrit
450	**5 000**	0.046 329	1030	1261.6	5.6148	2.2078	2.8164	1.62E−05	0.061 257	supercrit
475	**5 000**	0.049 12	1087.4	1333	5.7692	2.2989	2.896	1.68E−05	0.065 808	supercrit
500	**5 000**	0.051 886	1147	1406.5	5.9199	2.3912	2.9788	1.74E−05	0.070 509	supercrit
525	**5 000**	0.054 633	1208.8	1482	6.0672	2.4839	3.0637	1.80E−05	0.075 343	supercrit
550	**5 000**	0.057 364	1272.8	1559.7	6.2118	2.5765	3.1498	1.86E−05	0.080 295	supercrit
575	**5 000**	0.060 08	1339.1	1639.5	6.3537	2.6686	3.2364	1.92E−05	0.085 35	supercrit
600	**5 000**	0.062 785	1407.5	1721.5	6.4932	2.7599	3.3229	1.98E−05	0.090 498	supercrit
625	**5 000**	0.065 479	1478.2	1805.6	6.6306	2.8502	3.4091	2.03E−05	0.095 726	supercrit
275	**10 000**	0.011 267	623.81	736.49	3.7864	1.7927	3.3326	1.38E−05	0.044 21	supercrit
300	**10 000**	0.013 302	682.08	815.1	4.0602	1.8074	3.0023	1.39E−05	0.044 73	supercrit
325	**10 000**	0.015 148	736.63	888.11	4.294	1.8476	2.858	1.43E−05	0.046 6	supercrit
350	**10 000**	0.016 874	789.98	958.73	4.5034	1.9048	2.8021	1.47E−05	0.049 236	supercrit
375	**10 000**	0.018 52	843.42	1028.6	4.6963	1.9737	2.7953	1.52E−05	0.052 393	supercrit
400	**10 000**	0.020 106	897.68	1098.7	4.8773	2.0509	2.8187	1.58E−05	0.055 941	supercrit
425	**10 000**	0.021 648	953.23	1169.7	5.0494	2.1338	2.8618	1.63E−05	0.059 802	supercrit
450	**10 000**	0.023 156	1010.4	1241.9	5.2145	2.2206	2.918	1.69E−05	0.063 922	supercrit
475	**10 000**	0.024 637	1069.3	1315.7	5.374	2.31	2.9833	1.74E−05	0.068 259	supercrit
500	**10 000**	0.026 095	1130.2	1391.2	5.5288	2.4008	3.0548	1.80E−05	0.072 781	supercrit
525	**10 000**	0.027 535	1193.1	1468.5	5.6797	2.4923	3.1304	1.86E−05	0.077 463	supercrit
550	**10 000**	0.028 959	1258.1	1547.7	5.8271	2.5839	3.2089	1.91E−05	0.082 283	supercrit
575	**10 000**	0.030 371	1325.2	1628.9	5.9715	2.6752	3.2892	1.97E−05	0.087 223	supercrit
600	**10 000**	0.031 772	1394.4	1712.2	6.1132	2.7658	3.3704	2.02E−05	0.092 268	supercrit
625	**10 000**	0.033 163	1465.8	1797.4	6.2525	2.8554	3.452	2.08E−05	0.097 406	supercrit
275	**20 000**	0.005 431	534.36	642.97	3.174	1.84	3.8744	2.11E−05	0.066 04	supercrit
300	**20 000**	0.006 44	606.86	735.66	3.4968	1.8579	3.5478	1.93E−05	0.061 744	supercrit
325	**20 000**	0.007 429	672.54	821.11	3.7705	1.8927	3.3057	1.84E−05	0.060 136	supercrit
350	**20 000**	0.008 374	734.27	901.76	4.0097	1.9435	3.1601	1.81E−05	0.060 346	supercrit
375	**20 000**	0.009 279	794.13	979.7	4.2248	2.0069	3.0852	1.80E−05	0.061 777	supercrit
400	**20 000**	0.010 148	853.45	1056.4	4.4228	2.0795	3.0577	1.82E−05	0.064 057	supercrit
425	**20 000**	0.010 988	913.1	1132.9	4.6082	2.1587	3.0624	1.84E−05	0.066 955	supercrit
450	**20 000**	0.011 804	973.62	1209.7	4.7839	2.2425	3.089	1.88E−05	0.070 32	supercrit
475	**20 000**	0.012 601	1035.4	1287.4	4.952	2.3292	3.131	1.91E−05	0.074 052	supercrit
500	**20 000**	0.013 381	1098.7	1366.3	5.1139	2.4179	3.1838	1.95E−05	0.078 078	supercrit
525	**20 000**	0.014 148	1163.7	1446.7	5.2707	2.5075	3.2442	2.00E−05	0.082 346	supercrit
550	**20 000**	0.014 902	1230.6	1528.6	5.4231	2.5975	3.3101	2.04E−05	0.086 815	supercrit
575	**20 000**	0.015 647	1299.3	1612.2	5.5718	2.6874	3.3797	2.09E−05	0.091 454	supercrit
600	**20 000**	0.016 383	1369.9	1697.6	5.7171	2.7768	3.452	2.13E−05	0.096 237	supercrit
625	**20 000**	0.017 112	1442.6	1784.8	5.8595	2.8653	3.526	2.18E−05	0.101 15	supercrit

Source: Linstrom and Mallard (2017–2021).

A.8 Properties of Hydrogen (Gas)

The enthalpy of formation of hydrogen is $\Delta\hat{h}_f^0 = 0.0\ kJ/kmol$. The molar mass of hydrogen is $\hat{M} = 2.015\ 68\ kg/kmol$. These properties were derived from Linstrom and Mallard (2017–2021).

T (K)	p (kPa)	v (m^3/kg)	u (kJ/kg)	h (kJ/kg)	s (J/g*K)	c_v (J/g*K)	c_p (J/g*K)	μ (Pa*s)	k (W/m*K)	H_2 phase
200	20	41.249	1732.1	2557.1	54.487	9.4069	13.532	6.78E−06	0.132 32	vapor
210	20	43.312	1826.8	2693.1	55.15	9.5386	13.664	7.01E−06	0.138 24	vapor
220	20	45.374	1922.8	2830.3	55.789	9.6549	13.78	7.24E−06	0.143 96	vapor
230	20	47.436	2019.9	2968.6	56.404	9.7571	13.882	7.46E−06	0.149 59	vapor
240	20	49.499	2117.9	3107.9	56.996	9.8468	13.972	7.68E−06	0.155 14	vapor
250	20	51.561	2216.8	3248	57.568	9.9252	14.05	7.90E−06	0.160 49	vapor
260	20	53.624	2316.4	3388.9	58.121	9.9934	14.118	8.12E−06	0.165 75	vapor
270	20	55.686	2416.6	3530.3	58.655	10.053	14.177	8.33E−06	0.170 92	vapor
280	20	57.748	2517.4	3672.4	59.171	10.104	14.229	8.54E−06	0.175 9	vapor
290	20	59.81	2618.7	3814.9	59.671	10.148	14.272	8.75E−06	0.180 88	vapor
300	20	61.873	2720.4	3957.8	60.156	10.185	14.31	8.95E−06	0.185 68	vapor
310	20	63.935	2822.4	4101.1	60.626	10.218	14.342	9.15E−06	0.190 48	vapor
320	20	65.997	2924.7	4244.6	61.081	10.245	14.37	9.35E−06	0.195 09	vapor
330	20	68.06	3027.3	4388.5	61.524	10.268	14.393	9.55E−06	0.200 2	vapor
340	20	70.122	3130	4532.5	61.954	10.288	14.412	9.75E−06	0.205 22	vapor
350	20	72.184	3233	4676.7	62.372	10.304	14.429	9.94E−06	0.210 25	vapor
360	20	74.246	3336.1	4821	62.779	10.318	14.442	1.01E−05	0.215 18	vapor
370	20	76.308	3439.4	4965.5	63.175	10.329	14.454	1.03E−05	0.219 82	vapor
380	20	78.371	3542.7	5110.1	63.56	10.339	14.463	1.05E−05	0.224 67	vapor
390	20	80.433	3646.1	5254.8	63.936	10.347	14.471	1.07E−05	0.229 32	vapor
400	20	82.495	3749.6	5399.5	64.302	10.353	14.478	1.09E−05	0.233 98	vapor
200	50	16.503	1731.9	2557.1	50.707	9.4073	13.534	6.78E−06	0.132 37	vapor
210	50	17.328	1826.7	2693.1	51.371	9.539	13.666	7.01E−06	0.138 28	vapor
220	50	18.153	1922.7	2830.3	52.009	9.6552	13.782	7.24E−06	0.144	vapor
230	50	18.978	2019.7	2968.7	52.624	9.7575	13.884	7.46E−06	0.149 63	vapor
240	50	19.803	2117.8	3108	53.217	9.8471	13.973	7.68E−06	0.155 18	vapor
250	50	20.628	2216.7	3248.1	53.789	9.9255	14.051	7.90E−06	0.160 53	vapor
260	50	21.453	2316.3	3388.9	54.341	9.9937	14.12	8.12E−06	0.165 79	vapor
270	50	22.278	2416.5	3530.4	54.875	10.053	14.179	8.33E−06	0.170 96	vapor
280	50	23.103	2517.3	3672.5	55.392	10.104	14.23	8.54E−06	0.175 93	vapor
290	50	23.928	2618.6	3815	55.892	10.148	14.273	8.75E−06	0.180 92	vapor
300	50	24.753	2720.3	3957.9	56.376	10.186	14.311	8.95E−06	0.185 71	vapor
310	50	25.578	2822.3	4101.2	56.846	10.218	14.343	9.15E−06	0.190 51	vapor
320	50	26.403	2924.6	4244.8	57.302	10.245	14.37	9.36E−06	0.195 12	vapor

T (K)	p (kPa)	v (m^3/kg)	u (kJ/kg)	h (kJ/kg)	s (J/g*K)	c_v (J/g*K)	c_p (J/g*K)	μ (Pa*s)	k (W/m*K)	H_2 phase
330	**50**	27.228	3027.2	4388.6	57.745	10.268	14.393	9.55E−06	0.200 23	vapor
340	**50**	28.053	3130	4532.6	58.175	10.288	14.413	9.75E−06	0.205 25	vapor
350	**50**	28.878	3232.9	4676.9	58.593	10.304	14.429	9.95E−06	0.210 28	vapor
360	**50**	29.703	3336.1	4821.2	58.999	10.318	14.443	1.01E−05	0.215 21	vapor
370	**50**	30.528	3439.3	4965.7	59.395	10.329	14.454	1.03E−05	0.219 85	vapor
380	**50**	31.353	3542.6	5110.3	59.781	10.339	14.464	1.05E−05	0.224 7	vapor
390	**50**	32.178	3646.1	5255	60.157	10.347	14.472	1.07E−05	0.229 35	vapor
400	**50**	33.003	3749.6	5399.7	60.523	10.353	14.478	1.09E−05	0.234 01	vapor
200	**100**	8.254 3	1731.6	2557.1	47.847	9.4079	13.538	6.78E−06	0.132 44	vapor
210	**100**	8.667	1826.4	2693.1	48.51	9.5396	13.669	7.01E−06	0.138 34	vapor
220	**100**	9.079 6	1922.4	2830.4	49.149	9.6558	13.785	7.24E−06	0.144 07	vapor
230	**100**	9.492 3	2019.5	2968.7	49.764	9.758	13.886	7.46E−06	0.149 7	vapor
240	**100**	9.904 9	2117.6	3108.1	50.357	9.8477	13.976	7.68E−06	0.155 24	vapor
250	**100**	10.318	2216.5	3248.2	50.929	9.926	14.054	7.90E−06	0.160 59	vapor
260	**100**	10.73	2316.1	3389.1	51.482	9.9942	14.121	8.12E−06	0.165 84	vapor
270	**100**	11.143	2416.3	3530.6	52.016	10.053	14.18	8.33E−06	0.171 01	vapor
280	**100**	11.555	2517.2	3672.7	52.532	10.104	14.231	8.54E−06	0.175 99	vapor
290	**100**	11.968	2618.4	3815.2	53.033	10.148	14.275	8.75E−06	0.180 97	vapor
300	**100**	12.38	2720.1	3958.2	53.517	10.186	14.312	8.95E−06	0.185 76	vapor
310	**100**	12.793	2822.2	4101.4	53.987	10.218	14.344	9.16E−06	0.190 56	vapor
320	**100**	13.205	2924.5	4245	54.443	10.246	14.372	9.36E−06	0.195 17	vapor
330	**100**	13.618	3027.1	4388.9	54.886	10.269	14.395	9.56E−06	0.200 28	vapor
340	**100**	14.03	3129.9	4532.9	55.316	10.288	14.414	9.75E−06	0.205 3	vapor
350	**100**	14.443	3232.8	4677.1	55.734	10.305	14.43	9.95E−06	0.210 32	vapor
360	**100**	14.855	3336	4821.5	56.14	10.318	14.444	1.01E−05	0.215 26	vapor
370	**100**	15.268	3439.2	4966	56.536	10.33	14.455	1.03E−05	0.219 9	vapor
380	**100**	15.68	3542.6	5110.6	56.922	10.339	14.465	1.05E−05	0.224 74	vapor
390	**100**	16.093	3646	5255.3	57.298	10.347	14.472	1.07E−05	0.229 4	vapor
400	**100**	16.505	3749.5	5400	57.664	10.354	14.479	1.09E−05	0.234 06	vapor
200	**101.33**	8.146 4	1731.6	2557.1	47.792	9.408	13.538	6.78E−06	0.132 44	vapor
210	**101.33**	8.553 7	1826.4	2693.1	48.456	9.5396	13.669	7.01E−06	0.138 35	vapor
220	**101.33**	8.961	1922.4	2830.4	49.095	9.6558	13.785	7.24E−06	0.144 07	vapor
230	**101.33**	9.368 2	2019.5	2968.7	49.71	9.758	13.886	7.46E−06	0.149 7	vapor
240	**101.33**	9.775 5	2117.6	3108.1	50.303	9.8477	13.976	7.68E−06	0.155 24	vapor
250	**101.33**	10.183	2216.5	3248.2	50.875	9.926	14.054	7.90E−06	0.160 59	vapor
260	**101.33**	10.59	2316.1	3389.1	51.427	9.9942	14.121	8.12E−06	0.165 85	vapor
270	**101.33**	10.997	2416.3	3530.6	51.962	10.053	14.18	8.33E−06	0.171 01	vapor
280	**101.33**	11.404	2517.2	3672.7	52.478	10.104	14.231	8.54E−06	0.175 99	vapor
290	**101.33**	11.811	2618.4	3815.2	52.978	10.148	14.275	8.75E−06	0.180 97	vapor
300	**101.33**	12.219	2720.1	3958.2	53.463	10.186	14.312	8.95E−06	0.185 76	vapor

(Continued)

T (K)	p (kPa)	υ (m^3/kg)	u (kJ/kg)	h (kJ/kg)	s (J/g*K)	c_v (J/g*K)	c_p (J/g*K)	μ (Pa*s)	k (W/m*K)	H_2 phase
310	**101.33**	12.626	2822.2	4101.5	53.933	10.218	14.344	9.16E−06	0.190 56	vapor
320	**101.33**	13.033	2924.5	4245	54.389	10.246	14.372	9.36E−06	0.195 17	vapor
330	**101.33**	13.44	3027.1	4388.9	54.831	10.269	14.395	9.56E−06	0.200 28	vapor
340	**101.33**	13.847	3129.9	4532.9	55.261	10.288	14.414	9.75E−06	0.205 3	vapor
350	**101.33**	14.254	3232.8	4677.1	55.679	10.305	14.43	9.95E−06	0.210 32	vapor
360	**101.33**	14.661	3336	4821.5	56.086	10.318	14.444	1.01E−05	0.215 26	vapor
370	**101.33**	15.068	3439.2	4966	56.482	10.33	14.455	1.03E−05	0.219 9	vapor
380	**101.33**	15.475	3542.6	5110.6	56.868	10.339	14.465	1.05E−05	0.224 74	vapor
390	**101.33**	15.883	3646	5255.3	57.243	10.347	14.472	1.07E−05	0.229 4	vapor
400	**101.33**	16.29	3749.5	5400	57.61	10.354	14.479	1.09E−05	0.234 06	vapor
200	**200**	4.129 9	1731.1	2557.1	44.985	9.4092	13.545	6.78E−06	0.132 56	vapor
210	**200**	4.336 4	1825.9	2693.2	45.649	9.5407	13.675	7.01E−06	0.138 47	vapor
220	**200**	4.542 9	1921.9	2830.5	46.288	9.6569	13.79	7.24E−06	0.144 18	vapor
230	**200**	4.749 3	2019.1	2968.9	46.903	9.7591	13.891	7.47E−06	0.149 81	vapor
240	**200**	4.955 7	2117.1	3108.3	47.497	9.8487	13.98	7.69E−06	0.155 34	vapor
250	**200**	5.162 1	2216.1	3248.5	48.069	9.927	14.058	7.91E−06	0.160 69	vapor
260	**200**	5.368 4	2315.7	3389.4	48.622	9.9951	14.125	8.12E−06	0.165 95	vapor
270	**200**	5.574 8	2416	3531	49.156	10.054	14.184	8.33E−06	0.171 11	vapor
280	**200**	5.781 1	2516.8	3673.1	49.672	10.105	14.234	8.54E−06	0.176 08	vapor
290	**200**	5.987 5	2618.1	3815.6	50.173	10.149	14.278	8.75E−06	0.181 06	vapor
300	**200**	6.193 8	2719.8	3958.6	50.657	10.187	14.315	8.96E−06	0.185 85	vapor
310	**200**	6.400 1	2821.9	4101.9	51.127	10.219	14.347	9.16E−06	0.190 65	vapor
320	**200**	6.606 4	2924.2	4245.5	51.583	10.246	14.374	9.36E−06	0.195 26	vapor
330	**200**	6.812 7	3026.8	4389.4	52.026	10.269	14.397	9.56E−06	0.200 36	vapor
340	**200**	7.019	3129.6	4533.4	52.456	10.289	14.416	9.75E−06	0.205 38	vapor
350	**200**	7.225 3	3232.6	4677.7	52.874	10.305	14.432	9.95E−06	0.210 4	vapor
360	**200**	7.431 5	3335.8	4822.1	53.281	10.319	14.446	1.01E−05	0.215 34	vapor
370	**200**	7.637 8	3439	4966.6	53.677	10.33	14.457	1.03E−05	0.219 97	vapor
380	**200**	7.844 1	3542.4	5111.2	54.063	10.34	14.466	1.05E−05	0.224 82	vapor
390	**200**	8.050 3	3645.8	5255.9	54.438	10.348	14.474	1.07E−05	0.229 47	vapor
400	**200**	8.256 6	3749.4	5400.7	54.805	10.354	14.48	1.09E−05	0.234 13	vapor
200	**500**	1.655 3	1729.4	2557.1	41.198	9.413	13.565	6.79E−06	0.132 92	vapor
210	**500**	1.738 1	1824.3	2693.4	41.863	9.5443	13.693	7.02E−06	0.138 81	vapor
220	**500**	1.820 8	1920.5	2830.9	42.502	9.6603	13.807	7.25E−06	0.144 51	vapor
230	**500**	1.903 5	2017.7	2969.4	43.118	9.7623	13.906	7.48E−06	0.150 13	vapor
240	**500**	1.986 1	2115.9	3109	43.712	9.8518	13.994	7.70E−06	0.155 65	vapor
250	**500**	2.068 8	2214.9	3249.3	44.285	9.93	14.07	7.91E−06	0.160 99	vapor
260	**500**	2.151 4	2314.6	3390.3	44.838	9.998	14.137	8.13E−06	0.166 23	vapor
270	**500**	2.234 1	2415	3532	45.373	10.057	14.194	8.34E−06	0.171 39	vapor
280	**500**	2.316 7	2515.8	3674.2	45.89	10.108	14.244	8.55E−06	0.176 35	vapor
290	**500**	2.399 3	2617.2	3816.8	46.39	10.152	14.287	8.76E−06	0.181 33	vapor

T (K)	p (kPa)	υ (m^3/kg)	u (kJ/kg)	h (kJ/kg)	s (J/g*K)	c_v (J/g*K)	c_p (J/g*K)	μ (Pa*s)	k (W/m*K)	H_2 phase
300	**500**	2.481 9	2719	3959.9	46.875	10.189	14.323	8.96E−06	0.186 11	vapor
310	**500**	2.564 4	2821.1	4103.3	47.346	10.221	14.355	9.17E−06	0.190 9	vapor
320	**500**	2.647	2923.5	4247	47.802	10.249	14.381	9.37E−06	0.195 5	vapor
330	**500**	2.729 6	3026.1	4390.9	48.245	10.272	14.403	9.56E−06	0.200 6	vapor
340	**500**	2.812 1	3129	4535	48.675	10.291	14.422	9.76E−06	0.205 61	vapor
350	**500**	2.894 7	3232	4679.3	49.093	10.307	14.438	9.95E−06	0.210 63	vapor
360	**500**	2.977 2	3335.2	4823.8	49.5	10.321	14.451	1.01E−05	0.215 56	vapor
370	**500**	3.059 8	3438.5	4968.3	49.896	10.332	14.462	1.03E−05	0.220 19	vapor
380	**500**	3.142 3	3541.9	5113	50.282	10.342	14.471	1.05E−05	0.225 03	vapor
390	**500**	3.224 8	3645.3	5257.7	50.658	10.35	14.478	1.07E−05	0.229 68	vapor
400	**500**	3.307 4	3748.9	5402.6	51.025	10.356	14.484	1.09E−05	0.234 33	vapor
200	**1000**	0.830 47	1726.6	2557.1	38.325	9.4193	13.599	6.81E−06	0.133 5	vapor
210	**1000**	0.871 97	1821.8	2693.7	38.992	9.5503	13.723	7.04E−06	0.139 37	vapor
220	**1000**	0.913 44	1918.1	2831.5	39.633	9.666	13.834	7.27E−06	0.145 04	vapor
230	**1000**	0.954 89	2015.5	2970.4	40.25	9.7678	13.931	7.49E−06	0.150 64	vapor
240	**1000**	0.996 32	2113.8	3110.1	40.845	9.857	14.016	7.71E−06	0.156 14	vapor
250	**1000**	1.037 7	2212.9	3250.6	41.418	9.9349	14.091	7.93E−06	0.161 46	vapor
260	**1000**	1.079 1	2312.8	3391.9	41.972	10.003	14.155	8.14E−06	0.166 69	vapor
270	**1000**	1.120 5	2413.2	3533.7	42.508	10.062	14.211	8.35E−06	0.171 83	vapor
280	**1000**	1.161 9	2514.2	3676.1	43.025	10.112	14.26	8.56E−06	0.176 78	vapor
290	**1000**	1.203 2	2615.7	3818.9	43.526	10.156	14.301	8.77E−06	0.181 74	vapor
300	**1000**	1.244 6	2717.5	3962.1	44.012	10.193	14.337	8.97E−06	0.186 51	vapor
310	**1000**	1.285 9	2819.7	4105.6	44.483	10.225	14.367	9.18E−06	0.191 29	vapor
320	**1000**	1.327 2	2922.2	4249.4	44.939	10.252	14.393	9.38E−06	0.195 88	vapor
330	**1000**	1.368 5	3024.9	4393.5	45.382	10.275	14.414	9.57E−06	0.200 97	vapor
340	**1000**	1.409 8	3127.8	4537.7	45.813	10.295	14.432	9.77E−06	0.205 97	vapor
350	**1000**	1.451 2	3230.9	4682.1	46.231	10.311	14.447	9.96E−06	0.210 98	vapor
360	**1000**	1.492 5	3334.2	4826.6	46.639	10.324	14.46	1.02E−05	0.215 9	vapor
370	**1000**	1.533 7	3437.5	4971.3	47.035	10.336	14.47	1.03E−05	0.220 53	vapor
380	**1000**	1.575	3541	5116	47.421	10.345	14.479	1.05E−05	0.225 36	vapor
390	**1000**	1.616 3	3644.5	5260.8	47.797	10.353	14.486	1.07E−05	0.23	vapor
400	**1000**	1.657 6	3748.1	5405.7	48.164	10.359	14.491	1.09E−05	0.234 65	vapor
200	**1200**	0.693	1725.5	2557.1	37.568	9.4218	13.612	6.82E−06	0.133 73	vapor
210	**1200**	0.727 63	1820.7	2693.9	38.235	9.5527	13.735	7.05E−06	0.139 59	vapor
220	**1200**	0.762 23	1917.1	2831.8	38.877	9.6683	13.845	7.27E−06	0.145 26	vapor
230	**1200**	0.796 8	2014.6	2970.7	39.494	9.7699	13.941	7.50E−06	0.150 84	vapor
240	**1200**	0.831 35	2112.9	3110.6	40.089	9.859	14.025	7.72E−06	0.156 34	vapor
250	**1200**	0.865 88	2212.1	3251.2	40.663	9.9369	14.099	7.93E−06	0.161 65	vapor
260	**1200**	0.900 4	2312	3392.5	41.218	10.005	14.163	8.15E−06	0.166 87	vapor
270	**1200**	0.934 9	2412.5	3534.4	41.753	10.063	14.218	8.36E−06	0.172 01	vapor

(*Continued*)

T (K)	p (kPa)	v (m^3/kg)	u (kJ/kg)	h (kJ/kg)	s (J/g*K)	c_v (J/g*K)	c_p (J/g*K)	μ (Pa*s)	k (W/m*K)	H_2 phase
280	**1200**	0.969 39	2513.6	3676.8	42.271	10.114	14.266	8.57E−06	0.176 95	vapor
290	**1200**	1.003 9	2615.1	3819.7	42.772	10.158	14.307	8.78E−06	0.181 91	vapor
300	**1200**	1.038 3	2717	3963	43.258	10.195	14.342	8.98E−06	0.186 67	vapor
310	**1200**	1.072 8	2819.2	4106.5	43.729	10.227	14.372	9.18E−06	0.191 45	vapor
320	**1200**	1.107 3	2921.7	4250.4	44.186	10.254	14.397	9.38E−06	0.196 03	vapor
330	**1200**	1.141 7	3024.4	4394.5	44.629	10.277	14.418	9.58E−06	0.201 12	vapor
340	**1200**	1.176 1	3127.4	4538.8	45.06	10.296	14.436	9.77E−06	0.206 12	vapor
350	**1200**	1.210 6	3230.5	4683.2	45.478	10.312	14.451	9.97E−06	0.211 12	vapor
360	**1200**	1.245	3333.8	4827.8	45.886	10.326	14.463	1.02E−05	0.216 04	vapor
370	**1200**	1.279 4	3437.2	4972.4	46.282	10.337	14.473	1.04E−05	0.220 66	vapor
380	**1200**	1.313 8	3540.6	5117.2	46.668	10.346	14.482	1.05E−05	0.225 49	vapor
390	**1200**	1.348 2	3644.2	5262.1	47.044	10.354	14.489	1.07E−05	0.230 13	vapor
400	**1200**	1.382 7	3747.8	5407	47.411	10.361	14.494	1.09E−05	0.234 78	vapor
200	**2000**	0.418 09	1721.2	2557.3	35.439	9.4319	13.664	6.85E−06	0.134 65	supercrit
210	**2000**	0.438 96	1816.7	2694.6	36.109	9.5622	13.782	7.08E−06	0.140 46	supercrit
220	**2000**	0.459 81	1913.3	2832.9	36.753	9.6773	13.887	7.30E−06	0.146 1	supercrit
230	**2000**	0.480 63	2011	2972.3	37.372	9.7786	13.979	7.52E−06	0.151 65	supercrit
240	**2000**	0.501 43	2109.6	3112.5	37.969	9.8673	14.06	7.74E−06	0.157 11	supercrit
250	**2000**	0.522 21	2209	3253.4	38.544	9.9448	14.131	7.96E−06	0.162 39	supercrit
260	**2000**	0.542 98	2309.1	3395.1	39.1	10.012	14.192	8.17E−06	0.167 59	supercrit
270	**2000**	0.563 73	2409.8	3537.3	39.636	10.071	14.245	8.38E−06	0.172 7	supercrit
280	**2000**	0.584 47	2511	3679.9	40.155	10.121	14.291	8.59E−06	0.177 63	supercrit
290	**2000**	0.605 2	2612.7	3823.1	40.657	10.165	14.33	8.79E−06	0.182 56	supercrit
300	**2000**	0.625 92	2714.7	3966.5	41.144	10.202	14.363	9.00E−06	0.187 31	supercrit
310	**2000**	0.646 63	2817.1	4110.3	41.615	10.233	14.392	9.20E−06	0.192 06	supercrit
320	**2000**	0.667 33	2919.7	4254.3	42.072	10.26	14.416	9.40E−06	0.196 63	supercrit
330	**2000**	0.688 02	3022.6	4398.6	42.516	10.283	14.436	9.60E−06	0.201 7	supercrit
340	**2000**	0.708 71	3125.6	4543	42.948	10.302	14.452	9.79E−06	0.206 68	supercrit
350	**2000**	0.729 4	3228.8	4687.6	43.367	10.318	14.466	9.98E−06	0.211 68	supercrit
360	**2000**	0.750 07	3332.2	4832.4	43.774	10.331	14.477	1.02E−05	0.216 58	supercrit
370	**2000**	0.770 75	3435.7	4977.2	44.171	10.342	14.486	1.04E−05	0.221 19	supercrit
380	**2000**	0.791 42	3539.2	5122.1	44.558	10.351	14.494	1.06E−05	0.226 01	supercrit
390	**2000**	0.812 08	3642.9	5267	44.934	10.359	14.5	1.07E−05	0.230 63	supercrit
400	**2000**	0.832 74	3746.6	5412.1	45.301	10.365	14.505	1.09E−05	0.235 27	supercrit

Source: Linstrom and Mallard (2017–2021).

A.9 Properties of Propane (Liquid–Vapor)

The enthalpy of formation of propane is $\Delta\hat{h}_f^0 = -104\ 700\ kJ/kmol$. The molar mass of propane is $\hat{M} = 44.091\ 52\ kg/kmol$. These properties were derived from Linstrom and Mallard (2017–2021).

T (K)	p (kPa)	ρ (kg/m³)	u (kJ/kg)	h (kJ/kg)	s (J/g*K)	c_v (J/g*K)	c_p (J/g*K)	μ (Pa*s)	k (W/m*K)	C_3H_8 phase
200	2.0	0.053 092	451.9	489.57	3.012 4	1.0834	1.2729	5.54E−06	0.008 994	vapor
225	2.0	0.047 177	480.12	522.52	3.167 5	1.1746	1.3638	6.20E−06	0.011 096	vapor
250	2.0	0.042 451	510.71	557.83	3.316 2	1.2733	1.4623	6.87E−06	0.013 379	vapor
275	2.0	0.038 586	543.85	595.68	3.460 5	1.3785	1.5674	7.54E−06	0.015 845	vapor
300	2.0	0.035 367	579.69	636.24	3.601 5	1.4887	1.6775	8.20E−06	0.018 492	vapor
325	2.0	0.032 645	618.32	679.58	3.740 3	1.6019	1.7906	8.86E−06	0.021 322	vapor
350	2.0	0.030 311	659.8	725.78	3.877 1	1.7164	1.9051	9.52E−06	0.024 334	vapor
375	2.0	0.028 29	704.14	774.83	4.012 5	1.8306	2.0192	1.02E−05	0.027 528	vapor
400	2.0	0.026 521	751.32	826.73	4.146 4	1.9432	2.1319	1.08E−05	0.030 904	vapor
200	5.0	0.132 95	451.77	489.38	2.838 9	1.0844	1.2752	5.53E−06	0.008 99	vapor
225	5.0	0.118 08	480.02	522.37	2.994 3	1.1752	1.3653	6.20E−06	0.011 093	vapor
250	5.0	0.106 21	510.63	557.71	3.143 1	1.2737	1.4633	6.87E−06	0.013 378	vapor
275	5.0	0.096 524	543.79	595.59	3.287 5	1.3788	1.5681	7.54E−06	0.015 844	vapor
300	5.0	0.088 459	579.63	636.16	3.428 6	1.4888	1.678	8.20E−06	0.018 493	vapor
325	5.0	0.081 641	618.27	679.52	3.567 3	1.602	1.791	8.86E−06	0.021 323	vapor
350	5.0	0.075 8	659.76	725.72	3.704 3	1.7165	1.9054	9.52E−06	0.024 336	vapor
375	5.0	0.070 74	704.1	774.78	3.839 6	1.8306	2.0195	1.02E−05	0.027 53	vapor
400	5.0	0.066 314	751.28	826.68	3.973 5	1.9433	2.1321	1.08E−05	0.030 906	vapor
200	10	0.266 63	451.55	489.05	2.707 1	1.086	1.2791	5.53E−06	0.008 984	vapor
225	10	0.236 6	479.85	522.12	2.862 8	1.1762	1.3678	6.20E−06	0.011 089	vapor
250	10	0.212 71	510.5	557.51	3.011 9	1.2743	1.465	6.87E−06	0.013 376	vapor
275	10	0.193 24	543.68	595.43	3.156 4	1.3792	1.5693	7.54E−06	0.015 844	vapor
300	10	0.177 06	579.54	636.02	3.297 6	1.4892	1.6789	8.20E−06	0.018 494	vapor
325	10	0.163 38	618.19	679.4	3.436 4	1.6022	1.7917	8.86E−06	0.021 325	vapor
350	10	0.151 67	659.69	725.62	3.573 4	1.7166	1.906	9.52E−06	0.024 339	vapor
375	10	0.141 53	704.04	774.7	3.708 8	1.8308	2.0199	1.02E−05	0.027 534	vapor
400	10	0.132 67	751.23	826.61	3.842 7	1.9434	2.1325	1.08E−05	0.030 911	vapor
200	20	0.536 25	451.09	488.39	2.574 1	1.0894	1.2873	5.52E−06	0.008 971	vapor
225	20	0.475	479.51	521.62	2.730 6	1.1782	1.3729	6.19E−06	0.011 081	vapor
250	20	0.426 59	510.23	557.11	2.880 1	1.2756	1.4685	6.86E−06	0.013 372	vapor
275	20	0.387 27	543.46	595.1	3.024 9	1.3801	1.5718	7.53E−06	0.015 843	vapor
300	20	0.354 66	579.36	635.75	3.166 3	1.4898	1.6808	8.20E−06	0.018 496	vapor
325	20	0.327 16	618.04	679.17	3.305 2	1.6027	1.7932	8.86E−06	0.021 329	vapor

(Continued)

T (K)	p (kPa)	ρ (kg/m³)	u (kJ/kg)	h (kJ/kg)	s (J/g*K)	c_v (J/g*K)	c_p (J/g*K)	μ (Pa*s)	k (W/m*K)	C_3H_8 phase
350	**20**	0.303 64	659.55	725.42	3.442 3	1.717	1.9071	9.52E−06	0.024 345	vapor
375	**20**	0.283 29	703.92	774.52	3.577 7	1.831	2.0208	1.02E−05	0.027 542	vapor
400	**20**	0.265 5	751.12	826.45	3.711 8	1.9436	2.1332	1.08E−05	0.030 92	vapor
200	**50**	615.42	31.971	32.052	0.290 43	1.3912	2.129	0.000 288	0.147 84	liquid
216.18	**50**	597.73	66.883	66.967	0.458 27	1.4224	2.1885	0.000 235	0.138	liquid
216.18	**50**	1.254	468.07	507.94	2.498 1	1.1537	1.3613	5.93E−06	0.010 289	vapor
225	**50**	1.201 6	478.46	520.07	2.553 1	1.1847	1.3896	6.17E−06	0.011 059	vapor
250	**50**	1.075 4	509.41	555.91	2.704 1	1.2796	1.4794	6.85E−06	0.013 361	vapor
275	**50**	0.974 16	542.8	594.13	2.849 7	1.3828	1.5795	7.53E−06	0.015 841	vapor
300	**50**	0.890 81	578.81	634.94	2.991 7	1.4916	1.6864	8.20E−06	0.018 502	vapor
325	**50**	0.820 87	617.57	678.48	3.131	1.604	1.7974	8.86E−06	0.021 342	vapor
350	**50**	0.761 28	659.15	724.83	3.268 3	1.718	1.9104	9.52E−06	0.024 363	vapor
375	**50**	0.709 86	703.57	774	3.404	1.8318	2.0235	1.02E−05	0.027 565	vapor
400	**50**	0.665 01	750.81	825.99	3.538 2	1.9442	2.1354	1.08E−05	0.030 948	vapor
200	**100**	615.47	31.943	32.105	0.290 29	1.3912	2.1289	0.000 288	0.147 87	liquid
225	**100**	587.86	86.309	86.479	0.546 35	1.4421	2.2254	0.000 212	0.132 77	liquid
230.74	**100**	581.27	99.146	99.318	0.602 69	1.4559	2.2515	0.000 198	0.129 38	liquid
230.74	**100**	2.384 8	483.62	525.55	2.45	1.2166	1.4379	6.30E−06	0.011 542	vapor
250	**100**	2.182 1	508.00	553.83	2.567 6	1.287	1.4995	6.84E−06	0.013 345	vapor
275	**100**	1.968 9	541.67	592.46	2.714 9	1.3875	1.5931	7.52E−06	0.015 84	vapor
300	**100**	1.795 8	577.88	633.56	2.857 8	1.4949	1.6962	8.20E−06	0.018 513	vapor
325	**100**	1.651 9	616.78	677.32	2.997 9	1.6063	1.8048	8.86E−06	0.021 364	vapor
350	**100**	1.53	658.47	723.83	3.135 7	1.7197	1.9162	9.52E−06	0.024 394	vapor
375	**100**	1.425 2	702.97	773.13	3.271 7	1.8331	2.0281	1.02E−05	0.027 604	vapor
400	**100**	1.334 2	750.28	825.23	3.406 2	1.9452	2.1391	1.08E−05	0.030 995	vapor
200	**101.32**	615.47	31.942	32.107	0.290 28	1.3913	2.1289	0.000 288	0.147 87	liquid
225	**101.32**	587.86	86.308	86.481	0.546 34	1.4421	2.2254	0.000 212	0.132 77	liquid
231.03	**101.32**	580.93	99.816	99.99	0.605 59	1.4566	2.2529	0.000 197	0.129 21	liquid
231.03	**101.32**	2.414 2	483.94	525.91	2.449 1	1.218	1.4395	6.31E−06	0.011 568	vapor
250	**101.32**	2.211 9	507.96	553.77	2.565	1.2872	1.5001	6.83E−06	0.013 344	vapor
275	**101.32**	1.995 6	541.64	592.42	2.712 3	1.3876	1.5935	7.52E−06	0.015 84	vapor
300	**101.32**	1.82	577.85	633.53	2.855 3	1.495	1.6965	8.20E−06	0.018 513	vapor
325	**101.32**	1.674	616.76	677.29	2.995 3	1.6064	1.805	8.86E−06	0.021 365	vapor
350	**101.32**	1.550 4	658.45	723.8	3.133 2	1.7198	1.9163	9.53E−06	0.024 395	vapor
375	**101.32**	1.444 3	702.95	773.11	3.269 2	1.8331	2.0283	1.02E−05	0.027 605	vapor
400	**101.32**	1.352	750.26	825.21	3.403 6	1.9452	2.1392	1.08E−05	0.030 996	vapor
200	**200**	615.55	31.886	32.211	0.29	1.3913	2.1286	0.000 288	0.147 93	liquid
225	**200**	587.96	86.235	86.576	0.546 02	1.4422	2.225	0.000 212	0.132 84	liquid
247.7	**200**	561.21	137.95	138.31	0.764 98	1.5011	2.3382	0.000 164	0.119 68	liquid
247.7	**200**	4.553 2	501.92	545.84	2.410 3	1.2966	1.5424	6.74E−06	0.013 101	vapor
250	**200**	4.502 4	504.98	549.4	2.424 6	1.3044	1.5479	6.81E−06	0.013 324	vapor

T (K)	p (kPa)	ρ (kg/m^3)	u (kJ/kg)	h (kJ/kg)	s (J/g*K)	c_v (J/g*K)	c_p (J/g*K)	μ (Pa*s)	k (W/m*K)	C_3H_8 phase
275	**200**	4.026	539.32	589	2.575 5	1.3981	1.6239	7.51E−06	0.015 846	vapor
300	**200**	3.651 2	575.96	630.74	2.720 7	1.5018	1.7175	8.19E−06	0.018 542	vapor
325	**200**	3.345 6	615.17	674.95	2.862 2	1.6112	1.8205	8.87E−06	0.021 413	vapor
350	**200**	3.090 3	657.08	721.8	3.001	1.7233	1.9282	9.53E−06	0.024 46	vapor
375	**200**	2.873	701.76	771.37	3.137 8	1.8358	2.0376	1.02E−05	0.027 685	vapor
400	**200**	2.685 4	749.2	823.68	3.272 8	1.9473	2.1468	1.08E−05	0.031 089	vapor
200	**500**	615.8	31.716	32.528	0.289 15	1.3916	2.1277	0.000 289	0.148 11	liquid
225	**500**	588.29	86.014	86.864	0.545 03	1.4424	2.2235	0.000 212	0.133 04	liquid
250	**500**	558.81	143.05	143.95	0.785 49	1.5079	2.3489	0.000 161	0.118 62	liquid
274.88	**500**	526.31	203.4	204.35	1.015 7	1.5867	2.5158	0.000 123	0.105 11	liquid
274.88	**500**	10.883	531.07	577.01	2.371 5	1.4401	1.7566	7.50E−06	0.015 94	vapor
275	**500**	10.875	531.26	577.23	2.372 2	1.4405	1.7567	7.50E−06	0.015 953	vapor
300	**500**	9.640 3	569.7	621.57	2.526 5	1.5272	1.7988	8.21E−06	0.018 685	vapor
325	**500**	8.710 4	610.05	667.45	2.673 4	1.6276	1.8758	8.89E−06	0.021 597	vapor
350	**500**	7.970 2	652.76	715.49	2.815 7	1.7349	1.9687	9.57E−06	0.024 684	vapor
375	**500**	7.360 6	698.02	765.95	2.955	1.8443	2.0686	1.02E−05	0.027 946	vapor
400	**500**	6.846 3	745.91	818.94	3.091 7	1.9538	2.1714	1.09E−05	0.031 383	vapor
200	**1000**	616.21	31.435	33.058	0.287 74	1.392	2.1262	0.000 29	0.148 4	liquid
225	**1000**	588.82	85.648	87.346	0.543 4	1.4428	2.2211	0.000 213	0.133 38	liquid
250	**1000**	559.53	142.56	144.35	0.783 53	1.5083	2.3448	0.000 162	0.119 02	liquid
275	**1000**	527.17	203.02	204.92	1.014 3	1.5873	2.509	0.000 124	0.105 51	liquid
300	**1000**	489.49	268.36	270.4	1.242 1	1.6799	2.7481	9.53E−05	0.092 871	liquid
300.09	**1000**	489.35	268.6	270.64	1.242 9	1.6803	2.7492	9.52E−05	0.092 828	liquid
300.09	**1000**	21.675	556.95	603.09	2.350 7	1.591	2.0418	8.34E−06	0.019 249	vapor
325	**1000**	18.866	600.29	653.29	2.511 4	1.6632	2.0104	9.01E−06	0.022 083	vapor
350	**1000**	16.895	644.86	704.05	2.661 9	1.7582	2.0567	9.68E−06	0.025 168	vapor
375	**1000**	15.387	691.36	756.36	2.806 2	1.8603	2.1311	1.03E−05	0.028 45	vapor
400	**1000**	14.174	740.16	810.71	2.946 5	1.9654	2.2185	1.10E−05	0.031 915	vapor
200	**2000**	617.02	30.879	34.12	0.284 94	1.3929	2.1233	0.000 293	0.148 98	liquid
225	**2000**	589.87	84.926	88.316	0.540 17	1.4437	2.2164	0.000 215	0.134 05	liquid
250	**2000**	560.94	141.6	145.17	0.779 66	1.509	2.3369	0.000 163	0.119 79	liquid
275	**2000**	529.17	201.69	205.47	1.009 4	1.5876	2.4944	0.000 126	0.106 43	liquid
300	**2000**	492.65	266.35	270.41	1.235 3	1.6791	2.7156	9.74E−05	0.093 998	liquid
325	**2000**	446.55	338.11	342.59	1.466 2	1.7879	3.1109	7.31E−05	0.082 153	liquid
330.4	**2000**	434.07	355.16	359.77	1.518 6	1.8154	3.2632	6.80E−05	0.079 553	liquid
330.4	**2000**	46.234	583.24	626.5	2.325 9	1.7954	2.7006	9.84E−06	0.025 226	vapor
350	**2000**	39.405	625.01	675.77	2.470 9	1.8286	2.3971	1.02E−05	0.026 891	vapor
375	**2000**	34.212	675.96	734.42	2.632 8	1.9013	2.3224	1.08E−05	0.029 853	vapor
400	**2000**	30.655	727.42	792.66	2.783 1	1.9928	2.3456	1.14E−05	0.033 2	vapor
200	**4000**	618.62	29.789	36.255	0.279 43	1.3947	2.1178	0.000 298	0.150 13	liquid

(*Continued*)

T (K)	p (kPa)	ρ (kg/m³)	u (kJ/kg)	h (kJ/kg)	s (J/g*K)	c_v (J/g*K)	c_p (J/g*K)	μ (Pa*s)	k (W/m*K)	C_3H_8 phase
225	4000	591.92	83.519	90.277	0.533 84	1.4454	2.2077	0.000 22	0.135 38	liquid
250	4000	563.66	139.75	146.85	0.772 15	1.5105	2.3225	0.000 167	0.121 32	liquid
275	4000	532.97	199.16	206.66	1.000 1	1.5885	2.4687	0.000 13	0.108 2	liquid
300	4000	498.42	262.64	270.67	1.222 7	1.6784	2.6625	0.000 101	0.096 127	liquid
325	4000	456.98	331.83	340.59	1.446 4	1.7813	2.957	7.79E−05	0.084 973	liquid
350	4000	399	411.43	421.45	1.685 8	1.9108	3.6565	5.61E−05	0.073 954	liquid
366.49	4000	304.59	490.88	504.01	1.915 5	2.1882	12.89	3.44E−05	0.068 793	liquid
366.49	4000	139.55	577.85	606.51	2.195 2	2.3376	18.098	1.57E−05	0.056 757	vapor
375	4000	101.72	623.7	663.03	2.348	2.0826	4.28	1.36E−05	0.038 966	vapor
400	4000	76.078	693.82	746.4	2.563 6	2.0719	2.9156	1.30E−05	0.037 657	vapor
200	5000	619.4	29.254	37.327	0.276 71	1.3956	2.1152	0.0003	0.150 7	vapor
225	5000	592.92	82.833	91.266	0.530 74	1.4463	2.2036	0.000 222	0.136 03	vapor
250	5000	564.98	138.86	147.7	0.768 49	1.5112	2.3158	0.000 169	0.122 06	vapor
275	5000	534.78	197.95	207.3	0.995 58	1.589	2.4572	0.000 131	0.109 06	vapor
300	5000	501.07	260.93	270.91	1.216 9	1.6783	2.6404	0.000 103	0.097 139	vapor
325	5000	461.41	329.13	339.96	1.437 8	1.7793	2.9035	8.01E−05	0.086 245	vapor
350	5000	409.23	405.75	417.97	1.668 8	1.8993	3.4136	5.95E−05	0.075 885	vapor
375	5000	298.04	512.34	529.12	1.974 3	2.1538	7.8691	3.35E−05	0.064 887	supercritical
400	5000	112.21	668.97	713.53	2.454 2	2.1297	3.7059	1.49E−05	0.042 438	supercritical

Source: Linstrom and Mallard (2017–2021).

A.10 Properties of R32 (Liquid–Vapor)

The molar mass of R32 (CH_2F_2) is $\hat{M} = 52.022\ 08\ kg/kmol$. These properties were derived from Linstrom and Mallard (2017–2021).

T (K)	p (kPa)	υ (m³/kg)	u (kJ/kg)	h (kJ/kg)	s (J/g*K)	c_v (J/g*K)	c_p (J/g*K)	μ (Pa*s)	k (W/m*K)	R32 phase
200	10	3.176 3	454.93	486.7	2.697 0	0.559 37	0.726 77	8.37E−06	0.007 851	vapor
210	10	3.339 4	460.59	493.98	2.732 5	0.565 23	0.730 78	8.80E−06	0.008 142	vapor
220	10	3.501 7	466.3	501.32	2.766 6	0.572 9	0.737 15	9.23E−06	0.008 475	vapor
230	10	3.663 5	472.1	508.73	2.799 6	0.581 92	0.745 25	9.67E−06	0.008 849	vapor
240	10	3.824 9	477.98	516.23	2.831 5	0.592 07	0.754 7	1.01E−05	0.009 264	vapor
250	10	3.986	483.97	523.83	2.862 5	0.603 17	0.765 29	1.05E−05	0.009 721	vapor
260	10	4.146 8	490.07	531.54	2.892 8	0.6151	0.776 83	1.10E−05	0.010 218	vapor
270	10	4.307 5	496.29	539.37	2.922 3	0.627 78	0.789 2	1.14E−05	0.010 758	vapor

T (K)	p (kPa)	v (m^3/kg)	u (kJ/kg)	h (kJ/kg)	s (J/g*K)	c_v (J/g*K)	c_p (J/g*K)	μ (Pa*s)	k (W/m*K)	R32 phase
280	**10**	4.468 1	502.64	547.33	2.951 2	0.641 1	0.802 28	1.18E−05	0.011 338	vapor
290	**10**	4.628 5	509.13	555.42	2.979 6	0.655	0.815 99	1.23E−05	0.011 959	vapor
300	**10**	4.788 9	515.76	563.65	3.007 5	0.669 4	0.830 24	1.27E−05	0.012 622	vapor
310	**10**	4.949 2	522.53	572.02	3.035 0	0.684 24	0.844 94	1.31E−05	0.013 326	vapor
320	**10**	5.109 4	529.45	580.55	3.062 1	0.699 44	0.860 04	1.35E−05	0.014 072	vapor
330	**10**	5.269 6	536.53	589.22	3.088 8	0.714 94	0.875 46	1.40E−05	0.014 858	vapor
340	**10**	5.429 8	543.76	598.06	3.115 1	0.730 69	0.891 13	1.44E−05	0.015 686	vapor
350	**10**	5.589 9	551.15	607.05	3.141 2	0.746 62	0.907 01	1.48E−05	0.016 555	vapor
360	**10**	5.75	558.7	616.20	3.167	0.762 7	0.923 03	1.52E−05	0.017 466	vapor
370	**10**	5.91	566.41	625.51	3.192 5	0.778 86	0.939 14	1.56E−05	0.018 417	vapor
380	**10**	6.070 1	574.28	634.98	3.217 7	0.795 07	0.955 31	1.60E−05	0.019 41	vapor
390	**10**	6.230 1	582.31	644.61	3.242 7	0.811 28	0.971 49	1.64E−05	0.020 444	vapor
400	**10**	6.390 1	590.51	654.41	3.267 5	0.827 46	0.987 64	1.68E−05	0.021 519	vapor
410	**10**	6.550 1	598.87	664.37	3.292 1	0.843 58	1.003 70	1.72E−05	0.022 636	vapor
420	**10**	6.710 1	607.38	674.48	3.316 5	0.859 61	1.019 70	1.76E−05	0.023 793	vapor
430	**10**	6.870 1	616.06	684.76	3.340 7	0.875 53	1.035 6	1.80E−05	0.024 992	vapor
200	**20**	1.577 9	454.17	485.73	2.582 4	0.574 71	0.750 57	8.36E−06	0.007 859	vapor
210	**20**	1.661 2	459.99	493.21	2.618 9	0.575 51	0.747 13	8.79E−06	0.008 149	vapor
220	**20**	1.743 6	465.82	500.69	2.653 6	0.580 3	0.749 16	9.23E−06	0.008 481	vapor
230	**20**	1.825 5	471.69	508.2	2.687 1	0.587 45	0.754 38	9.66E−06	0.008 854	vapor
240	**20**	1.907	477.64	515.78	2.719 3	0.596 28	0.761 79	1.01E−05	0.009 269	vapor
250	**20**	1.988 2	483.68	523.45	2.750 6	0.606 43	0.770 89	1.05E−05	0.009 725	vapor
260	**20**	2.069 2	489.82	531.21	2.781	0.617 65	0.781 31	1.10E−05	0.010 222	vapor
270	**20**	2.15	496.08	539.08	2.810 7	0.629 79	0.792 84	1.14E−05	0.010 761	vapor
280	**20**	2.230 6	502.45	547.07	2.839 8	0.642 72	0.805 27	1.18E−05	0.011 341	vapor
290	**20**	2.311 1	508.96	555.18	2.868 3	0.656 3	0.818 47	1.23E−05	0.011 962	vapor
300	**20**	2.391 6	515.61	563.44	2.896 2	0.670 47	0.832 32	1.27E−05	0.012 624	vapor
310	**20**	2.472	522.39	571.83	2.923 8	0.685 11	0.846 71	1.31E−05	0.013 328	vapor
320	**20**	2.552 3	529.33	580.37	2.950 9	0.700 16	0.861 56	1.35E−05	0.014 073	vapor
330	**20**	2.632 6	536.41	589.06	2.977 6	0.715 55	0.876 77	1.40E−05	0.014 859	vapor
340	**20**	2.712 8	543.65	597.91	3.004	0.731 2	0.892 27	1.44E−05	0.015 687	vapor
350	**20**	2.793	551.05	606.91	3.030 1	0.747 06	0.908 01	1.48E−05	0.016 556	vapor
360	**20**	2.873 2	558.61	616.07	3.055 9	0.763 07	0.923 91	1.52E−05	0.017 466	vapor
370	**20**	2.953 4	566.32	625.39	3.081 5	0.779 18	0.939 93	1.56E−05	0.018 417	vapor
380	**20**	3.033 5	574.2	634.87	3.106 7	0.795 35	0.956 02	1.60E−05	0.019 409	vapor
390	**20**	3.113 6	582.24	644.51	3.131 8	0.811 52	0.972 12	1.64E−05	0.020 443	vapor
400	**20**	3.193 7	590.44	654.31	3.156 6	0.827 68	0.988 21	1.68E−05	0.021 518	vapor
410	**20**	3.273 8	598.8	664.27	3.181 2	0.843 77	1.004 3	1.72E−05	0.022 634	vapor
420	**20**	3.353 9	607.32	674.4	3.205 6	0.859 78	1.020 2	1.76E−05	0.023 792	vapor
430	**20**	3.433 9	616	684.68	3.229 8	0.875 69	1.036 1	1.80E−05	0.024 991	vapor

(*Continued*)

T (K)	p (kPa)	v (m^3/kg)	u (kJ/kg)	h (kJ/kg)	s (J/g*K)	c_v (J/g*K)	c_p (J/g*K)	μ (Pa*s)	k (W/m*K)	R32 phase
200	**50**	0.000 787	80.693	80.732	0.495 9	0.956 94	1.564 1	0.000 367	0.204 71	liquid
208.56	**50**	0.000 801	94.116	94.156	0.561 62	0.949 39	1.571	0.000 326	0.197 91	liquid
208.56	**50**	0.648 84	457.17	489.61	2.457 7	0.617 29	0.812 79	8.71E−06	0.008 129	vapor
210	**50**	0.653 9	458.08	490.78	2.463 3	0.614 42	0.808	8.78E−06	0.008 173	vapor
220	**50**	0.688 62	464.31	498.74	2.500 3	0.604 89	0.789 03	9.21E−06	0.008 502	vapor
230	**50**	0.722 65	470.47	506.6	2.535 3	0.604 84	0.783 28	9.65E−06	0.008 872	vapor
240	**50**	0.756 24	476.62	514.43	2.568 6	0.609 24	0.783 79	1.01E−05	0.009 285	vapor
250	**50**	0.789 5	482.81	522.29	2.600 7	0.616 35	0.788 08	1.05E−05	0.009 739	vapor
260	**50**	0.822 53	489.07	530.2	2.631 7	0.625 38	0.795 02	1.10E−05	0.010 234	vapor
270	**50**	0.855 38	495.43	538.19	2.661 9	0.635 9	0.803 92	1.14E−05	0.010 771	vapor
280	**50**	0.888 09	501.88	546.28	2.691 3	0.647 59	0.814 35	1.18E−05	0.011 35	vapor
290	**50**	0.920 69	508.45	554.49	2.720 1	0.660 24	0.826 01	1.23E−05	0.011 97	vapor
300	**50**	0.953 21	515.15	562.81	2.748 3	0.673 67	0.838 64	1.27E−05	0.012 631	vapor
310	**50**	0.985 65	521.98	571.26	2.776	0.687 74	0.852 07	1.31E−05	0.013 333	vapor
320	**50**	1.018	528.95	579.85	2.803 3	0.702 35	0.866 13	1.35E−05	0.014 077	vapor
330	**50**	1.050 4	536.07	588.59	2.830 1	0.717 37	0.880 72	1.40E−05	0.014 863	vapor
340	**50**	1.082 7	543.33	597.47	2.856 6	0.732 74	0.895 71	1.44E−05	0.015 689	vapor
350	**50**	1.114 9	550.75	606.5	2.882 8	0.748 37	0.911 03	1.48E−05	0.016 557	vapor
360	**50**	1.147 2	558.33	615.69	2.908 7	0.764 19	0.926 58	1.52E−05	0.017 466	vapor
370	**50**	1.179 4	566.06	625.03	2.934 3	0.780 15	0.942 3	1.56E−05	0.018 417	vapor
380	**50**	1.211 6	573.96	634.53	2.959 7	0.796 19	0.958 14	1.60E−05	0.019 408	vapor
390	**50**	1.243 7	582.01	644.2	2.984 7	0.812 26	0.974 04	1.64E−05	0.020 442	vapor
400	**50**	1.275 9	590.22	654.02	3.009 6	0.828 33	0.989 95	1.68E−05	0.021 516	vapor
410	**50**	1.308	598.59	663.99	3.034 3	0.844 35	1.005 8	1.72E−05	0.022 632	vapor
420	**50**	1.340 1	607.13	674.13	3.058 7	0.860 3	1.021 7	1.76E−05	0.023 788	vapor
430	**50**	1.372 2	615.82	684.43	3.082 9	0.876 15	1.037 4	1.80E−05	0.024 987	vapor
200	**100**	0.000 787	80.677	80.755	0.495 82	0.956 95	1.564	0.000 367	0.204 74	liquid
210	**100**	0.000 804	96.353	96.433	0.572 31	0.948 28	1.572 3	0.000 32	0.196 78	liquid
220	**100**	0.000 822	112.13	112.21	0.645 73	0.941 65	1.584 7	0.000 281	0.188 63	liquid
221.24	**100**	0.000 824	114.1	114.18	0.654 64	0.940 97	1.586 5	0.000 277	0.187 6	liquid
221.24	**100**	0.338 85	462.43	496.32	2.381 9	0.658 05	0.873 42	9.25E−06	0.008 586	vapor
230	**100**	0.354 82	468.32	503.8	2.415 1	0.638 94	0.84	9.63E−06	0.008 908	vapor
240	**100**	0.372 52	474.86	512.11	2.450 4	0.632 61	0.823 92	1.01E−05	0.009 315	vapor
250	**100**	0.389 85	481.33	520.32	2.483 9	0.633 6	0.818 45	1.05E−05	0.009 765	vapor
260	**100**	0.406 93	487.81	528.5	2.516	0.638 61	0.818 82	1.10E−05	0.010 257	vapor
270	**100**	0.423 82	494.32	536.71	2.547	0.646 26	0.823	1.14E−05	0.010 791	vapor
280	**100**	0.440 56	500.91	544.97	2.577	0.655 83	0.829 9	1.18E−05	0.011 367	vapor
290	**100**	0.457 19	507.59	553.31	2.606 3	0.666 87	0.838 85	1.23E−05	0.011 984	vapor
300	**100**	0.473 72	514.38	561.75	2.634 9	0.679 06	0.849 38	1.27E−05	0.012 643	vapor
310	**100**	0.490 19	521.28	570.3	2.663	0.692 17	0.861 14	1.31E−05	0.013 344	vapor
320	**100**	0.506 6	528.32	578.98	2.690 5	0.706 01	0.873 88	1.35E−05	0.014 086	vapor

T (K)	p (kPa)	υ (m³/kg)	u (kJ/kg)	h (kJ/kg)	s (J/g*K)	c_v (J/g*K)	c_p (J/g*K)	μ (Pa*s)	k (W/m*K)	R32 phase
330	100	0.522 95	535.49	587.78	2.717 6	0.720 44	0.887 39	1.40E−05	0.014 869	vapor
340	100	0.539 27	542.8	596.73	2.744 3	0.735 32	0.901 51	1.44E−05	0.015 694	vapor
350	100	0.555 55	550.26	605.81	2.770 6	0.750 56	0.916 11	1.48E−05	0.016 56	vapor
360	100	0.571 8	557.87	615.05	2.796 7	0.766 07	0.931 07	1.52E−05	0.017 468	vapor
370	100	0.588 03	565.63	624.44	2.822 4	0.781 77	0.946 29	1.56E−05	0.018 417	vapor
380	100	0.604 23	573.55	633.98	2.847 8	0.797 59	0.961 71	1.60E−05	0.019 408	vapor
390	100	0.620 42	581.63	643.67	2.873	0.813 49	0.977 25	1.64E−05	0.020 439	vapor
400	100	0.636 59	589.86	653.52	2.897 9	0.829 41	0.992 85	1.68E−05	0.021 513	vapor
410	100	0.652 75	598.25	663.53	2.922 6	0.845 31	1.008 5	1.72E−05	0.022 627	vapor
420	100	0.668 89	606.8	673.69	2.947 1	0.861 16	1.024 1	1.76E−05	0.023 783	vapor
430	100	0.685 02	615.51	684.01	2.971 4	0.876 92	1.039 6	1.80E−05	0.024 98	vapor
200	200	0.000 787	80.644	80.802	0.495 66	0.956 98	1.563 8	0.000 368	0.204 79	liquid
210	200	0.000 804	96.317	96.477	0.572 14	0.948 31	1.572	0.000 321	0.196 84	liquid
220	200	0.000 822	112.09	112.26	0.645 54	0.941 68	1.584 4	0.000 281	0.188 69	liquid
230	200	0.000 841	128.01	128.18	0.716 32	0.937 01	1.601 3	0.000 248	0.180 43	liquid
235.83	200	0.000 853	137.37	137.54	0.756 53	0.935 18	1.613 4	0.000 231	0.175 58	liquid
235.83	200	0.176 25	467.96	503.21	2.307 1	0.709 55	0.956 39	9.86E−06	0.009 227	vapor
240	200	0.180 35	471.06	507.13	2.323 6	0.692 76	0.927 6	1.01E−05	0.009 396	vapor
250	200	0.189 84	478.22	516.19	2.360 6	0.672 9	0.888 89	1.05E−05	0.009 833	vapor
260	200	0.199	485.18	524.98	2.395	0.667 04	0.871 25	1.09E−05	0.010 316	vapor
270	200	0.207 95	492.06	533.65	2.427 8	0.667 93	0.863 92	1.14E−05	0.010 841	vapor
280	200	0.216 73	498.93	542.28	2.459 1	0.672 82	0.862 75	1.18E−05	0.011 41	vapor
290	200	0.225 39	505.84	550.92	2.489 5	0.680 44	0.865 72	1.23E−05	0.012 021	vapor
300	200	0.233 95	512.81	559.6	2.518 9	0.690 04	0.871 69	1.27E−05	0.012 674	vapor
310	200	0.242 43	519.87	568.36	2.547 6	0.701 16	0.879 9	1.31E−05	0.013 37	vapor
320	200	0.250 86	527.04	577.21	2.575 7	0.713 44	0.889 83	1.36E−05	0.014 107	vapor
330	200	0.259 23	534.32	586.16	2.603 3	0.726 63	0.901 09	1.40E−05	0.014 886	vapor
340	200	0.267 56	541.72	595.23	2.630 3	0.740 53	0.913 39	1.44E−05	0.015 707	vapor
350	200	0.275 85	549.26	604.43	2.657	0.754 98	0.926 49	1.48E−05	0.016 57	vapor
360	200	0.284 12	556.94	613.76	2.683 3	0.769 85	0.940 21	1.52E−05	0.017 474	vapor
370	200	0.292 35	564.77	623.24	2.709 2	0.785 03	0.954 4	1.56E−05	0.018 42	vapor
380	200	0.300 57	572.74	632.85	2.734 9	0.800 42	0.968 95	1.60E−05	0.019 408	vapor
390	200	0.308 77	580.86	642.62	2.760 3	0.815 96	0.983 75	1.65E−05	0.020 437	vapor
400	200	0.316 95	589.14	652.53	2.785 3	0.831 59	0.998 73	1.69E−05	0.021 508	vapor
410	200	0.325 11	597.57	662.59	2.810 2	0.847 24	1.013 8	1.73E−05	0.022 62	vapor
420	200	0.333 26	606.15	672.81	2.834 8	0.862 88	1.028 9	1.76E−05	0.023 773	vapor
430	200	0.341 4	614.89	683.17	2.859 2	0.878 46	1.044 1	1.80E−05	0.024 969	vapor
200	500	0.000 787	80.547	80.941	0.495 17	0.957 07	1.563 2	0.000 368	0.204 95	liquid
210	500	0.000 803	96.208	96.61	0.571 62	0.948 39	1.571 3	0.000 321	0.197 01	liquid

(Continued)

$T (K)$	p (kPa)	υ (m^3/kg)	u (kJ/kg)	h (kJ/kg)	s $(J/g*K)$	c_v $(J/g*K)$	c_p $(J/g*K)$	μ $(Pa*s)$	k $(W/m*K)$	R32 phase
220	**500**	0.000 821	111.97	112.38	0.644 98	0.941 76	1.583 5	0.000 282	0.188 88	liquid
230	**500**	0.000 84	127.87	128.29	0.715 72	0.937 08	1.600 1	0.000 249	0.180 64	liquid
240	**500**	0.000 861	143.97	144.4	0.784 26	0.934 31	1.621 9	0.000 22	0.172 34	liquid
250	**500**	0.000 884	160.31	160.75	0.851 01	0.933 41	1.649 8	0.000 196	0.164 04	liquid
258.82	**500**	0.000 907	174.98	175.43	0.908 72	0.934 17	1.680 8	0.000 177	0.156 74	liquid
258.82	**500**	0.073 186	475.19	511.79	2.208 3	0.795 53	1.119 4	1.09E−05	0.010 572	vapor
260	**500**	0.073 724	476.24	513.1	2.213 4	0.787 77	1.104 7	1.09E−05	0.010 623	vapor
270	**500**	0.078 085	484.64	523.68	2.253 3	0.747 01	1.022 4	1.14E−05	0.011 094	vapor
280	**500**	0.082 194	492.58	533.68	2.289 7	0.730 3	0.981 01	1.18E−05	0.011 621	vapor
290	**500**	0.086 135	500.3	543.37	2.323 7	0.724 59	0.958 54	1.23E−05	0.012 2	vapor
300	**500**	0.089 956	507.91	552.89	2.355 9	0.725 02	0.946 77	1.27E−05	0.012 826	vapor
310	**500**	0.093 684	515.48	562.32	2.386 9	0.729 42	0.941 89	1.31E−05	0.013 498	vapor
320	**500**	0.097 34	523.07	571.74	2.416 8	0.736 6	0.941 82	1.36E−05	0.014 215	vapor
330	**500**	0.100 94	530.7	581.17	2.445 8	0.745 83	0.945 26	1.40E−05	0.014 976	vapor
340	**500**	0.104 49	538.41	590.65	2.474 1	0.756 61	0.951 33	1.44E−05	0.015 781	vapor
350	**500**	0.108	546.21	600.21	2.501 8	0.768 56	0.959 4	1.48E−05	0.016 629	vapor
360	**500**	0.111 48	554.11	609.85	2.529	0.781 42	0.969 02	1.53E−05	0.017 52	vapor
370	**500**	0.114 93	562.13	619.59	2.555 6	0.794 97	0.979 82	1.57E−05	0.018 453	vapor
380	**500**	0.118 36	570.27	629.45	2.581 9	0.809 03	0.991 54	1.61E−05	0.019 43	vapor
390	**500**	0.121 76	578.54	639.42	2.607 8	0.823 48	1.004	1.65E−05	0.020 448	vapor
400	**500**	0.125 15	586.95	649.53	2.633 4	0.838 19	1.016 9	1.69E−05	0.021 509	vapor
410	**500**	0.128 52	595.5	659.76	2.658 7	0.853 08	1.030 3	1.73E−05	0.022 612	vapor
420	**500**	0.131 88	604.19	670.13	2.683 7	0.868 08	1.043 9	1.77E−05	0.023 758	vapor
430	**500**	0.135 23	613.03	680.64	2.708 4	0.883 12	1.057 8	1.81E−05	0.024 945	vapor
200	**1000**	0.000 786	80.387	81.173	0.494 37	0.957 21	1.562 2	0.000 369	0.205 21	liquid
210	**1000**	0.000 803	96.028	96.831	0.570 76	0.948 53	1.57	0.000 322	0.197 3	liquid
220	**1000**	0.000 821	111.77	112.59	0.644 06	0.941 89	1.581 9	0.000 283	0.189 2	liquid
230	**1000**	0.000 84	127.64	128.48	0.714 72	0.937 21	1.598 2	0.000 249	0.180 99	liquid
240	**1000**	0.000 861	143.71	144.57	0.783 17	0.934 41	1.619 4	0.000 221	0.172 72	liquid
250	**1000**	0.000 883	160.01	160.89	0.849 8	0.933 48	1.646 7	0.000 197	0.164 46	liquid
260	**1000**	0.000 909	176.62	177.52	0.915 04	0.934 39	1.681 4	0.000 175	0.156 22	liquid
270	**1000**	0.000 937	193.61	194.55	0.979 29	0.937 24	1.726	0.000 156	0.148 02	liquid
279.77	**1000**	0.000 969	210.72	211.69	1.041 6	0.942 1	1.783 4	0.000 14	0.140 04	liquid
279.77	**1000**	0.036 713	479.6	516.31	2.130 5	0.877 4	1.324 8	1.18E−05	0.012 406	vapor
280	**1000**	0.036 775	479.84	516.61	2.131 5	0.875 49	1.320 4	1.18E−05	0.012 414	vapor
290	**1000**	0.039 342	489.71	529.06	2.175 2	0.819 29	1.184 5	1.23E−05	0.012 823	vapor
300	**1000**	0.041 696	498.82	540.51	2.214	0.793 82	1.113 3	1.27E−05	0.013 334	vapor
310	**1000**	0.043 912	507.51	551.42	2.249 8	0.782 42	1.071 6	1.32E−05	0.013 919	vapor
320	**1000**	0.046 028	515.97	562	2.283 4	0.778 83	1.046 4	1.36E−05	0.014 567	vapor
330	**1000**	0.048 069	524.31	572.38	2.315 4	0.780 2	1.031 6	1.40E−05	0.015 271	vapor

T (K)	p (kPa)	υ (m^3/kg)	u (kJ/kg)	h (kJ/kg)	s (J/g*K)	c_v (J/g*K)	c_p (J/g*K)	μ (Pa*s)	k (W/m*K)	R32 phase
340	**1000**	0.050 052	532.6	582.65	2.346	0.785 02	1.023 9	1.45E−05	0.016 028	vapor
350	**1000**	0.051 988	540.89	592.88	2.375 7	0.792 34	1.021 2	1.49E−05	0.016 835	vapor
360	**1000**	0.053 885	549.21	603.09	2.404 4	0.801 53	1.022 3	1.53E−05	0.017 689	vapor
370	**1000**	0.055 751	557.58	613.33	2.432 5	0.812 14	1.026 3	1.57E−05	0.018 591	vapor
380	**1000**	0.057 59	566.04	623.62	2.459 9	0.823 82	1.032 4	1.61E−05	0.019 538	vapor
390	**1000**	0.059 406	574.58	633.99	2.486 9	0.836 32	1.040 2	1.66E−05	0.020 531	vapor
400	**1000**	0.061 202	583.23	644.43	2.513 3	0.849 42	1.049 3	1.70E−05	0.021 568	vapor
410	**1000**	0.062 982	591.99	654.98	2.539 3	0.862 98	1.059 4	1.74E−05	0.022 65	vapor
420	**1000**	0.064 747	600.88	665.62	2.565	0.876 87	1.070 2	1.78E−05	0.023 775	vapor
430	**1000**	0.066 499	609.88	676.38	2.590 3	0.890 98	1.081 7	1.81E−05	0.024 944	vapor
200	**2000**	0.000 785	80.068	81.638	0.492 77	0.957 5	1.560 2	0.000 371	0.205 72	liquid
210	**2000**	0.000 802	95.67	97.274	0.569 05	0.948 82	1.567 6	0.000 324	0.197 87	liquid
220	**2000**	0.000 819	111.36	113	0.642 22	0.942 16	1.578 9	0.000 284	0.189 83	liquid
230	**2000**	0.000 838	127.19	128.87	0.712 73	0.937 46	1.594 4	0.000 251	0.181 68	liquid
240	**2000**	0.000 859	143.19	144.91	0.781	0.934 63	1.614 7	0.000 223	0.173 48	liquid
250	**2000**	0.000 881	159.42	161.18	0.847 43	0.933 63	1.640 6	0.000 198	0.165 28	liquid
260	**2000**	0.000 906	175.93	177.74	0.912 39	0.934 44	1.673 5	0.000 177	0.157 12	liquid
270	**2000**	0.000 934	192.81	194.68	0.976 29	0.937 11	1.715 3	0.000 158	0.149 01	liquid
280	**2000**	0.000 966	210.16	212.09	1.039 6	0.941 79	1.769 7	0.000 141	0.140 95	liquid
290	**2000**	0.001 003	228.13	230.14	1.102 9	0.948 88	1.843 4	0.000 126	0.132 91	liquid
300	**2000**	0.001 047	246.97	249.07	1.167 1	0.959 34	1.949 7	0.000 112	0.124 8	liquid
304.58	**2000**	0.001 072	256	258.14	1.197 1	0.965 82	2.016 9	0.000 105	0.121 02	liquid
304.58	**2000**	0.017 512	480.39	515.42	2.041 8	0.984 63	1.746 2	1.32E−05	0.016 355	vapor
310	**2000**	0.018 454	487.46	524.37	2.070 9	0.936 45	1.570 6	1.34E−05	0.016 22	vapor
320	**2000**	0.019 993	499.07	539.05	2.117 6	0.886 99	1.386 8	1.38E−05	0.016 32	vapor
330	**2000**	0.021 371	509.63	552.37	2.158 6	0.862 28	1.285 1	1.43E−05	0.016 667	vapor
340	**2000**	0.022 644	519.6	564.88	2.195 9	0.850 05	1.222 5	1.47E−05	0.017 165	vapor
350	**2000**	0.023 843	529.21	576.89	2.230 7	0.845 25	1.182 1	1.51E−05	0.017 773	vapor
360	**2000**	0.024 985	538.6	588.57	2.263 6	0.845 39	1.155 7	1.55E−05	0.018 468	vapor
370	**2000**	0.026 084	547.87	600.04	2.295 1	0.849 01	1.139	1.59E−05	0.019 237	vapor
380	**2000**	0.027 148	557.08	611.37	2.325 3	0.855 19	1.129	1.63E−05	0.020 072	vapor
390	**2000**	0.028 184	566.27	622.64	2.354 5	0.863 28	1.12 4	1.67E−05	0.020 969	vapor
400	**2000**	0.029 195	575.48	633.87	2.383	0.872 82	1.122 8	1.71E−05	0.021 922	vapor
410	**2000**	0.030 186	584.73	645.1	2.410 7	0.883 46	1.124 4	1.75E−05	0.022 929	vapor
420	**2000**	0.031 16	594.04	656.36	2.437 9	0.894 93	1.128 3	1.79E−05	0.023 988	vapor
430	**2000**	0.032 12	603.43	667.67	2.464 5	0.907 03	1.133 9	1.83E−05	0.025 097	vapor
200	**5000**	0.000 783	79.13	83.044	0.488 04	0.958 37	1.554 4	0.000 377	0.207 24	liquid
210	**5000**	0.000 799	94.623	98.617	0.564 02	0.949 68	1.560 7	0.000 329	0.199 56	liquid
220	**5000**	0.000 816	110.19	114.27	0.636 83	0.943	1.570 4	0.000 29	0.191 69	liquid

(Continued)

T (K)	p (kPa)	v (m³/kg)	u (kJ/kg)	h (kJ/kg)	s (J/g*K)	c_v (J/g*K)	c_p (J/g*K)	μ (Pa*s)	k (W/m*K)	R32 phase
230	5000	0.000 834	125.87	130.04	0.706 92	0.938 24	1.584 0	0.000 256	0.183 71	liquid
240	5000	0.000 854	141.69	145.96	0.774 69	0.935 32	1.601 7	0.000 227	0.175 7	liquid
250	5000	0.000 876	157.71	162.09	0.840 52	0.934 18	1.624 1	0.000 203	0.167 7	liquid
260	5000	0.000 899	173.97	178.46	0.904 74	0.934 76	1.652 2	0.000 181	0.159 75	liquid
270	5000	0.000 925	190.53	195.15	0.967 73	0.937 03	1.687 1	0.000 163	0.151 88	liquid
280	5000	0.000 955	207.46	212.24	1.029 8	0.941 03	1.731	0.000 146	0.144 1	liquid
290	5000	0.000 988	224.88	229.82	1.091 5	0.946 89	1.787 5	0.000 131	0.136 39	liquid
300	5000	0.001 027	242.91	248.05	1.153 3	0.954 92	1.862 8	0.000 117	0.128 72	liquid
310	5000	0.001 075	261.8	267.18	1.216 0	0.965 87	1.969 2	0.000 104	0.121 01	liquid
320	5000	0.001 134	281.95	287.63	1.281 0	0.981 5	2.134 7	9.19E−05	0.113 14	liquid
330	5000	0.001 218	304.24	310.32	1.350 8	1.006 7	2.443 4	7.95E−05	0.104 83	liquid
340	5000	0.001 362	331.45	338.26	1.434 1	1.062 8	3.367 8	6.52E−05	0.095 473	liquid
344.33	5000	0.001 501	348.49	356.00	1.485 9	1.134 4	5.428 2	5.64E−05	0.091 412	liquid
344.33	5000	0.004 819	452.17	476.27	1.835 2	1.264 5	7.606 9	1.85E−05	0.048 235	vapor
350	5000	0.005 914	473.3	502.87	1.911 9	1.120 6	3.392 9	1.76E−05	0.034 829	vapor
360	5000	0.007 049	494.5	529.74	1.987 7	1.030 6	2.245 7	1.74E−05	0.028 951	vapor
370	5000	0.007 901	510.49	550.00	2.043 2	0.989 74	1.856 2	1.75E−05	0.026 799	vapor
380	5000	0.008 626	524.34	567.46	2.089 8	0.967 82	1.655 5	1.77E−05	0.025 879	vapor
390	5000	0.009 273	537.00	583.37	2.131 1	0.956 16	1.534 2	1.79E−05	0.025 571	vapor
400	5000	0.009 867	548.95	598.29	2.168 9	0.950 97	1.454 8	1.82E−05	0.025 632	vapor
410	5000	0.010 423	560.43	612.55	2.204 1	0.950 24	1.400 5	1.86E−05	0.025 943	vapor
420	5000	0.010 95	571.60	626.35	2.237 4	0.952 72	1.362 6	1.89E−05	0.026 439	vapor
430	5000	0.011 453	582.57	639.84	2.269 1	0.957 57	1.336 0	1.92E−05	0.027 081	vapor

Source: Linstrom and Mallard (2017–2021).

A.11 Properties of R134a (Liquid–Vapor)

The molar mass of R134a is $\hat{M} = 102.028\ 48\ kg/kmol$. These properties were derived from Linstrom and Mallard (2017–2021).

T (K)	p (kPa)	v (m³/kg)	u (kJ/kg)	h (kJ/kg)	s (J/g*K)	c_v (J/g*K)	c_p (J/g*K)	μ (Pa*s)	k (W/m*K)	R134a phase
200	10	0.000 662	107.39	107.4	0.607 32	0.801 56	1.205 7	0.000 867	0.127 75	liquid
206.29	10	0.000 67	115	115.01	0.644 8	0.806 63	1.213 6	0.000 757	0.124 34	liquid
206.29	10	1.666 7	340.32	356.99	1.817 8	0.587 93	0.674 54	8.26E−06	0.006 004	vapor
210	10	1.697 7	342.52	359.5	1.829 8	0.593 73	0.679 77	8.41E−06	0.006 3	vapor
220	10	1.781 2	348.56	366.38	1.861 8	0.610 71	0.695 63	8.80E−06	0.007 101	vapor

T (K)	p (kPa)	v (m³/kg)	u (kJ/kg)	h (kJ/kg)	s (J/g*K)	c_v (J/g*K)	c_p (J/g*K)	μ (Pa*s)	k (W/m*K)	R134a phase
230	**10**	1.864 2	354.78	373.42	1.893 1	0.628 7	0.712 89	9.20E−06	0.007 902	vapor
240	**10**	1.946 9	361.17	380.63	1.923 8	0.647 08	0.730 75	9.59E−06	0.008 703	vapor
250	**10**	2.029 4	367.74	388.03	1.954	0.665 54	0.748 84	9.98E−06	0.009 503	vapor
260	**10**	2.111 7	374.49	395.61	1.983 8	0.683 95	0.766 96	1.04E−05	0.010 304	vapor
270	**10**	2.193 9	381.43	403.37	2.013	0.702 23	0.785 01	1.08E−05	0.011 105	vapor
280	**10**	2.27 6	388.55	411.31	2.041 9	0.720 34	0.802 94	1.11E−05	0.011 906	vapor
290	**10**	2.35 8	395.85	419.43	2.070 4	0.738 26	0.820 72	1.15E−05	0.012 707	vapor
300	**10**	2.439 9	403.33	427.73	2.098 5	0.756	0.838 33	1.19E−05	0.013 508	vapor
310	**10**	2.521 8	410.98	436.2	2.126 3	0.773 54	0.855 77	1.23E−05	0.014 309	vapor
320	**10**	2.603 6	418.8	444.84	2.153 7	0.790 89	0.873 04	1.27E−05	0.015 109	vapor
330	**10**	2.685 4	426.8	453.66	2.180 9	0.808 05	0.890 13	1.31E−05	0.015 91	vapor
340	**10**	2.767 2	434.97	462.64	2.207 7	0.825 03	0.907 05	1.34E−05	0.016 711	vapor
350	**10**	2.848 9	443.31	471.8	2.234 2	0.841 84	0.923 81	1.38E−05	0.017 512	vapor
360	**10**	2.930 7	451.81	481.12	2.260 5	0.858 48	0.940 4	1.42E−05	0.018 313	vapor
370	**10**	3.012 3	460.48	490.6	2.286 5	0.874 95	0.956 83	1.46E−05	0.019 114	vapor
380	**10**	3.094	469.31	500.25	2.312 2	0.891 27	0.973 11	1.50E−05	0.019 915	vapor
390	**10**	3.175 7	478.31	510.07	2.337 7	0.907 43	0.989 24	1.53E−05	0.020 716	vapor
400	**10**	3.257 3	487.47	520.04	2.362 9	0.923 44	1.005 2	1.57E−05	0.021 517	vapor
410	**10**	3.339	496.78	530.17	2.388	0.939 3	1.021 1	1.61E−05	0.022 318	vapor
420	**10**	3.420 6	506.25	540.46	2.412 8	0.955 03	1.036 8	1.64E−05	0.023 119	vapor
430	**10**	3.502 2	515.88	550.9	2.437 3	0.970 62	1.052 3	1.68E−05	0.023 92	vapor
440	**10**	3.583 8	525.67	561.51	2.461 7	0.986 08	1.067 8	1.72E−05	0.024 72	vapor
450	**10**	3.665 4	535.61	572.26	2.485 9	1.001 4	1.083 1	1.76E−05	0.025 521	vapor
200	**20**	0.000 662	107.39	107.4	0.607 31	0.801 56	1.205 7	0.000 867	0.127 75	liquid
210	**20**	0.000 674	119.51	119.52	0.666 44	0.809 87	1.218 6	0.000 702	0.122 37	liquid
216.74	**20**	0.000 683	127.75	127.77	0.705 09	0.816 13	1.228 2	0.000 618	0.118 84	liquid
216.74	**20**	0.870 81	346.17	363.58	1.793 1	0.612 56	0.701 99	8.66E−06	0.006 845	vapor
220	**20**	0.884 7	348.18	365.88	1.803 6	0.617 15	0.705 85	8.79E−06	0.007 106	vapor
230	**20**	0.927	354.46	373	1.835 3	0.632 92	0.719 97	9.18E−06	0.007 906	vapor
240	**20**	0.968 97	360.9	380.28	1.866 3	0.650 06	0.736 01	9.58E−06	0.008 707	vapor
250	**20**	1.010 7	367.51	387.73	1.896 7	0.667 77	0.752 93	9.97E−06	0.009 507	vapor
260	**20**	1.052 3	374.3	395.34	1.926 5	0.685 68	0.770 25	1.04E−05	0.010 308	vapor
270	**20**	1.093 7	381.26	403.13	1.955 9	0.703 61	0.787 72	1.08E−05	0.011 109	vapor
280	**20**	1.135 1	388.4	411.1	1.984 9	0.721 47	0.805 21	1.11E−05	0.011 909	vapor
290	**20**	1.176 4	395.71	419.24	2.013 4	0.739 2	0.822 64	1.15E−05	0.012 71	vapor
300	**20**	1.217 6	403.2	427.55	2.041 6	0.756 79	0.839 99	1.19E−05	0.013 511	vapor
310	**20**	1.258 7	410.86	436.04	2.069 4	0.774 21	0.857 2	1.23E−05	0.014 312	vapor
320	**20**	1.299 8	418.7	444.69	2.096 9	0.791 47	0.874 29	1.27E−05	0.015 112	vapor
330	**20**	1.340 9	426.7	453.52	2.124 1	0.808 55	0.891 23	1.31E−05	0.015 913	vapor
340	**20**	1.381 9	434.88	462.52	2.151	0.825 47	0.908 02	1.34E−05	0.016 714	vapor

(Continued)

T (K)	p (kPa)	υ (m^3/kg)	u (kJ/kg)	h (kJ/kg)	s (J/g*K)	c_v (J/g*K)	c_p (J/g*K)	μ (Pa*s)	k (W/m*K)	R134a phase
350	**20**	1.422 9	443.22	471.68	2.177 5	0.842 22	0.924 67	1.38E−05	0.017 515	vapor
360	**20**	1.463 9	451.73	481.01	2.203 8	0.858 82	0.941 17	1.42E−05	0.018 316	vapor
370	**20**	1.504 8	460.41	490.5	2.229 8	0.875 25	0.957 52	1.46E−05	0.019 116	vapor
380	**20**	1.545 7	469.25	500.16	2.255 6	0.891 53	0.973 73	1.50E−05	0.019 917	vapor
390	**20**	1.586 6	478.25	509.98	2.281 1	0.907 66	0.989 8	1.53E−05	0.020 718	vapor
400	**20**	1.627 5	487.41	519.96	2.306 3	0.923 64	1.005 7	1.57E−05	0.021 519	vapor
410	**20**	1.668 4	496.72	530.09	2.331 3	0.939 49	1.021 5	1.61E−05	0.022 32	vapor
420	**20**	1.709 3	506.2	540.39	2.356 1	0.955 19	1.037 2	1.64E−05	0.023 121	vapor
430	**20**	1.750 2	515.83	550.84	2.380 7	0.970 77	1.052 7	1.68E−05	0.023 922	vapor
440	**20**	1.791	525.62	561.44	2.405 1	0.986 21	1.068 1	1.72E−05	0.024 723	vapor
450	**20**	1.831 9	535.56	572.2	2.429 3	1.001 5	1.083 4	1.76E−05	0.025 523	vapor
200	**50**	0.000 662	107.38	107.41	0.607 27	0.801 57	1.205 7	0.000 868	0.127 76	liquid
210	**50**	0.000 674	119.5	119.53	0.666 4	0.809 88	1.218 6	0.000 703	0.122 38	liquid
220	**50**	0.000 687	131.76	131.79	0.723 42	0.819 31	1.233 1	0.000 582	0.117 18	liquid
230	**50**	0.000 701	144.17	144.2	0.778 58	0.829 5	1.249 2	0.000 491	0.112 15	liquid
232.7	**50**	0.000 705	147.54	147.57	0.793 17	0.832 35	1.253 9	0.000 47	0.110 81	liquid
232.7	**50**	0.369 25	355.25	373.71	1.765	0.651 44	0.747 62	9.25E−06	0.008 137	vapor
240	**50**	0.382 1	360.09	379.2	1.788 2	0.660 23	0.753 84	9.55E−06	0.008 72	vapor
250	**50**	0.399 45	366.82	386.79	1.819 2	0.674 93	0.766 13	9.95E−06	0.009 52	vapor
260	**50**	0.416 61	373.7	394.53	1.849 5	0.691 07	0.780 59	1.03E−05	0.010 32	vapor
270	**50**	0.433 64	380.73	402.41	1.879 3	0.707 85	0.796 12	1.07E−05	0.011 12	vapor
280	**50**	0.450 55	387.92	410.45	1.908 5	0.724 91	0.812 19	1.11E−05	0.011 92	vapor
290	**50**	0.467 38	395.28	418.65	1.937 3	0.742 05	0.828 54	1.15E−05	0.012 72	vapor
300	**50**	0.484 14	402.81	427.02	1.965 7	0.759 18	0.845 03	1.19E−05	0.013 52	vapor
310	**50**	0.500 84	410.51	435.55	1.993 6	0.776 25	0.861 57	1.23E−05	0.014 321	vapor
320	**50**	0.517 49	418.38	444.25	2.021 3	0.793 22	0.878 09	1.27E−05	0.015 121	vapor
330	**50**	0.534 11	426.41	453.12	2.048 5	0.810 06	0.894 56	1.31E−05	0.015 922	vapor
340	**50**	0.550 69	434.61	462.14	2.075 5	0.826 79	0.910 96	1.34E−05	0.016 722	vapor
350	**50**	0.567 24	442.97	471.34	2.102 1	0.843 37	0.927 28	1.38E−05	0.017 523	vapor
360	**50**	0.583 76	451.5	480.69	2.128 5	0.859 82	0.943 49	1.42E−05	0.018 323	vapor
370	**50**	0.600 27	460.19	490.2	2.154 5	0.876 14	0.959 6	1.46E−05	0.019 124	vapor
380	**50**	0.616 75	469.04	499.88	2.180 4	0.892 32	0.975 6	1.50E−05	0.019 925	vapor
390	**50**	0.633 22	478.06	509.72	2.205 9	0.908 36	0.991 49	1.53E−05	0.020 725	vapor
400	**50**	0.649 68	487.23	519.71	2.231 2	0.924 27	1.007 3	1.57E−05	0.021 526	vapor
410	**50**	0.666 12	496.56	529.86	2.256 3	0.940 04	1.022 9	1.61E−05	0.022 327	vapor
420	**50**	0.682 55	506.04	540.17	2.281 1	0.955 69	1.038 5	1.65E−05	0.023 127	vapor
430	**50**	0.698 97	515.68	550.63	2.305 7	0.971 22	1.053 9	1.68E−05	0.023 928	vapor
440	**50**	0.715 38	525.48	561.25	2.330 1	0.986 62	1.069 2	1.72E−05	0.024 729	vapor
450	**50**	0.731 78	535.42	572.01	2.354 3	1.001 9	1.084 4	1.76E−05	0.025 53	vapor
200	**100**	0.000 662	107.37	107.43	0.607 21	0.801 58	1.205 6	0.000 868	0.127 78	liquid

T (K)	p (kPa)	υ (m^3/kg)	u (kJ/kg)	h (kJ/kg)	s (J/g*K)	c_v (J/g*K)	c_p (J/g*K)	μ (Pa*s)	k (W/m*K)	R134a phase
210	**100**	0.000 674	119.49	119.55	0.666 34	0.809 89	1.218 5	0.000 703	0.122 4	liquid
220	**100**	0.000 687	131.74	131.81	0.723 35	0.819 32	1.233	0.000 583	0.117 2	liquid
230	**100**	0.000 701	144.15	144.22	0.778 51	0.829 51	1.249 1	0.000 492	0.112 17	liquid
240	**100**	0.000 715	156.73	156.8	0.832 04	0.840 27	1.266 9	0.000 42	0.107 28	liquid
246.79	**100**	0.000 726	165.37	165.44	0.867 56	0.847 84	1.28	0.000 38	0.104 04	liquid
246.79	**100**	0.192 56	363.34	382.6	1.747 5	0.687 64	0.793 19	9.77E−06	0.009 29	vapor
250	**100**	0.195 53	365.59	385.15	1.757 7	0.690 1	0.793 65	9.90E−06	0.009 546	vapor
260	**100**	0.204 61	372.65	393.11	1.789	0.701 37	0.800 36	1.03E−05	0.010 343	vapor
270	**100**	0.213 52	379.82	401.17	1.819 4	0.715 51	0.811 44	1.07E−05	0.011 141	vapor
280	**100**	0.222 31	387.12	409.35	1.849 1	0.730 94	0.824 6	1.11E−05	0.011 94	vapor
290	**100**	0.231 01	394.56	417.66	1.878 3	0.746 96	0.838 88	1.15E−05	0.012 739	vapor
300	**100**	0.239 63	402.16	426.13	1.907	0.763 27	0.853 79	1.19E−05	0.013 538	vapor
310	**100**	0.248 19	409.92	434.74	1.935 3	0.779 71	0.869 09	1.23E−05	0.014 337	vapor
320	**100**	0.256 7	417.84	443.51	1.963 1	0.796 17	0.884 61	1.27E−05	0.015 137	vapor
330	**100**	0.265 17	425.92	452.43	1.990 5	0.812 61	0.900 26	1.31E−05	0.015 937	vapor
340	**100**	0.273 61	434.15	461.51	2.017 7	0.829	0.915 98	1.34E−05	0.016 737	vapor
350	**100**	0.282 01	442.55	470.75	2.044 4	0.845 3	0.931 71	1.38E−05	0.017 537	vapor
360	**100**	0.290 39	451.11	480.15	2.070 9	0.861 52	0.947 44	1.42E−05	0.018 337	vapor
370	**100**	0.298 75	459.83	489.7	2.097 1	0.877 63	0.963 13	1.46E−05	0.019 137	vapor
380	**100**	0.307 09	468.7	499.41	2.123	0.893 64	0.978 77	1.50E−05	0.019 937	vapor
390	**100**	0.315 41	477.74	509.28	2.148 6	0.909 53	0.994 34	1.53E−05	0.020 737	vapor
400	**100**	0.323 72	486.93	519.3	2.174	0.925 31	1.009 8	1.57E−05	0.021 538	vapor
410	**100**	0.332 01	496.27	529.47	2.199 1	0.940 98	1.025 3	1.61E−05	0.022 338	vapor
420	**100**	0.340 29	505.77	539.8	2.224	0.956 53	1.040 6	1.65E−05	0.023 139	vapor
430	**100**	0.348 56	515.43	550.28	2.248 6	0.971 97	1.055 8	1.68E−05	0.023 939	vapor
440	**100**	0.356 83	525.24	560.92	2.273 1	0.987 29	1.071	1.72E−05	0.024 739	vapor
450	**100**	0.365 08	535.2	571.7	2.297 3	1.002 5	1.086	1.76E−05	0.025 54	vapor
200	**200**	0.000 662	107.34	107.48	0.607 09	0.801 61	1.205 5	0.000 869	0.127 82	liquid
210	**200**	0.000 674	119.46	119.59	0.666 21	0.809 92	1.218 4	0.000 704	0.122 44	liquid
220	**200**	0.000 687	131.71	131.85	0.723 22	0.819 34	1.232 9	0.000 583	0.117 24	liquid
230	**200**	0.000 701	144.12	144.26	0.778 37	0.829 53	1.248 9	0.000 492	0.112 21	liquid
240	**200**	0.000 715	156.69	156.83	0.831 89	0.840 28	1.266 6	0.000 421	0.107 33	liquid
250	**200**	0.000 731	169.45	169.6	0.883 99	0.851 49	1.286 2	0.000 364	0.102 58	liquid
260	**200**	0.000 748	182.41	182.56	0.934 85	0.863 13	1.308 1	0.000 317	0.097 935	liquid
263.07	**200**	0.000 753	186.45	186.6	0.950 27	0.866 79	1.315 4	0.000 304	0.096 526	liquid
263.07	**200**	0.099 877	372.64	392.62	1.733 4	0.732 01	0.854 04	1.04E−05	0.010 649	vapor
270	**200**	0.103 29	377.86	398.52	1.755 5	0.735 06	0.850 44	1.06E−05	0.011 196	vapor
280	**200**	0.108 07	385.42	407.03	1.786 5	0.744 87	0.853 74	1.11E−05	0.011 988	vapor
290	**200**	0.112 74	393.06	415.61	1.816 6	0.757 7	0.862 05	1.15E−05	0.012 783	vapor

(*Continued*)

T (K)	p (kPa)	υ (m^3/kg)	u (kJ/kg)	h (kJ/kg)	s (J/g*K)	c_v (J/g*K)	c_p (J/g*K)	μ (Pa*s)	k (W/m*K)	R134a phase
300	**200**	0.117 31	400.82	424.28	1.846	0.771 94	0.872 91	1.19E−05	0.013 579	vapor
310	**200**	0.121 82	408.71	433.07	1.874 8	0.786 91	0.885 22	1.23E−05	0.014 375	vapor
320	**200**	0.126 27	416.74	441.99	1.903 1	0.802 27	0.898 44	1.27E−05	0.015 173	vapor
330	**200**	0.130 68	424.91	451.04	1.931	0.817 84	0.912 24	1.30E−05	0.015 97	vapor
340	**200**	0.135 05	433.23	460.24	1.958 4	0.833 51	0.926 45	1.34E−05	0.016 769	vapor
350	**200**	0.139 38	441.7	469.57	1.985 5	0.849 23	0.940 93	1.38E−05	0.017 567	vapor
360	**200**	0.143 69	450.32	479.05	2.012 2	0.864 95	0.955 6	1.42E−05	0.018 366	vapor
370	**200**	0.147 98	459.09	488.68	2.038 6	0.880 65	0.970 4	1.46E−05	0.019 165	vapor
380	**200**	0.152 25	468.01	498.46	2.064 7	0.896 3	0.985 27	1.50E−05	0.019 964	vapor
390	**200**	0.156 5	477.09	508.39	2.090 5	0.911 89	1.000 2	1.54E−05	0.020 763	vapor
400	**200**	0.160 73	486.32	518.47	2.116	0.927 41	1.015 1	1.57E−05	0.021 563	vapor
410	**200**	0.164 95	495.7	528.69	2.141 2	0.942 85	1.03	1.61E−05	0.022 362	vapor
420	**200**	0.169 16	505.24	539.07	2.166 2	0.958 21	1.044 9	1.65E−05	0.023 162	vapor
430	**200**	0.173 36	514.92	549.59	2.191	0.973 48	1.059 8	1.69E−05	0.023 962	vapor
440	**200**	0.177 55	524.75	560.26	2.215 5	0.988 65	1.074 6	1.72E−05	0.024 762	vapor
450	**200**	0.181 73	534.74	571.08	2.239 8	1.003 7	1.089 4	1.76E−05	0.025 562	vapor
200	**500**	0.000 662	107.27	107.6	0.606 74	0.801 69	1.205 2	0.000 872	0.127 93	liquid
210	**500**	0.000 674	119.38	119.72	0.665 84	0.809 99	1.217 9	0.000 706	0.122 56	liquid
220	**500**	0.000 687	131.62	131.97	0.722 82	0.819 41	1.232 3	0.000 585	0.117 37	liquid
230	**500**	0.000 7	144.02	144.37	0.777 95	0.829 58	1.248 3	0.000 494	0.112 35	liquid
240	**500**	0.000 715	156.58	156.94	0.831 44	0.840 32	1.265 9	0.000 422	0.107 47	liquid
250	**500**	0.000 73	169.33	169.69	0.883 5	0.851 51	1.285 2	0.000 365	0.102 73	liquid
260	**500**	0.000 747	182.28	182.65	0.934 32	0.863 12	1.306 9	0.000 318	0.098 101	liquid
270	**500**	0.000 766	195.46	195.84	0.984 1	0.875 16	1.331 4	0.000 278	0.093 559	liquid
280	**500**	0.000 786	208.9	209.29	1.033	0.887 68	1.359 8	0.000 245	0.089 079	liquid
288.88	**500**	0.000 806	221.1	221.5	1.075 9	0.899 29	1.389 4	0.000 219	0.085 126	liquid
288.88	**500**	0.041 123	386.91	407.47	1.719 7	0.807 85	0.976 12	1.13E−05	0.012 93	vapor
290	**500**	0.041 386	387.87	408.56	1.723 4	0.806 87	0.972 28	1.14E−05	0.013 013	vapor
300	**500**	0.043 652	396.34	418.16	1.756	0.805 81	0.952 16	1.18E−05	0.013 772	vapor
310	**500**	0.045 798	404.75	427.65	1.787 1	0.812 54	0.946 8	1.22E−05	0.014 543	vapor
320	**500**	0.047 863	413.19	437.12	1.817 2	0.822 8	0.948 55	1.26E−05	0.015 322	vapor
330	**500**	0.049 865	421.7	446.63	1.846 5	0.834 87	0.954 19	1.30E−05	0.016 106	vapor
340	**500**	0.051 82	430.3	456.21	1.875 1	0.847 93	0.962 23	1.34E−05	0.016 893	vapor
350	**500**	0.053 735	439.02	465.88	1.903 1	0.861 62	0.971 86	1.38E−05	0.017 682	vapor
360	**500**	0.055 618	447.85	475.65	1.930 6	0.875 69	0.982 61	1.42E−05	0.018 474	vapor
370	**500**	0.057 474	456.8	485.54	1.957 7	0.890 03	0.994 16	1.46E−05	0.019 266	vapor
380	**500**	0.059 307	465.89	495.54	1.984 4	0.904 54	1.006 3	1.50E−05	0.020 06	vapor
390	**500**	0.061 12	475.11	505.67	2.010 7	0.919 16	1.019	1.54E−05	0.020 855	vapor
400	**500**	0.062 917	484.46	515.92	2.036 6	0.933 86	1.031 9	1.58E−05	0.021 65	vapor
410	**500**	0.064 698	493.96	526.31	2.062 3	0.948 59	1.045 2	1.62E−05	0.022 446	vapor
420	**500**	0.066 467	503.59	536.82	2.087 6	0.963 33	1.058 6	1.66E−05	0.023 242	vapor

T (K)	p (kPa)	v (m^3/kg)	u (kJ/kg)	h (kJ/kg)	s (J/g*K)	c_v (J/g*K)	c_p (J/g*K)	μ (Pa*s)	k (W/m*K)	R134a phase
430	**500**	0.068 225	513.37	547.48	2.112 7	0.978 07	1.072 2	1.69E−05	0.024 039	vapor
440	**500**	0.069 972	523.28	558.27	2.137 5	0.992 78	1.085 9	1.73E−05	0.024 836	vapor
450	**500**	0.071 711	533.34	569.2	2.162 1	1.007 5	1.099 7	1.77E−05	0.025 633	vapor
200	**1000**	0.000 661	107.15	107.82	0.606 14	0.801 83	1.204 6	0.000 877	0.128 11	liquid
210	**1000**	0.000 673	119.25	119.92	0.665 21	0.810 12	1.217 2	0.000 71	0.122 75	liquid
220	**1000**	0.000 686	131.48	132.17	0.722 16	0.819 52	1.231 5	0.000 588	0.117 57	liquid
230	**1000**	0.000 7	143.86	144.56	0.777 25	0.829 67	1.247 3	0.000 496	0.112 57	liquid
240	**1000**	0.000 714	156.4	157.12	0.830 69	0.840 38	1.264 6	0.000 425	0.107 71	liquid
250	**1000**	0.000 73	169.13	169.86	0.882 7	0.851 54	1.283 7	0.000 367	0.102 99	liquid
260	**1000**	0.000 746	182.05	182.8	0.933 45	0.863 1	1.304 9	0.000 32	0.098 376	liquid
270	**1000**	0.000 765	195.2	195.96	0.983 14	0.875 09	1.328 9	0.000 28	0.093 859	liquid
280	**1000**	0.000 785	208.6	209.39	1.032	0.887 53	1.356 5	0.000 247	0.089 409	liquid
290	**1000**	0.000 807	222.3	223.11	1.080 1	0.900 52	1.389 1	0.000 217	0.084 998	liquid
300	**1000**	0.000 832	236.36	237.19	1.127 8	0.914 2	1.428 7	0.000 192	0.080 587	liquid
310	**1000**	0.000 862	250.86	251.72	1.175 5	0.928 8	1.479 5	0.000 168	0.076 126	liquid
312.54	**1000**	0.000 87	254.63	255.5	1.187 6	0.932 7	1.494 8	0.000 163	0.074 978	liquid
312.54	**1000**	0.020 316	398.85	419.16	1.711 3	0.883 69	1.139 1	1.23E−05	0.015 374	vapor
320	**1000**	0.021 374	406.09	427.47	1.737 5	0.873 27	1.091 6	1.27E−05	0.015 84	vapor
330	**1000**	0.022 677	415.53	438.2	1.770 6	0.871 62	1.059 5	1.31E−05	0.016 525	vapor
340	**1000**	0.023 892	424.82	448.71	1.802	0.876 72	1.044 8	1.35E−05	0.017 246	vapor
350	**1000**	0.025 046	434.08	459.13	1.832 1	0.885 16	1.039 2	1.39E−05	0.017 989	vapor
360	**1000**	0.026 152	443.36	469.52	1.861 4	0.895 48	1.039	1.43E−05	0.018 746	vapor
370	**1000**	0.027 222	452.7	479.92	1.889 9	0.906 95	1.042 2	1.48E−05	0.019 512	vapor
380	**1000**	0.028 262	462.11	490.37	1.917 8	0.919 18	1.047 9	1.52E−05	0.020 284	vapor
390	**1000**	0.029 277	471.61	500.88	1.945 1	0.931 95	1.055 3	1.56E−05	0.021 061	vapor
400	**1000**	0.030 272	481.21	511.48	1.971 9	0.945 09	1.063 9	1.59E−05	0.021 842	vapor
410	**1000**	0.031 249	490.92	522.16	1.998 3	0.958 51	1.073 6	1.63E−05	0.022 626	vapor
420	**1000**	0.032 212	500.74	532.95	2.024 3	0.972 15	1.084	1.67E−05	0.023 412	vapor
430	**1000**	0.033 161	510.69	543.85	2.049 9	0.985 93	1.095	1.71E−05	0.024 2	vapor
440	**1000**	0.034 099	520.75	554.85	2.075 2	0.999 81	1.106 5	1.75E−05	0.024 989	vapor
450	**1000**	0.035 027	530.95	565.98	2.100 2	1.013 8	1.118 4	1.79E−05	0.025 779	vapor
200	**2000**	0.000 661	106.92	108.24	0.604 96	0.802 11	1.203 4	0.000 886	0.128 48	liquid
210	**2000**	0.000 673	118.99	120.34	0.663 97	0.810 37	1.215 9	0.000 717	0.123 14	liquid
220	**2000**	0.000 685	131.19	132.56	0.720 85	0.819 74	1.229 9	0.000 594	0.117 99	liquid
230	**2000**	0.000 699	143.54	144.94	0.775 86	0.829 86	1.245 3	0.000 501	0.113	liquid
240	**2000**	0.000 713	156.05	157.47	0.829 21	0.840 52	1.262 2	0.000 429	0.108 18	liquid
250	**2000**	0.000 728	168.73	170.19	0.881 1	0.851 61	1.280 7	0.000 371	0.103 49	liquid
260	**2000**	0.000 745	181.61	183.1	0.931 73	0.8631	1.301 2	0.000 324	0.098 919	liquid
270	**2000**	0.000 762	194.7	196.22	0.981 26	0.874 97	1.324 2	0.000 284	0.094 449	liquid
280	**2000**	0.000 782	208.03	209.59	1.029 9	0.887 27	1.350 3	0.000 25	0.090 056	liquid

(Continued)

T (K)	p (kPa)	v (m^3/kg)	u (kJ/kg)	h (kJ/kg)	s (J/g*K)	c_v (J/g*K)	c_p (J/g*K)	μ (Pa*s)	k (W/m*K)	R134a phase
290	**2000**	0.000 804	221.63	223.24	1.077 8	0.900 05	1.380 8	0.000 221	0.085 715	liquid
300	**2000**	0.000 828	235.57	237.22	1.125 2	0.913 44	1.417 2	0.000 195	0.081 395	liquid
310	**2000**	0.000 856	249.9	251.61	1.172 4	0.927 6	1.462 6	0.000 172	0.077 052	liquid
320	**2000**	0.000 889	264.74	266.52	1.219 7	0.942 84	1.522 1	0.000 151	0.072 631	liquid
330	**2000**	0.000 93	280.28	282.14	1.267 7	0.959 77	1.607 2	0.000 131	0.068 04	liquid
340	**2000**	0.000 985	296.88	298.85	1.317 6	0.98	1.748 2	0.000 112	0.063 122	liquid
340.63	**2000**	0.000 989	297.98	299.95	1.320 9	0.981 48	1.760 7	0.000 111	0.062 795	liquid
340.63	**2000**	0.009 292	409.7	428.28	1.697 6	0.987 82	1.538 9	1.42E−05	0.019 856	vapor
350	**2000**	0.010 254	421.11	441.62	1.736 2	0.958 23	1.339 1	1.45E−05	0.019 722	vapor
360	**2000**	0.011 114	432.26	454.49	1.772 5	0.948 78	1.246 3	1.49E−05	0.020 037	vapor
370	**2000**	0.011 878	442.93	466.69	1.805 9	0.948 8	1.198 5	1.53E−05	0.020 541	vapor
380	**2000**	0.012 58	453.37	478.53	1.837 5	0.953 48	1.171 8	1.56E−05	0.021 141	vapor
390	**2000**	0.013 237	463.69	490.16	1.867 7	0.960 82	1.156 9	1.60E−05	0.021 796	vapor
400	**2000**	0.013 86	473.97	501.69	1.896 9	0.969 83	1.149 2	1.64E−05	0.022 487	vapor
410	**2000**	0.014 457	484.25	513.16	1.925 2	0.979 96	1.146 5	1.68E−05	0.023 201	vapor
420	**2000**	0.015 033	494.56	524.63	1.952 9	0.990 92	1.147 1	1.72E−05	0.023 932	vapor
430	**2000**	0.015 591	504.93	536.11	1.979 9	1.002 5	1.150 3	1.76E−05	0.024 676	vapor
440	**2000**	0.016 134	515.37	547.64	2.006 4	1.014 5	1.155 4	1.80E−05	0.025 428	vapor
450	**2000**	0.016 665	525.9	559.23	2.032 4	1.026 9	1.161 9	1.83E−05	0.026 188	vapor
200	**4000**	0.000 659	106.46	109.1	0.602 63	0.802 68	1.201 2	0.000 905	0.129 2	liquid
210	**4000**	0.000 671	118.48	121.17	0.661 53	0.810 91	1.213 3	0.000 732	0.123 9	liquid
220	**4000**	0.000 684	130.63	133.37	0.718 27	0.820 22	1.226 8	0.000 607	0.118 8	liquid
230	**4000**	0.000 697	142.92	145.71	0.773 13	0.830 27	1.241 6	0.000 512	0.113 87	liquid
240	**4000**	0.000 711	155.36	158.2	0.826 3	0.840 83	1.257 7	0.000 438	0.109 1	liquid
250	**4000**	0.000 725	167.96	170.86	0.877 99	0.851 82	1.275 2	0.000 38	0.104 47	liquid
260	**4000**	0.000 741	180.74	183.71	0.928 37	0.863 16	1.294 3	0.000 332	0.099 976	liquid
270	**4000**	0.000 758	193.72	196.76	0.977 61	0.874 84	1.315 5	0.000 291	0.095 593	liquid
280	**4000**	0.000 777	206.92	210.03	1.025 9	0.886 89	1.339 2	0.000 257	0.091 303	liquid
290	**4000**	0.000 798	220.36	223.55	1.073 3	0.899 34	1.366 2	0.000 228	0.087 087	liquid
300	**4000**	0.000 82	234.09	237.37	1.120 2	0.912 26	1.397 6	0.000 202	0.082 921	liquid
310	**4000**	0.000 846	248.14	251.53	1.166 6	0.925 75	1.435 1	0.000 179	0.078 777	liquid
320	**4000**	0.000 876	262.59	266.1	1.212 8	0.939 97	1.481 3	0.000 159	0.074 619	liquid
330	**4000**	0.000 911	277.55	281.2	1.259 3	0.955 16	1.541 2	0.000 14	0.070 4	liquid
340	**4000**	0.000 955	293.18	297	1.306 5	0.971 81	1.624 5	0.000 122	0.066 052	liquid
350	**4000**	0.001 012	309.79	313.84	1.355 3	0.991 08	1.754 7	0.000 104	0.061 464	liquid
360	**4000**	0.001 095	328.11	332.49	1.407 8	1.016 9	2.012 7	8.66E−05	0.056 446	liquid
370	**4000**	0.001 27	351.22	356.3	1.473	1.074 8	3.145 5	6.46E−05	0.051 043	liquid
373.49	**4000**	0.001 58	369.25	375.57	1.524 7	1.185 5	26.326	4.55E−05	0.062 847	liquid
373.49	**4000**	0.002 563	395.13	405.38	1.604 6	1.226 1	37.632	2.64E−05	0.066 016	vapor
380	**4000**	0.003 97	422.92	438.79	1.693 5	1.088 7	2.600 9	2.01E−05	0.030 817	vapor
390	**4000**	0.004 791	440.48	459.65	1.747 7	1.047 1	1.783 7	1.91E−05	0.027 264	vapor

T (K)	p (kPa)	v (m^3/kg)	u (kJ/kg)	h (kJ/kg)	s (J/g*K)	c_v (J/g*K)	c_p (J/g*K)	μ (Pa*s)	k (W/m*K)	R134a phase
400	**4000**	0.005 385	454.57	476.11	1.789 4	1.034 3	1.543 4	1.89E−05	0.026 314	vapor
410	**4000**	0.005 881	467.39	490.91	1.826	1.031 6	1.428 9	1.90E−05	0.026 133	vapor
420	**4000**	0.006 32	479.57	504.85	1.859 6	1.033 8	1.364	1.91E−05	0.026 301	vapor
430	**4000**	0.006 722	491.39	518.28	1.891 1	1.038 9	1.324 3	1.93E−05	0.026 661	vapor
440	**4000**	0.007 095	503	531.38	1.921 3	1.046	1.299 1	1.96E−05	0.027 136	vapor
450	**4000**	0.007 447	514.5	544.29	1.950 3	1.054 4	1.283 3	1.99E−05	0.027 686	vapor

Source: Linstrom and Mallard (2017–2021).

A.12 Properties of R717 Ammonia (Liquid–Vapor)

The enthalpy of formation of ammonia is $\Delta \hat{h}_f^0 = -45\ 900\ kJ/kmol.$ The molar mass of ammonia is $\hat{M} = 17.029\ 95\ kg/kmol.$ These properties were derived from Linstrom and Mallard (2017–2021).

T (K)	p (kPa)	v (m^3/kg)	u (kJ/kg)	h (kJ/kg)	s (J/g*K)	c_v (J/g*K)	c_p (J/g*K)	μ (Pa*s)	k (W/m*K)	R717 phase
200	**10**	0.001 373	18.9849	18.9986	0.096 009	2.926 41	4.227 03	0.000 507	0.803 139	liquid
201.931	**10**	0.001 377	27.1558	27.1696	0.136 668	2.922 87	4.237 99	0.000 487	0.796 346	liquid
201.931	**10**	9.800 02	1398.45	1496.45	7.412 83	1.568 87	2.081 54	7.00E−06	0.019 711	vapor
210	**10**	10.202 7	1411.16	1513.18	7.494 09	1.559 55	2.067 03	7.22E−06	0.019 847	vapor
220	**10**	10.699 2	1426.8	1533.79	7.589 98	1.553 66	2.056 48	7.51E−06	0.020 095	vapor
230	**10**	11.193 8	1442.39	1554.33	7.681 27	1.552 44	2.051 96	7.81E−06	0.020 428	vapor
240	**10**	11.687	1457.98	1574.85	7.768 59	1.554 73	2.051 87	8.12E−06	0.020 845	vapor
250	**10**	12.179 2	1473.59	1595.38	7.852 41	1.559 75	2.055 12	8.45E−06	0.021 344	vapor
260	**10**	12.670 7	1489.25	1615.96	7.933 12	1.566 93	2.060 97	8.78E−06	0.021 923	vapor
270	**10**	13.161 5	1504.99	1636.61	8.011 04	1.575 88	2.068 91	9.12E−06	0.022 58	vapor
280	**10**	13.651 9	1520.82	1657.34	8.086 45	1.586 33	2.078 56	9.47E−06	0.023 313	vapor
290	**10**	14.142	1536.76	1678.18	8.159 58	1.598 04	2.089 65	9.82E−06	0.024 12	vapor
300	**10**	14.631 8	1552.82	1699.14	8.230 63	1.610 86	2.101 97	1.02E−05	0.024 999	vapor
310	**10**	15.121 4	1569.01	1720.23	8.299 77	1.624 64	2.115 36	1.06E−05	0.025 947	vapor
320	**10**	15.610 7	1585.34	1741.45	8.367 15	1.639 29	2.129 68	1.09E−05	0.026 962	vapor
330	**10**	16.1	1601.82	1762.82	8.432 91	1.654 72	2.144 83	1.13E−05	0.028 04	vapor
340	**10**	16.589 1	1618.46	1784.35	8.497 18	1.670 84	2.160 73	1.17E−05	0.029 18	vapor
350	**10**	17.078 1	1635.26	1806.04	8.560 05	1.687 59	2.177 3	1.21E−05	0.030 379	vapor
360	**10**	17.567	1652.23	1827.9	8.621 63	1.704 93	2.194 47	1.24E−05	0.031 633	vapor

(*Continued*)

T (K)	p (kPa)	υ (m^3/kg)	u (kJ/kg)	h (kJ/kg)	s (J/g*K)	c_v (J/g*K)	c_p (J/g*K)	μ (Pa*s)	k (W/m*K)	R717 phase
370	**10**	18.055 8	1669.37	1849.93	8.681 99	1.722 79	2.212 2	1.28E−05	0.032 939	vapor
380	**10**	18.544 6	1686.7	1872.14	8.741 23	1.741 14	2.230 43	1.32E−05	0.034 294	vapor
390	**10**	19.033 4	1704.21	1894.54	8.799 41	1.759 94	2.249 12	1.36E−05	0.035 695	vapor
400	**10**	19.522	1721.91	1917.13	8.856 59	1.779 14	2.268 24	1.40E−05	0.037 139	vapor
200	**20**	0.001 373	18.9808	19.0083	0.095 989	2.926 44	4.227 01	0.000 507	0.803 15	liquid
210	**20**	0.001 394	61.5376	61.5655	0.303 615	2.907 81	4.284 75	0.000 415	0.768 018	liquid
211.783	**20**	0.001 398	69.1854	69.2134	0.339 879	2.904 48	4.295 13	0.000 401	0.761 776	liquid
211.783	**20**	5.121 18	1412.02	1514.44	7.163 99	1.593 49	2.119 27	7.26E−06	0.019 901	vapor
220	**20**	5.328 55	1425.2	1531.77	7.244 26	1.581 08	2.099 16	7.50E−06	0.020 111	vapor
230	**20**	5.579 15	1441.1	1552.68	7.337 21	1.572 8	2.084 04	7.80E−06	0.020 443	vapor
240	**20**	5.828 32	1456.91	1573.48	7.425 72	1.570 08	2.076 39	8.12E−06	0.020 859	vapor
250	**20**	6.076 44	1472.7	1594.22	7.510 42	1.571 49	2.074 18	8.44E−06	0.021 357	vapor
260	**20**	6.323 8	1488.5	1614.97	7.591 8	1.576 04	2.076 02	8.77E−06	0.021 936	vapor
270	**20**	6.570 56	1504.34	1635.75	7.670 23	1.583 05	2.080 96	9.12E−06	0.022 592	vapor
280	**20**	6.816 88	1520.26	1656.6	7.746 04	1.592 04	2.088 34	9.47E−06	0.023 325	vapor
290	**20**	7.062 84	1536.27	1677.53	7.819 48	1.602 65	2.097 69	9.82E−06	0.024 132	vapor
300	**20**	7.308 52	1552.39	1698.56	7.890 77	1.614 61	2.108 65	1.02E−05	0.025 01	vapor
310	**20**	7.553 97	1568.63	1719.71	7.960 11	1.627 73	2.120 97	1.05E−05	0.025 958	vapor
320	**20**	7.799 25	1585	1740.98	8.027 66	1.641 86	2.134 43	1.09E−05	0.026 972	vapor
330	**20**	8.044 37	1601.51	1762.4	8.093 56	1.656 87	2.148 9	1.13E−05	0.028 05	vapor
340	**20**	8.289 37	1618.18	1783.96	8.157 93	1.672 66	2.164 23	1.17E−05	0.029 19	vapor
350	**20**	8.534 27	1635	1805.68	8.220 9	1.689 15	2.180 34	1.20E−05	0.030 388	vapor
360	**20**	8.779 08	1651.99	1827.57	8.282 56	1.706 26	2.197 13	1.24E−05	0.031 642	vapor
370	**20**	9.023 82	1669.15	1849.63	8.342 99	1.723 94	2.214 54	1.28E−05	0.032 948	vapor
380	**20**	9.268 49	1686.49	1871.86	8.402 29	1.742 14	2.232 5	1.32E−05	0.034 303	vapor
390	**20**	9.513 11	1704.02	1894.28	8.460 52	1.760 81	2.250 97	1.36E−05	0.035 704	vapor
400	**20**	9.757 68	1721.73	1916.89	8.517 74	1.779 91	2.269 89	1.40E−05	0.037 147	vapor
200	**50**	0.001 373	18.9687	19.0373	0.095 928	2.926 53	4.226 92	0.000 507	0.803 183	liquid
210	**50**	0.001 394	61.5239	61.5936	0.303 549	2.907 89	4.284 65	0.000 415	0.768 053	liquid
220	**50**	0.001 417	104.659	104.73	0.504 207	2.889 44	4.342 22	0.000 347	0.733 187	liquid
226.633	**50**	0.001 433	133.582	133.654	0.633 735	2.877 58	4.378 99	0.000 311	0.710 308	liquid
226.633	**50**	2.174 95	1431.47	1540.22	6.840 11	1.644 54	2.199 27	7.67E−06	0.020 367	vapor
230	**50**	2.209 91	1437.11	1547.6	6.872 43	1.635 87	2.184 89	7.78E−06	0.020 487	vapor
240	**50**	2.312 79	1453.64	1569.28	6.964 69	1.617 48	2.152 97	8.09E−06	0.020 902	vapor
250	**50**	2.414 57	1469.97	1590.7	7.052 15	1.607 63	2.133 32	8.42E−06	0.021 398	vapor
260	**50**	2.515 53	1486.2	1611.97	7.135 58	1.603 99	2.122 45	8.76E−06	0.021 975	vapor
270	**50**	2.615 88	1502.38	1633.17	7.215 58	1.604 96	2.117 97	9.10E−06	0.022 63	vapor
280	**50**	2.715 77	1518.56	1654.35	7.292 6	1.609 44	2.118 26	9.45E−06	0.023 361	vapor
290	**50**	2.815 28	1534.79	1675.55	7.366 99	1.616 64	2.122 19	9.81E−06	0.024 166	vapor
300	**50**	2.914 51	1551.08	1696.8	7.439 05	1.625 99	2.128 95	1.02E−05	0.025 044	vapor

T (K)	p (kPa)	υ (m³/kg)	u (kJ/kg)	h (kJ/kg)	s (J/g*K)	c_v (J/g*K)	c_p (J/g*K)	μ (Pa*s)	k (W/m*K)	R717 phase
310	**50**	3.013 52	1567.46	1718.14	7.509	1.637 08	2.137 97	1.05E−05	0.025 99	vapor
320	**50**	3.112 33	1583.95	1739.57	7.577 04	1.649 62	2.148 82	1.09E−05	0.027 003	vapor
330	**50**	3.211	1600.57	1761.12	7.643 35	1.663 37	2.161 18	1.13E−05	0.028 081	vapor
340	**50**	3.309 54	1617.32	1782.8	7.708 07	1.678 15	2.174 81	1.17E−05	0.029 22	vapor
350	**50**	3.407 98	1634.22	1804.62	7.771 32	1.693 81	2.189 51	1.20E−05	0.030 417	vapor
360	**50**	3.506 32	1651.27	1826.59	7.833 22	1.710 26	2.205 15	1.24E−05	0.031 67	vapor
370	**50**	3.604 6	1668.49	1848.72	7.893 86	1.727 4	2.221 59	1.28E−05	0.032 975	vapor
380	**50**	3.702 8	1685.88	1871.02	7.953 33	1.745 14	2.238 74	1.32E−05	0.034 329	vapor
390	**50**	3.800 96	1703.45	1893.5	8.011 71	1.763 43	2.256 52	1.36E−05	0.035 73	vapor
400	**50**	3.899 07	1721.2	1916.16	8.069 07	1.782 22	2.274 85	1.40E−05	0.037 172	vapor
200	**100**	0.001 373	18.9485	19.0858	0.095 827	2.926 69	4.226 77	0.000 507	0.803 238	liquid
210	**100**	0.001 394	61.5011	61.6405	0.303 441	2.908 02	4.284 49	0.000 415	0.768 111	liquid
220	**100**	0.001 417	104.633	104.775	0.504 091	2.889 55	4.342 04	0.000 347	0.733 249	liquid
230	**100**	0.001 441	148.328	148.472	0.698 325	2.871 77	4.396 93	0.000 295	0.698 854	liquid
239.562	**100**	0.001 466	190.607	190.753	0.878 43	2.855 27	4.446 55	0.000 256	0.666 556	liquid
239.562	**100**	1.138 09	1447.21	1561.01	6.598 3	1.703 16	2.294 67	8.05E−06	0.020 955	vapor
240	**100**	1.140 44	1447.98	1562.02	6.602 49	1.701 48	2.291 82	8.06E−06	0.020 974	vapor
250	**100**	1.193 61	1465.29	1584.66	6.694 9	1.671 07	2.238 97	8.39E−06	0.021 467	vapor
260	**100**	1.245 88	1482.27	1606.86	6.781 99	1.652 66	2.204 39	8.73E−06	0.022 041	vapor
270	**100**	1.297 5	1499.04	1628.79	6.864 75	1.642 85	2.182 64	9.08E−06	0.022 693	vapor
280	**100**	1.348 61	1515.68	1650.54	6.943 87	1.639 35	2.170 11	9.44E−06	0.023 422	vapor
290	**100**	1.399 35	1532.28	1672.21	7.019 91	1.640 56	2.164 37	9.80E−06	0.024 225	vapor
300	**100**	1.449 79	1548.87	1693.85	7.093 26	1.645 35	2.163 71	1.02E−05	0.025 1	vapor
310	**100**	1.499 99	1565.5	1715.5	7.164 25	1.652 94	2.166 96	1.05E−05	0.026 045	vapor
320	**100**	1.549 99	1582.2	1737.2	7.233 14	1.662 74	2.173 26	1.09E−05	0.027 056	vapor
330	**100**	1.599 85	1598.99	1758.97	7.300 14	1.674 32	2.181 99	1.13E−05	0.028 132	vapor
340	**100**	1.649 57	1615.89	1780.84	7.365 44	1.687 38	2.192 68	1.17E−05	0.029 269	vapor
350	**100**	1.699 19	1632.91	1802.83	7.429 17	1.701 66	2.204 99	1.20E−05	0.030 465	vapor
360	**100**	1.748 72	1650.08	1824.95	7.491 48	1.716 98	2.218 64	1.24E−05	0.031 717	vapor
370	**100**	1.798 18	1667.39	1847.21	7.552 47	1.733 19	2.233 44	1.28E−05	0.033 021	vapor
380	**100**	1.847 57	1684.86	1869.62	7.612 23	1.750 17	2.249 21	1.32E−05	0.034 374	vapor
390	**100**	1.8969	1702.5	1892.19	7.670 87	1.767 82	2.265 83	1.36E−05	0.035 773	vapor
400	**100**	1.946 19	1720.32	1914.94	7.728 46	1.786 07	2.283 17	1.40E−05	0.037 215	vapor
200	**200**	0.001 373	18.908	19.1827	0.095 625	2.927 01	4.226 47	0.000 508	0.803 348	liquid
210	**200**	0.001 394	61.4554	61.7343	0.303 223	2.908 28	4.284 17	0.000 415	0.768 228	liquid
220	**200**	0.001 417	104.582	104.865	0.503 858	2.889 77	4.341 69	0.000 347	0.733 372	liquid
230	**200**	0.001 441	148.271	148.559	0.698 075	2.871 96	4.396 54	0.000 295	0.698 984	liquid
240	**200**	0.001 467	192.492	192.786	0.886 294	2.854 7	4.448 33	0.000 255	0.665 227	liquid
250	**200**	0.001 494	237.22	237.519	1.068 9	2.837 8	4.498 16	0.000 223	0.632 21	liquid

(Continued)

T (K)	p (kPa)	υ (m^3/kg)	u (kJ/kg)	h (kJ/kg)	s (J/g*K)	c_v (J/g*K)	c_p (J/g*K)	μ (Pa*s)	k (W/m*K)	R717 phase
254.302	**200**	0.001 507	256.615	256.916	1.145 83	2.830 61	4.519 46	0.000 211	0.618 249	liquid
254.302	**200**	0.594 648	1463.47	1582.4	6.358 08	1.786 39	2.437 05	8.48E−06	0.021 846	vapor
260	**200**	0.610 586	1474.03	1596.14	6.411 52	1.758 74	2.387 9	8.68E−06	0.022 179	vapor
270	**200**	0.637 981	1492.09	1619.69	6.500 37	1.724 19	2.324 45	9.04E−06	0.022 825	vapor
280	**200**	0.664 812	1509.74	1642.7	6.584 09	1.702 77	2.281 96	9.40E−06	0.023 548	vapor
290	**200**	0.691 22	1527.13	1665.37	6.663 64	1.690 77	2.254 16	9.76E−06	0.024 346	vapor
300	**200**	0.717 302	1544.36	1687.82	6.739 75	1.685 65	2.236 93	1.01E−05	0.025 216	vapor
310	**200**	0.743 13	1561.51	1710.14	6.812 92	1.685 7	2.227 5	1.05E−05	0.026 157	vapor
320	**200**	0.768 756	1578.64	1732.39	6.883 57	1.689 68	2.223 94	1.09E−05	0.027 164	vapor
330	**200**	0.794 219	1595.79	1754.63	6.952 01	1.696 72	2.224 88	1.13E−05	0.028 236	vapor
340	**200**	0.819 549	1612.99	1776.9	7.018 49	1.706 17	2.229 35	1.16E−05	0.029 37	vapor
350	**200**	0.844 769	1630.27	1799.23	7.083 21	1.717 58	2.236 62	1.20E−05	0.030 563	vapor
360	**200**	0.869 897	1647.66	1821.64	7.146 35	1.730 57	2.246 15	1.24E−05	0.031 812	vapor
370	**200**	0.894 947	1665.17	1844.16	7.208 04	1.744 88	2.257 54	1.28E−05	0.033 113	vapor
380	**200**	0.919 931	1682.81	1866.8	7.268 42	1.760 3	2.270 47	1.32E−05	0.034 464	vapor
390	**200**	0.944 858	1700.6	1889.57	7.327 57	1.776 65	2.284 69	1.36E−05	0.035 861	vapor
400	**200**	0.969 737	1718.55	1912.49	7.385 61	1.793 82	2.3	1.40E−05	0.037 301	vapor
200	**500**	0.001 373	18.7869	19.4734	0.095 019	2.927 94	4.225 58	0.000 508	0.803 677	liquid
210	**500**	0.001 394	61.3187	62.0157	0.302 572	2.909 05	4.283 19	0.000 416	0.768 579	liquid
220	**500**	0.001 417	104.428	105.136	0.503 159	2.890 44	4.340 63	0.000 347	0.733 742	liquid
230	**500**	0.001 441	148.099	148.819	0.697 327	2.872 56	4.395 37	0.000 296	0.699 373	liquid
240	**500**	0.001 466	192.3	193.033	0.885 494	2.855 24	4.447 03	0.000 255	0.665 635	liquid
250	**500**	0.001 494	237.006	237.753	1.068 04	2.838 28	4.496 67	0.000 223	0.632 638	liquid
260	**500**	0.001 524	282.205	282.967	1.245 37	2.821 63	4.546 28	0.000 198	0.600 437	liquid
270	**500**	0.001 555	327.909	328.687	1.417 91	2.805 38	4.598 52	0.000 176	0.569 042	liquid
277.29	**500**	0.001 58	361.568	362.358	1.540 96	2.793 91	4.640 09	0.000 163	0.546 647	liquid
277.29	**500**	0.250 324	1484.53	1609.69	6.039 27	1.950 01	2.743 72	9.18E−06	0.023 762	vapor
280	**500**	0.253 689	1490.22	1617.07	6.065 74	1.927 45	2.700 58	9.29E−06	0.023 959	vapor
290	**500**	0.265 754	1510.54	1643.41	6.158 2	1.863 06	2.576 2	9.67E−06	0.024 736	vapor
300	**500**	0.277 39	1530.03	1668.72	6.244 01	1.820 56	2.491 15	1.01E−05	0.025 588	vapor
310	**500**	0.288 707	1548.97	1693.32	6.324 68	1.793 2	2.432 43	1.04E−05	0.026 513	vapor
320	**500**	0.299 781	1567.54	1717.43	6.401 23	1.776 66	2.392 06	1.08E−05	0.027 506	vapor
330	**500**	0.310 665	1585.87	1741.21	6.474 39	1.768 05	2.364 9	1.12E−05	0.028 566	vapor
340	**500**	0.321 397	1604.06	1764.76	6.544 71	1.765 4	2.347 5	1.16E−05	0.029 689	vapor
350	**500**	0.332 006	1622.18	1788.18	6.612 6	1.767 28	2.337 46	1.20E−05	0.030 871	vapor
360	**500**	0.342 515	1640.27	1811.53	6.678 37	1.772 7	2.333 09	1.24E−05	0.032 111	vapor
370	**500**	0.352 939	1658.39	1834.86	6.742 29	1.780 91	2.333 16	1.28E−05	0.033 404	vapor
380	**500**	0.363 293	1676.56	1858.21	6.804 55	1.791 36	2.336 76	1.32E−05	0.034 748	vapor
390	**500**	0.373 587	1694.81	1881.6	6.865 33	1.803 63	2.343 22	1.36E−05	0.036 139	vapor
400	**500**	0.383 829	1713.16	1905.08	6.924 76	1.817 4	2.352 02	1.39E−05	0.037 576	vapor

T (K)	p (kPa)	υ (m³/kg)	u (kJ/kg)	h (kJ/kg)	s (J/g*K)	c_v (J/g*K)	c_p (J/g*K)	μ (Pa*s)	k (W/m*K)	R717 phase
200	**1000**	0.001 373	18.5854	19.9582	0.094 01	2.929 5	4.224 11	0.000 509	0.804 226	liquid
210	**1000**	0.001 394	61.0913	62.485	0.301 488	2.910 34	4.281 58	0.000 417	0.769 162	liquid
220	**1000**	0.001 416	104.173	105.589	0.501 996	2.891 55	4.338 87	0.000 348	0.734 358	liquid
230	**1000**	0.001 44	147.813	149.253	0.696 083	2.873 54	4.393 44	0.000 296	0.700 021	liquid
240	**1000**	0.001 466	191.981	193.447	0.884 163	2.856 13	4.444 87	0.000 256	0.666 315	liquid
250	**1000**	0.001 493	236.65	238.144	1.066 62	2.839 1	4.494 2	0.000 224	0.633 35	liquid
260	**1000**	0.001 523	281.808	283.331	1.243 84	2.822 37	4.543 39	0.000 198	0.601 185	liquid
270	**1000**	0.001 555	327.465	329.02	1.416 26	2.806 04	4.595 07	0.000 177	0.569 829	liquid
280	**1000**	0.001 589	373.662	375.251	1.584 39	2.790 33	4.652 42	0.000 159	0.539 252	liquid
290	**1000**	0.001 626	420.474	422.1	1.748 78	2.775 59	4.719 35	0.000 143	0.509 389	liquid
298.045	**1000**	0.001 659	458.66	460.318	1.878 77	2.764 79	4.783 53	0.000 132	0.485 814	liquid
298.045	**1000**	0.128 497	1498.03	1626.52	5.791 61	2.129 81	3.133 02	9.83E−06	0.026 145	vapor
300	**1000**	0.129 868	1502.73	1632.6	5.811 93	2.105 99	3.082 32	9.91E−06	0.026 308	vapor
310	**1000**	0.136 61	1525.72	1662.33	5.909 46	2.010 12	2.878 96	1.03E−05	0.027 189	vapor
320	**1000**	0.143 004	1547.39	1690.39	5.998 55	1.945 9	2.741 01	1.07E−05	0.028 147	vapor
330	**1000**	0.149 144	1568.15	1717.29	6.081 33	1.902 92	2.644 82	1.11E−05	0.029 177	vapor
340	**1000**	0.155 091	1588.29	1743.38	6.159 22	1.874 78	2.576 8	1.15E−05	0.030 274	vapor
350	**1000**	0.160 888	1608.01	1768.89	6.233 18	1.857 37	2.528 6	1.19E−05	0.031 436	vapor
360	**1000**	0.166 566	1627.44	1794	6.303 91	1.847 88	2.494 76	1.23E−05	0.032 657	vapor
370	**1000**	0.172 147	1646.68	1818.83	6.371 93	1.844 4	2.471 62	1.27E−05	0.033 936	vapor
380	**1000**	0.177 648	1665.81	1843.46	6.437 62	1.845 53	2.456 62	1.31E−05	0.035 267	vapor
390	**1000**	0.183 082	1684.9	1867.98	6.501 31	1.850 29	2.447 95	1.35E−05	0.036 65	vapor
400	**1000**	0.188 46	1703.98	1892.44	6.563 23	1.857 92	2.444 28	1.39E−05	0.038 087	vapor
200	**2000**	0.001 372	18.1842	20.9285	0.092	2.932 59	4.221 16	0.000 511	0.805 322	liquid
210	**2000**	0.001 393	60.6386	63.4245	0.299 327	2.912 89	4.278 36	0.000 418	0.770 328	liquid
220	**2000**	0.001 415	103.664	106.495	0.499 68	2.893 75	4.335 38	0.000 35	0.735 588	liquid
230	**2000**	0.001 439	147.244	150.123	0.693 604	2.875 49	4.389 62	0.000 298	0.701 313	liquid
240	**2000**	0.001 465	191.347	194.276	0.881 512	2.857 9	4.440 6	0.000 257	0.667 67	liquid
250	**2000**	0.001 492	235.943	238.927	1.063 78	2.840 71	4.489 32	0.000 225	0.634 771	liquid
260	**2000**	0.001 521	281.019	284.061	1.240 79	2.823 85	4.537 7	0.000 199	0.602 676	liquid
270	**2000**	0.001 553	326.582	329.688	1.412 98	2.807 37	4.588 28	0.000 178	0.571 397	liquid
280	**2000**	0.001 587	372.671	375.844	1.580 84	2.791 48	4.644 13	0.000 159	0.540 909	liquid
290	**2000**	0.001 624	419.353	422.601	1.744 9	2.776 52	4.709	0.000 144	0.511 149	liquid
300	**2000**	0.001 664	466.742	470.07	1.905 83	2.762 96	4.787 64	0.000 13	0.482 023	liquid
310	**2000**	0.001 709	515.003	518.421	2.064 36	2.751 47	4.886 56	0.000 118	0.453 404	liquid
320	**2000**	0.001 759	564.382	567.9	2.221 44	2.742 99	5.015 39	0.000 107	0.425 117	liquid
322.501	**2000**	0.001 773	576.947	580.493	2.260 64	2.741 5	5.053 93	0.000 104	0.418 067	liquid
322.501	**2000**	0.064 453	1505.33	1634.24	5.528 05	2.375 79	3.801 05	1.07E−05	0.030 036	vapor
330	**2000**	0.067 526	1526.48	1661.53	5.611 72	2.257 14	3.496 37	1.10E−05	0.030 719	vapor

(Continued)

T (K)	p (kPa)	υ (m^3/kg)	u (kJ/kg)	h (kJ/kg)	s (J/g*K)	c_v (J/g*K)	c_p (J/g*K)	μ (Pa*s)	k (W/m*K)	R717 phase
340	**2000**	0.071 322	1552.37	1695.02	5.711 71	2.146 66	3.220 4	1.14E−05	0.031 718	vapor
350	**2000**	0.074 873	1576.49	1726.23	5.802 21	2.071 94	3.033 9	1.18E−05	0.032 805	vapor
360	**2000**	0.078 245	1599.39	1755.88	5.885 74	2.021 04	2.902 95	1.23E−05	0.033 969	vapor
370	**2000**	0.081 483	1621.45	1784.41	5.963 92	1.986 73	2.808 82	1.27E−05	0.035 204	vapor
380	**2000**	0.084 615	1642.91	1812.14	6.037 87	1.964 38	2.740 24	1.31E−05	0.036 505	vapor
390	**2000**	0.087 663	1663.95	1839.28	6.108 37	1.950 85	2.690 08	1.35E−05	0.037 872	vapor
400	**2000**	0.090 641	1684.71	1865.99	6.175 99	1.944 01	2.653 56	1.39E−05	0.039 336	vapor
200	**4000**	0.001 371	17.3889	22.8725	0.088 004	2.938 66	4.215 32	0.000 515	0.807 508	liquid
210	**4000**	0.001 392	59.7412	65.3074	0.295 033	2.917 91	4.271 99	0.000 422	0.772 652	liquid
220	**4000**	0.001 414	102.657	108.311	0.495 078	2.898 07	4.328 49	0.000 353	0.738 04	liquid
230	**4000**	0.001 437	146.118	151.867	0.688 682	2.879 34	4.382 08	0.000 3	0.703 887	liquid
240	**4000**	0.001 463	190.091	195.941	0.876 252	2.861 39	4.432 21	0.000 26	0.670 367	liquid
250	**4000**	0.001 49	234.544	240.503	1.058 15	2.843 93	4.479 78	0.000 227	0.637 594	liquid
260	**4000**	0.001 519	279.46	285.534	1.234 77	2.826 8	4.526 61	0.000 201	0.605 635	liquid
270	**4000**	0.001 55	324.842	331.04	1.406 5	2.810 04	4.575 1	0.000 179	0.574 507	liquid
280	**4000**	0.001 583	370.719	377.051	1.573 83	2.793 83	4.628 13	0.000 161	0.544 189	liquid
290	**4000**	0.001 619	417.152	423.629	1.737 27	2.778 46	4.689 13	0.000 145	0.514 627	liquid
300	**4000**	0.001 659	464.239	470.875	1.897 43	2.764 34	4.762 37	0.000 132	0.485 739	liquid
310	**4000**	0.001 703	512.125	518.936	2.055 01	2.752 05	4.853 46	0.000 12	0.457 413	liquid
320	**4000**	0.001 752	561.024	568.03	2.210 87	2.742 37	4.970 49	0.000 109	0.429 5	liquid
330	**4000**	0.001 807	611.245	618.473	2.366 08	2.736 47	5.126 04	9.86E−05	0.401 801	liquid
340	**4000**	0.001 872	663.261	670.75	2.522 13	2.736 3	5.341 87	8.93E−05	0.374 034	liquid
350	**4000**	0.001 95	717.848	725.649	2.681 25	2.745 43	5.660 96	8.06E−05	0.345 752	liquid
351.55	**4000**	0.001 964	726.615	734.472	2.706 4	2.748 07	5.724 75	7.92E−05	0.341 278	liquid
351.55	**4000**	0.030 707	1496.51	1619.34	5.223 43	2.718 38	5.234 91	1.19E−05	0.037 427	vapor
360	**4000**	0.033 006	1528.02	1660.05	5.337 91	2.519 21	4.471 62	1.22E−05	0.038 039	vapor
370	**4000**	0.035 389	1560.33	1701.88	5.452 57	2.364 38	3.940 47	1.27E−05	0.038 991	vapor
380	**4000**	0.037 538	1589.34	1739.5	5.552 9	2.261 27	3.605 5	1.31E−05	0.040 123	vapor
390	**4000**	0.039 53	1616.23	1774.35	5.643 43	2.190 73	3.378 26	1.35E−05	0.041 424	vapor
400	**4000**	0.041 406	1641.66	1807.28	5.726 82	2.142 08	3.216 92	1.39E−05	0.043 03	vapor

Source: Linstrom and Mallard (2017–2021).

A.13 Properties of Saturated Ammonia–Water Mixtures (Liquid–Vapor)

The composition variables are given in mass units (not mole units as presented in Chapters 6, 8, and 10 of this textbook): $x \doteq kg_{NH_3,liq}/\left(kg_{NH_3,liq} + kg_{H_2O,liq}\right)$ and $y \doteq kg_{NH_3,vap}/\left(kg_{NH_3,vap} + kg_{H_2O,vap}\right)$. The molar mass of ammonia is $\hat{M}_{NH_3} = 17.029\,95\ kg/kmol$ and the molar mass of water is $\hat{M}_{H_2O} = 18.01471\ kg/kmol$. (See Appendix A.14 for single-phase property values of $NH_3 + H_2O$ mixtures.)

T (K)	x (if bub) y (if dew)	Bubble point								Dew point							
		y (for x)	p_{bub} (kPa)	ρ_{liq} (kg/m^3)	ρ_{vap} (kg/m^3)	h_{liq} (kJ/kg)	h_{vap} (kJ/kg)	s_{liq} (kJ/kgK)	s_{vap} (kJ/kgK)	x (for y)	p_{dew} (kPa)	ρ_{liq} (kg/m^3)	ρ_{vap} (kg/m^3)	h_{liq} (kJ/kg)	h_{vap} (kJ/kg)	s_{liq} (kJ/kgK)	s_{vap} (kJ/kgK)
203.15	0.60	0.999 990	3.44	872.62	0.0347	−342.55	1500.90	−0.6543	7.9543	—	—	—	—	—	—	—	—
203.15	0.80	1.000 000	8.29	800.84	0.0840	−189.52	1499.50	−0.2843	7.5188	—	—	—	—	—	—	—	—
203.15	1.00	1.0	10.94	724.72	0.1110	32.33	1498.70	0.1622	7.3803	1.0	10.94	724.72	0.1110	32.33	1498.70	0.1622	7.3803
213.15	0.60	0.999 990	7.51	864.95	0.0724	−301.75	1520.30	−0.4583	7.6668	—	—	—	—	—	—	—	—
213.15	0.80	1.000 000	16.74	790.70	0.1621	−145.75	1518.10	−0.0741	7.2673	—	—	—	—	—	—	—	—
213.15	1.00	1.0	21.89	713.62	0.2125	75.08	1516.90	0.3675	7.1318	1.0	21.89	713.62	0.2125	75.08	1516.90	0.3675	7.1318
223.15	0.40	0.999 450	3.19	905.62	0.0293	−305.38	1542.00	−0.4300	8.1834	—	—	—	—	—	—	—	—
223.15	0.60	0.999 980	15.07	856.60	0.1391	−259.74	1539.30	−0.2657	7.4156	—	—	—	—	—	—	—	—
223.15	0.80	1.000 000	31.47	780.32	0.2923	−101.49	1536.20	0.1287	7.0449	—	—	—	—	—	—	—	—
223.15	1.00	1.0	40.84	702.09	0.3806	118.42	1534.30	0.5661	6.9112	1.0	40.84	702.09	0.3806	118.42	1534.30	0.5661	6.9112
233.15	0.40	0.999 170	6.57	901.79	0.0578	−268.24	1562.00	−0.2672	7.9194	—	—	—	—	—	—	—	—
233.15	0.60	0.999 996	28.13	847.74	0.2492	−216.42	1558.00	−0.0759	7.1949	—	—	—	—	—	—	—	—
233.15	0.80	1.000 000	55.63	769.71	0.4971	−56.75	1553.50	0.3247	6.8469	—	—	—	—	—	—	—	—
233.15	1.00	1.0	71.69	690.15	0.6438	162.32	1550.90	0.7583	6.7141	1.0	71.69	690.15	0.6438	162.32	1550.90	0.7583	6.7141
243.15	0.40	0.998 780	12.61	896.87	0.1066	−229.42	1582.00	−0.1042	7.6850	—	—	—	—	—	—	—	—
243.15	0.60	0.999 920	49.34	838.53	0.4207	−172.05	1576.20	0.1103	6.9999	—	—	—	—	—	—	—	—
243.15	0.80	0.999 990	93.22	758.89	0.8041	−11.54	1570.20	0.5144	6.6696	—	—	—	—	—	—	—	—
243.15	1.00	1.0	119.43	677.83	1.0374	206.75	1566.50	0.9446	6.5367	1.0	119.43	677.83	1.0374	206.75	1566.50	0.9446	6.5367
253.15	0.20	0.976 24	3.77	940.10	0.0306	−171.23	1623.00	−0.1134	8.4043	—	—	—	—	—	—	—	—

(Continued)

T (K)	x (if bub) y (if dew)	Bubble point								Dew point							
		y (for x)	p_{bub} (kPa)	ρ_{liq} (kg/m^3)	ρ_{vap} (kg/m^3)	h_{liq} (kJ/kg)	h_{vap} (kJ/kg)	s_{liq} (kJ/kgK)	s_{vap} (kJ/kgK)	x (for y)	p_{dew} (kPa)	ρ_{liq} (kg/m^3)	ρ_{vap} (kg/m^3)	h_{liq} (kJ/kg)	h_{vap} (kJ/kg)	s_{liq} (kJ/kgK)	s_{vap} (kJ/kgK)
253.15	0.40	0.998 25	22.81	891.07	0.1854	−189.05	1601.90	0.0584	7.4771	—	—	—	—	—	—	—	—
253.15	0.60	0.999 86	82.03	829.06	0.6752	−126.89	1593.90	0.2921	6.8266	—	—	—	—	—	—	—	—
253.15	0.80	0.999 98	149.09	747.84	1.2456	34.13	1585.90	0.6981	6.5100	—	—	—	—	—	—	—	—
253.15	1.00	1.0	190.08	665.14	1.6033	251.70	1580.80	1.1253	6.3757	1.0	190.08	665.14	1.6033	251.70	1580.80	1.1253	6.3757
263.15	0.20	0.969 77	6.87	938.21	0.0537	−129.22	1648.70	0.0494	8.1997	—	—	—	—	—	—	—	—
263.15	0.40	0.997 55	39.06	884.57	0.3062	−147.37	1621.80	0.2198	7.2923	—	—	—	—	—	—	—	—
263.15	0.60	0.999 77	130.22	819.41	1.0375	−81.15	1611.00	0.4691	6.6717	—	—	—	—	—	—	—	—
263.15	0.80	0.999 96	228.90	736.54	1.8586	80.24	1600.60	0.8763	6.3654	—	—	—	—	—	—	—	—
263.15	1.00	1.0	290.71	652.06	2.3906	297.16	1593.90	1.3009	6.2285	1.0	290.71	652.06	2.3906	297.16	1593.90	1.3009	6.2285
273.16	0.00	0.0	0.61	999.79	0.0049	0.00	2500.90	0.0000	9.1556	0.0	0.61	999.79	0.0049	0.00	2500.90	0.0000	9.1556
273.16	0.20	0.962 48	11.99	935.11	0.0903	−87.10	1675.00	0.2064	8.0142	0.008 92	0.77	998.01	0.0060	−6.98	2329.60	0.0153	9.3398
273.16	0.40	0.996 63	63.81	877.52	0.4832	−104.58	1641.50	0.3793	7.1279	0.020 25	1.02	994.85	0.0079	−14.97	2158.10	0.0287	9.3339
273.16	0.60	0.999 64	198.62	809.56	1.5358	−34.90	1627.40	0.6413	6.5325	0.038 26	1.51	988.86	0.0116	−26.37	1986.60	0.0471	9.1910
273.16	0.80	0.999 93	339.15	724.96	2.6856	126.84	1614.30	1.0496	6.2335	0.076 08	2.88	975.25	0.0218	−46.14	1815.10	0.0844	8.8391
273.16	1.00	1.0	429.55	638.56	3.4579	343.19	1605.40	1.4718	6.0925	1.0	429.55	638.56	3.4579	343.19	1605.40	1.4718	6.0925
283.15	0.00	0.0	1.23	999.65	0.0094	42.02	2519.20	0.1511	8.8998	0.0	1.23	999.65	0.0094	42.02	2519.20	0.1511	8.8998
283.15	0.20	0.954 38	20.07	931.06	0.1459	−44.88	1702.00	0.3582	7.8467	0.009 50	1.54	997.19	0.0116	34.85	2348.40	0.1680	9.0822
283.15	0.40	0.995 44	99.84	870.04	0.7319	−61.03	1661.20	0.5357	6.9817	0.021 81	2.04	993.22	0.0153	26.54	2177.30	0.1834	9.0746
283.15	0.60	0.999 43	292.15	799.57	2.1985	11.67	1643.10	0.8083	6.4072	0.041 66	3.02	986.05	0.0223	14.58	2006.20	0.2048	8.9303
283.15	0.80	0.999 88	486.44	713.11	3.7695	173.80	1626.80	1.2177	6.1128	0.083 65	5.73	970.31	0.0420	−6.34	1835.00	0.2473	8.5785
283.15	1.00	1.0	615.05	624.64	4.8679	389.71	1615.30	1.6380	5.9662	1.0	615.05	624.64	4.8679	389.71	1615.30	1.6380	5.9662

T (K)	x (if bub) y (if dew)	Bubble point								Dew point							
		y (for x)	p_{bub} (kPa)	ρ_{liq} (kg/m^3)	ρ_{vap} (kg/m^3)	h_{liq} (kJ/kg)	h_{vap} (kJ/kg)	s_{liq} (kJ/kgK)	s_{vap} (kJ/kgK)	x (for y)	p_{dew} (kPa)	ρ_{liq} (kg/m^3)	ρ_{vap} (kg/m^3)	h_{liq} (kJ/kg)	h_{vap} (kJ/kg)	s_{liq} (kJ/kgK)	s_{vap} (kJ/kgK)
293.15	0.00	0.0	2.34	998.16	0.0173	83.91	2537.50	0.2965	8.6661	0.0	2.34	998.16	0.0173	83.91	2537.50	0.2965	8.6661
293.15	0.20	0.945 41	32.34	926.22	0.2276	−2.35	1729.60	0.5057	7.6954	0.010 14	2.92	995.08	0.0214	76.54	2367.10	0.3152	8.8469
293.15	0.40	0.993 93	150.56	862.17	1.0706	−16.75	1680.70	0.6892	6.8512	0.023 52	3.88	990.34	0.0281	67.87	2196.50	0.3326	8.8379
293.15	0.60	0.999 13	416.63	789.37	3.0592	58.67	1658.20	0.9709	6.2938	0.045 37	5.72	981.96	0.0409	55.28	2025.80	0.3571	8.6926
293.15	0.80	0.999 80	678.61	700.92	5.1634	221.29	1638.00	1.3816	6.0016	0.091 66	10.83	964.07	0.0766	33.19	1854.90	0.4044	8.3412
293.15	1.00	1.0	857.48	610.20	6.7025	436.93	1623.30	1.8005	5.8475	1.0	857.48	610.20	6.7025	436.93	1623.30	1.8005	5.8475
303.15	0.00	0.0	4.25	995.61	0.0304	125.73	2555.60	0.4368	8.4520	0.0	4.25	995.61	0.0304	125.73	2555.60	0.4368	8.4520
303.15	0.20	0.935 56	50.37	920.72	0.3435	40.46	1757.80	0.6493	7.5587	0.010 83	5.30	991.95	0.0376	118.13	2385.70	0.4573	8.6317
303.15	0.40	0.992 04	219.63	853.96	1.5180	28.12	1700.20	0.8394	6.7346	0.025 38	7.03	986.42	0.0492	109.07	2215.60	0.4768	8.6215
303.15	0.60	0.998 70	577.83	778.95	4.1509	106.74	1672.50	1.1291	6.1908	0.049 37	10.35	976.80	0.0717	95.80	2045.30	0.5044	8.4755
303.15	0.80	0.999 67	923.28	688.36	6.9214	269.28	1648.00	1.5414	5.8985	0.100 04	19.48	956.72	0.1335	72.60	1874.80	0.5565	8.1246
303.15	1.00	1.0	1167.2	595.17	9.0533	484.90	1629.30	1.9597	5.7347	1.0	1167.2	595.17	9.0533	484.90	1629.30	1.9597	5.7347
313.15	0.00	0.0	7.38	992.18	0.0512	167.53	2573.50	0.5724	8.2556	0.0	7.38	992.18	0.0512	167.53	2573.50	0.5724	8.2556
313.15	0.20	0.924 82	76.01	914.65	0.5031	83.57	1786.60	0.7891	7.4351	0.011 58	9.22	987.96	0.0632	159.68	2404.20	0.5948	8.4343
313.15	0.40	0.989 69	311.06	845.44	2.0939	73.50	1719.60	0.9864	6.6303	0.027 39	12.21	981.61	0.0828	150.20	2234.50	0.6166	8.4232
313.15	0.60	0.998 09	781.75	768.29	5.5077	153.82	1686.10	1.2833	6.0972	0.053 64	17.94	970.71	0.1203	136.23	2064.70	0.6473	8.2766
313.15	0.80	0.999 05	1228.4	675.40	9.1037	317.83	1656.70	1.6975	5.8025	0.108 78	33.59	948.39	0.2231	112.01	1894.40	0.7041	7.9264
313.15	1.00	1.0	1555.4	579.44	12.034	533.78	1633.10	2.1161	5.6265	1.0	1555.4	579.44	12.0340	533.78	1633.10	2.1161	5.6265
323.15	0.00	0.0	12.35	988.00	0.0831	209.34	2591.30	0.7038	8.0749	0.0	12.35	988.00	0.0831	209.34	2591.30	0.7038	8.0749
323.15	0.20	0.913 16	111.48	908.07	0.7172	126.99	1815.80	0.9255	7.3233	0.012 38	15.41	983.21	0.1025	201.22	2422.50	0.7282	8.2528
323.15	0.40	0.986 81	429.13	836.62	2.8187	119.36	1738.90	1.1301	6.5369	0.029 54	20.39	976.02	0.1342	191.29	2253.40	0.7523	8.2410

(Continued)

T (K)	x (if bub) y (if dew)	Bubble point								Dew point							
		y (for x)	p_{bub} (kPa)	ρ_{liq} (kg/m^3)	ρ_{vap} (kg/m^3)	h_{liq} (kJ/kg)	h_{vap} (kJ/kg)	s_{liq} (kJ/kgK)	s_{vap} (kJ/kgK)	x (for y)	p_{dew} (kPa)	ρ_{liq} (kg/m^3)	ρ_{vap} (kg/m^3)	h_{liq} (kJ/kg)	h_{vap} (kJ/kg)	s_{liq} (kJ/kgK)	s_{vap} (kJ/kgK)
323.15	0.60	0.997 27	1034.30	757.35	7.1645	202.01	1699.00	1.4338	6.0118	0.058 17	29.90	963.77	0.1946	176.63	2084.00	0.7863	8.0941
323.15	0.80	0.999 17	1601.8	661.97	11.776	366.99	1664.00	1.8502	5.7127	0.117 88	55.71	939.19	0.3592	151.55	1913.90	0.8480	7.7444
323.15	1.00	1.000 00	2034.0	562.86	15.785	583.77	1634.20	2.2706	5.5213	1.0	2034.0	562.86	15.7850	583.77	1634.20	2.2706	5.5213
333.15	0.00	0.0	19.95	983.16	0.1304	251.18	2608.90	0.8313	7.9082	0.0	19.95	983.16	0.1304	251.18	2608.90	0.8313	7.9082
333.15	0.20	0.900 58	159.30	901.03	0.9979	170.76	1845.50	1.0586	7.2218	0.013 23	24.87	977.78	0.1608	242.78	2440.60	0.8577	8.0855
333.15	0.40	0.983 33	578.22	827.49	3.7130	165.67	1758.10	1.2706	6.4533	0.031 82	32.87	969.71	0.2102	232.41	2272.00	0.8843	8.0732
333.15	0.60	0.996 16	1341.4	746.09	9.1563	250.67	1711.20	1.5808	5.9339	0.062 98	48.12	956.06	0.3044	217.07	2103.00	0.9217	7.9259
333.15	0.80	0.998 73	2051.3	648.04	15.013	416.82	1670.00	2.0000	5.6282	0.127 35	89.24	929.16	0.5597	191.34	1933.00	0.9888	7.7568
333.15	1.00	1.000 00	2615.6	545.24	20.493	635.11	1632.40	2.4239	5.4174	1.0	2615.6	545.24	20.4930	635.11	1632.40	2.4239	5.4174
343.15	0.00	0.0	31.20	977.73	0.1984	293.07	2626.10	0.9551	7.7541	0.0	31.20	977.73	0.1984	293.07	2626.10	0.9551	7.7541
343.15	0.20	0.887 06	222.33	893.56	1.3583	214.79	1875.60	1.1887	7.1295	0.014 13	38.89	971.75	0.2445	284.39	2458.40	0.9837	7.9309
343.15	0.40	0.979 15	762.82	818.04	4.7979	212.43	1777.40	1.4083	6.3783	0.034 26	51.35	962.75	0.3194	273.57	2290.30	1.0129	7.9181
343.15	0.60	0.994 68	1708.3	734.47	11.518	299.84	1722.80	1.7248	5.8628	0.068 07	75.04	947.63	0.4620	257.60	2121.60	1.0540	7.7707
343.15	0.80	0.998 09	2584.3	633.52	18.894	467.42	1674.40	2.1472	5.5484	0.137 25	138.57	918.33	0.8466	231.47	1951.60	1.1269	7.4217
343.15	1.00	1.0	3313.5	526.31	26.407	688.19	1627.10	2.5770	5.3131	1.0	3313.5	526.31	26.4070	688.19	1627.10	2.5770	5.3131
353.15	0.00	0.0	47.41	971.77	0.2937	335.02	2643.00	1.0756	7.6111	0.0	47.41	971.77	0.2937	335.02	2643.00	1.0756	7.6111
353.15	0.20	0.872 61	303.73	885.68	1.8126	259.18	1905.90	1.3160	7.0452	0.015 09	59.07	965.15	0.3617	326.06	2475.90	1.1065	7.7876
353.15	0.40	0.974 19	987.39	808.27	6.0943	259.66	1796.60	1.5432	6.3109	0.036 85	77.94	955.17	0.4723	314.81	2308.20	1.1384	7.7745
353.15	0.60	0.992 75	2140.0	722.44	14.285	349.58	1733.80	1.8659	5.7979	0.073 46	113.72	938.51	0.6825	298.28	2139.90	1.1835	7.6267
353.15	0.80	0.997 19	3207.8	618.34	23.513	518.89	1677.40	2.2922	5.4726	0.147 64	209.20	906.72	1.2474	272.05	1969.70	1.2629	7.2776
353.15	1.00	1.0	4142.0	505.67	33.888	743.49	1617.50	2.7312	5.2060	1.0	4142.0	505.67	33.8880	743.49	1617.50	2.7312	5.2060

T (K)	x (if bub) y (if dew)	Bubble point								Dew point							
		y (for x)	p_{bub} (kPa)	ρ_{liq} (kg/m^3)	ρ_{vap} (kg/m^3)	h_{liq} (kJ/kg)	h_{vap} (kJ/kg)	s_{liq} (kJ/kgK)	s_{vap} (kJ/kgK)	x (for y)	p_{dew} (kPa)	ρ_{liq} (kg/m^3)	ρ_{vap} (kg/m^3)	h_{liq} (kJ/kg)	h_{vap} (kJ/kg)	s_{liq} (kJ/kgK)	s_{vap} (kJ/kgK)
363.15	0.00	0.0	70.18	965.30	0.4239	377.04	2659.60	1.1929	7.4782	0.0	70.18	965.30	0.4239	377.04	2659.60	1.1929	7.4782
363.15	0.20	0.857 23	406.92	877.39	2.3762	303.94	1936.40	1.4406	6.9679	0.016 11	87.40	958.02	0.5221	376.83	2493.00	1.2263	7.6544
363.15	0.40	0.968 35	1256.3	798.14	7.6235	307.38	1815.80	1.6755	6.2504	0.039 60	115.25	946.99	0.6814	356.17	2325.80	1.2613	7.6410
363.15	0.60	0.990 25	2641.1	709.95	17.491	399.95	1744.30	2.0046	5.7386	0.079 17	167.92	928.70	0.9839	339.17	2157.70	1.3106	7.4929
363.15	0.80	0.995 92	3927.9	602.43	28.974	571.35	1678.80	2.4354	5.4003	0.158 59	307.94	894.32	1.7955	313.19	1987.10	1.3972	7.1430
363.15	1.00	1.0	5116.7	482.75	43.484	801.75	1602.30	2.8884	5.0929	1.0	5116.7	482.75	43.4840	801.75	1602.30	2.8884	5.0929
373.15	0.00	0.0	101.42	958.35	0.5982	419.17	2675.60	1.3072	7.3542	0.0	101.42	958.35	0.5982	419.17	2675.60	1.3072	7.3542
373.15	0.20	0.840 94	535.61	868.72	3.0654	349.07	1967.10	1.5628	6.8964	0.017 19	126.27	950.37	0.7367	409.71	2509.60	1.3433	7.5302
373.15	0.40	0.961 52	1573.9	787.65	9.4073	355.63	1835.00	1.8054	6.1958	0.042 54	166.42	938.24	0.9615	397.69	2342.80	1.3816	7.5164
373.15	0.60	0.987 05	3215.2	696.95	21.172	451.03	1754.30	2.1412	5.6845	0.085 24	242.20	918.23	1.3878	380.33	2174.90	1.4357	7.3679
373.15	0.80	0.994 16	4750.2	585.67	35.399	624.96	1678.70	2.5772	5.3309	0.170 18	443.07	881.10	2.5314	354.99	2003.60	1.5305	7.0166
373.15	1.00	1.0	6255.3	456.63	56.117	864.15	1579.80	3.0513	4.9691	1.0	6255.3	456.63	56.1170	864.15	1579.80	3.0513	4.9691
383.15	0.00	0.0	143.38	950.95	0.8269	461.42	2691.10	1.4188	7.2381	0.0	143.38	950.95	0.8269	461.42	2691.10	1.4188	7.2381
383.15	0.20	0.823 76	693.71	859.65	3.8980	394.59	1997.60	1.6827	6.8300	0.018 35	178.48	942.23	1.0186	451.75	2527.70	1.4578	7.4141
383.15	0.40	0.953 61	1944.3	776.75	11.468	404.43	1854.30	1.9332	6.1465	0.045 67	235.15	928.91	1.3296	439.42	2359.20	1.4997	7.3999
383.15	0.60	0.983 02	3865.2	683.35	25.362	502.92	1763.80	2.2759	5.6351	0.091 70	341.97	907.79	1.9196	421.84	2191.40	1.5590	7.2506
383.15	0.80	0.991 70	5678.9	567.95	42.929	679.91	1676.90	2.7183	5.2639	0.182 53	624.54	867.02	3.5044	397.59	2019.10	1.6631	6.8972
383.15	1.00	1.0	7578.3	425.61	73.550	932.83	1546.20	3.2249	4.8258	1.0	7578.3	425.61	73.5500	932.83	1546.20	3.2249	4.8258
393.15	0.00	0.0	198.68	943.11	1.1221	503.82	2705.90	1.5279	7.1292	0.0	198.68	943.11	1.1221	503.82	2705.90	1.5279	7.1292

(Continued)

T (K)	x (if bub) y (if dew)	Bubble point								Dew point							
		y (for x)	p_{bub} (kPa)	ρ_{liq} (kg/m^3)	ρ_{vap} (kg/m^3)	h_{liq} (kJ/kg)	h_{vap} (kJ/kg)	s_{liq} (kJ/kgK)	s_{vap} (kJ/kgK)	x (for y)	p_{dew} (kPa)	ρ_{liq} (kg/m^3)	ρ_{vap} (kg/m^3)	h_{liq} (kJ/kg)	h_{vap} (kJ/kg)	s_{liq} (kJ/kgK)	s_{vap} (kJ/kgK)
393.15	0.20	0.805 70	885.35	850.18	4.8933	440.55	2028.10	1.8005	6.7678	0.019 58	247.31	933.59	1.3827	493.98	2541.10	1.5700	7.3050
393.15	0.40	0.944 51	2371.3	765.42	13.831	453.86	1873.50	2.0592	6.1017	0.049 02	325.77	919.01	1.8058	481.40	2374.90	1.6159	7.2903
393.15	0.60	0.977 98	4593.4	669.07	30.100	555.73	1773.00	2.4092	5.5898	0.098 58	473.56	895.25	2.6089	463.76	2207.10	1.6809	7.1400
393.15	0.80	0.988 41	6716.8	549.12	51.739	736.44	1673.20	2.8591	5.1987	0.195 74	864.23	852.01	4.7743	441.16	2033.30	1.7956	6.7834
393.15	1.00	1.0	9112.5	385.49	100.07	1013.1	1493.40	3.4218	4.6435	1.0	9112.5	385.49	100.0700	1013.10	1493.40	3.4218	4.6435
403.15	0.00	0.0	270.28	934.83	1.4970	546.39	2720.10	1.6347	7.0265	0.0	270.28	934.83	1.4970	546.39	2720.10	1.6347	7.0265
403.15	0.20	0.786 81	1114.9	840.31	6.0720	486.97	2058.20	1.9164	6.7090	0.020 89	336.48	924.47	1.8459	536.42	2555.80	1.6802	7.2022
403.15	0.40	0.934 13	2858.8	753.60	16.523	503.97	1892.70	2.1834	6.0606	0.052 60	443.26	908.52	2.4128	523.69	2389.80	1.7304	7.1868
403.15	0.60	0.971 73	5401.0	654.01	35.430	609.58	1781.70	2.5414	5.5481	0.105 95	644.32	882.70	3.4907	506.20	2221.70	1.8019	7.0352
403.15	0.80	0.983 91	7864.6	529.02	62.047	794.87	1667.60	3.0005	5.1346	0.210 00	1176.3	835.96	6.4156	485.86	2045.90	1.9286	6.6741
403.15	1.00	1.0	10 897.7	312.29	156.77	1135.2	1382.50	3.7153	4.3287	1.0	10 897.7	312.29	156.7700	1135.20	1382.50	3.7153	4.3287
413.15	0.00	0.0	361.54	926.13	1.9667	589.17	2733.50	1.7392	6.9294	0.0	361.54	926.13	1.9667	589.17	2733.50	1.7392	6.9294
413.15	0.20	0.767 13	1386.8	830.00	7.4573	533.90	2087.80	2.0306	6.6528	0.022 29	450.21	914.86	2.4273	579.13	2569.60	1.7885	7.1049
413.15	0.40	0.922 34	3410.3	741.25	19.573	554.84	1911.70	2.3062	6.0224	0.056 44	593.24	897.44	3.1768	566.34	2403.70	1.8436	7.0887
413.15	0.60	0.964 07	6288.5	638.05	41.403	664.63	1790.00	2.6729	5.5095	0.113 85	862.70	869.39	4.6061	549.24	2235.20	1.9221	6.9353
413.15	0.80	0.977 79	9120.2	507.40	74.154	855.66	1659.50	3.1435	5.0703	0.225 46	1577.8	818.75	8.5218	531.95	2056.70	2.0627	6.5679
413.15	0.90	0.982 53	10 563.0	426.74	102.77	988.84	1565.10	3.4357	4.8002	0.349 46	2786.5	764.19	15.624	541.39	1946.20	2.2300	6.1718

Source: Tillner-Roth and Friend (1998).

A.14 Properties of Single-Phase (Liquid, Gas) Ammonia–Water Mixtures

The composition variables are given in mass units (not mole units as presented in Chapters 6, 8, and 10 of this textbook). liquid property values are large for density and small for enthalpy and entropy; gas property values are opposite. Data gaps adjacent to liquid values but at lower temperatures indicate the occurrence of a solid phase. Data gaps at temperatures between liquid and gas phases indicate the occurrence of bubble point/dew point (saturation) phenomena for mixtures simultaneously having nonzero NH_3 and nonzero H_2O. (See Appendix A.13 for saturation property values of $NH_3 + H_2O$ mixtures.)

	z = 0.0 [H_2O]			z = 0.2			z = 0.4			z = 0.6			z = 0.8			z = 1.0 [NH_3]		
T (K)	ρ (kg/m³)	h (kJ/kg)	s (kJ/kgK)	ρ (kg/m³)	h (kJ/kg)	s (kJ/kgK)	ρ (kg/m³)	h (kJ/kg)	s (kJ/kgK)	ρ (kg/m³)	h (kJ/kg)	s (kJ/kgK)	ρ (kg/m³)	h (kJ/kg)	s (kJ/kgK)	ρ (kg/m³)	h (kJ/kg)	s (kJ/kgK)
$p = 100\ kPa$																		
248.15	—	—	—	940.54	−192.18	−0.1974	894.1	−209.34	−0.0229	833.83	−149.52	0.0218	—	—	—	0.8447	1580.50	6.6782
273.15	—	—	—	935.14	−87.06	0.2062	877.54	−104.59	0.3791	—	—	—	—	—	—	0.7612	1635.65	6.89
298.15	997.05	104.92	0.3672	923.57	19.07	0.578	—	—	—	—	—	—	—	—	—	0.6942	1689.84	7.0799
323.15	988.03	209.42	0.7038	—	—	—	—	—	—	—	—	—	—	—	—	0.6387	1744.04	7.2544
348.15	974.84	314.08	1.0157	—	—	—	—	—	—	—	—	—	0.5998	1964.71	7.6155	0.5917	1798.74	7.4175
373.15	0.5897	2675.79	7.3611	0.5814	2512.4	7.645	0.5736	2348.23	7.7682	0.5659	2183.79	7.8083	0.5585	2019.13	7.7665	0.5514	1854.24	7.5714
398.15	0.5503	2726.73	7.4933	0.5431	2563.94	7.7787	0.5362	2400.8	7.9046	0.5294	2237.54	7.9478	0.5227	2074.19	7.9093	0.5162	1910.71	7.7179
423.15	0.5164	2776.62	7.6148	0.5099	2615.13	7.9034	0.5036	2453.48	8.0329	0.4974	2291.79	8.0799	0.4913	2130.07	8.0454	0.4854	1968.27	7.8581
448.15	0.4866	2826.1	7.7284	0.4807	2666.34	8.0209	0.4749	2506.51	8.1547	0.4692	2346.7	8.206	0.4636	2186.88	8.1758	0.4581	2027.02	7.993
473.15	0.4603	2875.47	7.8356	0.4548	2717.77	8.1326	0.4494	2560.06	8.2709	0.4441	2402.38	8.3269	0.4388	2244.71	8.3014	0.4337	2087.02	8.1232
$p = 200\ kPa$																		

(Continued)

	$z = 0.0$ [H_2O]			$z = 0.2$			$z = 0.4$			$z = 0.6$			$z = 0.8$			$z = 1.0$ [NH_3]		
T (K)	ρ (kg/m^3)	h (kJ/kg)	s (kJ/kgK)	ρ (kg/m^3)	h (kJ/kg)	s (kJ/kgK)	ρ (kg/m^3)	h (kJ/kg)	s (kJ/kgK)	ρ (kg/m^3)	h (kJ/kg)	s (kJ/kgK)	ρ (kg/m^3)	h (kJ/kg)	s (kJ/kgK)	ρ (kg/m^3)	h (kJ/kg)	s (kJ/kgK)
248.15	—	—	—	940.58	−192.07	−0.1974	894.13	−209.25	−0.0230	833.87	−149.44	0.2016	753.43	11.3100	0.6068	671.56	229.20	1.0355
273.15	999.89	0.16	−0.0001	935.18	−86.96	0.2062	877.57	−104.50	0.3790	809.58	−34.94	0.6412	—	—	—	1.5468	1626.97	6.5272
298.15	997.09	105.01	0.3672	923.60	19.16	0.5779	858.11	5.63	0.7647	—	—	—	—	—	—	1.4035	1683.67	6.7259
323.15	988.08	209.51	0.7037	908.11	127.06	0.9254	—	—	—	—	—	—	—	—	—	1.2873	1739.39	6.9054
348.15	974.89	314.16	1.0157	—	—	—	—	—	—	—	—	—	—	—	—	1.1903	1795.08	7.0714
373.15	958.40	419.24	1.3071	—	—	—	—	—	—	1.1417	2177.60	7.4647	1.1243	2014.70	7.4229	1.1076	1851.27	7.2272
398.15	1.1138	2716.61	7.1531	1.0971	2555.98	7.4395	1.0812	2394.47	7.5651	1.0657	2232.65	7.6075	1.0507	2070.59	7.5679	1.0361	1908.23	7.3750
423.15	1.0418	2769.12	7.2811	1.0275	2608.97	7.5686	1.0136	2448.44	7.6966	1.0000	2287.80	7.7419	0.9866	2127.08	7.7055	0.9735	1966.17	7.5161
448.15	0.9798	2820.25	7.3985	0.9671	2661.39	7.6890	0.9546	2502.39	7.8205	0.9423	2343.37	7.8695	0.9301	2184.34	7.8370	0.9183	2025.21	7.6516
473.15	0.9255	2870.75	7.5081	0.9139	2713.69	7.8026	0.9024	2556.60	7.9382	0.8911	2399.55	7.9915	0.8800	2242.52	7.9633	0.8690	2085.44	7.7824
$p = 300\ kPa$																		
248.15	—	—	—	940.62	−191.97	−0.1974	894.16	−209.15	−0.023	833.90	−149.35	0.2015	753.48	11.39	0.6066	671.61	229.28	1.0352
273.15	999.94	0.26	−0.0001	935.22	−86.87	0.2062	877.61	−104.42	0.3789	809.62	−34.86	0.6410	—	—	—	2.3595	1617.88	6.3035
298.15	997.14	105.10	0.3671	923.64	19.24	0.5778	858.15	5.72	0.7646	—	—	—	—	—	—	2.1289	1677.34	6.5119
323.15	988.12	209.59	0.7037	908.15	127.14	0.9253	—	—	—	—	—	—	—	—	—	1.9464	1734.66	6.6965
348.15	974.93	314.24	1.0156	889.68	236.97	1.2526	—	—	—	—	—	—	—	—	—	1.7961	1791.38	6.8656
373.15	958.44	419.32	1.3071	—	—	—	—	—	—	—	—	—	1.6975	2010.19	7.2181	1.6690	1848.27	7.0234

T (K)	z = 0.0 [H_2O]			z = 0.2			z = 0.4			z = 0.6			z = 0.8			z = 1.0 [NH_3]		
	ρ (kg/m³)	h (kJ/kg)	s (kJ/kgK)	ρ (kg/m³)	h (kJ/kg)	s (kJ/kgK)	ρ (kg/m³)	h (kJ/kg)	s (kJ/kgK)	ρ (kg/m³)	h (kJ/kg)	s (kJ/kgK)	ρ (kg/m³)	h (kJ/kg)	s (kJ/kgK)	ρ (kg/m³)	h (kJ/kg)	s (kJ/kgK)
398.15	939.06	525.12	1.5815	—	—	—	1.6358	2387.92	7.3611	1.6095	2227.65	7.4045	1.5840	2066.95	7.3654	1.5597	1905.74	7.1275
423.15	1.5772	2761.21	7.0792	1.5534	2602.59	7.3677	1.5304	2443.28	7.4959	1.5079	2283.76	7.5412	1.4858	2124.06	7.5045	1.4645	1964.06	7.3145
448.15	1.4800	2814.19	7.2009	1.4595	2656.33	7.4911	1.4393	2498.19	7.6220	1.4193	2340.01	7.6703	1.3997	2181.79	7.6370	1.3805	2023.40	7.4507
473.15	1.3958	2865.91	7.3132	1.3774	2709.55	7.6067	1.3592	2553.10	7.7412	1.3412	2396.71	7.7934	1.3234	2240.33	7.7641	1.3060	2083.86	7.5820
$p = 500\ kPa$																		
248.15	—	—	—	940.70	−191.76	−0.1974	894.22	−208.96	−0.0232	833.97	−149.18	0.2012	753.58	11.56	0.6062	671.72	229.43	1.0347
273.15	1000.05	0.47	−0.0001	935.29	−86.68	0.2061	877.67	−104.24	0.3787	809.69	−34.70	0.6407	725.06	126.91	1.0491	638.62	343.19	1.4714
298.15	997.23	105.29	0.3671	923.72	19.42	0.5777	858.22	5.88	0.7644	784.19	82.31	1.0505	—	—	—	3.6329	1664.09	6.2287
323.15	988.21	209.76	0.7036	908.22	127.31	0.9251	836.65	119.41	1.1300		—	—	—	—	—	3.2979	1724.94	6.4246
348.15	975.02	314.40	1.0155	889.77	237.13	1.2524	—	—	—	—	—	—	—	—	—	3.0298	1783.85	6.6002
373.15	958.54	419.47	1.3069	—	—	—	—	—	—	—	—	—	—	—	—	2.8075	1842.20	6.7621
398.15	939.16	525.27	1.5813	—	—	—	—	—	—	2.7204	2217.34	7.1410	2.6677	2059.54	7.1049	2.6182	1900.72	6.9138
423.15	917.02	632.20	1.8418	2.6322	2589.13	7.1045	2.5856	2432.60	7.2355	2.5404	2275.49	7.2926	2.4967	2117.94	7.2472	2.4547	1959.82	7.0578
448.15	2.5032	2801.39	6.9428	2.4632	2645.82	7.2347	2.4242	2489.59	7.3664	2.3858	2333.19	7.4151	2.3481	2176.63	7.3819	2.3115	2019.75	7.1954
473.15	2.3528	2855.86	7.0611	2.3184	2701.03	7.3546	2.2845	2544.97	7.4888	2.2508	2390.94	7.5405	2.2174	2235.90	7.5106	2.1848	2080.68	7.3277
$p = 800\ kPa$																		
248.15	—	—	—	940.81	−191.44	−0.1974	894.31	−208.68	−0.0234	834.08	−148.92	0.2008	753.72	11.81	0.6057	671.88	229.67	1.0338
273.15	1000.20	0.77	−0.0001	935.40	−86.39	0.2060	877.77	−103.98	0.3784	809.81	−34.45	0.6402	725.24	127.15	1.0484	638.83	343.38	1.4703

(Continued)

T (K)	z = 0.0 [H_2O]			z = 0.2			z = 0.4			z = 0.6			z = 0.8			z = 1.0 [NH_3]		
	ρ (kg/m^3)	h (kJ/kg)	s (kJ/kgK)	ρ (kg/m^3)	h (kJ/kg)	s (kJ/kgK)	ρ (kg/m^3)	h (kJ/kg)	s (kJ/kgK)	ρ (kg/m^3)	h (kJ/kg)	s (kJ/kgK)	ρ (kg/m^3)	h (kJ/kg)	s (kJ/kgK)	ρ (kg/m^3)	h (kJ/kg)	s (kJ/kgK)
298.15	997.36	105.57	0.3670	923.83	19.69	0.5775	858.33	6.13	0.7640	784.33	82.54	1.0500	694.69	245.22	1.4619	6.0440	1642.56	5.9432
323.15	988.34	210.02	0.7035	908.34	127.56	0.9249	836.77	119.65	1.1296	—	—	—	—	—	—	5.4167	1709.67	6.1595
348.15	975.15	314.65	1.0153	889.89	237.36	1.2521	—	—	—	—	—	—	—	—	—	4.9401	1772.21	6.3459
373.15	958.68	419.69	1.3067	868.84	349.25	1.5625	—	—	—	—	—	—	—	—	—	4.5560	1832.91	6.5143
398.15	939.31	525.47	1.5810	—	—	—	—	—	—	—	—	—	4.3380	2048.09	6.8565	4.2355	1893.08	6.6704
423.15	917.19	632.38	1.8415	—	—	—	—	—	—	4.1342	2262.62	7.0353	4.0453	2108.57	7.0039	3.9619	1953.38	6.8173
448.15	4.1041	2780.06	6.6880	4.0213	2629.02	6.9863	3.9438	2476.13	7.1217	3.8683	2322.67	7.1732	3.7949	2168.77	7.1421	3.7244	2014.24	6.9570
473.15	3.8331	2839.77	6.8177	3.7675	2687.67	7.1136	3.7033	2534.95	7.2494	3.6396	2382.13	7.3023	3.5768	2229.19	7.2733	3.5157	2075.91	7.0909
$p = 1\,000\ kPa$																		
248.15	—	—	—	940.89	−191.23	−0.1974	894.38	−208.50	−0.0235	834.15	−148.75	0.2006	753.81	11.98	0.6053	671.99	229.83	1.0333
273.15	1000.30	0.98	−0.0001	935.48	−86.20	0.2059	877.84	−103.80	0.3782	809.89	−34.29	0.6399	725.35	127.30	1.0479	638.97	343.51	1.4697
298.15	997.45	105.75	0.3670	923.90	19.87	0.5774	858.41	6.30	0.7638	784.42	82.70	1.0497	694.84	245.36	1.4614	7.7778	1626.84	5.7927
323.15	988.43	210.20	0.7034	908.42	127.73	0.9247	836.85	119.80	1.1293	—	—	—	—	—	—	6.8984	1698.96	6.0252
348.15	975.24	314.81	1.0152	889.98	237.52	1.2519	813.26	236.08	1.4759	—	—	—	—	—	—	6.2568	1764.20	6.2198
373.15	958.77	419.84	1.3065	868.93	349.39	1.5622	—	—	—	—	—	—	—	—	—	5.7509	1826.59	6.3929
398.15	939.42	525.61	1.5809	845.30	463.70	1.8587	—	—	—	—	—		5.4838	2040.22	6.7340	5.3342	1887.91	6.5519
423.15	917.31	632.51	1.8412	—	—	—	—	—	—	5.2293	2253.70	6.9127	5.1006	2102.19	6.8849	4.9817	1949.05	6.7008
448.15	892.35	741.08	2.0905	5.0997	2617.06	6.8612	4.9874	2466.76	7.0003	4.8797	2315.47	7.0546	4.7761	2163.45	7.0256	4.6776	2010.53	6.8420
473.15	4.8538	2828.29	6.6956	4.7609	2678.35	6.9943	4.6712	2527.38	7.1319	4.5827	2376.14	7.1863	4.4957	2224.66	7.1585	4.4115	2072.69	6.9770

T (K)	z = 0.0 [H_2O]			z = 0.2			z = 0.4			z = 0.6			z = 0.8			z = 1.0 [NH_3]		
	ρ (kg/m³)	h (kJ/kg)	s (kJ/kgK)	ρ (kg/m³)	h (kJ/kg)	s (kJ/kgK)	ρ (kg/m³)	h (kJ/kg)	s (kJ/kgK)	ρ (kg/m³)	h (kJ/kg)	s (kJ/kgK)	ρ (kg/m³)	h (kJ/kg)	s (kJ/kgK)	ρ (kg/m³)	h (kJ/kg)	s (kJ/kgK)
$p = 1200\ kPa$																		
248.15	—	—	—	940.97	−191.02	−0.1974	894.44	−208.31	−0.0237	834.21	−148.58	0.2003	753.91	12.15	0.6049	672.10	229.99	1.0327
273.15	1000.40	1.18	−0.0001	935.55	−86.01	0.2058	877.91	−103.62	0.3781	809.97	−34.12	0.6396	725.47	127.45	1.0475	639.11	343.64	1.4690
298.15	997.54	105.94	0.3669	923.98	20.05	0.5772	858.48	6.47	0.7636	784.51	82.85	1.0493	694.98	245.49	1.4609	602.95	460.89	1.8796
323.15	988.51	210.37	0.7033	909.49	127.89	0.9246	836.93	119.96	1.1291	757.44	202.13	1.4334	—	—	—	8.4429	1687.77	5.9096
348.15	975.33	314.97	1.0150	890.06	237.67	1.2517	813.35	236.22	1.4756	—	—	—	—	—	—	7.6111	1755.98	6.1130
373.15	958.86	419.99	1.3064	869.02	349.54	1.5620	—	—	—	—	—	—	—	—	—	6.9704	1820.16	6.2911
398.15	939.52	525.75	1.5807	845.40	463.83	1.8584	—	—	—	—	—	—	—	—	—	6.4501	1882.67	6.4532
423.15	917.42	632.63	1.8410	—	—	—	—	—	—	—	—	—	6.1752	2095.69	6.7855	6.0138	1944.67	6.6043
448.15	892.48	741.19	2.0902	—	—	—	6.0579	2457.05	6.8975	5.9110	2308.10	6.9551	5.7713	2158.06	6.9287	5.6399	2006.8	6.7469
473.15	5.9053	2816.08	6.5909	5.7783	2668.68	6.8934	5.6579	2519.61	7.0334	5.5401	2370.75	7.0897	5.4250	2220.08	7.0633	5.3144	2069.47	6.8830
$p = 1500\ kPa$																		
248.15	—	—	—	941.09	−190.71	−0.1974	894.53	−208.03	−0.0239	834.32	−148.32	0.1999	754.05	12.41	0.6043	672.26	230.23	1.0319
273.15	1000.55	1.48	−0.0001	935.66	−85.72	0.2057	878.01	−103.36	0.3778	810.09	−33.88	0.6392	725.64	127.68	1.0468	639.32	343.84	1.4680
298.15	988.64	210.63	0.7031	908.61	128.14	0.9243	837.06	120.19	1.1287	757.60	202.33	1.4329	—	—	—	10.9000	1669.95	5.7577
323.15	975.46	315.21	1.0149	890.18	237.90	1.2514	813.49	236.44	1.4751	—	—	—	—	—	—	9.7198	1743.21	5.9763
348.15	959.00	420.22	1.3061	869.16	349.75	1.5616	—	—	—	—	—	—	—	—	—	8.8489	1810.30	6.1624
373.15	939.67	525.95	1.5804	845.56	464.02	1.8580	—	—	—	—	—	—	—	—	—	8.1576	1874.70	6.3295

(Continued)

T (K)	z = 0.0 [H_2O]			z = 0.2			z = 0.4			z = 0.6			z = 0.8			z = 1.0 [NH_3]		
	ρ (kg/m³)	h (kJ/kg)	s (kJ/kgK)	ρ (kg/m³)	h (kJ/kg)	s (kJ/kgK)	ρ (kg/m³)	h (kJ/kg)	s (kJ/kgK)	ρ (kg/m³)	h (kJ/kg)	s (kJ/kgK)	ρ (kg/m³)	h (kJ/kg)	s (kJ/kgK)	ρ (kg/m³)	h (kJ/kg)	s (kJ/kgK)
398.15	917.59	632.82	1.8407	—	—	—	—	—	—	—	—	—	7.8255	2085.71	6.6601	7.7561	1938.03	6.4838
423.15	892.68	741.35	2.0898	—	—	—	—	—	—	7.4976	2296.73	6.8294	7.2916	2149.83	6.8073	7.1019	2001.16	6.6288
448.15	7.5498	2796.01	6.4537	7.3539	2653.44	6.7640	7.1759	2507.57	6.9086	7.0046	2360.72	6.9683	6.8393	2213.14	6.9448	6.6820	2064.60	6.7665
473.15	7.5498	2796.01	6.4537	7.3539	2653.44	6.7640	7.1759	2507.57	6.9086	7.0046	2360.72	6.9683	6.8393	2213.14	6.9448	6.6820	2064.60	6.7665
$p = 2\,000\ kPa$																		
248.15	—	—	—	941.28	−190.18	−0.1975	894.68	−207.56	−0.0243	834.49	−147.89	0.1992	754.29	12.83	0.6033	672.53	230.62	1.0305
273.15	1000.81	1.99	0.0000	935.85	−85.24	0.2055	878.17	−102.92	0.3773	810.29	−33.46	0.6384	725.93	128.07	1.0457	639.67	344.16	1.4663
298.15	997.90	106.08	0.3667	924.27	20.76	0.5767	858.77	7.13	0.7627	784.88	83.46	1.0480	695.56	246.01	1.4588	603.71	461.22	1.8763
323.15	988.86	211.06	0.7029	908.80	128.56	0.9239	837.26	120.58	1.1280	757.87	202.68	1.4319	662.35	367.18	1.8490	15.45	1636.69	5.5357
348.15	975.69	315.62	1.0145	890.39	238.29	1.2509	813.72	236.79	1.4744	728.56	324.68	1.7955	—	—	—	13.47	1720.58	5.7861
373.15	959.24	420.60	1.3057	869.39	350.10	1.5611	787.88	355.89	1.8047	—	—	—	—	—	—	12.12	1793.22	5.9876
398.15	939.92	526.30	1.5799	845.82	464.33	1.8573	—	—	—	—	—	—	—	—	—	11.10	1861.07	6.1637
423.15	917.87	633.13	1.8401	819.43	581.53	2.1428	—	—	—	—	—	—	10.69	2068.43	6.4904	10.27	1926.77	6.3237
448.15	893.00	741.61	2.0892	—	—	—	—	—	—	10.26	2276.82	6.6581	9.9036	2135.74	6.6450	9.5858	1991.63	6.4727
473.15	865.00	852.46	2.3298	—	—	—	9.8185	2486.38	6.7383	9.5272	2344.62	6.8053	9.2533	2201.32	6.7874	8.9980	2056.39	6.6133
$p = 3000\ kPa$																		
248.15	—	—	—	941.67	−189.13	−0.1975	894.99	−206.62	−0.0250	834.83	−147.02	0.1978	754.76	13.67	0.6014	673.06	231.42	1.0277
273.15	1001.31	3.01	0.0000	936.22	−84.29	0.2050	878.51	−102.04	0.3764	810.68	−32.64	0.6369	726.50	128.84	1.0435	640.36	344.81	1.4630
298.15	998.35	107.60	0.3665	924.64	21.65	0.5761	859.13	7.97	0.7616	785.34	84.23	1.0463	696.27	246.67	1.4562	604.64	461.64	1.8721

T (K)	$z = 0.0$ [H_2O]			$z = 0.2$			$z = 0.4$			$z = 0.6$			$z = 0.8$			$z = 1.0$ [NH_3]		
	ρ (kg/m^3)	h (kJ/kg)	s (kJ/kgK)	ρ (kg/m^3)	h (kJ/kg)	s (kJ/kgK)	ρ (kg/m^3)	h (kJ/kg)	s (kJ/kgK)	ρ (kg/m^3)	h (kJ/kg)	s (kJ/kgK)	ρ (kg/m^3)	h (kJ/kg)	s (kJ/kgK)	ρ (kg/m^3)	h (kJ/kg)	s (kJ/kgK)
323.15	989.30	211.92	0.7024	909.19	129.40	0.9231	837.66	121.36	1.1268	758.42	203.37	1.4299	663.28	367.66	1.8458	564.19	583.74	2.2652
348.15	976.13	316.42	1.0139	890.80	239.07	1.2499	814.18	237.50	1.4729	729.24	325.25	1.7932	626.17	493.06	2.2195	22.18	1668.54	5.4706
373.15	959.71	421.35	1.3050	869.85	350.82	1.5599	788.43	356.51	1.8029	—	—	—	—	—	—	19.33	1756.19	5.7140
398.15	940.43	526.99	1.5790	846.34	464.96	1.8559	759.83	479.02	2.1207	—	—	—	—	—	—	17.39	1832.34	5.9117
423.15	918.44	633.75	1.8390	820.04	582.05	2.1411	—	—	—	—	—	—	—	—	—	15.93	1903.40	6.0848
448.15	893.65	742.14	2.0879	790.42	702.97	2.4187	—	—	—	—	—	—	15.4600	2105.97	6.3991	14.7500	1972.06	6.2424
473.15	865.76	852.86	2.3283	—	—	—	—	—	—	14.9300	2310.12	6.5560	14.3200	2176.71	6.5527	13.7800	2039.65	6.3892

Source: Tillner-Roth and Friend (1998).

Cited References

Linstrom, P.J. and Mallard, W.G., Eds.; (2017–2021); *NIST Chemistry WebBook, SRD 69; Thermophysical Properties of Fluid Systems.* National Institute of Standards and Technology, U.S. Dept. of Commerce; http://webbook.nist.gov/chemistry/fluid.

Tillner-Roth, R. and Friend, D.G.; (1998); "A Helmholtz free energy formulation of the thermodynamic properties of the mixture {water + ammonia}" *J. Phys. Chem. Ref. Data* **27**; 63; doi: 10.1063/1.556015.

Appendix B

Excel (VBA) Custom Functions

The underlying equations for the VBA (Visual Basic for Applications) procedures in this appendix were obtained from a variety of sources (see the Cited References). The VBA scripts were developed by the author and were validated at a range of conditions deemed relevant. For those who desire a deeper (yet still basic) treatment of VBA for engineers, see Chapra (2010).

VBA Instructions. To access VBA for the first time, open an Excel workbook, save it with file type **.xlsm* to enable ***macros*** (i.e. ***VBA procedures***), and navigate to ***File|Options|Customize Ribbon***. On the right side of the window, you will see a column heading "Customize the Ribbon" and just below that a drop-down menu that defaults to "Main Tabs," which is the correct selection. Slightly more than halfway down the list you will see an unchecked box at ***Developer***. When you check this box and click ***OK***, a new menu item called ***Developer*** will be available from your Main Ribbon. After the ***Developer*** tab is displayed, the Customize Ribbon step is no longer necessary for all future uses of VBA in the subject copy of Excel.

Click on Developer to access the Excel menu that contains Excel tools to facilitate interconnections between Excel and VBA. Only one of the tools in the Excel Developer tab is necessary to utilize the scripts in this appendix – the ***Visual Basic editor*** icon at the upper left. Click on this icon and you will leave the ***Excel Environment*** and enter the ***Visual Basic Environment***.

From the Visual Basic editor, navigate to ***Insert|Module*** and click to open a blank VBA Module. Then, simply copy/paste the entire VBA script from the desired appendix into the blank module to create the customized Excel functions outlined at the beginning of that appendix.

It is recommended to organize your VBA code in a modular fashion. In other words, one module should be inserted to hold all VBA instructions from any one of the appendices, and another module should be inserted to hold all VBA instructions for another appendix, and so on.

After creating modules and pasting VBA instructions, navigate to ***File|Save*** and then close the VBA Editor window to return back to the Excel Environment. In any cell of the workbook, test the availability and functionality of the new custom Excel function by entering the equals sign followed by the name of the custom function. The format is:

$$= Function_{name}(argment1, argument2, \ldots)$$

Many of the customized VBA functions available in this appendix utilize temperature $T\ (\doteq K)$ as one of the arguments, and most of the thermodynamic functions use density $\rho\ (\doteq kg/m^3)$ as the other argument. However, users should review the introductory materials in each relevant appendix to determine the correct ***arguments*** and their correct sequence in the ***argument list*** (i.e. the space between parentheses after ***Function_name***).

Thermal Systems Design: Fundamentals and Projects, Second Edition. Richard J. Martin.

Companion website: www.wiley.com\go\Martin\ThermalSystemsDesign2

Caution. Users should refrain from employing any VBA script without first performing their own ***validation*** of the output values ***for the conditions they are examining***. Usually, this means checking an appropriate range of output values against the NIST Chemistry Webbook (Linstrom and Mallard 2017–2021) or the tables given in Appendix A.

B.1 Chen Equation and Friction Factor (Chapter 2)

The VBA script below creates a user-defined Function named `ChenEq` (for friction factor f), which has two arguments `e_D` (relative roughness ϵ/D) and `Re_D` (Reynolds number $\rho VD/\mu$). The values returned by `ChenEq` are dimensionless, as are both arguments `e_D` and `Re_D`.

The governing equation is from Section 2.10 of this book (see Equation 2.19).

$$f = \left[-2\log_{10}\left(\frac{\varepsilon/D}{3.7065} - \frac{5.0452}{Re_D}\cdot \log_{10}\left[\frac{(\varepsilon/D)^{1.1098}}{2.8257} + \frac{5.8506}{Re_D^{0.8981}}\right]\right)\right]^{-2}$$

Source: Chen (1979)

```
Option Explicit
'Chen Equation function from Appendix B.1
'Source: Chen (1979) Cited in textbook Chapter 2
Dim e_D As Double
Dim Re_D As Double

Function ChenEq(e_D, Re_D)
Dim e1 As Double
Dim R1 As Double
Dim e2 As Double
Dim R2 As Double
Dim lgarg2 As Double
Dim lgarg1 As Double
Dim rslt As Double
e1 = e_D / 3.7065
R1 = 5.0452 / Re_D
e2 = (e_D ^ 1.1098) / 2.8257
R2 = 5.8506 / (Re_D ^ 0.8981)
lgarg2 = e2 + R2
lgarg1 = e1 - (R1 * Log(lgarg2) / Log(10))
rslt = (-2 * (Log(lgarg1) / Log(10))) ^ (-2)
ChenEq = rslt
End Function
```

B.2 Equilibrium Constant and Water Gas Shift Reaction (Chapter 4)

The VBA script below creates a user-defined Function named `watgas`, which has one argument `T`. The polynomial coefficients were determined by the author by performing a least-squares fit to the raw data from the JANAF tables (Chase 1998). The values returned by `watgas` are dimensionless, but the gaseous equilibrium equation it represents must use units of ***absolute pressure***. The units of **T** are *K*. See Equation 4.20.

$$K_p(T) = \exp\left(0.399\,437z^0 + 4.113\,18z^1 + 0.407\,999z^2 - 0.057\,381z^3\right)$$

$$z = \left(\frac{1000}{T(K)} - 1\right)$$

A second VBA script below creates a user-defined function named `richquadratic_a`, which has two arguments, `T_eq` and `Eqratio`. The coefficients for the quadratic equation are derived in Section 4.12 and are presented below. The units of `richquadratic_a` are $a \triangleq mol_{CO_2(watgas)}$, the units of `T_eq` are $T_{eq} \triangleq K$ and the units of `Eqratio` are:

$$\phi \triangleq \frac{\left(kmol_f/kmol_{O_2}\right)_{actual}}{\left(kmol_f/kmol_{O_2}\right)_{stoichiometric}}$$

The function returns the value computed by the quadratic formula for variable a, which is the solution to the following quadratic equation:

$$\underbrace{(1-K_p)}_{a_{Qu}}a^2 + \underbrace{\left[K_p\phi + (\phi-4)(1-K_p) + 2\phi\right]}_{b_{Qu}}a + \underbrace{K_p\phi(\phi-4)}_{c_{Qu}} = 0$$

The quadratic formula solved by VBA is Equation 4.21:

$$a = \frac{-b_{Qu} \pm \sqrt{b_{Qu}^2 - 4a_{Qu}c_{Qu}}}{2a_{Qu}}$$

The other three molar concentrations, $b \triangleq mol_{CO(watgas)}$, $c \triangleq mol_{H_2O(watgas)}$, and $d \triangleq mol_{H_2(watgas)}$, can be found manually (i.e. without using any VBA functions) by solving the following equations (again, from Section 4.12):

$$a + b = \phi \qquad 2a + b + c = 4 \qquad c + d = 2\phi$$

```
Option Explicit
'Water gas shift equilibrium function from Appendix B.2

'Thermodynamic data and formulas courtesy of the
'National Institute of Standards and Technology
'U.S. Department of Commerce. Not copyrightable in the United States.
'Source: K,p from Chase (1998)
Dim C_0(3) As Double
Dim a_ As Double
Dim a_Qu As Double
Dim b_Qu As Double
Dim c_Qu As Double
Dim K_p As Double
Dim Phi As Double
Dim Pz As Double
Dim T As Double
```

```
Dim T_eq As Double
Dim z As Double
Dim i As Integer

Function watgas(T)
C_0(0) = 0.399437
C_0(1) = 4.11318
C_0(2) = 0.407999
C_0(3) = -0.057381
Pz = 0
If T = 1000 Then T = 999.999
z = (1000 / T) - 1
For i = 0 To 3
Pz = Pz + C_0(i) * z ^ i
Next i
watgas = Exp(Pz)
End Function


Function Richquadratic_a(T_eq, Eqratio)
'From Section 4.12, find molar coefficient for CO2 using
'  Quadratic equation and Water-Gas-Shift equilibrium const
If (T_eq < 1100#) Or (T_eq > 2500#) Or (Eqratio <= 1#) Or (Eqratio > 2.5) Then
  Exit Function
End If
K_p = watgas(T_eq)
a_Qu = 1 - K_p
b_Qu = K_p * Eqratio + (Eqratio - 4#) * (1# - K_p) + 2# * Eqratio
c_Qu = K_p * Eqratio * (Eqratio - 4)

a_ = ((-b_Qu) + Sqr(b_Qu ^ 2 - 4 * a_Qu * c_Qu)) / (2 * a_Qu)
Richquadratic_a = a_
End Function
```

B.3 Enthalpy of Combustion Gases (Chapter 4)

The VBA scripts presented below create customized Excel Functions for enthalpy as a function of temperature for seven combustion gases – CO_2, CO, H_2O, H_2, CH_4, O_2, and N_2 – and *air*.

Caution

When solving combustion problems, the enthalpy values for *air* must not be used for combustion products because O_2 will be depleted from the products. Similarly, CH_4 should be used only as a combustion reactant, because it is not a product of either of the combustion chemistry formulations (i.e. ***fuel lean*** = complete combustion and ***fuel rich*** = water gas shift equilibrium) described in this textbook. The properties of *air* can be used to solve problems where its temperature changes only due to work (e.g. a compressor or turbine) or heat (e.g. noncombustion heat transfer or "pseudo-combustion" where the error introduced by assuming combustion products are 100 % *air* can be ignored), rather than chemistry. The sensible enthalpy of CH_4 may be important for problems where a fuel + air reactant mixture is preheated before combustion.

The enthalpy values are computed from the Shomate equations, with unique coefficients for each gas. The form of the Shomate equation is:

$$\left(\hat{h}^0 - \hat{h}^0_{298}\right) = \frac{AT^1}{1} + \frac{BT^2}{2} + \frac{CT^3}{3} + \frac{DT^4}{4} - ET^{-1} + FT^0 - H$$

where the left side is the enthalpy rise above the datum enthalpy at $T = 298.15\ K$ and has units of $(\doteq kJ/kmol)$ and the coefficients have appropriate units for consistency. The superscript zero on the enthalpy symbols indicates the enthalpy values are estimated for low density $(\rho \to 0)$ gases, which constitutes ideal gas behavior. For this reason, the Shomate equations used in this appendix should *not* be used to compute enthalpy for gases at high pressure or very close to the vapor dome. Hence, these VBA scripts are only recommended for combustion reactions at ambient temperature and higher. For computations involving the thermodynamics of high-pressure combustion, Kuo (2005) provides equations of state for gases and liquids above their critical points.

The source of data for the equations in this appendix is Linstrom and Mallard (2017–2021). When created, these custom Excel functions can be used to compute sensible enthalpies (see Section 4.7) and thereby determine adiabatic combustion temperatures for fuel-lean combustion (see Table 4.5) and in fuel-rich combustion (see Section 4.12).

The custom Excel functions are named in the form of `NISTH298_GAS(T)` where "`GAS`" is replaced by a two- or three-letter chemical formula for the compound whose sensible enthalpy is sought (e.g. `CO2, CO, H2O, H2, CH4, O2, N2, AIR`). The units of sensible enthalpy $\hat{h} - \hat{h}^0_{298}$ are $(\doteq kJ/kmol_{gas})$. The argument **T** is absolute temperature, $T(\doteq K)$.

The Shomate coefficients for seven of the gases are given in Table B.1. Enthalpy values for `AIR` are computed within its VBA function from the enthalpy values of O_2 and N_2 at their predefined molar ratio of 1 : 3.76 (see Chapter 4).

As shown in Table B.1, different coefficients are required for different temperature domains, and the VBA scripts account for those transitions.

One additional custom Excel function is included with this module that performs a pseudo reverse Shomate calculation. The function is `RevShomate_AIR(h-h298)` and it basically swaps output and argument values relative to the `NISTH298_AIR(T)` function. The correlation is not exact, because the reverse function was generated as a sixth-order polynomial least squares regression based on a set of approximately 75 temperature–enthalpy pairs generated by the forward function. Errors are less than 0.2 K for the range of temperatures $275 \leq T \leq 2100\ K$. This function can be used to explicitly determine the temperature of a blend of multiple air flows whose enthalpies are all known. This approach eliminates iteration but with the penalty of a small loss of accuracy.

It is recommended that all procedures in this appendix be kept together and pasted into one module in the Visual Basic Environment. Dividing them into separate modules may cause problems related to separation of the function's instructions from the data typing instructions (**Dim** statements) at the very beginning of the following script, which are only given once, even though some of the variables are used in multiple functions.

Table B.1 Shomate coefficients for six combustion gases at low density {Ideal Gas}. Different coefficients apply for different temperature domains, as indicated.

Gas	Min T (K)	Max T (K)	A	B	C	D	E	$F-H$
CO_2	298	1200	24.99 735	55.186 96	−33.691 37	7.948 387	−0.136 638	(−403.607 5) − (−393.522 4)
CO_2	1200	6000	58.166 39	2.720 074	−0.492 289	0.038 844	−6.447 293	(−425.918 6) − (−393.522 4)
CO	298	1300	25.567 59	6.096 13	4.054 656	−2.671 301	0.131 021	(−118.008 9) − (−110.527 1)
CO	1300	6000	35.1507	1.300 095	−0.205 921	0.013 55	−3.282 78	(−127.837 5) − (−110.527 1)
H_2O	298	1700	30.092	6.832 514	6.793 435	−2.534 48	0.082 139	(−250.881) − (−241.826 4)
H_2O	1700	6000	41.964 26	8.622 053	−1.499 78	0.098 119	−11.157 64	(−272.179 7) − (−241.826 4)
H_2	298	1000	33.066 178	−11.363 417	11.432 816	−2.772 874	−0.158 558	(−9.980 797) − (0)
H_2	1000	2500	18.563 083	12.257 357	−2.859 786	0.268 238	1.977 99	(−1.147 438) − (0)
CH_4	298	1300	−0.703 029	108.477 3	−42.521 57	5.862 788	0.678 565	(−76.843 76) − (−74.873 10)
CH_4	1300	6000	85.812 17	11.264 67	−2.114 146	0.138 190	−26.422 21	(−153.532 7) − (−74.873 10)
O_2	100	700	31.322 34	−20.235 31	57.866 44	−36.506 24	−0.007 374	(−8.903 471) − (0)
O_2	700	2000	30.032 35	8.772 972	−3.988 133	0.788 313	−0.741 599	(−11.324 68) − (0)
O_2	2000	6000	20.9111 1	10.720 71	−2.020 498	0.146 449	9.245 722	(5.337 651) − (0)
N_2	100	500	28.986 41	1.853 978	−9.647 459	16.635 37	0.000 117	(−8.671 914) − (0)
N_2	500	2000	19.505 83	19.887 05	−8.598 535	1.369 784	0.527 601	(−4.935 202) − (0)
N_2	2000	6000	35.518 72	1.128 728	−0.196 103	0.014 662	−4.553 76	(−18.970 91) − (0)

Source: Linstrom and Mallard (2017–2021).

```
Option Explicit
'Ideal Gas Enthalpy group of functions from Appendix B.3

'Underlying equations courtesy of the National Institute of
'Standards and Technology, U.S. Department of Commerce.
'Not copyrightable in the United States.
'Source: NIST Chemistry Webbook (Shomate Equations)
Dim T As Double
Dim t_1000 As Double
Dim Shomate As Double
Dim Sh_CO2(1, 5) As Double
Dim Sh_CO(1, 5) As Double
Dim Sh_H2O(1, 5) As Double
Dim Sh_H2(1, 5) As Double
Dim Sh_O2(2, 5) As Double
Dim Sh_N2(2, 5) As Double
Dim Sh_CH4(1,5) As Double
Dim Rangelo(2) As Double
Dim Rangehi(2) As Double
Dim i As Integer
Dim j As Integer

Function NISTH298_CO2(T)
'Underlying equations courtesy of the National Institute of
'Standards and Technology, U.S. Department of Commerce.
'Not copyrightable in the United States.
'Source: NIST Chemistry Webbook (Shomate Equations)
'Ideal Gas applications only (e.g., 1 atm combustion)
Rangehi(0) = 1200
Rangehi(1) = 6000
Rangelo(0) = 298
Rangelo(1) = Rangehi(0)
Sh_CO2(0, 0) = (-403.6075) - (-393.5224)
Sh_CO2(0, 1) = 24.99735
Sh_CO2(0, 2) = 55.18696
Sh_CO2(0, 3) = -33.69137
Sh_CO2(0, 4) = 7.948387
Sh_CO2(0, 5) = -0.136638
Sh_CO2(1, 0) = (-425.9186) - (-393.5224)
Sh_CO2(1, 1) = 58.16639
Sh_CO2(1, 2) = 2.720074
Sh_CO2(1, 3) = -0.492289
Sh_CO2(1, 4) = 0.038844
Sh_CO2(1, 5) = -6.447293
Shomate = 0
t_1000 = T / 1000
```

```
If T < Rangelo(0) Then
  i = 2
ElseIf T <= Rangehi(0) Then
  i = 0
ElseIf T <= Rangehi(1) Then
  i = 1
Else
  i = 2
End If

For j = 1 To 4
  Shomate = Shomate + Sh_CO2(i, j) * t_1000 ^ j / j
Next j

Shomate = Shomate - Sh_CO2(i, 5) / t_1000 + Sh_CO2(i, 0)

If i = 2 Then
  Exit Function
Else
  NISTH298_CO2 = Shomate * 1000
'Desired units are kJ/kmol not kJ/mol (as NIST uses)
End If
End Function

Function NISTH298_CO(T)
'Underlying equations courtesy of the National Institute of
'Standards and Technology, U.S. Department of Commerce.
'Not copyrightable in the United States.
'Source: NIST Chemistry Webbook (Shomate Equations)
'Ideal Gas applications only (e.g., 1 atm combustion)
Rangehi(0) = 1300
Rangehi(1) = 6000
Rangelo(0) = 298
Rangelo(1) = Rangehi(0)
Sh_CO(0, 0) = (-118.0089) - (-110.5271)
Sh_CO(0, 1) = 25.56759
Sh_CO(0, 2) = 6.09613
Sh_CO(0, 3) = 4.054656
Sh_CO(0, 4) = -2.671301
Sh_CO(0, 5) = 0.131021
Sh_CO(1, 0) = (-127.8375) - (-110.5271)
Sh_CO(1, 1) = 35.1507
Sh_CO(1, 2) = 1.300095
Sh_CO(1, 3) = -0.205921
Sh_CO(1, 4) = 0.01355
Sh_CO(1, 5) = -3.28278
Shomate = 0
```

```
t_1000 = T / 1000

If T < Rangelo(0) Then
  i = 2
ElseIf T <= Rangehi(0) Then
  i = 0
ElseIf T <= Rangehi(1) Then
  i = 1
Else
  i = 2
End If

For j = 1 To 4
  Shomate = Shomate + Sh_CO(i, j) * t_1000 ^ j / j
Next j

Shomate = Shomate - Sh_CO(i, 5) / t_1000 + Sh_CO(i, 0)

If i = 2 Then
  Exit Function
Else
  NISTH298_CO = Shomate * 1000
'Desired units are kJ/kmol not kJ/mol (as NIST uses)
End If
End Function

Function NISTH298_H2OV(T)
'Underlying equations courtesy of the National Institute of
'Standards and Technology, U.S. Department of Commerce.
'Not copyrightable in the United States.
'Source: NIST Chemistry Webbook (Shomate Equations)
'Ideal Gas applications only (e.g., 1 atm combustion)
Rangehi(0) = 1700
Rangehi(1) = 6000
Rangelo(0) = 298
'NIST claims Rangelo(0) should be 500K
Rangelo(1) = Rangehi(0)
Sh_H2O(0, 0) = (-250.881) - (-241.8264)
Sh_H2O(0, 1) = 30.092
Sh_H2O(0, 2) = 6.832514
Sh_H2O(0, 3) = 6.793435
Sh_H2O(0, 4) = -2.53448
Sh_H2O(0, 5) = 0.082139
Sh_H2O(1, 0) = (-272.1797) - (-241.8264)
Sh_H2O(1, 1) = 41.96426
Sh_H2O(1, 2) = 8.622053
Sh_H2O(1, 3) = -1.49978
```

```
Sh_H2O(1, 4) = 0.098119
Sh_H2O(1, 5) = -11.15764
Shomate = 0
t_1000 = T / 1000

If T < Rangelo(0) Then
  i = 2
ElseIf T <= Rangehi(0) Then
  i = 0
ElseIf T <= Rangehi(1) Then
  i = 1
Else
  i = 2
End If

For j = 1 To 4
  Shomate = Shomate + Sh_H2O(i, j) * t_1000 ^ j / j
Next j

Shomate = Shomate - Sh_H2O(i, 5) / t_1000 + Sh_H2O(i, 0)

If i = 2 Then
  Exit Function
Else
  NISTH298_H2OV = Shomate * 1000
'Desired units are kJ/kmol not kJ/mol (as NIST uses)
End If
End Function

Function NISTH298_H2(T)
'Underlying equations courtesy of the National Institute of
'Standards and Technology, U.S. Department of Commerce.
'Not copyrightable in the United States.
'Source: NIST Chemistry Webbook (Shomate Equations)
'Ideal Gas applications only (e.g., 1 atm combustion)
Rangehi(0) = 1000
Rangehi(1) = 2500
Rangelo(0) = 298
Rangelo(1) = Rangehi(0)
Sh_H2(0, 0) = (-9.980797) - (0)
Sh_H2(0, 1) = 33.066178
Sh_H2(0, 2) = -11.363417
Sh_H2(0, 3) = 11.432816
Sh_H2(0, 4) = -2.772874
Sh_H2(0, 5) = -0.158558
Sh_H2(1, 0) = (-1.147438) - (0)
Sh_H2(1, 1) = 18.563083
```

```
Sh_H2(1, 2) = 12.257357
Sh_H2(1, 3) = -2.859786
Sh_H2(1, 4) = 0.268238
Sh_H2(1, 5) = 1.97799
Shomate = 0
t_1000 = T / 1000

If T < Rangelo(0) Then
  i = 2
ElseIf T <= Rangehi(0) Then
  i = 0
ElseIf T <= Rangehi(1) Then
  i = 1
Else
  i = 2
End If

For j = 1 To 4
  Shomate = Shomate + Sh_H2(i, j) * t_1000 ^ j / j
Next j

Shomate = Shomate - Sh_H2(i, 5) / t_1000 + Sh_H2(i, 0)

If i = 2 Then
  Exit Function
Else
  NISTH298_H2 = Shomate * 1000
'Desired units are kJ/kmol not kJ/mol (as NIST uses)
End If
End Function

Function NISTH298_O2(T)
'Underlying equations courtesy of the National Institute of
'Standards and Technology, U.S. Department of Commerce.
'Not copyrightable in the United States.
'Source: NIST Chemistry Webbook (Shomate Equations)
'Ideal Gas applications only (e.g., 1 atm combustion)
Rangehi(0) = 700
Rangehi(1) = 2000
Rangehi(2) = 6000
Rangelo(0) = 100
Rangelo(1) = Rangehi(0)
Rangelo(2) = Rangehi(1)
Sh_O2(0, 0) = (-8.903471) - (0)
Sh_O2(0, 1) = 31.32234
Sh_O2(0, 2) = -20.23531
Sh_O2(0, 3) = 57.86644
```

```
Sh_O2(0, 4) = -36.50624
Sh_O2(0, 5) = -0.007374
Sh_O2(1, 0) = (-11.32468) - (0)
Sh_O2(1, 1) = 30.03235
Sh_O2(1, 2) = 8.772972
Sh_O2(1, 3) = -3.988133
Sh_O2(1, 4) = 0.788313
Sh_O2(1, 5) = -0.741599
Sh_O2(2, 0) = (5.337651) - (0)
Sh_O2(2, 1) = 20.91111
Sh_O2(2, 2) = 10.72071
Sh_O2(2, 3) = -2.020498
Sh_O2(2, 4) = 0.146449
Sh_O2(2, 5) = 9.245722
Shomate = 0
t_1000 = T / 1000

If T < Rangelo(0) Then
  i = 3
ElseIf T <= Rangehi(0) Then
  i = 0
ElseIf T <= Rangehi(1) Then
  i = 1
ElseIf T <= Rangehi(2) Then
  i = 2
Else
  i = 3
End If

For j = 1 To 4
  Shomate = Shomate + Sh_O2(i, j) * t_1000 ^ j / j
Next j

Shomate = Shomate - Sh_O2(i, 5) / t_1000 + Sh_O2(i, 0)

If i = 3 Then
  Exit Function
Else
  NISTH298_O2 = Shomate * 1000
'Desired units are kJ/kmol not kJ/mol (as NIST uses)
End If
End Function

Function NISTH298_N2(T)
'Underlying equations courtesy of the National Institute of
'Standards and Technology, U.S. Department of Commerce.
'Not copyrightable in the United States.
```

```
'Source: NIST Chemistry Webbook (Shomate Equations)
'Ideal Gas applications only (e.g., 1 atm combustion)
Rangehi(0) = 500
Rangehi(1) = 2000
Rangehi(2) = 6000
Rangelo(0) = 100
Rangelo(1) = Rangehi(0)
Rangelo(2) = Rangehi(1)
Sh_N2(0, 0) = (-8.671914) - (0)
Sh_N2(0, 1) = 28.98641
Sh_N2(0, 2) = 1.853978
Sh_N2(0, 3) = -9.647459
Sh_N2(0, 4) = 16.63537
Sh_N2(0, 5) = 0.000117
Sh_N2(1, 0) = (-4.935202) - (0)
Sh_N2(1, 1) = 19.50583
Sh_N2(1, 2) = 19.88705
Sh_N2(1, 3) = -8.598535
Sh_N2(1, 4) = 1.369784
Sh_N2(1, 5) = 0.527601
Sh_N2(2, 0) = (-18.97091) - (0)
Sh_N2(2, 1) = 35.51872
Sh_N2(2, 2) = 1.128728
Sh_N2(2, 3) = -0.196103
Sh_N2(2, 4) = 0.014662
Sh_N2(2, 5) = -4.55376
Shomate = 0
t_1000 = T / 1000

If T < Rangelo(0) Then
  i = 3
ElseIf T <= Rangehi(0) Then
  i = 0
ElseIf T <= Rangehi(1) Then
  i = 1
ElseIf T <= Rangehi(2) Then
  i = 2
Else
  i = 3
End If

For j = 1 To 4
  Shomate = Shomate + Sh_N2(i, j) * t_1000 ^ j / j
Next j

Shomate = Shomate - Sh_N2(i, 5) / t_1000 + Sh_N2(i, 0)
```

```
If i = 3 Then
  Exit Function
Else
  NISTH298_N2 = Shomate * 1000
'Desired units are kJ/kmol not kJ/mol (as NIST uses)
End If
End Function

Function NISTH298_AIR(T)
'Ideal Gas applications only (e.g., 1 atm combustion)

Dim H_O2 As Double
Dim H_N2 As Double
Dim Mol_O2 As Double
Dim Mol_N2 As Double

Mol_O2 = 1#
Mol_N2 = 3.76

H_O2 = NISTH298_O2(T)
H_N2 = NISTH298_N2(T)
NISTH298_AIR = (Mol_O2 * H_O2 + Mol_N2 * H_N2) / (Mol_O2 + Mol_N2)

End Function

Function NISTH298_CH4(T)
'Underlying equations courtesy of the National Institute of
'Standards and Technology, U.S. Department of Commerce.
'Not copyrightable in the United States.
'Source: NIST Chemistry Webbook (Shomate Equations)
'Ideal Gas applicaitons only (e.g., 1 atm combustion)
'Rangehi per NIST is excluded because sensible enthalpy
'  for CH4 should only be used for preheated reactants

Rangehi(0) = 1300
'Rangehi(1) = 6000
Rangelo(0) = 298
'Rangelo(1) = Rangehi(0)
Sh_CH4(0, 0) = (-76.84376) - (-74.8731)
Sh_CH4(0, 1) = -0.703029
Sh_CH4(0, 2) = 108.4773
Sh_CH4(0, 3) = -42.52157
Sh_CH4(0, 4) = 5.862788
Sh_CH4(0, 5) = 0.678565
'Sh_CH4(1, 0) = (-153.5327) - (-74.8731)
'Sh_CH4(1, 1) = 85.81217
'Sh_CH4(1, 2) = 11.26467
```

```
'Sh_CH4(1, 3) = -2.114146
'Sh_CH4(1, 4) = 0.138190
'Sh_CH4(1, 5) = -26.42221

Shomate = 0
t_1000 = T / 1000

If T < Rangelo(0) Then
  i = 2
ElseIf T <= Rangehi(0) Then
  i = 0
'ElseIf T <= Rangehi(1) Then
'  i = 1
Else
  i = 2
End If

For j = 1 To 4
  Shomate = Shomate + Sh_CH4(i, j) * t_1000 ^ j / j
Next j

Shomate = Shomate - Sh_CH4(i, 5) / t_1000 + Sh_CH4(i, 0)

If i = 2 Then
  Exit Function
Else
  NISTH298_CH4 = Shomate * 1000
'Desired units are kJ/kmol not kJ/mol (as NIST uses)
End If
End Function

Function RevShomate_AIR(hdum)
'By R.Martin, April 2021 6th order polynomial
'  curve fit to reverse Shomate Equation for air
'  where air is 1 mol O2 + 3.76 moles N2
'  argument is sensible enthalpy (h-h298) kJ/kmol
'  output is temperature K

Revsum = 0
RevSh_air(0) = 298.11: 'adjusted from 298.077
RevSh_air(1) = 0.03443632
RevSh_air(2) = -0.00000006105801
RevSh_air(3) = -4.047586E-12
RevSh_air(4) = 1.482415E-16
RevSh_air(5) = -2.03114E-21
RevSh_air(6) = 1.021676E-26
For k = 0 To 6
```

```
Revsum = Revsum + RevSh_air(k) * hdum ^ k
Next k
RevShomate_AIR = Revsum

End Function
```

B.4 VBA Functions for Water [Liquid + Vapor] (Chapter 6)

The VBA script presented below creates customized Excel functions for computing pressure ($p \doteq kPa$), specific enthalpy ($h \doteq kJ/kg$), specific entropy ($s \doteq kJ/kg \cdot K$), and quality ($X \doteq kg_{H_2O,vap}/kg_{H_2O,liq + vap}$) of water in single-phase or two-phase (vapor + liquid) embodiments as functions of temperature ($T \doteq K$) and density (i.e. reciprocal specific volume) ($v^{-1} = \rho \doteq kg/m^3$). The parameter Z is the dimensionless compressibility factor, which quantifies the embodiment's departure from ideal gas behavior. The parameter τ comprises a scaled variable representing reciprocal temperature. The parameter $Q(T, \rho)$ has units of reciprocal density but has no physical meaning. Q provides the mathematical framework for the curve-fit relationship of compressibility Z to temperature T and density ρ.

The parameter, ψ (i.e. Helmholtz free energy, which was associated with variable a in Chapter 6), while not directly necessary to compute pressure, enthalpy, or entropy is provided here to illustrate the interrelationship of all the thermodynamic parameters to each other. By starting with a universal equation for ψ (written in terms of two thermodynamic state variables T and ρ), the other three thermodynamic potentials (u, h, g) and the other two thermodynamic state variables (p, s) can be computed using Maxwell's relations (and some meticulous calculus and algebra). The following equations present the multiple-summation equations used in Keenan and adopted here.

$$p = Z\rho RT$$

$$h = RT\left[\rho\tau\left(\frac{\partial Q}{\partial \tau}\right)_{\rho} + Z\right] + \frac{d}{d\tau}(\psi_0\tau)$$

$$\tau = \frac{1000}{T}$$

$$Z = \left[1 + \rho Q + \rho^2\left(\frac{\partial Q}{\partial \rho}\right)_{\tau}\right]$$

$$Q(\rho,\tau) = (\tau - \tau_c)\sum_{j=1}^{7}(\tau - \tau_{aj})^{(j-2)} \cdot \left[\sum_{i=1}^{8} A_{ij}\left(\rho - \rho_{aj}\right)^{(i-1)} + e^{-E\rho}\sum_{i=9}^{10} A_{ij}\rho^{(i-9)}\right]$$

$$\left(\frac{\partial Q}{\partial \rho}\right)_{\tau} = (\tau - \tau_c)\sum_{j=1}^{7}(\tau - \tau_{aj})^{(j-2)}\left[\sum_{i=1}^{8} A_{ij}(i-1)\left(\rho - \rho_{aj}\right)^{(i-2)}\right.$$

$$\left. + (-E)e^{-E\rho}\sum_{i=9}^{10} A_{ij}\rho^{(i-9)} + e^{-E\rho}(i-9)\sum_{i=9}^{10} A_{ij}\rho^{(i-10)}\right]$$

$$\left(\frac{\partial Q}{\partial \tau}\right)_{\rho} = (\tau - \tau_c)\sum_{j=1}^{7}(j-2)(\tau - \tau_{aj})^{(j-3)}\left[\sum_{i=1}^{8} A_{ij}\left(\rho - \rho_{aj}\right)^{(i-1)} + e^{-E\rho}\sum_{i=9}^{10} A_{ij}\rho^{(i-9)}\right]$$

$$+ \sum_{j=1}^{7}(\tau - \tau_{aj})^{(j-2)} \cdot \left[\sum_{i=1}^{8} A_{ij}\left(\rho - \rho_{aj}\right)^{(i-1)} + e^{-E\rho}\sum_{i=9}^{10} A_{ij}\rho^{(i-9)}\right]$$

$$\psi = \psi_0(\tau) + RT[\ln \rho + \rho Q(\rho, \tau)]$$

$$\psi_0 = \sum_{k=1}^{6} C_k \tau^{-(k-1)} + C_7 \ln\left(\frac{1000}{\tau}\right) + C_8 \frac{\ln(1000/\tau)}{\tau}$$

$$\frac{d\psi_0}{d\tau} = \sum_{k=1}^{6} (-(k-1))C_k \tau^{(-(k-1)-1)} + (-1)C_7 \tau^{(-1)} + C_8 \tau^{(-2)}[\ln(\tau) - \ln(1000) - 1]$$

$$\frac{d(\psi_0 \tau)}{d\tau} = \tau \frac{d\psi_0}{d\tau} + \psi_0$$

$$s = -R\left[\ln\rho + \rho Q - \rho\tau\left(\frac{\partial Q}{\partial \tau}\right)_\rho\right] - \frac{d\psi_0}{dT}$$

$$\frac{d\psi_0}{dT} = \sum_{k=1}^{6} C_k(k-1)\frac{T^{(k-1)}}{1000^{(k-1)}} + C_7 T^{(-1)} + \frac{C_8}{1000}(\ln(T) + 1)$$

The source of data for these equations is Keenan et al. (1969) with modifications by Reynolds (1979). The thermodynamic property customized functions are named `H2O_p`, `H2O_h`, `H2O_s`, and `H2O_X` and each has T and ρ as its two arguments. Quality X is determined only when $\rho_{satvap} < \rho < \rho_{satliq}$ and X is equal to the mass fraction of water in the vapor phase of the two-phase mixture. The Keenan expressions for p, h, and s are not always reliable under the vapor dome, but the VBA scripts given here correct for those errors. The resolution for p is simple $p = p_{sat}$ everywhere under the dome. The resolution provided here for h and s under the dome is to apply the quality law to Keenan's accurate values of h and s at the saturated liquid and vapor endpoints, respectively.

In addition to the four `H2O_?` functions named above, three intermediate Excel functions `psatH2O`, `rhosatH2Ovap`, and `rhosatH2Oliq` are created by the VBA script below. These three intermediate functions each have one argument `TsatK`. The units of the values returned by p_{sat} and the two ρ_{sat} functions are kPa and kg/m^3, respectively, while the units of T are K. Like p_{sat}, the value for $\rho_{sat,\,liq}$ is computed from a multi-coefficient summation shown below. In contrast, the returned value for $\rho_{sat,\,vap}$ is computed by Newton–Raphson iteration on a dummy variable ρ_{dum} until $p(T, \rho_{dum}) \approx p_{sat}(T)$ is approached to within a very close tolerance of the exact value.

The governing equation for `psatH2O` is given below:

$$\ln\left(\frac{p_{sat}(T)}{p_c}\right) = \left[\left(\frac{T_c}{T}\right) - 1\right] \cdot \sum_{i=1}^{8} F_i[a(T - T_p)]^{(i-1)}$$

$$p_c = 22\,089\ kPa$$

$$T_c = 647.286\ K$$

$$F_1 = -7.419\,242\,0$$

$$F_2 = 2.972\,100\,0 \times 10^{-1}$$

$$F_3 = -1.155\,286\,0 \times 10^{-1}$$

$$F_4 = 8.685\,635\,0 \times 10^{-3}$$

$$F_5 = 1.094\,098\,0 \times 10^{-3}$$

$$F_6 = -4.399\,930\,0 \times 10^{-3}$$

$$F_7 = 2.520\,658\,0 \times 10^{-3}$$

$$F_8 = -5.218\,684\,0 \times 10^{-4}$$

$a = 0.01$

$T_p = 338.15\ K$

Here is the governing equation for **rhosatH2Oliq**:

$$\rho_{satliq} = \rho_c\left[1 + \sum_{j=1}^{8} D_j\left(1 - \frac{T_{sat}}{T_c}\right)^{\frac{j}{3}}\right]$$

$\rho_c = 317.0\ kg/m^3$

$T_c = 647.286\ K$

$D_1 = 3.671\ 125\ 7$

$D_2 = -2.851\ 239\ 6 \times 10^1$

$D_3 = 2.226\ 524\ 0 \times 10^2$

$D_4 = -8.824\ 385\ 2 \times 10^2$

$D_5 = 2.000\ 276\ 5 \times 10^3$

$D_6 = -2.612\ 255\ 7 \times 10^3$

$D_7 = 1.829\ 767\ 4 \times 10^3$

$D_8 = -5.335\ 052\ 0 \times 10^2$

Students may find it less than ideal to utilize a set of custom functions that always rely on T and ρ as the arguments needed to compute the other thermodynamic parameters. The author has found that T and v are the most natural to use as the independent variables for classical thermodynamic theory and analysis, and most engineers prefer to specify T and p when computing the other thermodynamic parameters. This is presumably because they are much more readily measurable than v and s. The obvious fix for this minor inconvenience is that students must use their own iterative methods (manual or built-in) in the Excel environment to find the correct input values of ρ (and T) that correlate with the desired value of p or s.

Caution

Importantly, we note here that the Keenan equations provided below are unsuitable for estimating pressure (as a function of T, ρ) in subcooled liquids because such liquids are nearly incompressible with pressure (especially at temperatures well below the critical point). In the same vein, quality is meaningless in both the superheated vapor and the subcooled liquid domains. To warn users of this defect, the custom functions return a value of (-1) for pressure and quality in the subcooled liquid regime and they return a value of (-9) for quality in the superheated vapor regime. These negative integer flags also have mnemonic value due to their shapes' correspondence to letter designations for liquid and gas (i.e. 1~l for liquid and 9~g for gas).

Users are also cautioned to refrain from using these customized Excel functions for conditions above the critical density of water ρ_c.

The coefficient matrices (C_{ij}, τ_{aj}, ρ_{aj}, and C_k) are given in Tables B.2 and B.3.

Table B.2 Coefficient matrix Aw_{ij} for water properties (liquid and vapor).

Aw_{ij}	$j = 1$	$j = 2$	$j = 3$	$j = 4$	$j = 5$	$j = 6$	$j = 7$
$i = 1$	$2.949\,293\,7 \times 10^{-2}$	$-5.198\,586 \times 10^{-3}$	$6.833\,535\,4 \times 10^{-3}$	$-1.564\,104 \times 10^{-4}$	$-6.397\,240\,5 \times 10^{-3}$	$-3.966\,140\,1 \times 10^{-3}$	$-6.904\,855\,4 \times 10^{-4}$
$i = 2$	$-1.321\,391\,7 \times 10^{-4}$	$7.777\,918\,2 \times 10^{-6}$	$-2.614\,975\,1 \times 10^{-5}$	$-7.254\,610\,8 \times 10^{-7}$	$2.640\,928\,2 \times 10^{-5}$	$1.545\,306\,1 \times 10^{-5}$	$2.740\,741\,6 \times 10^{-6}$
$i = 3$	$2.746\,463\,2 \times 10^{-7}$	$-3.330\,190\,2 \times 10^{-8}$	$6.532\,639\,6 \times 10^{-8}$	$-9.273\,428\,9 \times 10^{-9}$	$-4.774\,037\,4 \times 10^{-8}$	$-2.914\,247 \times 10^{-8}$	$-5.102\,807 \times 10^{-9}$
$i = 4$	$-3.609\,382\,8 \times 10^{-10}$	$-1.625\,462\,2 \times 10^{-11}$	$-2.618\,197\,8 \times 10^{-11}$	$4.312\,584 \times 10^{-12}$	$5.632\,313 \times 10^{-11}$	$2.956\,879\,6 \times 10^{-11}$	$3.963\,608\,5 \times 10^{-12}$
$i = 5$	$3.421\,843\,1 \times 10^{-13}$	$-1.773\,107\,4 \times 10^{-13}$	0.0	0.0	0.0	0.0	0.0
$i = 6$	$-2.445\,004\,2 \times 10^{-16}$	$1.27\,487\,42 \times 10^{-16}$	0.0	0.0	0.0	0.0	0.0
$i = 7$	$1.551\,853\,5 \times 10^{-19}$	$1.374\,615\,3 \times 10^{-19}$	0.0	0.0	0.0	0.0	0.0
$i = 8$	$5.972\,848\,7 \times 10^{-24}$	$1.559\,783\,6 \times 10^{-22}$	0.0	0.0	0.0	0.0	0.0
$i = 9$	$-4.103\,084\,8 \times 10^{-1}$	$3.373\,118 \times 10^{-1}$	$-1.374\,661\,8 \times 10^{-1}$	$6.787\,498\,3 \times 10^{-3}$	$1.368\,731\,7 \times 10^{-1}$	$7.984\,797 \times 10^{-2}$	$1.304\,125\,3 \times 10^{-2}$
$i = 10$	$-4.160\,586 \times 10^{-4}$	$-2.098\,886\,6 \times 10^{-4}$	$-7.339\,684\,8 \times 10^{-4}$	$1.040\,171\,7 \times 10^{-5}$	$6.458\,188 \times 10^{-4}$	$3.991\,757 \times 10^{-4}$	$7.153\,135\,3 \times 10^{-5}$

Table B.3 Additional coefficients τ_{aj}, ρ_{aj}, and C_k for water properties (liquid and vapor). For coefficient C_2, dual entries are given because the source had inconsistent datum values for entropy.

	τ_{aj}	ρ_{aj}	C_k
$j, k = 1$	T_a/T_c	634.	$1.857\ 065 \times 10^3$
$j, k = 2$	2.5	1000.	$3.229\ 12 \times 10^3$ p, h $4.160\ 5 \times 10^1$ s
$j, k = 3$	2.5	1000.	$-4.194\ 65 \times 10^2$
$j, k = 4$	2.5	1000.	$3.666\ 49 \times 10^1$
$j, k = 5$	2.5	1000.	$-2.055\ 16 \times 10^1$
$j, k = 6$	2.5	1000.	$4.852\ 33 \times 10^0$
$j, k = 7$	2.5	1000.	4.60×10^1
$k = 8$	n/a	n/a	$-1.011\ 249 \times 10^3$

...where, $T_a = 1000\ K$, and the critical point values for water are $T_c = 647.286\ K$, $\rho_c = 317.0\ kg/m^3$, and $p_c = 22\ 089\ kPa$. Two additional parameters needed for the aforementioned equations are:

$$R = \frac{\hat{R}}{\hat{M}_{H_2O}} = \frac{8.314\ 46}{18.014\ 71} = 0.461\ 537\ \frac{kJ}{kg \cdot K}$$

$$E = 4.8 \times 10^{-3}$$

It is recommended that all procedures in this appendix be kept together and pasted into a single module in the Visual Basic Environment. Dividing them into separate modules is likely to cause problems related to separation of the function's instructions from the data typing instructions (**Dim** statements) at the beginning of the script below, which are only given once, even though many of the variables are used in multiple functions.

```
Option Explicit
'Water Liquid+Vapor group of functions from Appendix B.4


Dim Aw(10, 7) As Double
Dim D_H2Oliq(8) As Double
Dim rho_aj(7) As Double
Dim Tau_aj(7) As Double
Dim Psi_C(8) As Double
Dim a_keenan As Double
Dim e_ As Double
Dim F_keenan(8) As Double
Dim p_c As Double
Dim Psidatum As Double
Dim Psizero As Double
Dim Psizero_prime_dT As Double
Dim Psizero_prime_dtau As Double
Dim Psizerotau_prime_dtau As Double
```

```
Dim Q_ As Double
Dim Q_prime_dtau As Double
Dim Q_prime_drho As Double
Dim R_H2O As Double
Dim rho_ As Double
Dim rho_c As Double
Dim s_ As Double
Dim Sumall As Double
Dim Sumall2 As Double
Dim sum17 As Double
Dim Sum18 As Double
Dim Sum910 As Double
Dim Sum110 As Double
Dim Sum17prime_drho As Double
Dim Sum17prime_dtau As Double
Dim Sum18prime_drho As Double
Dim Sum910prime_drho As Double
Dim Sum110prime_drho As Double
Dim T_c As Double
Dim T_a As Double
Dim Tp_keenan As Double
Dim T_0 As Double
Dim Tau_ As Double
Dim Tau_c As Double
Dim T_ As Double
Dim Zc_ As Double

Dim pdum As Double
Dim rhodum As Double
Dim dp As Double
Dim dpdrho As Double
Dim drhonxt As Double
Dim tol_dp As Double

Dim X_case As Double
Dim p_case As Double
Dim v_case As Double
Dim h_case As Double
Dim s_case As Double

Dim domain As Integer
Dim i As Integer
Dim j As Integer
Dim k As Integer
Dim n As Integer
Dim flag_s As Boolean
```

```
'VBA procedures below are from TPSI p.154 Water and
'   Equations Q-2, S-6, D-5, G-6 on p. 122-126

Function psat_H2O(T_)
'Equation of state for saturated water pressure as a function of T
'  Eq. S-6 from TPSI (orig Keenan)
F_keenan(1) = -7.419242:      F_keenan(2) = 0.29721
F_keenan(3) = -0.1155286:     F_keenan(4) = 0.008685635
F_keenan(5) = 0.001094098:    F_keenan(6) = -0.00439993
F_keenan(7) = 0.002520658:    F_keenan(8) = -0.0005218684
p_c = 22089
T_c = 647.286
rho_c = 317#
a_keenan = 0.01
Tp_keenan = 338.15
Sumall = 0

For i = 1 To 8
  Sumall = Sumall + F_keenan(i) * (a_keenan * (T_ - Tp_keenan)) ^ (i - 1)
Next i
psat_H2O = p_c * Exp(((T_c / T_) - 1) * Sumall)
End Function

Function rhosat_H2Oliq(T_)
'Equation of state for saturated density as a function of T from TPSI
'WARNING: Does not apply to nonzero quality (under the dome) or
'   supercritical conditions
'   Eq. D-5 from TPSI (orig Keenan)

D_H2Oliq(1) = 3.6711257:  D_H2Oliq(2) = -28.512396
D_H2Oliq(3) = 222.6524:   D_H2Oliq(4) = -882.43852
D_H2Oliq(5) = 2000.2765:  D_H2Oliq(6) = -2612.2557
D_H2Oliq(7) = 1829.7674:  D_H2Oliq(8) = -533.5052

Sumall2 = 0
For j = 1 To 8
  Sumall2 = Sumall2 + D_H2Oliq(j) * (1 - T_ / T_c) ^ (j / 3)
Next j
rhosat_H2Oliq = rho_c * (1 + Sumall2)
End Function

'WARNING: This function gives obviously erroneous results for pressure
'   under and to the left of the vapor dome. These errors are avoided by using
'   supervising functions H2O_? below.
Function DonotuseH2O_p(T_, rho_)
'p, rho, T from Equation 11 in Keenan (also Equation Q-2 in TPSI)
'Custom functions that begin "Donotuse" are necessary, and they are visible
```

```
' in Excel but are not meant to be used independently, i.e., outside the
' structure of this family of functions

e_ = 0.0048
p_c = 22089#
rho_c = 317#
R_H2O = 0.461537266
T_0 = 273.16
T_a = 1000#
T_c = 647.286
If T_ = 400 Then T_ = 400.000001:  'Avoids Zero to the Zeroth Power in Sum17
flag_s = False:  'flag_s is only true for entropy calculation
Call MultiSums(T_, rho_, flag_s, Psizero_prime_dT, Psizerotau_prime_dtau, Q_, _
   Q_prime_dtau, Zc_)
DonotuseH2O_p = rho_ * R_H2O * T_ * Zc_
End Function

'WARNING: This function gives erroneous values for enthalpy under the vapor dome.
Function DonotuseH2O_h(T_, rho_)
'h from Equation 14 in Keenan (using coefficients A,ij from TPSI)

e_ = 0.0048
p_c = 22089#
R_H2O = 0.461537266
rho_c = 317#
T_a = 1000#
T_c = 647.286
T_0 = 273.16
If T_ = 400 Then T_ = 400.000001:  'Avoids Zero to the Zeroth Power in Sum17
flag_s = False:  'flag_s is only true for entropy calculation
Call MultiSums(T_, rho_, flag_s, Psizero_prime_dT, Psizerotau_prime_dtau, _
   Q_, Q_prime_dtau, Zc_)
DonotuseH2O_h = R_H2O * T_ * (rho_ * Tau_ * Q_prime_dtau + Zc_) + _
   Psizerotau_prime_dtau
End Function

'WARNING: This function gives erroneous values for entropy under the vapor dome.
Function DonotuseH2O_s(T_, rho_)
's from Equation 13 in Keenan (using coefficients A,ij from TPSI and modified datum)

e_ = 0.0048
p_c = 22089#
R_H2O = 0.461537266
rho_c = 317#
T_a = 1000#
T_c = 647.286
T_0 = 273.16
```

```
If T_ = 400 Then T_ = 400.000001:  'Avoids Zero to the Zeroth Power in Sum17
flag_s = True:  'flag_s is only True for entropy calculation
Call MultiSums(T_, rho_, flag_s, Psizero_prime_dT, Psizerotau_prime_dtau, _
   Q_, Q_prime_dtau, Zc_)
DonotuseH2O_s = -R_H2O * (Log(rho_) + rho_ * Q_ - rho_ * Tau_ * Q_prime_dtau) - _
   Psizero_prime_dT
End Function

Function rhosat_H2Ovap(T_)
'Iterative function to determine vapor density for saturated vapor at T
'NewtonRaphson method with IdealGas approx for derivative: dp/drho = RT

n = 0
R_H2O = 0.461537266
tol_dp = 0.00000001
rhodum = 0.01
pdum = DonotuseH2O_p(T_, rhodum)
dp = psat_H2O(T_) - pdum
dpdrho = R_H2O * T_

drhonxt = 0

Do
  drhonxt = dp / dpdrho
  rhodum = rhodum + drhonxt
  pdum = DonotuseH2O_p(T_, rhodum)
  dp = psat_H2O(T_) - pdum
  If Abs(dp) < tol_dp Then Exit Do
  If Abs(dp) > 99000 Then Exit Function
  If n > 9999 Then Exit Function
  n = n + 1
Loop

rhosat_H2Ovap = rhodum
End Function

'WARNING: This Sub cannot be used independently - it must be called from
'  one of the H2O_? Functions below
Sub H2OPropAll(T_, rho_, X_case, p_case, v_case, h_case, s_case)
'Subroutine to apply appropriate equations for p, h at given T_
'  for 5 different rho_ domains
'Note: Quality of superheated water is assigned a value of -9 (for gas)
'      Quality of subcooled water is assigned a value of -1 (for liquid)
'      Pressure of subcooled water is assigned a value of -1 (for liquid)
If rho_ < 0.9999999 * rhosat_H2Ovap(T_) Then domain = 1
If rho_ <= 1.0000001 * rhosat_H2Ovap(T_) And rho_ >= 0.9999999 * _
   rhosat_H2Ovap(T_) Then domain = 2
```

```
If rho_ > 1.0000001 * rhosat_H2Ovap(T_) And rho_ < 0.9999999 * _
   rhosat_H2Oliq(T_) Then domain = 3
If rho_ <= 1.0000001 * rhosat_H2Oliq(T_) And rho_ >= 0.9999999 * _
   rhosat_H2Oliq(T_) Then domain = 4
If rho_ > 1.0000001 * rhosat_H2Oliq(T_) Then domain = 5

Select Case domain
  Case Is = 1:  'Superheated Vapor
    X_case = -9
    p_case = DonotuseH2O_p(T_, rho_)
    v_case = 1 / rho_
    h_case = DonotuseH2O_h(T_, rho_)
    s_case = DonotuseH2O_s(T_, rho_)
  Case Is = 2:  'Sat Vapor
    X_case = 1
    p_case = psat_H2O(T_)
    v_case = 1 / rho_
    h_case = DonotuseH2O_h(T_, rho_)
    s_case = DonotuseH2O_s(T_, rho_)
  Case Is = 3:  'Vapor Dome
    X_case = ((1 / rho_) - (1 / rhosat_H2Oliq(T_))) / ((1 / rhosat_H2Ovap(T_)) - _
   (1 / rhosat_H2Oliq(T_)))
    p_case = psat_H2O(T_)
    v_case = 1 / rho_
    h_case = DonotuseH2O_h(T_, rhosat_H2Oliq(T_)) + X_case * (DonotuseH2O_h(T_,
rhosat_H2Ovap(T_)) - DonotuseH2O_h(T_, rhosat_H2Oliq(T_)))
    s_case = DonotuseH2O_s(T_, rhosat_H2Oliq(T_)) + X_case * (DonotuseH2O_s(T_,
rhosat_H2Ovap(T_)) - DonotuseH2O_s(T_, rhosat_H2Oliq(T_)))
  Case Is = 4:  'Sat Liquid
    X_case = 0
    p_case = psat_H2O(T_)
    v_case = 1 / rho_
    h_case = DonotuseH2O_h(T_, rhosat_H2Oliq(T_))
    s_case = DonotuseH2O_s(T_, rhosat_H2Oliq(T_))
  Case Is = 5:  'Subcooled Liquid
    X_case = -1
    p_case = -1
    v_case = 1 / rho_
    h_case = DonotuseH2O_h(T_, rho_)
    s_case = DonotuseH2O_s(T_, rho_)
  End Select

End Sub

Function H2O_X(T_, rho_)
'Customized function to use in Excel to compute values for Quality X of water
Call H2OPropAll(T_, rho_, X_case, p_case, v_case, h_case, s_case)
```

```
H2O_X = X_case
End Function

Function H2O_p(T_, rho_)
'Customized function to use in Excel to compute values for pressure p of water
Call H2OPropAll(T_, rho_, X_case, p_case, v_case, h_case, s_case)
H2O_p = p_case
End Function

Function H2O_v(T_, rho_)
'Customized function to use in Excel to compute values for spec volume v of water
Call H2OPropAll(T_, rho_, X_case, p_case, v_case, h_case, s_case)
H2O_v = v_case
End Function

Function H2O_h(T_, rho_)
'Customized function to use in Excel to compute values for enthalpy h of water
Call H2OPropAll(T_, rho_, X_case, p_case, v_case, h_case, s_case)
H2O_h = h_case
End Function

Function H2O_s(T_, rho_)
'Customized function to use in Excel to compute values for entropy s of water
Call H2OPropAll(T_, rho_, X_case, p_case, v_case, h_case, s_case)
H2O_s = s_case
End Function

'The following Sub must be called from the p, h, or s functions above.
Sub MultiSums(T_, rho_, flag_s, dumm1, dumm2, dumm3, dumm4, dumm5)

e_ = 0.0048
p_c = 22089#
R_H2O = 0.461537266
rho_c = 317#
T_a = 1000#
T_c = 647.286
T_0 = 273.16
If T_ = 400 Then T_ = 400.000001:  'Avoids Zero to the Zeroth Power in Sum17
If rho_ = 1000 Then rho_ = 1000.000001:  'Avoids Zero to Zeroth in Sum18
If rho_ = 634 Then rho_ = 634.000001:    'Avoids Zero to Zeroth in Sum17
Tau_aj(1) = T_a / T_c:      rho_aj(1) = 634
Tau_aj(2) = 2.5:            rho_aj(2) = 1000
Tau_aj(3) = 2.5:            rho_aj(3) = 1000
Tau_aj(4) = 2.5:            rho_aj(4) = 1000
Tau_aj(5) = 2.5:            rho_aj(5) = 1000
Tau_aj(6) = 2.5:            rho_aj(6) = 1000
Tau_aj(7) = 2.5:            rho_aj(7) = 1000
```

```
Aw(1, 1) = 0.029492937:          Aw(1, 2) = -0.005198586
Aw(2, 1) = -0.00013213917:       Aw(2, 2) = 0.0000077779182
Aw(3, 1) = 0.00000027464632:     Aw(3, 2) = -0.000000033301902
Aw(4, 1) = -3.6093828E-10:       Aw(4, 2) = -1.6254622E-11
Aw(5, 1) = 3.4218431E-13:        Aw(5, 2) = -1.7731074E-13
Aw(6, 1) = -2.4450042E-16:       Aw(6, 2) = 1.2748742E-16
Aw(7, 1) = 1.5518535E-19:        Aw(7, 2) = 1.3746153E-19
Aw(8, 1) = 5.9728487E-24:        Aw(8, 2) = 1.5597836E-22
Aw(9, 1) = -0.41030848:          Aw(9, 2) = 0.3373118
Aw(10, 1) = -0.0004160586:       Aw(10, 2) = -0.00020988866
Aw(1, 3) = 0.0068335354:         Aw(1, 4) = -0.0001564104
Aw(2, 3) = -0.000026149751:      Aw(2, 4) = -0.00000072546108
Aw(3, 3) = 0.000000065326396:    Aw(3, 4) = -9.2734289E-09
Aw(4, 3) = -2.6181978E-11:       Aw(4, 4) = 4.312584E-12
Aw(5, 3) = 0#:                   Aw(5, 4) = 0#
Aw(6, 3) = 0#:                   Aw(6, 4) = 0#
Aw(7, 3) = 0#:                   Aw(7, 4) = 0#
Aw(8, 3) = 0#:                   Aw(8, 4) = 0#
Aw(9, 3) = -0.13746618:          Aw(9, 4) = 0.0067874983
Aw(10, 3) = -0.00073396848:      Aw(10, 4) = 0.000010401717
Aw(1, 5) = -0.0063972405:        Aw(1, 6) = -0.0039661401
Aw(2, 5) = 0.000026409282:       Aw(2, 6) = 0.000015453061
Aw(3, 5) = -0.000000047740374: Aw(3, 6) = -0.00000002914247
Aw(4, 5) = 5.632313E-11:         Aw(4, 6) = 2.9568796E-11
Aw(5, 5) = 0#:                   Aw(5, 6) = 0#
Aw(6, 5) = 0#:                   Aw(6, 6) = 0#
Aw(7, 5) = 0#:                   Aw(7, 6) = 0#
Aw(8, 5) = 0#:                   Aw(8, 6) = 0#
Aw(9, 5) = 0.13687317:           Aw(9, 6) = 0.07984797
Aw(10, 5) = 0.0006458188:        Aw(10, 6) = 0.0003991757
Aw(1, 7) = -0.00069048554:       Psi_C(1) = 1857.065
Aw(2, 7) = 0.0000027407416:      Psi_C(2) = 3229.12:
  'Psi_C(2) = 3229.12 has s datum problem, 41.605 has h, p datum problems
  'see flag_s below for fix
Aw(3, 7) = -0.000000005102807: Psi_C(3) = -419.465
Aw(4, 7) = 3.9636085E-12:        Psi_C(4) = 36.6649
Aw(5, 7) = 0#:                   Psi_C(5) = -20.5516
Aw(6, 7) = 0#:                   Psi_C(6) = 4.85233
Aw(7, 7) = 0#:                   Psi_C(7) = 46#
Aw(8, 7) = 0#:                   Psi_C(8) = -1011.249
Aw(9, 7) = 0.013041253
Aw(10, 7) = 0.000071531353

If flag_s = False Then
  Psi_C(2) = 3229.12
Else
  Psi_C(2) = 41.605
End If
```

```
Psizero = 0
Psizero_prime_dT = 0
Psizero_prime_dtau = 0
Psizerotau_prime_dtau = 0
Q_ = 0
Q_prime_dtau = 0
Q_prime_drho = 0
Tau_ = T_a / T_
Tau_c = T_a / T_c
sum17 = 0:  Sum17prime_drho = 0:  Sum17prime_dtau = 0
Sum18 = 0:  Sum18prime_drho = 0
Sum910 = 0: Sum910prime_drho = 0
Sum110 = 0: Sum110prime_drho = 0
Zc_ = 0

For j = 1 To 7
  Sum18 = 0: Sum18prime_drho = 0
  Sum910 = 0:  Sum910prime_drho = 0
  For i = 1 To 8
    Sum18 = Sum18 + Aw(i, j) * (rho_ - rho_aj(j)) ^ (i - 1)
    Sum18prime_drho = Sum18prime_drho + Aw(i, j) * (i - 1) * _
   (rho_ - rho_aj(j)) ^ (i - 2)
  Next i
  For i = 9 To 10
    Sum910 = Sum910 + Exp(-e_ * rho_) * Aw(i, j) * rho_ ^ (i - 9)
    Sum910prime_drho = Sum910prime_drho + (-e_) * Exp(-e_ * rho_) * Aw(i, j) * _
   rho_ ^ (i - 9)
    Sum910prime_drho = Sum910prime_drho + Exp(-e_ * rho_) * (i - 9) * Aw(i, j) * _
   rho_ ^ (i - 10)
  Next i
  sum17 = sum17 + (Tau_ - Tau_aj(j)) ^ (j - 2) * (Sum18 + Sum910)
  Sum17prime_drho = Sum17prime_drho + (Tau_ - Tau_aj(j)) ^ (j - 2) * _
   (Sum18prime_drho + Sum910prime_drho)
  Sum17prime_dtau = Sum17prime_dtau + (j - 2) * ((Tau_ - Tau_aj(j)) ^ (j - 3)) * _
   (Sum18 + Sum910)
Next j

Q_ = (Tau_ - Tau_c) * sum17
Q_prime_drho = (Tau_ - Tau_c) * Sum17prime_drho
Q_prime_dtau = (Tau_ - Tau_c) * Sum17prime_dtau + sum17
Zc_ = (1 + rho_ * Q_ + rho_ ^ 2 * Q_prime_drho)

For k = 1 To 6
  Psizero = Psizero + (Psi_C(k) * Tau_ ^ (-(k - 1)))
  Psizero_prime_dtau = Psizero_prime_dtau + (-(k - 1)) * Psi_C(k) * Tau_ ^ _
   (-(k - 1) - 1)
  Psizero_prime_dT = Psizero_prime_dT + (k - 1) * Psi_C(k) * T_ ^ (k - 2) / _
```

```
    (1000 ^ (k - 1))
Next k

Psizero = Psizero + Psi_C(7) * Log(T_) + Psi_C(8) * Log(T_) / Tau_
Psizero_prime_dtau = Psizero_prime_dtau + (-1) * Psi_C(7) * Tau_ ^ (-1)
Psizero_prime_dtau = Psizero_prime_dtau + Psi_C(8) * Tau_ ^ (-2) * (Log(Tau_) - _
    Log(1000) - 1)
Psizerotau_prime_dtau = Psizero + Psizero_prime_dtau * Tau_
Psizero_prime_dT = Psizero_prime_dT + Psi_C(7) * T_ ^ (-1) + (Psi_C(8) / 1000) * _
    (Log(T_) + 1)

End Sub
```

B.5 Thermodynamic Properties of Humid Air (Chapter 6)

The VBA script presented below creates customized Excel functions for thermodynamic properties of mixtures of air with water vapor.

For these computations, dry air is assumed to have the composition that was defined for combustion air in Chapter 4, namely, where O_2 and N_2 are always present at their predefined molar ratio of 1 : 3.76 and no other gases are present. The enthalpy values given here vary slightly from other sources, including ASHRAE (2013) which is the source of the curve fit formulas chosen as the basis for these VBA scripts, but the differences are modest ($|\epsilon| < 1\%$).

Ten unique functions (all but one of which require the same three arguments – T_{db}, p_∞, and ϕ) are created that provide computations for:

1) `Humair_psat_T`: Saturation pressure of water vapor in fully humid air. $p_{sat,w}(T) \doteq kPa$
2) `Humair_Chiw_Tprh`: Mole fraction of water vapor in humid air. $\chi_w(T,p,\phi) \doteq kmol_w/kmol_{ha}$
3) `Humair_gammaw_Tprh`: Specific humidity (also known as mass fraction) of water vapor in humid air. $\gamma_w(T,p,\phi) \doteq kg_w/kg_{ha}$
4) `Humair_Ww_Tprh`: Humidity ratio of water vapor in humid air (per unit mass of dry air). $W(T,p,\phi) \doteq kg_w/kg_{da}$
5) `Humair_mu_Tprh`: Saturation degree of water vapor in humid air (actual humidity as a function of saturated humidity), mass basis. $\mu(T,p,\phi) = W/W_s$ (a mass version of relative humidity)
6) `Humair_hhat_Tprh`: Molar enthalpy of humid air. $\hat{h}(T,p,\phi) \doteq kJ/kmol_{ha}$
7) `Humair_h_Tprh`: Specific enthalpy of humid air. $h(T,p,\phi) \doteq kJ/kg_{ha}$
8) `Humair_h_perda_Tprh`: Specific enthalpy of humid air per mass of dry air. $h_{/da}(T,p,\phi) \doteq kJ_{ha}/kg_{da}$
9) `Humair_tdp_Tprh`: Dew point temperature of humid air $T_{dp}(T,p,\phi) \doteq K$ (temperature at saturation if the mixture's humidity ratio remains constant, i.e. $W_{sat,\ tdp} = W$).
10) `Humair_twb_Tprh`: Wet bulb temperature of humid air $T_{wb}(T,p,\phi) \doteq K$(temperature at saturation if the mixture's enthalpy remains constant, i.e. $h_{sat,\ twb} = h$).

Also contained within the VBA script are other custom functions that compute intermediate values necessary to obtain the aforementioned nine results. Users may discover that these intermediate functions can be helpful to answer certain problems – but because they have no built-in constraints to prevent their misuse in domains where the formulas are not applicable, users should take extra precautions to validate every result they obtain. For the purposes of this textbook, only the aforementioned nine customized functions are recommended.

The arguments used in the formulas are itemized in the list below. Subscripts *w*, *da*, and *ha* may be applied, respectively, to the gas-specific parameters below for the *water vapor* or *dry air* portions or for the *humid air* in total.

- $p[\doteqdot kPa]$ humid air pressure
- $T[\doteqdot K]$ humid air temperature (i.e. dry bulb)
- $\phi=[\chi_w/\chi_{w,\ sat}]$ relative humidity (dimensionless)

The domain boundaries are taken to be:

- Relative humidity: $0.0 \leq \phi \leq 1.0$
- Dry bulb temperature: $273\ K \leq T \leq 351\ K$
- Ambient pressure: $80\ kPa \leq p \leq 110\ kPa$

Error messages or negative numbers will be displayed if these bounds are exceeded.

The governing formula for ***water vapor pressure*** (ASHRAE 2013) is given below (for $273.16 < T$ $(K) < 350$). Note: this equation is not as accurate as the water vapor equation from Keenan et al. (1969), but it is adequate for psychrometry analyses and it is mathematically simpler.

$$p_{sat,w}(T) = \frac{1}{1000} \exp\left[\sum_{i=1}^{6}\left(C_i T^{(i-2)}\right) + C_7 \ln T\right]$$

$$C_1 = -5.800\,220\,6 \times 10^3 \quad C_2 = 1.391\,499\,3 \quad C_3 = -4.864\,023\,9 \times 10^{-2}$$

$$C_4 = 4.176\,476\,8 \times 10^{-5} \quad C_5 = -1.445\,209\,3 \times 10^{-8} \quad C_6 = 0.0$$

$$C_7 = 6.545\,947\,3$$

The zeroth coefficient in the Shomate ***enthalpy*** formulas (Linstrom and Mallard 2017–2021) was altered to obtain consistent datum states for water, oxygen, and nitrogen at $T_0 = 273.16\ K \rightarrow h_{0,f} = 0.0\ kJ/kmol$ for liquid water at the triple point temperature (0.01 ° C).

$$\hat{h}_{H_2O,g} = 1000\left[Sh_{H_2O_0} + \sum_{j=1}^{4} Sh_{H_2O_j}\left(\frac{\left(\frac{T}{1000}\right)^j}{j}\right) + Sh_{H_2O_5}\left(\frac{T}{1000}\right)^{-1}\right]$$

$$Sh_{H_2O_0} = (-250.881) - (-286.628\,9) - (-0.837\,91)$$

$$Sh_{H_2O_1} = 30.092 \quad Sh_{H_2O_2} = 6.832\,514 \quad Sh_{H_2O_3} = 6.793\,435$$

$$Sh_{H_2O_4} = -2.534\,48 \quad Sh_{H_2O_5} = 0.082\,139\,0$$

$$\hat{h}_{O_2,g} = 1000\left[Sh_{O_{20}} + \sum_{j=1}^{4} Sh_{O_{2j}}\left(\frac{\left(\frac{T}{1000}\right)^j}{j}\right) + Sh_{O_{25}}\left(\frac{T}{1000}\right)^{-1}\right]$$

$$Sh_{O_{20}} = (-8.903\,471) - (0) - (-0.731\,954)$$

$$Sh_{O_{21}} = 31.322\,34 \quad Sh_{O_{22}} = -20.235\,31 \quad Sh_{O_{23}} = 57.866\,44$$

$$Sh_{O_{24}} = -36.506\,24 \quad Sh_{O_{25}} = -0.007\,374\,00$$

$$\hat{h}_{N_2,g} = 1000\left[Sh_{N_{20}} + \sum_{j=1}^{4} Sh_{N_{2j}}\left(\frac{\left(\frac{T}{1000}\right)^j}{j}\right) + Sh_{N_{25}}\left(\frac{T}{1000}\right)^{-1}\right]$$

$$Sh_{N_{20}} = (-8.671\,914) - (0) - (-0.727\,935)$$

$$Sh_{N_{21}} = 28.986\,41 \quad Sh_{N_{22}} = 1.853\,978 \quad Sh_{N_{23}} = -9.647\,459$$

$$Sh_{N_{24}} = 16.635\,37 \quad Sh_{N_{25}} = 0.000\,117\,00$$

$$\hat{h}_{da} = \frac{1 \cdot h_{O_2,g} + 3.76 \cdot h_{O_2,g}}{(1 + 3.76)}$$

The equations for the various psychrometric parameters computed are given below:

$$\chi_w = \phi \frac{p_{sat,w}(T)}{p} \qquad \gamma_w = \frac{\chi_w \hat{M}_{H_2O}}{(\chi_w \hat{M}_{H_2O} + (1-\chi_w)\hat{M}_{da})}$$

$$W = \frac{\chi_w \hat{M}_{H_2O}}{(1-\chi_w)\hat{M}_{da}} \qquad \mu = \phi \frac{(1-\chi_{w,sat})}{(1-\chi_w)}$$

$$\hat{h}_{ha} = \chi_w \hat{h}_{H_2O,g} + (1-\chi_w)\hat{h}_{da}$$

$$h_{ha} = \gamma_w \left(\frac{\hat{h}_{H_2O,g}}{\hat{M}_{H_2O}} \right) + (1-\gamma_w) \left(\frac{\hat{h}_{da}}{\hat{M}_{da}} \right)$$

$$h_{/da} = W \left(\frac{\hat{h}_{H_2O,g}}{\hat{M}_{H_2O}} \right) + 1 \left(\frac{\hat{h}_{da}}{\hat{M}_{da}} \right)$$

```
Option Explicit
'Humidity group of functions from Appendix B.5

'Use Ideal gas Shomate relations from module associated with Appendix B.3
' to compute enthalpy of humid air and other humidity parameters,
' while adjusting enthalpy datum states to 0.01 C at water triple point
Public Const Mhat_H2O As Double = 18.01471
Public Const Mhat_O2 As Double = 31.99806
Public Const Mhat_N2 As Double = 28.01286
Public Const Mhat_da As Double = 28.85009
Dim Sh_H2O(5) As Double
Dim Sh_O2(5) As Double
Dim Sh_N2(5) As Double
Dim Csat_ASHRAE(2, 7) As Double
Dim Csat_ASHpart(7) As Double
Dim Chi_w As Double
Dim Chi_wsat As Double
Dim dp As Double
Dim dpdrho As Double
Dim drhonxt As Double
Dim DTm As Double
Dim DW As Double
Dim DH As Double
Dim errtol As Double
Dim gamma_w As Double
Dim hhat_ As Double
Dim h_w As Double
```

```
Dim h_da As Double
Dim m_ As Double
Dim mu_ As Double
Dim p_ As Double
Dim p_c As Double
Dim pdum As Double
Dim phi_ As Double
Dim R_H2O As Double
Dim Rangelo As Double
Dim Rangehi As Double
Dim rhodum As Double
Dim Shomate As Double
Dim Sumall As Double
Dim Sumall2 As Double
Dim Sumall3 As Double
Dim Sumall4 As Double
Dim T_c As Double
Dim T_ As Double
Dim t_1000 As Double
Dim Tdum As Double
Dim tol_dp As Double
Dim W_w As Double
Dim i As Integer
Dim j As Integer
Dim n As Integer


Function Humair_psat_T(T_)
'Equation of state for saturated water p as a function of T
'   from ASHRAE(2013) Ch.1 (with indices modified by Martin
'   for programming consistency)
'   i = 1 is for liq+vap equil;  i = 2 is for sol+vap equil
'   p has units of kPa and T has units of K
Csat_ASHRAE(1, 1) = -5800.2206:          Csat_ASHRAE(1, 2) = 1.3914993
Csat_ASHRAE(1, 3) = -0.048640239:        Csat_ASHRAE(1, 4) = 0.000041764768
Csat_ASHRAE(1, 5) = -0.000000014452093: Csat_ASHRAE(1, 6) = 0#
Csat_ASHRAE(1, 7) = 6.5459473
Csat_ASHRAE(2, 1) = -5674.5359:          Csat_ASHRAE(2, 2) = 6.3925247
Csat_ASHRAE(2, 3) = -0.009677843:        Csat_ASHRAE(2, 4) = 0.00000062215701
Csat_ASHRAE(2, 5) = 2.0747825E-09:       Csat_ASHRAE(2, 6) = -9.484024E-13
Csat_ASHRAE(2, 7) = 4.1635019:
i = 0
j = 0

If T_ > 350.01 Then
  Humair_psat_T = -9
  Exit Function
```

```
ElseIf T_ < 233 Then
  Humair_psat_T = -1
  Exit Function
ElseIf T_ > 273.15 Then
  i = 1
Else
  i = 2
End If

p_c = 22089
T_c = 647.286
Sumall = 0

For j = 1 To 6
  Sumall = Sumall + Csat_ASHRAE(i, j) * T_ ^ (j - 2)
Next j
Sumall = Sumall + Csat_ASHRAE(i, 7) * Log(T_)
Humair_psat_T = Exp(Sumall) / 1000:   'convert to kPa from Pa
End Function

Function H3P_H2OV(T_)
'Underlying equations courtesy of the National Institute of
'Standards and Technology, U.S. Department of Commerce.
'Not copyrightable in the United States.
Rangehi = 1700
Rangelo = 273
'NIST claims Rangelo(0) should be 500K, but we reduced it to 273
'2nd range (T>1700) deleted for psychrometry
'Datum below at j=0 comprises 3 parts: constant, dHf,H2Ov,
'   adjust from 298 to 273 (Triple Pt 3P); values are per kmol
'   In Sh_H2O(0), (-241.8264) is gas (-286.6289) is liq datum
Sh_H2O(0) = (-250.881) - (-286.6289) - (-0.83791):
Sh_H2O(1) = 30.092
Sh_H2O(2) = 6.832514
Sh_H2O(3) = 6.793435
Sh_H2O(4) = -2.53448
Sh_H2O(5) = 0.082139
Shomate = 0
t_1000 = T_ / 1000

If T_ < Rangelo Then
    Exit Function
ElseIf T_ > Rangehi Then
    Exit Function
End If

For j = 1 To 4
```

```
  Shomate = Shomate + Sh_H2O(j) * t_1000 ^ j / j
Next j
Shomate = Shomate - Sh_H2O(5) / t_1000 + Sh_H2O(0)
H3P_H2OV = 1000 * Shomate
'Factor of 1000 is conversion from mol to kmol in denominator of H3P values

End Function

Function H3P_O2(T_)
'Underlying equations courtesy of the National Institute of
'Standards and Technology, U.S. Department of Commerce.
'Not copyrightable in the United States.
'2nd & 3rd ranges (T>1700) deleted for psychrometry
'Datum below at j=0 comprises 3 parts: constant, dHf,H2Ov,
'   adjust from 298 to 273 (Triple Pt 3P); values are per kmol
Rangehi = 700
Rangelo = 100
Sh_O2(0) = (-8.903471) - (0) - (-0.731954)
Sh_O2(1) = 31.32234
Sh_O2(2) = -20.23531
Sh_O2(3) = 57.86644
Sh_O2(4) = -36.50624
Sh_O2(5) = -0.007374
Shomate = 0
t_1000 = T_ / 1000

If T_ < Rangelo Then
  Exit Function
ElseIf T_ > Rangehi Then
  Exit Function
Else
End If

For j = 1 To 4
  Shomate = Shomate + Sh_O2(j) * t_1000 ^ j / j
Next j
Shomate = Shomate - Sh_O2(5) / t_1000 + Sh_O2(0)
H3P_O2 = 1000 * Shomate
'Factor of 1000 is conversion from mol to kmol in denominator of H3P values

End Function

Function H3P_N2(T_)
'Underlying equations courtesy of the National Institute of
'Standards and Technology, U.S. Department of Commerce.
'Not copyrightable in the United States.
'2nd & 3rd ranges (T>1700) deleted for psychrometry
```

```
'Datum below at j=0 comprises 3 parts: constant, dHf,H2Ov,
'   adjust from 298 to 273 (Triple Pt 3P); values are per kmol
Rangehi = 500
Rangelo = 100
Sh_N2(0) = (-8.671914) - (0) - (-0.727935)
Sh_N2(1) = 28.98641
Sh_N2(2) = 1.853978
Sh_N2(3) = -9.647459
Sh_N2(4) = 16.63537
Sh_N2(5) = 0.000117
Shomate = 0
t_1000 = T_ / 1000

If T_ < Rangelo Then
  Exit Function
ElseIf T_ > Rangehi Then
  Exit Function
Else
End If

For j = 1 To 4
  Shomate = Shomate + Sh_N2(j) * t_1000 ^ j / j
Next j
Shomate = Shomate - Sh_N2(5) / t_1000 + Sh_N2(0)
H3P_N2 = 1000 * Shomate
'Factor of 1000 is conversion from mol to kmol in denominator of H3P values

End Function

Function H3P_dryair(T_)
' Compute molar enthalpy of dry air (defined in Chapter 4)
'  from Shomate enthalpy values for O2, N2
Dim H_O2 As Double
Dim H_N2 As Double
Dim Mol_O2 As Double
Dim Mol_N2 As Double

Mol_O2 = 1#
Mol_N2 = 3.76

H_O2 = H3P_O2(T_)
H_N2 = H3P_N2(T_)
H3P_dryair = (Mol_O2 * H_O2 + Mol_N2 * H_N2) / (Mol_O2 + Mol_N2)
End Function

Function Humair_Chiw_Tprh(T_, p_, phi_)
'Compute mole fraction of water vapor in humid air from pressure,
```

```
'  temperature, and rel humidity
If (phi_ > 1 Or phi_ < 0 Or T_ < 273 Or T_ > 351 Or p_ > 110 Or p_ < 80) Then
  Humair_Chiw_Tprh = -1
  Exit Function
End If
Chi_w = phi_ * Humair_psat_T(T_) / p_
Humair_Chiw_Tprh = Chi_w
End Function

Function Humair_gammaw_Tprh(T_, p_, phi_)
'Compute mass fraction of water vapor in humid air from pressure,
'  temperature, and rel humidity
If (phi_ > 1 Or phi_ < 0 Or T_ < 273 Or T_ > 351 Or p_ > 110 Or p_ < 80) Then
  Humair_gammaw_Tprh = -1
  Exit Function
End If
Chi_w = Humair_Chiw_Tprh(T_, p_, phi_)
gamma_w = Chi_w * Mhat_H2O / (Chi_w * Mhat_H2O + (1 - Chi_w) * Mhat_da)
Humair_gammaw_Tprh = gamma_w
End Function

Function Humair_Ww_Tprh(T_, p_, phi_)
'Compute humidity ratio of humid air from pressure, temperature,
'  and rel humidity
If (phi_ > 1 Or phi_ < 0 Or T_ < 273 Or T_ > 351 Or p_ > 110 Or p_ < 80) Then
  Humair_Ww_Tprh = -1
  Exit Function
End If
Chi_w = Humair_Chiw_Tprh(T_, p_, phi_)
W_w = Chi_w * Mhat_H2O / ((1 - Chi_w) * Mhat_da)
Humair_Ww_Tprh = W_w
End Function

Function Humair_mu_Tprh(T_, p_, phi_)
'Compute saturation degree of humid air from pressure, temperature,
'  and rel humidity
If (phi_ > 1 Or phi_ < 0 Or T_ < 273 Or T_ > 351 Or p_ > 110 Or p_ < 80) Then
  Humair_mu_Tprh = -1
  Exit Function
End If
Chi_w = Humair_Chiw_Tprh(T_, p_, phi_)
Chi_wsat = Humair_psat_T(T_) / p_
mu_ = phi_ * (1 - Chi_wsat) / (1 - Chi_w)
Humair_mu_Tprh = mu_
End Function
```

```
Function Humair_hhat_Tprh(T_, p_, phi_)
'Compute molar enthalpy of humid air from pressure, temperature,
'  and rel humidity
If (phi_ > 1 Or phi_ < 0 Or T_ < 273 Or T_ > 351 Or p_ > 110 Or p_ < 80) Then
  Humair_hhat_Tprh = -1
  Exit Function
End If
Chi_w = Humair_Chiw_Tprh(T_, p_, phi_)
hhat_ = Chi_w * H3P_H2OV(T_) + (1 - Chi_w) * H3P_dryair(T_)
Humair_hhat_Tprh = hhat_
End Function

Function Humair_h_Tprh(T_, p_, phi_)
'Compute specific enthalpy of humid air from pressure, temperature,
'  and rel humidity
If (phi_ > 1 Or phi_ < 0 Or T_ < 273 Or T_ > 351 Or p_ > 110 Or p_ < 80) Then
  Humair_h_Tprh = -1
  Exit Function
End If
Chi_w = Humair_Chiw_Tprh(T_, p_, phi_)
gamma_w = Humair_gammaw_Tprh(T_, p_, phi_)
h_w = H3P_H2OV(T_) / Mhat_H2O
h_da = H3P_dryair(T_) / Mhat_da
Humair_h_Tprh = gamma_w * h_w + (1 - gamma_w) * h_da
End Function

Function Humair_h_perda_Tprh(T_, p_, phi_)
'Compute specific enthalpy of humid air (per kg dry air) from pressure,
'  temperature, and rel humidity
If (phi_ > 1 Or phi_ < 0 Or T_ < 273 Or T_ > 351 Or p_ > 110 Or p_ < 80) Then
  Humair_h_perda_Tprh = -1
  Exit Function
End If
Chi_w = Humair_Chiw_Tprh(T_, p_, phi_)
W_w = Chi_w * Mhat_H2O / ((1 - Chi_w) * Mhat_da)
h_w = H3P_H2OV(T_) / Mhat_H2O
h_da = H3P_dryair(T_) / Mhat_da
Humair_h_perda_Tprh = W_w * h_w + 1 * h_da
End Function

Function Humair_tdp_Tprh(T_, p_, phi_)
'Compute dew point temperature of humid air from pressure,
'  temperature, and rel humidity
If (phi_ > 1 Or phi_ < 0 Or T_ < 273 Or T_ > 351 Or p_ > 110 Or p_ < 80) Then
  Humair_tdp_Tprh = -1
  Exit Function
End If
```

```
W_w = Humair_Ww_Tprh(T_, p_, phi_)
errtol = 0.00000001
DW = 0.5
DTm = 0.001
Tdum = T_
n = 0
m_ = 0

Do
If DW < errtol Then Exit Do
  m_ = (Humair_Ww_Tprh(Tdum + DTm, p_, 1) - Humair_Ww_Tprh(Tdum, p_, 1)) / DTm
  DW = Humair_Ww_Tprh(Tdum, p_, 1) - Humair_Ww_Tprh(T_, p_, phi_)
  Tdum = Tdum - DW / m_
  n = n + 1
If n >= 1000 Then Exit Function
Loop
Humair_tdp_Tprh = Tdum
End Function

Function Humair_twb_Tprh(T_, p_, phi_)
'Compute wet bulb temperature of humid air from pressure,
'  temperature, and rel humidity
If (phi_ > 1 Or phi_ < 0 Or T_ < 273 Or T_ > 351 Or p_ > 110 Or p_ < 80) Then
  Humair_twb_Tprh = -1
  Exit Function
End If

errtol = 0.00001
DH = 0.1
DTm = 0.001
Tdum = T_
n = 0
m_ = 0

Do
If DH < errtol Then Exit Do
  m_ = (Humair_h_Tprh(Tdum + DTm, p_, 1) - Humair_h_Tprh(Tdum, p_, 1)) / DTm
  DH = Humair_h_Tprh(Tdum, p_, 1) - Humair_h_Tprh(T_, p_, phi_)
  Tdum = Tdum - DH / m_
  n = n + 1
If n >= 1000 Then Exit Function
Loop

Humair_twb_Tprh = Tdum
End Function
```

B.6 Thermodynamic Properties of Ammonia–Water Mixtures (Chapter 6)

The VBA script presented below creates customized Excel functions for thermodynamic properties of two-phase, ammonia/water binary mixtures. Five unique functions are given that provide property values for problems involving these mixtures (e.g. the chemical engineer's refrigerator from Section 10.10).

1) `T_px`: Mixture temperature as a function of mixture pressure and mole fraction of ammonia in the liquid phase $T(p,\hat{x})$
2) `T_py`: Temperature as a function of pressure and mole fraction of ammonia in the vapor phase $T(p,\hat{y})$.
3) `y_px`: Mole fraction of ammonia in the vapor phase as a function of pressure and mole fraction of ammonia in the liquid phase $\hat{y}(p,\hat{x})$
4) `hliq_Tx`: Specific enthalpy of the liquid phase as a function of temperature and mole fraction of ammonia in the liquid phase $h_{liq}(T,\hat{x})$
5) `hvap_Ty`: Specific enthalpy of the vapor phase as a function of temperature and mole fraction of ammonia in the vapor phase $h_{vap}(T,\hat{y})$

The parameters used in the finite sum equations are:

- $p[\doteq kPa]$ mixture pressure
- $T[\doteq K]$ mixture temperature
- $\hat{x}\left[\doteq kmol_{NH_3,liq}/kmol_{liq}\right]$ mole fraction of NH_3 in liquid phase
- $\hat{y}\left[\doteq kmol_{NH_3,vap}/kmol_{vap}\right]$ mole fraction of NH_3 in vapor phase
- $h_{liq}\left[\doteq kJ/kg_{liq}\right]$ specific enthalpy of liquid phase ($NH_3 + H_2O$)
- $h_{vap}\left[\doteq kJ/kg_{vap}\right]$ specific enthalpy of vapor phase ($NH_3 + H_2O$)
- $p_0[\doteq kPa]$ datum state for pressure in three equations below
- $T_0[\doteq K]$ datum state for temperature in two equations below

For the first three equations (excluding the two enthalpy equations), the two datum values are:

$$p_0 = 2000\ kPa, \qquad T_0 = 100\ K$$

For the two enthalpy equations only, four datum values are required:

$$h_0 = 100\ kJ/kg, \quad T_0 = 273.16\ K \quad \text{for liquid phase enthalpy}$$

$$h_0 = 1000\ kJ/kg, \quad T_0 = 324\ K \quad \text{for vapor phase enthalpy}$$

Users will note that even though the water triple point temperature is used for the liquid enthalpy datum, the liquid enthalpy value is identically zero only for the two endpoints – pure water and pure ammonia. Mixtures at $T = 0.01$ °C with values of $\hat{x}$ other than 0.0 or 1.0 result (perhaps counterintuitively) in negative liquid enthalpy values. This phenomenon is a consequence of the exothermicity of mixing ammonia into water. If the two components initially at the same temperature are mixed, and if the mixing process is adiabatic with respect to the surroundings, the mixture temperature (i.e. sensible enthalpy) will rise to account for the reduction in chemical enthalpy of the mixture. This phenomenon is illustrated in Figure B.1 (Patek and Klomfar 1995), which is part of a four-component mixture where only the liquid phase's enthalpy is being displayed.

Similarly, we also share a Patek-style plot of vapor phase enthalpy for the ammonia water system in Figure B.2 (Patek and Klomfar 1995). These curves are itemized by mixture pressure from 2 to

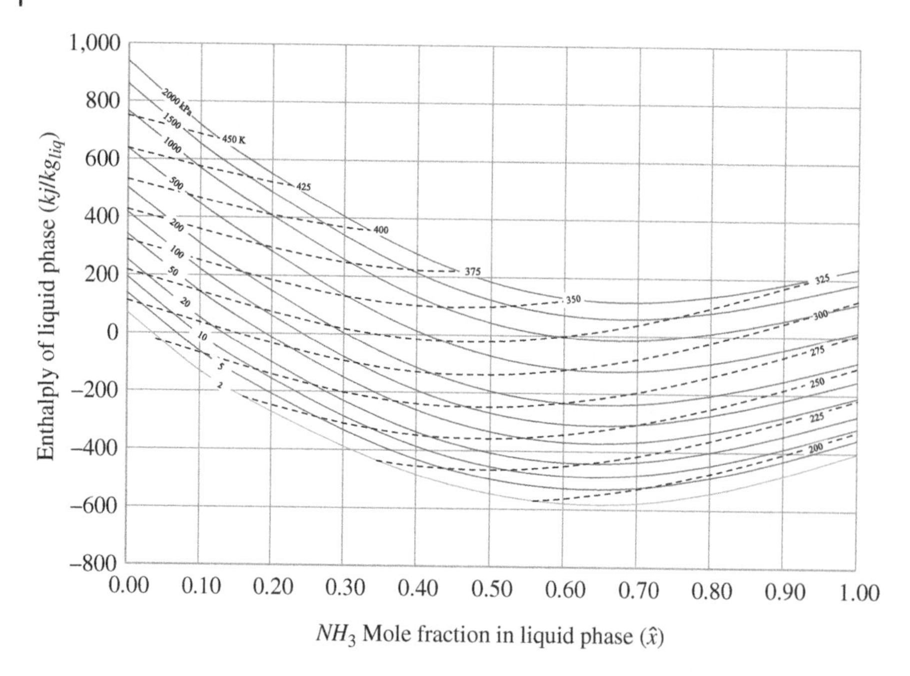

Figure B.1 Liquid phase enthalpy diagram for NH_3 + H_2O mixtures. The outer boundaries indicate the approximate domain of validity for the property equations given in this appendix. *Source:* Patek and Klomfar (1995).

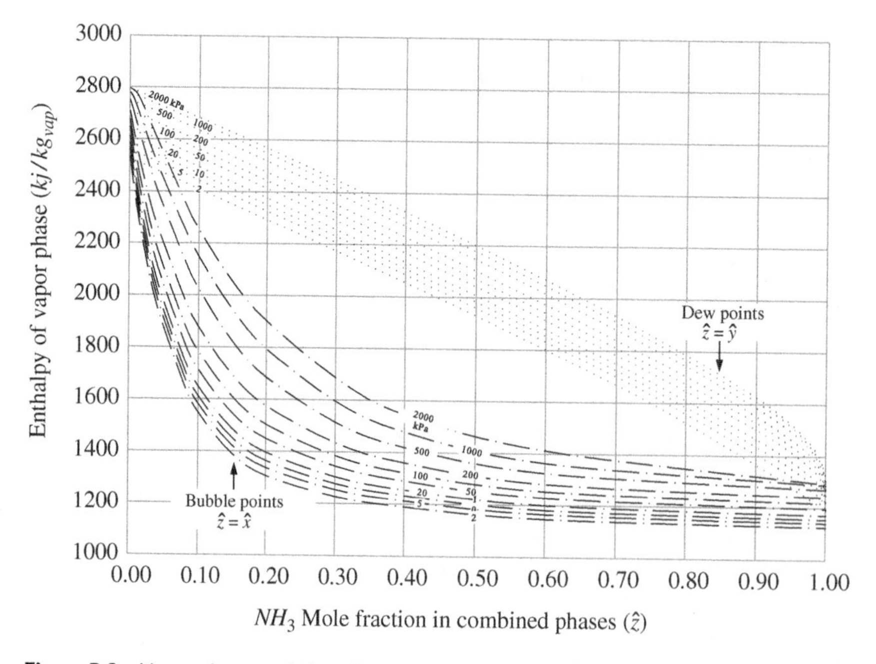

Figure B.2 Vapor phase enthalpy diagram for NH_3 + H_2O mixtures between bubble point and dew point. The outer boundaries indicate the approximate domain of validity for the property equations given in this appendix. *Source:* Patek and Klomfar (1995).

2000 kPa for bubble point and dew point. Again, all the enthalpy values reported are for saturated mixtures having all four components (liquid + vapor, $NH_3 + H_2O$, but where the reported enthalpy values apply just to the vapor phase.

All the customized functions introduced in this appendix are explained first and then a single VBA script is provided that contains all of the individual functions. A total of five customized functions become available when the VBA script is employed: $T(p,\hat{x})$, $T(p,\hat{y})$, $y(p,\hat{x})$, $h_{liq}(T,\hat{x})$, and $h_{vap}(T,\hat{y})$. The coefficients for each of the finite sum formulas are given in Tables B.4–B.8.

The VBA scripts were created by the author, and they carry the same warning as all other programming instructions included in this appendix – users *must* validate the output values over the entire domain of temperature and pressure desired before relying on the results for any purpose.

The VBA script at the end of this appendix creates a user-defined Function named `T_px`, which has two arguments $p, \hat{x}$. The units of the values returned by `T_px` are K, the units of p are kPa, and the units of $\hat{x}$ are $kmol_{NH_3,liq}/kmol_{liq}$.

$$T(p,\hat{x}) = T_0 \sum_i a_i (1-\hat{x})^{m_i} \left[\ln\left(\frac{p_0}{p}\right) \right]^{n_i}$$

$$p_0 = 2000\ kPa, \qquad T_0 = 100\ K$$

It is recommended that users incorporate all the VBA instructions in this appendix section together into a single VBA Module, because the Public Constant declarations and the comments are only introduced once at the beginning of the module and are not repeated for each new VBA function.

Table B.4 Mixture temperature T as a function of mixture pressure p and ammonia liquid mole fraction $\hat{x}$.

i	m_i	n_i	a_i
1	0	0	$3.223\ 02 \times 10^{0}$
2	0	1	$-3.842\ 06 \times 10^{-1}$
3	0	2	$4.609\ 65 \times 10^{-2}$
4	0	3	$-3.789\ 45 \times 10^{-3}$
5	0	4	$1.356\ 10 \times 10^{-4}$
6	1	0	$4.877\ 55 \times 10^{-1}$
7	1	1	$-1.201\ 08 \times 10^{-1}$
8	1	2	$1.061\ 54 \times 10^{-2}$
9	2	3	$-5.335\ 89 \times 10^{-4}$
10	4	0	$7.850\ 41 \times 10^{0}$
11	5	0	$-1.159\ 41 \times 10^{1}$
12	5	1	$-5.231\ 50 \times 10^{-2}$
13	6	0	$4.895\ 96 \times 10^{0}$
14	13	1	$4.210\ 59 \times 10^{-2}$

Table B.5 Mixture temperature T as a function of mixture pressure p and ammonia vapor mole fraction $\hat{y}$.

i	m_i	n_i	a_i
1	0	0	$3.240\,04 \times 10^{0}$
2	0	1	$-3.959\,20 \times 10^{-1}$
3	0	2	$4.356\,24 \times 10^{-2}$
4	0	3	$-2.189\,43 \times 10^{-3}$
5	1	0	$-1.435\,26 \times 10^{0}$
6	1	1	$1.052\,56 \times 10^{0}$
7	1	2	$-7.192\,81 \times 10^{-2}$
8	2	0	$1.223\,62 \times 10^{1}$
9	2	1	$-2.243\,68 \times 10^{0}$
10	3	0	$-2.017\,10 \times 10^{1}$
11	3	1	$1.108\,34 \times 10^{0}$
12	4	0	$1.453\,99 \times 10^{1}$
13	4	2	$6.443\,12 \times 10^{-1}$
14	5	0	$-2.212\,46 \times 10^{0}$
15	5	2	$-7.562\,66 \times 10^{-1}$
16	6	0	$-1.355\,29 \times 10^{0}$
17	7	2	$1.835\,41 \times 10^{-1}$

Table B.6 Ammonia vapor mole fraction $\hat{y}$ as a function of mixture pressure p and ammonia liquid mole fraction $\hat{x}$.

i	m_i	n_i	a_i
1	0	0	$1.980\,220\,17 \times 10^{1}$
2	0	1	$-1.180\,926\,69 \times 10^{1}$
3	0	6	$2.774\,799\,80 \times 10^{1}$
4	0	7	$-2.886\,342\,77 \times 10^{1}$
5	1	0	$-5.916\,166\,08 \times 10^{1}$
6	2	1	$5.780\,913\,05 \times 10^{2}$
7	2	2	$-6.217\,367\,43 \times 10^{0}$
8	3	2	$-3.421\,984\,02 \times 10^{3}$
9	4	3	$1.194\,031\,27 \times 10^{4}$
10	5	4	$-2.454\,137\,77 \times 10^{4}$
11	6	5	$2.915\,918\,65 \times 10^{4}$
12	7	6	$-1.847\,822\,90 \times 10^{4}$
13	7	7	$2.348\,194\,34 \times 10^{1}$
14	8	7	$4.803\,106\,17 \times 10^{3}$

Table B.7 Ammonia + water liquid enthalpy h_{liq} as a function of mixture pressure p and ammonia liquid fraction $\hat{x}$.

i	m_i	n_i	a_i
1	0	1	$-7.610\ 80 \times 10^0$
2	0	4	$2.569\ 05 \times 10^1$
3	0	8	$-2.470\ 92 \times 10^2$
4	0	9	$3.259\ 52 \times 10^2$
5	0	12	$-1.588\ 54 \times 10^2$
6	0	14	$6.190\ 84 \times 10^1$
7	1	0	$1.143\ 14 \times 10^1$
8	1	1	$1.181\ 57 \times 10^0$
9	2	1	$2.841\ 79 \times 10^0$
10	3	3	$7.416\ 09 \times 10^0$
11	5	3	$8.918\ 44 \times 10^2$
12	5	4	$-1.613\ 09 \times 10^3$
13	5	5	$6.221\ 06 \times 10^2$
14	6	2	$-2.075\ 88 \times 10^2$
15	6	4	$-6.873\ 93 \times 10^0$
16	8	0	$3.507\ 16 \times 10^0$

Table B.8 Ammonia + water vapor enthalpy h_{vap} as a function of mixture pressure p and ammonia vapor fraction $\hat{y}$.

i	m_i	n_i	a_i
1	0	0	$1.288\ 27 \times 10^0$
2	1	1	$1.252\ 47 \times 10^{-1}$
3	2	2	$-2.087\ 48 \times 10^0$
4	3	3	$2.176\ 96 \times 10^0$
5	0	4	$2.356\ 87 \times 10^0$
6	1	0	$-8.869\ 87 \times 10^0$
7	2	1	$1.026\ 35 \times 10^1$
8	3	2	$-2.374\ 40 \times 10^0$
9	0	3	$-6.701\ 55 \times 10^0$
10	1	0	$1.645\ 08 \times 10^1$
11	2	0	$-9.368\ 49 \times 10^0$
12	0	1	$8.422\ 54 \times 10^0$
13	1	0	$-8.588\ 07 \times 10^0$
14	0	1	$-2.770\ 49 \times 10^0$
15	4	0	$-9.612\ 48 \times 10^{-1}$
16	2	0	$9.880\ 09 \times 10^{-1}$
17	1	0	$3.084\ 82 \times 10^{-1}$

The VBA script at the end of this appendix creates a user-defined Function named `T_py`, which has two arguments $p, \hat{y}$. The units of the values returned by `T_py` are K, the units of p are kPa, and the units of $\hat{y}$ are $kmol_{NH_3,vap}/kmol_{vap}$.

$$T(p,\hat{y}) = T_0 \sum_i a_i (1-\hat{y})^{(m_i/4)} \left[\ln\left(\frac{p_0}{p}\right) \right]^{n_i}$$

$$p_0 = 2000\ kPa, \qquad T_0 = 100\ K$$

It is recommended that users incorporate all the VBA instructions in this appendix section together into a single VBA Module, because the Public Constant declarations and the comments are only introduced once at the beginning of the module and are not repeated for each new VBA function.

The VBA script at the end of this appendix creates a user-defined Function named `y_px`, which has two arguments $p, \hat{x}$. The units of the values returned by `y_px` are $kmol_{NH_3,vap}/kmol_{vap}$, the units of p are kPa, and the units of $\hat{x}$ are $kmol_{NH_3,liq}/kmol_{liq}$.

$$y(p,\hat{x}) = 1 - \exp\left[\ln(1-\hat{x}) \sum_i a_i \left(\frac{p}{p_0}\right)^{m_i} \hat{x}^{(n_i/3)} \right]$$

$$p_0 = 2000\ kPa$$

Users should be cautioned that use of first three functions should be thermodynamically consistent with each other, but small discrepancies can be found that may be unacceptable some users under some circumstances. For example, in the equation below, the nested set of expressions on the left *should* be precisely equal to the single expression on the right. By inspection, the temperatures are found to differ by several degrees Celsius for some conditions.

$$T(p, y(p,\hat{x})) = ?T(p,\hat{y})$$

It is recommended that users incorporate all the VBA instructions in this appendix section together into a single VBA Module, because the Public Constant declarations and the comments are only introduced once at the beginning of the module and are not repeated for each new VBA function.

The VBA script below creates a user-defined Function named `hliq_Tx`, which has two arguments $T, \hat{x}$. The units of the values returned by `hliq_Tx` are kJ/kg_{liq}, the units of T are K, and the units of $\hat{x}$ are $kmol_{NH_3,liq}/kmol_{liq}$.

$$h_{liq}(T,\hat{x}) = h_0 \sum_i a_i \left(\frac{T}{T_0} - 1\right)^{m_i} \hat{x}^{n_i}$$

$$h_0 = 100\ kJ/kg, \qquad T_0 = 273.16\ K$$

We note that even though the introductory content in Chapter 6 heavily utilized molar concentration values (e.g. $kmol, \hat{x}$, and $\dot{n}$), these enthalpy expressions given by Patek and Klomfar (1995) are reported on a mass basis (e.g. kg, x, and $\dot{m}$).

It is recommended that users incorporate all the VBA instructions in this appendix section together into a single VBA Module, because the Public Constant declarations and the comments are only introduced once at the beginning of the module and are not repeated for each new VBA function.

The VBA script below creates a user-defined Function named hvap_py, which has two arguments $p, \hat{y}$. The units of the values returned by hvap_px are kJ/kg_{vap}, the units of p are kPa, and the units of $\hat{y}$ are $kmol_{NH_3,vap}/kmol_{vap}$.

$$h_{vap}(T,\hat{y}) = h_0 \sum_i a_i \left(1 - \frac{T}{T_0}\right)^{m_i} (1-\hat{y})^{(n_i/4)}$$

$$h_0 = 1000\,kJ/kg, \qquad T_0 = 324\,K$$

It is recommended that users incorporate all the VBA instructions in this appendix section together into a single VBA Module, because the Public Constant declarations and the data types for variables are only assigned once at the beginning of the module and are not repeated for each new VBA function.

The VBA script at the end of this appendix creates a user-defined Function named x_py, which has two arguments $p, \hat{y}$. The units of the values returned by x_py are $kmol_{NH_3,liq}/kmol_{liq}$, the units of p are kPa, and the units of $\hat{y}$ are $kmol_{NH_3,vap}/kmol_{vap}$.

This function procedure gives results purely by iteration on the inverse function y_px described above. Iterating successfully was a challenge due to the shape of the family of curves $\hat{y}(p,\hat{x})$, which precluded the use of Newton–Raphson as an efficient iterator. Due to domain limits on the Patek fitting equations, iterations were deemed converged at slightly under 100 *ppm* error. Users may notice minor delays in the execution of this function for certain combinations of p and $\hat{x}$. This is because those cases require nearly 10^5 iterations to converge.

After creating these VBA scripts and comparing them to Patek's results, the author discovered another data source (Tillner-Roth and Friend 1998) for thermodynamic properties of ammonia–water mixtures. For several reasons, it is expected that the Tillner-Roth equations will prove to be superior to Patek, but in the time available for completion of the manuscript for this book, it was not possible to track down and translate two critical references (cited in Tillner-Roth) and to implement and cross-check their more rigorous (and extensive) state equations.

Selected property values (ρ, h, s) from the Tillner-Roth appendix table as presented in Appendices A.13 and A.14, and some of those results appear to differ somewhat from the corresponding values obtained by using the formulas given here. Therefore, we recommend against using the Patek formulas for any purpose other than academic problem solving, and instead we point users to Tillner-Roth for more accurate $NH_3 + H_2O$ thermodynamic property data.

```
Option Explicit
'Ammonia+Water group of functions from Appendix B.6

Dim i As Integer
Dim a_(17) As Double
Dim m_(17) As Double
Dim n_(17) As Double
Dim Sumall As Double
Dim Tdiffratio As Double
Dim Term1 As Double
Dim Term2 As Double
Dim Term3 As Double
Dim Term4 As Double
Dim Term5 As Double
Dim Tdatumvap As Double
```

```
Dim hdatumvap As Double
Dim Tdatumliq As Double
Dim hdatumliq As Double
Dim Tsat As Double
Dim p1 As Double
Dim T1 As Double
Dim x1 As Double
Dim y1 As Double
Dim deltay As Double
Dim xdum As Double
Dim ydum As Double
Dim yfactor As Double
Dim ytollo As Double
Dim ytolhi As Double
'p,sat(T) for water from Keenan (1969)
'p,sat units are kPa
Public Const p_0 As Double = 2000
Public Const T_0 As Double = 100
Public Const pcrit_H2O As Double = 22089:   ' Reynolds, 1981
Public Const Tcrit_H2O As Double = 647.286: ' Reynolds, 1981
Public Const pcrit_NH3 As Double = 11360:   ' Tillner-Roth 1998
Public Const Tcrit_NH3 As Double = 405.4:   ' Tillner-Roth 1998
Dim n As Integer


'VBA uses "Log" for natural logarithm, not "Ln"

Function T_px(p1, x1)
'T(p,x) NH3 Liquid Fraction
'Units of p are kPa, units of T are K.
'Units of x, y are mol,NH3/mol,phase.
'Source: Patek (1995)
If (p1 > 2000 Or p1 < 2 Or x1 < 0 Or x1 > 1) Then
  T_px = -9999
  Exit Function
End If
If x1 > 0.99999 Then x1 = 0.99999
'Solution becomes unstable at x = 1
m_(1) = 0: m_(2) = 0: m_(3) = 0: m_(4) = 0: m_(5) = 0
m_(6) = 1: m_(7) = 1: m_(8) = 1: m_(9) = 2: m_(10) = 4
m_(11) = 5: m_(12) = 5: m_(13) = 6: m_(14) = 13
n_(1) = 0: n_(2) = 1: n_(3) = 2: n_(4) = 3: n_(5) = 4
n_(6) = 0: n_(7) = 1: n_(8) = 2: n_(9) = 3: n_(10) = 0
n_(11) = 0: n_(12) = 1: n_(13) = 0: n_(14) = 1
a_(1) = 3.22302: a_(2) = -0.384206: a_(3) = 0.0460965
a_(4) = -0.00378945: a_(5) = 0.00013561: a_(6) = 0.487755
a_(7) = -0.120108: a_(8) = 0.0106154: a_(9) = -0.000533589
```

```
a_(10) = 7.85041: a_(11) = -11.5941: a_(12) = -0.052315
a_(13) = 4.89596: a_(14) = 0.0421059
Sumall = 0
Term1 = 0
For i = 1 To 14
  Term1 = a_(i) * (1 - x1) ^ m_(i) * (Log(p_0 / p1)) ^ n_(i)
  Sumall = Sumall + Term1
Next i
'NH3 triple point is 195.4; but T,min = 182 in Fig B.1
If T_0 * Sumall < 182 Then
  T_px = -9999
  Exit Function
End If
T_px = T_0 * Sumall
End Function

Function T_py(p1, y1)
'T(p,y) NH3 Vapor Fraction
'Source: Patek (1995)
If (p1 > 2000 Or p1 < 2 Or x1 < 0 Or x1 > 1) Then
  T_py = -9999
  Exit Function
End If
If y1 > 0.999999 Then y1 = 0.999999
'Solution becomes unstable at y = 1
m_(1) = 0: m_(2) = 0: m_(3) = 0: m_(4) = 0: m_(5) = 1
m_(6) = 1: m_(7) = 1: m_(8) = 2: m_(9) = 2: m_(10) = 3
m_(11) = 3: m_(12) = 4: m_(13) = 4: m_(14) = 5: m_(15) = 5
m_(16) = 6: m_(17) = 7
n_(1) = 0: n_(2) = 1: n_(3) = 2: n_(4) = 3: n_(5) = 0
n_(6) = 1: n_(7) = 2: n_(8) = 0: n_(9) = 1: n_(10) = 0
n_(11) = 1: n_(12) = 0: n_(13) = 2: n_(14) = 0: n_(15) = 2
n_(16) = 0: n_(17) = 2
a_(1) = 3.24004: a_(2) = -0.39592: a_(3) = 0.0435624
a_(4) = -0.00218943: a_(5) = -1.43526: a_(6) = 1.05256
a_(7) = -0.0719281: a_(8) = 12.2362: a_(9) = -2.24368
a_(10) = -20.178: a_(11) = 1.10834: a_(12) = 14.5399
a_(13) = 0.644312: a_(14) = -2.21246: a_(15) = -0.756266
a_(16) = -1.35529: a_(17) = 0.183541
Sumall = 0
Term2 = 0
For i = 1 To 17
  Term2 = a_(i) * (1 - y1) ^ (m_(i) / 4) * (Log(p_0 / p1)) ^ n_(i)
  Sumall = Sumall + Term2
Next i
T_py = T_0 * Sumall
End Function
```

```
Function y_px(p1, x1)
'y(p,x) Vapor Fraction vs Pressure and Liquid Fraction
'Source: Patek (1995)
If (p1 > 2000 Or p1 < 2 Or x1 < 0 Or x1 > 1) Then
  y_px = -9999
  Exit Function
End If
If x1 < 0.000001 Then x1 = 0.000001
If x1 > 0.999999 Then x1 = 0.999999
'Solution becomes unstable at x = 1 and x = 0
m_(1) = 0: m_(2) = 0: m_(3) = 0: m_(4) = 0: m_(5) = 1
m_(6) = 2: m_(7) = 2: m_(8) = 3: m_(9) = 4: m_(10) = 5
m_(11) = 6: m_(12) = 7: m_(13) = 7: m_(14) = 8
n_(1) = 0: n_(2) = 1: n_(3) = 6: n_(4) = 7: n_(5) = 0
n_(6) = 1: n_(7) = 2: n_(8) = 2: n_(9) = 3: n_(10) = 4
n_(11) = 5: n_(12) = 6: n_(13) = 7: n_(14) = 7:
a_(1) = 19.8022017: a_(2) = -11.8092669: a_(3) = 27.747998
a_(4) = -28.8634277: a_(5) = -59.1616608: a_(6) = 578.091305
a_(7) = -6.21736743: a_(8) = -3421.98402: a_(9) = 11940.3127
a_(10) = -24541.3777: a_(11) = 29159.1865: a_(12) = -18478.229
a_(13) = 23.4819434: a_(14) = 4803.10617
Sumall = 0
Term3 = 0
For i = 1 To 14
Term3 = a_(i) * (p1 / p_0) ^ m_(i) * x1 ^ (n_(i) / 3)
  Sumall = Sumall + Term3
Next i
'Set negative values to zero
If 1 - Exp(Log(1 - x1) * Sumall) < 0 Then
  y_px = 0
  Exit Function
End If
y_px = 1 - Exp(Log(1 - x1) * Sumall)
End Function

Function hliq_Tx(T1, x1)
'hliq(T,x) Enthalpy vs Temperature, NH3 Liquid Fraction
'Source: Patek (1995)
'h has units kJ/kg, x,y have units kmol,NH3/kmol,phase

If (T1 > 487 Or T1 < 182 Or x1 < 0 Or x1 > 1) Then
  hliq_Tx = -9999
  Exit Function
End If
If x1 < 0.000001 Then x1 = 0.000001
If T1 = 273.16 Then T1 = 273.160001
'Solution becomes unstable at x = 0, T = Tdatumliq
```

```
m_(1) = 0: m_(2) = 0: m_(3) = 0: m_(4) = 0: m_(5) = 0
m_(6) = 0: m_(7) = 1: m_(8) = 1: m_(9) = 2: m_(10) = 3
m_(11) = 5: m_(12) = 5: m_(13) = 5: m_(14) = 6: m_(15) = 6
m_(16) = 8
n_(1) = 1: n_(2) = 4: n_(3) = 8: n_(4) = 9: n_(5) = 12
n_(6) = 14: n_(7) = 0: n_(8) = 1: n_(9) = 1: n_(10) = 3
n_(11) = 3: n_(12) = 4: n_(13) = 5: n_(14) = 2: n_(15) = 4
n_(16) = 0
a_(1) = -7.6108: a_(2) = 25.6905: a_(3) = -247.092
a_(4) = 325.952: a_(5) = -158.854: a_(6) = 61.9084
a_(7) = 11.4314: a_(8) = 1.18157: a_(9) = 2.84179
a_(10) = 7.41609: a_(11) = 891.844: a_(12) = -1613.09
a_(13) = 622.106: a_(14) = -207.588: a_(15) = -6.87393
a_(16) = 3.50716
Tdatumliq = 273.16
hdatumliq = 100
Sumall = 0
Term4 = 0
For i = 1 To 16
  Term4 = a_(i) * ((T1 / Tdatumliq) - 1) ^ m_(i) * x1 ^ n_(i)
  Sumall = Sumall + Term4
Next i
hliq_Tx = hdatumliq * Sumall
'Patek NH3 datum prob, consider adding +340*y1 to RHS to equal Tillner-Roth
End Function

Function hvap_Ty(T1, y1)
'hvap(T,y) Enthalpy vs Temperature and Vapor Fraction of NH3
'Source: Patek (1995)
'h has units kJ/kg, x,y have units kmol,NH3/kmol,phase
If (T1 > 487 Or T1 < 182 Or y1 < 0 Or y1 > 1) Then
  hvap_Ty = -9999
  Exit Function
End If
If y1 > 0.999999 Then y1 = 0.999999
If T1 = 324# Then T1 = 324.000001
'Solution becomes unstable at y = 1, T = Tdatumvap
m_(1) = 0: m_(2) = 1: m_(3) = 2: m_(4) = 3: m_(5) = 0
m_(6) = 1: m_(7) = 2: m_(8) = 3: m_(9) = 0: m_(10) = 1
m_(11) = 2: m_(12) = 0: m_(13) = 1: m_(14) = 0: m_(15) = 4
m_(16) = 2: m_(17) = 1:
n_(1) = 0: n_(2) = 0: n_(3) = 0: n_(4) = 0: n_(5) = 2
n_(6) = 2: n_(7) = 2: n_(8) = 2: n_(9) = 3: n_(10) = 3
n_(11) = 3: n_(12) = 4: n_(13) = 4: n_(14) = 5: n_(15) = 6
n_(16) = 7: n_(17) = 10
a_(1) = 1.28827: a_(2) = 0.125247: a_(3) = -2.08748
a_(4) = 2.17696: a_(5) = 2.35687: a_(6) = -8.86987
```

```
a_(7) = 10.2635: a_(8) = -2.3744: a_(9) = -6.70155
a_(10) = 16.4508: a_(11) = -9.36849: a_(12) = 8.42254
a_(13) = -8.58807: a_(14) = -2.77049: a_(15) = -0.961248
a_(16) = 0.988009: a_(17) = 0.308482
Tdatumvap = 324
hdatumvap = 1000
Sumall = 0
Tdiffratio = 0
Term5 = 0
For i = 1 To 17
  Tdiffratio = (1 - (T1 / Tdatumvap))
  Term5 = a_(i) * Tdiffratio ^ m_(i) * (1 - y1) ^ (n_(i) / 4)
  Sumall = Sumall + Term5
Next i
hvap_Ty = hdatumvap * Sumall
'Patek NH3 datum prob, consider adding +340*y1 to RHS to equal Tillner-Roth
End Function

'R.J.Martin (2021) compute x_py by iteration.
'Note that values diverge slowly at very high x.
Function x_py(p1, y1)
'x(p,y) Liquid Fraction vs Pressure and Vapor Fraction
'Original source: Patek (1995)
If (p1 > 2000 Or p1 < 2 Or y1 < 0 Or y1 > 1) Then
  x_py = -9999
  Exit Function
End If
If y1 < 0.00002 Then y1 = 0.00002
If y1 > 0.9999995 Then y1 = 0.9999995

n = 0
xdum = 0.5
ydum = 0.8
ytollo = -0.0000005
ytolhi = 0.0000005
If y1 < 0.001 Then
  If p1 < 200 Then
    yfactor = 0.0001
  Else
    yfactor = 0.01
  End If
ElseIf y1 < 0.3 Then
  yfactor = 0.01
ElseIf y1 < 0.8 Then
  yfactor = 0.1
ElseIf y1 < 0.9999 Then
  yfactor = 0.5
```

```
Else
  yfactor = 2
End If

Do
ydum = y_px(p1, xdum)
deltay = ydum - y1
xdum = xdum - deltay * yfactor
n = n + 1
If deltay > ytollo And deltay < ytolhi Then Exit Do
If n > 99999 Then Exit Function
If xdum > 1 Or xdum < 0 Then Exit Function
Loop
x_py = xdum
End Function

'From Keenan (1968) via Reynolds (1979)
'p,sat(T) for H2O; units of T are K, units of p are kPa
Function psatH2O(T_sat)
Dim i As Integer
Dim F_keeH2O(8) As Double
Dim Sumall As Double
Const a_keeH2O As Double = 0.01
Const Tp_keeH2O As Double = 338.15

F_keeH2O(1) = -7.419242:   F_keeH2O(2) = 0.29721
F_keeH2O(3) = -0.1155286:    F_keeH2O(4) = 0.008685635
F_keeH2O(5) = 0.001094098:   F_keeH2O(6) = -0.00439993
F_keeH2O(7) = 0.002520658:   F_keeH2O(8) = -0.0005218684

For i = 1 To 8
Sumall = Sumall + F_keeH2O(i) * (a_keeH2O * _
(T_sat - Tp_keeH2O)) ^ (i - 1)
Next i

psatH2O = pcrit_H2O * Exp(((Tcrit_H2O / T_sat) - 1) * Sumall)
End Function

'From TPSI Thermodynamic Properties in SI (Reynolds, 1979)
'p,sat(T) for NH3; units of T are K, units of p are kPa
Function psatNH3(T_sat)
Dim i As Integer
Dim F_TPSINH3(8) As Double
Dim Sumall As Double
Const Tp_TPSINH3 As Double = 297.6823
F_TPSINH3(1) = -6.74405:   F_TPSINH3(2) = -0.0073895
F_TPSINH3(3) = -2.0331:     F_TPSINH3(4) = 0.23925
```

```
F_TPSINH3(5) = 0.3445:       F_TPSINH3(6) = -12.43895
F_TPSINH3(7) = -55.635:       F_TPSINH3(8) = -66.785

Sumall = 0
For i = 1 To 8
Sumall = Sumall + F_TPSINH3(i) * ((T_sat / Tp_TPSINH3) - 1) ^ (i - 1)
Next i
psatNH3 = pcrit_NH3 * Exp(((Tcrit_NH3 / T_sat) - 1) * Sumall)
End Function
```

B.7 Iterative Solution for Von Platen Mixer (Chapter 10)

This appendix describes a VBA tool that can greatly increase the efficiency of analyzing the Von Platen Mixer addressed briefly in Section 10.10. The mixer, which is analogous to the thermostatic expansion valve in the conventional refrigeration cycle) brings two streams together: ⑤ $NH_3 + H_2O$ subcooled liquid exiting from the condenser + ⑪ H_2 gas exiting from heat exchanger #1 and combines them together to become ⑥ $NH_3 + H_2O + H_2$ multiphase mixture just upstream of the evaporator.

The mixer is assumed adiabatic, but phase change and temperature drop occur as a consequence of the infusion of the hydrogen gas, which reduces the partial pressure of ammonia and water and which allows a substantial portion of the liquid ammonia to evaporate and causes the temperature of the three-component mixture to drop by several dozen kelvins.

A brute-force method can be applied successfully to solve the mixer's challenging thermodynamics, where Excel's *solver* is used in an alternating sequence to approach the correct answers for two unknown variables. To begin, we assign trial values to: (i) $\hat{x}_⑥$, the mixer-exit mole fraction of NH_3 in the liquid phase, and (ii) $p_{H_2⑥}$, the mixer-exit partial pressure of hydrogen. The successful calculation *will* converge to a single outcome for both values, and the other rules (e.g. conservation of enthalpy flow and conservation of species in the two phases combined) will be obeyed, if the accounting is performed correctly.

However, the two unknowns impose competing effects on the mixture. If we guess $p_{H_2⑥}$ to be too low while guessing $\hat{x}_⑥$ correctly, we will overpredict $T_⑥$ by a large amount and the ***error*** between outlet and inlet enthalpy flow $\Delta\dot{H}$ will be substantially positive (i.e. outlet too high compared to the inlet). On the contrary, if we guess $\hat{x}_⑥$ to be too low while guessing $p_{H_2⑥}$ correctly, the ***error*** between independent computations of χ_{H_2} the gas phase mole fraction of hydrogen, and $p_{H_2⑥}/p_{tot}$ the hydrogen partial pressure divided by total mixture pressure will be negative while the $\Delta\dot{H}$ ***error*** will be slightly positive. By using *solver* to tweak one trial value at a time (and then driving its counterpart's ***error*** to zero), the *one* correct solution can be found eventually (i.e. after dozens of alternating applications of *solver*). The counterpart to trial value $\hat{x}_⑥$ is $\Delta\dot{H}$ ***error*** and the counterpart to trial value $p_{H_2⑥}$ is χ_{H_2} ***error***.

The author notes that Excel's *in-cell iteration* method is more elegant and much faster than Solver when it works, but it is also more likely to diverge quickly, which can be difficult to undo even for skillful Excel practitioners. Students who are comfortable with *in-cell iteration* may find it a much more satisfactory way of proceeding, but unfamiliar students should first attempt the brute-force iteration using *solver* or use the customized VBA procedures presented here to bypass both *solver* and *in-cell iteration*.

Before we get into the iteration, we need to reintroduce the fourth ***preferred*** variable reported in the Von Platen and Munters (1926) patent. Recall the prior ***preferred*** variables were $\boxed{\hat{z}_{①} = \hat{x}_{strong} \approx 0.36}$, $\boxed{p_{overall} \approx 1000\ kPa}$, and $\boxed{\hat{z}_{④} = \hat{x}_{weak} = 0.25}$. The patent also cites a ***preferred*** mass flow ratio of hydrogen gas entering the mixer to that of the ammonia vapor generated in the boiler. The minimum ratio specified is $\boxed{\dot{m}_{H_2,gas⑪}/\dot{m}_{NH_3,vap③} > 0.40}$. Since the vapor flow that leaves the separator at state ③ contains some water along with the desired ammonia, a convenient way to enforce this fourth patent ***preference*** is to simply ***assign*** the hydrogen molar flow rate entering the mixer to be equal to a specific multiple of the combined (NH_3 + H_2O) molar flow rate at the condenser inlet (i.e. ***assign*** hydrogen mole flow at state ⑪ to be based on the *equality* $\dot{m}_{H_2,gas⑪}/\dot{m}_{NH_3+H_2,vap③} = 0.40$). With this ***assignment***, the actual value of the ***preferred*** mass ratio (above) will be greater than 0.4 because the denominator in the ***preferred*** ratio is smaller than the denominator in the ***assigned*** ratio.

$$\frac{\dot{m}_{H_2,gas⑪}}{\dot{m}_{NH_3+H_2O,vap③}} = \frac{\hat{M}_{H_2} \cdot \dot{n}_{H_2⑪}}{\hat{M}_{NH_3} \cdot \dot{n}_{NH_3,vap⑪} + \hat{M}_{H_2O} \cdot \dot{n}_{H_2O,vap③}} = 0.4$$

$$\dot{n}_{H_2⑪} = 0.4 \cdot \frac{\left(\hat{M}_{NH_3} \cdot \dot{n}_{NH_3,vap③} + \hat{M}_{H_2O} \cdot \dot{n}_{H_2O,vap③}\right)}{\hat{M}_{H_2}}$$

$$\rightarrow \quad \frac{\hat{M}_{H_2} \cdot \dot{n}_{H_2⑪}}{\hat{M}_{NH_3} \cdot \dot{n}_{NH_3,vap③}} > 0.4$$

Due to space constraints, the balance of this iterative (brute-force) solution is available on the companion website for this textbook, in the solution key for end-of-chapter Problem 10.9.

To use the VBA script below, copy and paste the instructions into a separate VBA module. This content relies on the customized functions created by the VBA script in Section B.6, so it is necessary to copy/paste those instructions into their own VBA module in the same Excel workbook.

The script contains several Sub procedures that are not visible in the Excel environment and will not be described here. A total of three customized Excel functions are created by these instructions, and users must *execute* all three in the same VBA *session*, in the sequence given below. The term *session* is not official VBA syntax, but its meaning should be apparent from context. A VBA session begins when any VBA procedure is first *executed* (i.e. by typing a VBA function into a cell in Excel and pressing *Enter*). The VBA session ends when an execution error (e.g. divide by zero) occurs, when the user presses the *Reset* button in the VBA environment, or the workbook is closed. If any of these events occur, the execution of the three functions in the correct sequence must be performed again. If the VBA session ends, all the *Public* variables stored when `Setzhat` was executed are lost and so are values for `CalcppH2_6` and `Calcxhat_6()`. They are easily recovered by simply clicking in the Formula Bar for the cell where `Setzhat` is placed and by pressing *Enter* again (followed by the same cell-clicking procedure for the other two functions).

The three function names are: `Setzhat(xh5, cf5, pt, t05, t11)`, `CalcppH2_6()`, and `Calcxhat_6()`. The arguments and values returned for these functions are:

`Setzhat` arguments are `xh5` = $\hat{x}_{⑤}\left(\doteq kmol_{H_3}/kmol_{liquid}\right)$ mole fraction of ammonia liquid at state 5; `cf5` = $\dot{n}_{⑤}\left(\doteq kmol_{liq}/s\right)$ condenser exit mole flow at state 5; `pt` = $p_{tot}\left(\doteq kPa\right)$ total pressure for all fluids (assumed constant for all states); `t05` = $T_{⑤}\left(\doteq K\right)$ temperature of liquid at state 5; and `t11` = $T_{⑪}\left(\doteq K\right)$ temperature of hydrogen gas at state 11. The value returned to the cell where `Setzhat` is executed is $\hat{z}_{⑤}\left(\doteq kmol_{H_3}/kmol_{comb}\right)$ mole fraction of ammonia in the combined mixture at state 5, which is equal to $\hat{x}_{⑤}$ because no vapor is present in the subcooled liquid leaving the condenser.

`CalcppH2_6()` has no arguments, but it nonetheless returns the value of $p_{H_2⑥}(\doteq kPa_{H_2})$, partial pressure of hydrogen gas after the mixer, which is one of the two trial variables being iterated.

`Calcxhat_6()` has no arguments, but it nonetheless returns the value of $\hat{x}_{⑥}(\doteq kmol_{NH_3}/kmol_{liquid})$, mole fraction of ammonia in the liquid phase after the mixer, which is one of the two trial variables being iterated. We note again that $\hat{x}_{⑤} = \hat{z}_{⑤}$, but $\hat{x}_{⑥} \neq \hat{z}_{⑥}$, because a substantial portion of the liquid NH_3 gets evaporated in the mixer. While this set of functions provides results that produce reasonably small errors for the conservation equations, the residual errors are slightly to high for the author to be comfortable recommending this method without qualification. The author suspects the higher residuals are related to a glitch in our implementation of the Newton–Raphson technique. Nonetheless, this packaged iterator can provide a good first guess for the final solution, and such a guess will reduce the effort needed to obtain a more accurate solution by manual iteration.

```
Option Explicit

'Double-iteration for Von Platen Mixer
' by R.J.Martin, 2021
' Sub procedures solve all intermediate values
' Function procedure arguments are T_5, T_11, p_all, xhat_5
' Function procedure outputs are Calcxhat_6, Calcph2_6
' Public variables, constants accessible everywhere
' If inlet is not subcooled, function gives error -91

'General variables
Public ptot As Double
Public zhat As Double
Public condflo_5 As Double
Public TK05 As Double
Public TK11 As Double
Public ndotH2 As Double
Public Hdot6_1stLaw As Double
'Group of trial variables
Public xhat6trial As Double
Public pHyd6trial As Double
Public Hdot6trial As Double
Public ChiH2g6trial As Double
Public ErrHdottrial As Double
Public ErrChiH2trial As Double
'Chemistry constants (..ratio from Von Platen patent)
Public Const Mhat_amm As Double = 17.02995
Public Const Mhat_wat As Double = 18.01471
Public Const Mhat_hyd As Double = 2.01568
Public Const H2ratiomass As Double = 0.4
'Factors and tolerances for Newton Raphson analysis
```

```
Public Const NRfactorHdot As Double = 0.6
Public Const NRfactorChiH2 As Double = 0.4
Public Const dxhat As Double = -0.0000001
Public Const dppH2 As Double = 0.0001
Public Const Tol_Hdot As Double = 0.0001
Public Const Tol_ChiH2 As Double = 0.0000001
'Intermediate variables used in different subs
Dim Hd_6_ As Double
Dim ChiH2_6_ As Double

Function Setzhat(xh5, cf5, pt, t05, t11)
'Because these parameters (zhat, etc) are critical parameters
'  that do not change throughout the entire mixer solving
'  process, users are asked to enter them here to avoid
'  errors. These are stored as Public variables accessible to
'  all functions and subs.
'  condflo is total NH3+H2O flow kmol/s out of condenser _5
'  Units for xh5 are kmol,NH3,liq5/kmol,NH3+H2O,liq5
'  Units for cf5: kmol/s. Use cf5=1.0 to obtain example soln.
'  Units for pt are kPa
'  Units for T05 and T11 are K
'  xhat6trial is trial val of mole fraction NH3 in liquid
'  pHyd6trial is trial val of partial pressure H2 after mixing
zhat = xh5: 'No vapor exists at condenser exit, so zhat=xhat
condflo_5 = cf5
ptot = pt
TK05 = t05
TK11 = t11
'Set initial guesses for trial variables
xhat6trial = xh5 - 0.15
pHyd6trial = 0.9 * ptot
'Set values for iteration procedure

If TK05 > T_px(ptot, zhat) Then
  MsgBox "T5 is not subooled, T5>Tbub"
  Setzhat = -91
  Exit Function
End If
Setzhat = zhat
End Function

Function NISTH298_H2ApxB7(TK__)
'Underlying equations courtesy of the National Institute of
'Standards and Technology, U.S. Department of Commerce.
'Not copyrightable in the United States.
'Source: NIST Chemistry Webbook (Shomate Equations)
'Ideal Gas applications only (e.g., 1 atm combustion)
```

```
Dim Rangehi(1) As Double
Dim Rangelo(1) As Double
Dim Sh_H2(1, 5) As Double
Dim Shomate As Double
Dim t_1000 As Double
Dim i As Integer
Dim j As Integer

Rangehi(0) = 1000
Rangehi(1) = 2500
Rangelo(0) = 196: 'NIST Eq only permits T>298 (extrap beware)
Rangelo(1) = Rangehi(0)
Sh_H2(0, 0) = (-9.980797) - (0)
Sh_H2(0, 1) = 33.066178
Sh_H2(0, 2) = -11.363417
Sh_H2(0, 3) = 11.432816
Sh_H2(0, 4) = -2.772874
Sh_H2(0, 5) = -0.158558
Sh_H2(1, 0) = (-1.147438) - (0)
Sh_H2(1, 1) = 18.563083
Sh_H2(1, 2) = 12.257357
Sh_H2(1, 3) = -2.859786
Sh_H2(1, 4) = 0.268238
Sh_H2(1, 5) = 1.97799
Shomate = 0
t_1000 = TK__ / 1000

If TK__ < Rangelo(0) Then
  i = 2
ElseIf TK__ <= Rangehi(0) Then
  i = 0
ElseIf TK__ <= Rangehi(1) Then
  i = 1
Else
  i = 2
End If

For j = 1 To 4
  Shomate = Shomate + Sh_H2(i, j) * t_1000 ^ j / j
Next j

Shomate = Shomate - Sh_H2(i, 5) / t_1000 + Sh_H2(i, 0)

If i = 2 Then
  Exit Function
Else
  NISTH298_H2ApxB7 = Shomate * 1000
```

```
'Desired units are kJ/kmol not kJ/mol (as NIST uses)
End If

End Function

Sub Hdot511(pt)
'Input argument is ptot, but it is a placeholder
'  to prevent the Sub from being visible as a macro
'Returned are <none> but two Public variables are
'  computed and stored here: ndotH2, Hdot6_1stLaw
'State 5 must be subcooled or saturated liquid
Dim xhat As Double
Dim yhat As Double
Dim X_hat As Double
Dim ndotliq_5 As Double
Dim Mhatliq_5 As Double
Dim hhatliq_5 As Double
Dim hhatH2_11 As Double
Dim Hdot05 As Double
Dim Hdot11 As Double
'For State 5, condenser exit the below apply
xhat = zhat
Mhatliq_5 = xhat * Mhat_amm + (1 - xhat) * Mhat_wat
ndotliq_5 = condflo_5
ndotH2 = ndotliq_5 * Mhatliq_5 * H2ratiomass / Mhat_hyd
hhatliq_5 = hliq_Tx(TK05, xhat) * Mhatliq_5
hhatH2_11 = NISTH298_H2ApxB7(TK11)
Hdot05 = ndotliq_5 * hhatliq_5
Hdot11 = ndotH2 * hhatH2_11
Hdot6_1stLaw = Hdot05 + Hdot11
End Sub

Sub Mixprop_6(xh_6, pH2_6, Hd_6_, ChiH2_6_)
'This sub computes actual mole flow rates a,b,c,d,e,f at _6
'  Arguments = xh_6, pH2_6
'  Returned = Hd_6_, ChiH2_6_
'This sub requires saturated mixture above Tbub, below Tdew
'  aw means ammonia+water (no H2)
'  Ensure zhat, others are pre-entered using Setzhat function

Dim ppawv As Double: 'awv ammonia water vapor (noH2, noliq)
Dim xhat As Double
Dim yhat As Double
Dim X_hat As Double
Dim ah6 As Double
Dim bh6 As Double
Dim ch6 As Double
```

```
Dim dh6 As Double
Dim eh6 As Double
Dim fh6 As Double
Dim Mhatliq_6 As Double
Dim Mhatvap_6 As Double
Dim Mhatgas_6 As Double
Dim TK06 As Double
Dim hhatliq_6 As Double
Dim hhatvap_6 As Double
Dim hhatH2_6 As Double

ppawv = ptot - pH2_6
xhat = xh_6
yhat = y_px(ppawv, xhat)
X_hat = (zhat - xhat) / (yhat - xhat)
'Variable values below (kmol/s actual)
eh6 = zhat * condflo_5
fh6 = (1 - zhat) * condflo_5
ch6 = X_hat * (eh6 + fh6) / (1 + (1 - yhat) / yhat)
dh6 = ch6 * (1 - yhat) / yhat
ah6 = eh6 - ch6
bh6 = fh6 - dh6
Mhatliq_6 = xhat * Mhat_amm + (1 - xhat) * Mhat_wat
Mhatvap_6 = yhat * Mhat_amm + (1 - yhat) * Mhat_wat: 'vapor only, no H2
Mhatgas_6 = (ch6 * Mhat_amm + dh6 * Mhat_wat + ndotH2 * Mhat_hyd) _
    / (ch6 + dh6 + ndotH2): 'ndotH2 is a Public variable

TK06 = T_px(ppawv, xhat)
hhatliq_6 = hliq_Tx(TK06, xhat) * Mhatliq_6
hhatvap_6 = hvap_Ty(TK06, yhat) * Mhatvap_6
hhatH2_6 = NISTH298_H2ApxB7(TK06)
Hd_6_ = (ah6 + bh6) * hhatliq_6 + (ch6 + dh6) * hhatvap_6 + ndotH2 * hhatH2_6
ChiH2_6_ = ndotH2 / (ch6 + dh6 + ndotH2)
End Sub

Sub NewtRaph_xhat(xhat_dxadj_)
'Sub that computes adjustment to trial value for xhat_6
'  while leaving ppH2 constant
'  by tweaking xhat to find slope on Err curve
'  then computing adjustment for xhat trial value that
'  moves Error closer to zero, per Newton Raphson method
Dim xhat_6tweak As Double
Dim ppH2_6tweak As Double
Dim Hdot_6tweak As Double
Dim ChiH2_6tweak As Double
Dim ErrHdot_6tweak As Double
Dim ErrChiH2_6tweak As Double
```

```
Dim dy_Hdot_6dev As Double
Dim dy_ChiH2_6dev As Double
Dim m_xhat_6dev As Double
Dim m_ppH2_6dev As Double
Dim dx_xhat_6dev As Double
Dim dx_ppH2_6dev As Double

'Tweak the first trial variable xhat
  xhat_6tweak = xhat6trial + dxhat
  Call Mixprop_6(xhat_6tweak, pHyd6trial, Hd_6_, ChiH2_6_)
  ErrHdot_6tweak = Hd_6_ - Hdot6_1stLaw
  dy_Hdot_6dev = ErrHdot_6tweak - ErrHdottrial
  m_xhat_6dev = dy_Hdot_6dev / dxhat
  dx_xhat_6dev = -NRfactorHdot * ErrHdottrial / m_xhat_6dev
'Return xhat_dxadjustment for next iteration
xhat_dxadj_ = dx_xhat_6dev
End Sub

Sub NewtRaph_ppH2(ppH2_dxadj_)
'Sub that finds the next trial values for xhat_6, ppH2_6
'  by tweaking one variable to find slope on Err curve
'  then computing adjustment for trial value that will
'  move Error closer to zero, per Newton Raphson method
Dim xhat_6tweak As Double
Dim ppH2_6tweak As Double
Dim Hdot_6tweak As Double
Dim ChiH2_6tweak As Double
Dim ErrHdot_6tweak As Double
Dim ErrChiH2_6tweak As Double
Dim dy_Hdot_6dev As Double
Dim dy_ChiH2_6dev As Double
Dim m_xhat_6dev As Double
Dim m_ppH2_6dev As Double
Dim dx_xhat_6dev As Double
Dim dx_ppH2_6dev As Double

'Tweak the second trial variable ppH2
  ppH2_6tweak = pHyd6trial + dppH2
  Call Mixprop_6(xhat6trial, ppH2_6tweak, Hd_6_, ChiH2_6_)
  ErrChiH2_6tweak = ChiH2_6_ - ppH2_6tweak / ptot
  ErrHdot_6tweak = Hd_6_ - Hdot6_1stLaw
  dy_ChiH2_6dev = ErrChiH2_6tweak - ErrChiH2trial
  m_ppH2_6dev = dy_ChiH2_6dev / dppH2
  dx_ppH2_6dev = -NRfactorChiH2 * ErrChiH2trial / m_ppH2_6dev
'Return ppH2_dxadjustment for next iteration
ppH2_dxadj_ = dx_ppH2_6dev
End Sub
```

```
Function Calcxhat_6()
Dim SetBool As Boolean
Dim Warn As String
Dim Hd_6_ As Double
Dim ChiH2_6_ As Double
Dim xhat_dxadj_ As Double
Dim ppH2_dxadj_ As Double
Dim n As Integer

SetBool = (zhat = 0 Or condflo_5 = 0 Or ptot = 0)
SetBool = (SetBool Or TK05 = 0 Or TK11 = 0)
Warn = "Place Setzhat first. Args: xhat5, condflo5, ptot, T5, T11"
If SetBool = True Then
  MsgBox Warn
  Calcxhat_6 = -999
  Exit Function
End If

n = 0
Call Hdot511(ptot)
Call Mixprop_6(xhat6trial, pHyd6trial, Hd_6_, ChiH2_6_)
  Hdot6trial = Hd_6_
  ChiH2g6trial = ChiH2_6_
  ErrHdottrial = Hdot6trial - Hdot6_1stLaw
  ErrChiH2trial = ChiH2g6trial - pHyd6trial / ptot
Do
  n = n + 1
  Call NewtRaph_ppH2(ppH2_dxadj_)
    pHyd6trial = pHyd6trial + ppH2_dxadj_
    xhat6trial = xhat6trial: 'xhat & ppH2 get updated every other iteration
  Call Mixprop_6(xhat6trial, pHyd6trial, Hd_6_, ChiH2_6_)
    Hdot6trial = Hd_6_
    ChiH2g6trial = ChiH2_6_
    ErrHdottrial = Hdot6trial - Hdot6_1stLaw
    ErrChiH2trial = ChiH2g6trial - pHyd6trial / ptot
Call NewtRaph_xhat(xhat_dxadj_)
    xhat6trial = xhat6trial + xhat_dxadj_
    pHyd6trial = pHyd6trial: 'xhat & ppH2 get updated every other iteration
  Call Mixprop_6(xhat6trial, pHyd6trial, Hd_6_, ChiH2_6_)
    Hdot6trial = Hd_6_
    ChiH2g6trial = ChiH2_6_
    ErrHdottrial = Hdot6trial - Hdot6_1stLaw
    ErrChiH2trial = ChiH2g6trial - pHyd6trial / ptot
  If (ErrHdottrial < Tol_Hdot And ErrChiH2trial < Tol_ChiH2) Then Exit Do
  If n > 500 Then Exit Do
Loop
```

```
Calcxhat_6 = xhat6trial
End Function

Function CalcppH2_6()
Dim SetBool As Boolean
Dim Warn As String
Dim Hd_6_ As Double
Dim ChiH2_6_ As Double
Dim xhat_dxadj_ As Double
Dim ppH2_dxadj_ As Double
Dim n As Integer

SetBool = (zhat = 0 Or condflo_5 = 0 Or ptot = 0)
SetBool = (SetBool Or TK05 = 0 Or TK11 = 0)
Warn = "Place Setzhat first. Args: xhat5, condflo5, ptot, T5, T11"
If SetBool = True Then
  MsgBox Warn
  Calcxhat_6 = -999
  Exit Function
End If

n = 0
Call Hdot511(ptot)
Call Mixprop_6(xhat6trial, pHyd6trial, Hd_6_, ChiH2_6_)
  Hdot6trial = Hd_6_
  ChiH2g6trial = ChiH2_6_
  ErrHdottrial = Hdot6trial - Hdot6_1stLaw
  ErrChiH2trial = ChiH2g6trial - pHyd6trial / ptot
Do
  n = n + 1
  Call NewtRaph_ppH2(ppH2_dxadj_)
    pHyd6trial = pHyd6trial + ppH2_dxadj_
    xhat6trial = xhat6trial: 'xhat & ppH2 get updated every other iteration
  Call Mixprop_6(xhat6trial, pHyd6trial, Hd_6_, ChiH2_6_)
    Hdot6trial = Hd_6_
    ChiH2g6trial = ChiH2_6_
    ErrHdottrial = Hdot6trial - Hdot6_1stLaw
    ErrChiH2trial = ChiH2g6trial - pHyd6trial / ptot
Call NewtRaph_xhat(xhat_dxadj_)
    xhat6trial = xhat6trial + xhat_dxadj_
    pHyd6trial = pHyd6trial: 'xhat & ppH2 get updated every other iteration
  Call Mixprop_6(xhat6trial, pHyd6trial, Hd_6_, ChiH2_6_)
    Hdot6trial = Hd_6_
    ChiH2g6trial = ChiH2_6_
    ErrHdottrial = Hdot6trial - Hdot6_1stLaw
    ErrChiH2trial = ChiH2g6trial - pHyd6trial / ptot
  If (ErrHdottrial < Tol_Hdot And ErrChiH2trial < Tol_ChiH2) Then Exit Do
```

```
  If n > 500 Then Exit Do
Loop

CalcppH2_6 = pHyd6trial
End Function
```

Cited References

Note: Thermodynamic data and formulas that are provided courtesy of the National Institute of Standards and Technology, US Department of Commerce, are not copyrightable in the United States.

ASHRAE; (2013); *ASHRAE Handbook—Fundamentals (SI)*; "Chapter 1: Psychrometrics"; American Society of Heating, Refrigerating, and Air Conditioning Engineers.

Chapra, S.C.; (2010); *Introduction to VBA for Excel* – ***2 Edition***; Upper Saddle River, NJ: Prentice Hall (Pearson Publishing).

Chase, M.W., Ed.; (1998); *NIST-JANAF Thermochemical Tables* – 4 Edition; Journal of Physical and Chemical Reference Date, Monograph 9; Gaithersburg, MD: National Institute of Standards and Technology; https://janaf.nist.gov.

Chen, N.H.; (1979); "An explicit equation for friction factor in pipe"; *Ind. Eng. Chem. Fund.;* **18**; pp.296–297

Keenan, J.H., Keyes, F.G., Hill, P.C., and Moore, J.G.; (1969); *Steam Tables*; New York: Wiley.

Kuo, K.K.; (2005); *Principles of Combustion* – 2 Edition; Hoboken: John Wiley & Sons.

Linstrom, P.J. and Mallard, W.G., Eds.; (2017–2021); NIST Chemistry WebBook, SRD 69; "Thermophysical properties of fluid systems"; National Institute of Standards and Technology, U.S. Dept. of Commerce; http://webbook.nist.gov/chemistry/fluid

Patek, J. and Klomfar, J.; (1995); "Simple functions for fast calculations of selected thermodynamic properties of the ammonia-water system"; *Int. J. Refrig.;* **18**; 228–234.

Reynolds, W.C.; (1979); *Thermodynamic Properties in SI*; Stanford University Mechanical Engineering.

Tillner-Roth, R. and Friend, D.G.; (1998); "A Helmholtz free energy formulation of the thermodynamic properties of the mixture {water + ammonia}"; *J. Phys. Chem. Ref. Data Monogr.;* **27**; 63; https://www.nist.gov/system/files/documents/srd/jpcrd537.pdf

Von Platen, B.C. and Munters, C.G. (1926). Refrigeration. U.S. Patent #1,609,334.

Index

Thermal Systems Design: Fundamentals and Projects, Second Edition. Richard J. Martin.

Companion website: www.wiley.com\go\Martin\ThermalSystemsDesign2

d

e

f

g

h

t

Printed and bound by CPI Group (UK) Ltd, Croydon, CR0 4YY